PEARSON ALWAYS LEARNING

LEAP Log Workbook

Third Edition

Edited by Susan Barbitta, Sue Brown, Anita Hollar, and Donna Lemons

Cover art designed by Pearson Learning Solutions. Back cover photograph courtesy of Jeffrey Vanhoutte.

Pearson Learning Solutions, 501 Boylston Street, Suite 900, Boston, MA 02116
A Pearson Education Company
www.pearsoned.com

Printed in the United States of America

3 4 5 6 7 8 9 10 V003 17 16 15 14 13

000200010271769744

BW/JG

ISBN 10: 1-269-31791-1
ISBN 13: 978-1-269-31791-7

Contents

DMA 010 Operations with Integers

DMA 020 Fractions and Decimals

DMA 030 Proportion/Ratio/Rates/Percents

DMA 040 Expressions, Linear Equations, and Inequalities

DMA 050 Graphs/Equations of Lines

DMA 060 Polynomials and Quadratic Applications

DMA 070 Rational Expressions/Equations

DMA 080 Radical Expressions/Equations

Acknowledgments

To Debbie, Billie, Gail, Mark, Stephanie, Andrew, Terri, Johnathan, Janie, Nazar, Derrick, Katie, Toi, Andrea, Andy, Brandi, Noel, Kelly, Jay, Monty, Belinda, Robert, Marianne, Paul, Kenyatta, Miles, Rodney, Anita C., Gwen, Avery, Dawn, Janet, Jen, Karey, Robin, Ken, Logan, Will, Tia, Jasmine, Tori, Ruth, Matt, Josh, Claire, Shalamar, Janine, Daniel, Tonya, Patti, Danny, June, Liz and the Developmental Math students of GTCC.

What a joy to work in such a great Developmental Math Department!

DMA 010

DMA 010

OPERATIONS WITH INTEGERS

This course provides a conceptual study of integers and integer operations. Topics include integers, absolute value, exponents, square roots, perimeter and area of basic geometric figures, Pythagorean Theorem, and use of the correct order of operations. Upon completion, students should be able to demonstrate an understanding of pertinent concepts and principles and apply this knowledge in the evaluation of expressions.

You will learn how to:

- Visually represent an integer and its opposite on the number line.
- Explain the concept of the absolute value of an integer.
- Demonstrate the conceptual understanding of operations with integers to solve application problems.
- Correctly apply commutative and associative properties to integer operations.
- Apply the proper use of exponents and calculate the principal square root of perfect squares.
- Simplify multi-step expressions using the rules for order of operations.
- Solve geometric application problems involving area and perimeter of rectangles and triangles, angles, and correctly apply the Pythagorean Theorem.

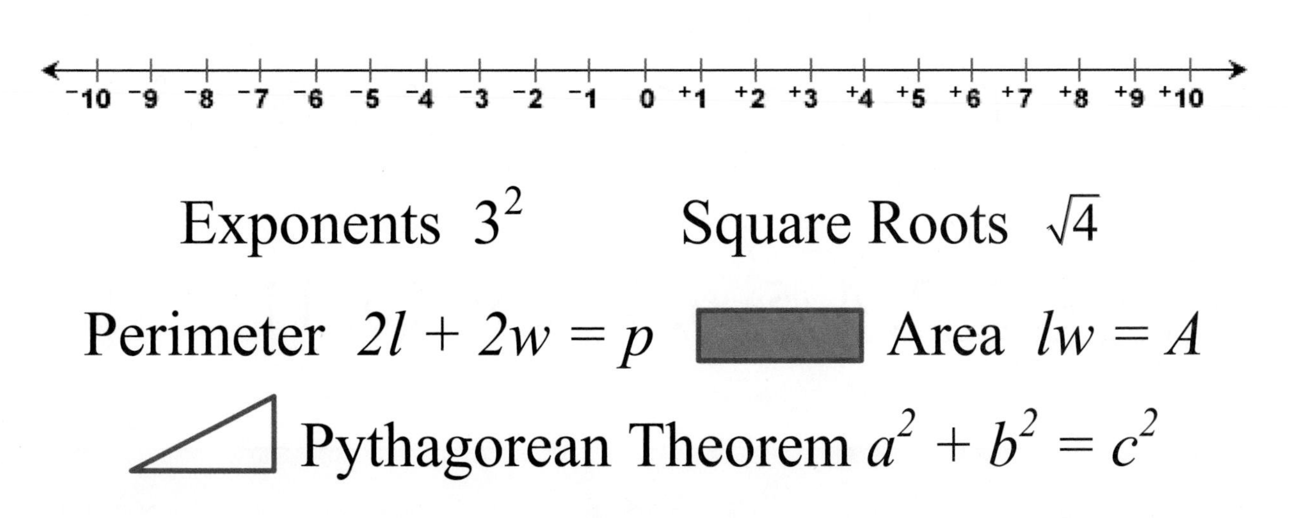

1.01 Introduction to Math 1.01

Make plans now to keep a Math notebook. Your work for every assignment should be organized and labeled. You will need these notes to study and to get help when needed.

Sample Notebook Work

HW 3.02 **(Reference the Assignment from MML.)**

1. $2x + 5 = x - 3$ **(Number the problem from MML.)**

$-x \quad -x$

$x + 5 = -3$ **(Show how your get your solution.)**

$-5 \quad -5$

$\boxed{x = -8}$ **(Box/circle your solution.)**

2. $-4x - 1 = 7$

$+1 \quad +1$

$-4x = 8$

$-4 \quad -4$

$\boxed{x = -2}$

Hints on a Quiz:
You can "X" out of a quiz and finish later. "Submit" your quiz after checking over your solutions.

Hints on a TEST:
You *cannot* "X" out of a test.
Please check over your solutions before you "Submit".

Why study math? Think of a day when you don't use math. You buy gas, food, pay rent. You might build or do repair in your home. You might even train for a profession in engineering, science, or business. To one extent or another, we use math every single day. So we begin with building the sets of numbers we use.

THE SET OF WHOLE NUMBERS

These are the counting numbers and zero. We write these in these brackets { } because they make up a set.

$$\{0, 1, 2, 3, 4, 5, 6, 7, 8, 9, 10, 11, \ldots\}$$

Complete MML HW Section 1.01 online and show your work in your notebook. This is a basic math review assignment.

1.02 An Introduction to Problem Solving 1.02

Problems involving applications of mathematics are usually written in sentence form. Most students like to refer to these as "word problems". As you read a word problem, look for **indicator words** that will help you determine which operation to use. In other words, will you need to add, subtract, multiply, or divide to solve the problems? Some applications may even use more than one operation to solve. Here's a chart that has some of the most common **indicator words**.

Addition	Subtraction	Multiplication	Division	Equality
sum plus increased by more than total added to longer than	difference less subtracted from less than decreased by minus fewer than	product times multiply of twice double/triple	quotient divided by divide equally per	equals is equal to is/was yields

Translate the problem into numbers and symbols. Look for the indicator words in these examples.

a. 35 increased by 6 is what number.

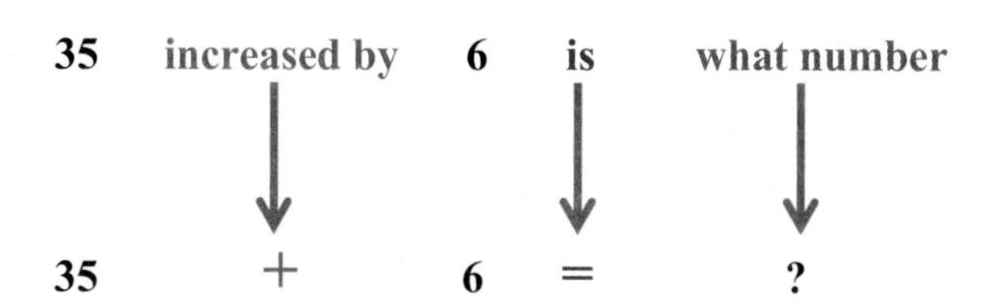

Then solve the problem by performing the calculation.

35 + 6 = 41

b. What is the quotient of 200 and 5 ?

quotient indicates division ⟶ **200 ÷ 5 = ?**

200 ÷ 5 = 40

c. What is the total of 24 and 6 ? ⟶ **24 + 6 = ?**

total indicates addition **24 + 6 = 30**

1.02 An Introduction to Problem Solving 1.02

EXAMPLE 1

The Hudson River in New York is 306 miles long. The Snake River, in the northwestern United States, is 732 miles longer than the Hudson River. How long is the Snake River?

Aurora Photos/Alamy

Step 1: Draw a picture.

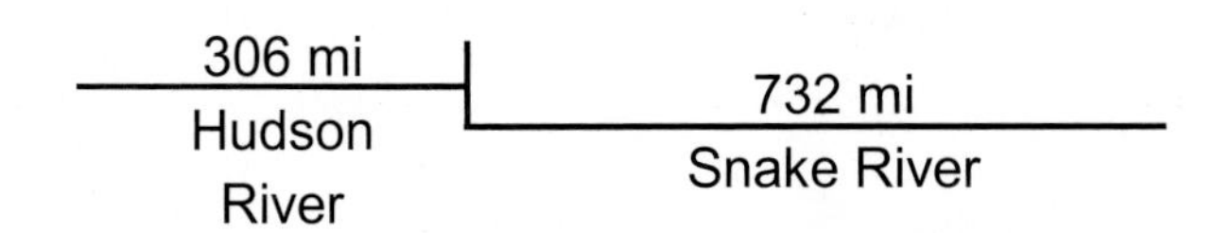

Step 2: Translate the phrase.

The phrase longer than means to add (+).
"Snake River is 732 miles longer than the Hudson River."

Translation: Snake River = 732 + 306

Step 3: Solve.

$$\begin{array}{r} 732 \\ +\ 306 \\ \hline 1{,}038 \end{array}$$

The Snake River is 1,038 miles long.

EXAMPLE 2

Zachary earns $800 per week. If he works 20 hours in one week, what is his hourly pay?

Hourly pay is weekly pay divided by hours worked in one week

Hourly pay = $800 ÷ 20 Write the equation.

Hourly pay = $40/hour Solve and answer.

EXAMPLE 3

Find the total cost of 3 hamburgers at $2.50 each and 4 fries at $1.25 each.

Total cost is # hamburgers multiplied by cost per burger plus # fries multiplied by cost per fries

Total cost = 3 • $2.50 + 4 • $1.25

Total cost = 3(2.50) + 4(1.25) Write the equation.

Total cost = 7.50 + 5 Solve.

Total cost = $12.50 Solve.

Boris Zaytsev/iStockphoto

1.02 An Introduction to Problem Solving 1.02

EXAMPLE 4

Sam is a landlord at a local apartment complex. Last month, he collected $725 in rent from each of his 12 tenants. After paying $3,120 in expenses, how much rent money did Sam have left?

Step 1: Find the total amount of rent money Sam collected. Multiply.

$725 from each of 12 tenants = 725 · 12 = $ 8,700 Write the equation.

Step 2: Subtract expenses from this amount.

$ 8,700 – 3,120 = $ 5,580 Solve.

Step 3: Solve. The amount of money remaining is $5,580. Answer.

EXAMPLE 5

Suppose the estimated 2008 population of a country was 213,900,000. During the summer months, 6 billion hot dogs are consumed. Approximately how many hot dogs were consumed per person during the summer? Round your answer to the nearest person (whole number).

Divide the number of hotdogs by the number of people who ate them.

# of hotdogs ÷ # of people

6,000,000,000 ÷ 213,900,000 = 28.0504 Write the equation, solve and answer.

28 hotdogs per person were consumed during the summer months

1.02 An Introduction to Problem Solving 1.02

YOUR TURN

Write a math sentence and solve.

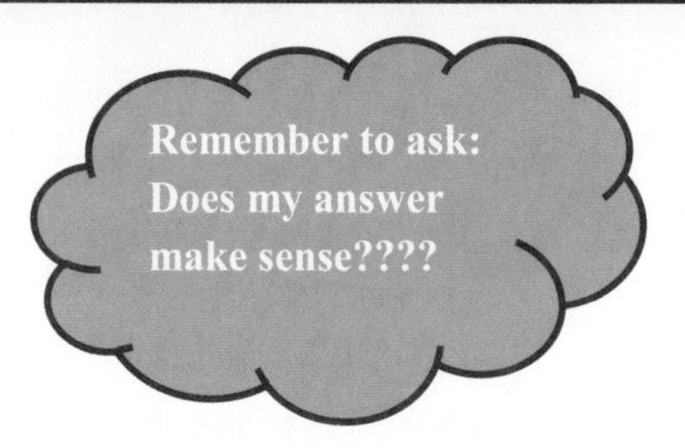

1. A family bought a house for $185,700 and later sold the house for $201,200. How much money did they make by selling the house?

Dennis M. Oschner/Getty

Equation:
Solve.

2. Alvin and John go to the student cafeteria on campus to eat dinner. Alvin orders a hamburger, onion rings, and a soda. John orders a hot dog, French fries, an apple, and a soda. Who orders the cheaper meal? By how much?

Cafeteria Menu	
Hamburger	$4
Hot Dog	$2
Onion Rings	$3
French fries	$2
Apple	$1
Soda	$1

Equation:
Solve.

3. A college student has $850 in her checking account. At the beginning of the semester she spends $314 to buy books and $40 for a parking permit. After receiving her paycheck, she deposits $550 in the same account. How much money is in the account now?

Equation:
Solve.

4. A whole pizza is 3624 calories. If the pizza is cut into 12 equal pieces, how many calories will each piece have?

Equation:
Solve.

Courtesy of Fotolia

1.02 An Introduction to Problem Solving 1.02

YOUR TURN Answers to Section 1.02.

1. \$15,500 2. John, \$2 cheaper 3. \$1,046 4. 302

Remember to use a Math notebook for all your homework. Write down every problem. Show your work.

Sample Notebook Work

HW 3.02 (Reference the Assignment from MML.)

1. $2x + 5 = x - 3$ (Number the problem from MML.)

$-x \quad -x$

$x + 5 = -3$ (Show how your get your solution.)

$-5 \quad -5$

$\boxed{x = -8}$ (Box/circle your solution.)

2. $-4x - 1 = 7$

$+1 \quad +1$

$\frac{-4x}{-4} = \frac{8}{-4}$

$\boxed{x = -2}$

Complete MML HW Section 1.02 online and show your work in your notebook.

1.03 Exponents, Square Roots, and Order of Operations 1.03

EXPONENTIAL NOTATION:

Exponents are shorthand for repeated multiplication.

For instance, if we wanted to multiply eight 5's together it would be time consuming to write it out.

$$5 \cdot 5 \cdot 5 \cdot 5 \cdot 5 \cdot 5 \cdot 5 \cdot 5$$

Suppose we wanted to multiply twenty 5's together. Who would want to write the number 5 twenty times? This is why we use exponential notation. An exponent indicates the number of times you want to multiply the number by itself. The base is the number you are multiplying.

Write $5 \cdot 5 \cdot 5 \cdot 5 \cdot 5 \cdot 5 \cdot 5 \cdot 5$ in exponential notation. This would be 5^8.

5^8 ----→ power or exponent (8)
----→ base (5)

The exponent tells you how many times the base gets multiplied by itself.

So, $3^5 = 3 \cdot 3 \cdot 3 \cdot 3 \cdot 3 = 243$

Courtesy of Fotolia

Don't forget that exponents are repeated multiplication.

$2^4 = 2 \cdot 2 \cdot 2 \cdot 2 = 16$ **NOT** ⟶ $2 \cdot 4 = 8$

EXAMPLE 1

Evaluate. (Evaluate means to find the answer.)

a. $8^3 = 8 \cdot 8 \cdot 8 = 512$

b. $2 \cdot 4^3 = 2 \cdot (4 \cdot 4 \cdot 4) = 2 \cdot (64) = 128$

YOUR TURN

Evaluate.

1. 6^4

2. $3 \cdot 2^4$

1.03 Exponents, Square Roots, and Order of Operations

There is a difference between writing a list of factors in exponential notation and evaluating. Look at the following two examples. ***Look closely at what the directions ask you to do.***

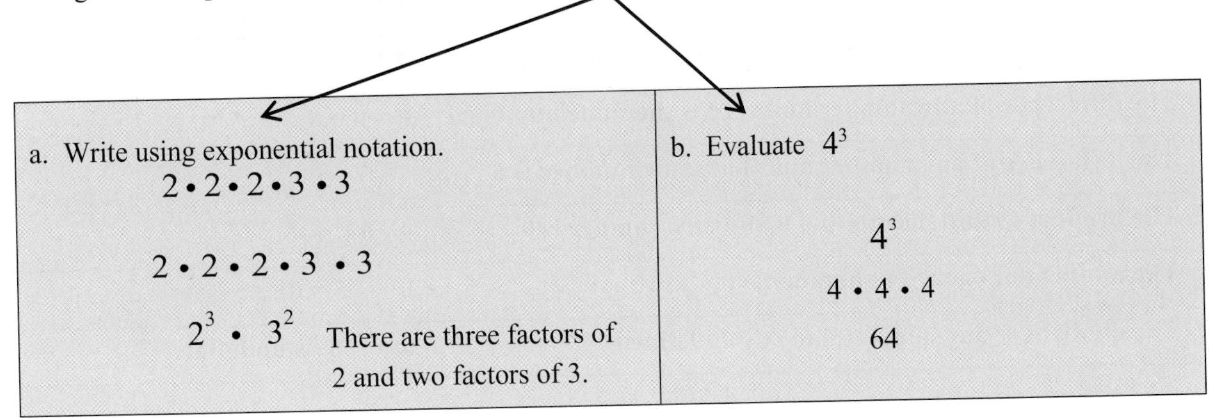

a. Write using exponential notation. $2 \cdot 2 \cdot 2 \cdot 3 \cdot 3$	b. Evaluate 4^3
$2 \cdot 2 \cdot 2 \cdot 3 \cdot 3$	4^3
$2^3 \cdot 3^2$ There are three factors of 2 and two factors of 3.	$4 \cdot 4 \cdot 4$ 64

Square Roots:

Radical Sign

$\sqrt{100}$ Read "the square root of 100."

$\sqrt{100}$ means "What number times itself is 100?"

$\sqrt{100} = 10$ because $10 \cdot 10 = 100$.

Avoiding Common Errors: Taking the square root of a number is not the same as dividing by 2. $\sqrt{100} = 10$ ~~$\sqrt{100} = 50$~~

Courtesy of Fotolia

0 is a very strange number. It is neither positive or negative. If you add or subtract zero to any number, that number stays the same. If you multiply any number by zero, you get zero . However, it gets even worse. You cannot divide a number by zero. But you can divide zero by a number and get zero.

1.03 Exponents, Square Roots, and Order of Operations

Look at these statements and examples!

Statement	Example
The sum of 0 and any number is the same number	**$4 + 0 = 4$**
The difference of any number and zero is the same number.	**$8 - 0 = 8$**
The difference of any number and that same number is 0	**$7 - 7 = 0$**
The product (multiplication) of 0 and any number is 0	**$4(0) = 0$**
The quotient of 0 and any number is 0	**$0 \div 5 = 0$**
The quotient of any number and 0 is undefined	**$5 \div 0$ is undefined**

Division with Zero: Remember that division is splitting into equal parts or groups.

There are 12 pieces of candy and 3 friends want to share them. How do they divide the candy?

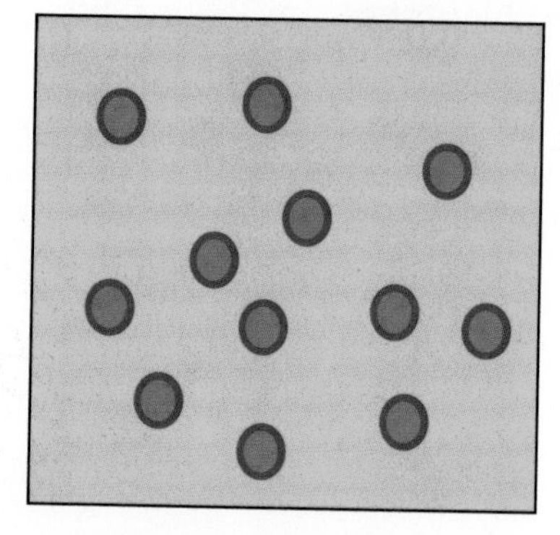

12 pieces of candy

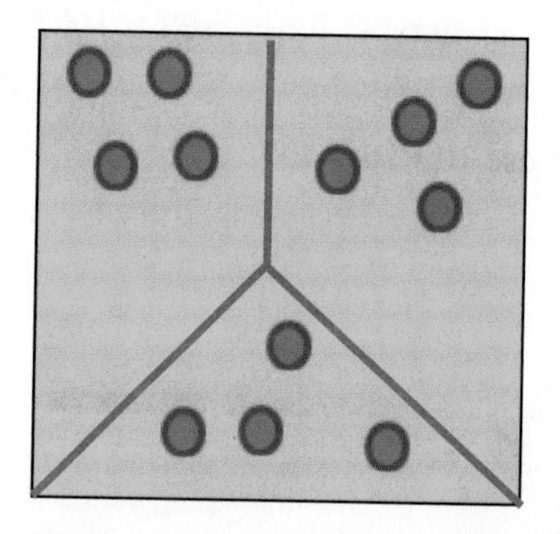

12 pieces of candy divided by 3
Each person would get 4 pieces of candy.

Now, let us try dividing 12 pieces of candy among zero people, how much does each friend get?

Does that question even make sense? NO!!

You can't share among zero friends, which means you can't divide by zero.

Here's another way to think of it. When you first learned to divide, you also probably thought a lot about multiplication.

12 divided by 3 is 4 since 4 times 3 is 12.

18 divided by 9 is 2 since 2 times 9 is 18.

What happens when we try this with 0?

5 divided by 0 is ? since ? times 0 is 5. ➡ But there is no number we can multiply by 0 and get 5.

Any number *times* 0 IS 0. Once again, dividing by zero doesn't make sense!

Division by zero is an operation for which we cannot find an answer, so it is not allowed. Or in mathematical words, **division by zero is UNDEFINED.**

Dividing by Zero	Zero Divided by a Number
$\frac{7}{0} \rightarrow$ **undefined**	$\frac{0}{7} = 0$

EXAMPLE 2

Simplify.

a. $\frac{3+6\cdot 2}{0} \Rightarrow \frac{3+12}{0} \Rightarrow \frac{15}{0} \Rightarrow$ **undefined** If zero is **under** the line, it is **undefined**.

b. $\frac{0}{4^2} \Rightarrow \frac{0}{16} = \mathbf{0}$ If zero is in the numerator, the entire fraction is zero.

1.03 Exponents, Square Roots, and Order of Operations 1.03

Order of Operations:

Two students are in a math class and have been given this problem to evaluate: $3 + 4 \cdot 2$. Here are their answers. Below is their work. Do you notice a difference?

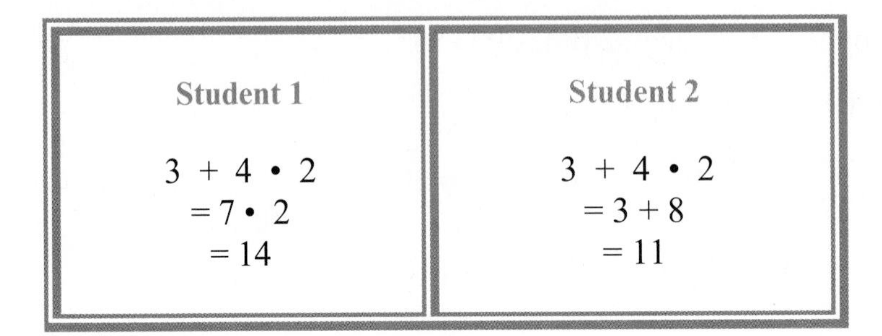

Each student worked the problem differently which resulted in two different answers. This can't be in math!!! **When performing arithmetic operations, there can be only one correct answer.**

We need a set of rules in order to avoid this confusion.

Mathematicians have agreed on a standard order of operations so that arithmetic problems can have only one correct answer.

Order of Operations

Parentheses or other grouping symbols.

Exponents or Square Roots

Multiply /Divide from LEFT to RIGHT

Add / Subtract from LEFT to RIGHT

EXAMPLE 3

Simplify. $16 \div 4 \cdot 2$

$4 \cdot 2$

8

Order of operations states to multiply or divide from LEFT to RIGHT. In this problem, division comes first, so we will divide before we multiply!!!

EXAMPLE 4

Simplify. $6 \cdot \sqrt{9} + 3 \cdot \sqrt{4}$

$6 \cdot \sqrt{9} + 3 \cdot \sqrt{4}$ → Evaluate square roots.

$6 \cdot 3 + 3 \cdot 2$ → Multiply.

$18 + 6$ → Add.

24

EXAMPLE 5

Simplify. $(10 - 7)^4 + 2 \cdot 3^2$

$(10 - 7)^4 + 2 \cdot 3^2$ → Parentheses first. $(10 - 7) = 3$

$(3)^4 + 2 \cdot 3^2$ → Exponents are next. $(3)^4 = 81$ *and* $(3)^2 = 9$

$81 + 2 \cdot 9$ → Multiply $2 \cdot 9 = 18$

$81 + 18$ → Add $81 + 18 = 99$

99

EXAMPLE 6

Simplify. $\dfrac{25 + 8 \cdot 2 - 3^3}{2\,(3 - 2)}$ ← The fraction bar is like a grouping symbol. We will evaluate above and below the fraction bar separately.

$\dfrac{25 + 8 \cdot 2 - 3^3}{2\,(3 - 2)}$ → Exponents above the fraction bar. $3^3 = 27$ / Parentheses below the fraction bar. $(3 - 2) = 1$

$\dfrac{25 + 8 \cdot 2 - 27}{2\,(1)}$ → Multiply above the fraction bar. $8 \cdot 2 = 16$ / Multiply below the bar. $2\,(1) = 2$

$\dfrac{25 + 16 - 27}{2}$ → Add/Subtract from Left to Right. Add comes first, so we will add first.

$\dfrac{41 - 27}{2}$ → Subtract.

$\dfrac{14}{2}$ → We have simplified above and below the fraction bar separately. After this, remember the fraction bar means to divide. So, we will divide as our last step.

7

1.03 Exponents, Square Roots, and Order of Operations 1.03

YOUR TURN

Simplify each expression by using the order of operations.

3. $48 \div 4 \bullet 2 - (7 - 5)$

4. $2^3(10 - 8) + 3 \bullet 5^2$

5. $\dfrac{7(9-6)+3}{3^3-3}$

6. $\dfrac{\sqrt{4}+4^2}{5(20-16)-3^2-5}$

1.03 Square Roots and the Pythagorean Theorem 1.03

The Pythagorean Theorem is used to find **missing SIDE lengths** of right triangles (which contain a 90 degree angle).

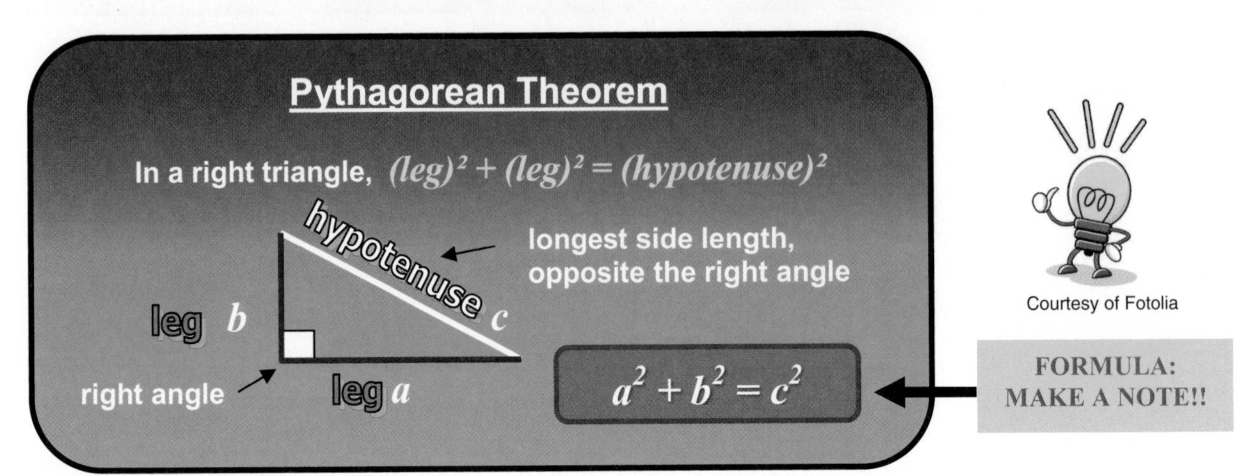

Courtesy of Fotolia

FORMULA: MAKE A NOTE!!

EXAMPLE 7

Find the missing side length of the right triangle.

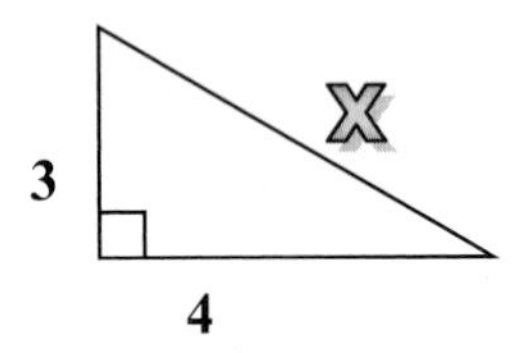

In the right triangle above, we are missing the *hypotenuse.*

$$a^2 + b^2 = c^2$$

$3^2 + 4^2 = x^2$ *(leg)² + (leg)² = (hypotenuse)²*

$9 + 16 = x^2$ Write the equation.

$25 = x^2$ Solve.

$\sqrt{25} = \sqrt{x^2}$ Take the positive square root of both sides.

$\boxed{5 = x}$

EXAMPLE 8

The opening in my wall unit for a television is 36" by 48". What is the largest size television that will fit in that spot? **Remember that when purchasing a television the size represents the diagonal length of the screen.**

$(48)^2 + (36)^2 = c^2$ Equation

$2304 + 1296 = c^2$ Solve.

$3600 = c^2$ Take the positive root.

$60 = c$

A 60" television is the largest that would fit. (Answer.)

Courtesy of Fotolia

1.03 Square Roots and the Pythagorean Theorem 1.03

EXAMPLE 9

Insert <, >, or = between the given pair of numbers to make a true statement.

a. $\|-1\| \quad \|15\|$	b. $\|-32\| \quad -(32)$	c. $-\|-2\| \quad -(-2)$
$\|-1\| \quad \|15\|$	$\|-32\| \quad -(32)$	$-\|-2\| \quad -(-2)$
$1 \quad 15$	$32 \quad 32$	$-2 \quad 2$
$1 < 15$	$32 = 32$	$-2 < 2$

YOUR TURN

Write an equation and answer each question.

7. Sketch the right triangle and find the length of the side not given.

leg = 6, leg = 8

Formula: ______________________

Equation:
Solve.

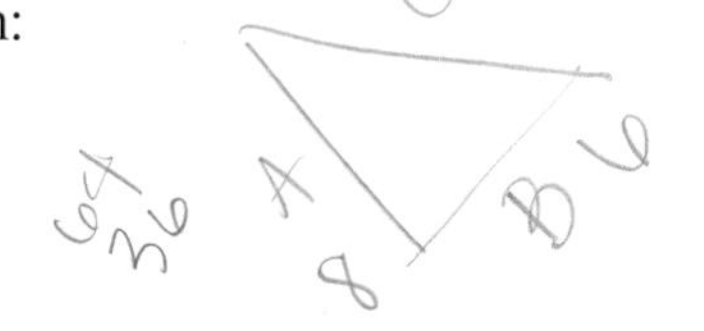

8. A ladder is leaning against a wall. The distance from the base of the ladder to the wall is 12 m and the ladder reaches 9 m up the wall. How long is the ladder? Draw a picture. Round your answer to the nearest thousandth if necessary.

Formula: ______________________

Equation:
Solve.

9. The opening in my wall unit for a new television is 27" by 36". What is the largest size television I can put in that spot? Remember that when purchasing a television, the size represents the diagonal length of the screen. Draw a picture.

Formula: ______________________

Equation:
Solve.

YOUR TURN Answers to Section 1.03.

1. 1296	**6. 3**
2. 48	**7. 10**
3. 22	**8. 15 m**
4. 83	**9. 45 in**
5. 1	

Complete MML HW Section 1.03 online and show your work in your notebook.

Equality and Inequality Symbols

$=$	is equal to	$5 = 5$
$\neq$	is not equal to	$5 \neq 6$
$<$	**is** less than	$5 < 6$
$\leq$	**is** less than or equal to	$9 \leq 9$ (must be true for $9 < 9$ OR for $9 = 9$)
$>$	**is** greater than	$7 > 6$
$\geq$	**is** greater than or equal to	$6 \geq 6$ (must be true for $6 > 6$ OR for $6 = 6$

Number Line and Order

Remember: On a number line the farther to the right a number is, the greater that number is. The inequality sign points to the lesser number.

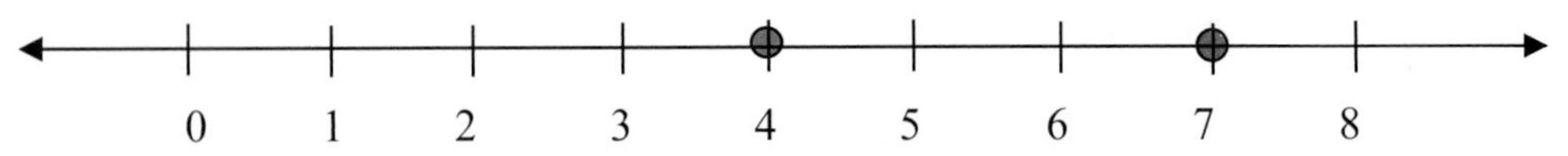

0 1 2 3 4 5 6 7 8

7 is further right than 4, therefore, $7 > 4$ (seven is greater than four) or $4 < 7$ (four is less than seven)

EXAMPLE 1

True or false: $8 < 10$

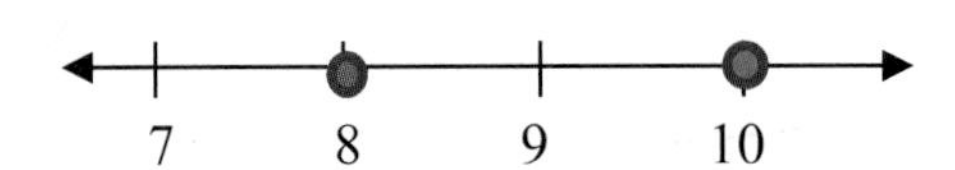

10 is further to the right of 8 on the number line, therefore 10 is the greater number; thus **8 is less than** 10. The statement is **true**.

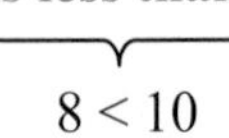

$8 < 10$

EXAMPLE 2

True or False: $4 \leq 4$

$4 \leq 4$ is true if $4 < 4$ OR if $4 = 4$.
Is $4 < 4$ true? Of course not!
Is $4 = 4$ true? Of course!
Since one of the statements is true then $4 \leq 4$ is true.

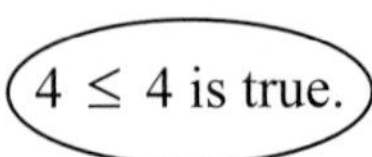

EXAMPLE 3

Use < or > to make true statements.

a. 24 __>__ 12 Since 24 is to the right of 12 on the number line **24 > 12**.
b. 3 __<__ 5 Since 3 is to the left of 5 on the number line **3 < 5**.

Notice that the inequality sign always points to the lesser number.

1.04 Introduction to Integers

> Check this out! $2 < 5$ is the same as $5 > 2$.
>
> If you need to change the order then you must change the direction of the inequality sign. This will come in handy when graphing inequalities.

YOUR TURN

Tell whether these statements are true or false. Use the number line, if necessary.

1. $5 > 3$ X
2. $19 < 5$ F
3. $2 \leq 6$ X
4. $2 \geq 2$ X

Use $<$ or $>$ to make a true statement.

5. $3 \underline{<} 45$
6. $12 \underline{>} 4$

Up to this point, we have only looked at numbers greater than or equal to 0. However, some situations exist that cannot be represented by a number greater than 0.

THINK ABOUT IT!

- How would we express a temperature 5 degrees below zero?
- What number would represent an overdrawn banking account?

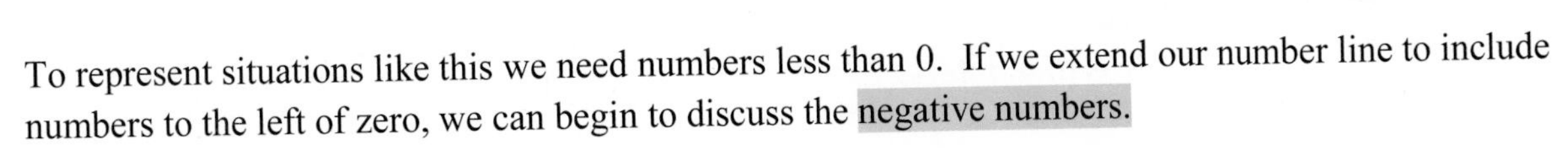

To represent situations like this we need numbers less than 0. If we extend our number line to include numbers to the left of zero, we can begin to discuss the negative numbers.

1.04 Introduction to Integers 1.04

Let's look at the number line again this time including negative numbers.

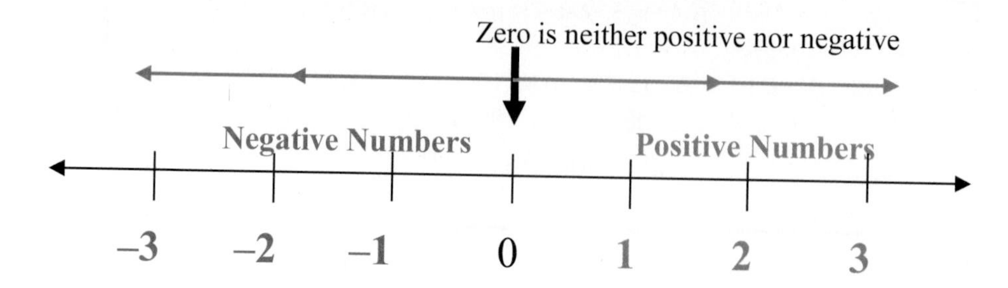

The group of numbers that include negative numbers, zero, and the positive numbers is called the integers.

Integers: { . . . , -5, -4, -3, -2, -1, 0, 1, 2, 3, 4, 5, ...}

EXAMPLE 4

We can use integers to represent real world situations.

a. Three feet below sea level can be represented by – 3.

b. **Digging a hole 6 feet below ground level** can be represented by – 6.

c. **Withdrawal of $59** from a checking account can be represented by – 59.

d. **8 steps backwards** can be represented by – 8.

e. **A checking account balance of 125** can be represented by +125.

EXAMPLE 5

Represent each quantity as an integer on a number line.

a. **10 degress below zero.**

b. **A gain of 5 points.**

c. **A loss of 4 dolars.**

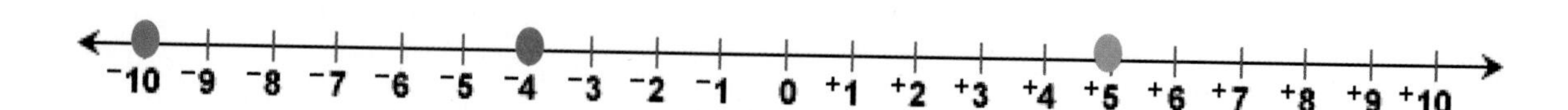

YOUR TURN

7. Represent each quantity with an integer.

a. A loss of 3 dollars. -3 ________

b. Temperature fell 8 degrees. -8 ________

c. A deposit of 4 dollars. +4 ________

d. Checking account balance overdrawn by 10 dollars. -10 ________

8. Represent each quantity above as an integer on a number line.

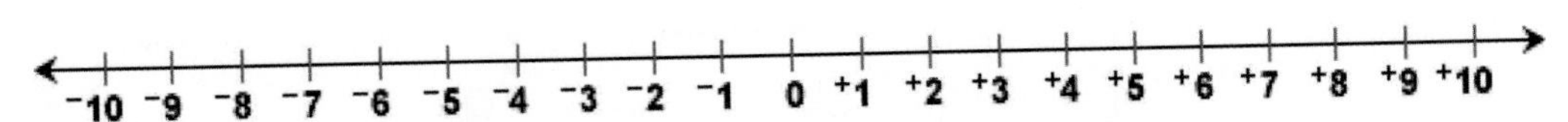

Comparing integers

We have already looked at how to compare two positive numbers and learned that a number to the right on a number line is larger than a number to the left.

Remember: The farther to the right the number is the greater the number. In other words, integers get smaller in value as you move to the left on the number line, and larger as you move to the right on the number line.

EXAMPLE 6

A meteorologist recorded temperatures in four cities around the world. List these cities in order from least to greatest according to their temperatures.

Recorded Temperatures	
City	Temp
Albany	$^{+}5°$
Anchorage	-6°
Buffalo	-7°
Reno	$^{+}10°$

Since the temperatures are given as integers, a number line will help us solve this problem. The farther to the right the number is the greater the number is.

Let's graph the four integers on the number line and then compare.

1.04 Introduction to Integers

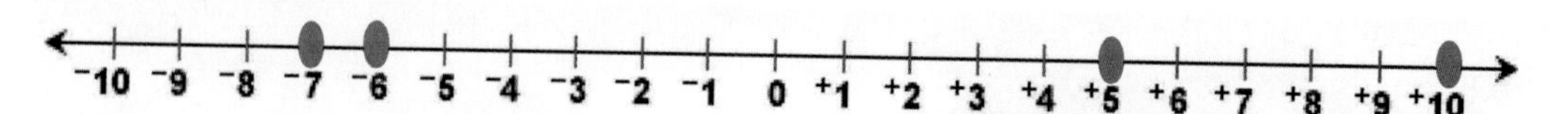

Reading the graph from left to right we can determine the order of the integers.
-7 is the smallest, then – 6, then +5 and the largest is + 10.

Recorded Temperatures From Least to Greatest	
City	Temp
Buffalo	- 7°
Anchorage	- 6°
Albany	+5°
Reno	+10°

YOUR TURN

Tell whether these statements are true or false.

9. -5 < -3 ___f___
10. 0 < -5 ___t___
11. -2 ≤ -6 ___t___
12 -2 ≥ -2 ___t___

Use < or > to make a true statement.

13. – 3 ___<___ – 45
14. – 12 ___>___ 4

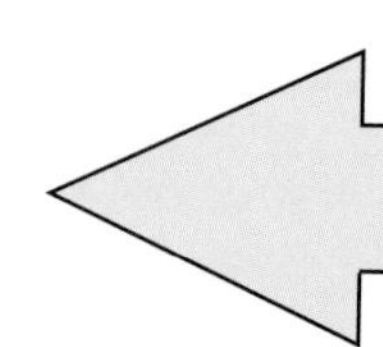

ABSOLUTE VALUE AND OPPOSITES

Problem: Jack and Jill were at Sam's house. Jill rode her bicycle 3 miles west of Sam's house, and Jack rode his bicycle 3 miles east of Sam's house. Who traveled a greater distance from Sam's house, Jill or Jack?

Solution: Jill and Jack both traveled the same distance from Sam's house since each traveled 3 miles (in opposite directions).

The problem above can be solved using integers.
Traveling 3 miles west can be represented by –3
Traveling 3 miles east can be represented by +3.
Sam's house can be represented by the integer 0.

Two integers that are the ***same distance from zero*** in opposite directions are called **opposites**.

1.04 Introduction to Integers 1.04

The integers +3 and –3 are opposites since they are each 3 units from zero.

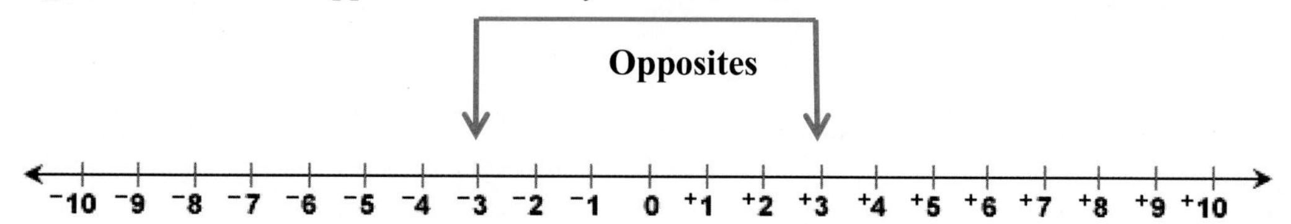

The absolute value of an integer is its distance from zero on the number line. The absolute value of +3 is 3, and the absolute value of – 3 is 3. Opposite integers have the same absolute value. Mathematicians use the symbol | | for absolute value. Let's look at some examples.

EXAMPLE 7

Find the absolute value of + 6, - 6, + 2, - 4 . Use the number line given.

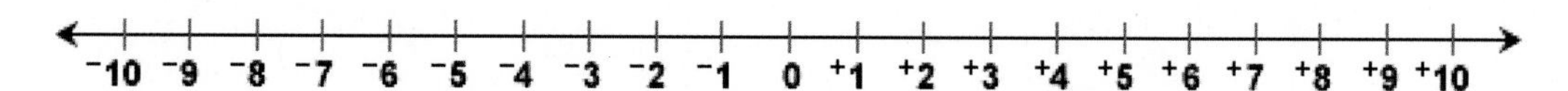

Solution $|+6| = 6$ **since + 6 is 6 units from 0.**

$|-6| = 6$ **since - 6 is 6 units from 0.**

$|+2| = 2$ **since + 2 is 2 units from 0.**

$|-4| = 4$ **since - 4 is 4 units from 0.**

Remember that when we find the absolute value of an integer, we are finding its *distance from 0* on the number line. Opposite integers have the same absolute value since they are both the same distance from 0.

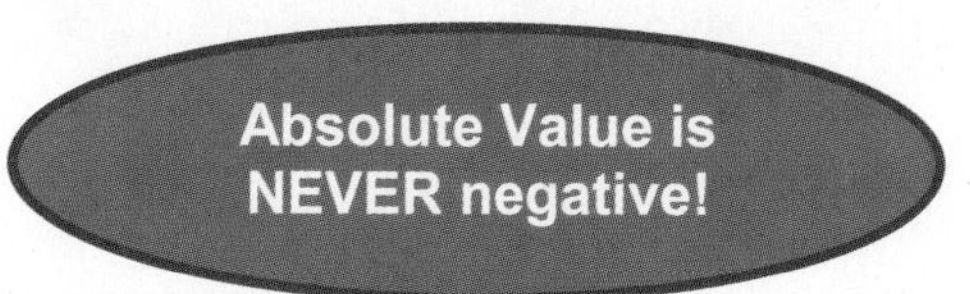

In understanding absolute value, we used the concept of opposites. Above sea level is the opposite of below sea level. Forward is the opposite of backward. An increase is the opposite of a decrease. As we see above, a positive is the opposite of a negative.

The symbol for the opposite of a number is " – ". The opposite of 3 is written – (3).

The opposite of 4 is – 4 or in symbols $-(4) = -4$

The opposite of – 4 is 4 or in symbols $-(-4) = 4$

To find the opposite of a number just change its sign.

1.04 Introduction to Integers 1.04

EXAMPLE 8

Evaluate.

a. $-(-8)$ b. $-|-4|$ c. $-|9|$

Solutions:

a. $-(-8)$
The opposite of – 8 is 8.

b. $-|-4| = -4$
The opposite of the absolute value of – 4.

- First find the absolute value of – 4 which is 4.
- Now find the opposite of 4 which is – 4.

The answer is – 4.

c. $-|9| = -9$ The opposite of the absolute value of 9.

- The absolute value of 9 is 9.

The opposite of 9 is – 9.

YOUR TURN

Simplify.

15. $|-12|$ 16. $-(-12)$ 17. $-|8|$ 18. $-|-34|$

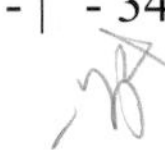

YOUR TURN Answers to Section 1.04.

1. True 2. False 3. True 4. True 5. < 6. >

7. a. – 3 b. – 8 c. + 4 d. - 10 8. Number line below.

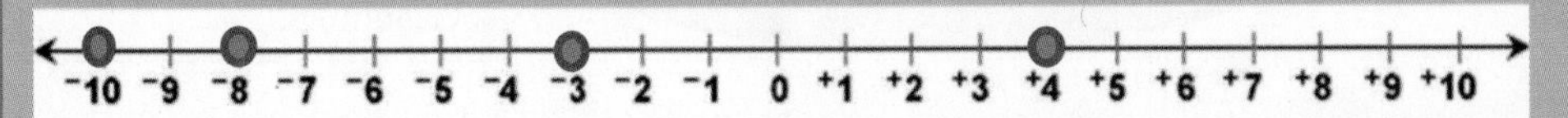

9. True 10 False 11. False 12. True 13. > 14. <

15. 12 16. 12 17. – 8 18. - 34

Complete MML HW Section 1.04 online and show your work in your notebook.

1.05 Adding Integers 1.05

ADDING NUMBERS THAT HAVE THE SAME SIGNS

Add (the absolute values) and keep the **common sign.**

ADDING NUMBERS THAT HAVE DIFFERENT SIGNS

Subtract (the absolute values) and use the **sign of the "greater" number**. *(It is not really the greater number but actually the number with the greater absolute value).*

EXAMPLE 1

Simplify.

a. $-12+8$
-4

Signs are different.
Subtract and use the sign of the greater number.

b. $-2+(-3)$
-5

Signs are the same.
Add and use the common sign.

c. $5+(-2)$
3

Signs are different.
Subtract and use the sign of the greater number.

d. $-2+(-6)+7$
$-8+7$
-1

Signs are the same.
Add and use the common sign.
Then signs are different.
Subtract and use the sign of the greater number.

YOUR TURN

Simplify.

1. $-3+(-13)$
-16

2. $5+(-8)$
-3

3. Find the sum of -9 and 12.

4. $16+(-16)$
0

5. $-5+(-2)+8+(-3+9)$

Additive Inverse

The **opposite** of a number is also called the additive inverse of the number because when you add a number and its additive inverse you get 0.

$-3+3=0$ $5+-5=0$

EXAMPLE 2

Let a = 5 and b = - 2. Are the following statements True or False?

$a > -(a)$

This question is read,
Is "a" greater than the opposite of "a"?

Since a = 5 , the question becomes
Is 5 greater than the opposite of 5?

The opposite of 5 is – 5.

Is 5 > - 5 ?

TRUE

$-(b) + b = 0$

This question is read,
Is the opposite of "b" added to "b" equal to 0?

Since b = - 2 , the question becomes
Is the opposite of – 2 added to - 2 = 0?

The opposite of - 2 is 2.

Is 2 + - 2 = 0 ?

TRUE

YOUR TURN

Let a = 3 and b = - 7. Are the following statements True or False? (Remember to use parentheses when you substitute a negative number.)

6. $-b < b$

-7 < 7
f

7. $a + (-a) = 0$

3 + (-3)
0
t

1.05 Adding Integers 1.05

EVALUATING EXPRESSIONS

Evaluating expressions is very much like order of operations. The expressions we will work with will contain variables. Variables are representations for unknown numbers. When evaluating expressions you are given an expression with variables and values for the variables. Substitute, or replace, the value for the variable for that variable by placing parentheses around the number. Then follow the order of operations.

Make sure to put parentheses around the number you are substituting.

EXAMPLE 3

Evaluate when a. $x = 3$ and b. $y = -5$. Remember to add parentheses when substituting numbers.

a. $2x + y$ Substitute.

$2(3) + (-5)$ Multiply.

$6 + (-5)$ Now add.

1

b. $|x| + |y|$ Substitute.

$|3| + |-5|$ Find absolute value.

$3 + 5$ Add.

8

YOUR TURN

Evaluate. Remember to add parentheses when substituting numbers.

8. $4x + y$ when $x = 3$ and $y = -2$

9. $|x| + |y|$ when $x = 7$, $y = -5$

1.05 Adding Integers 1.05

Applications of Integers

EXAMPLE 4

a. In Whistler, B.C. the record extreme high temperature is 102° F. If you decrease this temperature by 111°, the result is the extreme low temperature.
Find the new temperature.

extreme low temperature is **extreme high temperature** + **decrease of 111°**

↓ ↓ ↓

$x \quad = \quad 102 \quad + \quad -111$

$x = 102 + (-111)$

$x = -9°$

The extreme low temperature is 9° below zero.

b. My checking account was overdrawn by $13. I then wrote a check for $87. Later I deposited $75. What is the current balance in my account?
(NOTE: Checks are negative integers and Deposits are positive integers)

account overdrawn by $13		I wrote a check for $87		I deposited $75		balance in my account
-13	+	(-87)	+	75	=	x

$-13 + (-87) + 75 = x$

$-100 + 75 = x$

$-25 = x$ Yikes! **The checking account is overdrawn by $25!**

YOUR TURN

10. A negative net income (a loss) results when a company spends more money than it brings in. In 2009, the Smith Company had net income as follows:

Quarter	Income
1st Quarter	$2.5 million
2nd Quarter	$10 million
3rd Quarter	−$14.2 million
4th Quarter	−$4 million

a. Was the net income a gain or loss for 2009?

b. What was the amount of gain or loss?

1.05 Adding Integers 1.05

Properties of Integers

Property	What's Happening	Examples
Commutative Property of Addition $a + b = b + a$ **Commutative Property of Multiplication** $ab = ba$	The order in which you add or multiply numbers can change. This is NOT true for subtraction and division! CHANGE ORDER OF NUMBERS	$3+5=5+3$ $8=8$ $(-7)(5)=(5)(-7)$ $-35=-35$
Associative Property of Addition $a + (b + c) = (a + b) + c$ **Associative Property of Multiplication** $a(bc) = (ab)c$	The parentheses shift to include different numbers while the order stays the same. You cannot change the parentheses if you have subtraction and/or division. MOVE PARENTHESES	$4+(5+6)=(4+5)+6$ $4+11=9+6$ $15=15$ $6(2\cdot8)=(6\cdot2)\cdot8$ $6(16)=(12)\cdot8$ $96=96$
Additive Identity **Multiplicative Identity**	When you add 0 to a number that number stays the same. If you multiply a number by 1 then the number stays the same.	$124 + 0 = 124$ $725 \cdot 1 = 725$ Sort of a no-brainer!

YOUR TURN

11. Complete the number sentence below using the commutative property of multilication.

$7 \cdot 8 =$ 8·7

12. Complete the number sentence below using the associative property of multilication.

$7(5 \cdot 2) =$ ____________

YOUR TURN **Answers to Section 1.05.**

1. -16 **2.** -3 **3.** 3 **4.** 0 **5.** 7 **6.** False **7.** True **8.** 10

9. 12 **10.** a loss of 5.7 million **11.** $7 \cdot 8 = 8 \cdot 7$ **12.** $7(5 \cdot 2) = (7 \cdot 5) \cdot 2$

Complete MML HW Section 1.05 online and show your work in your notebook.

1.06 Subtracting Integers 1.06

SUBTRACTING REAL NUMBERS

Numbers with Same signs → add and use the common sign.
Numbers with Different signs → subtract and use the sign of the "greater" number.

The temperature in Anchorage, Alaska was 8°F in the morning and dropped to –5°F in the evening. What is the **difference** between these temperatures?

Courtesy of Fotolia

Solution: We can solve this problem using integers. Using the number line below, the distance from +8 to 0 is 8, and the distance from 0 to –5 is 5, for a total of 13.

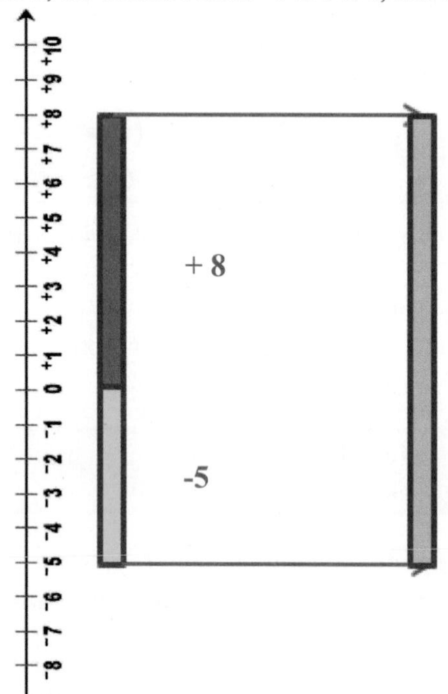

+ 13

$+8-(-5) = +13.$
The difference is 13 degrees.

When solving integer problems with large numbers it is not always practical to use a number line. If you had been asked to find the difference of – 282 and 546 , drawing a number line would be very time consuming and not necessary as we will learn. As we look at subtraction, we can begin to see, that subtracting an integer is exactly the same as adding the opposite of a number. Let's look at an example.

$$6-2=4 \qquad 6+(-2)=4$$

1.06 Subtracting Integers 1.06

EXAMPLE 1

Simplify.

$30 - 58$	• Rewrite the subtraction.	$-34 - 10$	• Rewrite the subtraction.
$30 + (-58)$	• Keep the 30, change subtract to add and 58 to – 58.	$-34 + (-10)$	• Keep the –34, change subtract to add and 10 to – 10.
-28	• Follow the addition rules. Signs are different so subtract and use the sign of the "greater" number.	-44	• Follow the addition rules. Signs are the same so add and use the common sign.
$40 - (-4)$	• Rewrite the subtraction.	$-2 - (-10)$	• Rewrite the subtraction.
$40 + 4$	• Keep the 40, change subtract to add and – 4 to 4.	$-2 + 10$	• Keep the –2 , change subtract to add –10 to 10.
44	• Follow the addition rules. Signs are the same so add and use the common sign.	8	• Follow the addition rules. Signs are different so subtract and use the sign of the "greater" number.

Subtract –8 from – 5

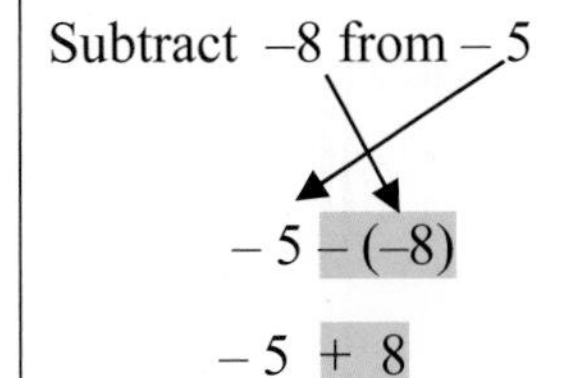

$-5 - (-8)$

$-5 + 8$

3

- Remember that "**subtract from**" means that you will switch the numbers – what you see first goes last and what is last will be first!
- Now rewrite the subtraction. Keep the first number and change subtraction to addition and – 8 to positive 8.
- Now add; since signs are different, subtract and use the sign of the 8.

EXAMPLE 2

Simplify.

$-10 - 5 + (-6) - (-3)$

$-10 + (-5) + (-6) + 3$

$-15 + (-6) + 3$

$-21 + 3$

-18

- When adding and subtracting several signed numbers, rewrite everything as addition. Make sure that you do not rewrite anything that is already addition.
- Notice that the plus – 6 is still plus – 6, but the subtract 5 and subtract – 3 have changed.
- Now add from left to right.

1.06 Subtracting Integers 1.06

Courtesy of Fotolia

Be careful with subtraction! It makes a difference what order you subtract!

Look closely at the following example. Notice the order you write the numbers when you subtract changes your answer.

Subtract 25 from 17.

"subtract from" tells us to switch the numbers—what you see first goes last and what you see last will be first.

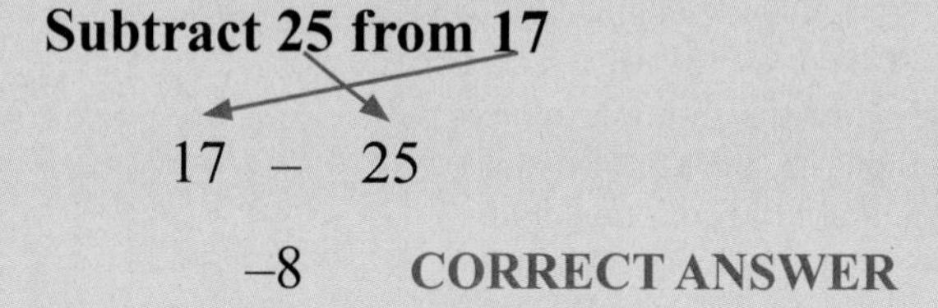

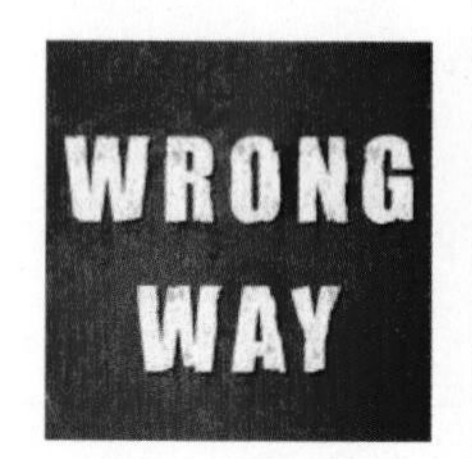

You will **not get the correct answer** if you write the numbers in the order they are given!

Subtract 25 from 17

25 – 17

8 **WRONG ANSWER !!!**

" Less than" is exactly like "subtract from". You must switch the order of the numbers.

12 less than 10

10 – 12

–2

1.06 Subtracting Integers 1.06

Let's look at one more word that indicates subtraction.

Courtesy of Fotolia

Find the difference of – 16 and 21.

– 16 – 21

– 37

(Notice the order doesn't change when the word "difference" is used!)

Courtesy of Fotolia

Make a note about subtraction!

Indicator Word	Example	Solution
Subtract from (switch order)	Subtract –2 from –20	–20 – (–2) = –18
Less than (switch order)	6 less than –5	–5 – 6 = –11
Difference (keep order)	difference of –16 and –21	–16 – 21 = –37

YOUR TURN

Simplify.

1. $11 - 15$
2. $11 - (-15)$
3. $-11 - (-15)$
4. -1 subtracted from 5
5. $12 - 19 + 5 - (-32)$
6. $10 + [-8 + (-10 - 2)]$

Evaluating Expressions Again!

EXAMPLE 3

Find the values of the expression when $x = 2$ and $y = -4$.

$x - y$

$(2) - (-4)$

$2 + 4$

6

- Use parentheses around what you are substituting, especially when substituting negative numbers!
- Change the subtraction to add the opposite; so subtract -4 becomes add 4.

Applications of Integers, Again!

EXAMPLE 4

A football team **gains 8 yards** on the first play of their drive, then **loses 4 yards** on the second play. They **lose 6 more yards** on the third play. What are the total yards lost or gained?

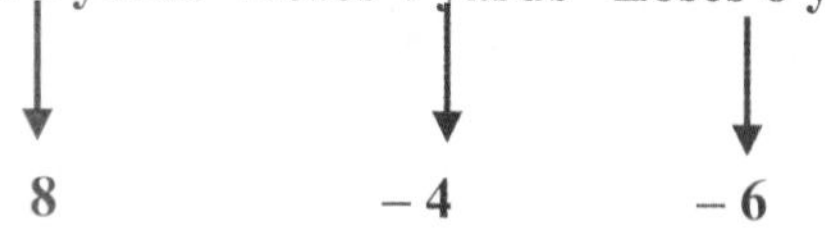

$8 - 4 - 6$ Rewrite each subtraction as add the opposite.

$8 + (-4) + (-6)$ Now add the first two numbers.
(Different signs: subtract and use the sign of the "greater" number.)

$4 + (-6)$ Now subtract 4 and -6. (Different signs: subtract, keep sign of "greater" number.)

-2 Since the 2 is negative and negative means a loss, then *the team lost 2 yards during the 3 downs*.

1.06 Subtracting Integers 1.06

YOUR TURN

Write an equation and solve.

7. A scuba diver makes a dive of 120 feet to view a sunken ship. He then dives another 40 feet. Represent the depth of the diver.

Eq.

Solve.

8. Alex has \$225 in his checking account. He writes a check for \$117 , makes a deposit of \$45, and then writes another check for \$167. Find the balance in his account. (Write the number as an integer.)

Eq.

Solve.

9. The temperature on a March afternoon is -4° Celsius at 1 pm. If the temperature drops 3 degrees by 4 pm, rises 4 degrees between 4 pm and 6 pm, and then drops 7 degrees between 6 pm and 9 pm. What is the temperature at 9 pm?

Eq.

Solve.

YOUR TURN Answers to Section 1.06.

1. - 4

2. 26

3. 4

4. 6

5. 30

6. - 10

7. 160 feet below sea level

8. – 14 dollars

9. – 10 ° C

Complete MML HW Section 1.06 online and show your work in your notebook.

1.07 Multiplying and Dividing Integers 1.07

MULTIPLYING AND DIVIDING TWO NUMBERS THAT HAVE THE

SAME SIGNS

The product or quotient of two positive or of two negative numbers is **POSITIVE.**

MULTIPLYING AND DIVIDING TWO NUMBERS THAT HAVE

DIFFERENT SIGNS

The product or quotient of one positive number and one negative number is **NEGATIVE.**

Multiplying and Dividing Integers is similar to multiplying and dividing whole numbers. The only difference is we will need to know whether the answer will be positive or negative. Let's look at an example.

EXAMPLE 1

Alice owes $6 to each of 4 friends. How much money does she owe? If Alice owes 4 friends $6 each then we realize that she owes a total of $24. Now let's think about this problem using integers.

Owing $6 can be represented by **- 6** .

If she owes 4 friends $6 then the problem now would be (- 6) (4) .

She owes a total of $24. So (- 6) (4) = **- 24**

In this example, we see that a negative multiplied by a positive gives us a negative answer. In order to solve integer problems it would be helpful to know the rules for determining whether a product or quotient will be positive or negative.

1.07 Multiplying and Dividing Integers 1.07

A little trick for you to try!

Read in any direction (across, up & down, or diagonal) to determine the sign of your answer when multiplying & dividing.

+	–	–
–	+	–
–	–	+

For example:

- If I read down the left most column, I can see that + • – = –
- If I read down the right most column, I see that – • – = +
- If I read across the middle row, I see that – • + = –

It works the same way for division, as well! Pretty cool, huh?

Courtesy of Fotolia

OR REMEMBER!

If signs are the SAME, the result is always POSITIVE. (+)(+)=(+) or (-)(-)=(+)

If signs are the DIFFERENT, the result is always NEGATIVE. (+)(-)=(-) or (-)(+)=(-)

EXAMPLE 2

Multiply or divide.

$(-8) \cdot (4) = -32$	Negative times positive is negative.
$(8) \cdot (-4) = -32$	Positive times negative is negative.
$(-8) \cdot (-4) = 32$	Negative times negative is positive.
$(8) \cdot (4) = 32$	Positive times positive is positive.
$(16) \div -2 = -8$	Positive divided by negative is negative.

Be very careful when 0 is involved in division.

Dividing with

Courtesy of Fotolia

$\frac{0}{-23} = 0$ Zero divided by any number is 0.

$\frac{-9}{0}$ ⇒ Undefined ⇒ Division by zero is undefined. So if 0 is **under** the fraction bar then the fraction is **undefined**.

EXPONENTS OF NEGATIVE INTEGERS

When we multiply negative integers together, we must utilize parentheses !!!

$(-2)(-2)(-2)(-2)=(-2)^4$

Courtesy of Fotolia

BEWARE!! -2^4 is NOT the same as $(-2)^4$.

The missing parentheses mean that -2^4 will multiply **four 2's** together first (by order of operations), and then **take the negative of that answer.**

$-2^4 = -(2)(2)(2)(2) = -(16) = -16$

Let's look at one more example .

$(-5)^2=(-5)(-5)=25$ *compared to* $-5^2=-(5)(5)=-(25)=-25$

The negative number must be enclosed by parentheses to have the exponent apply to the negative term.

Calculator Tip

If you are planning on using a calculator, then you must learn whether your calculator puts the number raised to a power in parentheses automatically or not!

Evaluating Expressions

EXAMPLE 3

Evaluate the following expressions for $x = -12$ and $y = 6$. Remember to add parentheses if needed.

a. $2xy$

$2(-12)(6)$	Substitute in values.
$-24(6)$	Multiply the first two numbers.
-144	Multiply again!

b. $\dfrac{x}{y}$

$\dfrac{(-12)}{6}$	Substitute in values.
-2	Divide.

YOUR TURN

Evaluate the following expressions for $x = -8$ and $y = -2$. Remember to add parentheses.

1. $2xy$

2. $\frac{x}{y}$

Applications of Integers, Again!

EXAMPLE 4

The temperature of a chemical compound was – 5° at 4:20 pm. During a chemical reaction the temperature increased by 2° C per minute until 4:52. What was the temperature at 4:24?

The reaction began at 4:20. The questions ask for the temperature at 4:24. So, we will need to calculate the increase for 4 minutes. (4:24 – 4:20 = 4 minutes).

The temperature increased 2° C per minute. So the total increase in temperature would be

$4 (2) = 8$ degrees.

The compound began at – 5° and we now know it increased by 8°. The present temperature of the compound would be – 5° + 8° = **3°**.

A chemist doing this experiment would probably want to record the temperature at several different points in the experiment. We could organize the information in a chart to help keep track of the current temperature at a given time.

Current Time	Minutes past 4:20	Change in Temp (Multiply minutes by 2.)	Current Temp (Add Change in Temp to beginning Temp of – 5 °)
4:23	3	6	1°
4:35	15	30	25°
4:40	20	40	35°
4:44	24	48	43°

1.07 Multiplying and Dividing Integers 1.07

YOUR TURN

Complete # 3- 6 using the information from the previous example.

Current Time	Minutes past 4:20	Change in Temp (Multiply minutes by 2.)	Current Temp (Add Change in Temp to beginning Temp of – 5 °)
3. 4:46			
4. 4:48			
5. 4:50			
6. 4:52			

Translating Algebraic Expressions

EXAMPLE 5

Translate each phrase into an expression. Use x to represent "a number".

a. The product of 5 and a number → $5x$ → product indicates multiplication

b. The difference of a number and 9 → $x - 9$ → difference indicates subtraction

c. Divide a number by – 3 → $\frac{x}{-3}$ → divide indicates a quotient

YOUR TURN

Translate.

7. The product of a number and – 2

8. – 29 increased by a number

9. Subtract a number from 5

10. The quotient of – 8 and a number

1.07 Multiplying and Dividing Integers 1.07

EXAMPLE 6

One year, a total of 98 million music cassettes were shipped. Four years, later, this number had dropped to 30 million music cassettes.

a. Find the change in the number of music cassettes shipped over the four years.

b. Find the average change per year in the number of cassettes shipped over this period.

Courtesy of Fotolia

Solution:

a. Find the change in the number of music cassettes shipped over the four years.

Since the number of cassettes dropped we know that that answer will be represented by a negative number. The difference between 98 and 30 is 68. ($98 - 30 = 68$).

The change in the number of music cassettes would be – 68 over the four year period.

b. Find the average change per year in the number of cassettes shipped over this period.

The cassettes were shipped over a four year period. So, to find the average change per year we will divide the above answer by four.

–68 divided by 4 is – 17.

The average change per year in the number of cassettes shipped is – 17.

YOUR TURN **Answers to Section 1.07.**

1. 32 **2.** 4 **3.** 26, 52, 47 **4.** 28, 56, 51 **5.** 30, 60, 55 **6.** 32, 64, 59 **7.** $-2x$ **8.** $-29+x$ **9.** $5-x$ **10.** $\frac{-8}{x}$

Complete MML HW Section 1.07 online and show your work in your notebook.

1.08 Order of Operations 1.08

Now we can look at integer problems that involve add, subtract, multiply, and divide. Let's summarize what we know about these operations.

Add (the absolute values)
Keep the **common sign.**

$5 + 6 = 11$

$-5 + -6 = -11$

ADDING/ DIFFERENT SIGNS

Subtract (the absolute values)
Take the **sign of the "greater" number**.

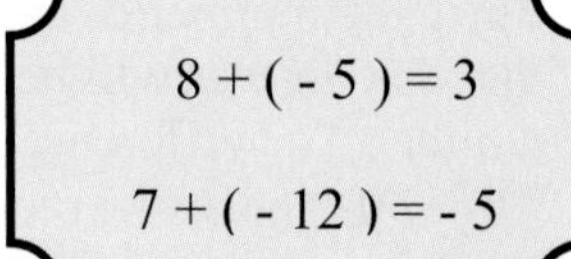

SUBTRACTING

To **subtract** an integer **add it's opposite.**

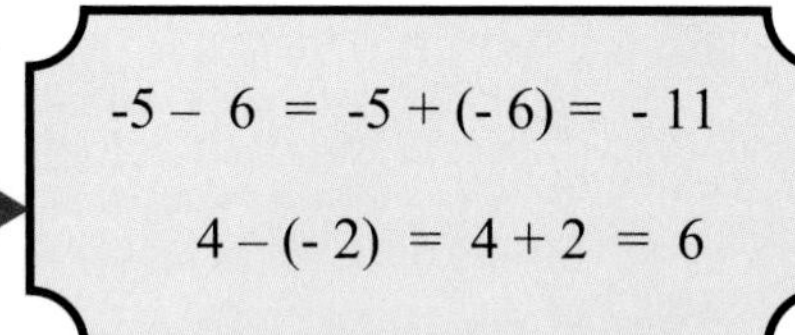

MULTIPLYING AND DIVIDING
SAME SIGNS

Product or quotient is always POSITIVE.

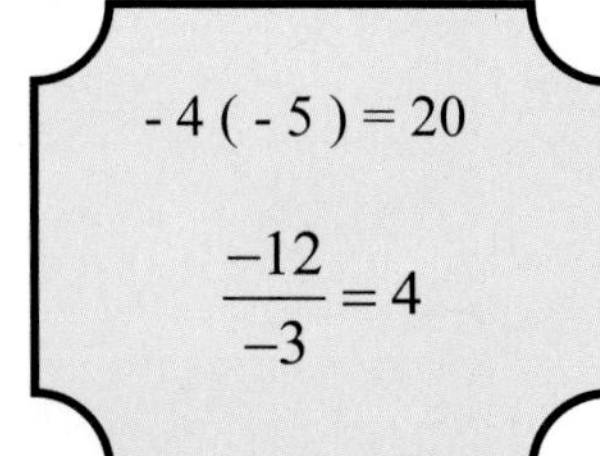

MULTIPLYING AND DIVIDING
DIFFERENT SIGNS

Product or quotient is NEGATIVE.

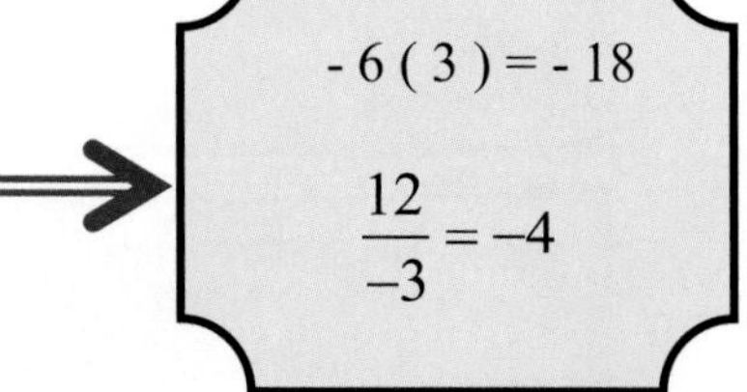

Order of Operations with Integers

Parentheses or other grouping symbols.

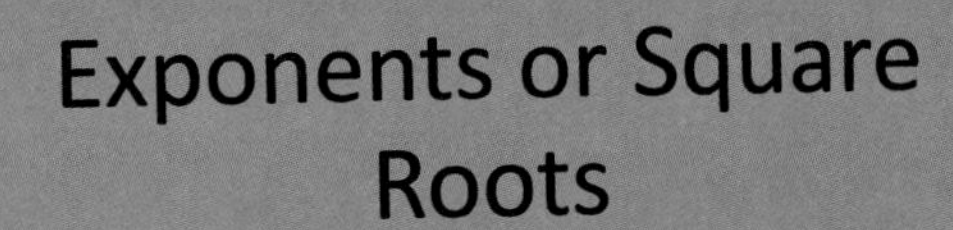

Multiply /Divide from LEFT to RIGHT

Add / Subtract from LEFT to RIGHT

EXAMPLE 1

Simplify.

a. $-30 \div 6 \cdot 3$ — Multiply or divide first?? Division comes first so you do it first!

$-5 \cdot 3$ — Now multiply.

-15

b. $8 + 2(4 - 12)$ — This time we have parentheses. So subtract 4 and 12 in the ( ).

$8 + 2 \cdot -8$ — When a number is written directly beside a "(" that means to multiply. Now, should we add or multiply next? → Multiply!

$8 + -16$ — Now add.

-8

EXAMPLE 2

Simplify using the order of operations.

$(-2)^2+3(-3-2)-|4-6|$ Follow the order of operations; parentheses first!

$(-2)^2+3(-5)-|4-6|$ Absolute value is like a grouping symbol, so evaluate it next.

$(-2)^2+3(-5)-|-2|$ Now take care of the exponent.

$4+3(-5)-|-2|$ Ok, look at what is left. We have addition, multiplication and subtraction. What comes first?

$4+(-15)-|-2|$ Change the absolute value of -2 to 2.

$4+(-15)-(2)$ Now add or subtract from left to right.

$-11-(2)$

$-11+(-2)$

-13

EXAMPLE 3

Simplify using the order of operations.

$\frac{-3^2+|9-6|-4}{10-5}$ Absolute value bars serve as a grouping symbol. So, perform the operation inside the absolute value.

$\frac{-3^2+|3|-4}{10-5}$ Now we have an exponent. Raise 3 to the second power. Remember only square the 3 and not the negative sign since there are no ( ). Then take the absolute value of the second 3. You can also subtract 10 and 5 since that is in the denominator.

$\frac{-9+3-4}{5}$ Now what should we do? Now add from left to right.

$\frac{-10}{5}$ Now divide.

-2

EXAMPLE 4

Evaluate $10-x^2$ if $x = -5$.

$10-x^2$ Substitute – 5 in for x. Put (-5) in parentheses.

$10-(-5)^2$ Evaluate exponent first. $(-5)^2 = 25$

$10-(25)$ Subtract.

-15

1.08 Order of Operations 1.08

YOUR TURN

Simplify.

1. $\dfrac{2(-7)^2-10}{\lvert -5\rvert+3}$	2. $3^2-2[(5-7)-(-7-4)]$
3. $\dfrac{-3^4-5^2}{\lvert -7+2\rvert-4}$	4. Evaluate x^2-y if $x=-3$ and $y=-7$.

YOUR TURN Answers to Section 1.08.

1. 11 **2. – 9** **3. – 106** **4. 16**

Complete MML HW Section 1.08 online and show your work in your notebook.

1.09 Introduction to Geometry 1.09

This section corresponds to the eBook Section 1.01.

Identify Lines, Line Segments, Rays, and Angles

Vocabulary	*Definition*	*Diagram*	*Name/Notation*
Point	A location.	A	A point is represented by a dot and labeled with a capital letter.
Line	A set of points that extend in two opposite directions. **A line uses an arrow on the ends.*	A B	To name a line use any two points on the line. Write these as capital letters with a line above them. $\overleftrightarrow{AB}$
Line Segment	A segment (piece) of a line with two endpoints.	**A** **B**	To name a line segment, use the two endpoints with a small line segment drawn above the two capital letters. $\overline{\mathbf{AB}}$
Plane	A flat surface with no thickness that extends in all directions.	A plane A	To name a plane use a single capital letter or at least three points that lie in the plane.
Ray	Part of a line with one endpoint. **A ray looks like an arrow.*	A B	To name a ray use the endpoint first and any other point on the ray with an arrow drawn above it. $\overrightarrow{AB}$
Angle	An angle is formed by two rays with the same end-point. The common end-point is called the vertex.	C Angle b B A *Angles are measured in degrees.*	To name an angle, use the vertex with an angle sign drawn in front. You can name it three different ways: 1. ∠ABC or ∠CBA 2. ∠B 3. ∠b *The vertex must be in the middle.*

1.09 Introduction to Geometry 1.09

Classifying Angles

We classify angles into four different categories based on their measure (number of degrees). The first question you will want to ask yourself is, "What is the measure of the angle?"

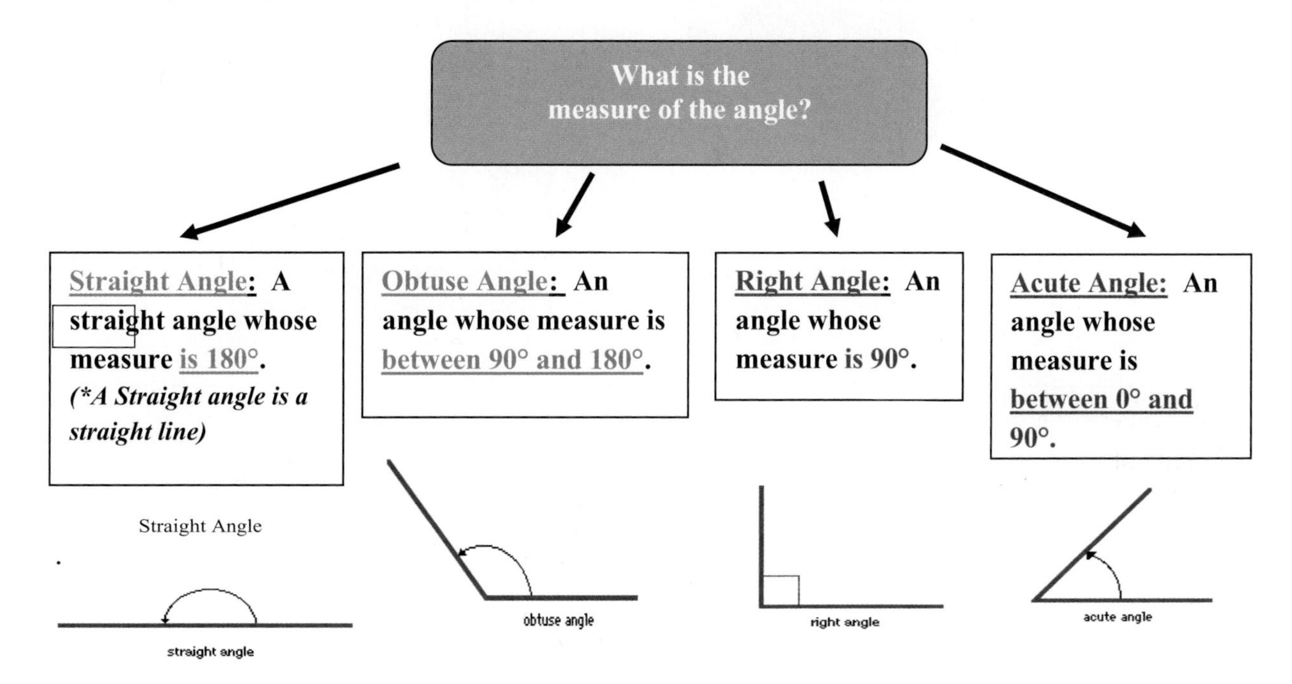

YOUR TURN

Classify each angle as acute, right, obtuse, or straight.

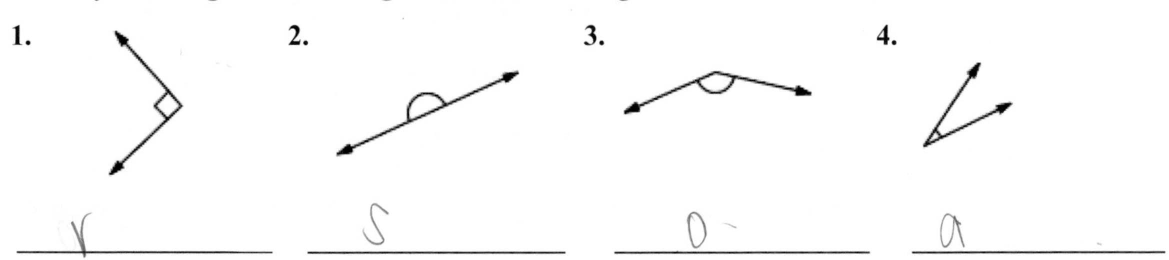

1. r
2. s
3. o
4. a

5.

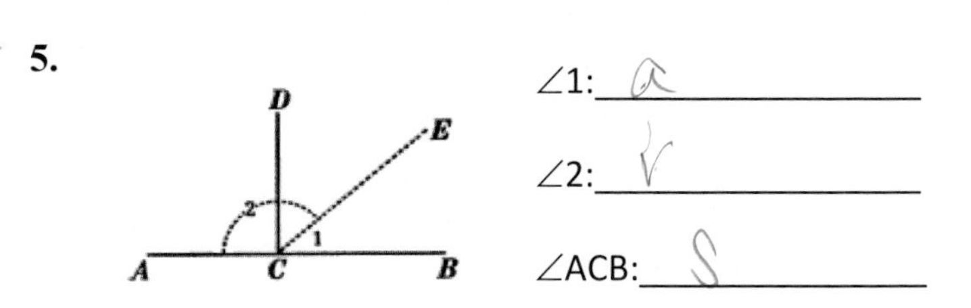

1.09 Introduction to Geometry 1.09

Complementary Angles: sum is 90°

*We say that each angle is a complement of the other.

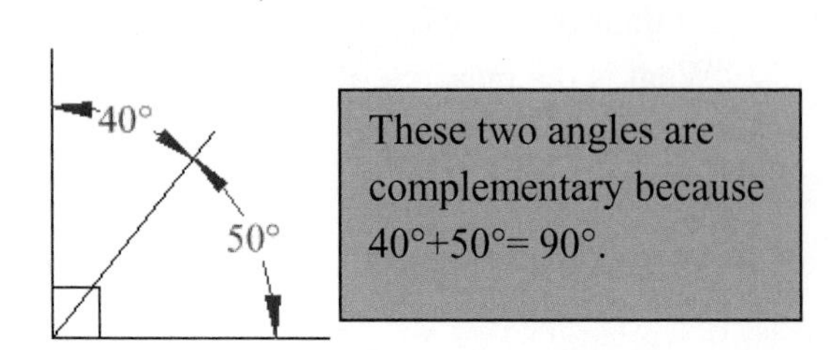

Supplementary Angles: sum is 180°

*We say that each angle is a supplement of the other.

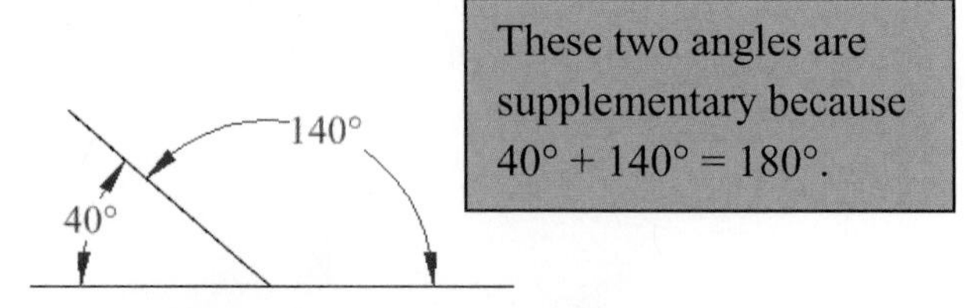

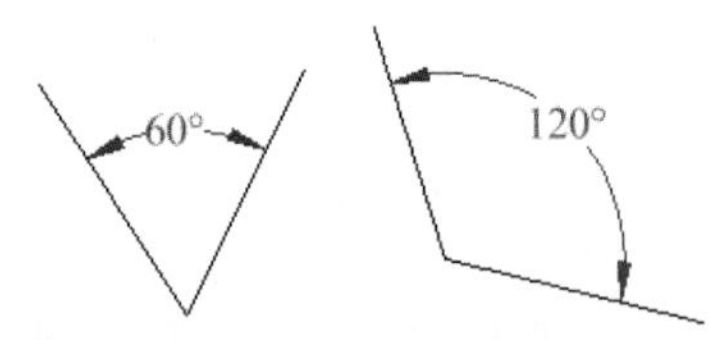

These two angles are supplementary because 60° + 120° = 180°.

 **Hints:**

Complementary think Corner.

A corner makes a 90° angle.

Supplementary think Straight.

A straight line is 180°.

EXAMPLE 1

a. Find the complement of a 35° angle.

Two angles that have a sum of 90° are complementary. Therefore, the complement of a 35° angle can be found by subtracting the angle from 90°.

90° – 35° = 55°

b. Find the supplement of a 127° angle.

Two angles that have a sum of 180° are supplementary. Therefore, the supplement of a 127° angle can be found by subtracting the angle from 180°.

180° - 127° = 53°

 **YOUR TURN**

Write an equation and solve.

6. Find the complement of a 14° angle.

____

7. Find the supplement of an 83° angle.

____

Finding Angle Measures

Adjacent angles: adjacent angles share a common vertex and ray with no common interior points.

EXAMPLE 2

Find the measure of angle x. Then classify it as acute, obtuse, right, or straight.

Given m∠ ADC = 125° and m∠ BDC = 87°.

In the example below, ∠ADB and ∠ BDC are adjacent angles.

*** m∠ ADB is the same as angle x.**

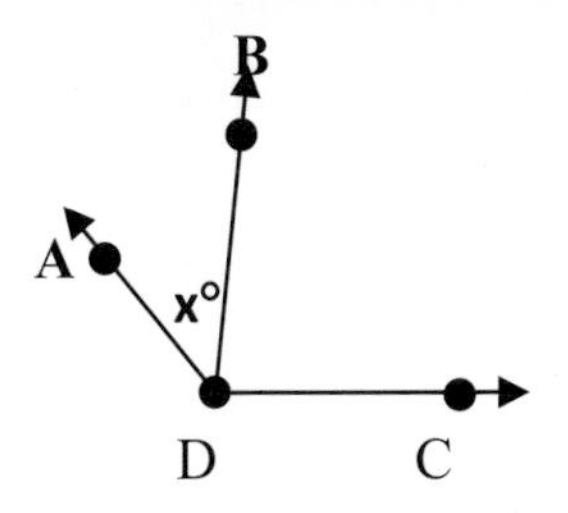

m∠ ADB + m∠ BDC = m∠ ADC

x° + 87° = 125° **Plug in what we know.**

- 87° - 87° **Subtract 87 from both sides to solve for x.**

x° = 38°

acute angle

Vertical Angles: two angles that are opposite of one another when two lines intersect. Vertical angles are congruent, which means they have the same measure.

EXAMPLE 3

If the measure of ∠ 1 = 42°, find the measure of m∠ 2, m∠ 3, and m∠ 4.

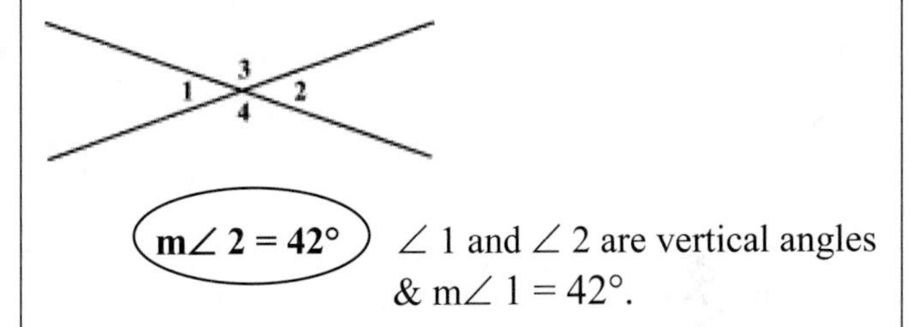

m∠ 2 = 42° ∠ 1 and ∠ 2 are vertical angles & m∠ 1 = 42°.

∠ 2 and ∠ 3 are on a straight line, which means they add up to 180°.

42° + m∠ 3 = 180°

- 42° - 42° **Subtract 42 from both sides.**

m∠ 3 = 138°

∠3 and ∠4 are vertical angles which means they have the same measure.

Therefore, m∠ 4 = 138°

YOUR TURN

8. Find the measure of angle b.

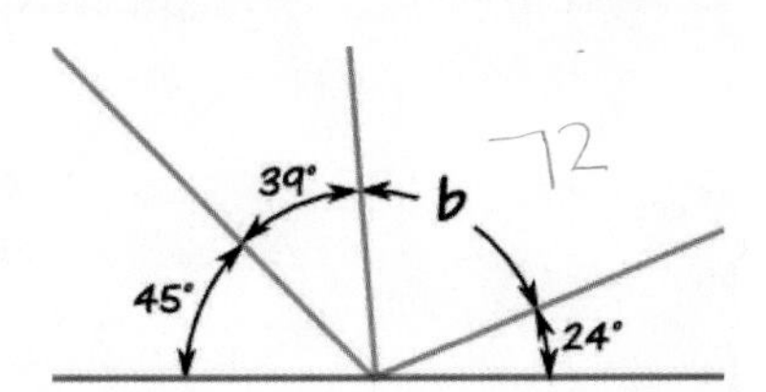

9. If the measure of ∠ BCP is 65°, what is the measure of ∠ OCA and ∠ ACP?

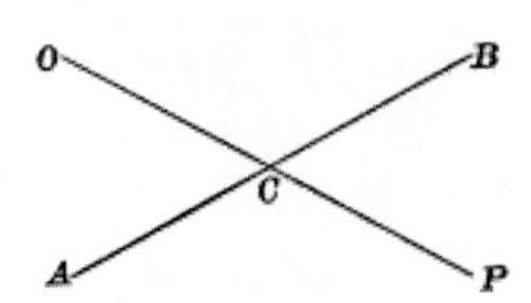

1.09 Introduction to Geometry 1.09

Parallel and Perpendicular Lines and Their Angles

Vocabulary	*Definition*	*Diagram*	*Name/Notation*
Parallel Lines	Two lines that lie in the same plane and do not intersect.	a b	The symbol for "is parallel to" is ‖. **line a ‖ line b* **sometimes arrows are drawn on the lines to show they are parallel to one another.*
Perpendicular Lines	Two lines that intersect and form a right angle at the point of intersection.	a b	The symbol for perpendicular looks like an upside down T. line a ⊥ line b
Transversal	A line that intersects two or more lines at different points.	E G A B H C D F	In the figure to the left, line EF is the transversal.

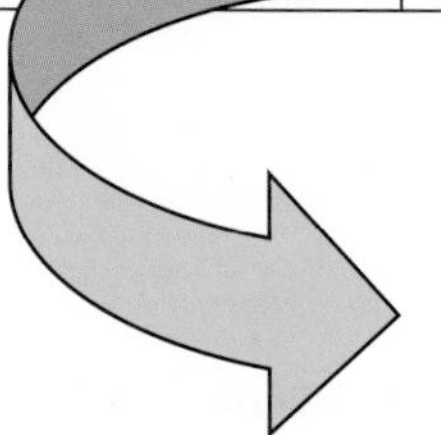

If two parallel lines are cut by a transversal:

- **corresponding angles have equal measure**
- **alternate interior angles have equal measure.**

EXAMPLE 4

In the example below line *j* and line *k* are parallel.

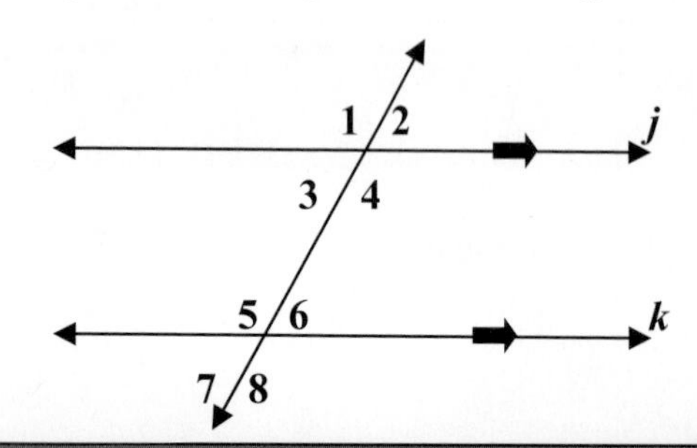

Corresponding Angles:
Angles on the same side of the transversal, one is outside the parallel lines, one is inside.
In the example on the left,

$\angle 1$ and $\angle 5$

$\angle 3$ and $\angle 7$

$\angle 2$ and $\angle 6$

$\angle 4$ and $\angle 8$

Alternate Interior Angles:
Angles on opposite sides of the transversal inside the parallel lines. In the example,

$\angle 3$ and $\angle 6$

$\angle 4$ and $\angle 5$

EXAMPLE 5

Line m and line n are parallel. The measure of $\angle 3 = 88°$.

Find the measure of $\angle 2$.

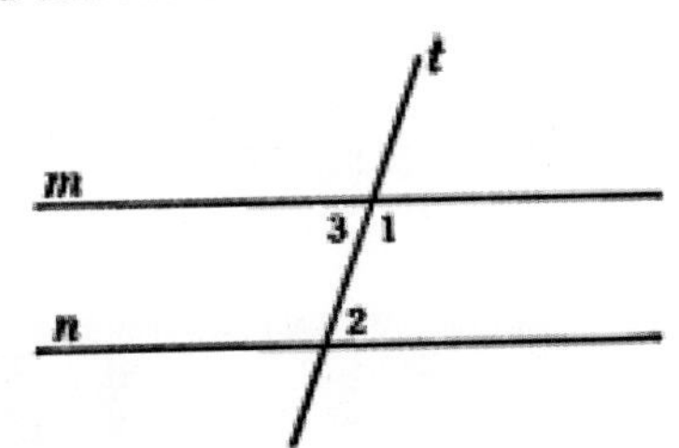

$\angle 3$ and $\angle 2$ are alternate interior angles.
Alternate interior angles have the same measure.

Therefore, $m\angle 2 = 88°$.

EXAMPLE 6

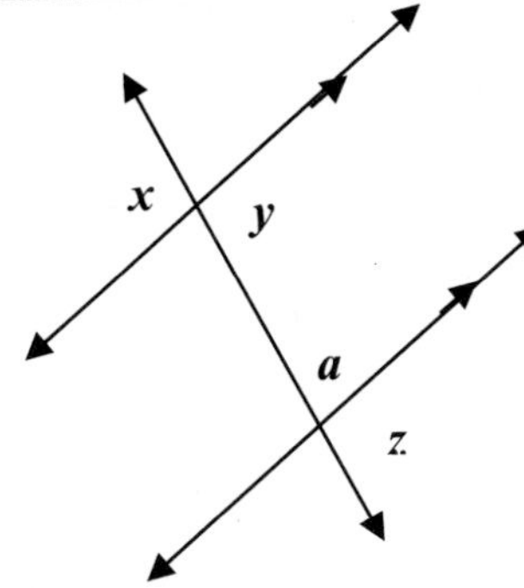

If the measure of $\angle y$ is 91° find the measures of $\angle x$, $\angle z$, and $\angle a$.

$\angle y$ and $\angle z$ are corresponding angles. **Therefore,**

$m\angle z = 91°$.

$\angle x$ and $\angle y$ are vertical angles.

Therefore, $m\angle x = 91°$.

$\angle a$ and $\angle z$ are on a straight line which means they sum to 180°.

$180° - 91° = m\angle a$

$89° = m\angle a$

YOUR TURN

10. If the measure of $\angle 2 = 45°$, find the remaining angles.

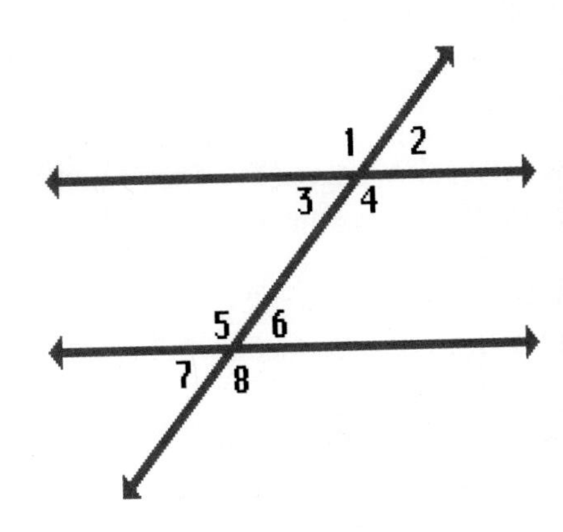

Angle	Measure	Angle	Measure
$\angle 1$		$\angle 5$	
$\angle 2$		$\angle 6$	
$\angle 3$		$\angle 7$	
$\angle 4$		$\angle 8$	

Perimeter and Area

Perimeter is a measure of distance around a shape

For example, someone might want to figure out the perimeter around their garden before buying material to make a fence so that they know how much material to buy.

Area is a measure of the amount of surface inside a shape.

For example, someone might want to know how much space their garden takes up.

Geometric Figure	Picture	Perimeter Formulas	Area Formulas
Rectangle	width, length	$P = 2l + 2w$ l = length, w = width	*Area = length · width* $A = l \cdot w$
Square	side, side	$P = 4s$ s = side All sides - same length.	*Area = side · side* $A = s^2$
Triangle	height, base	$P = a + b + c$ *Where a,b,c are the sides of the triangle.* Add up the sides of the triangle.	$Area = \frac{1}{2} \cdot base \cdot height$ $A = \frac{1}{2} \cdot b \cdot h$

Perimeter

Perimeter simply measures the distance around a shape. It can be measured in inches, feet, yards, miles, centimeters, meters, kilometers, and so on (any standard distance measurement). You can measure the perimeter of nearly any shape, you just add together the measure of each of its sides.

EXAMPLE 7

Perimeter of a Square

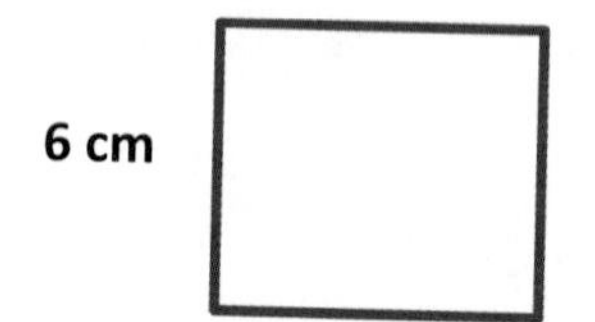

In order to calculate perimeter, add together the lengths of all four sides of the square. The length of one side is given. Remember, all sides of a square are equal.
So the perimeter of this square would be $6 + 6 + 6 + 6 = 24$.
We could also use the *formula* $P = 4s$ to find the perimeter.

$P = 4s$	Write an equation.
$P = 4(6)$	Solve.
$P = 24$ cm	Answer.

1.09 Introduction to Geometry 1.09

EXAMPLE 8

Perimeter of a Rectangle

8 in

4 in

In order to calculate perimeter, you need to add together the lengths of all four sides of the rectangle. You are given the length of one side and the width of one side. Remember, opposite sides of a rectangle are equal.

So the perimeter of this rectangle is 8 + 8 + 4 + 4 = 24 inches.

We could also use the formula $P = 2l + 2w$ to find the perimeter.

$\boldsymbol{P = 2l + 2w}$	Formula.
$P = 2(8) + 2(4)$	Write an equation.
$P = 16 + 8$	Solve.
$P = 24$ inches	Answer.

YOUR TURN

Find the perimeter. Find the formula. Write an equation. Solve and answer.

11.

14 cm

23 cm

Formula: ____________________

Equation:

Solve.

12. Find the perimeter of a rectangular football which is 53 yards wide and 120 yards long.

Formula: ____________________

Equation:

Solve.

13. Find the perimeter of a square ceramic tile with a side of length 3 inches.

3 in

Formula: ____________________

Equation:

Solve.

14. Find the perimeter.

Equation:

Solve.

13 in

? 6 in

13 in

?

30 in

1.09 Introduction to Geometry 1.09

Area

Area is the measure of the amount of surface inside a shape. Area formulas are given in the chart above for a square, rectangle, and a triangle. When calculating area, you will take the units given in the problem (feet, yards, etc.) and square them, so your unit measure would be in square feet (ft.^{2}) or whatever units are given.

EXAMPLE 9

Area of a Square

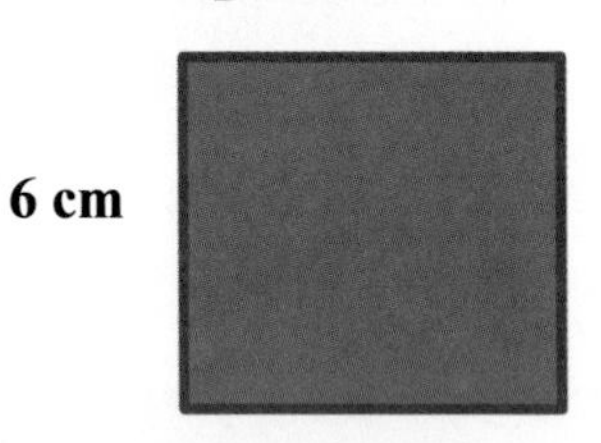

In order to calculate the area of a square, we multiply side *times* side. So, we have 6 x 6, which is 36. Therefore, the area is 36 square centimeters.

Or we can use the formula:

$A = s^2$	Formula.
$A = 6^2$	Write an equation.
$A = 36\text{ cm}^2$	Solve and answer.

EXAMPLE 10

Area of a Rectangle

In order to calculate the area of a rectangle, we multiply the length times the width. So, we have 6 x 4, which is 24. Therefore, the area is 24 square inches.

$A = l \cdot w$	Formula.
$A = 6 \cdot 4$	Write an equation.
$A = 24\text{ in}^2$	Solve and answer.

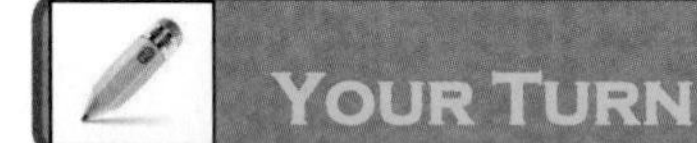

YOUR TURN

Find the area. Find the formula. Write an equation. Solve and answer.

15.

Formula: ______________________

Equation:

Solve.

16. The floor of Jane's room is 24 feet by 10 feet. Find the area of the room.

Formula: ______________________

Equation:

Solve.

1.09 Introduction to Geometry

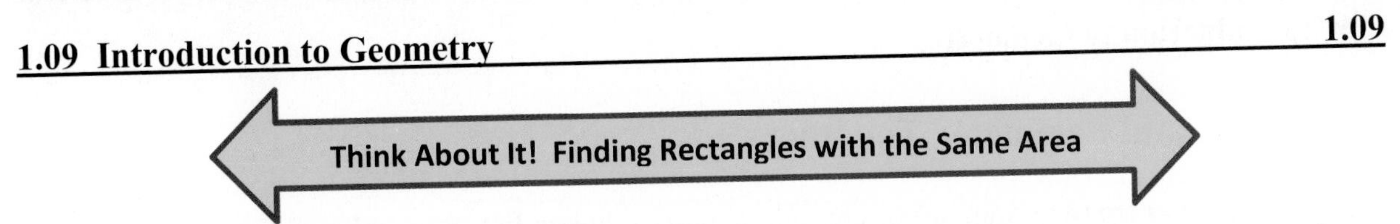

The area of a rectangle below is 24 in^2. Can you think of any other rectangles that we could draw that have an area of 24 in^2? Remember that area is the product of length *times* width. So, if the area is 24 in^2, then I am looking for other pairs of numbers that will multiply to give me 24. Let's list possible lengths and width for a rectangle that has an area of 24 in^2.

Length	Width	Area = $l \cdot w$
1	24	24 in^2
2	12	24 in^2
3	8	24 in^2
4	6	24 in^2

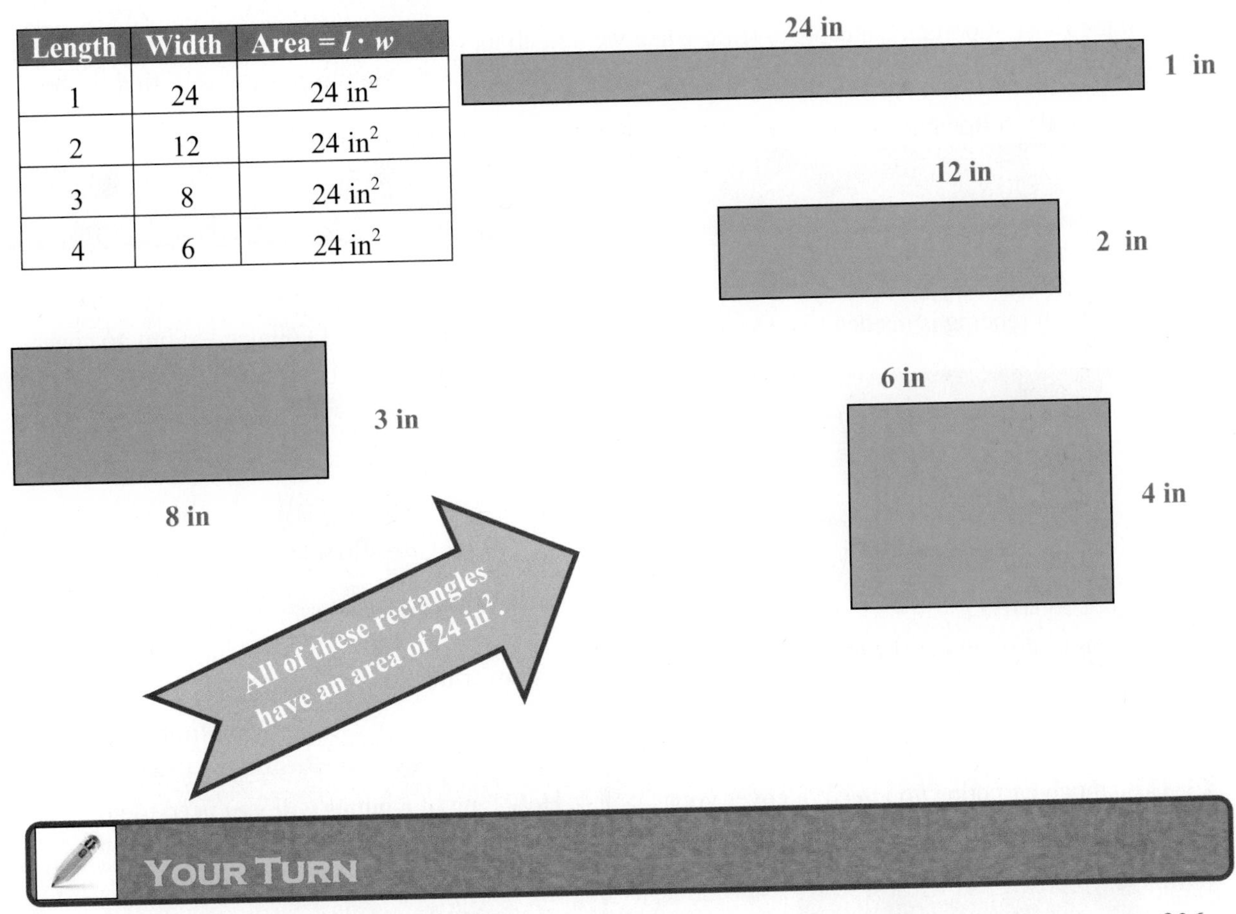

YOUR TURN

17. Draw rectangles with a different length and width that are possible for a rectangle with an area of 36 in^2. Hint: You should have 5 rectangles. Remember a square is also a rectangle.

18. Find the **perimeter** of each rectangle you drew.

Using Area and Perimeter Together

Area and perimeter are two examples of measurement used in real life. Remember perimeter is the distance around the outside of a shape. Area is the measure of the amount of surface covered by something. Think about it like this: **Perimeter goes *around the outside* edge, add (+) up all the sides.**

Area is the *inside*, multiply (·) the length times width.

One of the most important concepts to learn when we talk about perimeter and area is to understand when an application problem is asking us to find the perimeter or when a problem is asking us to find the area.

Think about the following situations and determine if we will need to find perimeter or area.

YOUR TURN

Fill in the blanks.

1. How much fencing is needed to go around a pool?

Courtesy of Fotolia

Fencing goes around the edges of the pool, so we will find ____________ by adding up the sides.

2. How many tiles will you need to buy to cover the floor of your patio?

Courtesy of Fotolia

Tiles cover the inside region of the patio so we will find the ___________ by multiplying.

3. How much carpeting you need to cover your bedroom floor?

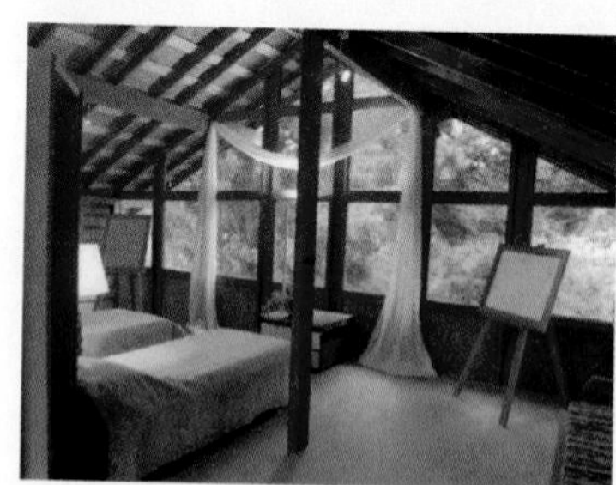

Courtesy of Fotolia

Carpeting covers the inside region of the room so we will find the __________ by multiplying.

4. How long of a gutter will you need to go around the roof of your house?

Courtesy of Fotolia

Guttering goes around the edges of a house, so we will find ____________by adding up the sides.

(Fill in the blank answers: 1. Perimeter 2. Area 3. Area 4. Perimeter)

YOUR TURN

Determine if you would find the perimeter or area to solve the following problems.

19. a. How much wood you would use to make a picture frame? ________________

    b. How much paint you will need to paint a wall ? __________________

    c. How much wall border you will need around your bedroom wall? ______________

    d. How much flooring you will need to remodel a kitchen? ________________

EXAMPLE 11

Ben is building a fenced area for his dogs. He wants to make a rectangular yard for the dogs that is 12 feet by 8 feet.

a. How much fencing will he need?

Since fencing goes around the outside edges, we know we need to find the perimeter.

$P = 2l + 2w$	Formula.
$= 2(12) + 2(8)$	Write an equation.
$= 24 + 16$	Solve.
$= 40$ feet	

Ben will need 40 feet of fencing.

Courtesy of Fotolia

b. If the fencing cost $12 a foot, How much will it cost Ben to fence in this area?

Ben needs 40 feet of fencing. The fence cost $12 per foot. To find his cost, multiply!
$40(12) = 480$

It will cost Ben $480 to build the fence.

1.09 Introduction to Geometry 1.09

EXAMPLE 12

A hotel is going to recarpet their lobby. The lobby is in the shape of a square that measures 45 meters on each side.

a. How much carpet will the hotel need to purchase?

Courtesy of Fotolia

Carpet covers the inside of an area so we will need to find the area of the square.

$\mathbf{A = s^2}$	Formula
$A = (45)^2$	Write an equation and solve.
$A = 2{,}025$ square meters	

The hotel will need 2,025 square meters of carpet.

b. If the carpet cost $10 per square meter, how much will it cost to recarpet the lobby?

The hotel needs 2,025 square meters at a cost of $10 per square meter. To find the cost we will multiply.

2,025 (10) = $ 20,250

The hotel will pay $20,250 for the carpet.

YOUR TURN

Write an equation and solve. Answer the question using words.

20. Isabella is making a display board for the school elections. The display board is 10 ft. by 6 ft. She needs to add a ribbon border around the entire display board.
a. What is the length of the ribbon that she needs?

Eq: ______________________________

Solve.

b. If the cost of the ribbon is $2 per foot, how much will Isabella pay?

Eq: ______________________________

Solve.

21. Paul wants to paint his 4 bedroom walls. Each wall measures 12 ft. by 10 ft.
a. How many square feet of paint will he need? (Hint: Don't forget to multiply your answer by 4 walls.)

Eq: ______________________________

Solve.

b. If one gallon covers 60 square feet, how many gallons will Paul need?

Eq: ______________________________

Solve.

1.09 Introduction to Geometry 1.09

YOUR TURN **Answers to Section 1.09.**

1. right 2. straight 3. obtuse 4. acute

5. $\angle$1: acute, $\angle$2: obtuse, $\angle$ACB: straight

6. 76° 7. 97° 8. 72° 9. $\angle$OCA = 65°, $\angle$ACP = 115°

10. m$\angle$1 = 135°, m$\angle$2 = 45°, m$\angle$3 = 45°, m$\angle$4 = 135°

m$\angle$5 = 135°, m$\angle$6 = 45°, m$\angle$7 = 45°, m$\angle$8 = 135°

11. 74 cm 12. 346 yards 13. 12 in

14. 86 in 15. 529 square cm 16. 240 square feet

17. 5 rectangles:
1 in by 36 in, 2 in by 18 in, 3 in by 12 in, 4 in by 9 in, and 6 in by 6 in

18. 74 in, 40 in, 30 in, 26 in, 24 in 19. a. Perimeter b. Area c. Perimeter d. Area

20. 32 feet, \$64 21. 480 ft^2, 8 gallons

Complete MML HW Section 1.09 online and show your work in your notebook.

DMA 010 TEST: **Review your notes, key concepts, formulas, and MML work.** ☺

Student Name: ______________________________ **Date:** ____________

Instructor Signature: ______________________________

DMA 020

DMA 020

FRACTIONS AND DECIMALS

This course provides a conceptual study of the relationship between fractions and decimals and covers related problems. Topics include application of operations and solving contextual application problems, including determining the circumference and area of circles with the concept of pi. Upon completion, students should be able to demonstrate an understanding of the connections between fractions and decimals.

<u>You will learn how to</u>:

- ➢ Solve contextual application problems involving operations with fractions and decimals.
- ➢ Visually represent fractions and their decimal equivalents.
- ➢ Simplify fractions.
- ➢ Find the lowest common denominator of two fractions.
- ➢ Correctly perform arithmetic operations on fractions.
- ➢ Explain the relationship between a number and its reciprocal.
- ➢ Correctly order fractions and decimals on a number line.
- ➢ Convert decimals between standard notation and word form.
- ➢ Round decimals to a specific place value.
- ➢ Estimate sums, differences, products, and quotients with decimals.
- ➢ Demonstrate an understanding of the connection between fractions and decimals.
- ➢ Convert between standard notation and scientific notation.
- ➢ Solve geometric applications involving the circumference and area of circles.

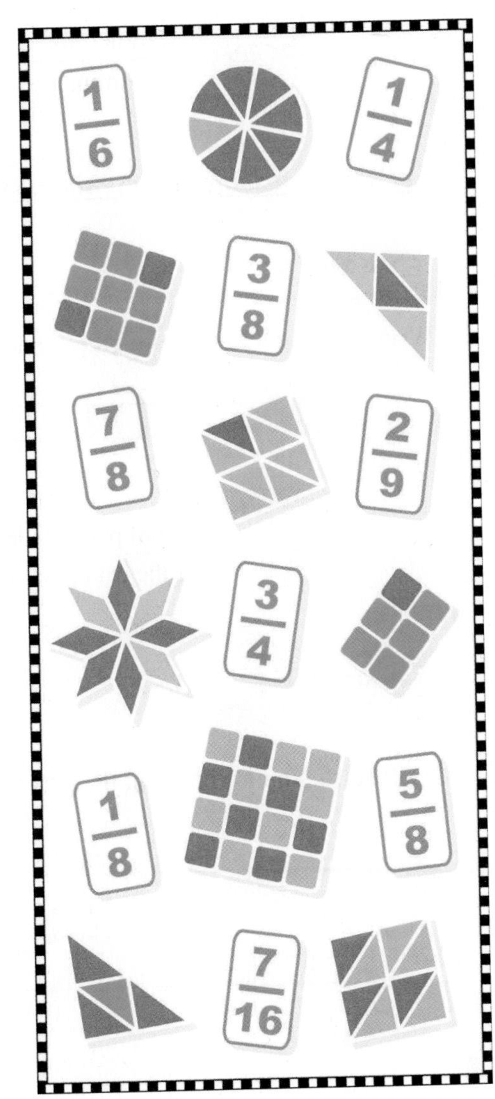

Courtesy of Fotolia

2.01 Intro to Fractions and Mixed Numbers 2.01

Whole numbers are used to count whole things as units, such as: cars, houses, dollars and people. A fraction is used to represent the parts of a whole thing or a group.

Fractions can be thought of as a:

- Part of a whole – the fraction $\frac{75}{83}$ indicates that the whole is divided into 83 parts and 75 parts are "special."
- Quotient – the fraction $\frac{7}{2}$ can be thought of as $7 \div 2$.
- Ratio – so the fraction $\frac{3}{5}$ can be thought of as the ratio of 3 to 5.

Each part of a fraction has a name with a specific purpose, for example:

$$\frac{4}{7} \Rightarrow \frac{\text{Numerator}}{\text{Denominator}} \Rightarrow \frac{\text{Number of parts considered}}{\text{Number of equal parts in the whole}}$$

Let's look at some representations of fractions.

EXAMPLE 1

Write a fraction to represent the shaded region of the figure.

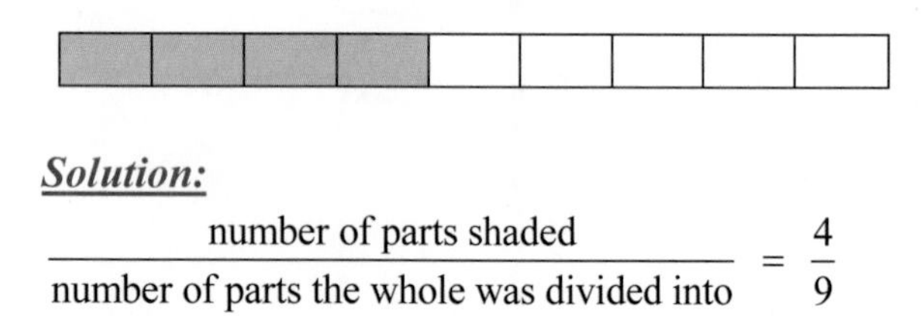

Solution:

$$\frac{\text{number of parts shaded}}{\text{number of parts the whole was divided into}} = \frac{4}{9}$$

EXAMPLE 2

Write a fraction to represent the part of the whole shown.

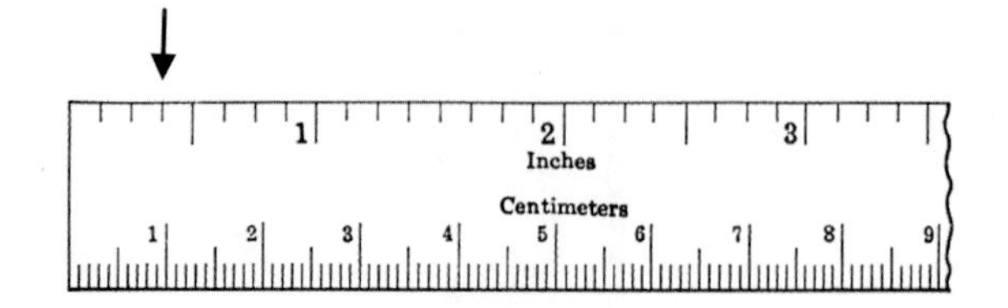

Solution:
Count the number of spaces between the end of the ruler and the 1. Since there are 8 spaces, 8 is the denominator. The number of spaces between the end of the ruler and the indicated point is 3. Therefore the fraction represented here is $\frac{3}{8}$.

2.01 Intro to Fractions and Mixed Numbers 2.01

EXAMPLE 3

Out of 40 students in a math class there are 30 students that are passing. What fraction of the class is passing?

Solution:

$$\frac{\text{Number of parts considered}}{\text{Number of equal parts in the whole}} = \frac{30}{40}$$

More Vocabulary: Proper, Improper and Mixed

- A proper fraction is a fraction less than 1. The numerator is smaller than the denominator.
- An improper fraction is a fraction greater than 1 or equal to 1. The numerator is either greater than or equal to the denominator.

EXAMPLE 4

What improper fraction is represented by this fraction model? What mixed number is represented?

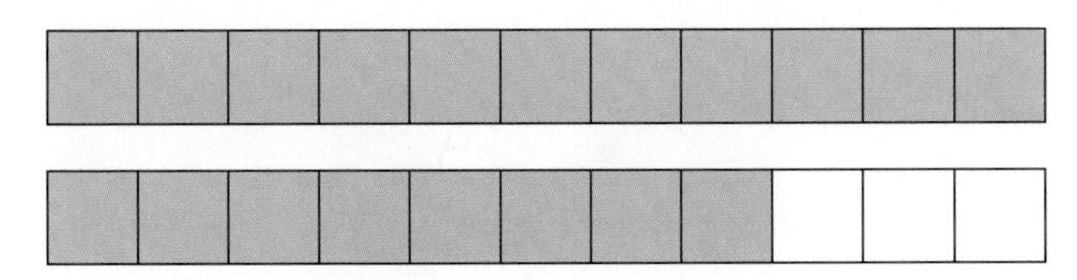

What improper fraction represents the shaded region?

Each whole has been divided into 11 parts. So each block is $\frac{1}{11}$ of the whole.

$\frac{1}{11}$	$\frac{1}{11}$	$\frac{1}{11}$	$\frac{1}{11}$	$\frac{1}{11}$	$\frac{1}{11}$	$\frac{1}{11}$	$\frac{1}{11}$	$\frac{1}{11}$	$\frac{1}{11}$	$\frac{1}{11}$

$\frac{1}{11}$	$\frac{1}{11}$	$\frac{1}{11}$	$\frac{1}{11}$	$\frac{1}{11}$	$\frac{1}{11}$	$\frac{1}{11}$	$\frac{1}{11}$			

We have 19 parts shaded. The improper fraction represented is $\frac{19}{11}$. You could also count each shaded block as $\frac{1}{11}, \frac{2}{11}, \frac{3}{11}, \ldots\ldots, \frac{19}{11}$

2.01 Intro to Fractions and Mixed Numbers 2.01

What mixed number represents the shaded region?

ONE WHOLE										
$\frac{1}{11}$	$\frac{1}{11}$	$\frac{1}{11}$	$\frac{1}{11}$	$\frac{1}{11}$	$\frac{1}{11}$	$\frac{1}{11}$	$\frac{1}{11}$			

We have 1 whole shaded and 8 parts shaded out of 11 parts in another whole. This 8 parts out of 11 is represented by $\frac{8}{11}$. Put these two together to get the mixed number, $1\frac{8}{11}$.

YOUR TURN

Answer each question.

1. What fraction of the group is yellow?

2. What fraction of the pizza is represented by one slice? What fraction of the pizza is left after one slice is removed?

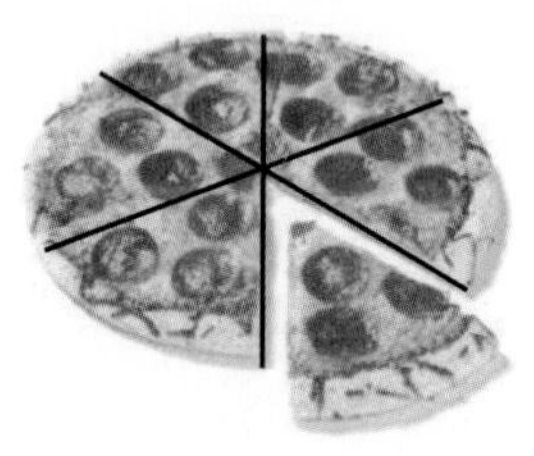

3. In a class of 27 students, 4 are male. What fraction of the students is male?

4. Use the box to draw a fraction model that represents $\frac{6}{8}$.

5. Write the improper fraction and a mixed number represented below.

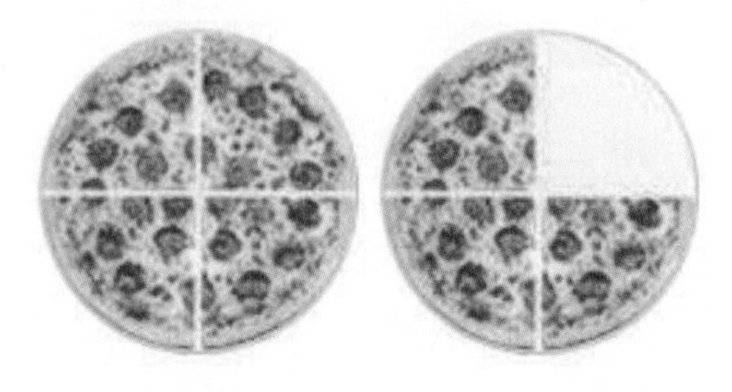

6. Write an improper fraction and a mixed number to represent the point indicated.

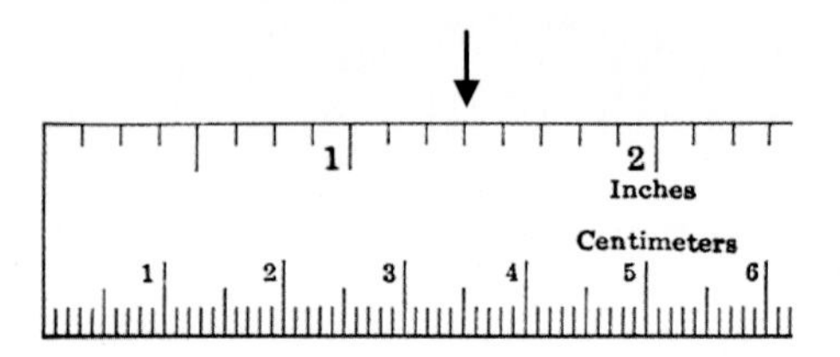

2.01 Intro to Fractions and Mixed Numbers

YOUR TURN

7. There are 15 children on a T-ball team. Eleven of the children are boys.

   a. What fraction of the team is boys?

   b. How many girls are on the team?

   c. What fraction of the team is girls?

8. Represent $1\frac{3}{5}$ by shading the fraction model.

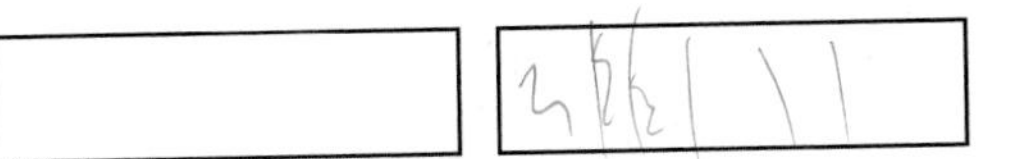

9. If the numerators of two fractions are different but the denominators are the same, which fraction is larger? Prove your answer by using fraction models.

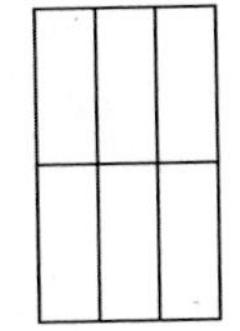
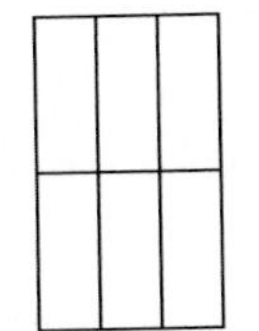
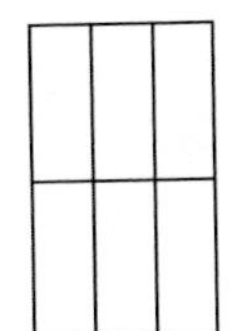
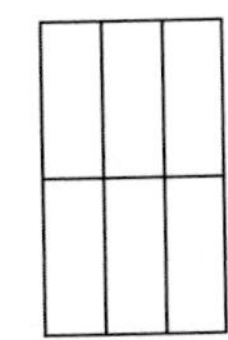
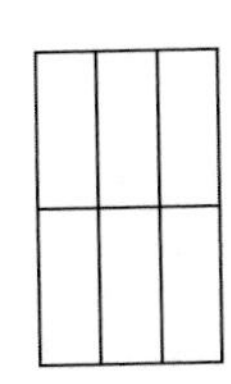

10. If the denominators of two fractions are different but the numerators are the same, which fraction is larger? Prove your answer by using fraction models.

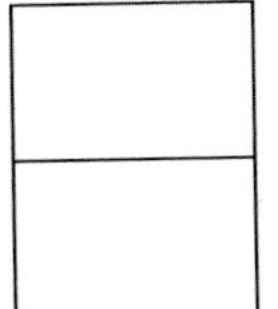
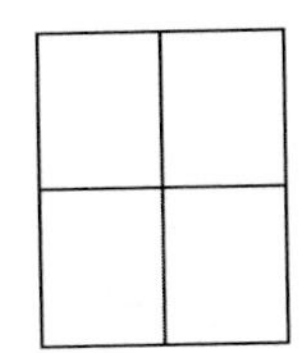
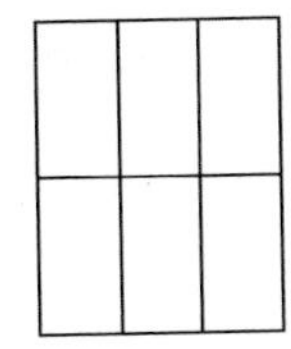
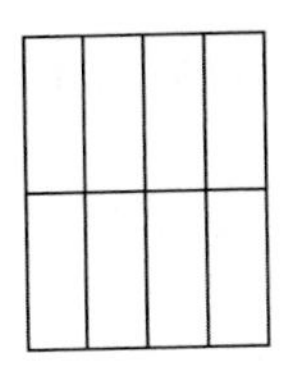

2.01 Intro to Fractions and Mixed Numbers 2.01

YOUR TURN

Graphing Fractions on the Number Line

EXAMPLE 5

Another way to visualize fractions is by looking at them on a number line. Think of 1 unit as the whole, exactly like reading the ruler. For example, to graph $\frac{4}{7}$, divide the distance from 0 to 1 into 7 parts, then count 4 parts (spaces) to the right.

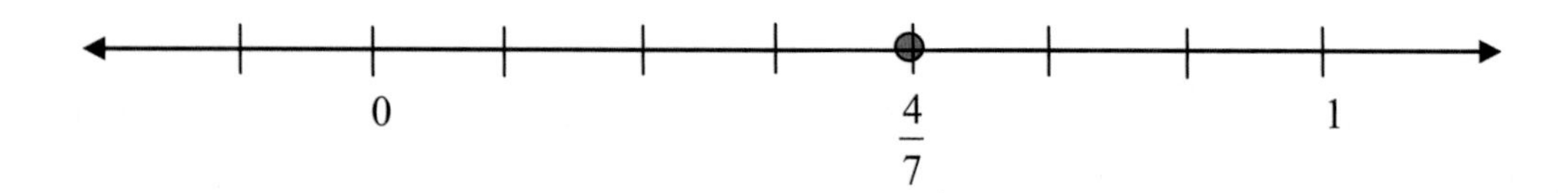

YOUR TURN

11. Graph $\frac{7}{4}$ on the number line below. Label the number line with 0, 1, and 2 and then graph the point.

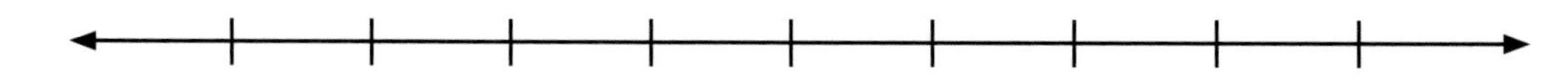

12. Graph $\frac{3}{5}$ on the number line below. Label the number line with 0 and 1 and then graph the point.

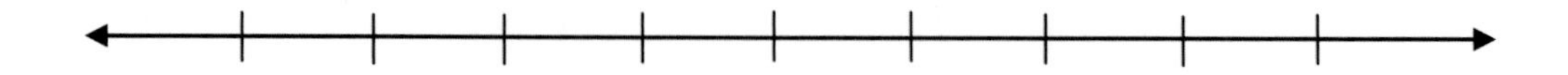

2.01 Intro to Fractions and Mixed Numbers

YOUR TURN **Answers to Section 2.01.**

1. $\frac{3}{4}$ 2. $\frac{1}{6}, \frac{5}{6}$ 3. $\frac{4}{27}$ 4. 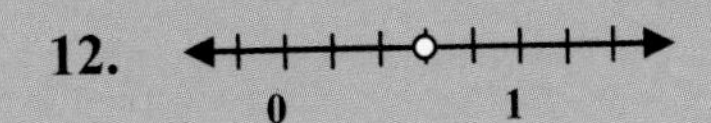 5. $\frac{7}{4}, 1\frac{3}{4}$ 6. $\frac{11}{8}, 1\frac{3}{8}$ 7a. $\frac{11}{15}$ 7b. 4 7c. $\frac{4}{15}$

8. $1\frac{3}{5}$

9. If denominators are the same, the fraction with the larger numerator is the greater fraction.

10. If numerators are the same, the fraction with the smaller denominator is the greater fraction.

11. 0 1 2

12. 0 1

Complete MyMathLab Section 2.01 Homework in your Homework notebook.

2.02 Fractions and Simplest Form 2.02

Looking at the factors of numbers will help us simplify, or reduce fractions. Factors also help us to find common denominators.

- **Factors** are numbers multiplied together to give a product. The whole number factors of a number divide that number evenly (there is no remainder).
- Factors help to determine if a number is **prime** or **composite**.
- A **prime number** is a whole number greater than one that is evenly divisible by only 1 and itself.
  Ex: 2, 3, 5, 7, 11, 13, 17, 19, 23, 29, 31 . . .
- A **composite number** is a number that is not prime. It is a number with more than two factors.
  Ex: 4, 6, 8, 9, 10, 12, 14, 15, 16 . . .
- 1 is neither prime nor composite

Divisibility Tests

Divisibility tests are helpful when looking for factors of numbers.

A number is divisible by	*if*
2	the # is ***even*** (ends in 0, 2, 4, 6, or 8). Ex: 97,538 ends in 8 (even #), thus 97,538 **is divisible by 2.**
3	the ***sum of its digits*** add up to a number that is divisible by 3. Ex. 159; since its digits add up to 15 (**1 + 5 + 9 = 15**) and 15 is **divisible by 3**, then 159 is divisible by 3.
4	if the ***last two digits*** of the number are ***divisible by 4***. Ex: 2016 is divisible by 4 since the **last two digits**, 16, **are divisible by 4.**
5	the number **ends in 0** or **5**. Ex: 3515 is divisible by 5 since it ends in 5.
6	it is *divisible by 2 and 3*. Ex: 324 is divisible by 2 since it is even. It is also divisible by 3 since 3 + 2 + 4 = 9, which is divisible by 3. So 324 is **divisible by 2 and 3** therefore it is **divisible by 6.**
9	the *sum of its digits* add up to a number that is divisible by 9. Ex. 1458; since its digits add up to 18 (**1 + 4 + 5 + 8 = 18**) and 18 is divisible by 9, then 1458 is divisible by 9.
10	it ends in **0**. Ex: 2340 is divisible by 10 because its last digit is 0.

2.02 Fractions and Simplest Form 2.02

Prime Factorization

The **prime factorization** of a number means writing the number as a product of its prime factors.

EXAMPLE 1

Write the prime factorization of 60.

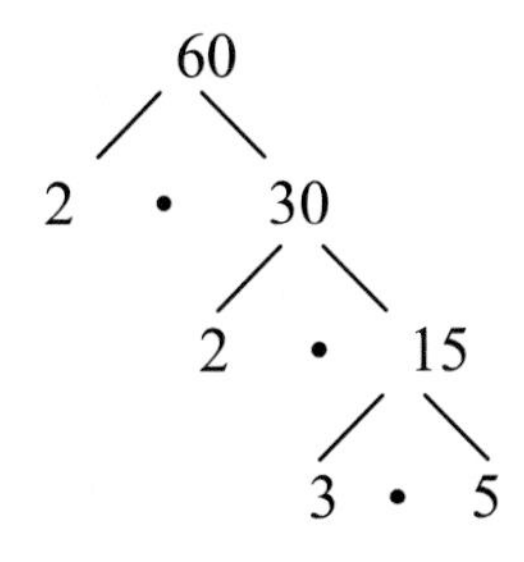

- Pick any 2 factors. It does not matter which two you choose.
- Continue factoring the composite numbers until you are left with only prime numbers.
- Express any repeated factors using exponents.
- The order that you write the factors in does not matter since multiplication is commutative.

So, the prime factorization of 60 is $2^2 \bullet 3 \bullet 5$

EXAMPLE 2

Write the prime factorization of 105.

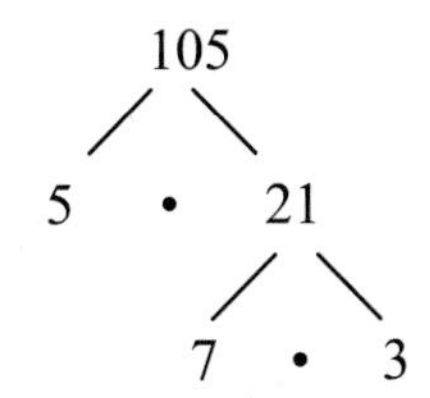

The prime factorization of 105 is $3 \cdot 5 \cdot 7$

YOUR TURN

(Use the divisibility rules to help you with the following.)

1. Write the prime factorization of 72.

2. Write the prime factorization of 297.

2.02 Fractions and Simplest Form 2.02

Equivalent Fractions

Fractions that represent the same part of a whole are called ***equivalent fractions.***

Example

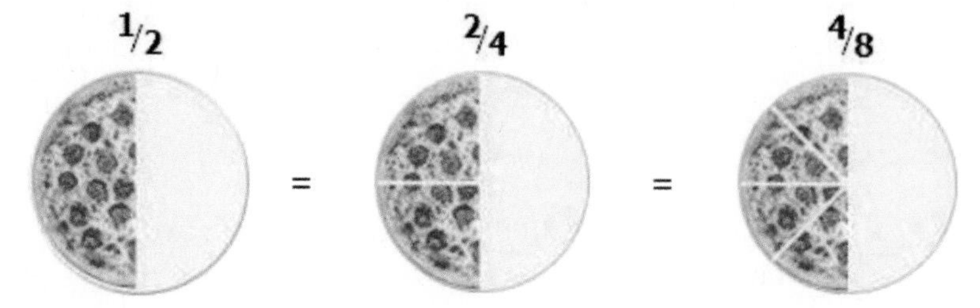

There are many other fractions that are equal to ½. But ½ is the simplest form for the numbers represented here.

A fraction is in simplest form when the numerator and denominator have no common factors. We can use prime factorization to help us simplify fractions. If both the numerator and denominator have common prime factors, then we can divide (cancel) these factors out.

Simplest Form

EXAMPLE 3

Write $\frac{42}{66}$ in simplest form.

$\frac{42}{66}=\frac{2\cdot3\cdot7}{2\cdot3\cdot11}$ Step 1: Write the prime factorization for 42 and 66.

$\frac{42}{66}=\frac{\overset{1}{\cancel{2}}\cdot\overset{1}{\cancel{3}}\cdot7}{\underset{1}{\cancel{2}}\cdot\underset{1}{\cancel{3}}\cdot11}$ Step 2: Divide out the common factors in the numerator and the denominator.

$\frac{42}{66}=\frac{7}{11}$ Step 3: Clean it up! Finished!

2.02 Fractions and Simplest Form

EXAMPLE 4

Write $-\frac{72}{26}$ in simplest form.

$$-\frac{72}{26} = -\frac{2 \cdot 2 \cdot 2 \cdot 3 \cdot 3}{2 \cdot 13}$$

Step 1: Write the prime factorization for 72 and 26.

$$-\frac{72}{26} = -\frac{\overset{1}{\cancel{2}} \cdot 2 \cdot 2 \cdot 3 \cdot 3}{\underset{1}{\cancel{2}} \cdot 13}$$

Step 2: Divide out the common factors in the numerator and the denominator.

$$-\frac{72}{26} = -\frac{2 \cdot 2 \cdot 3 \cdot 3}{13}$$

Step 3: Multipy the remaining factors.

$$-\frac{72}{26} = -\frac{36}{13}$$

Finished!

EXAMPLE 5

Forty of the states in the US have at least one four-star hotel.

a. What fraction of the states has at least one 4-star hotel?

b. How many states do not have a 4-star hotel?

c. What fraction of the states does not have a 4-star hotel?

Solution

a. What fraction of the states has at least one 4-star hotel?

There are 50 states in the US. So our task is to represent 40 out of 50 as a simplified fraction.

40 ← States with a 4-star hotel
50 ← Total number of states in the US

$\frac{40}{50} = \frac{2 \cdot 2 \cdot 2 \cdot 5}{2 \cdot 5 \cdot 5} = \frac{4}{5}$ of the states in the US have at least one 4-star hotel.

b. How many states do not have a 4-star hotel?

Total # of states − number of states with a 4-star hotel = number of states without a 4-star hotel
50 − 40 = 10

Thus, 10 states do not have a 4-star hotel. I wonder which ones they are?

c. What fraction of the states does not have a 4-star hotel?

10 ← Number of states without a 4-star hotel
50 ← Total number of states in the US

$\frac{10}{50} = \frac{2 \cdot 5}{2 \cdot 5 \cdot 5} = \frac{1}{5}$ of the states in the US have no 4-star hotels.

YOUR TURN

Write in simplest form.

3. $\frac{64}{20}$

4. $-\frac{98}{126}$

5. Jennifer purchased a new car for $30,000. She got $12,000 for her trade-in. How much of the price of her new car was NOT covered by her trade-in? What fraction, in simplest form, of the purchase price was not covered by the trade-in?

ARE THESE FRACTIONS EQUIVALENT???

One way to determine whether fractions are equivalent (equal) is to simplify each fraction and then compare.

EXAMPLE 6

Are $\frac{20}{30}$ and $\frac{26}{39}$ equal?

Simplify each fraction and compare the results.

$\frac{20}{30} = \frac{2 \cdot 2 \cdot 5}{2 \cdot 3 \cdot 5} = \frac{2}{3}$

$\frac{26}{39} = \frac{2 \cdot 13}{3 \cdot 13} = \frac{2}{3}$

Since these simplified fractions are equal, then $\frac{20}{30} = \frac{26}{39}$.

2.02 Fractions and Simplest Form 2.02

Another way to determine whether two fractions are equal is to use a property of proportions that allows us to cross multiply.

EXAMPLE 7

Are $\frac{6}{21}$ and $\frac{14}{35}$ equal?

210 294

$$\frac{6}{21} = \frac{14}{35}$$

Since $210 \neq 294$, $\frac{6}{21} \neq \frac{14}{35}$.

YOUR TURN

Determine whether the fractions are equivalent.

6. $\frac{16}{20}$ and $\frac{9}{12}$

7. $\frac{63}{72}$ and $\frac{147}{168}$

8. $\frac{27}{89}$ and $-\frac{27}{89}$

YOUR TURN Answers to Section 2.02

1. $2^3 \cdot 3^2$ 2. $3^3 \cdot 11$ 3. $\frac{16}{5}$ 4. $-\frac{7}{9}$ 5. \$18,000; $\frac{3}{5}$ 6. $\neq$ 7. $=$ 8. $\neq$

2.02 Fractions and Simplest Form 2.02

Calculator Tip

Scientific calculators generally have a fraction key such as [a b/c], that allows you to easily perform operations on fractions.

For example, to simplify $\frac{432}{810}$, enter [432] [a b/c] [810] [=]. You should see a simplified fraction, $\frac{8}{15}$.

Graphing Calculators also allow you to simplify fractions. The fraction option is usually under the [MATH] menu. To simplify $\frac{432}{810}$, enter $432 \div 810$ [MATH] [ENTER] [ENTER] . You should see a simplified fraction, $\frac{8}{15}$.

Complete MyMathLab Section 2.02 Homework in your Homework notebook.

2.03 Multiplying and Dividing Fractions

MULTIPLYING FRACTIONS

How do we multiply fractions? Let's look at fraction models again to discover how to find $\frac{1}{2}$ of $\frac{3}{5}$.

Here we have $\frac{3}{5}$ shaded.

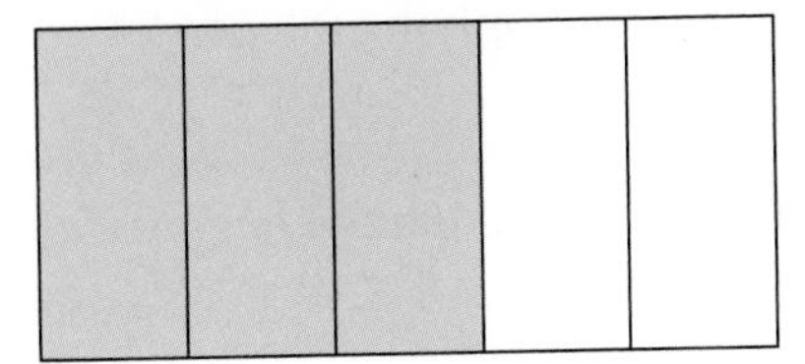

Now let's divide this area into halves.

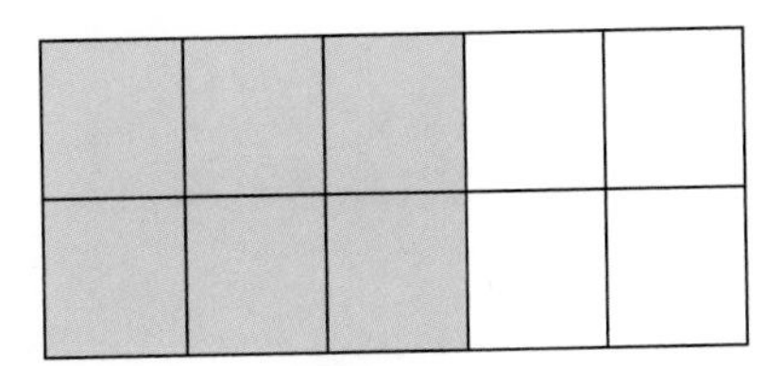

To find $\frac{1}{2}$ of $\frac{3}{5}$ we will shade $\frac{1}{2}$ of this area with a different color.

Now look at $\frac{1}{2}$ of the shaded area.

It is $\frac{3}{10}$ of the model. This means that $\frac{1}{2}$ of $\frac{3}{5}$ is $\frac{3}{10}$.

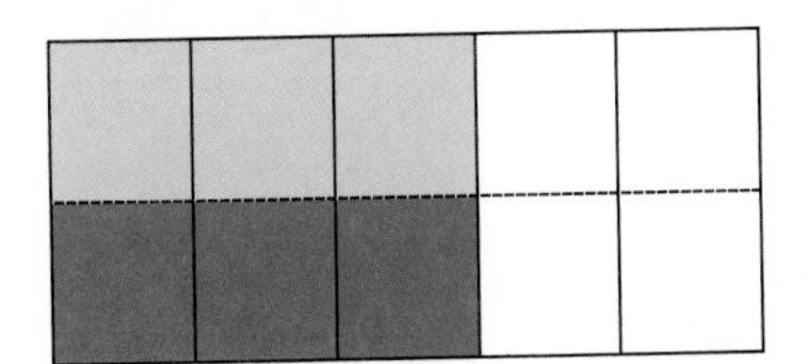

This leads us to the fact that $\frac{1}{2} \bullet \frac{3}{5} = \frac{1 \bullet 3}{2 \bullet 5} = \frac{3}{10}$

MULTIPLYING FRACTIONS

To multiply fractions, multiply the numerators together and then multiply the denominators together. Simplify your answer.

EXAMPLE 1

Multiply and simplify your answer. Prime factor the numerator and denominator to simplify.

$$\frac{4}{27} \bullet \frac{3}{8} = \frac{4 \bullet 3}{27 \bullet 8} = \frac{\overset{1}{\cancel{2}} \bullet \overset{1}{\cancel{2}} \bullet \overset{1}{\cancel{3}}}{\underset{1}{\cancel{3}} \bullet 3 \bullet 3 \bullet \underset{1}{\cancel{2}} \bullet \underset{1}{\cancel{2}} \bullet 2} = \frac{1}{18}$$

2.03 Multiplying and Dividing Fractions 2.03

EXAMPLE 2

Multiply and simplify. Remember the rules for multiplying integers!

$$-\frac{7}{8}\bullet\frac{5}{11}=-\frac{7\bullet 5}{8\bullet 11}=-\frac{35}{88}$$ ***Recall that a positive times a negative equals a negative!***

EXAMPLE 3

Multiply and simplify.

$$\left(-\frac{3}{4}\right)^2=\left(-\frac{3}{4}\right)\bullet\left(-\frac{3}{4}\right)=\frac{3\bullet 3}{4\bullet 4}=\frac{9}{16}$$ ***Recall that a negative times a negative equals a positive!***

EXAMPLE 4

Multiply and express the answer as a simplified fraction.

$$-\frac{1}{12}\bullet\frac{3}{25}\bullet\frac{15}{7}$$

You can cancel out common factors anywhere in the multiplication as long as one of the factors is in the numerator and one factor is in the denominator of any of the fractions.

$$-\frac{1}{12}\bullet\frac{3}{25}\bullet\frac{15}{7}=-\frac{1}{2\bullet 2\bullet \overset{}{\underset{1}{\cancel{3}}}}\bullet\frac{\overset{1}{\cancel{3}}}{5\bullet \underset{1}{\cancel{5}}}\bullet\frac{3\bullet \overset{1}{\cancel{5}}}{7}=-\frac{3}{140}$$ *Don't forget that you are multiplying by that negative number!*

2.03 Multiplying and Dividing Fractions

EXAMPLE 5

Multiply and express your answer as a simplified fraction.

$\frac{2}{3} \bullet \left(-\frac{1}{7}\right)^2$ Be sure to work the exponent first. Remember that negative times negative is positive!

$$\frac{2}{3} \bullet \left(-\frac{1}{7}\right) \bullet \left(-\frac{1}{7}\right) = \frac{2 \bullet -1 \bullet -1}{3 \bullet 7 \bullet 7} = \frac{2}{147}$$

YOUR TURN

Multiply and express your answer as a simplified fraction.

1. $\frac{3}{2} \bullet \frac{8}{27}$

2. $\left(-\frac{5}{6}\right)\left(-\frac{8}{15}\right)$

3. $\left(-\frac{2}{5}\right)^3$

4. $-\frac{1}{3} \bullet \left(-\frac{3}{4}\right)^2$

2.03 Multiplying and Dividing Fractions 2.03

DIVIDING FRACTIONS

You may recall that division of fractions requires that we change the division to multiplication and the second number to its reciprocal. Why do we change it to multiplication? Consider the fraction model for 2/3. How many 1/6's are in 2/3 or in other words what is 2/3 divided by 1/6?

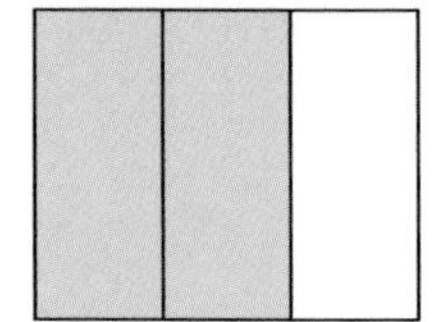

This fraction model represents $\frac{2}{3}$. We will create a fraction model the same size with the same shading as this one and divide it into sixths.

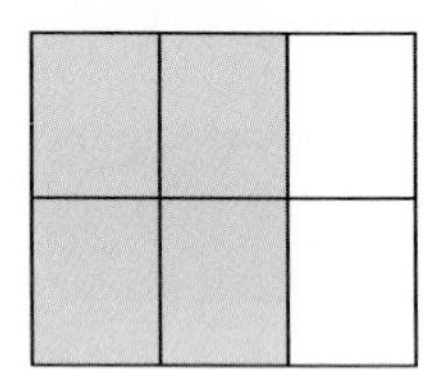

You can see that there are 4 of the sixths in the shaded part $\left(\frac{2}{3}\right)$. So $\frac{2}{3} \div \frac{1}{6} = 4$.

Now look at $\frac{2}{3} \div \frac{1}{6} = \frac{2}{3} \bullet \frac{6}{1} = \frac{12}{3} = 4$. Same answer! Amazing, right?

Fractions are fun!

> **DIVIDING FRACTIONS**
>
> To divide two fractions, multiply the first fraction by the reciprocal of the second fraction. In other words, keep the first, change the operation sign and flip the second fraction.

Reciprocals

When we "flip" that fraction we are changing it to its ***reciprocal***. Reciprocals are two numbers whose product is 1. Let's look at some numbers and their reciprocals.

Number	Reciprocal
$\frac{2}{3}$	$\frac{3}{2}$ because $\frac{2}{3} \cdot \frac{3}{2} = \frac{6}{6} = 1$.
$-\frac{1}{7}$	$-\frac{7}{1} = -7$ If a number is negative, then its reciprocal is negative. $-\frac{1}{7} \cdot \left(-\frac{7}{1}\right) = \frac{7}{7} = 1$.
0	Zero has no reciprocal since there is no number that when multiplied by 0 will give us 1.

2.03 Multiplying and Dividing Fractions 2.03

Dividing Fractions Rule

KEEP, CHANGE, FLIP

To divide 2 fractions,

- Keep the first one as it is.
- Change division to multiplication.
- Flip the second fraction.
- Now multiply.
- Simplify (or reduce) if possible.

EXAMPLE 6

Divide and express your answer as a simplified fraction.

$\frac{2}{15} \div \frac{5}{7}$

$\frac{2}{15}$ Keep the first fraction as it is.

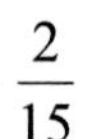

$\frac{2}{15} \bullet$ Change division to multiplication.

$\frac{2}{15} \bullet \frac{7}{5}$ "Flip" the second fraction.

$\frac{2}{15} \bullet \frac{7}{5} = \frac{14}{75}$ Now multiply.

EXAMPLE 7

Divide: $\frac{9}{10} \div \left(-\frac{15}{4}\right) = \frac{9}{10} \bullet \left(-\frac{4}{15}\right) = -\frac{\overset{3}{\cancel{9}} \bullet \overset{2}{\cancel{4}}}{\underset{5}{\cancel{10}} \bullet \underset{5}{\cancel{15}}} = -\frac{6}{25}$

Recall that negative times positive is negative!

This example shows how we can divide out common factors without prime factoring all of the numbers.

- Here we notice that 9, in the numerator, and 15 in the denominator have a common factor of **3.** Divide 9 and 15 by 3 to factor out that common factor. We get the 3 and 5 in red above.

- There is also a factor common to 4 and 10. So we divide 4 and 10 by the common factor of 2. We get the 2 and 5 in blue above.

Of course you can always prime factor all of the numbers and simplify.

2.03 Multiplying and Dividing Fractions 2.03

YOUR TURN

Divide and express your answer in simplest form. Watch out for signs!

5. $\frac{3}{2} \div \frac{11}{5}$

$\frac{3}{2} \cdot \frac{5}{11}$ $\frac{15}{22}$

6. $\left(-\frac{2}{11}\right) \div \left(-\frac{7}{22}\right)$

$\frac{-2}{11} \cdot \frac{-22}{7}$ $-\frac{44}{77}$

$-\frac{4}{7}$

7. $\frac{7}{8} \div (-2)$ Remember that $-2 = -\frac{2}{1}$.

$\frac{7}{8} \cdot \frac{-1}{2}$ $\frac{-7}{16}$

8. $\left(\frac{3}{4}\right)^2 \div \left(-\frac{3}{4}\right)$ Be sure to follow order of operations!

$\frac{9}{16} \cdot -\frac{4}{3} = \frac{-36}{48}$

$-\frac{3}{4}$

APPLICATIONS

One key word to look for when solving fraction applications is the word **"of"**. When you see the phrase "one half **of** the money", mathematically that means one-half **times** the amount of money.

It is also important to know the **"per"** indicates division. For example, **miles per gallon** would mean to **divide the number of miles by the number of gallons.**

AS ALWAYS, REMEMBER TO CHECK TO MAKE SURE YOUR ANSWER MAKES SENSE!

2.03 Multiplying and Dividing Fractions

EXAMPLE 8

A cruise to the Bahamas is being discounted. It is advertised as being $\frac{2}{3}$ of the regular price, which is $4500. Find the sale price of the trip.

$\frac{2}{3}$ of the regular price

⬍ ⬍ ⬍

$\frac{2}{3} \quad \bullet \quad 4500$

$$\frac{2}{3} \bullet 4500 = \frac{2}{3} \bullet \frac{4500}{1} = \frac{2 \bullet \overset{1500}{\cancel{4500}}}{\underset{1}{\cancel{3}} \bullet 1} = \frac{3000}{1} = 3000$$

<u>The sale price of the trip is $3000</u>. Does the answer make sense???

EXAMPLE 9

Find the radius of a circle if the diameter of the circle is $\frac{3}{4}$ of an inch.

Remember that the radius of a circle is one-half of its diameter and that "of" indicates multiplication.

$\frac{1}{2} \bullet \frac{3}{4} = \frac{1 \bullet 3}{2 \bullet 4} = \frac{3}{8}$ **The radius of the circle is $\frac{3}{8}$ of an inch**. Does the answer make sense?

EXAMPLE 10

Christina is making shorts for the cheerleaders. She has 12 yards of fabric. Each pair of shorts requires $\frac{3}{4}$ of a yard of fabric. How many pairs of shorts can she make with the fabric she has?

Total amount of fabric ÷ fabric needed for one pair

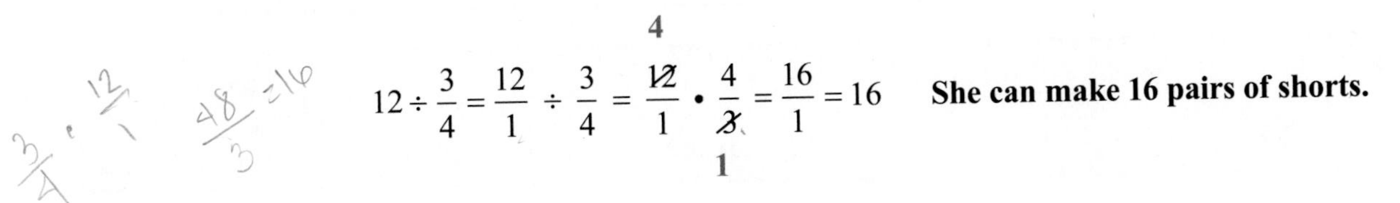

$$12 \div \frac{3}{4} = \frac{12}{1} \div \frac{3}{4} = \frac{\overset{4}{\cancel{12}}}{1} \bullet \frac{4}{\underset{1}{\cancel{3}}} = \frac{16}{1} = 16$$ **She can make 16 pairs of shorts.**

2.03 Multiplying and Dividing Fractions 2.03

YOUR TURN

Answer each question. **Remember to check to see if your answer makes sense. Draw a picture, if it helps.**

9. Find the diameter of a circle whose radius is $\frac{5}{16}$ of an inch.

10. Anthony has a carpet remnant that is 2 yards wide and 3 yards long. He is making scratch posts for cats that require $\frac{3}{4}$ of a square yard of carpet each to sell at the flea market. How many of the posts can he make? *(Hint: First find the area of the carpet he has!)*

11. The sugar bowls at the church each hold $\frac{3}{4}$ cup of sugar. How many sugar bowls can be filled with 9 cups of sugar? (Make sure that your answer makes sense.)

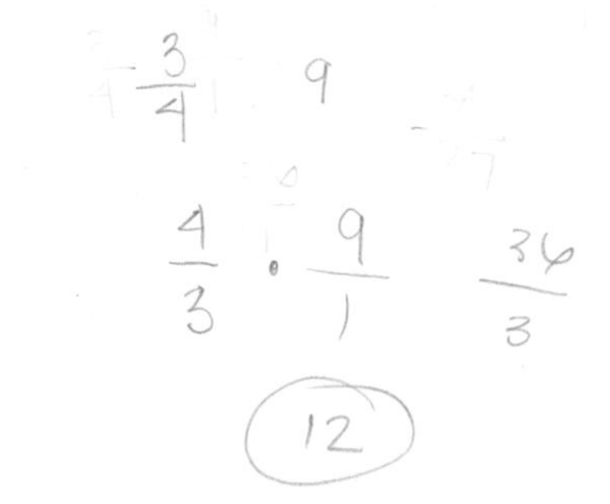

YOUR TURN Answers to Section 2.03

1. $\frac{4}{9}$ 2. $\frac{4}{9}$ 3. $-\frac{8}{125}$ 4. $-\frac{3}{16}$

5. $\frac{15}{22}$ 6. $\frac{4}{7}$ 7. $-\frac{7}{16}$ 8. $-\frac{3}{4}$

9. $\frac{5}{8}$ 10. 8 posts 11. 12 sugar bowls

Calculator Tip

You can use your calculator to multiply and divide fractions.

Enter the first fraction using the *a b/c* key, then the operation sign, and then the second number.

Hit enter and you should have your answer! This works the same way with a graphing calculator .

Use the same process as discussed in section 2.02.

Complete MyMathLab Section 2.03 Homework in your Homework notebook.

2.04 Adding and Subtracting Fractions, Least Common Denominator 2.04

The next step in developing your understanding of fractions is learning how to add and subtract fractions. We will start by adding and subtracting like fractions, that is fractions with the same denominator.

Let's take a look at adding two like fractions using the fraction models.

Let's add $\frac{1}{5}+\frac{3}{5}$

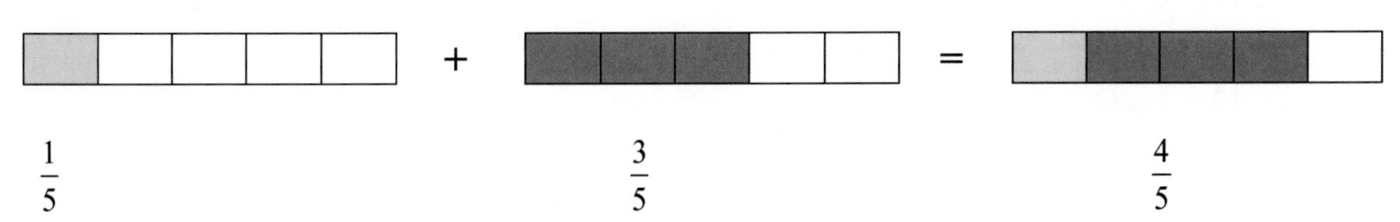

$\frac{1}{5}$ $\frac{3}{5}$ $\frac{4}{5}$

We can see that when we add fractions with like denominators, we simply add the numerators and keep the denominator.

Adding and Subtracting Fractions with Common Denominators

Add or subtract the numerators and write the sum or difference over the common denominator.

EXAMPLE 1

Add: $\frac{5}{13}+\frac{3}{13}$ $\frac{\textit{add} \text{ the numerators}}{\textit{leave} \text{ denominator the same}}$ $\frac{5}{13}+\frac{3}{13}=\frac{5+3}{13}=\frac{8}{13}$

EXAMPLE 2

Add and simplify. Remember that when you add a positive number and a negative number, you actually subtract the numbers and use the sign of the "greater" number.

$$-\frac{3}{8}+\frac{5}{8}=\frac{-3+5}{8}=\frac{2}{8}=\frac{1}{4}$$

2.04 Adding and Subtracting Fractions, Least Common Denominator 2.04

EXAMPLE 3

Subtract and simplify. Remember to change subtraction to add the opposite!

$$-\frac{11}{15}-\left(-\frac{7}{15}\right)$$

$$-\frac{11}{15}+\left(+\frac{7}{15}\right)=\frac{-11+7}{15}=\frac{-4}{15}=-\frac{4}{15}$$

Remember that a fraction can be thought of as division. Here we have a negative divided by a positive and then a positive divided by a negative. Both of these make a negative quotient. The negative sign needs to be written outside the fraction.

$$\frac{-4}{15}=\frac{4}{-15}=-\frac{4}{15}$$

YOUR TURN

Add or subtract and simplify.

1. $\frac{4}{11}+\frac{3}{11}+\frac{1}{11}$

2. $\frac{17}{18}-\frac{7}{18}$

3. $\frac{1}{8}-\frac{7}{8}$

4. $-\frac{9}{10}+\frac{9}{10}$

5. $-\frac{3}{10}-\left(-\frac{7}{10}\right)$

6. $\frac{7}{15}-\frac{2}{15}+\left(-\frac{5}{15}\right)$ Follow order of operations!

2.04 Adding and Subtracting Fractions, Least Common Denominator 2.04

WRITING EQUIVALENT FRACTIONS

In the next sections we will learn to add fractions with unlike denominators. In order to do that we must be able to write equivalent fractions.

Back to the fraction models!

EXAMPLE 4

How would we change $\frac{1}{2}$ to a fraction whose denominator is 8? Look at the figures below.

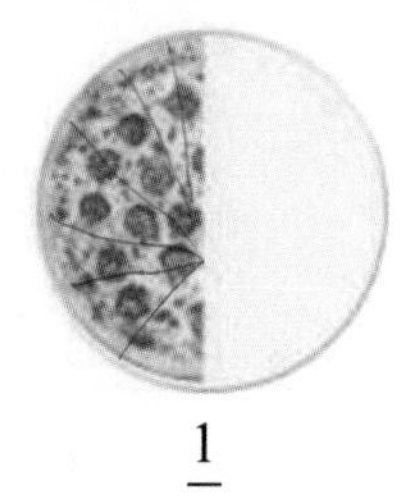

$\frac{1}{2}$

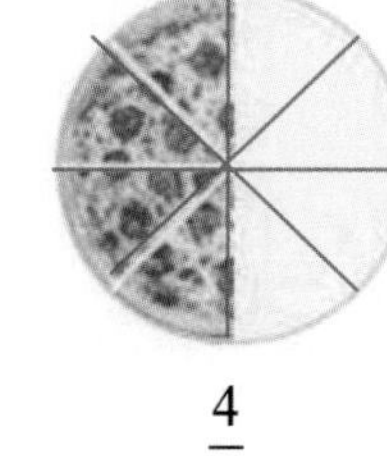

$\frac{4}{8}$

You see two representations of the same amount of pizza. How do we change halves to eighths? We would divide each half into four pieces, right? We can conclude that

$$\frac{1}{2} = \frac{4}{8}.$$

How do we do this with the fraction models? We ask ourselves how we change the denominator 2 into the denominator, 8; remember that the only number we can multiply a fraction by and not change its value is 1. This time 1 will be in the form of $\frac{4}{4}$ since $8 \div 2 = 4$.

$$\frac{1}{2} = \frac{\ }{8}$$

$$\frac{1}{2} \bullet \frac{4}{4} = \frac{4}{8}$$

EXAMPLE 5

Write $\frac{7}{20}$ as an equivalent fraction with the given denominator.

$$\frac{7}{20} = \frac{\quad}{100}$$

$$\frac{7}{20} \bullet \frac{5}{5} = \frac{35}{100}$$

2.04 Adding and Subtracting Fractions, Least Common Denominator 2.04

EXAMPLE 6

Write 3 as an equivalent fraction with the given denominator.

$$3 = \frac{?}{11}$$

$$\frac{3}{1} \bullet \frac{11}{11} = \frac{33}{11}$$

Remember that $3 = \frac{3}{1}$

YOUR TURN

Write as an equivalent fraction.

7. $\frac{3}{5} = \frac{}{75}$

8. $-\frac{8}{11} = -\frac{}{121}$

YOUR TURN Answers to Section 2.04

1. $\frac{8}{11}$ 2. $\frac{5}{9}$ 3. $-\frac{3}{4}$ 4. 0 5. $\frac{2}{5}$ 6. 0 7. $\frac{45}{75}$ 8. $-\frac{88}{121}$

Complete MyMathLab Section 2.04 Homework in your Homework notebook.

2.05 Adding and Subtracting Unlike Fractions 2.05

In order to add fractions with unlike denominators in the next section, we must also be able to find common denominators.

The least common denominator (LCD) is the same as the least common multiple of the denominators. It is the smallest number that is a multiple of the denominators. In other words, it is the smallest number that the denominators will divide into with no remainder.

FINDING THE LEAST COMMON DENOMINATOR

Let's find the least common denominator for $\frac{1}{4}$ and $\frac{1}{3}$

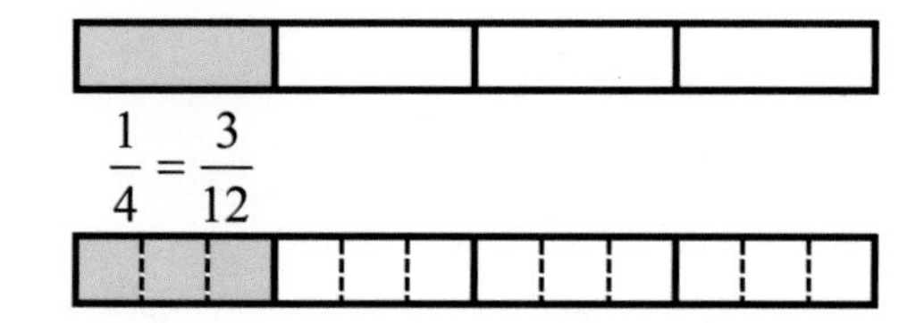

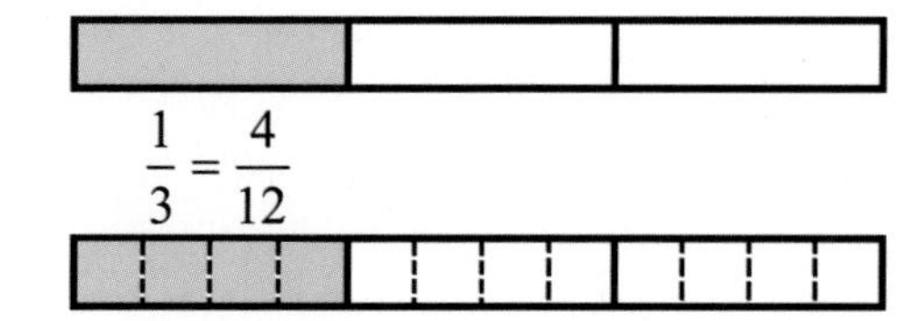

The common denominator is 12. That is because 12 is divisible by both 3 and 4.

How do we find the least common denominator (LCD) without using the fraction models?

Sometimes the LCD will just come to you. When it doesn't just come to you try using the prime factorizations of the denominators to help you find the LCD.

EXAMPLE 1

Find the LCD of $\frac{1}{72}$ and $\frac{9}{60}$.

$72 = 2 \cdot 2 \cdot 2 \cdot 3 \cdot 3$
$60 = 2 \cdot 2 \cdot 3 \cdot 5$

Prime factor each of the denominators.

$72 = 2 \cdot 2 \cdot 2 \cdot 3 \cdot 3$

$60 = 2 \cdot 2 \cdot 3 \cdot 5$

The LCD will have all of the factors within each of the denominators. Since there are two factors of two in 60 but three factors of two in 72, we must include three factors of 2 in the LCD so that all of the factors of 72 are in the LCD.

There also must be two factors of 3 in the LCD in order for all the factors of 72 to be in the LCD.

$\text{LCD} = 2 \cdot 2 \cdot 2 \cdot 3 \cdot 3 \cdot 5$

The LCD must include three factors of 2, two factors of 3 and one factor of 5.

$\text{LCD} = 360$

The LCD is 360.

2.05 Adding and Subtracting Unlike Fractions 2.05

You should always check to see if one of the denominators is a multiple of the other denominator. If it is, then the bigger denominator is a common multiple of the smaller one. The larger denominator is the LCD.

EXAMPLE 2

Find the LCD of $\frac{1}{28}$ and $\frac{3}{4}$.

Since 28 is a multiple of 4 (4 goes into 28 with no remainder), the LCD is 28.

EXAMPLE 3

Find the LCD of $\frac{8}{15}, \frac{5}{12}, \frac{9}{10}$.

$15 = 3 \cdot 5$
$12 = 2 \cdot 2 \cdot 3$
$10 = 2 \cdot 5$

Prime factor each of the denominators.

$15 = \textcircled{3} \cdot 5$
$12 = \textcircled{2 \cdot 2} \cdot 3$
$10 = 2 \cdot \textcircled{5}$

The LCD will have all of the factors within each of the denominators.
Circle all of the unique factors in the LCD. Since there are two factors of 2 in 12, then the LCD contains two factors of 2.

$\text{LCD} = 2 \cdot 2 \cdot 3 \cdot 5$
$\text{LCD} = 60$

Multiply the circled factors.
You can check by dividing each of the denominators into the LCD.

YOUR TURN

Find the LCD of each list of fractions.

1. $\frac{1}{7}, \frac{9}{28}$

2. $\frac{23}{18}, \frac{1}{21}$

YOUR TURN

Find the LCD of each list of fractions.

3. $-\frac{3}{25}, \frac{5}{6}$

Ignore the sign when finding the LCD.

4. $\frac{9}{14}, \frac{13}{15}, \frac{1}{12}$

ADDING AND SUBTRACTING UNLIKE FRACTIONS

Let's look at $\frac{1}{4} + \frac{1}{3}$.

$\frac{1}{4}$

$\frac{1}{3}$

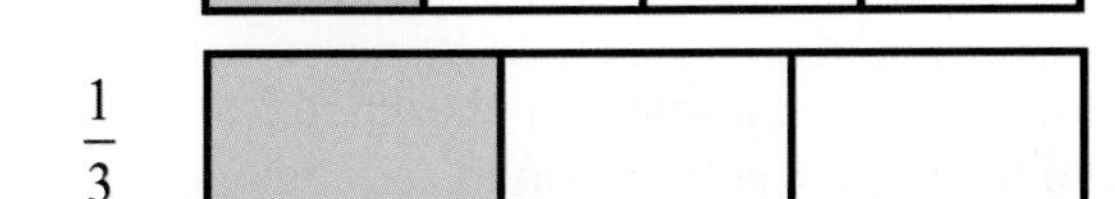

We cannot add these together since they are not divided into the same number of parts.

We can convert $\frac{1}{4}$ and $\frac{1}{3}$ to 12^{ths}.

$\frac{1}{4} = \frac{3}{12}$

$\frac{1}{3} = \frac{4}{12}$

$$\frac{1}{4} + \frac{1}{3}$$

$$\frac{1}{4} \bullet \frac{3}{3} + \frac{1}{3} \bullet \frac{4}{4}$$

$$\frac{3}{12} + \frac{4}{12} = \mathbf{\frac{7}{12}}$$

2.05 Adding and Subtracting Unlike Fractions

Method for Adding/Subtracting Unlike Fractions

1. Find the LCD of the denominators.
2. Rewrite both fractions to make the same LCD for each fraction. These are equivalent fractions.
3. Add or subtract the numerators, keep the denominators.
4. Simplify if possible.

EXAMPLE 4

Subtract and simplify: $\frac{1}{6} - \frac{5}{9}$

$\frac{1}{6} - \frac{5}{9}$

Find the LCD of 6 and 9.

$6 = 2 \bullet 3$

$9 = 3 \bullet 3$

$\text{LCD} = 2 \bullet 3 \bullet 3 = 18$

$\frac{1}{6} \bullet \frac{3}{3} - \frac{5}{9} \bullet \frac{2}{2}$

Rewrite both fractions so that each has a denominator of 18.

$\frac{3}{18} - \frac{10}{18}$

$\frac{3}{18} + \left(-\frac{10}{18}\right)$

Remember that if you are subtracting a small number minus (–) a large number, then you need to rewrite the subtraction to add the opposite.

$\frac{3 + (-10)}{18} = \frac{-7}{18} = -\frac{7}{18}$

When adding a positive and a negative, you actually subtract and use the sign of the "greater" number.

EXAMPLE 5

$2 - \frac{9}{11}$

Change 2 to a fraction.

$\frac{2}{1} - \frac{9}{11}$

Find the LCD.

$\frac{2}{1} \bullet \frac{11}{11} - \frac{9}{11}$

Rewrite the fractions so that each has the LCD as its denominator.

$\frac{22}{11} - \frac{9}{11} = \frac{22 - 9}{11} = \frac{13}{11}$

Subtract the numerators; keep the denominator!

We will look at changing improper fractions to mixed numbers later.

2.05 Adding and Subtracting Unlike Fractions

YOUR TURN

Add or subtract. Simplify your answer.

5. $\frac{1}{4} + \frac{1}{8}$

6. $\frac{2}{3} - \frac{1}{4}$

7. $-\frac{1}{6} - \frac{1}{3}$

8. $\frac{2}{5} + \frac{1}{3} + \frac{1}{10}$

COMPARING UNLIKE FRACTIONS

We have learned that if two fractions have the same denominators then the fraction with the greater numerator is the greater fraction.

EXAMPLE 6

Insert < or > to form a true statement.

$\frac{7}{12}$ ___ $\frac{8}{15}$

$\frac{7}{12} \bullet \frac{5}{5}$ ___ $\frac{8}{15} \bullet \frac{4}{4}$ Change each fraction so that each has the LCD as the denominator.

$\frac{35}{60} > \frac{32}{60}$ Compare the numerators. Since the denominators are the same, the fraction with the greater numerator is the greater fraction.

$\frac{7}{12} > \frac{8}{15}$

2.05 Adding and Subtracting Unlike Fractions 2.05

YOUR TURN

Be careful with the last one.

Insert $<$ or $>$ to form a true statement.

9. $\frac{5}{9}$ ___ $\frac{6}{11}$

10. $-\frac{7}{8}$ ___ $\frac{5}{6}$

11. $-\frac{3}{4}$ ___ $-\frac{11}{12}$

12. The gutter on a roof overflows whenever more than $\frac{1}{2}$ inch of rain falls. Saturday it rained $\frac{7}{10}$ inch. Did the gutter overflow? How do you know?

APPLICATIONS

Krystal has completed typing $\frac{5}{8}$ of her paper. What part of the paper does she have left to type?

The "*part does she have left*" indicates that we will subtract.

Whole paper – part already typed = part left to type

$1 - \frac{5}{8}$ = part left to type

$$\frac{8}{8} - \frac{5}{8} = \frac{3}{8}$$

Remember that 1 can be expressed as $\frac{8}{8}$.

Krystal has $\frac{3}{8}$ of the paper left to type.

2.05 Adding and Subtracting Unlike Fractions 2.05

YOUR TURN

Answer each question. Show your work.

13. Allie is making barbecue sauce that calls for $\frac{3}{4}$ cup of molasses. She has $\frac{1}{2}$ cup of molasses.

    How much more of the molasses does she need?

Courtesy of Fotolia

14. Reggie and three friends stopped at a pizza parlor Saturday night and shared a large pizza equally. The next day Reggie and seven friends stopped at the same pizza parlor for a snack. This time the eight friends shared a large pizza equally among them and then ordered a second large pizza and shared it equally. Did Reggie eat more pizza on Saturday or on Sunday? Or did he eat the same amount each day? ***Draw a diagram to prove your answer.***

YOUR TURN Answers to Section 2.05

1. 28 2. 126 3. 150 4. 420 5. $\frac{3}{8}$ 6. $\frac{5}{12}$ 7. $-\frac{1}{2}$ 8. $\frac{5}{6}$

9. > 10. < 11. > 12. Yes, since $\frac{1}{2} < \frac{7}{10}$ 13. $\frac{1}{4}$ cup 14. He ate the same amount on both days.

Complete MyMathLab Section 2.05 Homework in your Homework notebook.

2.06 Order of Operations and Mixed Numbers 2.06

Let's review order of operations.

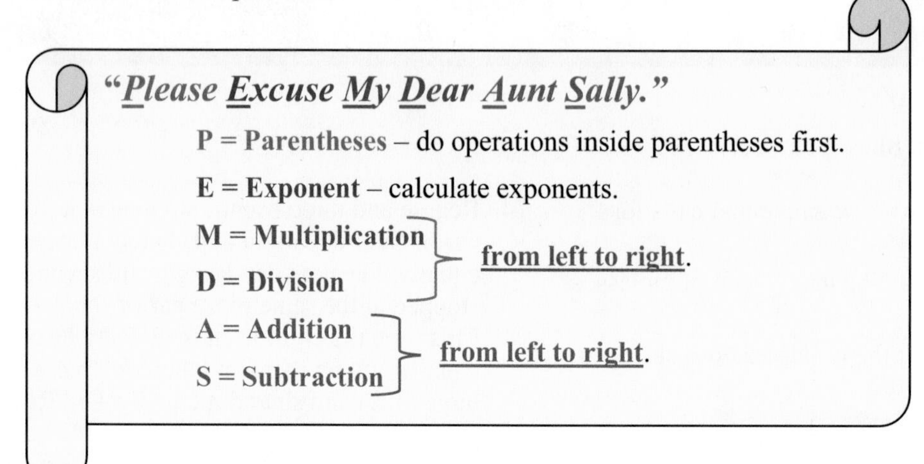

EXAMPLE 1

Simplify using the order of operations.

$\frac{2}{3} \div \frac{4}{5} \bullet \frac{3}{5}$ Multiply or divide from left to right, so do the division first.

$\frac{2}{3} \bullet \frac{5}{4} \bullet \frac{3}{5}$ Change $\div$ to multiply and $\frac{4}{5}$ to its reciprocal, $\frac{5}{4}$.

$\frac{\cancel{2}^{1}}{\cancel{3}_{1}} \bullet \frac{\cancel{5}^{1}}{\cancel{4}_{2}} \bullet \frac{\cancel{3}^{1}}{\cancel{5}_{1}} = \frac{1}{2}$ Now we have all multiplication. We can divide out the common factors.

Make sure that you are looking for common factors in a numerator and a denominator, not two numerators or two denominators.

Fractions are fun!

2.06 Order of Operations and Mixed Numbers

EXAMPLE 2

Simplify.

$$\left(\frac{3}{4}-\frac{1}{12}\right)\div\left(\frac{3}{4}\right)^2$$

$$\left(\frac{3}{4}\bullet\frac{3}{3}-\frac{1}{12}\right)\div\left(\frac{3}{4}\right)^2$$ Simplify the parentheses.

$$\left(\frac{9}{12}-\frac{1}{12}\right)\div\left(\frac{3}{4}\right)^2$$ **ADD/SUBTRACT FRACTIONS**  **LCD**

$$\frac{8}{12}\div\left(\frac{3}{4}\right)^2$$

$$\frac{8}{12}\div\frac{9}{16}$$ Next exponents. Remember that $\left(\frac{3}{4}\right)^2=\frac{3}{4}\bullet\frac{3}{4}$.

$$\frac{8}{12}\bullet\frac{16}{9}$$ Now division. 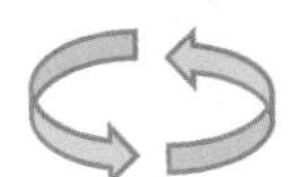 **KEEP, CHANGE, FLIP.**

$$\frac{8}{\not{12}_{3}}\bullet\frac{\not{16}^{4}}{9}=\frac{\mathbf{32}}{\mathbf{27}}$$ Simplify.

EXAMPLE 3

Evaluate $2yz-x$ when $x=-\frac{1}{3}$, $y=\frac{2}{5}$, $z=\frac{5}{6}$.

Substitute the values for the variables into the expression. It is a good idea to use parentheses around the number you are substituting.

$$2yz-x$$

$$2\left(\frac{2}{5}\right)\left(\frac{5}{6}\right)-\left(-\frac{1}{3}\right)$$

Be careful substituting negative numbers after a subtraction sign.

Courtesy of Fotolia

$$\frac{\not{2}^{1}}{1}\bullet\frac{2}{\not{5}_{1}}\bullet\frac{\not{5}^{1}}{\not{6}_{3}}-\left(-\frac{1}{3}\right)$$ Take care of the multiplication.

$$\frac{2}{3}-\left(-\frac{1}{3}\right)$$ Subtract a negative means to add a positive!!

$$\frac{2}{3}+\left(+\frac{1}{3}\right)=\frac{3}{3}=\mathbf{1}$$ Add and simplify.

2.06 Order of Operations and Mixed Numbers 2.06

FRACTION REMINDERS

Multiplying – Express each number as a fraction and multiply across.

Division – Keep, Change, Flip

Addition and Subtraction – Find a common denominator before adding or subtracting.

YOUR TURN

Simplify each expression using order of operations.

1. $\frac{1}{2} + \frac{3}{4} \cdot \frac{1}{3}$

2. $\left(\frac{4}{3}\right)^3 \div \left(2 - \frac{2}{3}\right)$

3. $\frac{1}{2} \cdot \left(-\frac{1}{3}\right)^2$

4. Evaluate when $x = -\frac{1}{3}$ and $y = \frac{2}{5}$.

   $y - x$

5. Evaluate when $x = -\frac{1}{3}$, $y = \frac{2}{5}$, $z = \frac{5}{6}$.

   $yz^2 + 2x$

MIXED NUMBERS AND IMPROPER FRACTIONS

We have looked at improper fractions and mixed numbers before. Here we can express the fraction represented in two ways: the mixed number $1\frac{11}{12}$ and the improper fraction $\frac{23}{12}$.

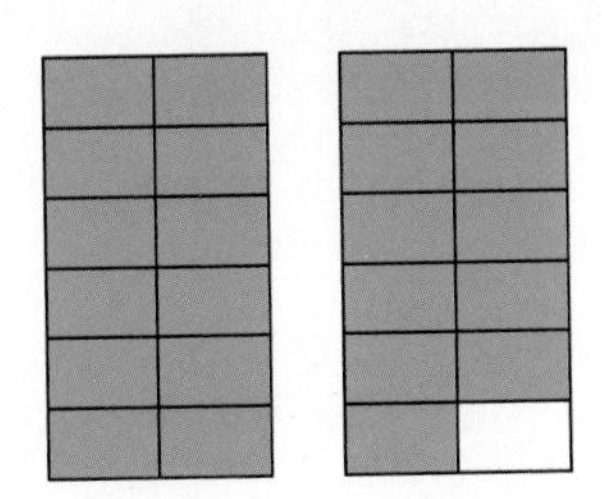

A mixed number is a number greater than 1 that has a whole number part and a fraction part.

An improper fraction is a fraction where the numerator is greater than or equal to its denominator.

How do we change a mixed number into an improper fraction?

Suppose we need to convert $1\frac{3}{4}$ to an improper fraction. We need to change the whole number part into fourths since the fraction part of the mixed number is fourths.

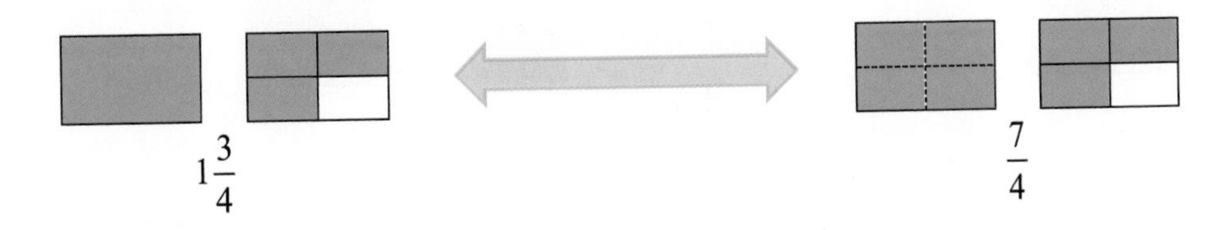

How do we do this without using fraction models?

EXAMPLE 4

Change $2\frac{1}{3}$ to an improper fraction. We have 2 wholes that need changing to thirds and then added to one-third.

$$2\frac{1}{3} \rightarrow \frac{3 \cdot 2 + 1}{3} = \frac{7}{3}$$

Since the denominator was 3 in the mixed number then it is 3 in the improper fraction.

2.06 Order of Operations and Mixed Numbers 2.06

Converting a Mixed Number to an Improper Fraction

- **Multiply the denominator times the whole number.**
- **Add the numerator.**
- **Write this answer over the denominator.**

EXAMPLE 5

Change $4\frac{6}{7}$ to an improper fraction.

$$\mathbf{4}\frac{6}{7} = \frac{7 \cdot \mathbf{4} + 6}{7} = \frac{34}{7}$$

Now let's look at how to convert an improper fraction to a mixed number. Looking at the model we can see that $\frac{7}{4} = 1\frac{3}{4}$.

Converting an Improper Fraction to a Whole or Mixed Number

- **Divide the denominator into the numerator. Answer is the whole number.**
- **The remainder is the numerator.**
- **The denominator is the same as the denominator in the improper fraction.**

$$\text{improper fraction} = \text{quotient}\ \frac{\text{remainder}}{\text{original denominator}}$$

2.06 Order of Operations and Mixed Numbers 2.06

EXAMPLE 6

Change $\frac{17}{5}$ to a whole or mixed number.

$$5\overline{)17} \text{ quotient } 3,\quad 17-15=2 \quad\rightarrow\quad 3\frac{2}{5}$$

$$\text{quotient}\ \frac{\text{remainder}}{\text{original denominator}}$$

1. Divide the denominator into the numerator. This answer is the whole number part.
2. Our numerator is the remainder.
3. Our denominator is the denominator of the original fraction.

EXAMPLE 7

Convert $\frac{182}{52}$ to a whole or mixed number.

$$52\overline{)182} \text{ quotient } 3,\quad 182-156=26 \quad\rightarrow\quad 3\frac{26}{52}=3\frac{1}{2}$$

$$\text{quotient}\ \frac{\text{remainder}}{\text{original denominator}}$$

1. Divide the denominator into the numerator. This answer is the whole number part.
2. Our numerator is the remainder.
3. Our denominator is the denominator of the original fraction.
4. Simplify the fraction.

EXAMPLE 8

Convert $\frac{96}{12}$ to a whole or mixed number.

$$12\overline{)96} \text{ quotient } 8,\quad 96-96=0 \quad\rightarrow\quad 8\frac{0}{12}=8$$

$$\text{quotient}\ \frac{\text{remainder}}{\text{original denominator}}$$

1. Divide the denominator into the numerator. This answer is the whole number part.
2. Our numerator is the remainder.
3. Our denominator is the denominator of the original fraction.
4. Since the remainder is 0, then there is no fraction part. This time the improper fraction gives us a whole number.

2.06 Order of Operations and Mixed Numbers 2.06

YOUR TURN

Change to improper fractions.

6. $8\frac{5}{9}$

7. $12\frac{2}{5}$

Change to a whole or mixed number.

8. $\frac{64}{9}$

9. $\frac{225}{15}$

10. $\frac{111}{6}$

YOUR TURN **Answers to Section 2.06.**

1. $\frac{3}{4}$ 2. $\frac{16}{9}$ 3. $\frac{1}{18}$ 4. $\frac{11}{15}$ 5. $-\frac{7}{18}$

6. $\frac{77}{9}$ 7. $\frac{62}{5}$ 8. $7\frac{1}{9}$ 9. 15 10. $18\frac{1}{2}$

Complete MyMathLab Section 2.06 Homework in your Homework notebook.

GRAPHING FRACTIONS AND MIXED NUMBERS

Let's look at graphing fractions on the number line again. Each unit has been divided into fourths.

Remember that when we look at the negative side, we have to kind of think "backwards".

Notice that $-\frac{3}{4}$ is closer to -1 than it is to 0. Likewise $\frac{3}{4}$ is closer to 1 than it is to 0 since $\frac{3}{4}$ is more than $\frac{1}{2}$ way from 0 to 1.

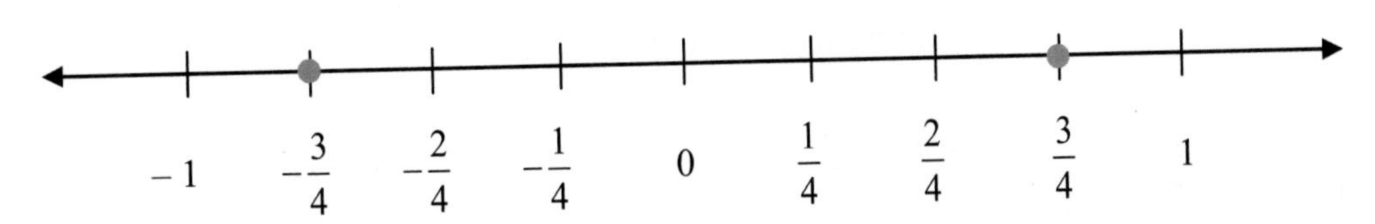

When placing the graph (dot) on the number line, think about how you must divide the space between the two integers. The number of ***spaces*** (not hash marks) between the 2 integers must be the same as the denominator of the fraction you are trying to graph. So if you are trying to graph $1\frac{3}{7}$, then you would think about dividing the space between 1 and 2 into 7 ***spaces***. Then $1\frac{3}{7}$ would be on the third mark after the 1 on the number line.

EXAMPLE 1

Graph and label each point on the number line below. $A:\frac{1}{2}, B:-\frac{2}{3}, C:-2\frac{1}{4}, D:3\frac{3}{5}, E:-3\frac{1}{8}$

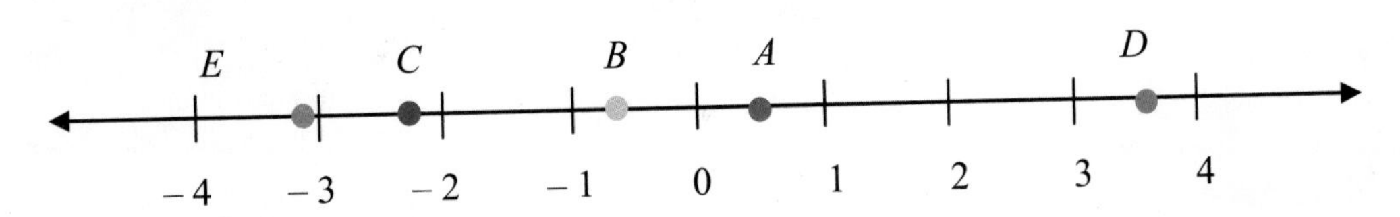

2.07 Operations on Mixed Numbers

When working with mixed numbers it is a good idea to estimate the answer to see if your answer is reasonable. Let's look at rounding mixed numbers first.

Courtesy of Fotolia

When estimating, you must round each number first and then perform the operation.

ROUNDING MIXED NUMBERS

Rounding mixed numbers is similar to rounding whole numbers or decimals.

- If the fraction part of the mixed number is greater than or equal to one-half, the whole number part will increase by one.
- If the fraction part of the mixed number is less than one-half, the whole number part will stay the same.
- One way to decide if the fraction is less than or greater than one-half is to think about whether the numerator is less than or greater than half of the denominator.

Another way to look at it!

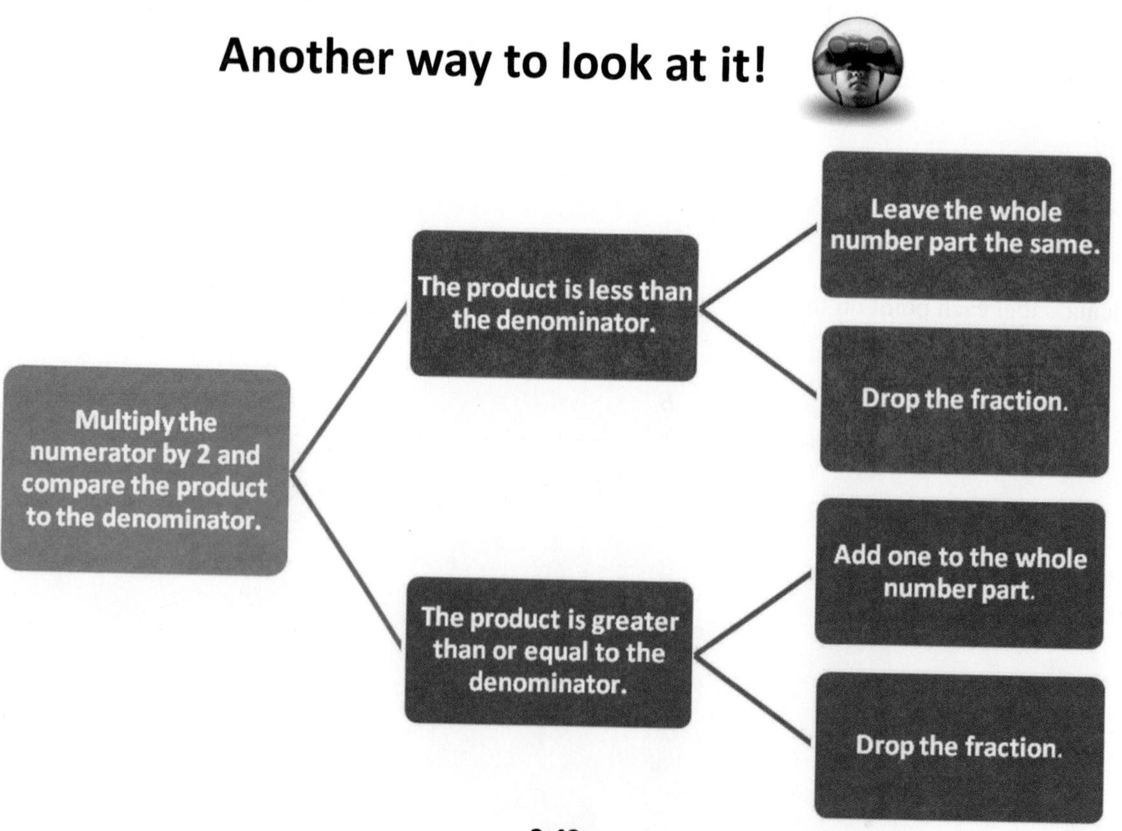

2.07 Operations on Mixed Numbers

EXAMPLE 2

Round $3\frac{1}{5}$ to the nearest whole number.

Solution:

Multiply the numerator (1) by 2. $\mathbf{1 \times 2 = 2}$

Is this product less than or greater than the denominator? $\mathbf{2 < 5}$

Since $2 < 5$, keep the whole number part, 3, and drop the fraction.

$$\mathbf{3\frac{1}{5} \approx 3}$$

EXAMPLE 3

Round $4\frac{8}{15}$ to the nearest whole number.

Solution:

Multiply the numerator, 8, by 2. $\mathbf{8 \times 2 = 16}$

Is this product less than or greater than the denominator, 15? $\mathbf{16 > 15}$

Since $16 > 15$, add 1 to the whole number part, 4, and drop the fraction.

$$\mathbf{4\frac{8}{15} \approx 5}$$

MULTIPLYING AND DIVIDING MIXED NUMBERS

Remember that when we are multiplying and dividing fractions, each of our numbers must be in fraction form. If we have a mixed number, then we must first change it to an improper fraction before multiplying or dividing.

EXAMPLE 4

Multiply $2\frac{2}{3} \bullet 3\frac{1}{4}$. First find the exact answer.

$$2\frac{2}{3} \bullet 3\frac{1}{4}$$

$$\frac{3 \bullet 2 + 2}{3} \bullet \frac{4 \bullet 3 + 1}{4}$$ Change each mixed number to an improper fraction.

$$\frac{\overset{2}{\cancel{8}}}{3} \bullet \frac{13}{\underset{1}{\cancel{4}}} = \frac{26}{3} = \mathbf{8\frac{2}{3}}$$ Divide out the common factors in the numerator and denominator. Then multiply across.

Change the improper fraction to a mixed number.

Let's estimate the answer. (Round the first two numbers and then multiply to find an estimate.)

Correct	Incorrect
$2\frac{2}{3} \approx 3$ **since** $2 \bullet 2 = 4$ **and** $4 > 3$ $3\frac{1}{4} \approx 3$ **since** $2 \bullet 1 = 2$ **and** $2 < 4$ Thus our answer is reasonable since $2\frac{2}{3} \bullet 3\frac{1}{4} \approx 3 \bullet 3 \approx 9$	$2\frac{2}{3} \bullet 3\frac{1}{4} = 8\frac{2}{3} \approx 9$ You must round each of the numbers ***BEFORE*** you multiply.

2.07 Operations on Mixed Numbers 2.07

EXAMPLE 5

Divide $8 \div 1\frac{3}{5}$. Find an exact solution.

$8 \div 1\frac{3}{5}$

$\frac{8}{1} \div \frac{8}{5}$ Change the whole number and the mixed number to improper fractions.

$\frac{\cancel{8}^{1}}{1} \bullet \frac{5}{\cancel{8}_{1}} = \frac{5}{1} = 5$ Remember how to divided fractions: **Keep, Change, Flip.**

Divide out the common factors in the numerator and denominator and multiply across. Change the improper fraction to a whole number. This is the exact solution.

Find the estimate.

The number 8 does not need to be rounded. Then note how to round the second number.

$1\frac{3}{5} \approx 2$ since **2 • 3 = 6 and 6 > 5.**

$8 \div 1\frac{3}{5} \approx 8 \div 2 \approx 4$ our estimate is reasonable (close enough).

YOUR TURN

1. Graph and label these points on the number line below.

$$A:\frac{1}{4}, B:-\frac{1}{3}, C:-3\frac{3}{4}, D:2\frac{1}{5}, E:-1\frac{7}{8}$$

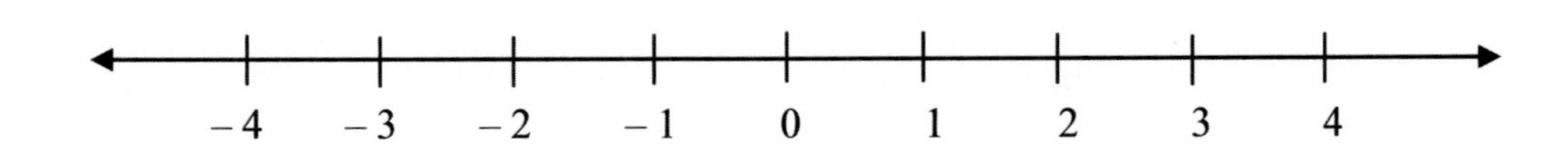

2. Multiply $6\frac{1}{4} \bullet 1\frac{1}{5}$.

Estimate:_______

Exact answer_______

3. Divide $4\frac{1}{5} \div \frac{5}{9}$.

Estimate:_______

Exact answer_______

ADDING MIXED NUMBERS

We will align mixed numbers vertically to add them. This is just one method which is shown here. (Another way to add mixed numbers is to change them to improper fractions first and then add .)

EXAMPLE 6

Add: $7\frac{1}{4} + 3\frac{1}{3}$

$$7\frac{1}{4} \cdot \frac{3}{3} = 7\frac{3}{12}$$

Find the LCD and change each fraction so that it has the LCD as its denominator.

$$+\ 3\frac{1}{3} \cdot \frac{4}{4} = 3\frac{4}{12}$$

Add the whole numbers together.

$$10\frac{7}{12}$$

Add the fractions together.

What if the fraction part of the answer is an improper fraction? Change that improper fraction to a mixed number. Remember the mixed number $3\frac{15}{8}$ can be thought of as $3 + \frac{15}{8}$.

$$3\frac{15}{8} = 3 + \frac{15}{8}$$

$$3 + 1\frac{7}{8}$$

Change the improper fraction to a mixed number.

Now add the whole numbers together and we have a simplified answer.

$$3\frac{15}{8} = 4\frac{7}{8}$$

2.07 Operations on Mixed Numbers

Refer back to the instructions for rounding mixed numbers on page 2–42 to help you find the estimate.

EXAMPLE 7

Add. Find an exact sum and an estimated sum. $3\frac{7}{8}+4\frac{5}{8}$

$$\begin{array}{r} 3\frac{7}{8} \\ +\,4\frac{5}{8} \\ \hline 7\frac{12}{8} \end{array}$$

$$7\frac{12}{8} = 7 + 1\frac{4}{8} = 8\frac{4}{8} = 8\frac{1}{2}$$

No need to find an LCD this time! ☺

We add and see that we cannot leave the answer with the improper fraction. Change $\frac{12}{8}=1\frac{4}{8}$ then add 7 to $1\frac{4}{8}$. This is the exam sum.

Estimate: $3\frac{7}{8} + 4\frac{5}{8} \approx 4 + 5 \approx 9$ so the answer is reasonable.

SUBTRACTING MIXED NUMBERS

We will align mixed numbers vertically to subtract them. This is just one method which is shown here. (Another way to add mixed numbers is to change them to improper fractions first and then subtract.)

EXAMPLE 8

Subtract. Find an exact difference and an estimated difference. $8\frac{1}{6}-4\frac{5}{6}$

Courtesy of Fotolia

Always subtract the fraction part before subtracting the whole number part, just in case you need to borrow. This is the case in this example.

$$\begin{array}{rcl} \overset{7}{\cancel{8}}\frac{\overset{11}{\cancel{1}}}{6} & = & 7\frac{11}{6} \\ -\,4\frac{5}{6} & = & 4\frac{5}{6} \\ \hline \end{array}$$

A BIG mistake that students make in borrowing is to just borrow 1 from 8 and make the numerator 10 more than it was.

WRONG!

2.07 Operations on Mixed Numbers 2.07

Look at the correct way to borrow.

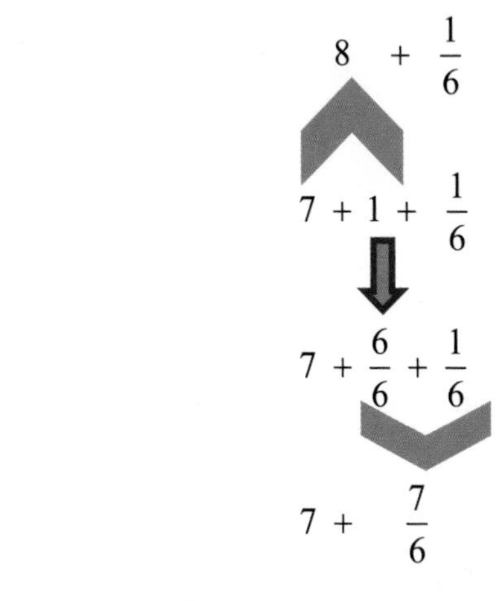

Recall: $8\frac{1}{6} = 8 + \frac{1}{6}$.

$8 = 7 + 1$

Change 1 to sixths since that is the denominator of the fraction.

Combine the fractions. Make sense? When we borrow 1 from the 8 we are actually borrowing 6 sixths. We already had 1 sixth, so we now have $\frac{7}{6}$ as the fraction part.

$$\begin{array}{rcr} 8\frac{1}{6} & = & 7\frac{7}{6} \\ -\ 4\frac{5}{6} & = & 4\frac{5}{6} \\ \hline 3\frac{2}{6} & = & 3\frac{1}{3} \end{array}$$

Correct solution!

Estimate: $8\frac{1}{6} - 4\frac{5}{6} \approx 8 - 5 \approx 3$. (Remember to round the first two numbers and then subtract.)

EXAMPLE 9

Subtract: $4\frac{1}{2} - 3\frac{3}{8}$. Always find the common denominator *before* borrowing.

Courtesy of Fotolia

$$\begin{array}{rcr} 4\frac{1}{2} & = & 4\frac{4}{8} \\ -3\frac{3}{8} & = & 3\frac{3}{8} \\ \hline & & 1\frac{1}{8} \end{array}$$

Change fractions so that they have the same LCD.

Subtract fractions, and then subtract whole numbers. No need to borrow! ☺

FINISHED!

2.07 Operations on Mixed Numbers

EXAMPLE 10

Subtract: $9\frac{2}{3} - 8\frac{3}{4}$

$$\begin{array}{rcccr} 9\frac{2}{3} & = & 9\frac{8}{12} & = & 8\frac{20}{12} \\ -\,8\frac{3}{4} & = & 8\frac{9}{12} & = & 8\frac{9}{12} \\ \hline & & & & \frac{11}{12} \end{array}$$

Find the LCD and make the necessary changes.

This time we do need to borrow. Borrow 1 from 9 and change it to $\frac{12}{12}$. Add this $\frac{12}{12}$ to $\frac{8}{12}$ that we already had.

Subtract. FINISHED!

EXAMPLE 11

Subtract: $14 - 3\frac{2}{15}$

$$\begin{array}{rcr} 14 & = & 13\frac{15}{15} \\ -\,3\frac{2}{15} & = & -3\frac{2}{15} \\ \hline & & 10\frac{13}{15} \end{array}$$

Borrow 1 from the 14 and change it to $\frac{15}{15}$.

Subtract. Don't forget to simplify, if possible.

FINISHED!

Fractions are even MORE fun now!

2.07 Operations on Mixed Numbers 2.07

YOUR TURN

Perform the indicated operation and simplify the answer. Give an estimate and an exact answer.

4. $11\frac{4}{5} + 1\frac{1}{5}$

Estimate:________

Exact:__________

5. $9 + 3\frac{1}{2}$

Estimate:________

Exact:__________

6. $4\frac{2}{3} + 1\frac{5}{6} + 3\frac{1}{4}$

Estimate:________

Exact:__________

7. $9\frac{7}{15} - 1\frac{11}{15}$

Estimate:________

Exact:__________

8. $3\frac{4}{5} - 1\frac{1}{3}$

Estimate:________

Exact:__________

9. $8 - 3\frac{11}{12}$

Estimate:________

Exact:__________

APPLICATIONS AGAIN!

EXAMPLE 12

Latisha drove $301\frac{3}{4}$ miles on $15\frac{1}{3}$ gallons of gas. What were her miles per gallon?

$301\frac{3}{4} \div 15\frac{1}{3}$ — Change mixed numbers to improper fractions,

$\frac{1207}{4} \div \frac{46}{3}$ — ***KEEP, CHANGE, FLIP***

$\frac{1207}{4} \times \frac{3}{46} = \frac{3621}{184} = 19\frac{125}{184}$ miles per gallon — Then multiply across the top; multiply across the bottom.
Change the improper fraction to a mixed number.

2.07 Operations on Mixed Numbers 2.07

EXAMPLE 13

Courtesy of Fotolia

A plane ride from Greensboro to Dallas requires $3\frac{1}{2}$ hours. If the plane has been flying $1\frac{3}{4}$ hours, find **how much time remains** before landing.

"How much time remains" is the same as how much time is left!

$$\begin{array}{r} 3\frac{1}{2} = 3\frac{1}{2} \cdot \frac{2}{2} = 3\frac{2}{4} = 2\frac{6}{4} \\ -1\frac{3}{4} = 1\frac{3}{4} \qquad\quad = 1\frac{3}{4} = 1\frac{3}{4} \\ \hline 1\frac{3}{4} \text{ hours} \end{array}$$

How much time is left indicates **subtraction.**

Determine the LCD (4), and rewrite both fractions

Subtract like fractions. You will need to borrow!

Simplify, if needed.

EXAMPLE 14

I am going to bake some sugar cookies that require $2\frac{1}{3}$ cups of flour. If I make $3\frac{1}{2}$ recipes, how much flour is needed?

Think about the answer. Will it be more than $2\frac{1}{3}$ cups of flour or less?

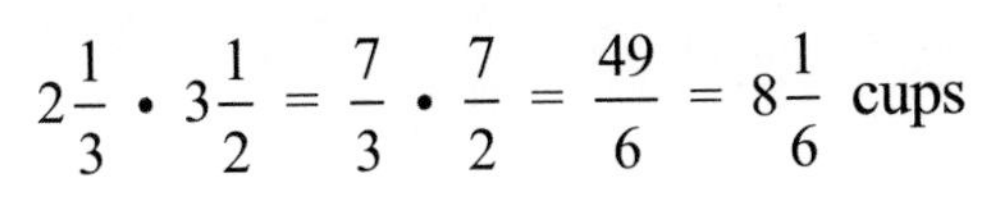

$$2\frac{1}{3} \cdot 3\frac{1}{2} = \frac{7}{3} \cdot \frac{7}{2} = \frac{49}{6} = 8\frac{1}{6} \text{ cups}$$

We are making ***more than 3 times the recipe***, right? Thus, we multiply! No common denominator required!

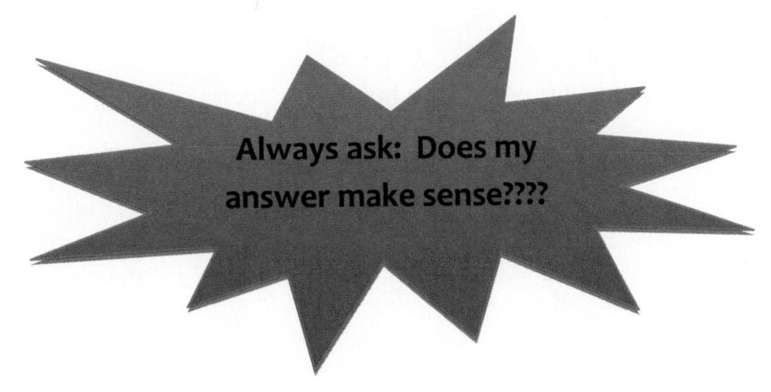

EXAMPLE 15

Jonathan had a 20 ft pipe to use for home improvements. He has already used a $3\frac{1}{2}$ foot piece of pipe and a $6\frac{3}{4}$ foot piece of pipe on other projects. He needs 10 ft of the pipe for his current project. Does he need to buy another pipe or does he have enough?

Estimate the answer.

$3\frac{1}{2} \approx 4$ and $6\frac{3}{4} \approx 7$. If he has used 11 (4 + 7) feet of the 20 ft pipe, does he have 10 ft left????

No, since 20 – 11 = 9. Now let's work out the exact amount he has left.

Let's find how much he has already used.

$$\begin{array}{rl} 3\frac{1}{2} = 3\frac{1}{2} \cdot \frac{2}{2} & = 3\frac{2}{4} \\ +\,6\frac{3}{4} = 6\frac{3}{4} \quad & = 6\frac{3}{4} \\ \hline & 9\frac{5}{4} = 10\frac{1}{4} \end{array}$$

ft already used.

Since this amount has already been "taken away" from the pipe then we will subtract that amount from 20 to find out how much is left for the new project.

$$\begin{array}{rl} 20 & = 19\frac{4}{4} \\ -10\frac{1}{4} & = 10\frac{1}{4} \\ \hline & 9\frac{3}{4} \text{ ft} \end{array}$$

Since he needs 10 ft he does not have enough of the pipe for his new project.

YOUR TURN

Answer each question. Show your work.

10. An average building can last 60 years. If a building which is considered temporary lasts $\frac{1}{14}$ as long, how long will it last? Round to the nearest year.

Courtesy of Fotolia

(continue to next page)

2.07 Operations on Mixed Numbers

YOUR TURN

11. You buy $1\frac{1}{2}$ lb of fudge and $\frac{3}{4}$ lb of taffy. What is the total amount of candy purchased?

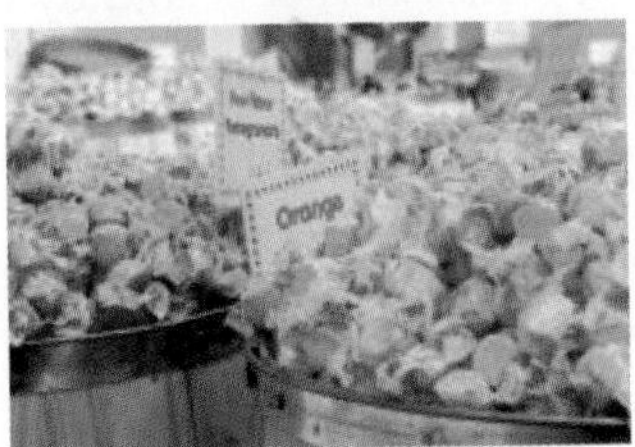

Courtesy of Fotolia

12. Cierra is going to bake some sugar cookies that require $2\frac{1}{4}$ cups of flour. If she makes $4\frac{1}{2}$ recipes, how much flour is needed?

Courtesy of Fotolia

13. The area of a rectangle is $93\frac{1}{2}$ square inches. Find the width if the length is 11 inches.

14. Jonathan had a 15 ft board to use for home improvements. He has already used a 2 foot piece of the board and a $6\frac{3}{4}$ foot piece on other projects. He needs a 5 ft of the board for his current project. Does he have enough or does he need to buy more? If so, how much will be left over? If not, how much more will be need to buy?

(continue to next page)

2.07 Operations on Mixed Numbers 2.07

YOUR TURN

15. Gail is working on sprucing up her home. She plans to put down new carpet and add a border to one of the rooms in her home. The wallpaper border is sold in 15 ft rolls. Carpet is sold by the square yard (1 square yard = 9 square feet). How much of each will she need if the room measures $17\frac{1}{2}$ ft by $18\frac{1}{2}$ ft.

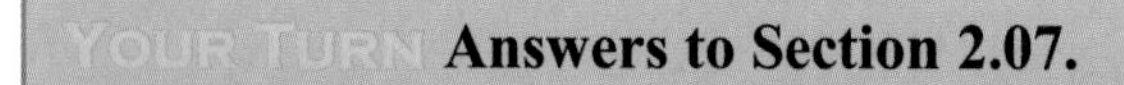

YOUR TURN **Answers to Section 2.07.**

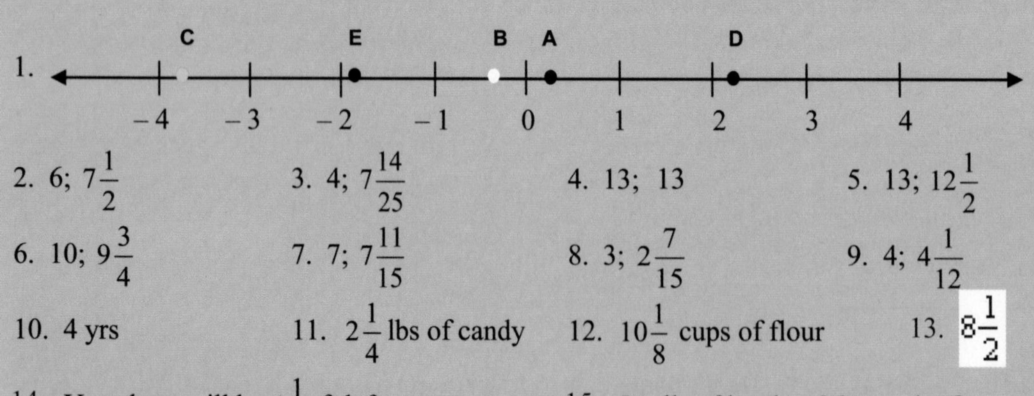

1\.

2\. 6; $7\frac{1}{2}$ 3. 4; $7\frac{14}{25}$ 4. 13; 13 5. 13; $12\frac{1}{2}$

6\. 10; $9\frac{3}{4}$ 7. 7; $7\frac{11}{15}$ 8. 3; $2\frac{7}{15}$ 9. 4; $4\frac{1}{12}$

10\. 4 yrs 11. $2\frac{1}{4}$ lbs of candy 12. $10\frac{1}{8}$ cups of flour 13. $8\frac{1}{2}$

14\. Yes, there will be $1\frac{1}{4}$ ft leftover 15. 5 rolls of border; 36 sq. yd. of carpet

Complete MyMathLab Section 2.07 Homework in your Homework notebook.

2.08 An Introduction to Decimals 2.08

Decimal notation is used to represent part of a whole – like fractions. A decimal number contains a decimal point. The number to the left of the decimal point is a whole number and the number to the right of the decimal is the decimal part.

Here is a chart that gives place value for whole numbers and decimals.

Notice that place values to the left of the red line (whole numbers) end in "s". Place values to the right of the red line (decimal part) end in "ths".

Hundred-thousands	Ten-thousands	Thousands	Hundreds	Tens	Ones	Tenths	Hundredths	Thousandths	Ten-thousandths	Hundred-thousandths
100,000	10,000	1000	100	10	1	$\frac{1}{10}$	$\frac{1}{100}$	$\frac{1}{1000}$	$\frac{1}{10,000}$	$\frac{1}{100,000}$

TRANSLATING DECIMAL NUMBERS TO WORDS

To translate a decimal into words use the following steps:

- **Write the whole number part in words.**
- **Write "and" for the decimal point.**
- **Write the decimal part in words followed by the place value of the last digit.**

EXAMPLE 1

Write 2.45 in words.

Read the number before the decimal.	Read the decimal point.	Read the number after the decimal.	Count the number of decimal places and convert that to the correct decimal word.
2	.	45	two places to left
two	and	forty-five	hundredths

2.45 ⟷ two and forty-five hundredths

2.08 An Introduction to Decimals 2.08

EXAMPLE 2

Write 700.009 in words.

Read the number before the decimal.	Read the decimal point.	Read the number after the decimal.	Count the number of decimal places and convert that to the correct decimal word.
700	**.**	**9**	**three places to left**
Seven hundred	**and**	**nine**	**Thousandths**

700.009 ⟷ seven hundred and nine thousandths

WRITING DECIMALS IN STANDARD FORM

EXAMPLE 3

Write three and forty five hundredths in standard form.

Write the number before the decimal.	Write the decimal point.	Write the number after the decimal.	Convert the decimal "word" to the correct number of decimal places.
three	**and**	**forty-five**	**Hundredths**
3	**.**	**45**	**two places to the left**

three and forty-five hundredths ⟷ 3.45

EXAMPLE 4

Write fifteen and six thousandths in standard form.

Write the number before the decimal.	Write the decimal point.	Write the number after the decimal.	Convert the decimal "word" to the correct number of decimal places.
fifteen	**and**	**six**	**Thousandths**
15	**.**	**6**	**three places to the left**

This time we need three decimal places. However, the number after the decimal, 6, is only one digit. This 6, the last digit of the number after the decimal, must be in the thousandths place (third place to the right of the decimal).

15.__ __ 6 Fill in the blanks **15.006**

fifteen and six thousandths ⟷ 15.006

2.08 An Introduction to Decimals 2.08

YOUR TURN

Write the following decimals in words.

1. 12.036

2. 1. 0002

Write in standard form.

3. Three and twenty-six thousandths

4. twenty-two and five ten-thousandths

WRITING DECIMALS AS FRACTIONS

Since decimal numbers represent part of a whole, let's look at fraction models again.

Look at the decimal number 1.5; this is one whole and five tenths of another whole thing. The fraction model for this may look like this:

Think about another name for this model. How about $1\frac{5}{10}$ or $1\frac{1}{2}$?

Look at the table below.

Decimal	In Words	Simplified Fraction or Mixed Number
0.3	Three **tenths**	$\frac{3}{10}$
0.08	Eight **hundredths**	$\frac{8}{100}=\frac{2}{25}$
0.006	Six **thousandths**	$\frac{6}{1000}=\frac{3}{500}$
9.05	Nine and five **hundredths**	$9\frac{5}{100}=9\frac{1}{20}$

Check this out!

one decimal place
one zero in the denominator

two decimal places
two zeros in the denominator

three decimal places
three zeros in the denominator

two decimal places
two zeros in the denominator

Get it????

2.08 An Introduction to Decimals 2.08

EXAMPLE 5

Write 12.025 as a simplified fraction or mixed number.

$$12.025 = 12\frac{25}{1000} = 12\frac{1}{40}$$

3 decimal places — three zeros

EXAMPLE 6

Write 5.12 as a simplified fraction or mixed number.

$$5.12 = 5\frac{12}{100} = 5\frac{3}{25}$$

2 decimal places — two zeros

YOUR TURN

Complete the chart.

	Decimal	In Words	Simplified Fraction or Mixed Number
5.		seven tenths	
6.	0.015		
7.	23.09		
8.		six thousandths	

9. Use the fraction model to represent 0.25

2.08 An Introduction to Decimals 2.08

COMPARING DECIMAL NUMBERS

Recall that when we compared fractions we needed to find common denominators. Comparing decimals works pretty much the same way. If the fraction representations of the decimals have the same denominators then the decimal numbers will be easy to compare.

Let's look at putting 0.06, 0.006, and 0.6 in order from smallest to largest using their fraction representations.

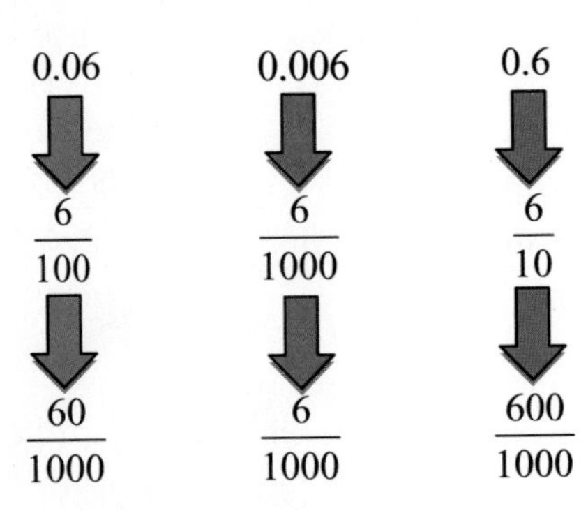

Express each decimal as a fraction.

We see that the denominators are not the same. We need to change these fractions so that they have the same denominators.

Now compare the numerators.

middle **smallest** **largest**

The correct order would be 0.006, 0.06, 0.6.

There is an easier way to do this!

From above we see that $0.06 = \frac{60}{1000}$ which is sixty thousandths or 0.060.

Likewise, we see that $0.6 = \frac{600}{1000}$ which is six hundred thousandths or 0.600.

Look at the numbers we started with again.

$0.06 = 0.060$

$0.006 = 0.006$

$0.6 = 0.600$

From this we see that we can add zeros to the end of a decimal number and not change its value! This helps when comparing decimals.

EXAMPLE 7

Use <, > or = between the pair of numbers to form a true statement.

0.42 ____ 0.419 Do these numbers have the same number of decimal places (same denominator)?

$0.420 > 0.419$ Add 0's to make the numbers have the same denominators. Ignore the decimal and compare the numbers: 420 ___ 419

$0.42 > 0.419$ If $0.420 > 0.419$, then $0.42 > 0.419$.

2.08 An Introduction to Decimals 2.08

Let's look at negative decimals

EXAMPLE 8

Use $<$, $>$ or $=$ between the pair of numbers to form a true statement.

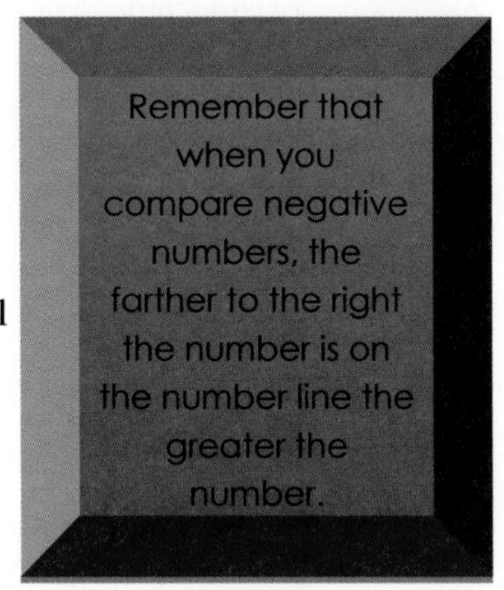

-8.0019 ___ -8.019	Do these numbers have the same number of decimal places?
$-8.0019 > -8.0190$	Add 0's to make the numbers have the same number of decimal places. Ignore the decimal and compare the numbers: -80019 ____ -80190.
$-8.0019 > -8.019$	If $-8.0019 > -8.0190$, then $-8.0019 > -8.019$.

EXAMPLE 9

Use $<$, $>$ or $=$ between the pair of numbers to form a true statement.

0.359 ____ 0.361

These numbers have the same number of decimal places so we don't need to add 0's.

$0.359 < 0.361$

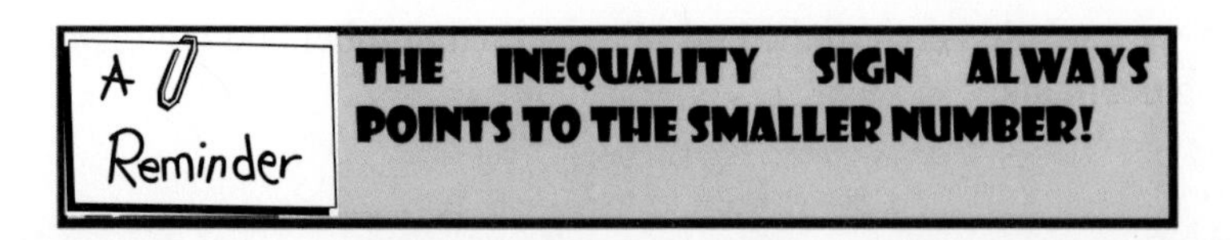

YOUR TURN

Use $<$, $>$ or $=$ between the pair of numbers to form a true statement.

10. 0.017 ___ 0.018 11. 0.098 ___ 0.1 12. -1.062 ___ -1.07 13. 54.090 ___ 54.09

ROUNDING DECIMAL NUMBERS

We round decimal numbers very much like we do whole numbers. The only difference is that the digits to the right of the place we are rounding to are dropped instead of being replaced with 0's.

- Locate the digit in the requested place .
- Look at the number to the right of this digit.
- If the number to the right is 5 or greater, add 1 to the digit in the given place value. If it is less than 5, keep the digit in the given place value.
- Drop off all digits to the right of the given place value.

EXAMPLE 10

Round 9.4732 to the nearest thousandth.

9 . 4 7 3 | 2 Locate the digit in the **thousandths** place. Draw a line after this digit.

thousandths

9 . 4 7 3 | 2 ← Look at the number to the right of the line. Since the number is less than 5, the underlined digit stays the same.

$9.4732 \approx 9.473$ Drop off all numbers to the right of the thousandths place (right of the line).

EXAMPLE 11

Round 234.572 to the nearest tenth.

2 3 4 . 5 | 7 2 Locate the digit in the **tenths** place.

2 3 4 . 5 | 7 2 Look at the number to the right of the underlined place value. Since the number is greater than 5, add 1 to the underlined digit.

$234.572 \approx 234.6$ Drop off all numbers to the right of the tenth place.

2.08 An Introduction to Decimals 2.08

EXAMPLE 12

Round 29.7113 to the nearest whole number.

2 9. 7 1 1 3 — Rounding to the nearest whole number is the same as rounding to the ones place. Locate the digit in the **ones** place.

2 9. 7 1 1 3 — Look at the number to the right of the underlined place value. Since the number is greater than 5, add 1 to the underlined digit. This will also change the digit to the left since we are actually adding 1 to 29.

29.7113 $\approx$ 30 — Drop off all numbers to the right of the ones place.

> **Helpful Hint: $$$** If the number represents money, then
>
> rounding to the nearest cent = rounding to the nearest hundredths
>
> rounding to the nearest dollar = rounding to the nearest ones

EXAMPLE 13

Round $0.894 to the nearest cent.

0 . 8 9 4 — Rounding to the nearest cent is the same as rounding to the hundredths place. Locate the digit in the **hundredths** place.

0 . 8 9 4 — Look at the number to the right of the underlined place value. Since the number is less than 5, the underlined digit remains the same.

$0.894 $\approx$ $0.89 — Drop off all numbers to the right of the hundredths place.

EXAMPLE 14

Round $92.51 to the nearest dollar.

9 2 . **5 1** — Rounding to the nearest dollar is the same as rounding to the ones place. Locate the digit in the **ones** place.

9 2 . **5 1** — Look at the number to the right of the underlined place value. Since the number is 5, add 1 to the underlined digit.

$92.51 $\approx$ $93 — Drop off all numbers to the right of the ones place.

2.08 An Introduction to Decimals

YOUR TURN

Round to the given place value.

14. 0.561 to nearest tenth

15. 0.234 to the nearest hundredth

16. 12.2996 to the nearest thousandth

17. 1.45 to the nearest whole number

18. $ 0.061 to the nearest cent

19. $ 345.62 to the nearest dollar

Answers to Section 2.08.

1. twelve and thirty-six thousandths 2. One and two ten-thousandths 3. 3.026 4. 22.0005

5. 0.7, $\frac{7}{10}$ 6. fifteen thousandths; $\frac{3}{200}$ 7. Twenty-three and nine hundredths; $23\frac{9}{100}$

8. 0.006; $\frac{3}{500}$

9.

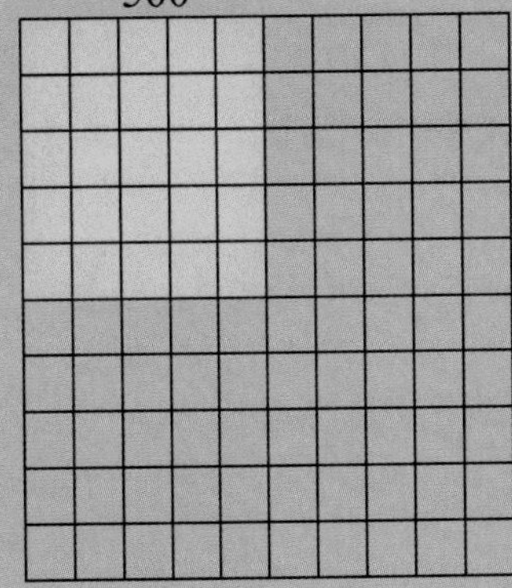

10. < 11. < 12. > 13. = 14. 0.6 15. 0.23

16. 12.300 17. 1 18. $0.06 19. $346

Complete MyMathLab Section 2.08 Homework in your Homework notebook.

2.09 Adding and Subtracting Decimals 2.09

Let's review adding and subtracting decimal numbers

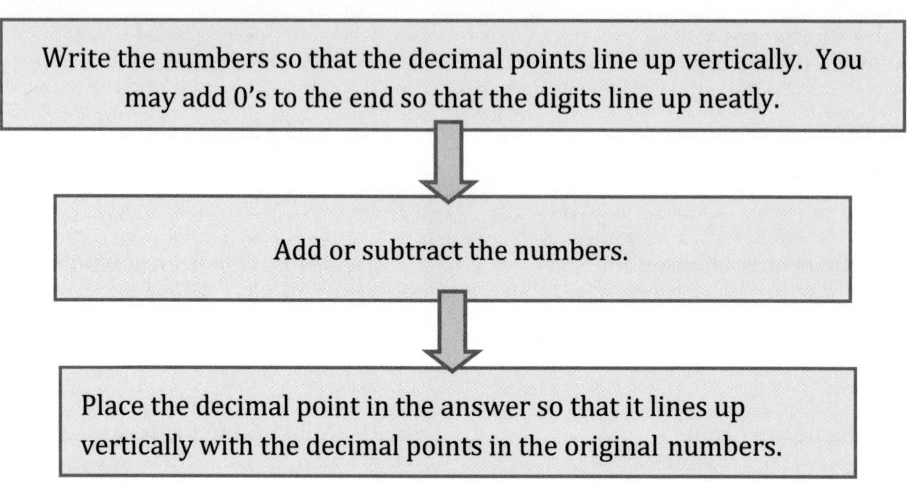

We will also estimate the answer to check to be sure our answer is reasonable. Remember that estimating is just a matter of rounding each of the numbers and then performing the operation on the rounded numbers.

ESTIMATING IS ESPECIALLY IMPORTANT IF YOU PLAN ON USING A CALCULATOR TO WORK THE PROBLEM.

EXAMPLE 1

Add. Estimate the answer. 12.27 + 3.457

Step 1: Line up decimal points vertically. Insert a 0 so digits line up neatly.

$$\begin{array}{r} 12.27\mathbf{0} \\ +\ \ 3.457 \\ \hline \end{array}$$

Step 2: Add the numbers from right to left as you do with whole numbers.

$$\begin{array}{r} 12.270 \\ +\ \ 3.457 \\ \hline 15.727 \end{array}$$

Place the decimal point in the sum so that all decimal points line up.

Estimate the sum to see if our answer is reasonable.

$$\begin{array}{rcr} 12.27 & \approx & 12 \\ +\ \ 3.457 & \approx & 3 \\ \hline & & 15 \end{array}$$

The answer we got is reasonable.

2.09 Adding and Subtracting Decimals

EXAMPLE 2

Subtract 0.5068 from 3.529. Estimate the answer.

Step 1: Line up decimal points vertically. Insert a 0 so digits line up neatly.

$$\begin{array}{r} 3.529\mathbf{0} \\ -\ 0.5068 \\ \hline \end{array}$$

Step 2: Subtract the numbers from right to left as you do with whole numbers.

$$\begin{array}{r} {\scriptstyle 8\ 10} \\ 3.52\not{9}\not{0} \\ -\ 0.5068 \\ \hline \mathbf{3.0222} \end{array}$$

Does the answer make sense? Estimate: 4 – 1 = 3

Yes, the answer is reasonable.

EXAMPLE 3

Subtract: $-3.56-(-5.07)$. Remember the rules for subtracting signed numbers. (Change to addition; change the second number to its opposite and then follow addition rules.)

$-3.56-(-5.07)$

$-3.56 + 5.07 = \mathbf{1.51}$

YOUR TURN

Add or subtract. Estimate the answer. (Remember, this means to round the numbers first, then add or subtract.)

1. 18.14 + 2.396

   Exact:__________

   Estimate:________

2. 35.9 – 4.06

   Exact:__________

   Estimate:________

3. Subtract 3.2 from 4.025

   Exact:__________

   Estimate:________

4. – 3.67 – 1.456

   Exact:__________

   Estimate:________

2.09 Adding and Subtracting Decimals 2.09

APPLICATIONS

EXAMPLE 4

Find the total monthly cost of owning and maintaining a car given the information below.

Monthly car payment	$324.65
Monthly insurance cost	$65.00
Cost of gasoline per month	$76.80
Average maintenance cost per month	$32.45

Courtesy of Fotolia

The phrase "total monthly cost" tells us to add. Remember, to add decimal numbers we line up the decimals vertically and then add as we would whole numbers.

$$\begin{array}{r} \$324.65 \\ \$65.00 \\ \$76.80 \\ +\ \$32.45 \\ \hline \$498.90 \end{array}$$

The total monthly cost is $498.90.

EXAMPLE 5

US paper currency measures 6.14 inches by 2.61nches. Anthony wants to make a frame so that he can frame the first $100. The framing material is sold by the inch. How many inches of the material does he need to by?

Since we want the distance around the edge of the dollar bill then we need to find the perimeter.

We find the perimeter by adding all the sides together

6.14 inches

Courtesy of Fotolia

Add all four sides to find the perimeter: 6. 14 + 2.61 + 6.14 + 2.61 = 17.50 inches
(Round up since the number to the right of the 7, the nearest one inch, is 5.)

Anthony will need to buy 18 inches of material.

By the way, every number has a decimal; sometimes you just don't see it.
125 = 125.

YOUR TURN

Answer each question. Show your work.

5. Gasoline was $4.139 per gallon one week and $4.287 per gallon the next. By how much did the price change?

6. Rebecca wanted to buy the following items: A DVD player for $49.95, a DVD holder for $19.95 and a personal stereo for $21.95. Does Rebecca have enough money to buy all three items if she has $90 with her?

7. The following table shows the batting averages for the top 8 hitters in MLB in 2012.

Player	Batting Average
Beltre	0.3212
Braun	0.3193
Cabrera	0.3296
Jeter	0.3163
Mauer	0.3193
McCutchen	0.3272
Posey	0.3358
Trout	0.3260

a. Which of the players had the highest batting average?

b. Which of the eight players had the lowest batting average?

c. List the players in order from highest batting average to lowest.

d. What is the difference between the highest average and the lowest average?

YOUR TURN Answers to Section 2.09

1. 20.536; 20
2. 31.84; 32
3. 0.825; 1
4. – 5.126; – 5
5. $0.148
6. no

7a. Posey
7b. Jeter

7c. Posey, Cabrera, McCutchen, Trout, Beltre, Braun, Mauer, Jeter

7d. 0.0195

Complete MyMathLab Section 2.09 Homework in your Homework notebook.

2.10 Multiplying Decimals and Circumference of a Circle 2.10

Let's review how to multiply decimals.

Multiply the numbers ignoring the decimal points.

THEN, put the decimal point in the answer. The answer will have as many decimal places as the two original numbers combined.

Let's look at why it works this way.

Let's multiply 0.7×1.9 by changing the numbers to fraction form.

$$\frac{7}{10} \times 1\frac{9}{10} = \frac{7}{10} \times \frac{19}{10} = \frac{133}{100} = 1\frac{33}{100}$$

Now change $1\frac{33}{100}$ to a decimal. $1\frac{33}{100} = 1.33$

This means that $0.7 \times 1.9 = 1.33$.

1 decimal place | 1 decimal place | 2 decimal places

The first number had one decimal place.

The second number had one decimal place.

The answer has 2 decimal places.

EXAMPLE 1

Multiply the numbers by changing to fraction form.

$$0.7 \times 0.009 = \frac{7}{10} \times \frac{9}{1000} = \frac{63}{10{,}000} = 0.0063$$

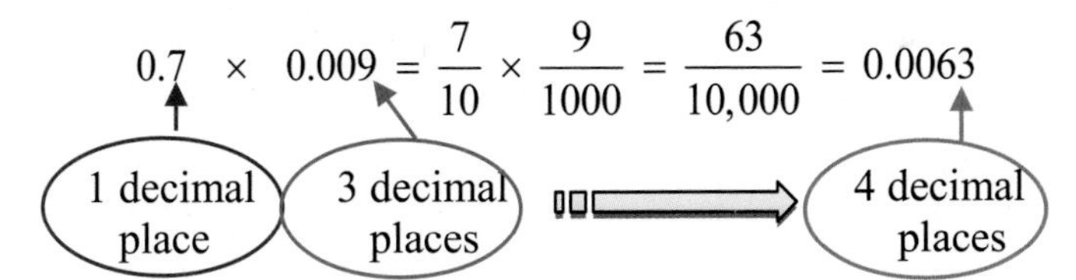

2.10 Multiplying Decimals and Circumference of a Circle 2.10

One more time!

ESTIMATING IS ESPECIALLY IMPORTANT IF YOU PLAN ON USING A CALCULATOR TO WORK THE PROBLEM.

EXAMPLE 2

Multiply 20.86 x 4.9 and estimate to see if your answer is reasonable.

Exact	**Estimate 1**	**Estimate**
20.86 × 4.9	(20.86 is rounded to the ones.)	(20.86 is rounded to the tens.)
18774	21 × 5	20 × 5
8344		
102.214	105	100

You do have choices when estimating. Either estimate indicates that the exact answer is reasonable.

YOUR TURN

Give an exact answer and an estimate.

1. 23.6 x 0.78

   Exact:__________

   Estimate:________

2. 12.1 x 2.1986

   Exact:__________

   Estimate:________

3. -0.9 x 12.1

   Exact:__________

   Estimate:________

4. (-7.83)(-1.9)

   Exact:__________

   Estimate:________

MULTIPLYING BY POWERS OF 10

Take a look at this:

12.345 x 10 = 123.45

12.345 x 100 = 1234.5

12.345 x 1000 = 12345.

Notice what is happening to the decimal as we multiply by powers of 10.

What conclusion can we make????

Move the decimal point to the right the same number of places as there are **zeros** in the power of 10.

EXAMPLE 3

Multiply.

a. 342.85 x 10 = 3428.5 — **1 zero** in the power of ten ⇔move decimal **1 place** to the right

b. 0.1234 x 1000 = 123.4 — **3 zeros** in the power of ten ⇔move decimal **3 places** to the right

c. 8.2 x 100 = 820. — **2 zeros** in the power of ten ⇔move decimal **2 places** to the right

We must add a zero here.

EXAMPLE 4

Many times we hear on the news or read in the newspaper numbers like 12.5 million. Let's look at what this number looks like in standard notation.

12.5 million = 12.5 x 1,000,000 = 12,500,000

6 zeros ⇔ move decimal to the right 6 places

2.10 Multiplying Decimals and Circumference of a Circle 2.10

YOUR TURN

Multiply.

5. 741.2 x 100

6. 10,000 x 0.9514

7. 0.861 x 10

Express these numbers in standard form.

8. 1.05 billion

9. 0.75 million

10. 375.2 billion

Circumference of a circle is the equivalent of 'perimeter' for a circle. In other words, if you took a circle and unrolled it, you would have its circumference. You can use either of the formulas below to find circumference of a circle. One formula uses the radius of a circle; the other formula uses the diameter.

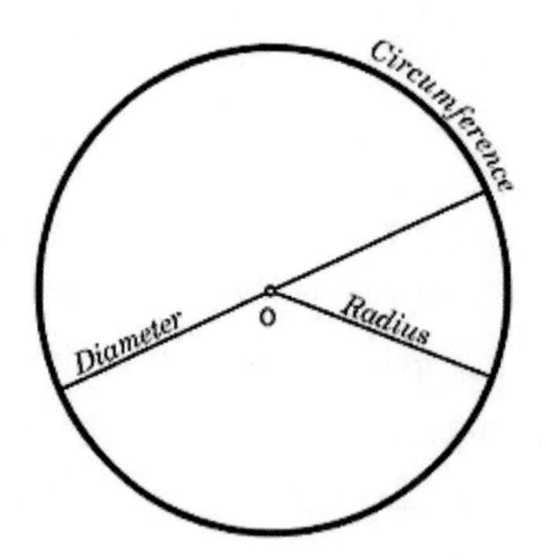

If you know the radius

Circumference $= 2 \cdot \pi \cdot$ radius

Circumference $= 2\pi r$

If you know the diameter

Circumference $= \pi \cdot$ diameter

Circumference $= \pi d$

2.10 Multiplying Decimals and Circumference of a Circle 2.10

EXAMPLE 5

Find the circumference of the circles below in terms of π. Then use the approximation 3.14 for π to approximate the circumference to the nearest hundredth.

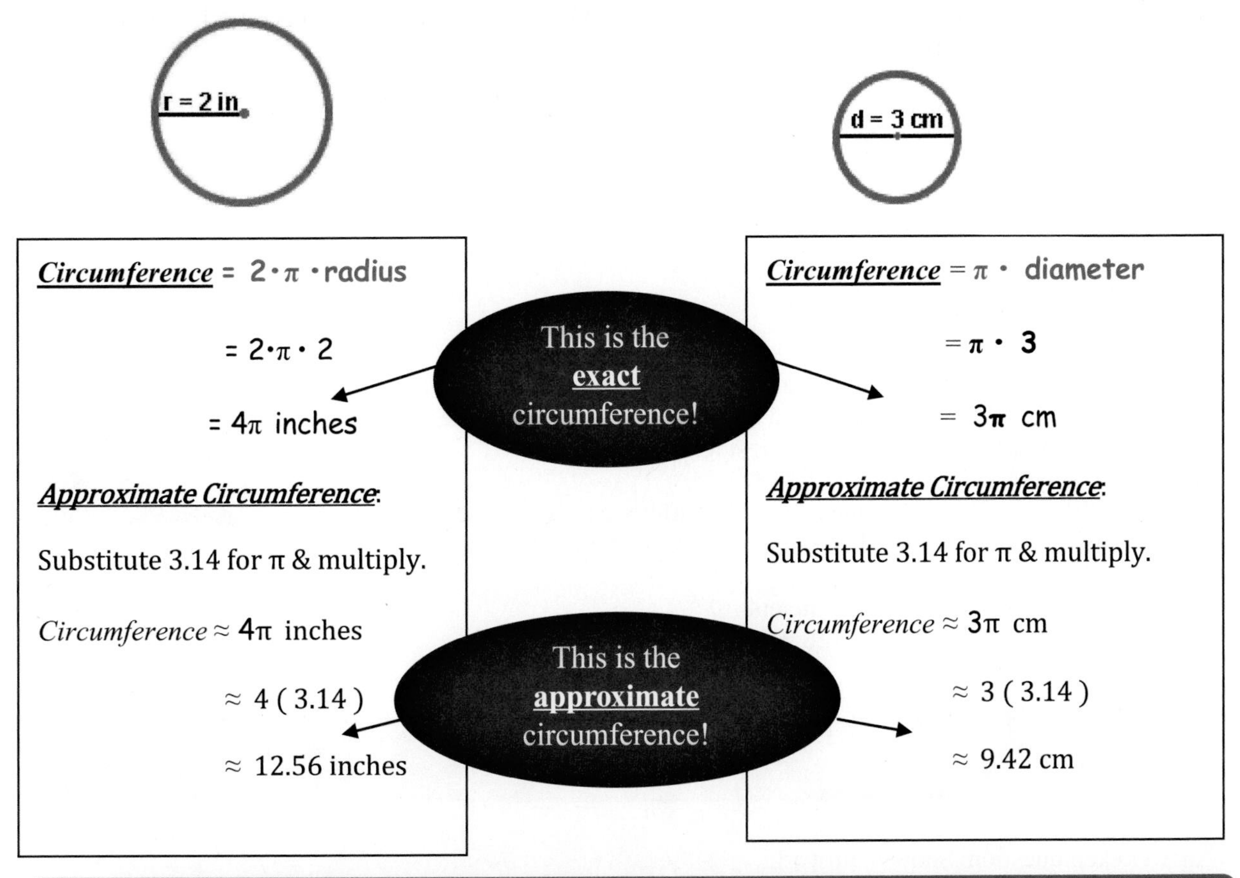

YOUR TURN

Find the circumference of the circles below in terms of π. Then use the approximation 3.14 for π to approximate the circumference to the nearest hundredth.

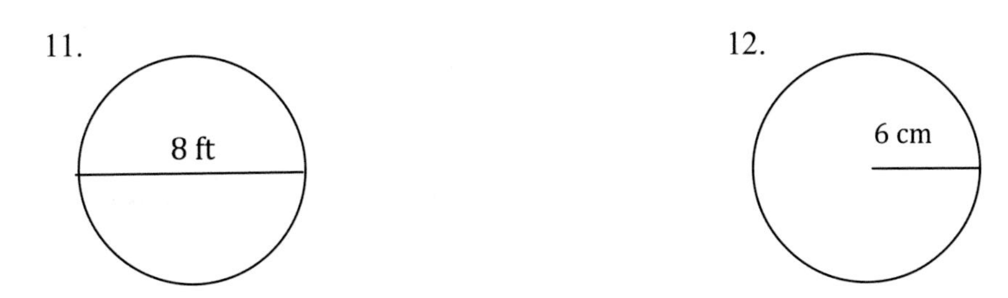

2.10 Multiplying Decimals and Circumference of a Circle 2.10

APPLICATIONS!

EXAMPLE 6

Courtesy of Fotolia

If Alan earns \$13.75 an hour, find his weekly pay (before taxes) if he works 40 hours a week.

Total pay = hourly rate x number of hours worked

Total pay = 13.75 x 40 = \$550 per week

EXAMPLE 7

The diameter of a ferris wheel at Cedar Point in Ohio is 126 meters. If a girl rides the ferris wheel for one revolution, find how far she travels. Find the exact distance and the approximate distance to the nearest tenth.

Courtesy of Fotolia

One revolution would be the same as the circumference, right? Use the formula for the circumference of a circle when you know the diameter. $C = \pi d$

Exact answer: $C = \pi(126) = 126\pi$ meters
Approximate answer: $C = 3.14 \cdot 126 = 395.64 \approx 395.6$ meters

YOUR TURN

Answer each question. Show your work.

13. The Apple iPod nano measures 3.6 inches by 1.5 inches. Find the area of the face of the iPod nano.

14. What is the cost in dollars of 16.3 gallons of gasoline at \$3.29 per gallon? Round to the nearest cent.

(continue to next page)

2.10 Multiplying Decimals and Circumference of a Circle 2.10

YOUR TURN

15. The radius of an US quarter is 12.13 mm. Find the circumference of the quarter. Give an exact answer and an approximation using 3.14 for π. Round to the nearest hundredth.

Your Turn **Answers to Section 2.10.**

1. 18.408; 24 or 20	2. 26.60306; 24 or 20	3. −10.89; -12
4. 14.877; 16	5. 74,120	6. 9514
7. 8.61	8. 1,050,000,000	9. 750,000
10. 375,200,000,000	11. 8π ft; 25.12 ft	12. 12π cm; 37.68 cm
13. 5.4 square inches	14. $53.63	15. 24.26π mm; 76.18 mm

Complete MyMathLab Section 2.10 Homework in your Homework notebook.

2.11 Dividing Decimals 2.11

First let's review some vocabulary about division.

divisor : number you are dividing by **dividend**: number you are dividing into **quotient**: the answer	**dividend ÷ divisor = quotient** $\text{divisor}\overline{)\text{dividend}}$ with **quotient** on top

DIVIDING BY A WHOLE NUMBER

When dividing decimals, it is important to remember that your divisor (the number on the outside #⌐) must be a whole number. If the divisor is a whole number, you divide as normal and place the decimal in the quotient (the answer) directly above the decimal in the dividend (number you're dividing into).

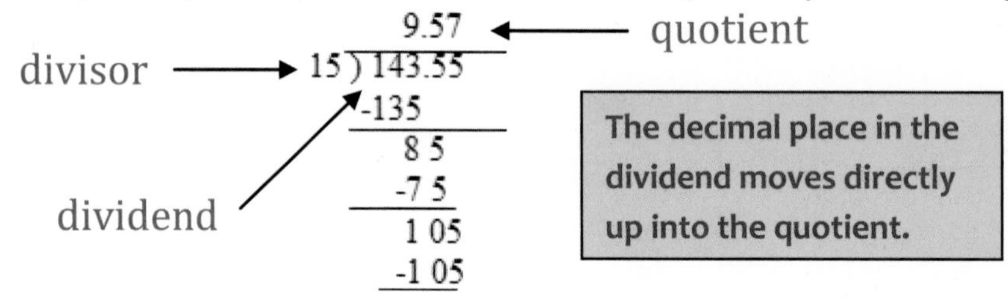

DIVIDING BY A DECIMAL NUMBER

What if the divisor is not a whole number? Look at $1.25 \div 0.005$.

Remember that division can be represented by a fraction. So we can express this division as a fraction:

$$1.25 \div 0.005 = \frac{1.25}{0.005}$$

The divisor (0.005) must be a whole number. How could we get a whole number in the denominator?

If we multiply the denominator by 1000, would we then have a whole number in the denominator?

Yes! $0.005 \times 1000 = 5$

Remember, we cannot just multiply the denominator of a fraction without multiplying the numerator by the same number. $\frac{1.25}{0.005} = \frac{1.25}{0.005} \times \frac{\mathbf{1000}}{\mathbf{1000}} = \frac{1250}{5}$

Now we have a whole number divisor and can complete the division.

$$0.005\overline{)1.25} \Rightarrow 5\overline{)1250.} \Rightarrow \begin{array}{r}250.\\ 5\overline{)1250}\end{array}$$

Remember that multiplying by powers of ten simply moves the decimal in a number. We use this process to change these decimal divisors into whole number divisors. But we now see that we must move that decimal in the dividend as well as in the divisor.

2.11 Dividing Decimals

EXAMPLE 1

Divide: $2.176 \div 0.34$

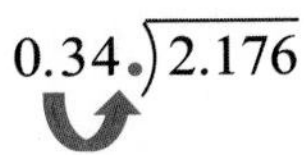

Move the decimal point in the divisor to the right until the divisor is a whole number.

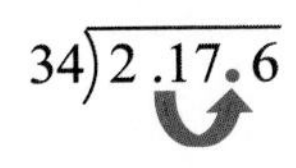

Move the decimal point in the dividend the same number of places as the decimal point in the divisor was moved, above.

$34\overline{)217.6}$

Place the decimal point in the quotient directly above the decimal point in the dividend.

$$\begin{array}{r} 6.4 \\ 34\overline{)217.6} \\ \underline{204} \\ 13\,6 \\ \underline{13\,6} \\ 0 \end{array}$$

Divide as you would with whole numbers.

Calculator Tip

Make sure that you are inputting the numbers correctly when using the calculator. If the problems are written this way $1.5\overline{)0.0036}$, then you will enter this "$0.0036 \div 1.5 =$ "

Last time!

ESTIMATING IS ESPECIALLY IMPORTANT IF YOU PLAN ON USING A CALCULATOR TO WORK THE PROBLEM.

EXAMPLE 2

Divide $6.3\overline{)54.08}$. Give an exact answer and an estimate.

Exact: $54.08 \div 6.3 = 8.6$ (using the calculator!)(rounded to the nearest tenth)

Estimate: Round the numbers first, then divide these numbers.

$54.08 \approx 54$

$6.3 \approx 6$

$54 \div 6 = 9$

2.11 Dividing Decimals 2.11

YOUR TURN

Divide and estimate.

1. $4\overline{)283.4}$
2. $0.04\overline{)0.112}$
3. $36 \div (-0.18)$
4. $(-93.5) \div (-100)$

Exact:__________ Exact:__________ Exact:__________ Exact:__________

Estimate:________ Estimate:________ Estimate:________ Estimate:________

DIVIDING BY POWERS OF 10

Take a look at this:

$$12.345 \div 10 = 1.2345$$

$$12.345 \div 100 = 0.12345$$

$$12.345 \div 1000 = 0.012345$$

Notice what is happening to the decimal as we divide by powers of 10.

What conclusion can we make????

> Move the decimal point to the left the same number of places as there are **zeros** in the power of 10.

2.11 Dividing Decimals 2.11

EXAMPLE 3

Divide.

a. 342.85 ÷ 100 = **3.4285** **2 zeros** in the power of ten ⇔ move decimal **2 places** to the left.

b. 8.2 ÷ 10 = **0.82** **1 zero** in the power of ten ⇔ move decimal **1 place** to the left.

c. 12.34 ÷ 1000 = **0.01234** **3 zeros** in the power of ten ⇔ move decimal **3 places** to the left.

We need to fill in the empty space with a zero.

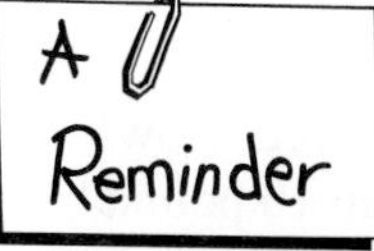

We put a zero in front of the decimal when there is not a whole number part to the decimal. It is not wrong to leave that zero out. It just makes the decimal point more visible when you include that zero in front.

YOUR TURN

Divide.

5. 54.23 ÷ 100

6. $-\dfrac{4.567}{1000}$

7. 263.12 ÷ 10

APPLICATIONS

KEY WORDS THAT INDICATE DIVISION

average *each*

per *part of*

Shared equally

KEEP IN MIND THE FOLLOWING

total ÷ size of part = number of equal parts

total ÷ number of equal parts = size of each part

2.11 Dividing Decimals 2.11

EXAMPLE 4

Will drives a Tahoe. He recently checked his gas mileage. He drove 335.9 miles on 22.7 gallons of gas. How many miles (per) gallon did he get? Round the answer to the nearest tenth.

Key word indicates division

miles per gallon

$335.9 \div 22.7 = 14.79735683 \approx 14.8$ miles per gallon (ugh!)

EXAMPLE 5

A pound of fertilizer covers 39 square feet of lawn. Kayla's lawn measures 4465.8 square feet. How much fertilizer does she need to the nearest hundredth of a pound? If the fertilizer is sold in 10 pound bags, how many bags will she need to buy?

Let's find out how many pounds of fertilizer she needs to make her grass grow.

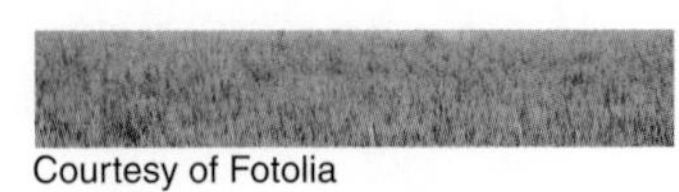
Courtesy of Fotolia

She needs enough to cover 4465.8 square feet. Each pound only covers 39 square feet so we know she needs several pounds. We will need to divide to know how many she needs. But which way will we divide?

Which one makes sense???

$4465.8 \div 39 = 114.5076923 \approx 114.51$ (circled) total ÷ size of equal parts

$39 \div 4465.8 = 0.008733038 \approx 0.01$ ~~size of equal parts ÷ total~~

The first one makes sense since we know that she will need more than one pound! She needs 114.51 pounds of fertilizer.

Now let's find out how many bags she needs. We know that each bag contains 10 pounds. We always divide the total (amount of fertilizer needed) by the size of the equal parts (10 pound bags).

Total	÷ size of equal parts	= number of equal parts
total number of pounds needed ÷	size of bag	= number of bags needed
114.51 ÷	10	= 11.451 (Try not to use your calculator!)

She needs 11.451 bags of fertilizer. She cannot buy part of a bag, right? **She will need to buy 12 bags.**

Always check to see if your answer makes sense!!

2.11 Dividing Decimals

EXAMPLE 6

Logan buys five organic chickens. The chickens weigh 2.9 pounds, 3.1 pounds, 3 pounds, 2.85 pounds, and 3.05 pounds. How much less than the average weight of the five chickens was the weight of the heaviest one?

What do we want to find?	How much less than the average weight was the weight of the heaviest?
What operation is indicated by How much less than?	How much less than means that we need to subtract.
What do we subtract?	Weight of heaviest chicken – average weight of the chickens
We have a problem: we don't know the average weight.	The heaviest chicken weighs 3.1 pounds (refer back to section 2.08 for help with this). Average weight of the chickens is ?.
How do I find the ***average*** weight??	Add the weight of the chickens and divide by 5, since there are 5 chickens. $\frac{2.9+3.1+3+2.85+3.05}{5}=\frac{14.9}{5}=2.98$ pounds
Now, we can find the difference.	**Weight of heaviest chicken – average weight of the chickens** **3.1 – 2.98** $3.1-2.98=0.12$ pounds

2.11 Dividing Decimals 2.11

YOUR TURN

Answer each question. Show your work!!

8. You need to put some gas in your car. Regular gasoline is $3.26 per gallon. You only have a $ 20 bill on you. How many gallons can you buy? Round your answer to the nearest tenth.

Courtesy of Fotolia

9. Jeff is painting the walls of a room. The walls have a total area of 678 square feet. A quart of paint covers 45 square feet. How many quarts does he need? Remember, you cannot buy part of a quart.

10. Mohammed has these scores on his tests in MyMathLab: 98.5, 92.35, 85, and 87.11. How much less than his average is his lowest test score? (Compare his average to his lowest test score.)

11. Matthew purchased new school supplies. He bought 3 pens for $11. How much did he have to pay for 1 pen? Round your answer to the nearest cent. (Remember the nearest cent is the same as the nearest hundredths.)

YOUR TURN Answers to Section 2.11

1. 70.85; 70 2. 2.8; 3 3. –200; –180 4. 0.935; 1 5. 0.5423 6. –0.004567
7. 26.312 8. 6.1 gallons 9. 16 quarts 10. 5.74 11. $3.67

Complete MyMathLab Section 2.11 Homework in your Homework notebook.

2.12 Fractions, Decimals and Order of Operations 2.12

Sometimes we need to change fractions to decimals. Recall that the fraction bar means division! We will use this fact to make these changes.

CHANGING FRACTIONS TO DECIMALS

To write a fraction as a decimal, divide the numerator by the denominator.

EXAMPLE 1

Write $\frac{3}{8}$ as a decimal. (This means to divide 3 by 8.)

You can use long division or use your calculator, just make sure that you are putting it in your calculator correctly!

$$\frac{3}{8} = 3 \div 8 = 0.375$$

You will know that you have entered the division in the calculator wrong if you start with a proper fraction and end up with a decimal number that is great than one!

EXAMPLE 2

Write $\frac{7}{18}$ as a decimal.

$$\frac{7}{18} = 7 \div 18 = 0.3888\ldots = 0.3\overline{8}$$

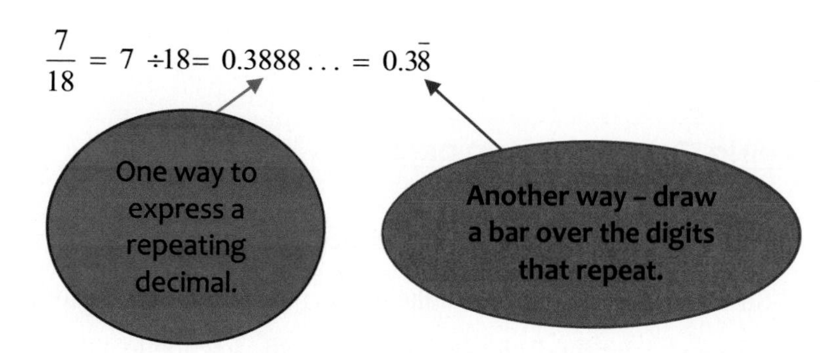

2.12 Fractions, Decimals and Order of Operations 2.12

Let's look at changing a mixed number to a decimal.

EXAMPLE 3

Change $4\frac{5}{16}$ to a decimal.

Method 1

In this method we just work on the fraction part. We know that the whole number part will remain as 4 when we convert the fraction to the decimal.

$$\frac{5}{16} = 5 \div 16 = 0.3125$$

So $4\frac{5}{16} = 4.3125$

Method 2

Here we change the mixed number to an improper fraction first and then express that fraction as a decimal.

$$4\frac{5}{16} = \frac{69}{16} = 69 \div 16 = 4.3125$$

YOUR TURN

Write these numbers as decimals.

1. $-\frac{2}{5}$
2. $7\frac{9}{40}$
3. $5\frac{1}{12}$

COMPARING FRACTIONS AND DECIMALS

How will we compare $\frac{1}{8}$ and 0.112? It is difficult to do if one number is a fraction and the other is a decimal. So we need to change one of the numbers to the form of the other one and then compare. Since it is easier to compare decimals, let's change the fraction to a decimal.

2.12 Fractions, Decimals and Order of Operations 2.12

EXAMPLE 4

Use <, >, or = to make a true statement.

$\frac{1}{8}$ ___ 0.122 — Change $\frac{1}{8}$ to a decimal number. $\frac{1}{8} = 1 \div 8 = 0.125$

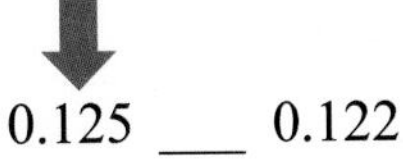

0.125 ___ 0.122 — Compare the decimal numbers. Since the two numbers have the same number of decimal places, ignore the decimal and compare 125 and 122.

$0.125 > 0.122$

$\frac{1}{8} > 0.122$

EXAMPLE 5

Use <, >, or = to make a true statement.

$\frac{2}{3}$ ___ 0.67

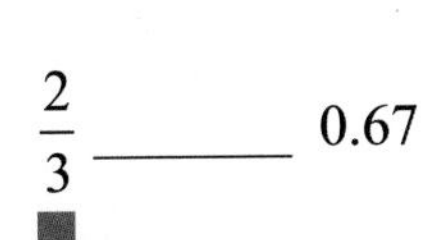

$\frac{2}{3}$ _______ 0.67 — Change $\frac{2}{3}$ to a decimal number. $\frac{2}{3} = 2 \div 3 = 0.666666667$. This is the number you will see on your calculator.

0.666666667 ___ 0.67

0.666666667 ___ 0.670000000 — Compare the decimal numbers. Add zeros to 0.67 so that it has as many decimal places as 0.6666666667.

$0.666666667 < 0.67$

$\frac{2}{3} < 0.67$

EXAMPLE 6

Write in order from smallest to largest: $2\frac{3}{11}$, 2.27, –2. 27

$2\frac{3}{11} = 2.\overline{27}$

$3 \div 11 = 0.272727273$ — Change $2\frac{3}{11}$ to a decimal.

Compare the decimals now.
Line up the numbers and add zeroes where necessary.
The negative is the smallest because the other two are positive.

2.2727
2.2700
–2.27

The correct order is: –2. 27, 2.27, $2\frac{3}{11}$.

2.12 Fractions, Decimals and Order of Operations 2.12

YOUR TURN

Use <, > , or = to make a true statement. **SHOW YOUR WORK!**

4. $0.7 \_\_ \frac{7}{9}$

5. $0.403 \_\_ \frac{2}{5}$

6. Write these numbers in order from smallest to largest.

   $3\frac{7}{12}$, 3.583, 3.5

Your Turn **Answers to Section 2.12.**

1. –0.4
2. 7.225
3. $5.08\overline{3}$
4. <
5. >
6. 3.5, 3.583, $3\frac{7}{12}$

Complete MyMathLab Section 2.12 Homework in your Homework notebook.

2.13 Square Roots and the Pythagorean Theorem

When we take the square root of a number we are answering the question, "What number do I square (multiply by itself) to get the number under the radical sign $\left(\sqrt{\ }\right)$?"

Memorize these square roots.

Square Root of a Number

A square root of a number, "a," is a number, "b," such that $b^2 = a$.
We use the radical sign to name square roots $\sqrt{\ }$.

$$\sqrt{a} = b \text{ if } b^2 = a$$

	$\sqrt{1} = 1$	$\sqrt{36} = 6$	$\sqrt{121} = 11$
$\sqrt{4} = 2$		$\sqrt{49} = 7$	$\sqrt{144} = 12$
$\sqrt{9} = 3$		$\sqrt{64} = 8$	$\sqrt{169} = 13$
$\sqrt{16} = 4$		$\sqrt{81} = 9$	$\sqrt{196} = 14$
$\sqrt{25} = 5$		$\sqrt{100} = 10$	$\sqrt{225} = 15$

FINDING SQUARE ROOTS OF PERFECT SQUARES

Perfect squares have exact values. The square root is a whole number, a decimal which terminates or a fraction.

EXAMPLE 1

Find each square root. (These examples are perfect squares.)

a. $\sqrt{100} = 10$ Read "the square root of 100".

This is 10 because $10^2 = 100$.

b. $\sqrt{3249} = 57$

c. $\sqrt{6.25} = 2.5$

d. $\sqrt{\frac{36}{49}} = \frac{\sqrt{36}}{\sqrt{49}} = \frac{6}{7}$

When finding the square root of a fraction, find the square root of the numerator separately from the square root of the denominator.

2.13 Square Roots and the Pythagorean Theorem 2.13

Approximating square roots

Use a calculator to approximate square roots of numbers that are not perfect squares. On some calculators you would enter the number and then the square root key. On others you enter the square root key and then the number.

Know thy calculator.

EXAMPLE 2

a. Use a calculator to approximate $\sqrt{7}$ to the nearest tenth.

$\sqrt{7} = 2.645751311$

$\sqrt{7} \approx 2.6$

When you put $\sqrt{7}$ in your calculator, you get 2.645751311. If your calculator had space for a million or more digits, you would not see this decimal repeat because $\sqrt{7}$ is an irrational number. Round 2.645751311 to the nearest tenth.

b. Use a calculator to approximate $\sqrt{21}$ to the nearest thousandth.

$\sqrt{21} = 4.582575695$

$\sqrt{21} \approx 4.583$

YOUR TURN

Find the square root of each number.

1. $\sqrt{144}$

2. $\sqrt{0.7396}$

3. $\sqrt{\frac{81}{196}}$

Approximate each square root to the given place.

4. $\sqrt{156}$ to the nearest tenth

5. $\sqrt{89}$ to the nearest hundredth

2.13 Square Roots and the Pythagorean Theorem 2.13

USING THE PYTHAGOREAN THEOREM

The Pythagorean Theorem is used to find missing **SIDE lengths** on right triangles (which contain a 90 degree angles).

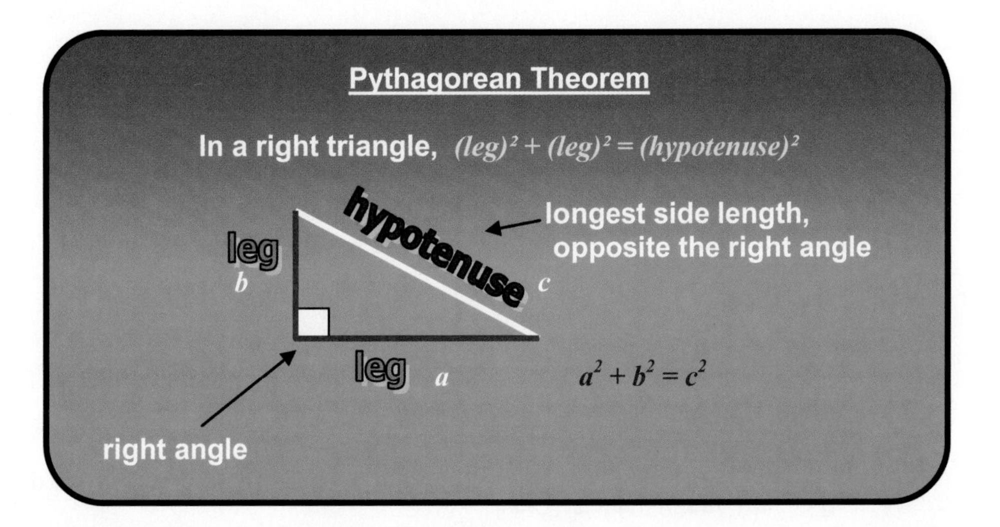

Courtesy of Fotolia

Be very careful when substituting the side lengths into the formula.

Before you plug into the formula make sure you know which sides are the legs and which is the hypotenuse. The hypotenuse is the longest side; it is opposite the right angle.

EXAMPLE 3

Find the missing side length of the right triangle.

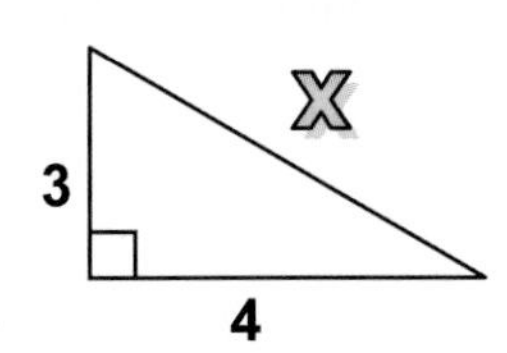

$(\text{leg})^2 + (\text{leg})^2 = (\text{hypotenuse})^2$ Determine which sides are the legs and which is the hypotenuse.

$(3)^2 + (4)^2 = (x)^2$ The legs form the right angle.

$9 + 16 = x^2$

$25 = x^2$ We see that $x^2 = 25$.

$x = \sqrt{25} = 5$ To find x we take the square root of 25.

EXAMPLE 4

The hypotenuse of a right triangle is 7 inches long. One of the legs is 4 inches long. Find the length of the other leg to a two decimal place approximation.

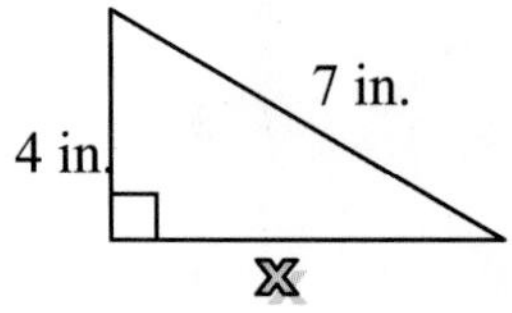

Draw the triangle and label the sides.

Plug the info into the formula.
This time we are missing the length of one of the legs.

$$(\text{leg})^2 + (\text{leg})^2 = (\text{hypotenuse})^2$$

$$(4)^2 + (x)^2 = (7)^2$$

$$16 + x^2 = 49$$

Subtract 16 from both sides before we find the square root.

$$x^2 = 49 - 16$$

$$x^2 = 33$$

Round to the nearest hundredths (two decimal places).

$$x = \sqrt{33} = 5.744562647 \approx \mathbf{5.74}$$

EXAMPLE 5

The distance between the bases on a baseball diamond is 90 feet. (The diamond is actually a square.)

a. Find how far the catcher must throw the ball from home plate to second base in order to throw out a runner trying to steal second base. Round the answer to the nearest whole number. (Draw a sketch of the baseball diamond.)

b. How far must a runner run if he hits a homerun? (He must run all the way around the diamond).

a.

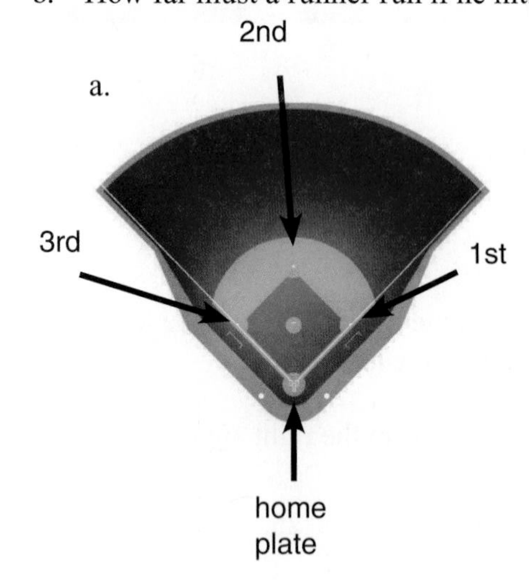

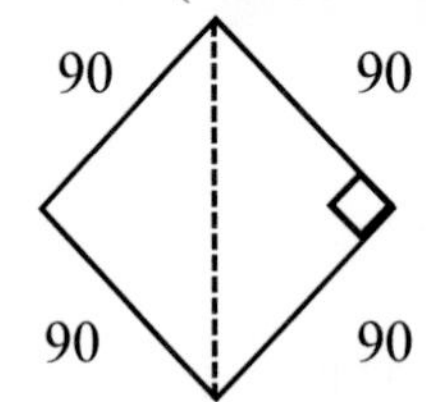

So we have a right triangle!

$$(\text{leg})^2 + (\text{leg})^2 = (\text{hypotenuse})^2$$

$$(90)^2 + (90)^2 = (x)^2 \quad \text{Write an equation.}$$

$$8100 + 8100 = x^2 \quad \text{Solve.}$$

$$16{,}200 = x^2$$

$$x = \sqrt{16{,}200} = 127.279 \approx 127$$

The catcher must throw the ball approximately 127 feet.

2.13 Square Roots and the Pythagorean Theorem 2.13

b. The homerun hitter runs around the diamond.

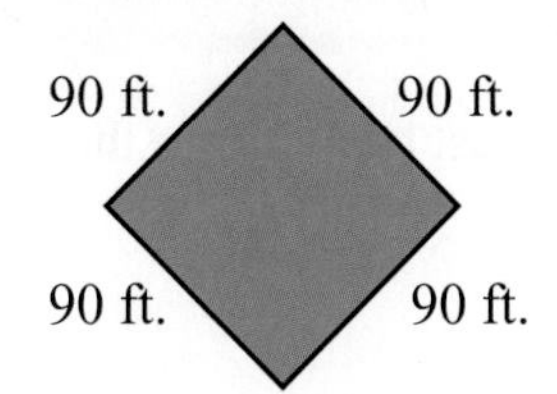

Distance around – Sound familiar???

We need the perimeter of the diamond. So we simply add up the length of the sides of the diamond.

$90 + 90 + 90 + 90 = 360$ ft.

The home run hitter must run 360 feet.

YOUR TURN

Answer each question. Show your work.

6. Sketch a right triangle and find the length of the missing side.

   leg = 5; leg = 12

   Write an equation.

   Solve.

7. Sketch a right triangle and find the length of the missing side. Round to the nearest thousandths, if needed.

   leg = 6; hypotenuse = 20

   Write an equation.

   Solve.

2.13 Square Roots and the Pythagorean Theorem 2.13

YOUR TURN

8. How far up a wall will an 11m ladder reach, if the foot of the ladder must be 4m from the base of the wall? Round answer to the nearest tenth.

   Write an equation.

   Solve.

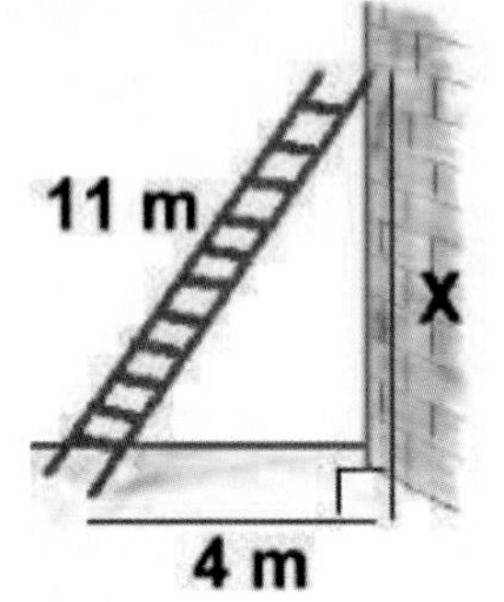

YOUR TURN **Answers to Section 2.13**

1. 12 2. 0.86 3. $\frac{9}{14}$ 4. 12.5 5. 9.43 6. 13 7. 19.079 8. 10.2 m

Complete MyMathLab Section 2.13 Homework in your Homework notebook.

2.14 More Work with Perimeter and Area

PERIMETER

Perimeter: perimeter represents the distance **around** a polygon.

The perimeter of a circle is called its circumference.

★ **To find perimeter, add up all the side lengths.**

Perimeter Formulas		
Shape	**Formula**	**Picture**
Rectangle	$P = 2l + 2w$ * l = length, w = width	Length (l), width (w)
Square	$P = 4s$ *s = side ★ All sides have the same length.	side (s)
Triangle	$P = a + b + c$ where a, b, and c are side lengths	a, b, c
Circle	$C = 2\pi r$ or $C = \pi d$ C = circumference r = radius d = diameter	Circumference, Diameter, Radius

EXAMPLE 1

Find the perimeter of the rectangle.

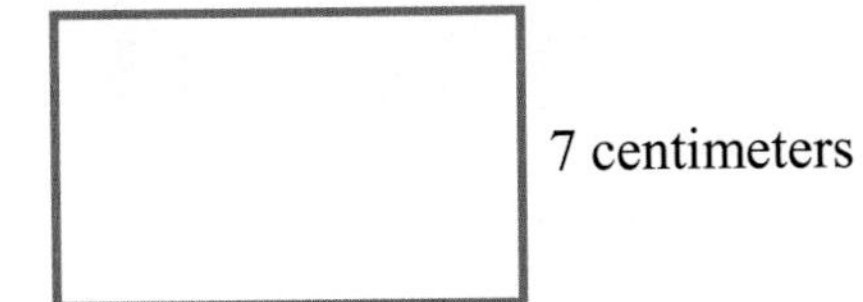

The perimeter of a rectangle can be found by adding up all four sides or by using the formula.

$$P = 2l + 2w.$$

$P = 2(10) + 2(7)$	Write an equation.
$P = 20 + 14$	Solve.
P = 34 centimeters	

2.14 More Work with Perimeter and Area 2.14

EXAMPLE 2

Find the perimeter of the trapezoid.

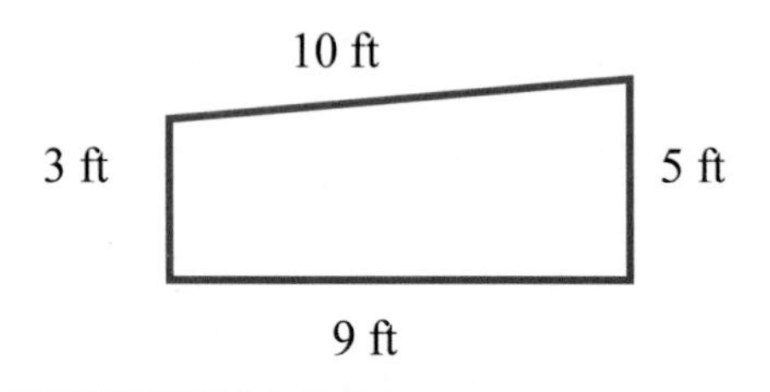

Solution: To find the perimeter of the trapezoid, add up all four sides.

$P = 10 + 5 + 9 + 3$ Write an equation.

P = 27 ft Solve.

EXAMPLE 3

How much would it cost to put a stone border around a rectangular garden measuring 7 yards by 5 yards if the stone border costs \$4 per yard?

Solution:

Think about what they are asking. To put a stone border around a garden means they are talking about the perimeter. Sometimes it helps to sketch a picture so that you remember there are four sides to a rectangle!

7 yards

5 yards

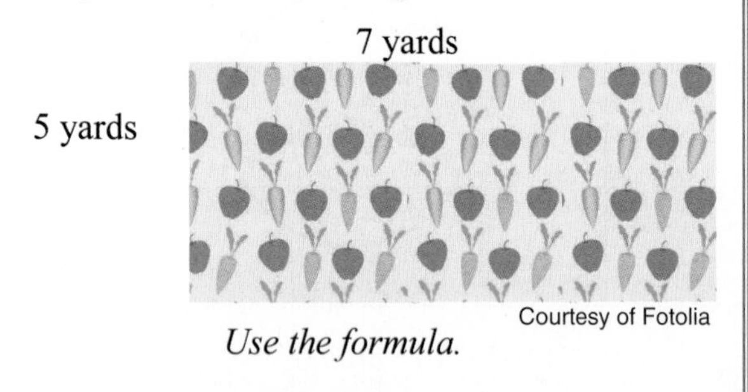

Courtesy of Fotolia

Use the formula.

length = 7 yards $\quad \boldsymbol{P = 2l + 2w}$

width = 5 yards $\quad \boldsymbol{P} = 2(7) + 2(5)$

$\boldsymbol{P} = 14 + 10$

$\boldsymbol{P} = 24$ yards

*Be careful of the question asked. They asked how much it would cost to build a stone border that costs \$4 per yard. We have 24 yards! We multiply to find the cost.

Total cost = \$4(24) = **\$96**

YOUR TURN

Write an equation. Solve and answer the question.

1. Find the perimeter of rectangle whose length is $15\frac{1}{2}$ ft and whose width is 11.75 ft.

   Draw a picture.

(continue to next page)

2.14 More Work with Perimeter and Area 2.14

YOUR TURN

2. How much fencing is required to enclose a circular garden whose radius is 20 meters? Give an exact answer and an approximation. Use 3.14 to approximate. Draw a picture.

   Write an equation.

   Solve.

AREA

Area: A measure of the amount of surface of a region, (the inside space).
The units for area are squared. For example, cm • cm = cm^2.

Geometric Figure	Picture	Area Formulas
Rectangle	width, length	***Area = length · width*** $A = l \cdot w$
Square	side, side	***Area = side · side*** $A = s \cdot s$ or $A = s^2$
Triangle	height, base	$Area = \frac{1}{2} \cdot base \cdot height$ $A = \frac{1}{2} \cdot b \cdot h$
Parallelogram	height, base	$Area = base \cdot height$ $A = b \cdot h$
Trapezoid	one base or b, height, other base or B	**Area =** $\frac{1}{2}(base + other\ base)height$ $A = \frac{1}{2}(b + B)h$
Circle	radius	$Area = \pi(radius)^2$ $A = \pi r^2$ ***The radius is half the diameter.***

EXAMPLE 4

Find the area of the triangle.

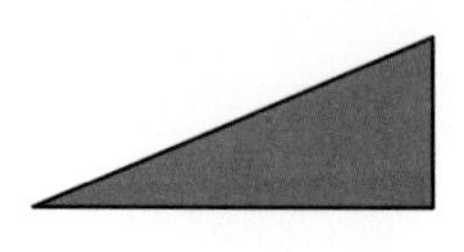

9 cm

20 cm

Since this is a right triangle, the base and height are actually the legs!

Use the Area of aTriangle formula.

$A = \frac{1}{2}bh$

$A = \frac{1}{2} \bullet 20 \bullet 9$ Write an equation. **Follow the order of operations!**

$A = 10 \bullet 9$ Solve.

$A = 90 \text{ cm}^2$

EXAMPLE 5

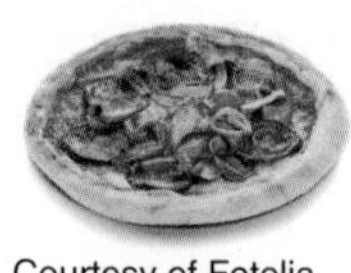

Courtesy of Fotolia

Which one of the following is the better buy?

1 large pizza (14 inch diameter) for \$ 17.99

1 medium pizza (12 inch diameter) for \$13.99

Courtesy of Fotolia

Area of the 14 inch pizza	We need to know how much pizza there is in a 14 in. and a 12 in. pizza (in square inches). The area of each pizza will give us that information.	Area of the 12 inch pizza
$A = \pi r^2$ $A \approx (3.14)(7)^2$ $A \approx (3.14)(49)$ $A \approx 153.86$		$A = \pi r^2$ $A \approx (3.14)(6)^2$ $A \approx (3.14)(36)$ $A \approx 113.04$
cost per square inch $17.99 \div 153.86$ $0.1169 \approx \$0.117$ per square inch	Now find the cost per square inch. Recall that "per" means to divide.	**cost per square inch** $13.99 \div 113.04$ $0.1237 \approx \$0.124$ per square inch

Finally, we compare the two costs per square inch. **Since \$0.117 < \$ 0.124 the large pizza is the better buy.**

2.14 More Work with Perimeter and Area

EXAMPLE 6

Janine wants to install a 3 feet wide walk around her swimming pool. The pool is 15 feet wide and 30 feet long. What is the area of the sidewalk? (Draw a picture!)

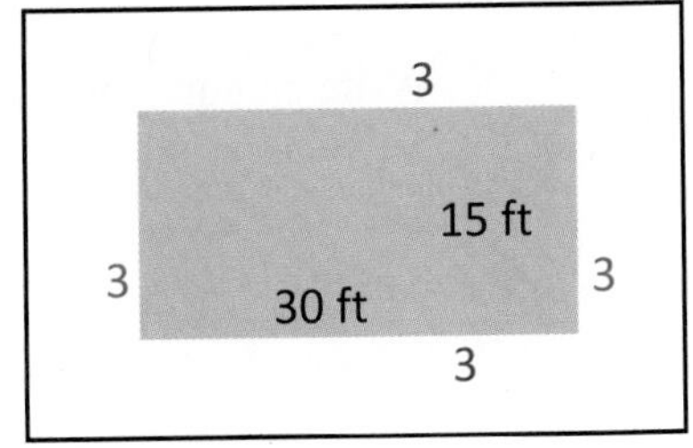

To find the area of the sidewalk, think about the big rectangle.

If we take away the "pool", or small rectangle, we would be left with the sidewalk, right?

Since we are talking about area of the sidewalk we will need to find the area of the big rectangle and the area of the "pool".

But we don't have the dimensions of the big rectangle!

Look at the diagram again. Look from left to right and you will see that the length of the big rectangle can be found by adding the width of the walk, length of the pool, and width of the walk again since the walk is on all sides of the pool.

Length of big rectangle = **3** + 30 + **3** = 36 ft

Back to the diagram to find the width of the big rectangle. Look from top to bottom and you will see that the width can be found by adding the 2 widths of the walk to the width of the pool.

Width of the big rectangle = **3** + 15 + **3** = 21

Now that we have the dimensions of the big rectangle and the pool we can find the area of the sidewalk.

36

21

15

30

Area of big rectangle = $36 \times 21 = 756$ square feet

Area of pool = $30 \times 15 = 450$ square feet

Area of walk = area of big rectangle – area of pool
= $756 - 450 = 306$ square feet

FINISHED!

2.14 More Work with Perimeter and Area 2.14

YOUR TURN

3. Find the area of a triangle with a base of 2.8 inches and a height of $3\frac{1}{4}$ inches. Draw a picture.

   Write an equation.
   Solve.

4. A softball diamond has 4 sides with each side length 60 feet. If a player hits a homerun, how far does the player run (from home plate, around the bases and then back to home plate)? Draw a picture.

   Write an equation.
   Solve.

Courtesy of Fotolia

5. Anita wants to put a trim around a rectangular tablecloth she is making. The tablecloth is $1\frac{7}{8}$ feet wide and $2\frac{1}{2}$ feet long. How much trim does she need to buy? Draw a picture.

Courtesy of Fotolia

Write an equation and solve.

2.14 More Work with Perimeter and Area

YOUR TURN

Write an equation. Solve.

6. Susan needs a new cover for her round pool. It is 20 feet across the pool. Use $3.14 \approx \pi$.

a. How many square feet of canvas is needed to make the cover?

b. If the canvas is sold in 11 foot wide bolts how many bolts will she need to purchase?

c. The cover needs to have an elastic border. How many feet of elastic will Susan need? DRAW A PICTURE.

7. Carli wants to install a 3 yard walk around a circular swimming pool. The diameter of the pool is 15 yards. In order to buy the concrete, she needs to know the area of the walk. What is the area of
the walk? Use 3.14 for π. Round the area to the nearest tenth. [Be sure to draw a picture.]

Write an equation.

Solve.

YOUR TURN **Answers to Section 2.14**

1. 54.5 ft or $54\frac{1}{2}$ft 2. 40π m; 125.6 m 3. 4.55 square inches 4. 240 ft 5. 8.75 ft or, $8\frac{3}{4}$ft

6a. 314 square feet 6b. 29 bolts 6c. 62.8 ft 7. 169.6 square feet

Complete MyMathLab Section 2.14 Homework in your Homework notebook.

2.15 Scientific Notation 2.15

When we are working with very large or very small numbers scientific notation provides a way of writing them.

For example, in 2006 the world population was about 6, 539,000,000 people. This number could be written as 6.539×10^9 in scientific notation.

The mass of a hydrogen atom is 0.0000000000000000000000017. This number could be written as 1.7×10^{-24}. Each number written in scientific notation is written as a number greater than or equal to 1 but less than 10 multiplied by some power of 10.

Writing Numbers in Scientific Notation

Step 1: Move the decimal until you have one digit to the left of the decimal. This will give you a number between 1 and 9 and then the decimal point.

1 digit between 1 & 9. → #._ _ _ _ X 10⁻

Step 2: Count the number of places you moved the decimal to get the number in step 1. If the original # was greater than or equal to 1, then the exponent is positive. If the original # was less than 1 then the exponent is negative.

Step 3: Multiply the number in step 1 by 10 raised to the power equal to the count found in step 2.

(# step 1) X $10^{(\text{# step 2})}$

EXAMPLE 1

Write 18,300,000 in scientific notation.

Step 1: Move decimal to the right of first nonzero digit. [Between the 1 and the 8.]

18 , 300 , 000. The number is now **1.83**

Step 2: The decimal moved to the **left** 7 places, so the exponent is **positive**.

Step 3: Put it all together. 1.83×10^7

2.15 Scientific Notation

EXAMPLE 2

Write 0. 0000052 in scientific notation.

Step 1: Move decimal to the right of first nonzero digit. [Between the 5 and the 2]

0 . 0000052 The number is now **5.2.**

Step 2: The decimal moved to the **right** 6 places so the exponent is **negative.**

Step 3: Put it all together! 5.2×10^{-6}

An easy way to remember whether the exponent should be positive or negative is to think that very **large numbers** (56,000,000,000) should be paired with **positive exponents** while very **small numbers** (0.000006) should be paired with **negative exponents**.

YOUR TURN

Write each number in scientific notation.

1. 5,300,000,000
2. 0.00000575

SCIENTIFIC NOTATION TO STANDARD FORM

A number written in scientific notation can be rewritten in standard form. This is going in reverse of what we just did.

To write a scientific notation number in standard form:

- Move the decimal point the same number of places as the exponent on 10.
- If the exponent is **negative**, move the decimal point to the **left**.
  ("Small" exponent then "small" number.)
- If the exponent is **positive**, move the decimal point to the **right.**
  ("Big" exponent then "big" number.)

EXAMPLE 3

Write 2.03×10^5 in standard notation.

2.03×10^5 move decimal 5 places to right → 2 0 3 , 0 0 0

$2.03 \times 10^5 = 203,000$

...10^{-3} in standard notation.

9.4×10^{-3} move decimal 3 places to left 0. 0 0 9 4

$9.4 \times 10^{-3} = 0.0094$

YOUR TURN

Write each number in standard form.

3. 1.25×10^{-3}

4. 8.79×10^{8}

YOUR TURN **Answers to Section 2.15**

1. 5.3×10^{9} 2. 5.75×10^{-6} 3. 0.00125 4. 879,000,000

Complete MyMathLab Section 2.15 Homework in your Homework notebook.

DMA 020 TEST: Review your notes, key concepts, formulas, and MML work. ☺

Student Name: ______________________________ **Date:** ______________

Instructor Signature: ______________________________

DMA 030

DMA 030

PROPORTION/RATIO/RATES/PERCENTS

This course provides a conceptual study of the problems that are represented by rates, ratios, percent, and proportions. Topics include rates, ratios, percent, proportion, conversion of English and metric units, and applications of the geometry of similar triangles. Upon completion, students should be able to use their understanding to solve conceptual application problems.

You will learn how to:

- Demonstrate an understanding of the concepts of ratios, rates, proportions, and percents in the context of application problems.
- Write a ratio using a variety of notations.
- Distinguish between events in a problem that should be represented by a ratio or a rate.
- Calculate a unit rate.
- Convert measurements using the U.S. customary and metric system using unit analysis.
- Represent percent as "parts of 100".
- Correctly convert between fractions, decimals, and percents.
- Solve application problems using ratios, rates, proportions, and percents.
- Recognize that two triangles are similar and solve for unknown sides using proportions in contextual applications.

Courtesy of Fotolia

3.01 Ratios 3.01

A ratio compares two quantities. You can compare two numbers, such as 5 and 10, or you can compare two measurements that have the same type of units, such as 3 *weeks* and 7 *weeks*.

A ratio can be written in 3 ways.

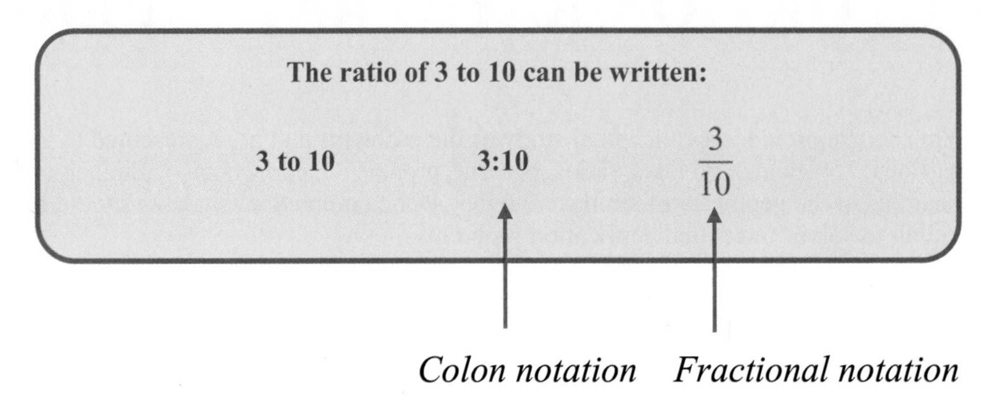

Writing a ratio as a fraction is the most common method. If the fraction happens to be an improper fraction, do not write the fraction as a mixed number.

Courtesy of Fotolia

If the fraction happens to be an improper fraction, do not write the fraction as a mixed number. A ratio is the comparison of two quantities.

A ratio of $\frac{3}{2}$ cannot be written as $1\frac{1}{2}$.

Writing a Ratio as a Fraction

The order of the quantities is important when writing ratios. To write a ratio as a fraction, write the 1st number as the *numerator* and the 2nd number as the *denominator*.

Simplifying a Ratio

To simplify a ratio, write the fraction in simplest form. In other words, reduce your fraction to lowest terms. Common factors and common units can be divided out.

3.01 Ratios 3.01

EXAMPLE 1

Write each ratio using fractional notation.

a. The ratio of 5 to 8.

5 is the *numerator* ⟷ 8 is the *denominator*

The ratio is written $\frac{5}{8}$.

Note: The order is important when writing ratios. The ratio $\frac{5}{8}$ is *NOT* the same as $\frac{8}{5}$.

b. The ratio of the length to the width.

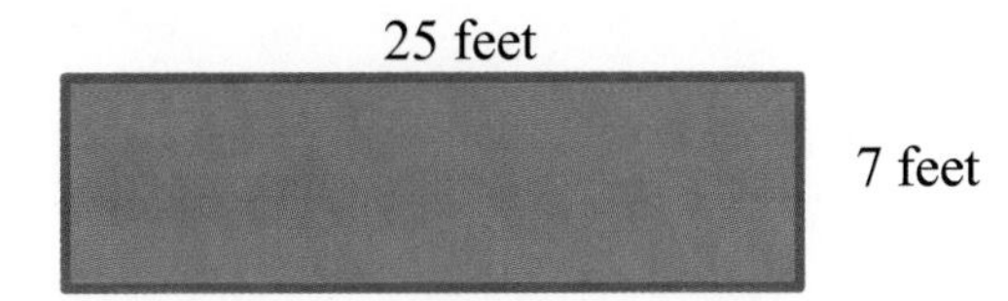

$$\frac{length}{width} = \frac{25 \cancel{feet}}{7 \cancel{feet}} = \frac{25}{7}$$

You can cancel *common units* just like you reduce common factors when writing fractions in lowest terms.

c. 40 days to 20 days.

$$\frac{40\ days}{20\ days} = \frac{40}{20} = \frac{2}{1}$$

Courtesy of Fotolia

In the Module on fractions you would have rewritten $\frac{2}{1}$ as 2.

But a **ratio** compares **two quantities**, so keep both parts of the ratio and write it as $\frac{2}{1}$.

EXAMPLE 2

Write the ratio as a ratio of whole numbers using fractional notation. Write the fraction in simplest form.

$3\frac{1}{2}$ *to* $2\frac{3}{8}$.

Remember that we can write a ratio as a fraction. $\dfrac{3\frac{1}{2}}{2\frac{3}{8}}$

Like any other fraction the line separating the numerator from the denominator is another way to represent division. So, let's divide.

Rewrite the numbers into a division problem written horizontally. $3\frac{1}{2} \div 2\frac{3}{8}$

Change the mixed numbers to improper fractions. $\frac{7}{2} \div \frac{19}{8}$

Take the reciprocal of the second fraction and multiply. $\frac{7}{2} \bullet \frac{8}{19} = \frac{56}{38}$

Reduce the ratio. $\frac{56}{38} = \frac{28}{19}$

The ratio $3\frac{1}{2}$ *to* $2\frac{3}{8}$ in simplest form is $\frac{28}{19}$

3.01 Ratios 3.01

YOUR TURN

1. Write each ratio as a fraction in lowest terms.

a. 4 to 3	b. 10 weeks to 2 weeks	c. \$20 to \$35

EXAMPLE 3

Find the ratio of the length of the swimming pool to the perimeter of the pool.

25 feet

15 feet

First find the perimeter of the pool. Remember perimeter means to find the distance *around the outside* of the pool. This can be done by adding the length of the 4 sides of the rectangular pool.

Perimeter = 25 + 25 + 15 + 15 = 80 feet.

$$\frac{length}{perimeter} = \frac{25\ feet}{80\ feet} = \frac{25}{80} = \frac{5}{16}$$

EXAMPLE 4

The table shows the number of greeting cards sold for various occasions. Write each ratio described as a fraction in lowest terms.

Holiday	*Cards Sold*
Valentine's Day	*900 million*
Mother's Day	*150 million*
Graduation	*60 million*
Father's Day	*95 million*

Courtesy of Fotolia

a. The ratio of Valentine's Day cards to Graduation Cards.

$$\frac{Valentine's\ Day\ cards}{Graduation\ cards} = \frac{900\ million}{60\ million} = \frac{900}{60} = \frac{15}{1}$$

b. The ratio of Father's Day cards to total cards sold.

First find the total number of cards sold by adding all the cards together.

900 + 150 + 60 + 95 = 1205 million

Now, find the ratio.

$$\frac{Father's\ Day\ cards}{total\ number\ of\ cards} = \frac{95\ million}{1205\ million} = \frac{95}{1205} = \frac{19}{241}$$

EXAMPLE 5

The price of Jane's dream car increased from \$25,000 to \$40,000. Find the ratio of the increase in price to the original price.

First, subtract to find the increase in price: 40,000 – 25,000 = 15,000. The car increased in price \$15,000.

$$\frac{increase\ in\ price}{original\ price} = \frac{\$15,000}{\$25,000} = \frac{3}{5}$$

YOUR TURN

2. Last week, Shane worked 45 hours. This week he worked 50 hours. Find the ratio of the increase in hours to the previous week of hours.

3. Find the ratio of the base of the triangle to the perimeter.

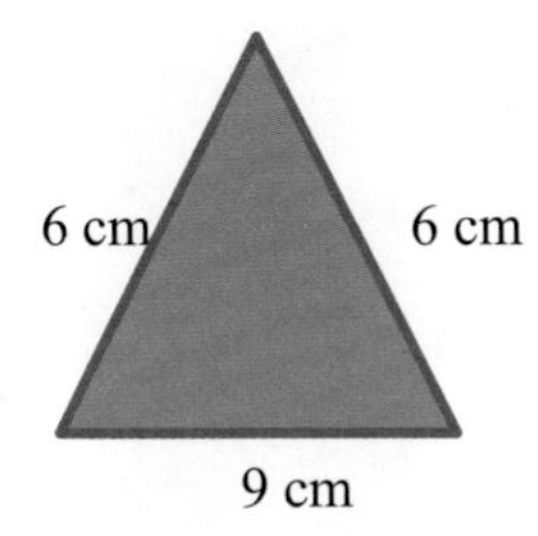

Use the table for numbers 4 and 5.

Holiday	*Cards Sold*
Valentine's Day	*900 million*
Mother's Day	*150 million*
Graduation	*60 million*
Father's Day	*95 million*

4. Write the ratio of graduation cards sold to Father's Day cards sold.

5. Write the ratio of Mother's Day cards sold to total number of cards sold.

3.01 Ratios 3.01

A ratio compares two measurements or quantities with the same type of units, such as 9 feet to 12 feet. But many comparisons we make use measurements with different types of unit. Two examples are given below.

500 miles on 20 gallons (miles to gallons)

200 dollars for 10 hours (money to time)

This type of comparison is called a **rate.**

EXAMPLE 6

Write each rate as a fraction in simplest form.

When comparing quantities with different units, write the units as part of the answer. They do not cancel out as with ratios!!!!

a. 2225 miles on 75 gallons of gas.

$$\frac{2225\,miles}{75\,gallons} = \frac{89\,miles}{3\,gallons}$$

b. $1500 in 10 weeks.

$$\frac{1500\,dollars}{10\,weeks} = \frac{150\,dollars}{1\,week}$$

A Reminder

Be sure to write the units in a rate!!

YOUR TURN

Write each rate as a fraction in simplest form.

6. 4 teachers for 90 students

7. 18 miles in 4 hours

Unit Rates

When the denominator of a rate is 1, it is called a unit rate.

Unit rates are used frequently. Here are two examples:

Sally earns $9 for 1 hour of work. = $ 9 <u>per</u> 1 hour = $\frac{9\ dollars}{1\ hour}$ = $9 /hour

Fred drove 28 miles on 1 gallon of gas = 28 miles <u>per</u> 1 gallon = $\frac{28\ miles}{1\ gallon}$ = 28 miles/gallon

We can use the word "per" to mean division.

Writing a Rate as a Unit Rate

To write a rate as a unit rate, divide the numerator by the denominator.

EXAMPLE 7

Write as a unit rate. $60 in 5 hours

$$\frac{60\ dollars}{5\ hours} = \frac{12\ dollars}{1\ hour} = \$12\ /\ hour$$

Divide the numerator by the denominator

60 ÷ 5 = 12

EXAMPLE 8

Charlie can assemble 250 computer boards in an 8-hour shift while Robert can assemble 402 computer boards in a 12-hour shift.

a. Find the unit rate of Charlie.

$$\frac{250\ boards}{8\ hours} = \frac{31.25\ boards}{1\ hour} = 31.25\ \ boards/hour$$

Divide the numerator by the denominator: 250 ÷ 8 = 31.25

Courtesy of Fotolia

b. Find the unit rate of Robert.

$$\frac{402\ boards}{12\ hours} = \frac{33.5\ boards}{1\ hour} = 33.5\ \ boards/hour$$

Divide the numerator by the denominator: 402 ÷ 12 = 33.5

c. Who can assemble computer boards faster, Charlie or Robert?

Charlie can assemble 31.25 boards per hour.
Robert can assemble 33.5 boards per hour.

Robert can assemble computer boards faster!!!

YOUR TURN

8. Write as a unit rate: On average it costs $1,495,000 to build 25 Habitat for Humanity houses in the United States.

9. A Toyota Corolla can travel 420 miles on 12 gallons of gas. A Chevrolet Cobalt requires 8 gallons of gas to travel 296 miles.

a. Find the unit rate (how many miles per 1 gallon of gas) for the Corolla.

b. Find the unit rate (how many miles per 1 gallon of gas) for the Cobalt.

c. Which car gets the best gas mileage?

Finding Unit Prices

The "Unit Price" (or "unit cost") tells you how much you pay per item.

To find the unit price just divide the cost by the quantity.

EXAMPLE 9

A store charges $4.20 for a bag of 12 apples. What is the unit price for an apple?

Unit price: $\dfrac{\$4.20}{12\ apples} = \dfrac{\$0.35}{1 apple}$ $or\ \ \$0.35\ per\ apple$

Divide the numerator by the denominator: 4.20 ÷ 12 = 0.35

3.01 Ratios

When shopping for groceries or other household items, you will find many brands and sizes from which to choose. Sometimes foods packed in the "giant" or "family" size may seem like the best buy. You may think that buying one large container will not cost as much as two or three smaller packages. But larger containers do not always end up costing less than smaller ones. It is important to look at the cost per unit and compare these costs to find the better buy.

Courtesy of Fotolia

You can save money by finding the lowest cost per unit. The **"best buy"** is the item with the lowest cost per unit or the lowest unit price.

EXAMPLE 10

Find each unit price and decide which one is the better buy. To compare, round each unit price to three decimal places.

Peanut butter

Courtesy of Fotolia

12 ounces for $1.29

Courtesy of Fotolia

28 ounces for $3.39

Unit Price of 12 ounces =

$$\frac{price}{number\ of\ units} = \frac{\$1.29}{12\ ounces} = \frac{\$0.108}{1\ ounce} = \$0.108\ per\ ounce$$

Unit Price of 28 ounces =

$$\frac{price}{number\ of\ units} = \frac{\$3.39}{28\ ounces} = \frac{\$0.121}{1\ ounce} = \$0.121\ per\ ounce$$

The lowest cost per ounce is $0.108, so the 12-ounce Peanut Butter is the best buy.

Calculator Tip

When using a calculator to find unit prices, remember that division is not commutative. So, regardless of the order the information is given, always divide the cost by the number of units.

$$\frac{price}{number\ of\ units} = price \div number\ of\ units$$

YOUR TURN

Find each unit price and decide which one is the better buy. To compare, round each unit price to three decimal places.

10. Soup: 2 cans for \$2.18 or
5 cans for \$5.29

Unit Price for 2 cans for \$2.18

$$unit\ price = \frac{price}{number\ of\ units}$$

= ________________

= \$ __________ *per can*

Unit Price for 5 cans for \$5.29

$$unit\ price = \frac{price}{number\ of\ units}$$

= ________________

= \$ __________ *per can*

Better Buy: ____________________________

11. Cereal: 14 ounces for \$2.89 or
18 ounces for \$3.96

Unit Price for 14 ounces for \$2.89

$$unit\ price = \frac{price}{number\ of\ units}$$

= ________________

= \$ __________ *per ounce*

Unit Price for 18 ounces for \$3.96

$$unit\ price = \frac{price}{number\ of\ units}$$

= ________________

= \$ __________ *per ounce*

Better Buy: ____________________________

3.01 Ratios

YOUR TURN **Answers to Section 3.01.**

1. a. $\frac{4}{3}$ **b.** $\frac{10}{2} = \frac{5}{1}$ **c.** $\frac{20}{15} = \frac{4}{3}$

2. $\frac{5}{45} = \frac{1}{9}$ **3.** $\frac{9}{21} = \frac{3}{7}$ **4.** $\frac{60}{95} = \frac{12}{19}$

5. $\frac{150}{1205} = \frac{30}{241}$ **6.** $\frac{2 \text{ teachers}}{45 \text{ students}}$ **7.** $\frac{9 \text{ miles}}{2 \text{ hours}}$

8. \$59,800/house **9.** a. 35 miles/gal b. 37 miles/gal c. Cobalt

10. \$1.09/can

\$1.058/ can

5 cans is better buy

11. \$0.206/ounce

\$0.22/ounce

14 ounces is better buy

Complete MyMathLab Section 3.01 Homework in your Homework notebook.

3.02 Proportions 3.02

PROPORTIONS: → **A proportion is a statement that 2 ratios or rates are equal. Proportions use "=" between two ratios.**

$\frac{4}{5} = \frac{20}{25}$ is a proportion and is read, "4 is to 5 as 20 is to 25."

EXAMPLE 1

Write as a proportion.

4 water bottles is to 5 players as 20 water bottles is to 25 players

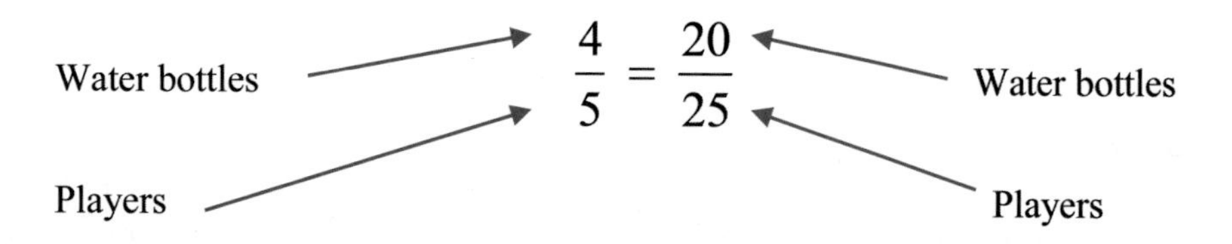

Courtesy of Fotolia

In the proportion above, the numerators contain the same units as well as the denominators. Proportions must be written in this way. **Order is very important!!!**

YOUR TURN

Write each sentence as a proportion.

1. 4 songs are to 1 CD as 20 songs are to 5 CDs.	2. 2 cups of milk is to 1 recipe as 12 cups of milk is to 6 recipes

3.02 Proportions 3.02

A proportion may be either true or false. In order to tell whether a proportion is true, simply cross multiply to see if the cross products are equal to each other. If they are, this is a *true proportion*. If the cross products are not equal, the proportion is *false*.

Using Cross Products to Determine if Proportions are True or False

$$\frac{2}{5} = \frac{4}{10}$$

5 · 4 = 20

2 · 10 = 20

Cross products are equal.

In this case, the **cross products** are both 20. When the cross products are ***equal***, the proportion is ***true***. If the cross products are **not equal**, the proportion is **false**.

EXAMPLE 2

a. Is $\frac{8}{24} = \frac{3}{9}$ **a true proportion?**

Look at the cross products. Does $8 \cdot 9 = 24 \cdot 3$?

$72 = 72$ ✓ Since the cross products are equal, the proportion is true.

b. Is $\frac{0.8}{0.3} = \frac{0.2}{0.6}$ **a true proportion?**

Look at the cross products. Does $0.8\,(0.6) = 0.2\,(0.3)$?

$0.48 \neq 0.06$ ✗ Since the cross products are not equal, the proportion is false.

YOUR TURN

Fill in the blanks below to determine whether each proportion is true or false.

3. $\frac{5}{6} = \frac{9}{7}$	4. $\frac{9}{36} = \frac{2}{8}$	5. $\frac{3.6}{6} = \frac{5.4}{8}$
____ • ____ = ____ • ____	____ • ____ = ____ • ____	____ • ____ = ____ • ____
____ = ____	____ = ____	____ = ____
Proportion is ____________.	Proportion is ____________.	Proportion is ____________.

Finding an Unknown Value in a Proportion

Four numbers are used in a proportion. If any three of these numbers are known, the fourth can be found.

Find the unknown number that will make this proportion true.

$$\frac{3}{5} = \frac{n}{40}$$

If this proportion is true, the cross products must be equal.

Set the cross products equal. ⟶ $5 \cdot n = 3 \cdot 40$

$$5 \cdot n = 120$$

To solve for n, divide the number by itself by the number beside the n. $n = \frac{120}{5}$

$$n = 24$$

Steps to Find an Unknown Value in a Proportion

Step 1: Set the cross products equal to each other.

Step 2: To solve for the unknown, "n", divide the number by itself by the number multiplied by n.

EXAMPLE 3

Find the unknown number *n*.

$$\frac{7}{n} = \frac{6}{5}$$

Step 1: Set the cross products equal. → $7 \cdot 5 = 6 \cdot n$

$$35 = 6 \cdot n$$

Step 2: Divide the number by itself by the number multiplied by n: $\frac{35}{6} = n$

$$5\frac{5}{6} = n$$

EXAMPLE 4

Find the unknown number *n*.

$$\frac{0.05}{12} = \frac{n}{0.6}$$

Step 1: Set the cross products equal. → $0.05 \cdot 0.6 = 12 \cdot n$

$$0.03 = 12 \cdot n$$

Step 2: Divide the number by itself by the number multiplied by n: $\frac{0.03}{12} = n$

$$0.0025 = n$$

YOUR TURN

Find the unknown number in each proportion.

6. $\frac{n}{5} = \frac{6}{10}$	7. $\frac{12}{10} = \frac{n}{16}$	8. $\frac{7.8}{1.3} = \frac{n}{2.6}$
____ • n = ____ • ____	____ • n = ____ • ____	____ • n = ____ • ____
$n =$	$n =$	$n =$

YOUR TURN **Answers to Section 3.02.**

1. $\frac{4\,songs}{1\,CD} = \frac{20\,songs}{5\,CD's}$

2. $\frac{2\text{ cups of milk}}{1\text{ recipe}} = \frac{12\text{ cups of milk}}{6\text{ recipes}}$

3. False 4. True 5. False

6. 3 7. 19.2 8. 15.6

Complete MyMathLab Section 3.02 Homework in your Homework notebook.

Solving Problems by Writing Proportions

Proportions can be used to solve a variety of problems. We will look at problems in which you are given a ratio or a rate and then will be asked to find part of a corresponding ratio or rate. Given a ratio or rate, a proportion can be used to determine the unknown quantity.

EXAMPLE 1

The scale on a map is 2 inches equals 5 miles. What is the distance between two points on the map that are 8 inches apart on the map?

This example compares inches to **miles**. Write a proportion using the two rates. Be sure that both rates compare inches to miles in the same order. In other words, inches is in both numerators and miles is in both denominators.

This rate compares **inches** to **miles**.

$$\frac{2 \ inches}{5 \ miles} = \frac{8 \ inches}{n \ miles}$$

This rate compares **inches** to **miles.**

$$2 \cdot n = 5 \cdot 8$$

$$2 \cdot n = 40$$

$$n = \frac{40}{2}$$

$$n = 20$$

Courtesy of Fotolia

When setting up a proportion be consistent in the rates.

This rate compares **inches to miles.**

$$\frac{2 \ inches}{5 \ miles} \neq \frac{n \ miles}{8 \ inches}$$

This rate compares **miles to inches** .

These rates do not compare things in the same order and cannot be set up as a true proportion.

3.03 Proportions and Problem Solving 3.03

EXAMPLE 2

A bag of grass seed covers 3000 square feet of lawn. Find how many bags of grass seed should be purchased to cover a rectangular lawn 120 by 210 feet.

Since grass seed covers the square footage of the lawn, we must first find the area of the lawn. A rectangle 120 by 210 feet has an area of 25,200 square feet. [multiply *length* · *width*]

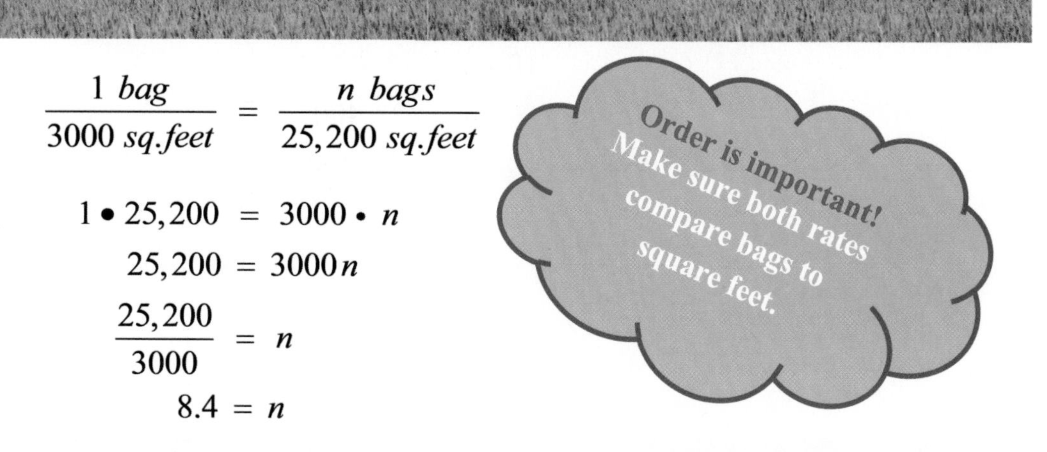

$$\frac{1\ bag}{3000\ sq.feet} = \frac{n\ bags}{25{,}200\ sq.feet}$$

$$1 \bullet 25{,}200 = 3000 \bullet n$$

$$25{,}200 = 3000n$$

$$\frac{25{,}200}{3000} = n$$

$$8.4 = n$$

We will need 8.4 bags of fertilizer, but we realize we cannot buy a portion of a bag of fertilizer. We must buy whole bags of fertilizer. In order to have enough, we will need **9 bags of fertilizer**.

> NOTICE: We did not round the answer like we normally do for decimals. Because we must purchase whole items to have enough to do the job, you will always need to purchase more than you actually need to do the job! So, we will always **round up** to determine your answer.

EXAMPLE 3

Susan decides to lose a few pounds. There is a new diet that claims you can lose 10 pounds in 7 days. If she wants to lose 14 pounds, how long should it take her?

$$\frac{10\ pounds}{7\ days} = \frac{14\ pounds}{n\ days}$$

$$10 \bullet n = 7 \bullet 14$$

$$10n = 98$$

$$n = \frac{98}{10}$$

$$n = 9.8\ days$$

> Susan can lose 14 pounds in 9.8 days!

EXAMPLE 4

A newspaper report says that 7 out of 10 people prefer Coke to Pepsi. At a convention of 3200 people, how many would you expect to prefer Coke?

People who prefer Coke. ⟶ $\frac{7}{10} = \frac{n}{3200}$ ⟵ People who prefer Coke.

Total surveyed. ⟶ (denominators) ⟵ Total people.

$$10 \bullet n = 7 \bullet 3200$$
$$10n = 22{,}400$$
$$n = \frac{22{,}400}{10}$$
$$n = 2240$$

You would expect 2240 people to prefer Coke.

Courtesy of Fotolia

Always ask yourself, "Is my answer reasonable?" If it is not, look at the way your proportion is set up. Be sure you have matching units in the numerators and matching units in the denominators.
For example, suppose you had set up the last proportion *incorrectly* like below:

$$\frac{7}{10} = \frac{3200}{n}$$
$$7 \bullet n = 3200 \bullet 10$$
$$7n = 32{,}000$$
$$n = 4571 \ (\textit{rounded to nearest person})$$

This answer is *unreasonable* because there are only 3200 people total at the convention. It is not possible that out of 3200 people 4571 people would prefer Coke.

EXAMPLE 5

An alloy consists of copper, zinc, and tin in the ratio of 2:3:5, representing the three amounts, respectively. Find the amount of each metal in 72 kilograms of the alloy.

Given that the alloy consists of three metals in the ratio 2:3:5, we know that there are a total of 10 units: $(2+3+5=10)$

Amount of copper would be 2 parts out of 10 which written as a ratio would be: $\frac{2}{10}$

Amount of zinc would be 3 parts out of 10 which written as a ratio would be: $\frac{3}{10}$

Amount of tin would be 5 parts out of 10 which written as a ratio would be: $\frac{5}{10}$

Now we will need to find the amount of each metal present in 72 kilograms of the alloy.

Amount of Copper

$$\frac{2}{10}=\frac{n}{72}$$
$$10n = 2 \cdot 72$$
$$10n = 144$$
$$n = \frac{144}{10}$$
$$n = 14.4 \text{ kilograms}$$

Amount of Zinc

$$\frac{3}{10}=\frac{n}{72}$$
$$10n = 3 \cdot 72$$
$$10n = 216$$
$$n = \frac{216}{10}$$
$$n = 21.6 \text{ kilograms}$$

Amount of Tin

$$\frac{5}{10}=\frac{n}{72}$$
$$10n = 5 \cdot 72$$
$$10n = 360$$
$$n = \frac{360}{10}$$
$$n = 36 \text{ kilograms}$$

3.03 Proportions and Problem Solving 3.03

YOUR TURN

Set up and solve a proportion for each problem.

1. It takes a word processor 35 minutes to spell check 10 pages. Find how long it takes to word process 62 pages.

$$\frac{\square \text{ min}}{\square \text{ pages}} = \frac{\square \text{ min}}{\square \text{ pages}}$$

2. A gallon of paint covers 120 square feet of a wall. Find how many gallons of paint are needed to paint a wall 20 feet by 32 feet.

$$\frac{\square \text{ gal}}{\square \text{ sq ft}} = \frac{\square \text{ gal}}{\square \text{ sq ft}}$$

3. A Honda Civic averages 420 miles on 12 gallons of gas. Shane is planning a trip to New York which is 2170 miles from his home. Find how many gallons of gas he will need.

$$\frac{\square \text{ miles}}{\square \text{ gal}} = \frac{\square \text{ miles}}{\square \text{ gal}}$$

4. In one state, 3 out of 5 college students receive financial aid. At this rate, how many of the 4500 students at Central Community College receive financial aid?

$$\frac{\square \text{ fin. aid}}{\square \text{ students}} = \frac{\square \text{ fin. aid}}{\square \text{ students}}$$

3.03 Proportions and Problem Solving 3.03

YOUR TURN

5. The scale on a map is 2 inches equals 15 miles. What is the distance between two points on the map that are 8 inches apart?

$$\frac{\square \text{ inches}}{\square \text{ miles}} = \frac{\square \text{ inches}}{\square \text{ miles}}$$

6. An alloy consists of copper, nickel, and tin in the ratio of 1:3:5, representing the three amounts, respectively. Find the amount of each metal in 72 kilograms of the alloy.

$$\frac{\square \text{ copper}}{\square \text{ alloy}} = \frac{\square \text{ kg of copper}}{\square \text{ kg of alloy}}$$

$$\frac{\square \text{ nickel}}{\square \text{ alloy}} = \frac{\square \text{ kg of nickel}}{\square \text{ kg of alloy}}$$

$$\frac{\square \text{ tin}}{\square \text{ alloy}} = \frac{\square \text{ kg of tin}}{\square \text{ kg of alloy}}$$

3.03 Proportions and Problem Solving 3.03

YOUR TURN **Answers to Section 3.03.**

1. 217 minutes
2. 6 gallons
3. 62 gallons
4. 2700 students
5. 60 miles
6. 8 kg copper, 24 kg nickel, 40 kg tin

Complete MyMathLab Section 3.03 Homework in your Homework notebook.

3.04 Congruent and Similar Triangles 3.04

Similar triangles are the same shape but not the same size. You can think of it as "zooming in" or "zooming out" and making the triangle bigger or smaller, but keeping its basic shape. All of the angles are exactly the same size, so the object looks exactly like the original, only the side lengths are larger or smaller.

The triangles below are "similar" – although they are different sizes, their shapes are the same. The triangles are not the same size but all of the angles are exactly the same size, so the triangles look exactly like each other, only larger or smaller.

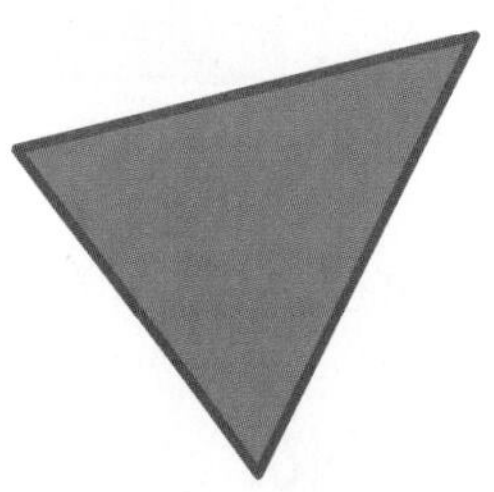

Similar Triangles

Corresponding angles are equal.
Corresponding sides are in proportion.

If two triangles are similar, then their corresponding angles have the same measure and their corresponding sides are proportional. To illustrate, if triangle ABC is similar to triangle RST, then angles A and R have the same measure, angles B and S have the same measure, and angles C and T have the same measure. In addition, the corresponding sides are proportional.

$$\Delta ABC \sim \Delta RST$$

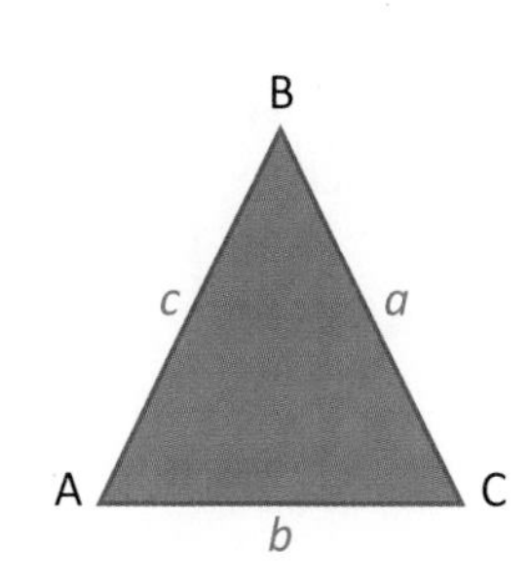

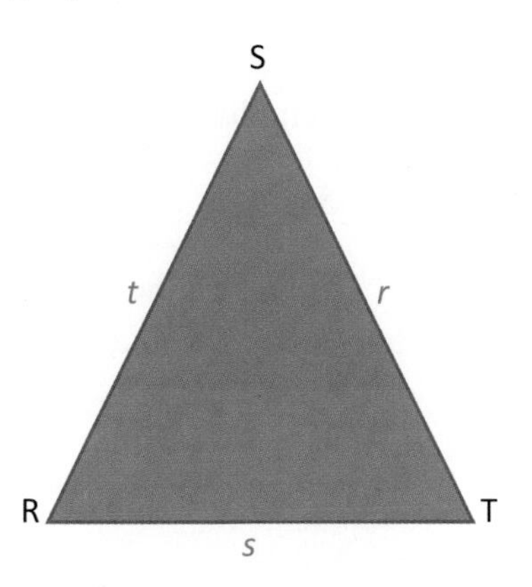

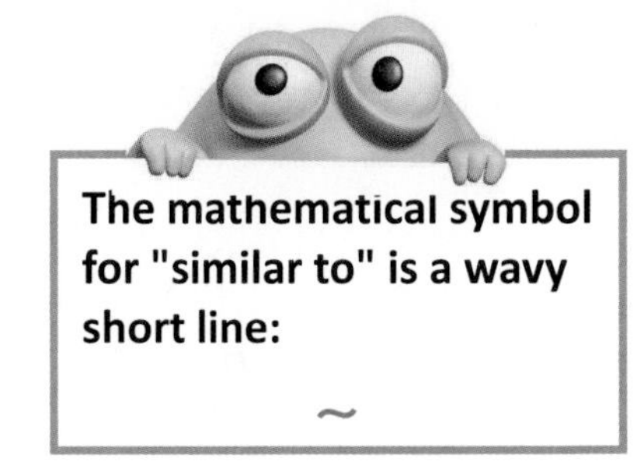

$$\frac{a}{r} = \frac{b}{s} = \frac{c}{t}$$

3.04 Congruent and Similar Triangles 3.04

EXAMPLE 1

Find the ratio of corresponding sides for the similar triangles below.

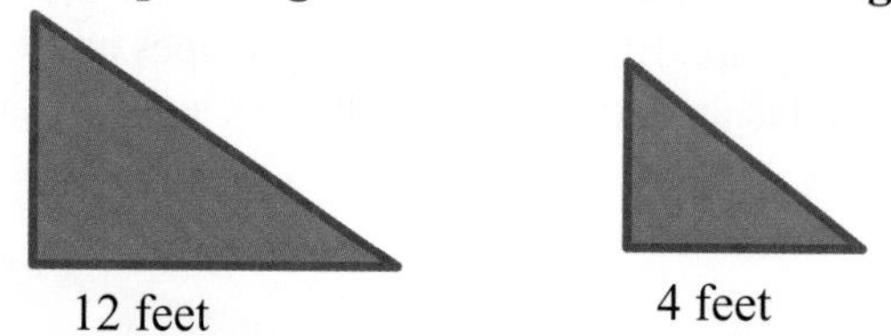

We are given the lengths of two corresponding sides. Their ratio is

$$\frac{12 \text{ feet}}{4 \text{ feet}} = \frac{3}{1}$$

If we know the ratio and the side measures of one triangle, we can easily find the side measures of the other triangle.

EXAMPLE 2

Given that the two triangles are similar, find the missing length *n*.

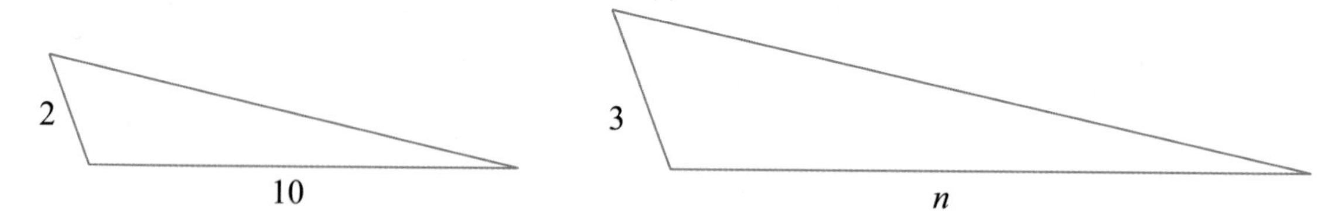

Since the triangles are similar, corresponding sides are in proportion. Therefore, the ratio of the side lengths 2 to 3 is the same as the ratio of the side lengths 10 to n. Let's write this as a proportion and then solve for n.

$$\frac{2}{3} = \frac{10}{n}$$ ⟶ To find the unknown length n, we set cross products equal and solve.

$$2 \bullet n = 3 \bullet 10$$
$$2n = 30$$
$$n = \frac{30}{2}$$
$$n = 15$$ ⟶ The missing length is 15 units.

EXAMPLE 3

Samantha, a 5-foot-tall park ranger needs to know the height of a tree. She notices that when the shadow of the tree is 24 feet long, her shadow is 8 feet long. Find the height of the tree.

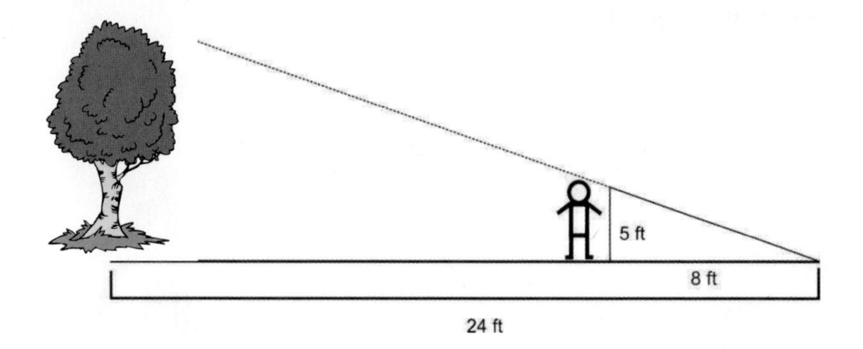

We will set up a proportion and solve.

$$\frac{\text{Samantha's height}}{\text{tree's height}} \longrightarrow \frac{5}{n} = \frac{8}{24} \longleftarrow \frac{\text{length of Samantha's shadow}}{\text{length of tree's shadow}}$$

$$5 \cdot 24 = 8 \cdot n$$

$$120 = 8n$$

$$\frac{120}{8} = n$$

$$15 = n$$

The height of the tree is 15 feet.

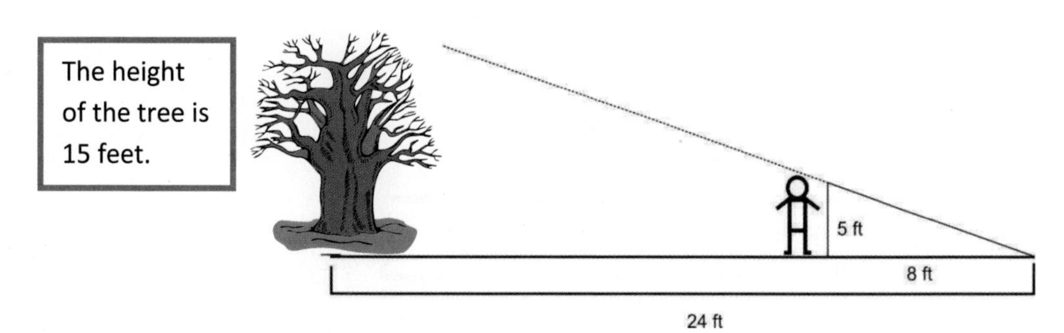

EXAMPLE 4

Samuel, a firefighter, needs to estimate the height of a burning building. He estimates the length of his shadow to be 9 feet long and the length of the building's shadow to be 75 feet long. Find the approximate height of the building if he is 6 feet tall.

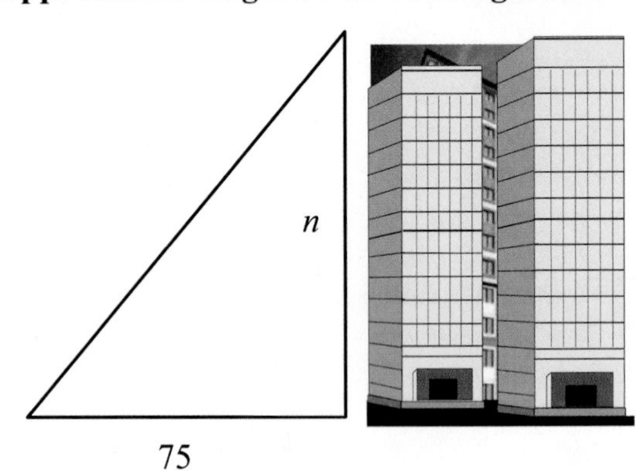

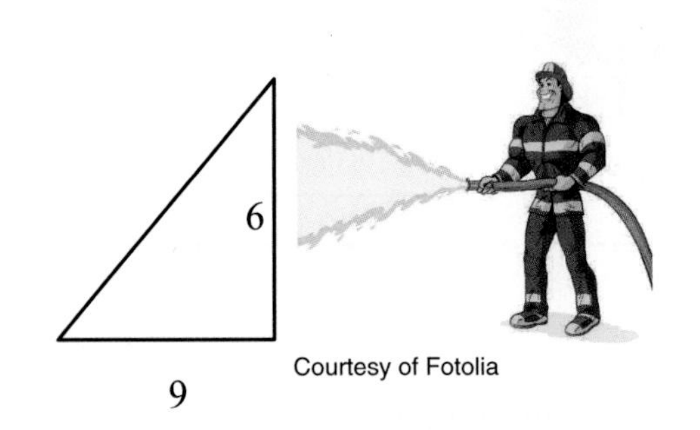

Courtesy of Fotolia

$$\frac{\text{firefighter's height}}{\text{building's height}} \longrightarrow \frac{6}{n} = \frac{9}{75} \longleftarrow \frac{\text{length of firefigher's shadow}}{\text{length of building's shadow}}$$

$$\frac{6}{n} = \frac{9}{75}$$

$$6 \bullet 75 = 9 \bullet n$$

$$450 = 9n$$

$$\frac{450}{9} = n$$

$$50 = n$$

The height of the building is 50 feet.

3.04 Congruent and Similar Triangles 3.04

EXAMPLE 5

In the diagram below, triangle $\Delta ADF \sim \Delta BCF$.

The crosswalk at point A is about 40 yds long. A bridge across a pond will be built, from point B to point C. What will the length of the bridge be?

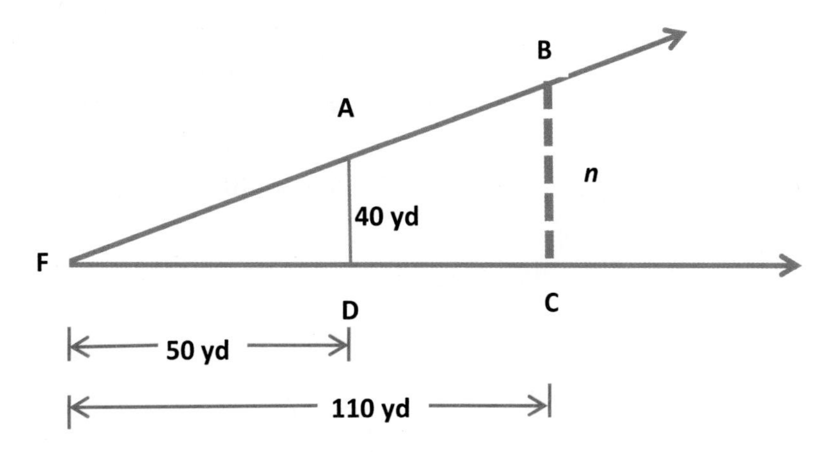

Since the two triangles are similar, we will set up a proportion using corresponding sides.

$$\frac{50}{110} = \frac{40}{n}$$

$$50 \cdot n = 110 \cdot 40$$

$$50n = 4400$$

$$n = 88$$

The length of the bridge will be 88 yds.

YOUR TURN

1. Find the length of the unknown side.

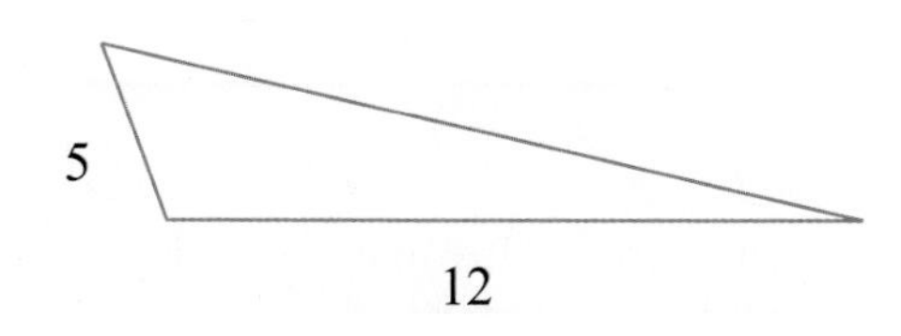

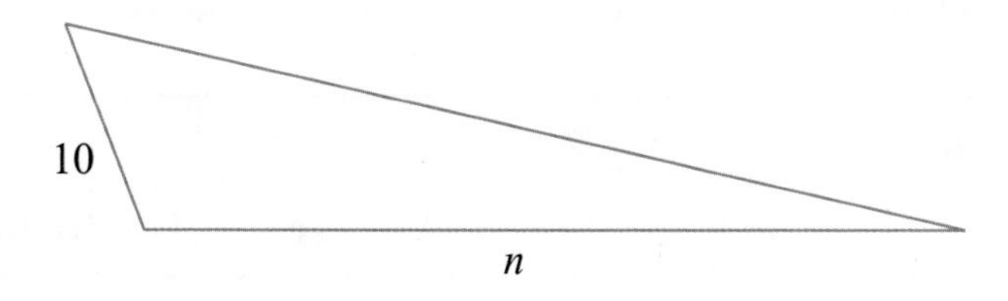

YOUR TURN

Write the proportion and solve to answer each question.

2. If a 30-foot tree casts an 18-foot shadow, find the length of the shadow cast by a 24-foot tree.

3. Marcus and Steve stood side-by-side on a sunny day and noticed that their shadows were different lengths. Marcus measured Steve's shadow and determined it was 9.5 feet. Steve measured Marcus' shadow and determined it was 10.25 feet. If Steve is 6 feet tall, how tall is Marcus? Round to the nearest thousandth.

YOUR TURN **Answers to Section 3.04.**

1. 24 units
2. 14.4 feet
3. 6.474 feet

Complete MyMathLab Section 3.04 Homework in your Homework notebook.

3.05 Length: U.S. and Metric Systems of Measurements 3.05

In the United States, we commonly use one of two measurement systems known as the United States or English System and the Metric System. We are familiar with using both systems in our daily lives. We speak of our cars going 55 miles per hour which uses the English System. When we go to the grocery store, we buy a 2-liter drink which uses the Metric System.

We use *linear measurement* to see how many inches our children grow each year, to complete household projects such as hanging curtains or installing new baseboards, and to install sprinkler pipes in the yard. These measurements include inches, centimeters, feet, yards, meters and even miles. We will begin by looking at the U.S. system for length.

U.S. Units of Length

1 mile (mi) = 5280 feet (ft)
1 yard (yd) = 3 feet (ft)
1 yard (yd) = 36 inches (in)
1 foot (ft) = 12 inches (in)

We can visualize these units as follows:

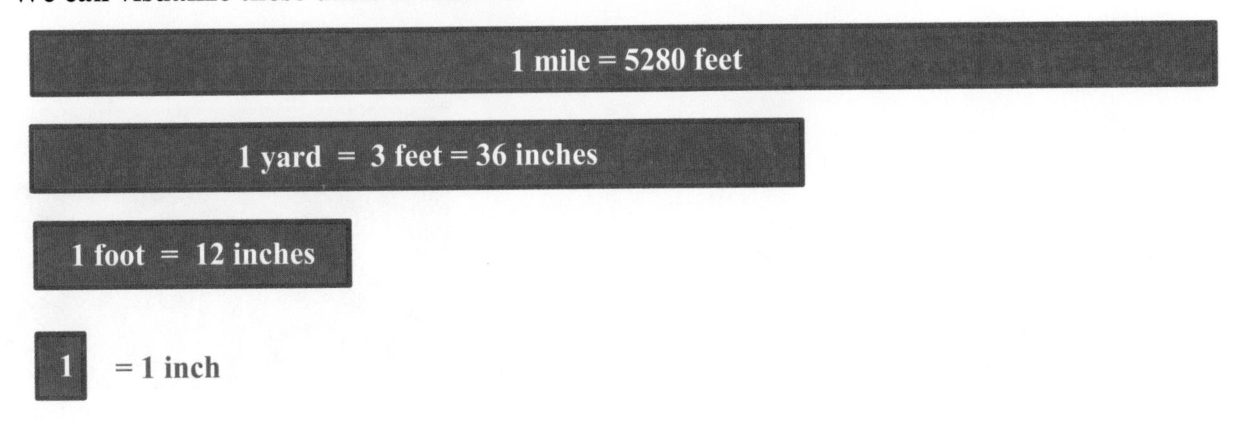

Linear measurement is used to measure the length or width of objects, the distance between points, or the distance around figures (perimeter). The most common units of linear measurement are inches, feet, yards, and miles. When measuring, we choose the unit that is best for the situation at hand. If we are measuring curtains, we would choose inches. If we are measuring the distance between cities, we would choose miles.

3.05 Length: U.S. and Metric Systems of Measurements 3.05

YOUR TURN

1. Write inch, foot, yard, or mile after each item listed below to show which unit of linear measurement would be most appropriate.

a. length of a piece of paper ________	g. flight across the country _________
b. distance between suburbs ________	h. length of a fish _________
c. fabric needed to sew a dress ________	i. length of a whale _________
d. height of a ceiling ________	j. depth of a swimming pool ________
e. perimeter of a baseball field ________	k. distance on a football field _______
f. length of a paper clip ________	l. size of a window _________

***Conversion Factors* are used to convert from one unit of measure to another in order to analyze comparisons.**

Courtesy of Fotolia

For example, is it faster to travel 60 miles per hour or 60 feet per second?

We could not answer this immediately unless we compared the same unit of measure.
This would require a conversion.
When using conversion factors you are really multiplying by a value of 1.
This sounds simple, however recognizing what "1" means or looks like can be tricky.

3.05 Length: U.S. and Metric Systems of Measurements 3.05

You will use fractions to convert from one unit of measure to another.

For instance, $\frac{60 \text{ seconds}}{1 \text{ minute}}$ *is really "1", because 60 seconds is the same as one minute!*

What "conversion factor of 1" is needed?

Conversions

1. **Begin with the original problem in the numerator.** (Remember, all whole numbers can be written over 1.)
2. **Multiply by the necessary conversion factor.** *** Remember the *numerator units of the 1st fraction* should match the *denominator units of the 2nd fraction.*
3. **Cancel the common units.**
4. **Simplify the mathematics.**

EXAMPLE 1

Convert 10 yards to feet.

Begin with the original 10 yards. Your goal is to make the yards cross out and in the final answer have feet. To do this, you will need to multiply by "1". The numerator (in this case yards) of the original 1st fraction will need to be the same units as the denominator of the 2nd conversion "1" fraction.

$$\frac{10 \, yards}{1} \cdot \frac{3 \, feet}{1 \, yard}$$

Pick the conversion factor that will cause the "yards" to cancel. The unit you want to end up with should always be in the numerator—in this problem "feet".

$$\frac{10 \, \cancel{yards}}{1} \cdot \frac{3 \, feet}{1 \, \cancel{yard}}$$

The yards cancel.

Conversion Factors

$\frac{3 \text{ feet}}{1 \text{ yard}}$ or $\frac{1 \text{ yard}}{3 \text{ feet}}$

$$\frac{10}{1} \cdot \frac{3 \, feet}{1}$$

Now, do the math.

30 feet

EXAMPLE 2

Convert 75 inches to feet.

Begin with the original 75 inches. Your goal is to make the inches cross out and in the final answer have feet. To do this, you will need to multiply by "1". The numerator (in this case inches) of the original 1st fraction will need to be the same units as the denominator of the 2nd conversion "1" fraction.

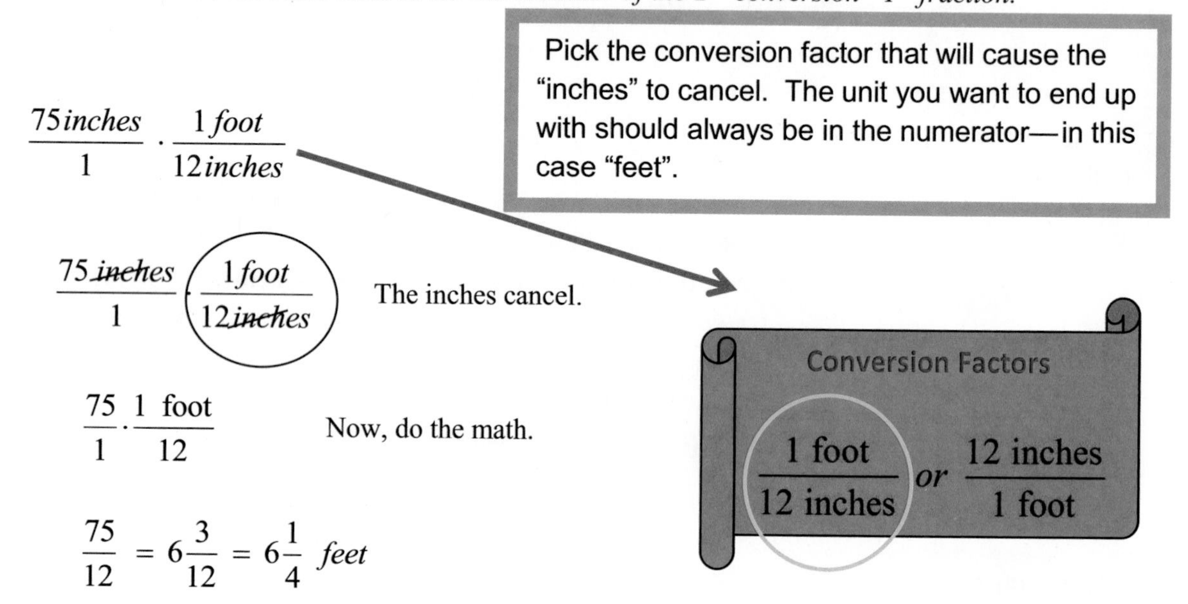

EXAMPLE 3

Convert 23,760 feet to miles.

We choose a symbol for 1 with "miles" in the top and "feet" in the bottom.

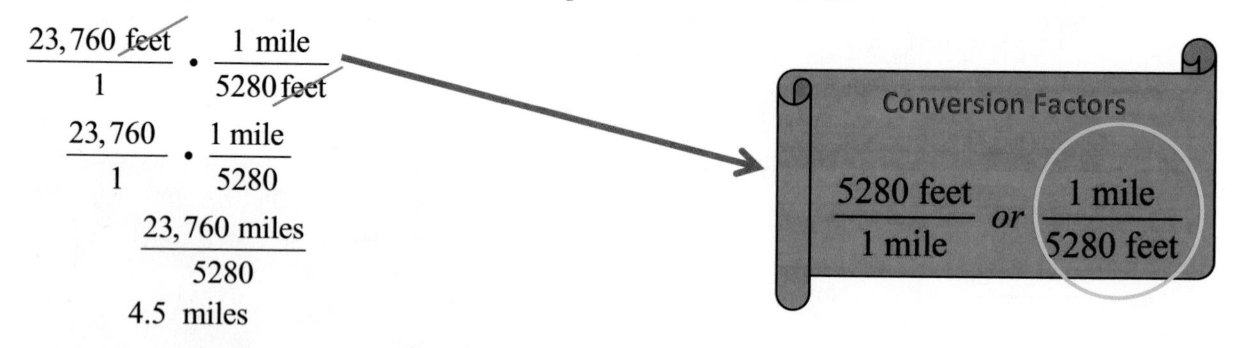

3.05 Length: U.S. and Metric Systems of Measurements 3.05

YOUR TURN

Convert the measurement as indicated. Circle the conversion factor you would use to convert and then follow steps to convert.

2. 24 yds to feet Conversion Factors: $\frac{3\,ft}{1\,yd}$ *or* $\frac{1\,yd}{3\,ft}$ $\frac{24\,yds}{1} \bullet$ ______ $24\,yds =$ __________ ft	3. 11.5 feet to inches Conversion Factors: $\frac{1\,ft}{12\,in}$ *or* $\frac{12\,in}{1\,ft}$ $\frac{11.5\,ft}{1} \bullet$ ______ $11.5\,ft =$ __________ in
4. 4 miles to feet Conversion Factors: $\frac{5280\,ft}{1\,mi}$ *or* $\frac{1\,mi}{5280\,ft}$ $\frac{4\,mi}{1} \bullet$ ______ $4\,mi =$ __________ ft	5. 90 inches to yds Conversion Factors: $\frac{36\,in}{1\,yd}$ *or* $\frac{1\,yd}{36\,in}$ $\frac{90\,in}{1} \bullet$ ______ $90\,in =$ __________ yd

Using Mixed Units of Length – U.S. System

Sometimes it will be necessary to convert a measurement of length with mixed units such as feet and inches.

EXAMPLE 4

Convert: 7 feet = _______yds _______ft

We know that 1 yard is equal to 3 feet or in other words, it takes 3 feet to make 1 yard. So the question becomes, if you have 7 feet, how many yards can you make? We will also consider if there are any "feet" remaining. Let's draw a picture.

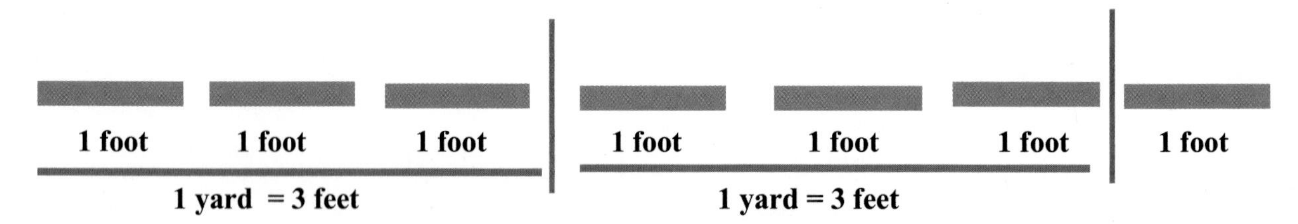

If we had 7 feet, we could make 2 yards with 1 foot remaining.

So, 7 feet = 2 yds 1 ft

EXAMPLE 5

Convert: 142 inches = _________ft _________in

It takes 12 inches to make 1 foot. So, we need to find out how many groups of 12 we can make out of 142 and if we have any inches remaining. Drawing a picture would be very time consuming since you would have to draw 142 inch blocks. Just realize that when we talk about how many groups we can make we are using **division**. A much faster way to complete this problem would be to divide 142 by 12.

$$\begin{array}{r} 11 \\ 12\,\overline{)\,142} \\ -12 \\ \hline 22 \\ -12 \\ \hline 10 \end{array}$$

We can make 11 groups of 12 with 10 remaining.

142 inches = 11 feet 10 inches

YOUR TURN

Convert the measurement as indicated.

6. 112 in = ______ ft _______ in Since there are 12 inches in 1 ft, we need to determine how many groups of 12 we can make out of 112. We will divide. $12\overline{)112}$ 112 in = ______ ft _______ in	7. 70 ft = ________ yd ________ ft Since there are 3 ft in 1 yd, we need to determine how many groups of 3 we can make out of 70. We will divide. $3\overline{)70}$ 70 ft = ________yd ________ft

EXAMPLE 6

Convert 4 feet 7 inches to inches.

Remember that 1 foot = 12 inches. So, for every foot we have 12 inches. In this problem, we have 4 feet plus 7 inches. Let's draw a picture.

1 foot	**1 foot**	**1 foot**	**1 foot**	
12 inches	12 inches	12 inches	12 inches	7 in.

Now, let's add up how many inches we have: 12 + 12 + 12 + 12 + 7 = 55 inches.

4 feet 7 inches = 55 inches

EXAMPLE 7

Convert: 3 yds 1 ft = ________ ft.

Remember that 1 yd = 3 ft. So for each yard we have 3 feet. In this problem, we have 3 yards and 1 ft. Let's draw a picture.

1 yard **1 yard** **1 yard**

Now, let's add up how many feet we have: 3 + 3 + 3 + 1 = 10 feet

3 yards 1 foot = 10 feet

YOUR TURN

Convert the measurement as indicated.

8. 5 yds 2 ft = ___________ ft.	9. 7 ft 8 in = _______________ inches

3.05 Length: U.S. and Metric Systems of Measurements 3.05

EXAMPLE 8

a. Add 5 ft 9 in and 7 ft 10 in.

Begin by stacking the original problem, lining up the like units.

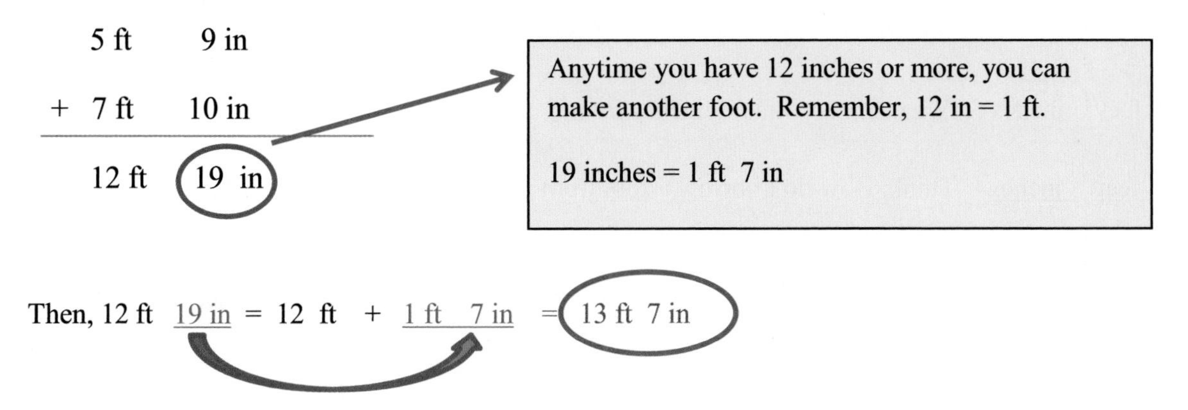

Remove a foot from the 19 inches: Subtract 12 inches from the 19 inches and add it (now as 1 foot) to 12 feet in front.

b. A rope that is 5 yds 1 ft long is attached to a rope that is 4 yds 2 ft long. How long is the entire piece of rope?

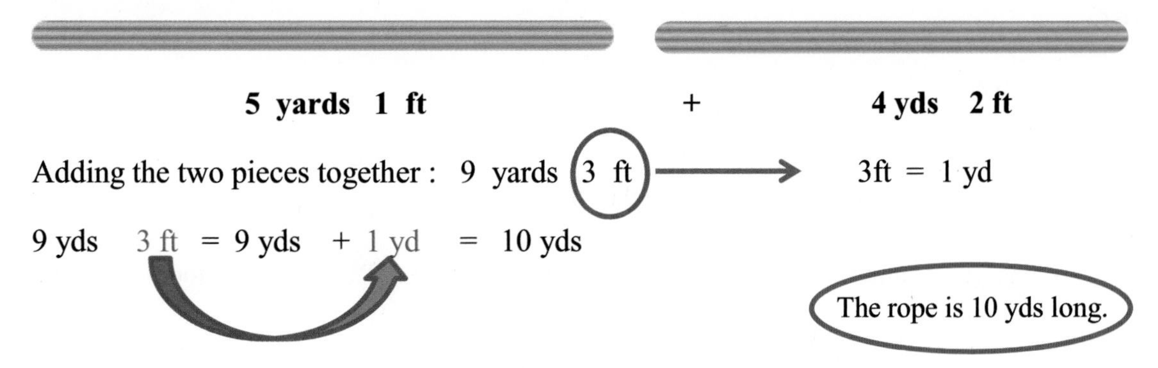

EXAMPLE 9

a. Subtract: 9 feet 5 inches - 4 feet 8 inches

Begin by stacking the original problem, lining up the like units.

9 feet 5 inches	We cannot subtract 8 inches from 5 inches.
- 4 feet 8 inches	Think: How do I borrow a foot from the 9 feet? Remember: 1 foot = 12 inches
8 17 ~~9~~ feet ~~5~~ inches	12 inches (1 foot) are borrowed from 9 feet leaving 8 feet and then added to the 5 inches giving us a total of 17 inches.
- 4 feet 8 inches	
4 feet 9 inches	Now subtract like units for the solution.

b. Jim is 6 ft 2 inches tall while John is 5 ft 10 inches tall. How much taller is Jim than John?

The indicator words " how much taller" tell us to subtract.

6 feet 2 inches	We cannot subtract 10 inches from 2 inches.
- 5 feet 10 inches	Think: How do I borrow a foot from the 6 feet. Remember: 1 foot = 12 inches
5 14 ~~6~~ feet ~~2~~ inches	12 inches (1 foot) are borrowed from 6 feet leaving 5 feet and then added to the 2 inches giving us a total of 14 inches.
- 5 feet 10 inches	
0 feet 4 inches	Now subtract like units for the solution.

Jim is 4 inches taller than John.

SUBTRACTION WITH BORROWING

Courtesy of Shutterstock

10 ft	6 in
– 2 ft	8 in

When we begin subtracting, we soon realize that we cannot subtract 8 inches FROM 6 inches. But there is an extra supply of inches right next door in the feet.

So we go over and borrow a foot!!

10 ft 6 in = ~~10~~ ft 6 in (9 ft written above the crossed-out 10)

We need to borrow 1 ft.

1 ft = 12 inches

So, we have borrowed 12 inches, leaving 9 ft rather than 10 ft.

We began with 10 ft 6 in.

Then we borrowed 12 inches which we will put with our other inches.
12 inches added to 6 inches = 18 inches.

Now we have 9 ft 18 in.

10 ft 6 in = 9 ft 18 in

Now we can subtract.

9 ft	18 in
– 2 ft	8 in
7 ft	10 in

3.05 Length: U.S. and Metric Systems of Measurements 3.05

EXAMPLE 10

a. Multiply 7 ft 9 in by 3.

Multiply each measurement by 3: $3 \bullet (7\ ft \quad 9\ inches) = 21\ ft \quad 27\ inches$

We have enough inches to make some feet. For every 12 inches, we can make 1 foot.

27 inches = 2 feet 3 inches

To simplify: 21 feet 27 inches = 21 feet + 2 feet 3 inches = 23 feet 3 inches

b. The Acme Microwave Company stacks its microwave ovens in a distribution warehouse. Each microwave is 1 ft 9 in wide. How far up the wall would 4 of these microwaves extend?

Courtesy of Fotolia

Each microwave is 1 ft 9 in. So we will need to multiply each number by 4.

4 (1 ft 9 in) = 4 ft 36 in ⟶ 36 in = 3 ft

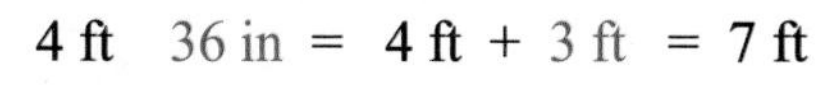

4 ft 36 in = 4 ft + 3 ft = 7 ft

The stack will be 7 ft tall.

EXAMPLE 11

a. Divide: 8 ft 6 in ÷ 2

Divide each measurement by 2: $\frac{8\text{ ft}}{2} \quad \frac{6\text{ in}}{2} = 4\text{ ft } 3\text{ in}$

b. A carpenter needs to cut a board into thirds. If the board is 18 ft 6 in long originally, how long will each cut piece be?

18 ft 6 in

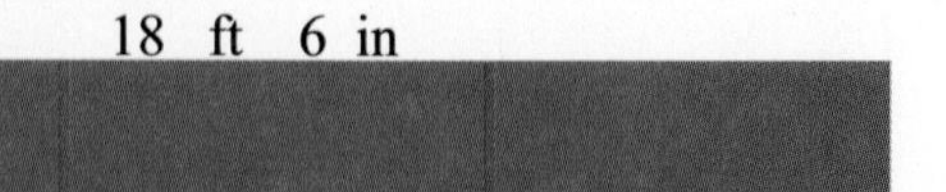

Divide the board into 3 pieces.

$\frac{18\text{ ft } 6\text{ in}}{3} = 6\text{ ft } 2\text{ in}$

Each piece will be 6 ft 2 in

3.05 Length: U.S. and Metric Systems of Measurements 3.05

YOUR TURN

Solve. Write in each blank what operation you will need to use to solve: add/subtract/multiply/divide. Make sure and fill in your final answer.

10. Mr. Jones hired a carpenter to extend a window in his kitchen. The original window measured 6 feet 9 inches across the top. He asked the carpenter to add 1 foot 9 inches and install a greenhouse shelf. What was the width of the new greenhouse window?

Operation needed to solve: _____________

The width of the window is ______ ft ______ in

11. A piece of rose that is 12 feet 8 inches long is cut in half. How long is each piece?

Operation needed to solve: _____________

The rope is ________ft _______ in long.

12. The Department of Transportation is installing concrete sound barriers along a highway. Each barrier is 1 yard 2 feet long. Find the total length of 24 barriers placed end to end.

Operation needed to solve: _____________

The total length of the barriers is _____ yd _____ ft

13. A 3 yard long cord has become frayed at both ends so that 3 inches is trimmed from each end. How long is the remaining cord?

Operation needed to solve: _____________

The remaining cord is ____ yd _____ in long.

3.05 Length: U.S. and Metric Systems of Measurements 3.05

Although the Metric System is used in most countries of the world, it is used very little in the United States. The Metric System does not use inches, feet, and miles. An advantage of the metric system is that it is very easy to convert from one unit to another because the entire system is based on powers of 10.

The basic unit of length in the metric system is the **meter**. A meter is a little longer than a yard. If you recall a yard is 36 inches, then we can compare a meter which is 39.37 inches.

1 yard = 36 inches

1 meter = 39.37 inches

All units of length in the Metric System are based on the meter.

Conversions Factors: Length
1 kilometer (km) = 1000 meters
1 hectometer (hm) = 100 meters
1 dekameter (dam) = 10 meters
1 meter (m) = 1 meter
10 decimeters (dm) = 1 meter
100 centimeters (cm) = 1 meter
1000 millimeters (mm) = 1 meter

Before we begin to work with the Metric System, let's familiarize ourselves with Metric units.

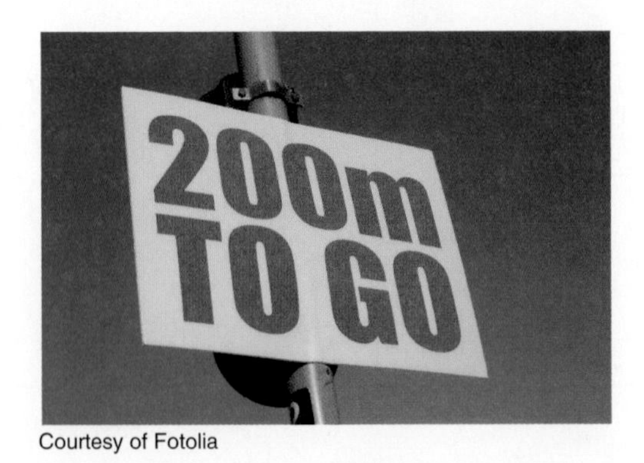

Courtesy of Fotolia

3.05 Length: U.S. and Metric Systems of Measurements 3.05

Millimeter

Courtesy of Fotolia

- Small units of length are called **millimeters**.
- A millimeter is about the **thickness of a plastic ID card** (or credit card).
- Or about the thickness of 10 sheets of paper on top of each other.
- This is a very small measurement!

Centimeter

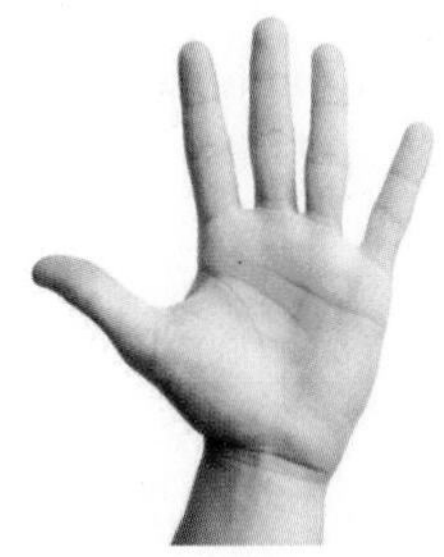

Courtesy of Fotolia

- 1 centimeter = 10 millimeters
- A fingernail is about **one centimeter wide**.
- 1 centimeter is about:
    - As long as a staple.
    - The width of a highlighter.
    - The diameter of a belly button.
    - The width of 5 CD's stacked on top of each other.

You might use millimeters or centimeters to measure how tall you are, or how wide a table is, but you would not use them to measure the length of a football field. In order to do that, you can switch to meters.

Meter

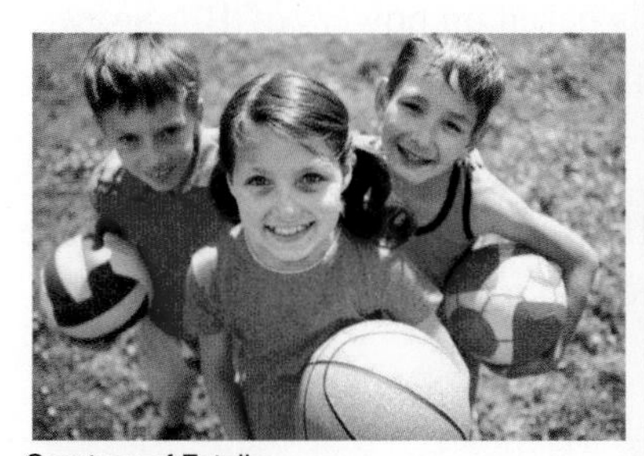

Courtesy of Fotolia

- A **meter** is equal to 100 centimeters.
    - An elementary school student is about one meter tall.

A meter (m) is about:

- a little more than a yard.
- the width of a doorway.
- four rungs up a ladder.
- five steps up a staircase.
- the depth of the shallow end of a swimming pool.
- the width of a dining table.
- waist high on an adult.

3.05 Length: U.S. and Metric Systems of Measurements 3.05

Kilometer

Courtesy of Fotolia

A **kilometer** is equal to 1000 meters.

- When you need to get from one place to another, you measure the distance using **kilometers**.
- The distance from one city to another or how far a plane travels would be measured using kilometers.

YOUR TURN

14. Write millimeter (mm), centimeter (cm), meter (m), or kilometer (km) after each item listed below to show which unit of linear measurement would be most appropriate.

a. distance between towns ________	e. Thickness of a book ________
b. fabric needed to sew a dress ________	f. length of a whale ________
c. thickness of a credit card ________	g. depth of a swimming pool ________
d. distance from New York to San Diego ________	h. width of a paper clip ________

CONVERTING METRIC SYSTEM UNITS OF LENGTH

There are two ways to convert within the Metric System.

- The first method is to continue using our unit fractions as we did with the U.S. System.
- The second method is based on the fact that the Metric System is based on powers of 10. So, we can convert from one unit to another by simply moving the decimal place the correct number of places to the right or left.

Look at the following table. Each place in the table has a value that is 10 times the value of the one to the left. Thus, moving one place in the table corresponds to moving one decimal place.

km	hm	dam	m	dm	cm	mm
kilometer	hectometer	dekameter	meter	decimeter	centimeter	millimeter

We will use this table to convert by looking at how many "decimal places" we move to get from one unit to another.

3.05 Length: U.S. and Metric Systems of Measurements 3.05

In the following examples, we will refer to the method using unit fractions as Method A and the method using the table as Method B.

EXAMPLE 12

Method A: Convert 3 km to meters.

Begin with the original 3 km. Your goal is to make the kilometers cross out and in the final answer have meters. To do this, multiply by "1". The 1st fraction's numerator units, km, will need to be the same as the 2nd fraction's denominator units. The 2nd fraction is the conversion fraction.

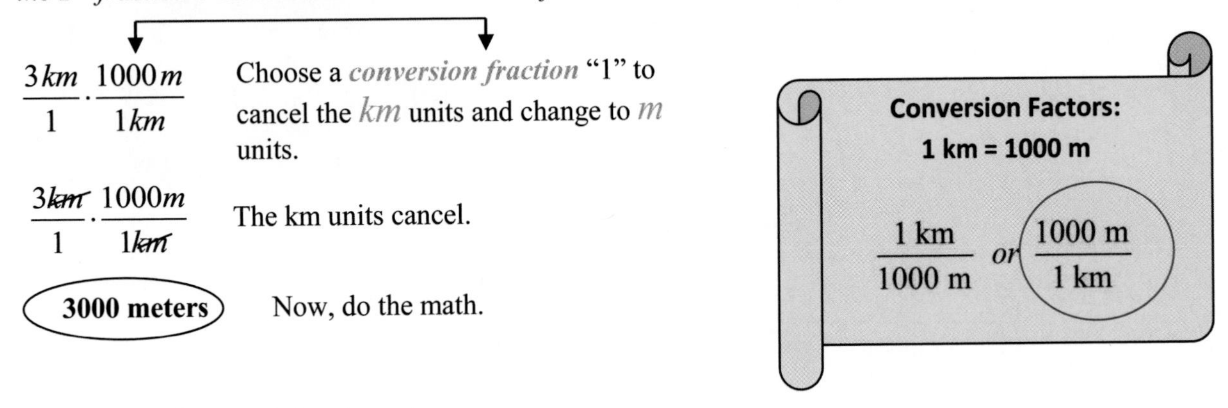

$\frac{3km}{1} \cdot \frac{1000m}{1km}$ Choose a *conversion fraction* "1" to cancel the *km* units and change to *m* units.

$\frac{3\cancel{km}}{1} \cdot \frac{1000m}{1\cancel{km}}$ The km units cancel.

3000 meters Now, do the math.

Method B: Convert 3 km to meters.

Think: To go from km to m in the table is a move three places to the right. Thus we move the decimal point three places to the right.

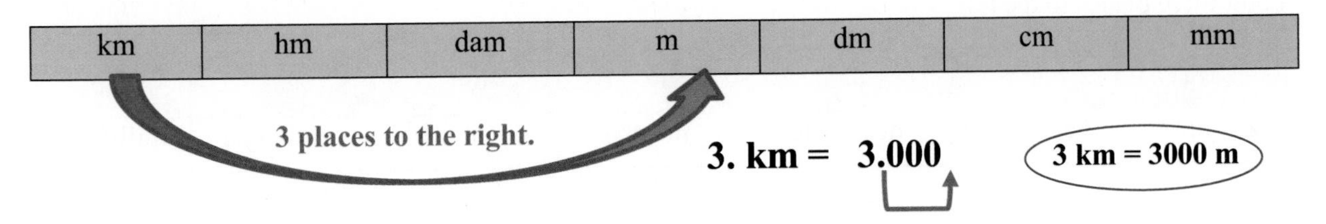

km	hm	dam	m	dm	cm	mm

3. km = 3.000

3 km = 3000 m

HERE'S A REALLY COOL WAY TO REMEMBER THE ORDER OF THESE METRIC UNITS!!!

Courtesy of Fotolia

THE BEGINNING LETTER OF EACH WORD IS THE BEGINNING LETTER OF THE METRIC UNIT!!!

Courtesy of Fotolia

Kangaroos	Hop	Down	Mountains	Drinking	Chocolate	Milk
Km	Hm	Dam	M	Dm	Cm	Mm

3.05 Length: U.S. and Metric Systems of Measurements 3.05

EXAMPLE 13

Method A: **Convert 450,000 mm to meters.**

We choose a symbol for 1 with "meters" in the top and "millimeters" in the bottom.

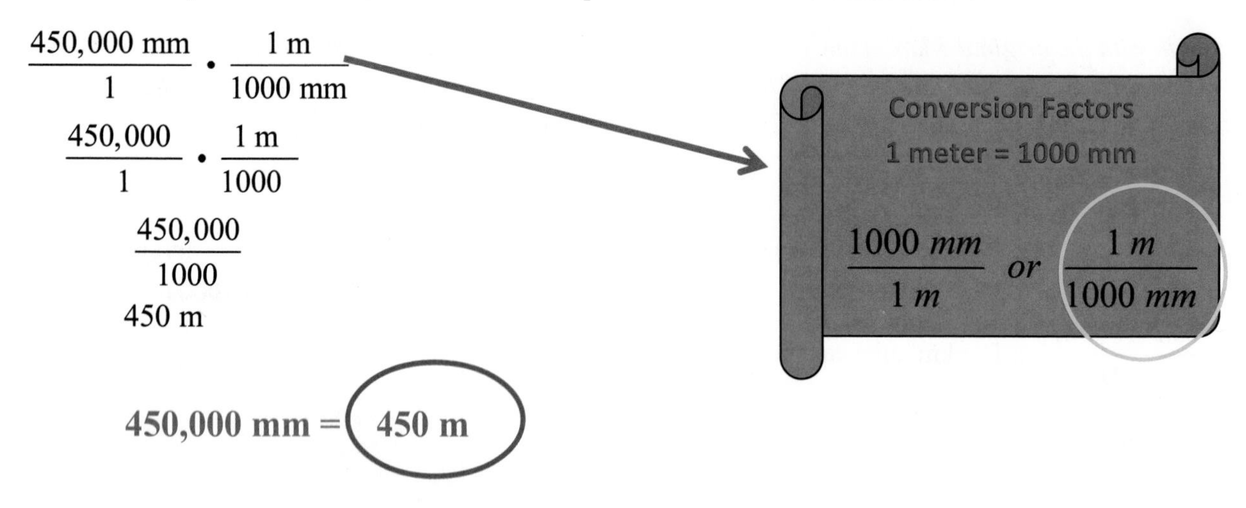

Method B. **Convert 450,000 mm to meters.**

Think: To go from mm to m in the table is a move three places to the left. Thus we move the decimal point three places to the left.

km	hm	dam	m	dm	cm	mm
kilometer	hectometer	dekameter	meter	decimeter	centimeter	millimeter

3 places to the left.

450,000 mm = 450,000. = 450 m

The fact that conversions can be done so easily is an important advantage of the Metric System. The most commonly used metric units of length are km, m, cm, and mm. In the exercises, you will notice these used more often than the other units.

3.05 Length: U.S. and Metric Systems of Measurements 3.05

YOUR TURN

Underneath the table given draw an arrow showing how you should move the decimal. Then convert each measurement.

km	hm	dam	m	dm	cm	mm

15. 3500 m = ______ km	16. 2.5 m = ________ mm
17. 14 m = ________ cm	18. 150 mm = ________ cm
19. 4.6 km = __________ m	20. 6.9 cm = ________ mm

3.05 Length: U.S. and Metric Systems of Measurements 3.05

YOUR TURN **Answers to Section 3.05**

1.	
a. length of a piece of paper inch	g. flight across the country miles
b. distance between suburb miles	h. length of a fish inch
c. fabric needed to sew a dress yards	i. length of a whale feet
d. height of a ceiling feet	j. depth of a swimming pool feet
e. perimeter of a baseball field feet	k. distances on a football field yards
f. length of a paper clip inch	l. size of a window feet or inches

2. 72 ft
3. 138 in
4. 21120 feet
5. 2.5 yd
6. 9 ft 4 in
7. 23 yd 1 ft
8. 17 feet
9. 92 in
10. 8 ft 6 in
11. 6 ft 4 in
12. 40 yds

10. 8 ft 6 in
11. 6 ft 4 in
12. 40 yds

13. 2 yds 30 inches

14.	
a. distance between towns kilometers	e. thickness of a book centimeters
b. fabric needed to sew a dress meters	f. length of a whale meters
c. thickness of a credit card millimeters	g. depth of a swimming pool meters
d. distance from New York to San Diego kilometers	h. width of a paper clip millimeters

15. 3.5 km
16. 2500 mm
17. 1400 cm
18. 15 cm
19. 4600 m
20. 69 mm

Complete MyMathLab Section 3.05 Homework in your Homework notebook.

3.06 Weight and Mass: U.S. and Metric Systems of Measurements 3.06

We measure the mass of an object by weighing it. We use weight when we shop at the grocery store and read how many ounces of cereal are in the box, weighing ourselves at the gym or doctor's office, and reading how many grams of fat or carbohydrates a certain food contains. The most common units of weight in the U.S. Standard System are the **ounce**, the **pound**, and the **ton**. Gram is the basic unit of mass in the Metric System.

24 ounces

Courtesy of Fotolia

65 pounds

Courtesy of Fotolia

10 tons

U.S. Units of Weight	
1 pound (lb) = 16 ounces (oz)	1 ton (T) = 2000 pounds (lb)

YOUR TURN

1. If you were describing the weight of the following, which type of unit would you use: ounce, pound, or ton? Write the correct unit in the blank beside each item.

a. elephant ___________	b. teenage boy ___________
c. paper clip ___________	d. car ___________
e. ladder ___________	f. bag of chips ___________

Unit fractions that equal 1 are used to convert between units of weight just like we did in the previous section with length. When converting, remember that the numerator should contain the unit we are converting to and the denominator should contain the original unit.

$$\frac{\text{unit to convert to}}{\text{original unit}}$$

3.06 Weight and Mass: U.S. and Metric Systems of Measurements 3.06

EXAMPLE 1

Convert 7000 pounds to tons.

Let's think about this. It takes 2000 pounds to make a ton. So, the question is really how many groups of 2000 can we make out of 7000.

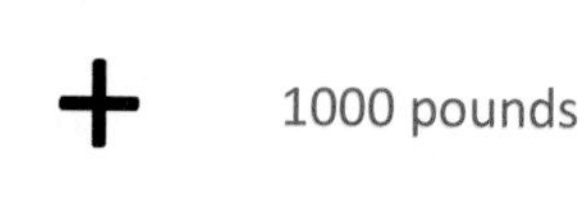

7000 pounds = 3 tons with 1000 pounds remaining. Since 1000 pounds is half of a ton, we can say that 7000 pounds = 3.5 tons.

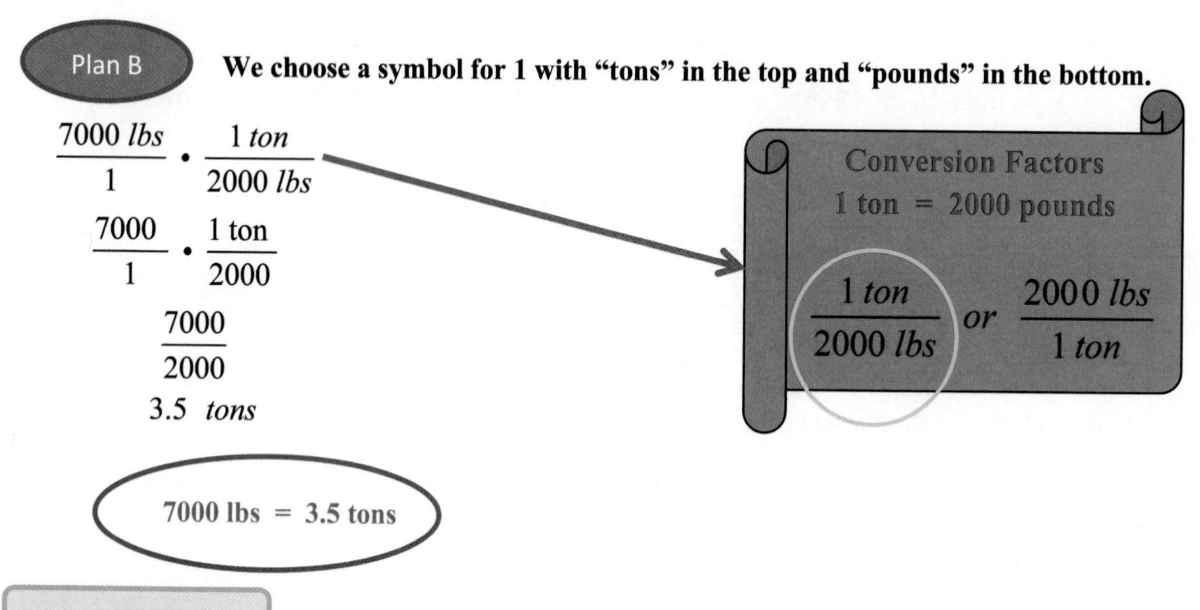

We choose a symbol for 1 with "tons" in the top and "pounds" in the bottom.

$$\frac{7000\ lbs}{1} \cdot \frac{1\ ton}{2000\ lbs}$$

$$\frac{7000}{1} \cdot \frac{1\text{ ton}}{2000}$$

$$\frac{7000}{2000}$$

$$3.5\ tons$$

7000 lbs = 3.5 tons

EXAMPLE 2

Convert 3 pounds to ounces.

For every pound, we have 16 ounces. So the question is really, if you have 3 pounds how many ounces do you have?

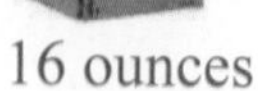

3.06 Weight and Mass: U.S. and Metric Systems of Measurements 3.06

Plan B **Convert 3 pounds to ounces.**

Choose a symbol for 1 with "ounces" in the top and "pounds" in the bottom.

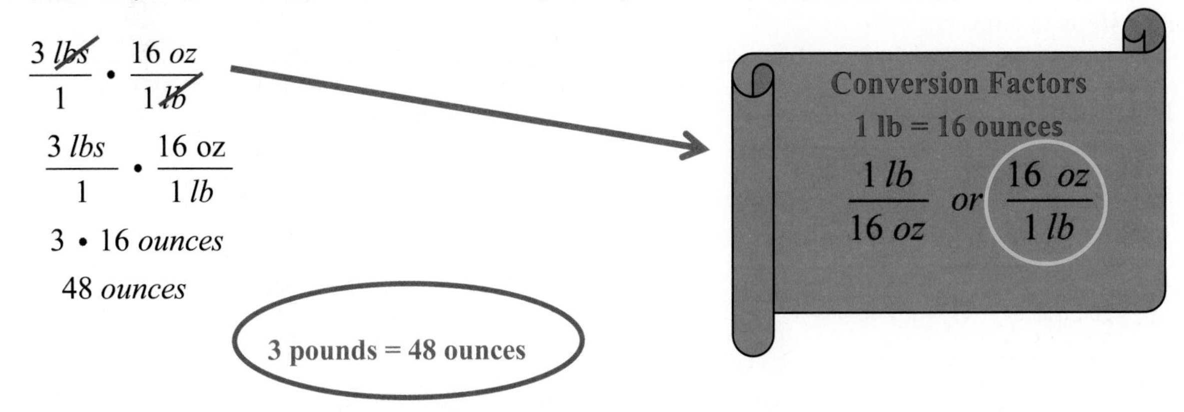

$$\frac{3\ lbs}{1} \bullet \frac{16\ oz}{1\ lb}$$

$$\frac{3\ lbs}{1} \bullet \frac{16\ \text{oz}}{1\ lb}$$

$$3 \bullet 16\ ounces$$

$$48\ ounces$$

3 pounds = 48 ounces

EXAMPLE 3

Convert 47 ounces = _______ lb ________ oz

Remember it takes 16 ounces to make a pound. So, we need to find out how many groups of 16 can be made out of 47 and if there are any ounces remaining. Drawing a picture would be very time consuming since you would have to draw 47 ounce blocks. Realize that when we talk about how many groups we can make we are using **division**. A much faster way to complete this problem would be to divide 47 by 16.

```
      2
16 | 47
    -32
     15
```

We can make 2 groups of 16 with 15 remaining.

47 ounces = 2 lbs 15 oz

EXAMPLE 4

Convert $3\frac{3}{4}$ pounds to ounces.

We choose a symbol for 1 with "ounces" in the top and "pounds" in the bottom.

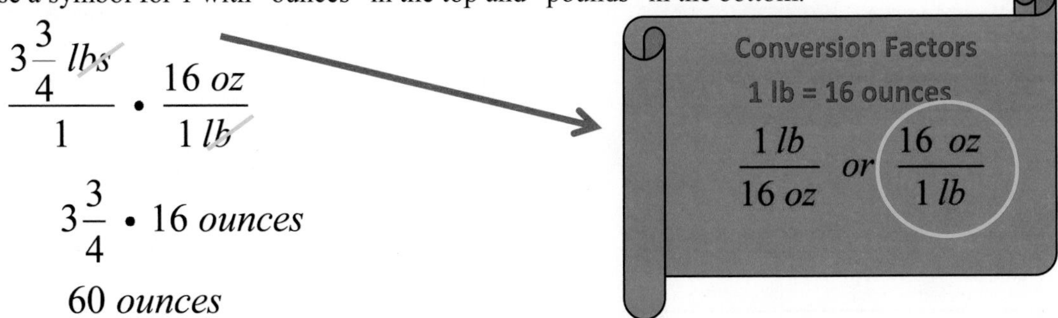

$$\frac{3\frac{3}{4}\ lbs}{1} \bullet \frac{16\ oz}{1\ lb}$$

$$3\frac{3}{4} \bullet 16\ ounces$$

$$60\ ounces$$

3.06 Weight and Mass: U.S. and Metric Systems of Measurements 3.06

YOUR TURN

Convert the measurement as indicated. Circle the conversion factor you would use to convert and then follow steps to convert.

2. 8 pounds to ounces Conversion Factors: $\frac{16\ oz}{1\ lb}$ *or* $\frac{1\ lb}{16\ oz}$ $\frac{8\ lb}{1} \bullet$ ______ $8\ lbs =$ __________ oz	3. 108 ounces to pounds Conversion Factors: $\frac{16\ oz}{1\ lb}$ *or* $\frac{1\ lb}{16\ oz}$ $\frac{108\ oz}{1} \bullet$ ______ $108\ oz =$ __________ lb
4. 2 ½ tons to pounds Conversion Factors: $\frac{1\ ton}{2000\ lb}$ *or* $\frac{2000\ lb}{1\ ton}$ $\frac{2\frac{1}{2}\ tons}{1} \bullet$ ______ $2\frac{1}{2}\ tons =$ __________ lb	5. 100 oz = _____ lb ______ oz Since there are 16 ounces in 1 lb, we need to determine how many groups of 16 we can make out of 100. We will divide. $16\overline{)100}$ 100 oz = ______ lb ______ oz
6. $5\frac{3}{4}$ pounds to ounces Conversion Factors: $\frac{16\ oz}{1\ lb}$ *or* $\frac{1\ lb}{16\ oz}$ $\frac{5\frac{3}{4}\ lb}{1} \bullet$ ______ $5\frac{3}{4}\ lbs =$ __________ oz	7. 17,000 pounds to tons Conversion Factors: $\frac{1\ ton}{2000\ lb}$ *or* $\frac{2000\ lb}{1\ ton}$ $\frac{17{,}000\ lbs}{1} \bullet$ ______ $17{,}000\ lbs =$ __________ $tons$

3.06 Weight and Mass: U.S. and Metric Systems of Measurements 3.06

In the Metric System, the basic unit for measuring mass is the **gram**. The standard for the gram was created by measuring a cubic centimeter of pure water in a vacuum. That mass was named one gram, so it is also related to the meter. Small things are weighed in grams, and larger things are weighed in kilograms. A gram is about the weight of a sugar cube. A kilogram is 2.2 pounds so it is similar to holding two boxes of butter.

The milligram, the gram, and the kilogram are the three most commonly used units of mass in the Metric System.

Milligram Small masses such as medicine and vitamins may be measured in milligrams.

A tablet contains 200 milligrams of ibuprofen.

Courtesy of Fotolia

Grams

A paperclip weighs about 1 gram or 1 raisin weighs about a gram.

Hold one small paperclip in your hand. Does that weigh a lot? No! A gram is very light. That is why you often see things measured in hundreds of grams.

Courtesy of Fotolia

Kilograms

The kilogram is used for larger items. When you weigh yourself on a scale, you would use kilograms.

A book weights about 1 kilogram. A kilogram is a little more than 2 pounds.

Courtesy of Fotolia

YOUR TURN

8. Determine whether the measurement in each statement is reasonable. Write yes or no.

a. A professor weighs less than 150 g. ____________

b. The doctor prescribed a pill containing 250 kg of medication. ____________

c. Samuel weighs 75 kg. ____________

d. A cat weighs 12 g. ____________

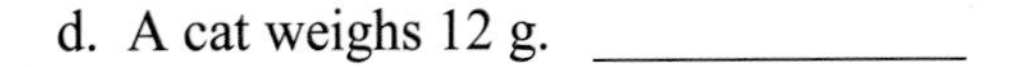

Courtesy of Fotolia

Converting within the Metric System

If you remember, to convert in the Metric System we can use one of two methods. We can use the unit fraction Method A or we can use the table Method B and simply move the decimal. For the following examples we will use Method B.

It is easy to go from one metric unit to another because it means multiplying or dividing by powers of ten. Remember the table is used to convert with length. We will see that same table used here.

.

kg	hg	dag	g	dg	cg	mg
kilogram	hectogram	dekagram	gram	decigram	centigram	milligram

EXAMPLE 5

Convert 2.3 mg to g.

Think: To go from mg to g in the table is a move three places to the left. Thus we move the decimal point three places to the left.

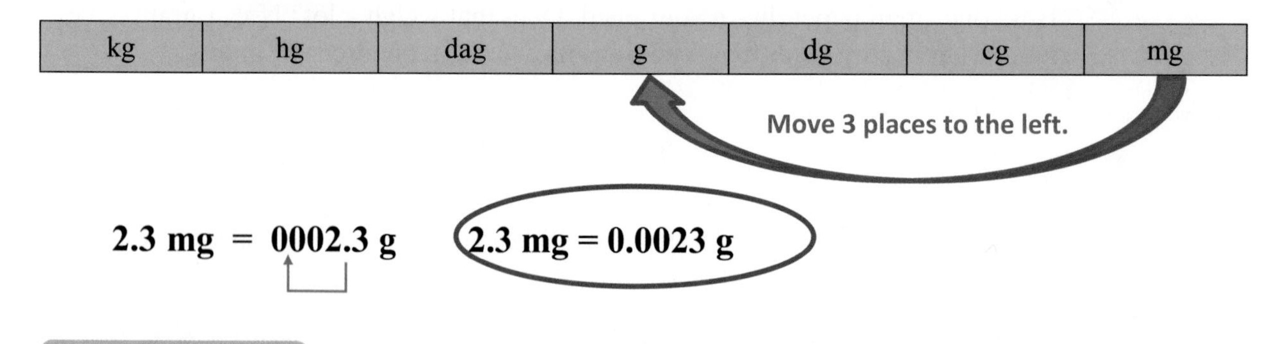

kg	hg	dag	g	dg	cg	mg

EXAMPLE 6

Convert 4 kg to g.

Think: To go from kg to g in the table is a move three places to the right. Thus we move the decimal point three places to the right.

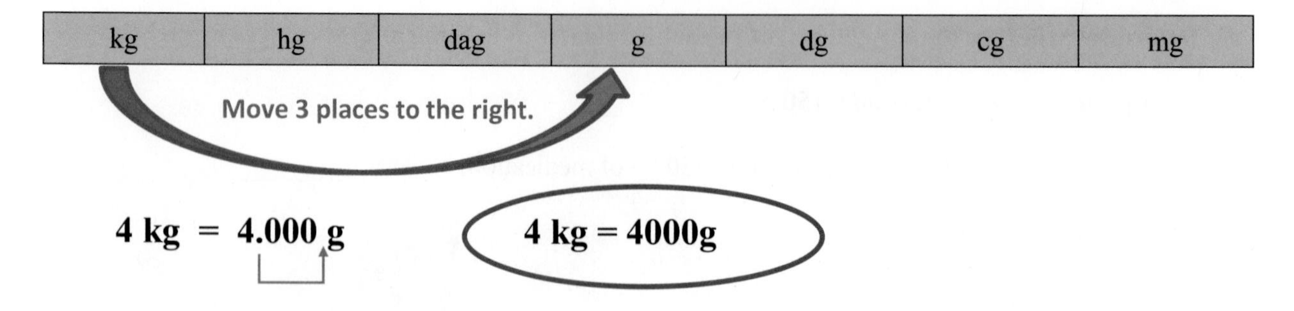

kg	hg	dag	g	dg	cg	mg

YOUR TURN

Underneath the table given draw an arrow showing how you should move the decimal. Then convert each measurement.

kg	hg	dag	g	dg	cg	mg

9. 23.6 mg to g	10. 1200 g to kg	11. 6.4 kg to g

YOUR TURN **Answers to Section 3.06.**

1. a. ton b. pound c. ounce d. ton e. pound f. ounce

2. 128 ounces 3. 6.75 pounds 4. 5000 pounds

5. 6 pounds 4 ounces 6. 92 ounces 7. 8 ½ tons

8. a. no b. no c. yes d. no

9. 0.0236 g 10. 1.2 kg 11. 6400 g

Complete MyMathLab Section 3.06 Homework in your Homework notebook.

3.07 Capacity: U.S. and Metric Systems of Measurements 3.07

We use liquid measurements when shopping for household items such as dish soap or soft drinks, when following a recipe, when mixing concentrates with water to make new products for cleaning or sprays for the indoor and outdoor plants, and when we put gasoline in the car. These measurements include ounces, cups, pints, quarts, and gallons.

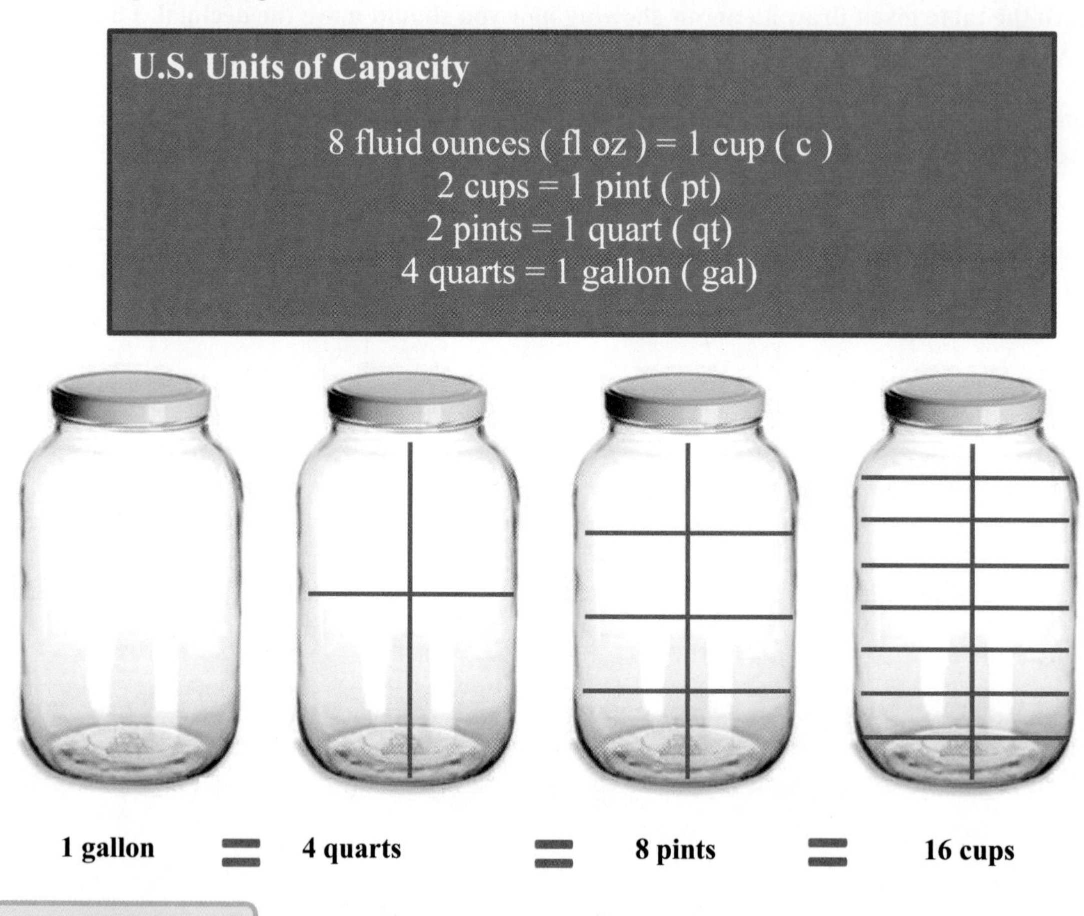

1 gallon = 4 quarts = 8 pints = 16 cups

EXAMPLE 1

Convert 9 quarts to gallons.

As in the previous sections, we can use unit fractions to convert or we can use our visual to convert. If it takes 4 quarts to make a gallon, we could make 2 gallons out of 9 quarts with one quart left over.

So, 9 quarts = $2\frac{1}{4}$ gallons.

Using unit fractions:

$$\frac{9 \text{ qt}}{1} \cdot \frac{1 \text{ gal}}{4 \text{ qt}}$$

$$\frac{9 \text{ gal}}{4}$$

$$2\frac{1}{4} \text{ gal}$$

3.07 Capacity: U.S. and Metric Systems of Measurements 3.07

Here is a great way to remember the relationship between all these units. Looking at the letters below, you can see that inside 1 gallon is 4 quarts. Then inside each quart is 2 pints and inside each pint is 2 cups. So, you can easily look at this diagram and see that there are 8 pints in a gallon or 16 cups in a gallon.

If you were asked how many quarts were in 5 gallons, you can reason that if there are 4 quarts in one gallon, that there would be 20 quarts if you had 5 of these "G" or gallons.

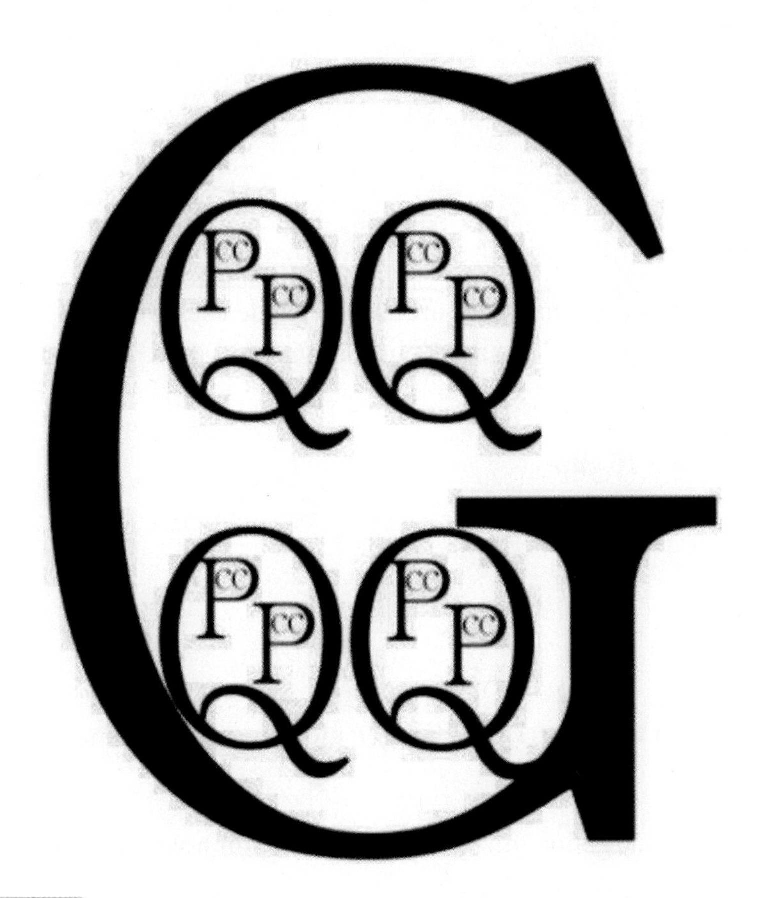

EXAMPLE 2

Convert: 7 pints = ________ quarts ________ pints

Looking at the diagram above, we can see that it takes 2 pints to make 1 quart. So, if we have 7 pints, we could make 3 quarts, which would be 6 of the pints, and have one pint leftover.

7 pints = 3 quarts 1 pint

EXAMPLE 3

Convert 40 pints to gallons.

Looking at the diagram, you can count that there are 8 "Ps" inside the "G". So, it takes 8 pints to make a gallon. We need to know how many groups of 8 we can make out of 40.

Given 40 pints, we could make 5 groups of 8.

40 pints = 5 gallons

EXAMPLE 4

Convert 7 ½ pints to cups.

Looking at the diagram, we can see that there are 2 cups inside each pint. So, if you had 7 pints you would actually have 14 cups. If there are 2 cups in one pint, then half of a pint would have 1 cup.

7 ½ pints = 14 cups + 1 cup = **15 cups.**

EXAMPLE 5

Can 3 qt 1 pt of fruit punch and 5 qt 1 pt of ginger ale be poured into a 2-gallon container without it overflowing?

Since there are 4 quarts in every gallon, a 2-gallon container can hold 8 quarts.

Add up how much liquid we are going to pour into the container.

3 qt 1 pt + 5 qt 1 pt = 8 qts 2 pts.

Our container can hold 8 quarts, but we have 8 qts 2 pts of liquid. So, the container could not hold the liquid without overflowing.

3.07 Capacity: U.S. and Metric Systems of Measurements 3.07

YOUR TURN

Convert each measurement.

1. 14 quarts to gallons	2. 5 quarts to pints
3. 23 pints = ________ quarts ______ pints	4. 8 ½ quarts to pints

5. Can 3 quarts 2 pints of fruit punch and 4 quarts 1 pint of ginger ale be poured into a 2-gallon container without it overflowing?

Courtesy of Fotolia

3.07 Capacity: U.S. and Metric Systems of Measurements 3.07

The two most used units for capacity within the Metric System are the liter and the milliliter. The basic unit for measuring liquid capacity is the **liter**. A liter is a little more than a quart. Most soft drinks are packaged in 2-liter bottles. Another common measure of capacity is the milliliter. It would fill a little box about the size of a sugar cube or take up the space of a sugar cube. Milliliters are used in medicine and to measure small amounts of liquids or small volumes.

Courtesy of Fotolia

A **milliliter** is a very small amount of liquid.
Here is a milliliter of medicine in a teaspoon.
It doesn't even fill the teaspoon!

20 drops of water

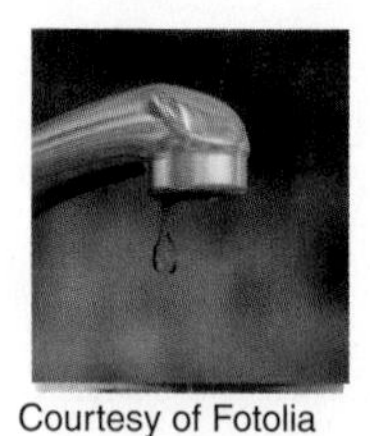

Courtesy of Fotolia

make about 1 milliliter.

1 full teaspoon of liquid

Courtesy of Fotolia

is about 5 milliliters.

Courtesy of Fotolia

1000 milliliters makes up 1 liter.

This pitcher has exactly 1 liter of water in it.

Milliliters are used to measure small amounts of liquid. Remember 1 teaspoon is about 5 milliliters. Liters are used to measure larger amounts. Soda comes in 2-liter bottles.

Would it be reasonable for Jackie to take a dose of 2 liters of cough medicine? NO!!
Would it be reasonable to take 20 milliliters of cough medicine? YES!!

Could you take a bath in 300 ml of water?

NO!!!

Courtesy of Fotolia

3.07 Capacity: U.S. and Metric Systems of Measurements 3.07

Converting from one unit of capacity to another is done exactly the same way as for length and weight. We have two methods from which to choose from. Unit fractions may be used or we can continue to use our chart and move the decimal place left or right as needed. Here is that chart once again with liters.

kl	hl	dal	L	dl	cl	ml
kiloliter	hectoliter	dekaliter	liter	deciliter	centiliter	milliliter

EXAMPLE 6

Convert 3240 ml to liters

Think: To go from ml to l in the table is a move three places to the left. Thus we move the decimal point three places to the left.

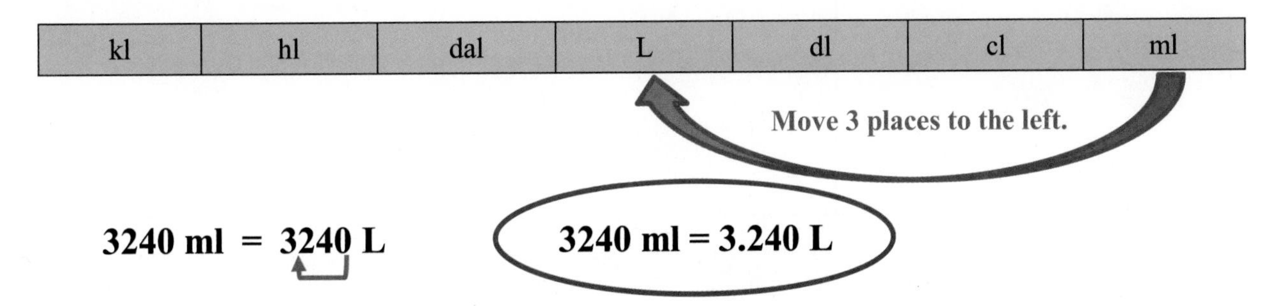

EXAMPLE 7

Convert 4.6 liters to milliliters

Think: To go from liters to milliliters in the table is a move three places to the right. Thus we move the decimal point three places to the right.

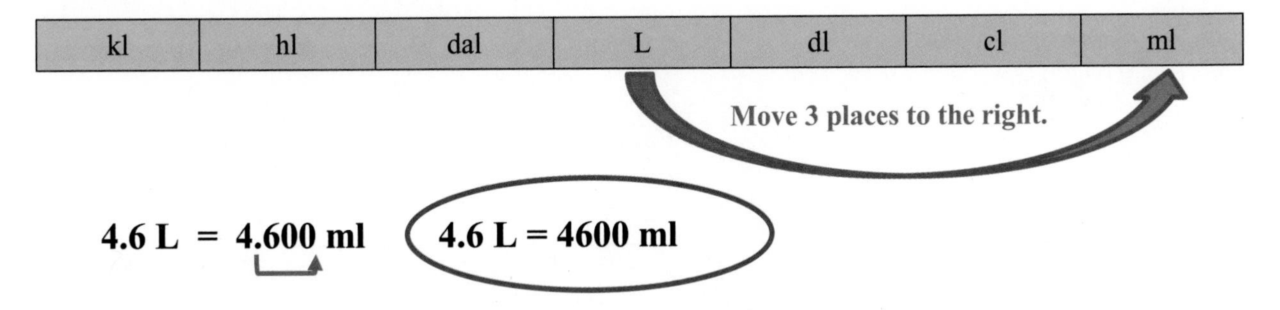

3.07 Capacity: U.S. and Metric Systems of Measurements 3.07

YOUR TURN

Underneath the table given draw an arrow showing how you should move the decimal. Then convert each measurement.

kl	hl	dal	l	dl	cl	ml

6. 6.5 L to ml	7. 346 ml to L	8. Is it reasonable to drink 30 ml of antacid? About how many teaspoons would this be?

YOUR TURN Answers to Section 3.07.

1. 3 ½ gallons
2. 10 pts
3. 11 qts 1 pt
4. 17 pts
5. No
6. 6500 ml
7. 0.346 L
8. Yes , 6 tsp

Complete MyMathLab Section 3.07 Homework in your Homework notebook.

3.08 Conversions between the U.S. and Metric System 3.08

To convert between the U.S. and the Metric System we will continue to use unit fractions. Remember when choosing the correct unit fraction, the unit you are converting to is in the numerator and the unit you are converting from is in the denominator.

When making these conversions, use the conversion factors in the charts below. Notice the use of the approximate symbol: $\approx$. Converting between these two systems are approximations. There are no "exact" conversions. You will usually be asked to round to two decimal places.

LENGTH
1 m ≈ 1.09 yd
1 m ≈ 3.28 ft
1 km ≈ 0.62 mile
1 in ≈ 2.54 cm
1 ft ≈ 0.30 m
1 mi ≈ 1.61 km

CAPACITY
1 L ≈ 1.06 qt
1 L ≈ 0.26 gal
1 gal ≈ 3.79 L
1 qt ≈ 0.95 L
1 fl oz ≈ 29.57 ml

WEIGHT
1 kg ≈ 2.20 lb
1 g ≈ 0.04 oz
1 mg ≈ 0.000035274 oz
1 lb ≈ 0.45 kg
1 oz ≈ 28.35 g

EXAMPLE 1

A balance beam is 6 centimeters wide. Convert this width to inches.

From the chart above we know that 1 inch $\approx$ 2.54 centimeters. From this fact we have the following two unit fractions from which to choose from: $\frac{1\text{ in}}{2.54\text{ cm}}$ or $\frac{2.54\text{ cm}}{1\text{ in}}$.

Since we are converting to inches, we will choose the fraction that has inches in the "top."

$$\frac{6\text{ cm}}{1} \bullet \frac{1\text{ in}}{2.54\text{ cm}}$$
$$\frac{6}{1} \bullet \frac{1\text{ in}}{2.54}$$
$$\frac{6\text{ in}}{2.54}$$
$$\approx 2.36\ in$$

EXAMPLE 2

The speed limit is 40 miles per hour. Convert this to kilometers per hour.

Given that 1 km = 0.62 mi, we will convert this using a unit fraction that has km in the "top."

$$\frac{40\text{ mi}}{1} \bullet \frac{1\text{ km}}{0.62\text{ mi}} = \frac{40}{1} \bullet \frac{1\text{ km}}{0.62} = \frac{40\text{ km}}{0.62} \approx 64.52\text{ km per hour}$$

EXAMPLE 3

Ibuprofen comes in 100 mg tablets. Convert this to ounces. Round to four decimal places.

Courtesy of Fotolia

$1\ mg \approx 0.000035274\ oz$. We will use a unit fraction with ounces in the "top."

$$\frac{100\ mg}{1} \bullet \frac{0.000035274\ oz}{1\ mg} = \frac{100}{1} \bullet \frac{0.000035274\ oz}{1} = \frac{100\ oz \bullet 0.000035274}{1} = 0.0035274\ oz$$

1 mg = 0.0035 ounces

EXAMPLE 4

Ibuprofen comes in 175 milligram tablets. Convert this to ounces. Round to three decimal places if needed.

Courtesy of Shutterstock

When we look at the conversion chart, we realize we do not have a conversion factor to convert milligrams to ounces. However, there is a conversion factor to convert grams to ounces.

We will need to convert 175 milligrams to grams and then convert to ounces.

STEP 1: convert milligrams to grams

175 mg = 0.175 grams ⟶ move 3 places to the left to convert **mg** to **grams.**

STEP 2: convert grams to ounces.

0.175 grams = ________ ounces.

The conversion factor taken from the chart is ***28.35 g = 1 oz***

$$\frac{0.175\ g}{1} \bullet \frac{1\ oz}{28.35\ g} = \frac{0.175\ oz}{28.35} = 0.006$$

Rounded to three decimal places.

175 mg = 0.006 ounces

3.08 Conversions between the U.S. and Metric System 3.08

YOUR TURN

Convert the measurement as indicated. Circle the conversion factor you would use to convert and then follow steps to convert. Round to the nearest hundredth.

1. 86 miles to kilometers Conversion Factors: $\frac{0.62\ mi}{1\ km}$ or $\frac{1\ km}{0.62\ mi}$ $\frac{86\ mi}{1} \cdot$ ___ 86 *mi* = __________ *km*	2. 86 inches to centimeters Conversion Factors: $\frac{2.54\ cm}{1\ in}$ or $\frac{1\ in}{2.54\ cm}$ $\frac{86\ in}{1} \cdot$ ___ 86 *in* = __________ *cm*
3. 1000 grams to ounces Conversion Factors: $\frac{0.04\ oz}{1\ g}$ or $\frac{1\ g}{0.04\ oz}$ $\frac{1000\ g}{1} \cdot$ ___ 1000 *g* = __________ *oz*	4. 756 milliliters to fluid ounces Conversion Factors: $\frac{29.57\ ml}{1\ fl\ oz}$ or $\frac{1\ fl\ oz}{29.57\ ml}$ $\frac{756\ ml}{1} \cdot$ ___ 756 *ml* = __________ *fl oz*
5. A Boeing 747 has a cruising speed of about 980 kilometers per hour. Convert this to miles per hour. Conversion Factors: $\frac{0.62\ mi}{1\ km}$ or $\frac{1\ km}{0.62\ mi}$ $\frac{980\ km}{1} \cdot$ ___ 980 *km* = __________ *mi*	6. A doctor orders a dosage of 7 ml of medicine to be taken twice a day. How many fluid ounces of medicine should be purchased for one day? Conversion Factors: $\frac{29.57\ ml}{1\ fl\ oz}$ or $\frac{1\ fl\ oz}{29.57\ ml}$ $\frac{7\ ml}{1} \cdot$ ___ *7 ml taken twice a day* = __________ *fl oz*

EXAMPLE 5

You are mixing some concrete for a home project and you've calculated according to the directions that you need six gallons of water for your mix. However, your bucket isn't calibrated, so you don't know how much it holds. You just finished a two-liter bottle of soda so you can use it to measure the water. How many times will you need to fill the soda bottle?

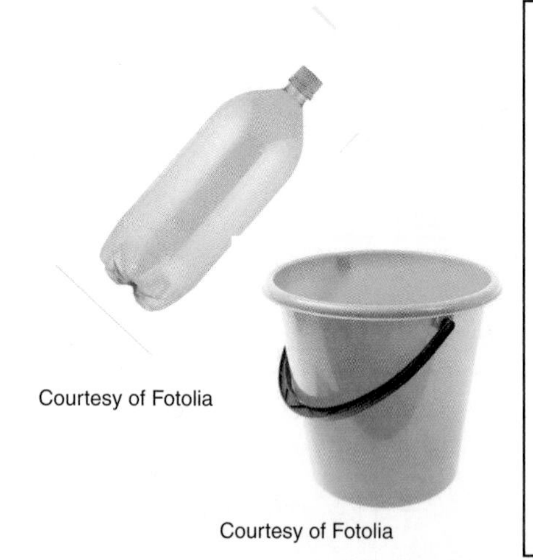

Courtesy of Fotolia

Courtesy of Fotolia

We will need to find out how many 2 liters are in 6 gallons.

1 gallon = 3.79 L.

If we have 6 gallons, then we have $6(3.79) = 22.74$ liters.

If we have a 2 liter bottle, and we need 22.74 liters of water, then we will divide to see how many times we will need to fill up the 2 liter bottle to make a total of 22.74 liters.

$\frac{22.74}{2} = 11.37$ **We will need to fill the bottle 12 times.**

YOUR TURN

7. You are mixing some concrete for a home project and you've calculated according to the directions that you need 8 gallons of water for your mix. However, your bucket isn't calibrated, so you don't know how much it holds. You just finished a two-liter bottle of soda so you can use it to measure the water. How many times will you need to fill the soda bottle?

YOUR TURN **Answers to Section 3.08.**

1. 138.46 km or 138.71 km
2. 218.44 cm
3. 35.27 ounces or 40 ounces
4. 25.57 fl oz
5. 608.70 mi/hr or 607.50 mi/hr
6. 0.47 fl oz
7. 16 times

Complete MyMathLab Section 3.08 Homework in your Homework notebook.

3.09 Percent, Decimals, and Fractions 3.09

Percents are **fractions and decimals** in disguise. The word percent means, " per 100." The denominator of a percent fraction is always 100, and we use the symbol, "%" to denote percent. Think about the symbol % as being made of the fraction bar and the two zeros in 100. So, this module will build on the previous module of fractions and decimals. 26% means 26 out of 100. A percent is just a ratio with 100 as the denominator.

EXAMPLE 1

Look at the following 10 by 10 grids—total of 100 squares. We will shade to represent the given percentages.

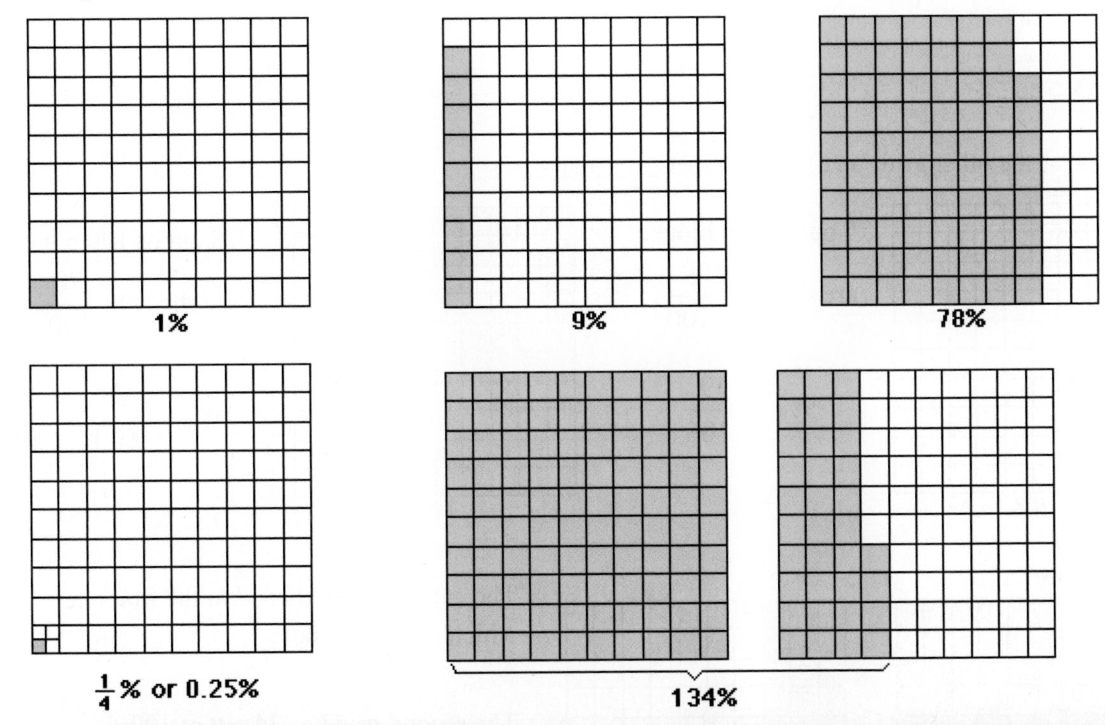

YOUR TURN

Shade in the model below to represent the given percentage.

1. 5 %

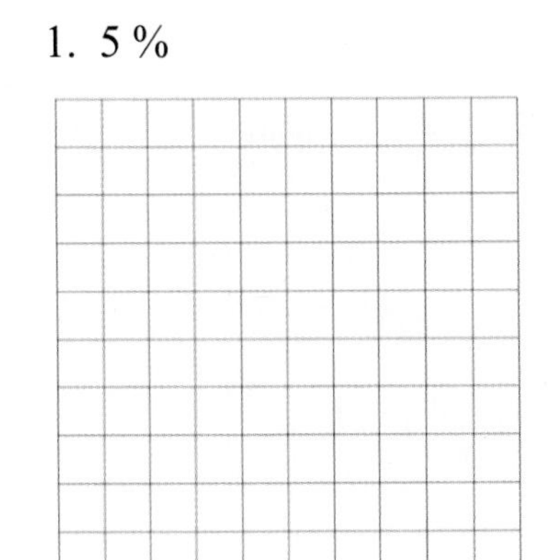

2. 25%

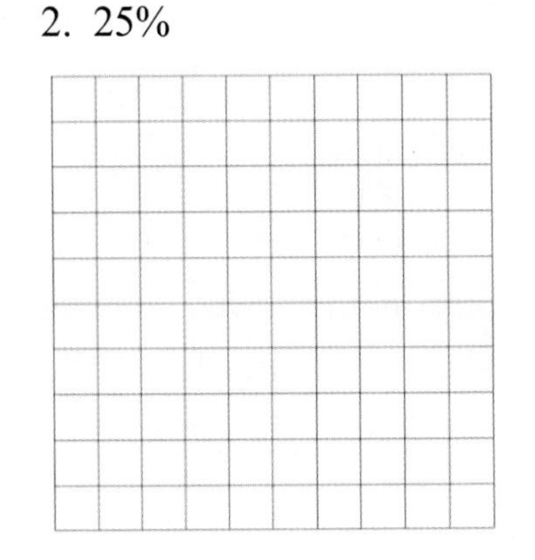

3. 52%

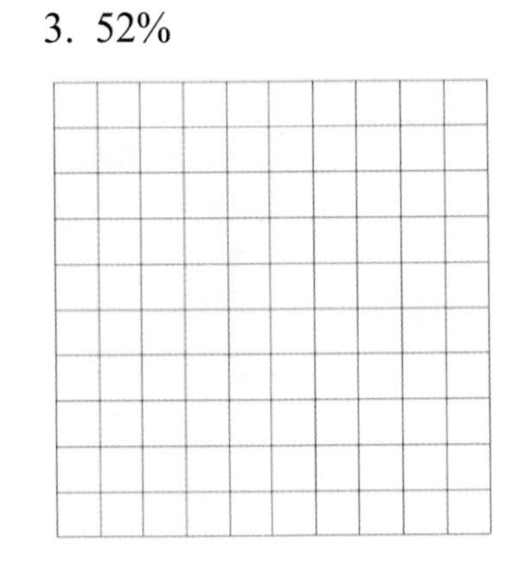

3.09 Percent, Decimals, and Fractions 3.09

We can also think about a percent as a fraction or a mixed number. Using the idea that a percent is "per hundred", we can rewrite a percent to look like a fraction that means the same thing.

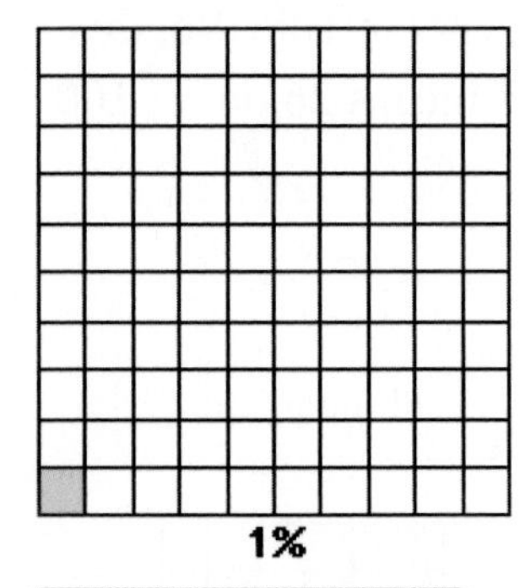

Looking at the grid, we have 1 block out of 100 shaded. We have already stated that this is equivalent to 1% but we can also think about this as 1 out of 100 or $\frac{1}{100}$. So, 1 % = $\frac{1}{100}$.

EXAMPLE 2

Let's look at the other grids and find the fraction which is it's equivalent.

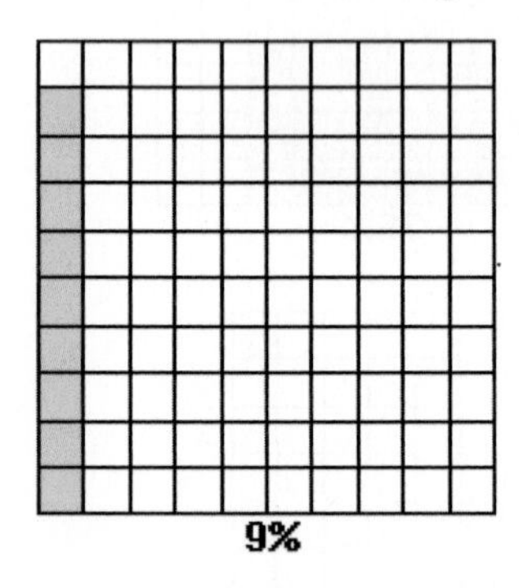

9 out of 100 blocks are shaded: $\frac{9}{100}$.

$$9\% = \frac{9}{100}$$

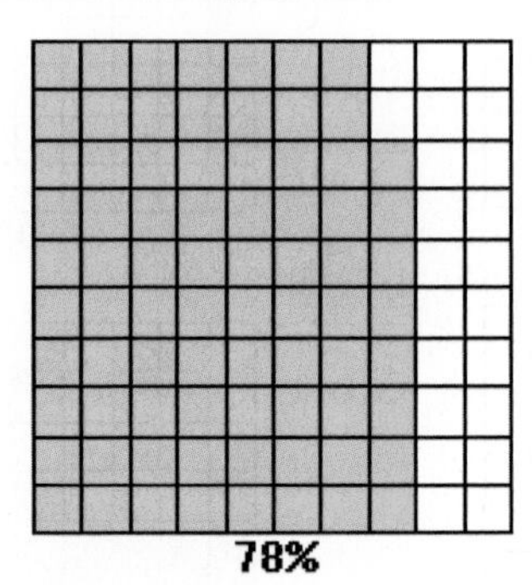

78 out of 100 blocks are shaded: $\frac{78}{100}$.

$$78\% = \frac{78}{100} = \frac{39}{50}$$

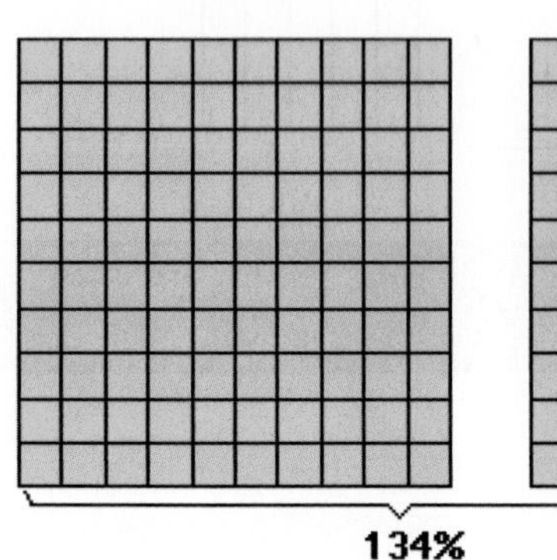

The entire grid is shaded in the first one which means 1 whole.

The second grid has 34 out of 100: $\frac{34}{100}$.

$$134\% = 1\frac{34}{100} = 1\frac{17}{50}$$

Percents are used everywhere in everyday life.

- The sales tax rate for North Carolina is 7.75%.
- 80% of homes in America have computers.
- Black Friday sales were as much as 80% off regular prices.
- 8% of people worldwide drive cars.

3.09 Percent, Decimals, and Fractions 3.09

YOUR TURN

Write each shaded region as a percent and as a fraction or a mixed number.

4.

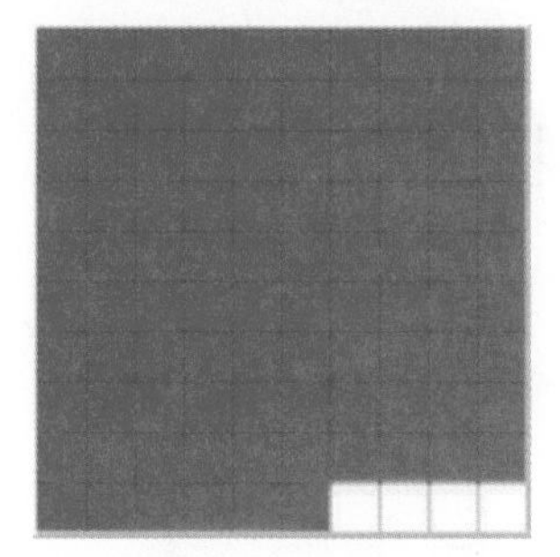

Percent = ___________

Fraction = ___________

5.

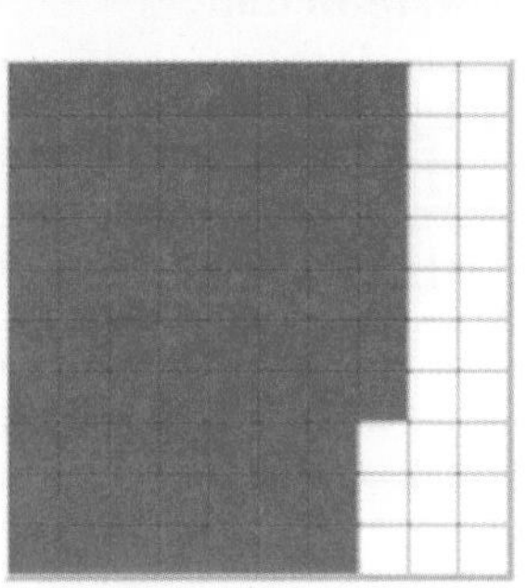

Percent = ___________

Fraction = ___________

6.

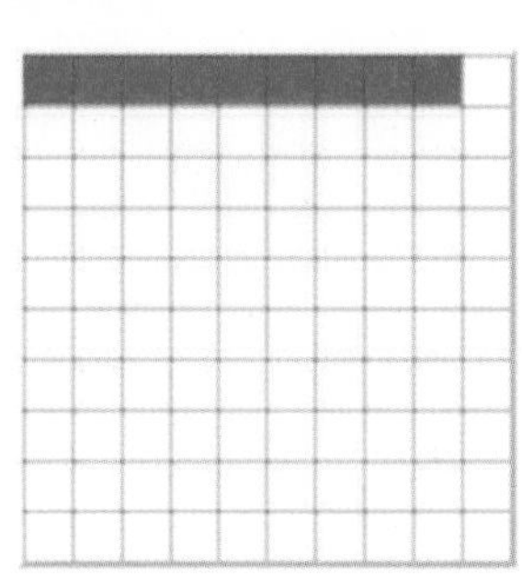

Percent = ___________

Fraction = ___________

7.

Percent = ___________ Fraction = ___________

If you are asked to rewrite a percent as a fraction, it would be time consuming to draw a grid each time. If we apply the thought that the percent sign, "%" , means to write a ratio with 100 as the denominator, we can simply put the number over 100 in front of the percent sign and reduce.

> **To write a percent as a fraction**, write the number over a denominator of 100, dropping the percent sign, and then reduce.

EXAMPLE 3

A library tracks the percent of each type of book checked out. The grid below shows the percents from last week. What percent of the books were fiction?

Solution:

We need to count the blocks that are shaded blue.

39 out of 100 blocks are blue.

This represents 39%.

Fiction

Nonfiction

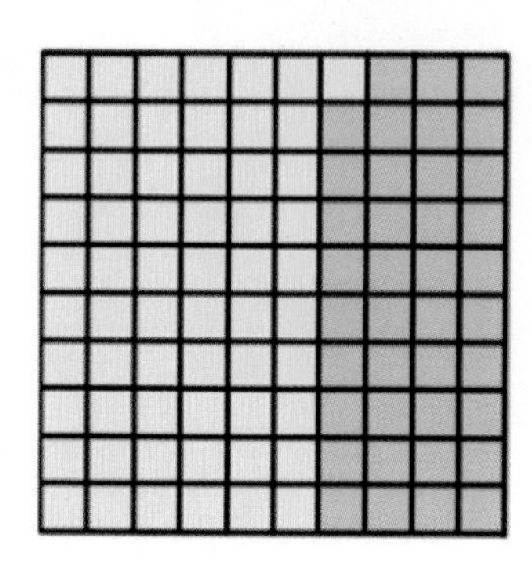

39% of the books were fiction.

YOUR TURN

7a. Using the grid above, determine what percent of the books checked out were nonfiction.

__________% of the books were nonfiction.

EXAMPLE 4

Write the fraction or mixed number that is equivalent to each percent.

a. $15\% = \frac{15}{100} = \frac{3}{20}$

b. $125\% = \frac{125}{100} = 1\frac{1}{4}$

c. $75\% = \frac{75}{100} = \frac{3}{4}$

To divide a fraction, multiply by its reciprocal.

Dividing by 100 is the same as multiplying by $\frac{1}{100}$

d. $12\frac{1}{2}\% = \frac{12\frac{1}{2}}{100} = 12\frac{1}{2} \div 100 = 12\frac{1}{2} \bullet \frac{1}{100} = \frac{25}{2} \bullet \frac{1}{100} = \frac{25}{200} = \frac{1}{8}$

Remember that the fraction bar means division. So, having a 100 in the denominator of a fraction means to divide by 100.

e. $33\frac{1}{3}\% = \frac{33\frac{1}{3}}{100} = 33\frac{1}{3} \div 100 = 33\frac{1}{3} \bullet \frac{1}{100} = \frac{100}{3} \bullet \frac{1}{100} = \frac{100}{300} = \frac{1}{3}$

To write a percent as a decimal:

- Drop the % symbol.
- Divide by 100 **OR** Multiply by 0.01.

Calculator Option: Putting the number over 100 is the same as dividing by 100. With your calculator, divide by 100 using your fraction key instead of the divide key.

20 % = 20 a b/c 100 = $\frac{1}{5}$

NOTE: Using the ÷ key will give you a decimal . Since you want a fraction you must use the fraction key!!!

3.09 Percent, Decimals, and Fractions 3.09

YOUR TURN

Write each percent as a fraction or mixed number in simplest form.

8. 35 %	9. 250 %	10. 37 ½ %	11. ¼ %

Converting a Percent to a Decimal

First, let's review decimals. You saw a chart similar to this one in the previous module. We can see from this place value chart that $\frac{1}{100}$ = 0.01 (one hundredth).

Thousands	Hundreds	Tens	Ones	Tenths	Hundredths	Thousandths
1000	100	10	1	$\frac{1}{10}$	$\frac{1}{100}$	$\frac{1}{1000}$
1000	100	10	1	0.1	**0.01**	0.001

Since percent means per one hundred, 17 % would be the same as $\frac{17}{100}$. From your knowledge of decimal place value (chart above), we see that $\frac{17}{100}$ = 0.17 (seventeen hundredths). Therefore, 17 % must equal 0.17.

3.09 Percent, Decimals, and Fractions 3.09

Let's look at some more examples.

EXAMPLE 5

Write the percent as a decimal.

a. $25\% = \dfrac{25}{100} = 0.25$

b. $55\% = \dfrac{55}{100} = 0.55$

c. $90\ \% = \dfrac{90}{100} = 0.90$

To write a percent as a decimal:

- Drop the % symbol.
- Divide by 100 **OR** Multiply by 0.01.

EXAMPLE 6

Write the percent as a decimal.

a. $120\ \% = 120 \bullet 0.01 = 1.20$

b. $3\ \% = 3 \bullet 0.01 = 0.03$

c. $0.72\ \% = 0.72 \bullet 0.01 = 0.0072$

Multiplying by 0.01 is the equivalent to dividing by 100.

So, another option would be to use your calculator and divide the decimal by 100 and then drop the % sign.

$29\ \% = 29 \div 100 = 0.29$

Since our number system is based on powers of 10 (look back at the place value chart), dividing by 100 is equivalent to moving through two powers of 10 – two place values. So, another way to divide by 100, is simply to move the decimal two places to the left.

3.09 Percent, Decimals, and Fractions 3.09

EXAMPLE 7

Write each percent as a decimal.

29% = 29. = 0.29 Simply remove % sign and move decimal 2 places left.

4.6% = 4.6 = 0 .046 Remove % sign and move 2 places. Add a zero as needed.

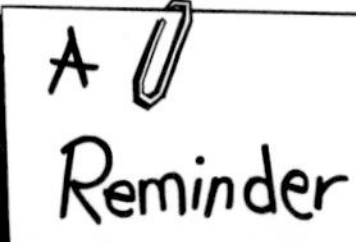

Remember where a decimal point is located if you don't already see it --- always at the end of the number.

For example, the number 23 is considered to have a decimal point at the end of the number 23.

To write a percent as a decimal:

- Drop the % symbol.
- Move decimal point two places to the left. (divide by 100)

YOUR TURN

Write each percent as a decimal.

12. 26 %	13. 250 %	14. 3.8 %	15. 0.4 %

3.09 Percent, Decimals, and Fractions 3.09

EXAMPLE 8

Let's take another look at those 10 X 10 grids. Keep in mind that a percent is just another way to write a fraction or a decimal.

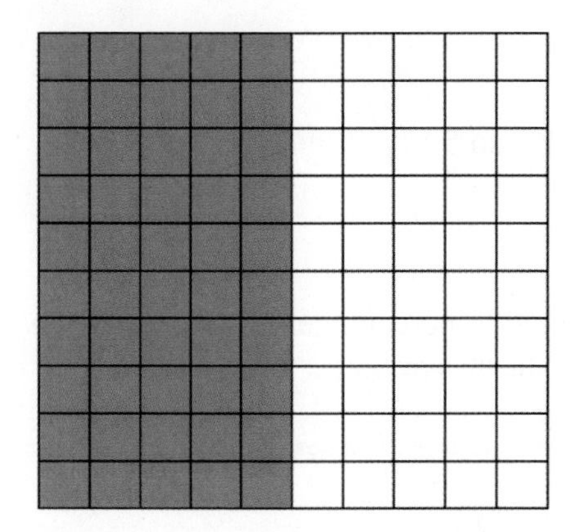

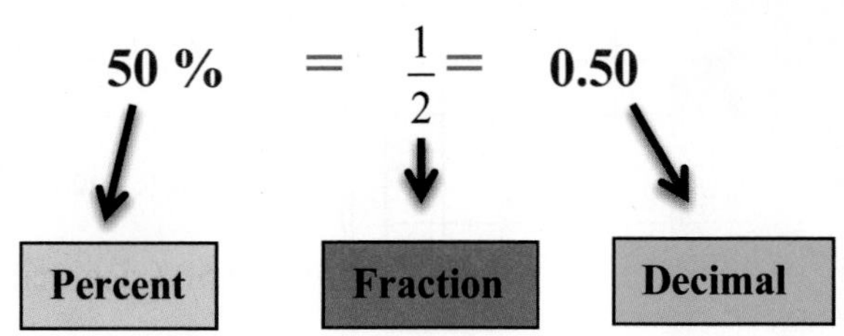

YOUR TURN

Fill in the chart by converting each percent to a fraction and then a decimal. Make sure to reduce your fractions.

Percent	Fraction	Decimal
16. 40 %		
17. 2 %		
18. 12 ½ %		
19. 8.4 %		

3.09 Percent, Decimals, and Fractions 3.09

Converting a Decimal to a Percent

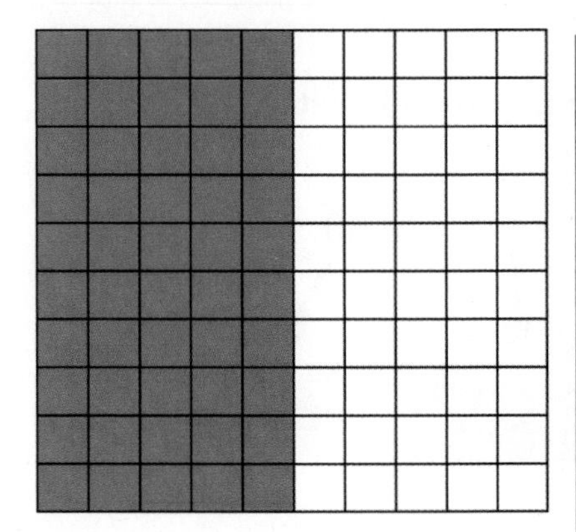

In this grid, 50 of the 100 squares are blue, so we know from the previous section that this would be represented as

$\frac{50}{100}$ or 0.50 as a decimal.

Percentage mean "per 100" so 50 % of the box is blue.

So, 0.50 (decimal number) = 50 %

So, to convert from decimal to percentage, just **multiply the decimal by 100** and add the % symbol.

EXAMPLE 9

Write each decimal as a percent.

a. 0.24 = 0.24 (100) = 24 %

b. 2 = 2 (100) = 200 %

c. 0.04 = 0.04 (100) = 4 %

d. 0.6 = 0.6 (100) = 60 %

To convert from decimal to percentage, just **multiply the decimal by 100** and add the % symbol.

Since our number system is based on powers of 10 (look back at the place value chart), multiplying by 100 is equivalent to moving through two powers of 10 – two place values. So, another way to multiply by 100, is simply to move the decimal two places to the right.

To write a decimal as a percent:

- Move decimal point two places to the right. (multiply by 100)
- Add the % symbol.

3.09 Percent, Decimals, and Fractions 3.09

EXAMPLE 10

Write each decimal as a percent.

a. 0.15 = 15.% = 15% Move the decimal 2 places right and add % sign.

b. 1.25 = 125.% = 125 %

c. 0.4 = 04 . = 40 %

YOUR TURN

Write each decimal as a percent.

20. 0.28	21. 0.09	22. 1.35	23. 0.8

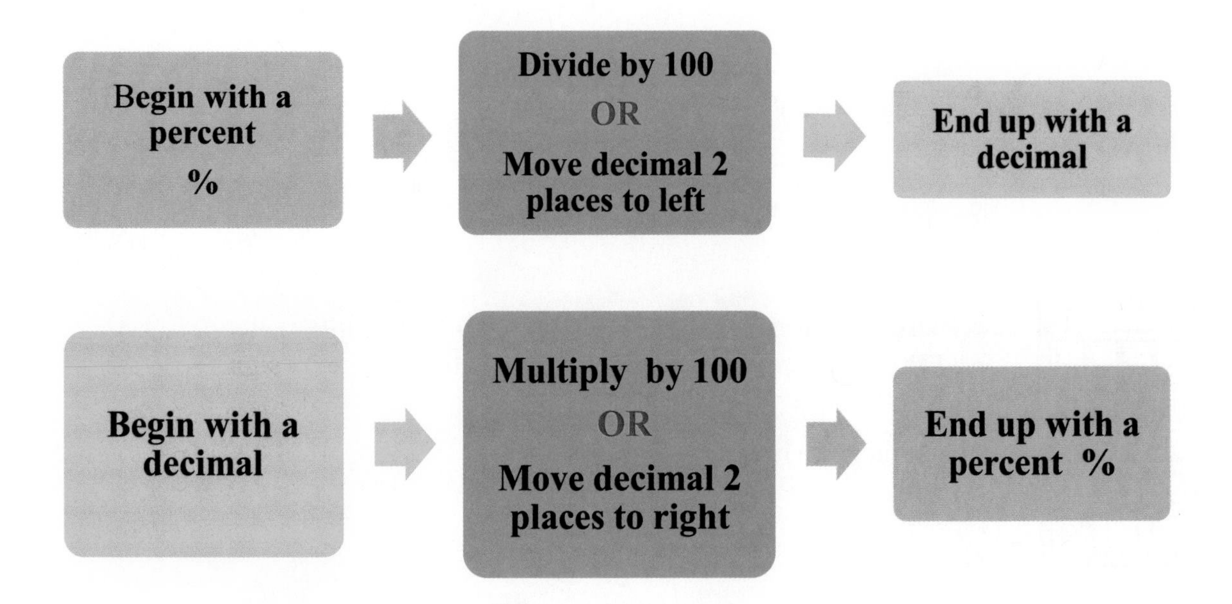

3.09 Percent, Decimals, and Fractions 3.09

Converting a Fraction to a Percent

We have already looked at how to change a percent into a fraction.

> To write a percent as a fraction:
>
> - Drop the % symbol.
> - Divide by 100 **OR** Multiply by $\frac{1}{100}$

To change a fraction into a percent, you will just reverse the above process. Instead of dividing by 100 (writing over a denominator of 100), we will multiply the fraction by 100 and then add the % sign.

$27\% = \frac{27}{100}$ ⟶ percent to fraction – divide by 100

$\frac{27}{100} = \frac{27}{100} \bullet \frac{100}{1} = 27\%$ ⟶ fraction into percent - multiply by 100

EXAMPLE 11

Write each fraction as a percent.

a. $\frac{1}{8} = \frac{1}{8} \bullet 100 = \frac{1}{8} \bullet \frac{100}{1} = \frac{100}{8} = \frac{25}{2} = 12\frac{1}{2}\ \%$

b. $\frac{7}{40} = \frac{7}{40} \bullet 100 = \frac{7}{40} \bullet \frac{100}{1} = \frac{700}{40} = \frac{35}{2} = 17\frac{1}{2}\ \%$

c. $2\frac{1}{2} = \frac{5}{2} \bullet 100 = \frac{5}{2} \bullet \frac{100}{1} = \frac{500}{2} = 250\ \%$

Calculator Option: Type in the fraction using your fraction key and then multiply by 100. Don't forget to add the % sign.

Example: Write $\frac{1}{5}$ as a percent.

1 a b/c 5 × 100 = 20 %

3.09 Percent, Decimals, and Fractions 3.09

YOUR TURN

Write each fraction or mixed number as a percent.

24. $\frac{5}{8}$	25. $\frac{4}{5}$	26. $1\frac{3}{5}$	27. $\frac{1}{3}$

Reference Chart: Percent, Decimals, Fractions

Operation	Explanation
Decimal to Percent	Move the decimal point 2 places to the right and add % sign. → OR Multiply by 100 using your calculator and add % sign.
Percent to Decimal	Move the decimal point 2 places to the left and drop % sign. ← OR Divide by 100 using your calculator.
Fraction to a Percent	Multiply the fraction by 100 and add % sign.
Percent to a Fraction	Put the number over a denominator of 100 and reduce. OR Divide by 100 using the fraction key on your calculator.

3.09 Percent, Decimals, and Fractions 3.09

Decimals, fractions and percentages are just different ways of showing the same value. We can change any value into another form that is equal. For example, we can change a percent to an equivalent fraction, or a decimal to an equivalent percent, or a fraction into an equivalent decimal. In other words, given a percent, fraction, or decimal, we can rewrite it as a percent, fraction, or decimal.

EXAMPLE 12

Complete the table.

Percent	Decimal	Fraction
35 %		
	0.6	
		$\frac{1}{8}$

35%

Decimal: Move 2 places left.

$35\,\% = 0.35$

Fraction: Divide by 100.

$35\% = \frac{35}{100} = \frac{7}{20}$

0.6

Percent: Move 2 places right.

$0.6 = 60\,\%$

Fraction: Divide by 100.

$60\% = \frac{60}{100} = \frac{3}{5}$

$\frac{1}{8}$

Percent: Multiply by 100.

$\frac{1}{8} = \frac{1}{8} \bullet \frac{100}{1} = \frac{100}{8} = 12\frac{1}{2}\%$

Decimal: move 2 places left.

$12\frac{1}{2}\% = 12.5\% = 0.125$

3.09 Percent, Decimals, and Fractions 3.09

YOUR TURN

28. Complete the table:

Percent	Decimal	Fraction or Mixed Number
80 %		
	2.5	
		$\frac{5}{8}$

YOUR TURN **Answers to Section 3.09.**

1. shade 5 blocks
2. Shade 25 blocks
3. Shade 52 blocks
4. 96%, $\frac{96}{100} = \frac{24}{25}$
5. 77%, $\frac{77}{100}$
6. 9%, $\frac{9}{100}$
7. 125%, $\frac{5}{4} = 1\frac{1}{4}$ 7a. 61%
8. $\frac{7}{20}$
9. $2\frac{1}{2} = \frac{5}{2}$
10. $\frac{3}{8}$
11. $\frac{1}{400}$
12. 0.26
13. 2.5
14. 0.038
15. 0.004
16. $\frac{2}{5} = 0.4$
17. $\frac{1}{50} = 0.02$
18. $\frac{1}{8} = 0.125$
19. $\frac{21}{250} = 0.084$
20. 28%
21. 9 %

(continued on next page)

3.09 Percent, Decimals, and Fractions 3.09

YOUR TURN **Answers to Section 3.09 continued.**

22. 135 %
23. 80 %
24. 62 ½ %
25. 80 %
26. 160 %
27. $33\frac{1}{3}\%$
28.

Percent	Decimal	Fraction or Mixed Number
80 %	0.8	$\frac{4}{5}$
250 %	2.5	$\frac{5}{2} = 2\frac{1}{2}$
$62\frac{1}{2}\%$	0.625	$\frac{5}{8}$

Complete MyMathLab Section 3.09 Homework in your Homework notebook.

3.10 Solving Percent Problems Using Equations 3.10

Before we start solving percent problems, let's look at those 10 by 10 grids again and see how we can use them to help us understand what a percentage is.

Recall that percent means "per 100" which means to divide by 100.

EXAMPLE 1

The entire unit square represents 400 people which would be 100%.

If 400 is shared equally among the 100 squares, then each square would represent 4 people.
(400 divided by 100 = 4)

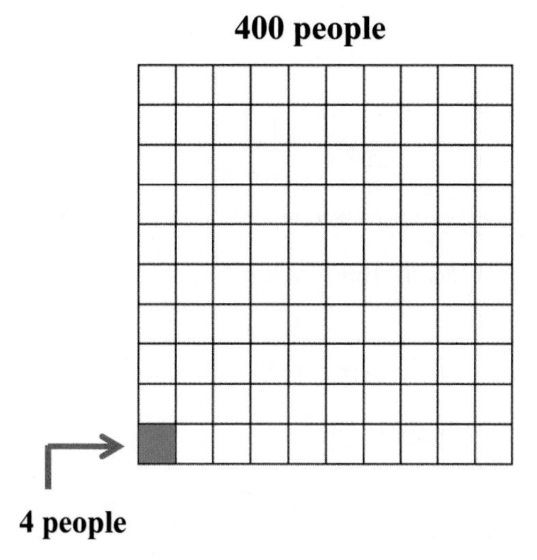

If 400 people represent 100% (the entire 100 blocks), then 4 people would represent 1% - (one block out of 100).

From this let's look at some other questions we could answer about percentages.

a. How many people would be represented by 20 small squares ?

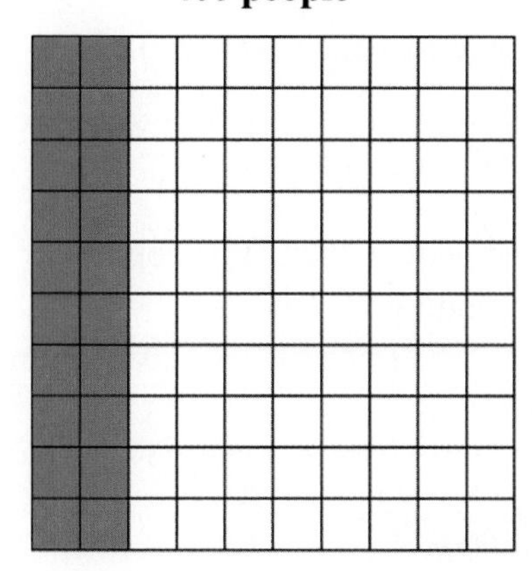

The question: **20% of 400 is what?**

If 4 people are one square, then 20 squares would be 20 (4) =80 people.

20 % of 400 is 80.

3.10 Solving Percent Problems Using Equations 3.10

b. How many people would be represented by 50 small squares?

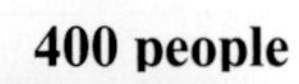

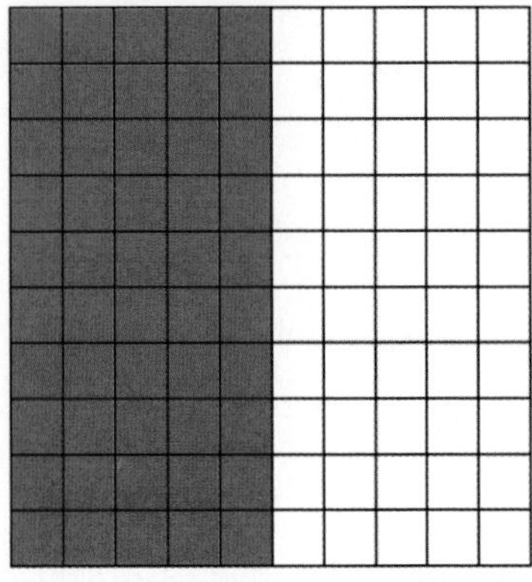

The question: **50% of 400 is what?**

If 4 people are one square, then 50 squares would be 50 (4) =200 people.

50% of 400 is 200.

c. How many people would be represented by one half of a small square?

400 people

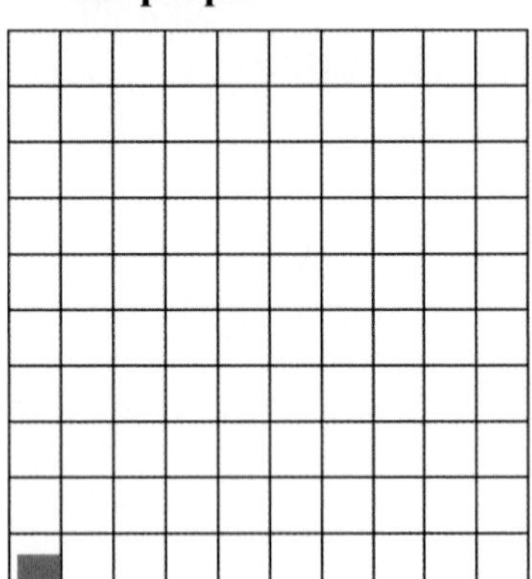

The question: $\frac{1}{2}$ **% of 400 is what?**

If 4 people are one square then one half of a square would be $\frac{1}{2}$ (4) = 2 people.

$\frac{1}{2}$ **% of 400 is 2.**

Let's look at this from another view. What part of the unit square would represent 100 people?

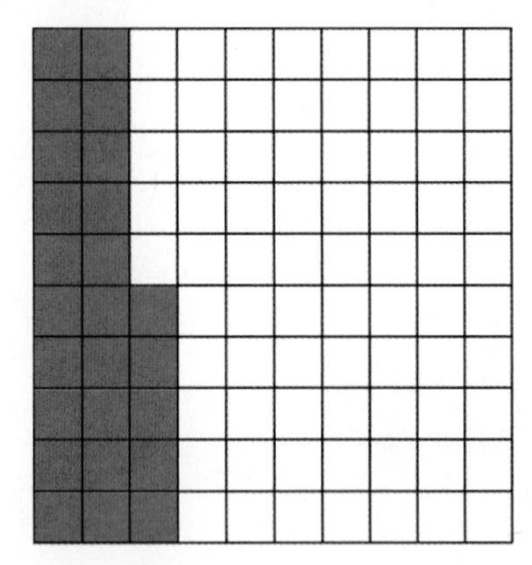

The question is, "What percent of 400 is 100?"

If each square represents 4 people, then 100 people would be represented by shading 25 squares.
(100 /4 = 25)

25% of 400 is 100.

3.10 Solving Percent Problems Using Equations

YOUR TURN

1. In Duke County, there are 300 schools in total. Represent 1% of the schools in a unit grid by shading and labeling one square.

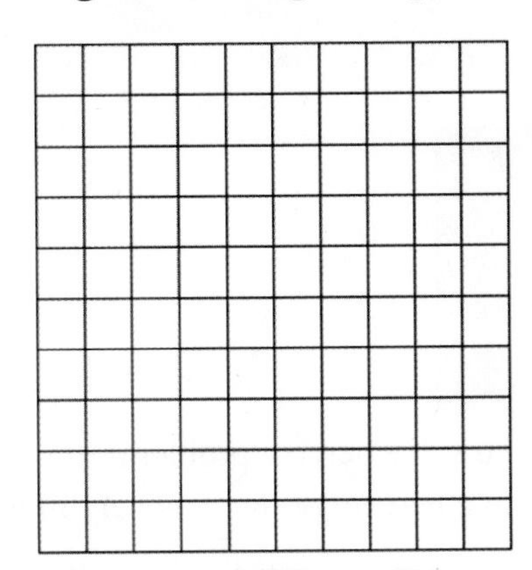

a. How many schools does 1 square represent? ______________

b. 60 of the schools have a teacher-to-student ratio that meets or exceeds the requirements for accreditation. Represent 60 schools on the grid by shading the appropriate number of squares.

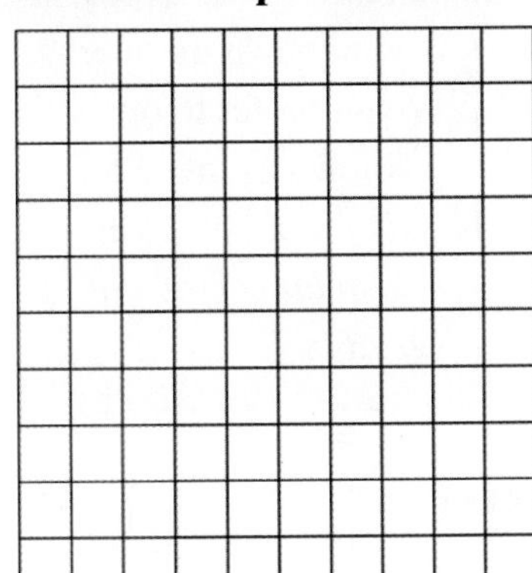

Recall from the grid above what one square represents.

b. What percent of the schools meet or exceed the requirements? In other words, what percent of 300 is 60?

____

c. An adjacent county has 40 schools that meet or exceed the requirement for accreditation. This represents 20% of the total number of schools. Can you use a grid to show how many schools this county has in total? (Hint: Shade in 20% of the grid and find how many schools is represented by one square if 20% is 40 schools. After finding the value of one square, find the value of the unit square which is 100 small squares.)

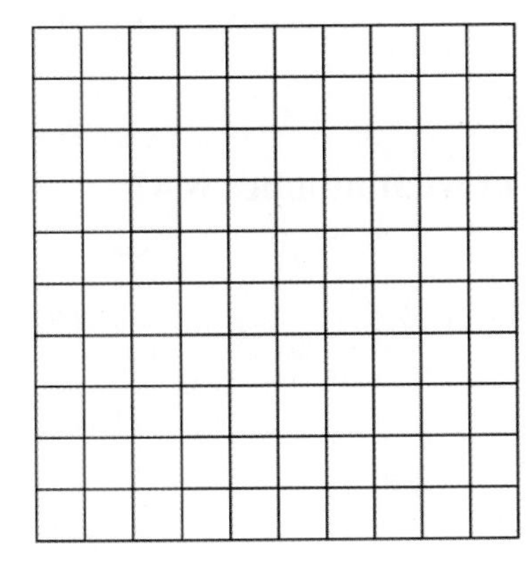

c. 20% of how many schools is represented by 40 schools? In other words, 20% of what is 40?

____

3.10 Solving Percent Problems Using Equations 3.10

In the previous examples, we worked percent problems using the unit square. Now, we want to look at another way to work these same problems.

Every statement of percent can be expressed verbally as: "***One number is some percent of another number.***" Percent statements will always involve three numbers.

____ is ____ % of ____

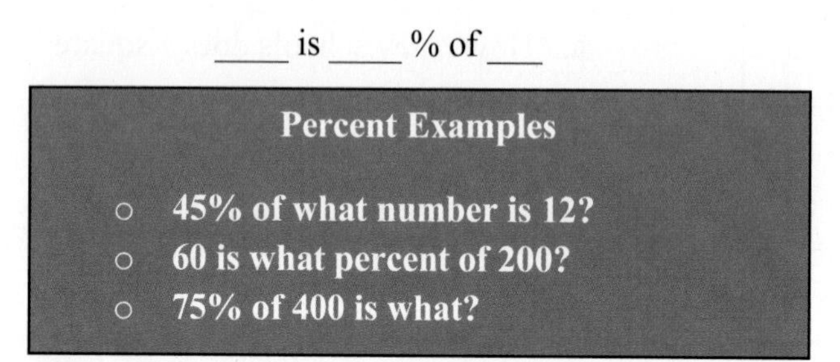

An equation is a mathematical statement that contains an equal sign. To solve percent problems, translate these problems into mathematical equations. Recognizing key words in a percent problem is helpful in writing the problem as an equation.

Key Words

of means **multiplication** $(\cdot)$

is means **equals** $(=)$

what number means **the unknown number** n

Any letter of the alphabet can be used to represent the unknown number. We will refer to our unknown as the letter ***n***. ***This unknown number n is also known as the variable.***

EXAMPLE 2

Translate to an equation using the key words above. 6 is what percent of 36 ?

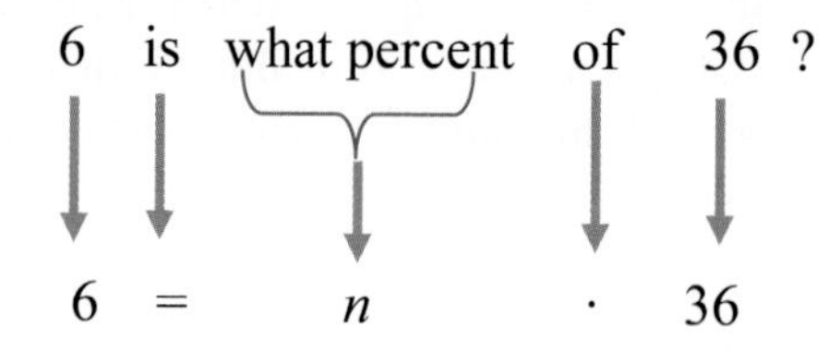

We will use the dot (·) to mean multiplication.

EXAMPLE 3

Translate to an equation using the key words above. 25% of what number is 4?

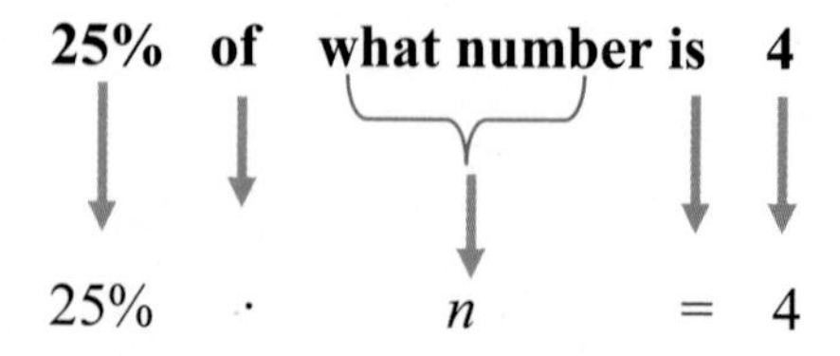

3.10 Solving Percent Problems Using Equations 3.10

YOUR TURN

Translate to an equation. Do not solve.

2. 30% of what number is 90?	3. 78% of 68 is what number?	4. What number is 5% of 120?

Solving Percent Problems

After translating the percent problem into an equation, we use the equation to find the unknown number.

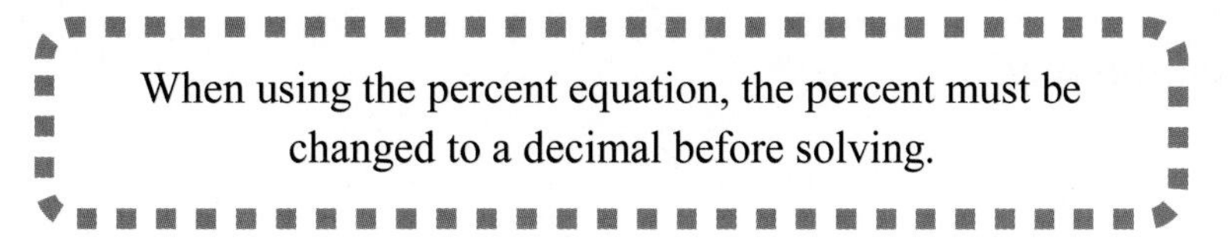

When using the percent equation, the percent must be changed to a decimal before solving.

EXAMPLE 4

What number is 25% of 80?

$n = 25\% \cdot 80$ ⟶ Translate using the key words.

$n = 0.25 \cdot 80$ ⟶ Write the percent as a decimal.

$n = 20$

20 is 25% of 80

EXAMPLE 5

16 ½% of 150 is what number?

$16\frac{1}{2}\% \cdot 150 = n$ ⟶ Translate using the key words.

$0.165 \cdot 150 = n$ ⟶ Write the percent as a decimal.

$24.75 = n$

16 ½% of 150 is 24.75

3.10 Solving Percent Problems Using Equations

EXAMPLE 6

12% of what number is 0.6?

$12\% \cdot n = 0.6$ ⟶ Translate.

$0.12 \cdot n = 0.6$ ⟶ Change the percent to a decimal.

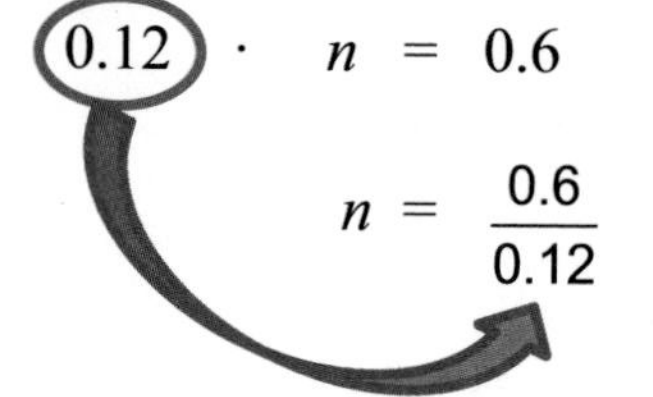

$n = \frac{0.6}{0.12}$ ⟶ To solve for n, we always divide the number by itself by the number in front of the n. In other words, the number in front of the n should always go in the denominator.

$n = 5$

12% of 5 is 0.6

EXAMPLE 7

135 is 150% of what number?

$135 = 150\% \cdot n$ ⟶ Translate.

$135 = 1.5 \cdot n$ ⟶ Change the percent to a decimal.

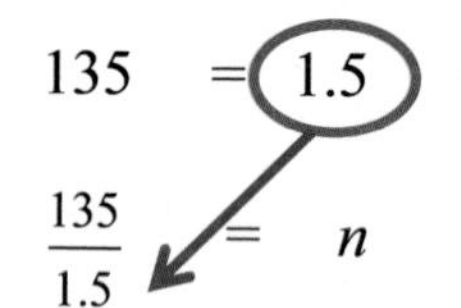

$\frac{135}{1.5} = n$ ⟶ To solve for n, we always divide the number by itself by the number in front of the n. In other words, the number in front of the n should always go in the denominator.

$90 = n$

135 is 150% of 90

EXAMPLE 8

What percent of 16 is 12?

$n \cdot 16 = 12$ ⟶ Translate.

$n = \frac{12}{16}$ ⟶ To solve for n, divide the number by itself by the number sitting beside n.

$n = 0.75$ ⟶ Notice that 0.75 is a decimal but the question asked for a percent. To change a decimal to a percent, move the decimal 2 places to the right.

$n = 75\%$

75% of 16 is 12

3.10 Solving Percent Problems Using Equations 3.10

EXAMPLE 9

35 is what percent of 25?

$35 = n \cdot 25$ → Translate the question into an equation and then solve.

$\frac{35}{25} = n$ → To solve for n, divide the number by itself by the number sitting beside n.

$1.4 = n$ → Don't forget to change the decimal answer to a percent.

$140\% = n$

35 is 140% of 25

YOUR TURN

Write an equation and then solve to answer each question. Circle your solution.

5. 25% of 60 is what number? Equation: ______	6. 110% of what number is 44? Equation: ______	7. What percent of 50 is 20? Don't forget to change your answer to a %. Equation: ______
8. 2.4% of 26 is what number? Equation: ______	9. 13 is $6\frac{1}{2}\%$ of what number? Equation: ______	10. 9 is what percent of 12? Don't forget to change your answer to a %. Equation: ______

3.10 Solving Percent Problems Using Equations 3.10

YOUR TURN **Answers to Section 3.10**

1. a. shade one square, 3 schools

   b. shade 20 squares , 20 %

   c. shade 20 squares, 200 students

2. $30\% \cdot \boldsymbol{n} = 90$

3. $78\% \cdot 68 = \boldsymbol{n}$

4. $\boldsymbol{n} = 5\% \cdot 120$

5. 15

6. 40

7. 40 %

8. 0.624

9. 200

10. 75 %

Complete MyMathLab Section 3.10 Homework in your Homework notebook.

3.11 Solving Percent Problems Using Equations 3.11

Percent has many applications in our daily lives. This section discusses percents as they apply to different real-life situations. In the word problems, look for the key sentence that asks the question. Remember, all percent problems relate three numbers. Always look to fill in the blanks of this percent statement: "____ is ____ % of ____." Your task is to fill in two of those blanks and solve for the third one.

EXAMPLE 1

An inspector found 18 defective light bulbs during an inspection. If this is 1.5% of the total number of bulbs inspected, how many bulbs were inspected?

- Find the key sentence/words. If you see a percent symbol, look there first.

  Key words: **"If this is 1.5 % of the total number."**

- Decide where the other number you are given goes and what is your unknown.

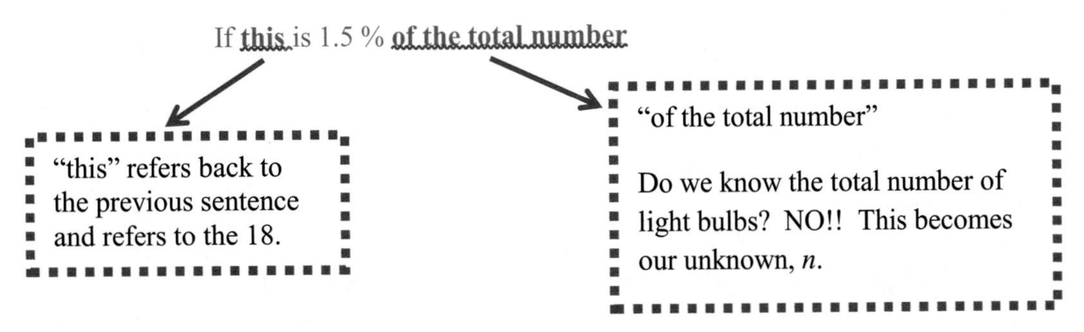

- Plugging in 18 and n in the right blanks, we have:

  18 is 1.5% of what number?

- Now, let's translate. Recall "is" means =, "of" means multiply, and "what" is our unknown, n.

  18 is 1.5% of what number?

$$18 = 0.015 \cdot n$$

$$\frac{18}{0.015} = n$$

$$1200 = n$$

Courtesy of Fotolia

There were 1200 light bulbs inspected.

3.11 Solving Percent Problems Using Equations 3.11

EXAMPLE 2

There are 115 Criminal Justice schools in the state of Florida. Of these schools, 65 offer BA degrees in Criminal Justice. What percent of Criminal Justice schools in Florida offer BA degrees? Round your answer to the nearest whole percent.

Key Words: What percent **of Criminal Justice schools** in Florida offer BA degrees?

"of" refers to the number of Criminal Justice schools. Do we know that number? YES!!
The first sentence states there are 115 Criminal Justice schools.

" BA degrees": Do we know that number? YES!!
The second sentence states there are 65 Criminal Justice schools that offer BA degrees.

Courtesy of Fotolia

Now, let's translate this into a percent equation.

What % of 115 is 65?

$$65 = n \cdot 115$$

$$\frac{65}{115} = n$$

$0.57 \approx n$ ⟶ Round to two decimal places.

57% ⟶ Change our decimal answer into a % by moving the decimal 2 places to the right.

EXAMPLE 3

A woman earns $2700 per month and budgets $540 per month for food. What percent of her monthly income is spent on food?

"of her monthly income" refers back to $2700.
"is spent on food" refers back to $540.

Percent equation: What percent "of" her monthly income "is" spent on food?
What percent of 2700 is 540 ?

$$n \cdot 2700 = 540$$

$$n = \frac{540}{2700}$$

$$n = 0.2$$

20% Remember to change your answer to a percent.

3.11 Solving Percent Problems Using Equations

EXAMPLE 4

Find the percent of total calories from fat.

Nutrition Facts	
Serving Size 5 oz	
Servings Per Container 4	
Amount Per Serving	
Calories 90	Calories from Fat 30
	% Daily Value*
Total Fat 3g	5%
Saturated Fat 0g	0%
Cholesterol 0mg	0%
Sodium 440mg	19%
Total Carbohydrate 13g	4%
Dietary Fiber 3g	4%
Sugars 3g	
Protein 3g	
Vitamin A 60% •	Vitamin C 60%
Calcium 4% •	Iron 4%

* Percent Daily Values are based on a 2,000 calorie diet. Your daily values may be higher or depending on your calorie needs:

Find the percent of total calories from fat.

"of" refers to the total number of calories. Do we know that?? YES!! The label states there are 90 total calories.

"from fat " refers to fat calories. Do we know that number?? YES!! The label states there are 30 total fat calories.

Translate into a percent equation.

What percent of total calories is from fat?

What percent of 90 is 30?

$$n = \frac{30}{90}$$

$$n = \frac{1}{3}$$

$$n = 33\frac{1}{3}\%$$

$33\frac{1}{3}$% *calories are from fat*

EXAMPLE 5

Approximately 41% of films are rated PG. If 950 films were recently rated, how many films were rated PG? Round to the nearest whole number.

41% of films are rated PG

"of" refers to the number of films. Do we know how many films? YES!! 950 films

"rated PG" refers to the number of films rated PG. Do we know how many films? NO!! This is our unknown.

Translate into a percent equation: 41% of 950 is what?

$$0.41(950) = n$$

$$389.5 = n$$

390 films are rated PG

EXAMPLE 6

Las Vegas, Nevada increased in population 19% from the 2000 census to the 2010 census. In 2000, the population of Las Vegas was about 478,500.

a. Find the increase in population from 2000 to 2010.
b. Find the population for Las Vegas in 2010.

Courtesy of Fotolia

What number is 19% of 478,500?

$$n = 19\% \cdot 478{,}500$$
$$n = 0.19 \cdot 478{,}500$$
$$n = 90{,}915$$

a. The population increased by 90,915 people between 2000 and 2010.

b. To find the population for Las Vegas in 2010, we will need to add 90,915 to the original population in 2000.

$$\textit{new population} = 90{,}915 + 478{,}500$$
$$\textit{new population} = 569{,}415$$

3.11 Solving Percent Problems Using Equations 3.11

YOUR TURN

Write an equation, solve and answer each question.

1. In one country, approximately 133,500 of 890,000 restaurants are pizza restaurants. What percent of the restaurants are pizza restaurants? What percent of ________ is ________? Equation: ______________________ Solve:	2. Approximately 58% of films are rated R. If 830 films were recently rated, how many are rated R? Round to nearest integer as needed. What percent of ________ is ________? Equation: ______________________ Solve:
3. Honda announced a 5.5% increase in the cost of a certain model car. This year the price is \$22,016. Find the increase and the new price. Equation: ______________________ The increase in price is: _______________ The new price is: _______________	4. A family pays \$25,900 as a down payment for a home. If this represents 14% of the price of the home, find the price of the home. What percent of ________ is ________? Equation: ______________________ Solve:

3.11 Solving Percent Problems Using Equations 3.11

Finding Percent Increase and Percent Decrease

Percent increase and percent decrease are measures of **percent change**, which is the extent to which something gains or loses value.

When a quantity shrinks (gets smaller), then we can compute its PERCENT DECREASE.

When a quantity grows (gets bigger), then we can compute its PERCENT INCREASE.

Courtesy of Fotolia

FORMULA: MAKE A NOTE!!

To Find Percent Increase or Decrease: Decimal form

$$percent\ increase = \frac{amount\ of\ increase\ \{difference\}}{original\ amount}$$

$$percent\ decrease = \frac{amount\ of\ decrease\ \{difference\}}{original\ amount}$$

A Reminder

The answer from the formula above will be a decimal. Don't forget to change your answer to a percent by moving the decimal 2 places to the right.

We can write this formula with a 100% at the end to emphasize that since it is a **percent** increase/decrease, it should be reported as a percent.

To Find Percent Increase or Decrease: Percent Form

$$percent\ increase = \frac{amount\ of\ increase\ \{difference\}}{original\ amount} \cdot 100\%$$

$$percent\ decrease = \frac{amount\ of\ decrease\ \{difference\}}{original\ amount} \cdot 100\%$$

EXAMPLE 7

Find percent increase.

Burning Man is a play on Broadway. One of its characters is a mechanical spider. On opening night in 2008, the play saw 49,500 people walk through its doors. In 2010, the play saw its numbers rise to 51,454 attendees. Find the percent increase in attendance. Round your answer to the nearest whole percent.

$$\text{percent increase} = \frac{\text{amount of increase \{difference\}}}{\text{original amount}} \bullet 100\%$$

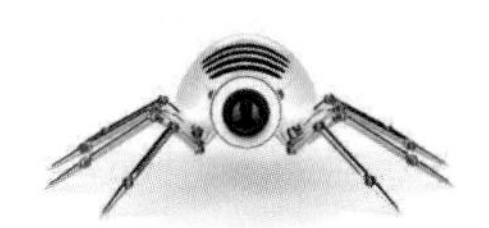
Courtesy of Fotolia

$$\text{percent increase} = \frac{(51{,}454 - 49{,}500)}{49{,}500} \bullet 100\%$$

$$\text{percent increase} = \frac{1954}{49{,}500} \bullet 100\%$$

$$\text{percent increase} = 0.04 \bullet 100\%$$

$$\text{percent increase} = 4\%$$

Note that when you compute percent increase or decrease, you always compare how much a quantity has changed from the original amount.

EXAMPLE 8

There are 110 calories in a cup of whole milk and only 64 calories in a cup of skim milk. In switching to skim milk, find the percent decrease in number of calories. Round to the nearest tenth, if needed.

$$\text{percent increase} = \frac{\text{amount of increase \{difference\}}}{\text{original amount}} \bullet 100\%$$

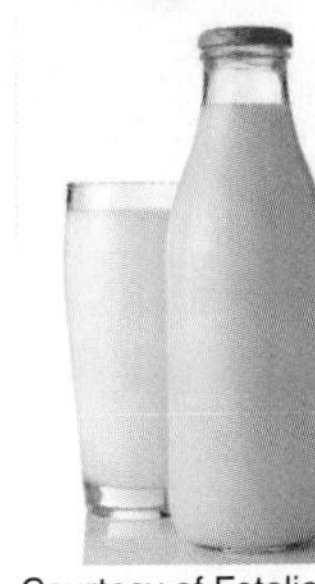
Courtesy of Fotolia

$$\text{percent increase} = \frac{(110 - 64)}{110} \bullet 100\%$$

$$\text{percent increase} = \frac{46}{110} \bullet 100\%$$

$$\text{percent increase} = 0.418 \bullet 100\%$$

$$\text{percent increase} = 41.8\%$$

3.11 Solving Percent Problems Using Equations 3.11

YOUR TURN

Write an equation and solve.

5. By changing how she drives, Martha increased her car's rate of miles per gallon from 19.5 to 23.7. Find the percent of increase. Round to the nearest tenth if needed. *percent increase* $= \frac{\text{amount of increase \{difference\}}}{\text{original amount}} \cdot 100\ \%$	6. The number of cable TV channels recently decreased from 1085 to 1070. Find the percent decrease. Round to the nearest tenth if needed. *percent decrease* $= \frac{\text{amount of decrease \{difference\}}}{\text{original amount}} \cdot 100\ \%$

YOUR TURN Answers to Section 3.11.

1. 15 %
2. 481 films
3. Increase: $1210.88, new price: $23,226.88
4. $185,000
5. 21.5 %
6. 1.4 %

Complete MyMathLab Section 3.11 Homework in your Homework notebook.

3.12 Sales Tax, Commission, and Discounts 3.12

SALES TAX

Percents are often used in the retail industry. Almost all states charge a tax on purchases made. This tax is called a sales tax and differs from state to state. Sales tax is given in terms of a percent of the purchase price.

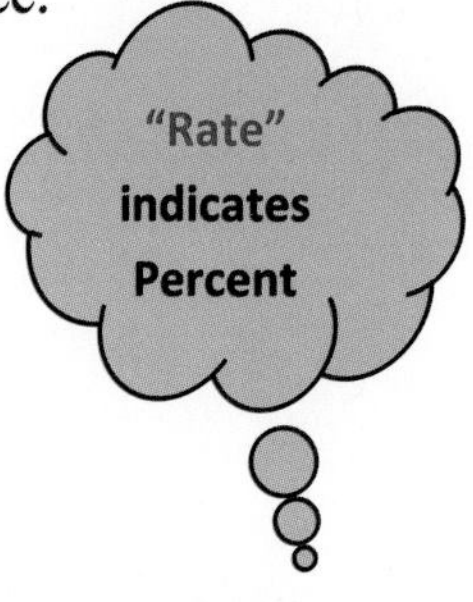

Sales Tax and Total Price

sales tax = tax rate • purchase price

total price = purchase price + sales tax

EXAMPLE 1

If the sales tax rate is 7.75% in North Carolina, what is the *sales tax* and the *total amount due* on a $75.99 leather jacket? Round your answer to the nearest cent.

sales tax = tax rate • purchase price	Start with the formula.
sales tax $= .0775 \cdot 75.99$	Plug in what you know.
sales tax $= \$5.89$	Multiply.
total price = purchase price + sales tax	Start with the formula.
total price $= \$75.99 + \$5.89$	Plug in what you know.
total price $= \$81.88$	Add.

Courtesy of Fotolia

EXAMPLE 2

The sales tax on a $7,500 motorcycle is $487.50. Find the sales tax rate.

In this problem, we are given the sales tax and purchase price. We are asked to find the rate!

Courtesy of Fotolia

sales tax = ***tax rate*** • *purchase price*	Start with the formula.
$487.50 = n \cdot 7500$	Plug in what you know.
$\frac{487.50}{7500} = n$	Divide both sides by 7500 to solve for n%.
$0.065 = n$	Convert the decimal to a percent.

$$\textit{tax rate} = 6.5\%$$

3.12 Sales Tax, Commission, and Discounts 3.12

YOUR TURN

1. What is the sales tax on a pair of shoes that cost $150 if the sales tax rate if 5%?

sales tax = tax rate • purchase price

sales tax = ________ • ____________________

sales tax = ____________________

Courtesy of Shutterstock

2. The sales tax on a $450 TV is $20.25. Find the sales tax rate.

sales tax = tax rate • purchase price

____________ = *rate* • ____________

= *rate*

tax rate = _____ %

3. A washer has a purchase price of $426. What is the total price if the sales tax rate is 8%?

sales tax = tax rate • purchase price

sales tax = ________ • ____________________

sales tax = ____________________

total price = purchase price + sales tax

total price = ____________ + ____________

4. Brian went shopping and bought a pair of sandals for $57 and sunglasses for $33. What is the total sales tax and the total of the purchases? The tax rate in Brian's city is 6%.

sales tax = tax rate • purchase price

sales tax = ________ • ____________________

sales tax = ____________________

total price = purchase price + sales tax

total price = ____________ + ____________

Courtesy of Shutterstock

3.12 Sales Tax, Commission, and Discounts 3.12

COMMISSION

A wage is defined as a payment for performing work. Hourly wages, commissions and salaries are all examples of a wage. People who work in the sales business are sometimes paid a commission. This is a payment of a percentage of total sales.

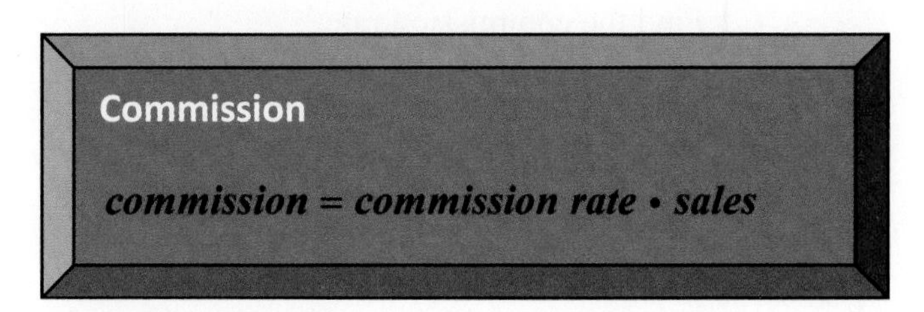

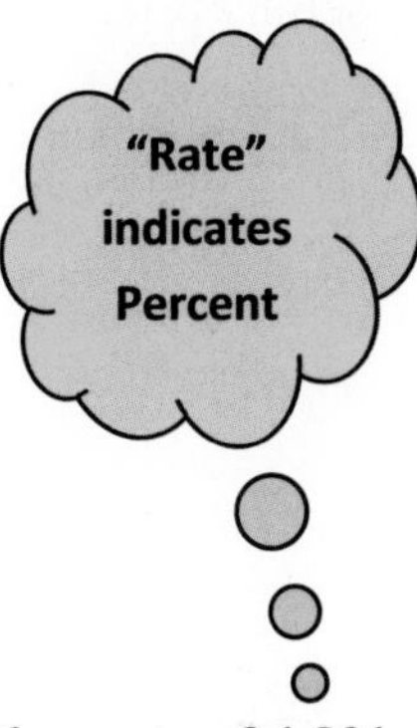

EXAMPLE 3

Emma Smith is a photographer for a fashion magazine. She is paid a commission rate of 4.8% per magazine sold. Find her commission if she sold $1,638 in magazines last month. Round your answer to the nearest cent.

commission = commission rate • sales Start with the formula.

commission = 4.8% • 1638 Plug in what you know

commission = 0.048 • 1638 Multiply

commission = $78.62

Courtesy of Fotolia

EXAMPLE 4

A SEARS salesperson earns $1,290 for selling $8,600 worth of LCD TV's. Find the rate of commission.

commission = ***commission rate*** • *sales*

In this problem, we are given the commission and the sales. We are asked to find the rate!

1290 = ***commission rate*** • 8600

$1290 = n \cdot 8600$

$\frac{1290}{8600} = n$

$0.15 = n$ Convert the decimal to a percent.

commission rate = 15%

Courtesy of Fotolia

YOUR TURN

5. How much commission will Jack make on the sale of a $161,500 house if he receives 3.1% of the selling price?	6. A salesperson earned a commission of $3520 for selling $32,000 worth of vacuum cleaners. Find the commission rate.
commission = rate • sales	*commission = rate • sales*
commission = ______________ • ______________	______________ *= rate •* ______________
commission = ______________.	*= rate*
	commission rate = _____ %

DISCOUNT

Suppose a television that normally sells for $600 is on sale at 25% off. This means the item's original price of $600 is discounted by 25% of $600, or $150. 25% is known as the discount rate. $150 is the amount of discount {in dollars}. And the sale price is $600 – $150 or $450.

Discount and Sale Price

amount of discount = discount rate • original price

sale price = original price – amount of discount

EXAMPLE 5

A Sony DVD Player was advertised on sale for 15% off the regular price of $70. Find the *discount* and the *sale price*.

Courtesy of Fotolia

$\textit{amount of discount} = \textit{discount rate} \bullet \textit{original price}$ Start with the formula.

$\textit{amount of discount} = 0.15 \bullet 70$ Plug in what you know, changing the % to a decimal.

$\textit{amount of discount} = \$10.50$ Multiply.

$\textit{sale price} = \textit{original price} - \textit{amount of discount}$ Start with the formula.

$\textit{sale price} = \$70 - \$10.50$ Plug in what you know.

$\textit{sale price} = \$59.50$ Subtract.

YOUR TURN

7. A $800 computer is on sale for 15% off. Find the amount of discount and the sale price.

discount = rate • price

discount = __________ • __________

discount = ______________

sale price = original price – discount

= __________ – __________

= ______________

8. Jane bought a pair of pants for $50, a blouse for $35, and a pair of shoes for $60 at Kohls. She had a coupon for 30% off. What was the final price of her purchase?

discount = rate • price

discount = __________ • __________

discount = ______________

sale price = original price – discount

= __________ – __________

= ______________

EXAMPLE 6

"Shoes For Less" is having a big sale this weekend. You want to buy two pairs of shoes. Each pair of shoes cost \$49. One offer states, "Buy one pair of shoes, get the second pair for free." Another promotion states, "Buy two pairs of shoes at 40% off for each pair." Which offer is the best buy?

Courtesy of Fotolia

Offer 1:

"Buy one pair of shoes, get the second pair for free."

With this offer, you will get two pairs for the price of one. So, your total cost for two pairs of shoes would be \$49.

Offer 2:

"Buy two pairs of shoes at 40% off for each pair"

With this offer, you get 40% off each pair. So, figure the cost of one pair and then multiply by 2 to get the cost for two pairs.

$\textit{amount of discount} = \textit{discount rate} \cdot \textit{original price}$

$\textit{amount of discount} = 40\% \cdot 49$

$\textit{amount of discount} = 0.40 \cdot 49$

$\textit{amount of discount} = \$19.60$

$\textit{sale price} = \textit{original price} - \textit{amount of discount}$

$\textit{sale price} = \$49 - \$19.60$

$\textit{sale price} = \$29.40$ → cost of 2 pairs = 2 (29.40)

= \$58.80

Offer 1: 2 pairs cost \$49 **Offer 2: 2 pairs cost \$58.80**

"Buy one pair of shoes, get the second pair for free" is the best buy.

EXAMPLE 7

Sally is going shopping to buy a coffeemaker that retails for $90. She looked in the sales papers and found the coffeemaker discounted 60% at Target. The same coffeemaker is discounted 35% at Kohls and she received a coupon for Kohls in the mail for an additional 30%. Where should she go to buy her coffeemaker?

Courtesy of Fotolia

Target: 60% Discount = 0.60 ($90) = 54.00

Sale Price = $90 - $54 = $36

Kohls: Discount of 35% = 0.35 ($90) = $31.50

Price = $90 - $31.50 = $58.50

Additional Discount Coupon 30% = 0.30 (58.50) = $ 17.55

Final Price = $58.50 - $17.55 = $40.95

Sally should buy the coffeemaker at Target.

YOUR TURN

9. "Shoes For Less" is having a big sale this weekend. You want to buy two pairs of shoes. Each pair of shoes costs $70. One offer states, "Buy one pair of shoes, get the second pair at half price." Another promotion states, "Buy two pairs of shoes and receive 30% off each pair." Which offer is the best buy?

Offer 1:

"Buy one pair of shoes, get the second pair for half price."

Cost of first pair: ___________________

Cost of second pair is one-half the cost of the first pair: ___________

Total cost of both pairs: ____________________

YOUR TURN

Offer 2:

"Buy two pairs of shoes at 30% off each pair"

Figure the cost of one pair and then multiply by 2 to get the cost for two pairs.

amount of discount = discount rate • original price

amount of discount = ________ **•** ______________

amount of discount =

sale price = original price – amount of discount

sale price =

cost of two pairs = 2 • $ _______________ = _______________

Offer 1: 2 pairs cost $ __________ **Offer 2: 2 pairs cost $** __________

________________ **is the best buy.**

YOUR TURN **Answers to Section 3.12.**

1. sales tax: $7.50
2. rate = 4.5 %
3. sales tax = $34.08, total price = $460.08
4. sales tax = $5.40, total price = $95.40
5. $5,006.50
6. 11 %
7. discount: $120 sale price: $680
8. $101.50
9. Offer 1: $105, Offer 2: $98, "Buy two pairs of shoes at 30% off each pair."

Complete MyMathLab Section 3.12 Homework in your Homework notebook.

3.13 Percent and Problem Solving: Percent 3.13

Interest is money charged for using other people's money. You are *paid* interest when you invest money. You *pay* interest when you borrow money. The original amount borrowed, loaned or invested is known as the principal. The interest rate is the rate at which the interest is calculated. Unless the problem tells you otherwise, the rate is understood to be *per year*. Simple interest is interest that is computed on the original principal and follows the formula stated below:

Simple Interest = Principal • Rate • Time

Simple Interest = P • R • T

EXAMPLE 1

Find the simple interest after 4 years on $1050 at an interest rate of 7.5%.

$\textit{Simple Interest} = P \cdot R \cdot T$	Start with the formula.
$\textit{Simple Interest} = \$1050 \cdot 0.075 \cdot 4$	Be sure to change the % to a decimal.
$I = \$315$	Multiply.

EXAMPLE 2

John Blue borrowed $1800 for 9 months at a simple interest rate of 18%. How much interest did he pay?

$\textit{Simple Interest} = P \cdot R \cdot T$

$\textit{Simple Interest} = 1800 \cdot 0.18 \cdot \frac{9}{12}$

$\textit{Simple Interest} = \$243$

Since there are 12 months in a year, we must find what part of a year 9 months is.

$$9 \text{ months} = \frac{9}{12} \text{ year}$$

Don't worry about reducing the fraction since we are going to use our calculator!

Don't forget you can use the fraction key. Simply multiply

1800 X 0.18 X 9 a b/c 12 = *answer.* 243

YOUR TURN

1. A money market fund advertises a simple interest rate of 5%. Find the interest earned on an investment of $3000 for 5 months.	2. A company borrows $120,000 for 6 years at a simple interest rate of 11.5%. Find the interest paid on the loan.
If time is given in months, don't forget to put the number over 12.	
$Interest = p \bullet r \bullet t$	$Interest = p \bullet r \bullet t$
= ________ • ________ • ________	= ________ • ________ • ________
= ____________	= ____________

When someone borrows money, they *pay* back the original money borrowed, the principal, plus any interest. When someone invests money, they *receive* the original amount invested, the principal, plus any interest. The total amount is the sum of the principal and the interest.

Total Amount = principal + interest

EXAMPLE 3

If $5000 is borrowed at a simple interest rate of 15% for 8 months, find the interest and the total amount paid.

Simple Interest = ***P • R • T*** — Start with the interest formula.

$$\textit{Simple Interest} = \$5000 \bullet 0.15 \bullet \frac{8}{12}$$

Put the time over 12 since it is given in months.

Simple Interest = $500

Total Amount = principal + interest

Total Amount = $5000 + $500 — Add the interest to the original amount borrowed.

Total Amount = $5500

EXAMPLE 4

A money market fund advertises a simple interest rate of 9%. Find the total amount received on an investment of $7000 for 5 years.

Simple Interest = ***P*** • ***R*** • ***T*** — Start with the interest formula.

Simple Interest = $7000 • 0.09 • 5

Simple Interest = $3150

If time is given in years, we plug that number in the formula. Put the number over 12 only when time is given in months.

Total Amount = principal + interest

Total Amount = $7000 + $3150

Total Amount = $10,150

YOUR TURN

Courtesy of Fotolia

Courtesy of Fotolia

3. Janet borrowed $65,000 as a down payment on a house. If the simple interest rate on a 30-year loan is 10.25%, find the total amount paid on the loan.	4. Piedmont Bank advertises a money market account at a simple interest rate of 9%. Find the total amount received if $5,000 is invested for 15 months. (If time is given in months, don't forget to put the number over 12.)
Interest = $p \cdot r \cdot t$	*Interest* = $p \cdot r \cdot t$
= ________ • ________ • ________	= ________ • ________ • ________
= ____________	= ____________
Total Amount = **principal** + **interest**	Total Amount = **principal** + **interest**
= ________ + ________	= ________ + ________
= ____________	= ____________

3.13 Percent and Problem Solving: Percent 3.13

When you borrow money from a bank, you do not pay the total amount back at one time. The bank will divide your payment into monthly payments to make the amount more manageable. To calculate a monthly payment, a bank figures the total amount you are going to owe and then divides by how many months you are going to borrow the money.

$$\text{monthly payment} = \frac{\text{principal} + \text{interest}}{\text{total number of payments}}$$

EXAMPLE 5

A college student borrows $12,000 for 4 years to pay for school. If the interest is $1928, find the monthly payment. Round to the nearest cent.

First, determine how many payments this student will have. The loan is for 4 years. Since there are 12 months per year, the number of payments will be 4 (12) or 48 payments.

$$\text{monthly payment} = \frac{\text{principal} + \text{interest}}{\text{total number of payments}}$$

$$\text{monthly payment} = \frac{\$12{,}000 + \$1928}{48}$$

$$\text{monthly payment} = \frac{\$13{,}928}{48}$$

$$\text{monthly payment} \approx \$290.17$$

The monthly payment is $290.17

YOUR TURN

5. Find the monthly payment on a $3000 loan for 3 years if the interest on the loan is $1123.58. Round to the nearest cent.	6. A college student borrows $1500 for 6 months to pay for a semester of school. If the interest is $61.88, find the monthly payment. Round to the nearest cent.

3.13 Percent and Problem Solving: Percent **3.13**

YOUR TURN **Answers to Section 3.13.**

1. **$62.50**
2. **$82,800**
3. **$264,875**
4. **$5562.50**
5. **$114.54**
6. **$260.31**

Complete MyMathLab Section 3.13 Homework in your Homework notebook.

DMA 030 TEST: Review your notes, key concepts, formulas, and MML work. ☺

Student Name: ______________________________ **Date:** ______________

Instructor Signature: ______________________________

DMA 040

DMA 040

EXPRESSIONS, LINEAR EQUATIONS, AND INEQUALITIES

This course provides a conceptual study of problems involving linear expressions, equations, and inequalities. Emphasis is placed on solving contextual application problems. Upon completion, students should be able to distinguish between simplifying expressions and solving equations and apply this knowledge to problems involving linear expressions, equations, and inequalities.

<u>You will learn how to</u>

- Differentiate between expressions, equations, and inequalities
- Simplify and evaluate, when appropriate, expressions, equations, and inequalities
- Effectively apply algebraic properties of equality
- Correctly represent the solution to an inequality on the number line
- Represent the structure of application problems pictorially and algebraically
- Apply effective problem solving strategies to contextual application problems
- Demonstrate conceptual knowledge by modeling and solving applications using linear equations and inequalities

Courtesy of Fotolia

<u>Application Types - Problems finding:</u>
Unknown Numbers
Board/String Piece lengths
Complementary/Supplementary Angle measures
Consecutive Numbers
Perimeter, Circumference and Area
Cost, Discount, Mark-up Percent Increases

We will answer the question,
"How do I balance an equation to
find unknown values?"

Courtesy of Shutterstock

4.01 Introduction to Variables, Algebraic Expressions, and Equations 4.01

VARIABLES

Variables are what make *algebra* different from other kinds of mathematics like arithmetic.

In *arithmetic*, we deal with *numbers*, and each number has a value, which is itself. And if we want to be more complicated, we can make **expressions**, like: $5 + 2$

What is the value of this expression? (This isn't a trick question.) Answer: 7.

So what is a variable? Below is one simple ***definition***.

A variable is something which has a value, but we haven't decided yet what the value is going to be.

Because we haven't yet decided what the value of a variable is, we can't write down the value, so we have to write the variable some other way. A **variable** is a symbol, usually a letter such as x, y, or z, used to represent an unknown number in an expression or an equation. The value of this number can vary or change. For this reason, we cannot use numbers to represent variables because the value of a number does not change.

EXAMPLE 1

Cara is working on her algebra homework and is trying to write an expression. The problem concerns interest and she needs a variable to represent the number of years. Which of the following cannot be used to represent the variable in this problem and why?

$X, x, y, n, 7, A, 4$

Cara could use all of the letters, but not the numbers 4 and 7. We cannot use numbers to represent variables because the value of a number does not change.

Algebraic Expression

An algebraic **expression** is a combination of numbers and/or variables which represent known or unknown values.

Here is an example of an expression containing the variable x:

$x + 6$

So what is the value of $x + 6$?

The basic answer is that it is 6 more than whatever x is. But since we haven't yet decided what the value of x is, we can't know what the value of $x + 6$ is.

4.01 Introduction to Variables, Algebraic Expressions, and Equations 4.01

Algebraic expressions can contain one or more variables which *are connected by plus and minus signs*. Note that an expression is only a phrase, not the whole sentence, so it doesn't include an equal sign! Here are a few examples of algebraic expressions:

$$x+5, \quad 2m-9, \quad 2x^2+3y, \quad 5(k+7), \quad mn$$

If two variables or a number and a variable are next to each other, with no operation sign between them, the operation is multiplication.

$2m-9 \;=\; 2 \bullet m - 9$ = the product of 2 and m subtract 9

$2x^2+3y \;=\; 2 \bullet x^2 + 3 \bullet y$ = the product of 2 and x^2 plus the product of 3 and y

$5(\boldsymbol{k}+7) \;=\; 5 \bullet (\boldsymbol{k}+7)$ = the product of 5 and $\boldsymbol{k}+7$

$mn \;=\; m \bullet n$ = the product of m and n

If a variable has an exponent attached to it, the variable is a base while the exponent tells us how many bases to multiply together.

$4^3 = 4 \bullet 4 \bullet 4$

$2^3 \bullet 5^4 = 2 \bullet 2 \bullet 2 \bullet 5 \bullet 5 \bullet 5 \bullet 5$

$\boldsymbol{x}^3 = \boldsymbol{x} \bullet \boldsymbol{x} \bullet \boldsymbol{x}$

$\boldsymbol{x}^3 \boldsymbol{y}^5 = \boldsymbol{x} \bullet \boldsymbol{x} \bullet \boldsymbol{x} \bullet \boldsymbol{y} \bullet \boldsymbol{y} \bullet \boldsymbol{y} \bullet \boldsymbol{y} \bullet \boldsymbol{y}$

Evaluating Expressions

Evaluating expressions is very much like order of operations. The expressions we will work with will contain variables. Remember variables are representations for unknown numbers. When evaluating expressions, you are given an expression with variables and values for the variables. Substitute, or replace, the value in place of that variable and then follow the order of operations. You will be more successful if you put parentheses around the number you are substituting.

> Important note: You cannot solve expressions. You can add, subtract, multiply, divide, factor, and evaluate expressions, but not solve them.

EXAMPLE 2

Find the value of each algebraic expression if $x = 5$. Another way to say this is to evaluate the expressions for the given value. (This means plug-in/replace the variable with the number. Then simplify.

a. $7x$	b. $2x^2$	c. $6x - 2$
$= 7(5)$ $= 35$	$= 2 \cdot (x)^2$ $= 2 \cdot (5)^2$ $= 2 \cdot 25$ $= 50$	$= 6(5) - 2$ $= 30 - 2$ $= 28$

EXAMPLE 3

Evaluate $\frac{x + 2y}{z}$ **for $x = 2, y = 3$, and $z = 2$.**

$$\frac{x + 2y}{z} = \frac{2 + 2(3)}{2} = \frac{2 + 6}{2} = \frac{8}{2} = 4$$

Evaluate **means "plug in the number" for the variable.**

EXAMPLE 4

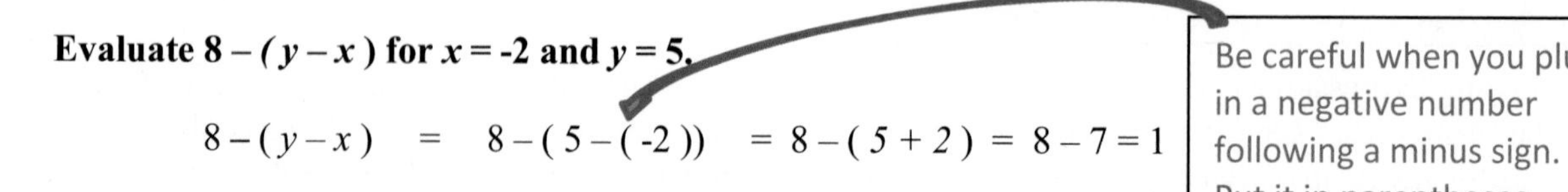

Evaluate $8 - (y - x)$ for $x = $ -2 and $y = 5$.

$$8 - (y - x) = 8 - (5 - (-2)) = 8 - (5 + 2) = 8 - 7 = 1$$

Be careful when you plug in a negative number following a minus sign. Put it in parentheses.

EXAMPLE 5

Evaluate $2m^2 + 6n$ **for $m = 7$ and $n = 3$.**

$$2 \cdot (m)^2 + 6(n) = 2 \cdot (7)^2 + 6(3) = 2 \cdot (49) + 18 = 98 + 18 = 116$$

YOUR TURN

Evaluate the following expressions for $x = 7, y = -4$, and $z = 2$. Plug in the number for the variable.

1. $2(x - y)$	**2.** $4x^2 - 2z$	**3.** $\frac{6x + 5y}{z}$

4.01 Introduction to Variables, Algebraic Expressions, and Equations 4.01

EQUATIONS EQUATIONS EQUATIONS EQUATIONS

An equation is a mathematical statement containing an equal sign that states that two expressions are equal. It is extremely important that you know the difference between an equation and an expression. Remember, we do not solve expressions. We *simplify expressions* and **solve equations**!

Expressions: $2x - 7$ or $7 + 2(3x - 6)$

Equations: $\mathbf{3x = 12}$ or $\mathbf{2x - 7 = 7 + 2(3x - 6)}$

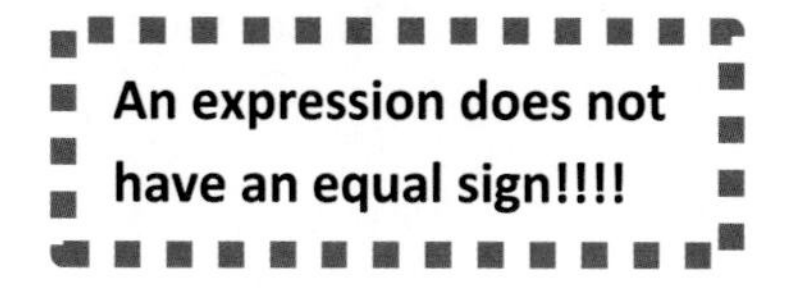

An equation has an equal sign.

First, let's learn how to see if a number is a solution of an equation.

Deciding if a number is a solution of an equation is very similar to evaluating expressions. Simply substitute the proposed solution for the variable and follow order of operations. Then check to see if each side of the equation simplifies to the same thing. Let's try it!

Is $y = 7$ a solution to the equation $y - 5 = 12$?

$y - 5 = 12$

$(7) - 5 = 12$

$2 = 12$

Replace y with 7 and then simplify the left side of the equation.

Since $2 \neq 12$, then **7 is not a solution of the equation.**

Determine if $x = 5$ is a solution of $4x + 5 = 6x - 5$.

$4x + 5 = 6x - 5$	Replace the x with 5 on both sides of the equation.
$4(5) + 5 = 6(5) - 5$	Now follow order of operation to simplify each side. Do the multiplications first.
$20 + 5 = 30 - 5$	Now subtract.
$25 = 25$ YES!	**Thus 5 is a solution of the equation.**

EXAMPLE 6

Determine which numbers in the set are solutions of the equations. $2(n-3)=8$; $\{5, 7, 12\}$

We will need to "plug-in" each number into the equation and see if we get a true statement.

Is 5 a solution?	Is 7 a solution?	Is 12 a solution?
$2(n-3)=8$ $2(5-3)=8$ $2(2)=8$ $4=8$	$2(n-3)=8$ $2(7-3)=8$ $2(4)=8$ $8=8$	$2(n-3)=8$ $2(12-3)=8$ $2(9)=8$ $18=8$
False Statement 5 is NOT a solution.	TRUE Statement 7 is a solution.	False Statement 12 is NOT a solution

YOUR TURN

Decide whether the number is a solution of the given equation.

4. Is 20 a solution for $2n-30=10$?	5. Is 1 a solution for $6x+3 = 4x + 17$?
6. Determine which numbers in the set are solutions of the equations. $5(n-1)=40$; $\{3, 5, 9\}$	

Translating English into Algebra

When translating from English into ***algebraic expressions***, the list of key words below may come in handy.

Add	**Subtract**	**Multiply**	**Divide**	**Equals**
Sum	***Difference***	***Product***	***Quotient***	=
Total	***Less than***	Twice	Half	Is
More than	***Subtracted from***	Half	Divided by	Is the same as
Increase	Decrease	Times	***Divided into***	Yields
Increased by	Decreased by	% of	Ratio	Gives
Plus	Minus	of	Per	Results in

- Be sure to translate word for word!
- When you see **sum, difference, product or quotient**, you will translate the "and" as the proper operation sign.
- The phrases that are highlighted above require that you switch the numbers around! (See example 7c and 7e which follow.)
- Use a variable for the unknown number. You get to choose the variable to use!

EXAMPLE 7

Translate into an algebraic expression.

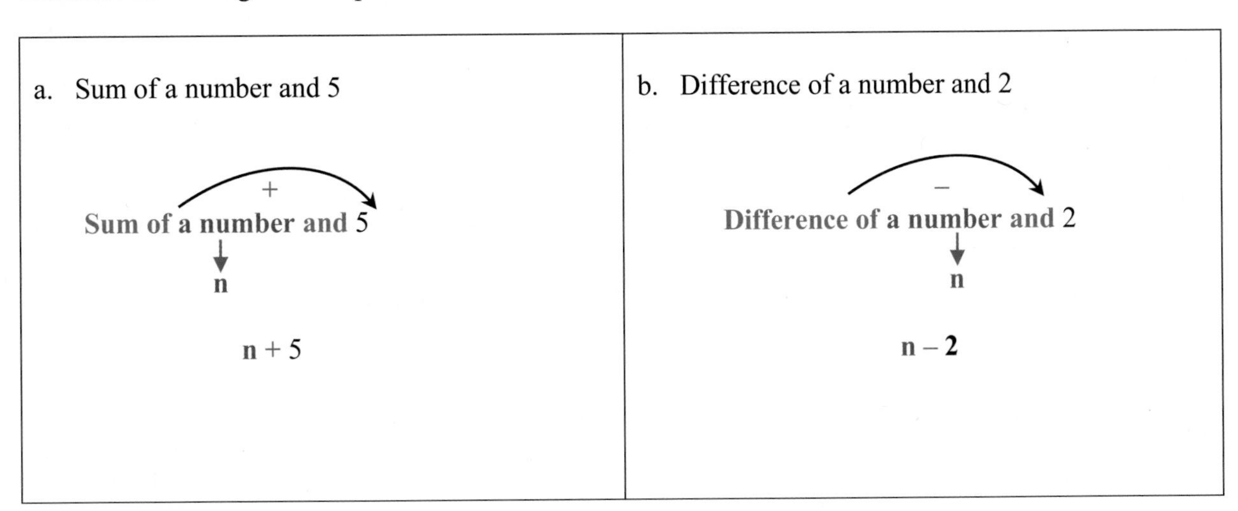

4.01 Introduction to Variables, Algebraic Expressions, and Equations 4.01

EXAMPLE 7 CONTINUED

c. Three less than a number

Remember that "**less than**" means to **subtract** and to **switch the numbers** around.

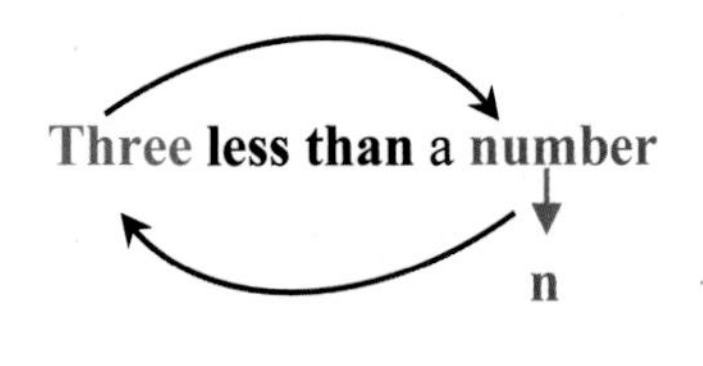

n – 3

d. **Twice** a number, **increased by** 1

$2 \quad \bullet \quad n \qquad + \qquad 1$

$2 \bullet n + 1$

$2n + 1$

e. 15 **subtracted from** a number

Treat "subtracted from" just like you do "less than".

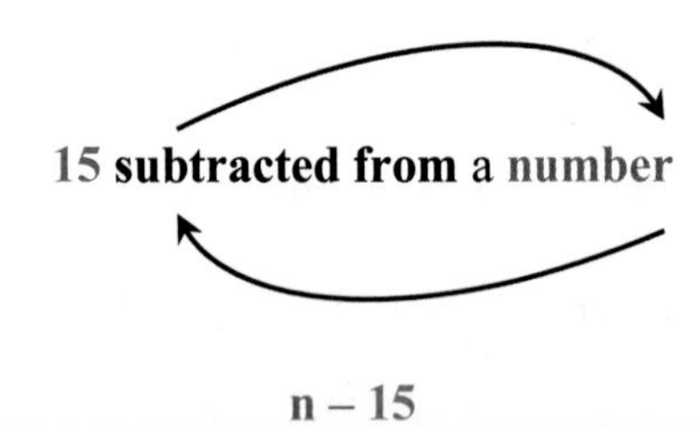

n – 15

f. Product of nine and a number

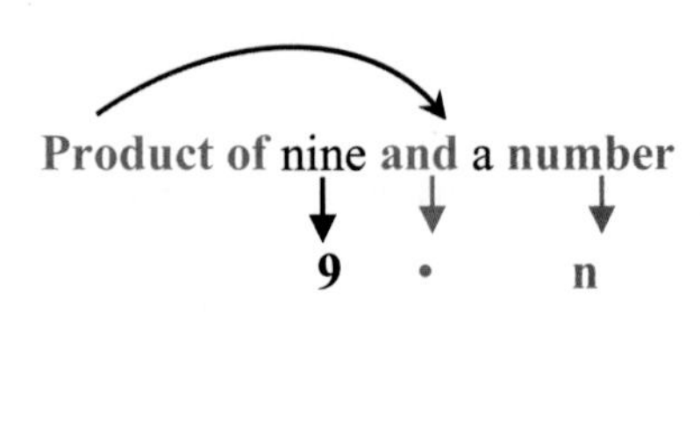

9n

YOUR TURN

Translate into an algebraic expression using *x* as the variable.

7. Twice a number	8. Six less than a number
9. Sum of twice a number and three	10. Difference of seven and a number
11. Four subtracted from three times a number	12. Ten added to three times a number

4.01 Introduction to Variables, Algebraic Expressions, and Equations 4.01

YOUR TURN **Answers to Section 4.01.**

1. 22	2. 192	3. 11	4. Yes	5. No
6. {9}	7. 2x	8. x – 6	9. 2x + 3	10. 7 – x
11. 3x – 4	12. 3x + 10			

Complete MyMathLab Section 4.01 Homework in your Homework notebook.

4.02 Properties of Real Numbers 4.02

In this section, we will look at the properties of real numbers. In working with numbers, you may have noticed some patterns in the way the numbers behave. For example, if you add $2 + 3$, you will get the same answer if you add $3 + 2$. Although, we don't know the name of the property that tells us we can do this, we know it is true. Let's look at some of these properties and call them by name.

COMMUTATIVE PROPERTIES

The word "*commute to*" means to go back and forth. Students commute to school. If a student travels from home to school and follows the same route from school to home, he will travel the same distance each time. The commutative properties say that if two numbers are added or multiplied in any order, they give the same results.

Courtesy of Fotolia

The Commutative Properties of Addition and Multiplication

$a + b = b + a$ *and* $ab = ba$

The Commutative Property states that changing the **ORDER** of two numbers which are added or multiplied does NOT change the value of the answer.

Notice that subtraction and division are not included in this property. Is subtraction commutative? If we change the order of the two numbers we are subtracting, will we get the same answer?

Is $7 - 2 = 2 - 7$? NO!!!

Therefore, subtraction is NOT commutative!

The same is true for division: $10 \div 5 \neq 5 \div 10$. Division is NOT commutative!!

EXAMPLE 1

Use the commutative property to rewrite each statement.

a. $x + 5 =$ _____________ b. $3 \bullet x =$ _________

Answer: $x + 5 = 5 + x$ *Answer:* $3 \bullet x = x \bullet 3$

The commutative property changes the ORDER.

4.02 Properties of Real Numbers 4.02

ASSOCIATIVE PROPERTIES

When we *associate* one object to another, we tend to think of those objects as being grouped together. A group of 5 players *associate* together and make one basketball team. The associative property states that when we add or multiply three numbers, we can group them in any way and get the same answer. Or in other words, the addition or multiplication of a set of numbers is the same regardless of how the numbers are grouped. Parentheses indicate the numbers that are considered one unit. Hence, the numbers are '*associated*' together.

The Associative Properties of Addition and Multiplication

$(a+b)+c = a+(b+c)$ *or* $(a \cdot b) \cdot c = a \cdot (b \cdot c)$

The Associative property states that changing the **GROUPING** of numbers which are added or multiplied does NOT change the value of the answer.

Both the commutative and associative properties can be used at the same time, but they do NOT work for subtraction or division!

EXAMPLE 2

Use the associative property to rewrite each statement and simplify if possible.

a. $(x+5)+9 =$ ____________

$(x+5)+9 = x+(5+9) = x+14$

b. $3 \cdot (4 \cdot x) =$ ________

$3 \cdot (4 \cdot x) = (3 \cdot 4) \cdot x = 12x$

The associative property changes the GROUPING.

YOUR TURN

Determine whether each statement is true by the commutative property or the associative property. Hint: Look at what changes from the left side of the equal sign to the right side of the equal sign. If the **order** changed, it is the commutative property. If **grouping** changed, it is the associative property.

1. $(2+4)+5 = 2+(4+5)$	2. $5(4 \cdot 2) = 5(2 \cdot 4)$
3. $(7+10)+4 = (10+7)+4$	4. $(8+1)+7 = 8+(7+1)$

4.02 Properties of Real Numbers 4.02

IDENTITY PROPERTIES

The Identity Property for Addition tells us that when you add 0 to a number its value does not change. The number 0 is called the identity element for addition.

The Identity Property for Multiplication tells us that when you multiply by 1 the number stays the same. The number 1 is called the identity element for multiplication.

The Identity Properties

0 is the Identity for Addition: $a + 0 = a$ and $0 + a = a$

1 is the Identity for Multiplication: $a \bullet 1 = a$ *and* $1 \bullet a = a$

INVERSE PROPERTIES

The Inverse Properties of Addition and Multiplication are connected to the identity properties mentioned above. If you add a number to another number and end up with the additive identity which is 0, then you have added the additive inverse of that number. If you multiply and end up with the multiplicative inverse which is 1, then you have multiplied by the multiplicative inverse of that number.

The Inverse Properties

Inverse for Addition: $a + -a = 0$ and $-a + a = 0$ (Opposites sum to 0.)

Inverse for Multiplication: $a \bullet \frac{1}{a} = 1$ *and* $\frac{1}{a} \bullet a = 1$ (Reciprocals multiply to 1.)

When we add the additive inverse of a number we must end up the additive identity which is 0. So, what could you add to 3 to end up with 0? In other words: $3 + \_\_\_\_ = 0$. (This number would be the additive inverse of 3.) $3 + -3 = 0$. -3 is the ADDITIVE INVERSE of 3.

What could you add to -6 and get 0, or $-6 + \_\_\_\_\_ = 0$. Of course, we realize $-6 + 6 = 0$. So, 6 is the ADDITIVE INVERSE of -6.

In general, the opposite of a number is its additive inverse. So, often we will interchange the words "additive inverse" or "opposite of a number ". These two mean the same thing.

EXAMPLE 3

What is the relationship between the numbers which sum to zero?

$$4 + -4 = 0$$

Additive Inverses add up to 0.

$$\frac{1}{2} + -\frac{1}{2} = 0$$

Additive Inverses are opposites and have different signs.

Recall the definition of reciprocals in Module 2.

Two numbers whose product is 1 are called reciprocals.

$$3 \bullet \frac{1}{3} = 1 \qquad \frac{3}{4} \bullet \frac{4}{3} = 1$$

Notice that when you multiply reciprocals together, the product is the identity number 1. We have already stated that if you multiply and end up with the multiplicative inverse which is 1, then you have multiplied by the multiplicative inverse of that number.

So, the reciprocal of a number is its multiplicative inverse. Often we will interchange the words "reciprocal" or "multiplicative inverse ". These two mean the same thing.

EXAMPLE 4

Find the multiplicative inverse of each number.

a. 4

$$4 \bullet \frac{1}{4} = 1$$

$\frac{1}{4}$ is the multiplicative inverse

b. $-\frac{5}{6}$

Multiplicative Inverses are reciprocals and have the same signs.

$$-\frac{5}{6} \bullet -\frac{6}{5} = 1$$

$-\frac{6}{5}$ is the multiplicative inverse

Multiplicative Inverses multiply to give 1.

YOUR TURN

Find the additive inverse or multiplicative inverse.

5. Find the additive inverse of 8.	6. Find the multiplicative inverse of -5.
7. Find the additive inverse of $\frac{1}{2}$.	8. Find the multiplicative inverse of $\frac{4}{5}$.

DISTRIBUTIVE PROPERTY

The meaning of the word "distribute" means to give out to several. An important property of real numbers involves this idea. Look at the following two expressions:

$2(3 + 8) = 2(11) = 22$

$2(3) + 2(8) = 6 + 16 = 22$

Both expressions equal 22

Since both expressions equal 22, $2(3 + 8) = 2(3) + 2(8)$.

This result is an example of the distributive property. In the example, 2 is distributed to the 3 and 8. The distributive property says that multiplying a number by a sum of numbers, gives the same answer as multiplying and then adding the two products.

Distributive Property

$a(b + c) = ab + ac$

Multiply the number outside the parentheses by each of the numbers inside the parentheses and then add them together!

EXAMPLE 5

Use the distributive property to simplify each expression.

a. $\mathbf{2}(x+2)$ Multiply both terms inside the parentheses by 2.

$(\mathbf{2})x + (\mathbf{2})2$

$\mathbf{2x + 4}$

b. $\mathbf{-2}(3v + 4)$ Multiply every term in the parentheses by **–2.**

$(\mathbf{-2})\,3v + (\mathbf{-2})\,4$

$(-6v) + (\mathbf{-8})$ **This should change the signs of the terms inside the parentheses.**

$\mathbf{-6v - 8}$

c. $\mathbf{-}(5z^2 + 3z - 6)$ The "-" out front represents "-1." Multiply every term in the parentheses by **–1.**

$(\mathbf{-1})\ 5z^2 + (\mathbf{-1})\ 3z - (\mathbf{-1})\ (6)$

$\mathbf{-5z^2 - 3z + 6}$ **This should change the signs of the terms inside the parentheses.**

d. $-\frac{3}{4}(16a-12c)$ Multiply the fraction by every term in parentheses.

$\left(-\frac{3}{4}\right)\cdot 16a - \left(-\frac{3}{4}\right)\cdot 12c$

$\left(-\frac{3}{\cancel{4}}\right)\cdot \overset{4}{\cancel{16}}a - \left(-\frac{3}{\cancel{4}}\right)\cdot \overset{3}{\cancel{12}}c$ Remember, you can divide out common factors (reduce).

$(-12a) - (-9c)$ Simplify the result.

$\mathbf{-12a + 9c}$

YOUR TURN

Use the distributive property to write each expression without parentheses.

9. $2(x-7)$	10. $-5(3y-2)$	11. $-(3m-6)$
12. $-5(-3+2z)$	13. $-\frac{5}{6}(12a-18b+6c)$	14. $-3(x-3y+4z)$

YOUR TURN Answers to Section 4.02.

1. Associative 2. Commutative 3. Commutative

4. Commutative and Associative 5. -8 6. $-\frac{1}{5}$ 7. $-\frac{1}{2}$ 8. $\frac{5}{4}$

9. 2x – 14 10. -15y + 10 11. – 3m + 6 12. 15 – 10 z

13. -10a + 15b -5c 14. – 3x + 9y – 12z

Complete MyMathLab Section 4.02 Homework in your Homework notebook.

4.03 Simplifying Expressions 4.03

Before we look at simplifying expressions, we will need to look at some vocabulary we will be using.

Term

number, variable, or the product of a number and variable

Examples
3, $-7x$, $-4x^2y^5$

Like terms

terms with the same variable to the same power

Examples
$2x$ and $8x$
$5xy^2$ and $-3xy^2$
$5n^2m^3$ and $-n^2m^3$

Numerical Coefficient

the number being multiplied by the varialble.

Examples
Coefficient of 6x is 6.
Coefficient of x is 1.

YOUR TURN

1. Fill in the chart below.

Terms	Coefficients	Like or Unlike. Explain why if they are unlike.
$7w$, $4w$		
$5xy$, $5x$		
$2ab$, $-ab$		
$3x^2$ and $-3y^2$		

To simplify an expression , we will combine like terms using the distributive property. Remember, like terms are terms with the same variable and the same exponent. The coefficient can be different but the variable(s) and the exponents must be the same.

EXAMPLE 1

Simplify the following expressions by combining like terms.

a. $2u + 3u = (2+3)u = 5u$

b. $10y^2 + y^2 = (10+1)y^2 = 11y^2$

c. $7x^2 + 2x - 3x = 7x^2 + (2-3)x = 7x^2 - 1x$ or $7x^2 - x$

d. $-3k - 5k + 8 = (-3-5)k + 8 = -8k + 8$

4.03 Simplifying Expressions 4.03

After studying the previous examples, we soon realize that we can combine like terms without having to show the middle step. Let's look at two more examples and work them a little differently. When we look at an expression that needs to be simplified, we will circle the like terms making sure to circle the sign in front of the term as part of the term.

EXAMPLE 2

Simplify the following expressions.

a. $5x - 5y - 3x - 16y$

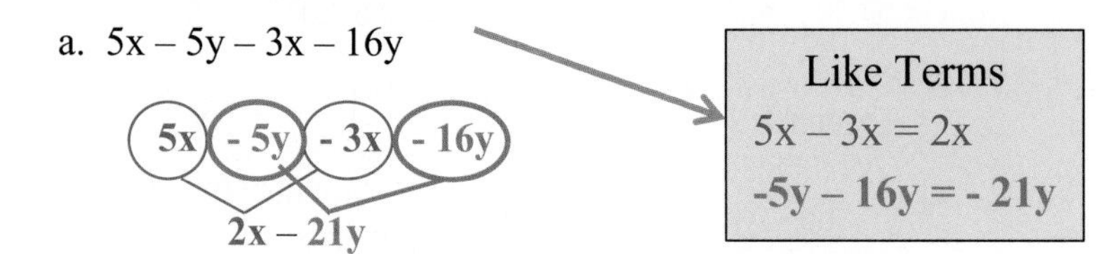

b. $-5a \; -3 \; + \; a \; + \; 2$

$-5a \; -3 \; + a \; + 2$

$-4a \; - 1$

Like Terms

$-5a + a = -4a$

$-3 + 2 = -1$

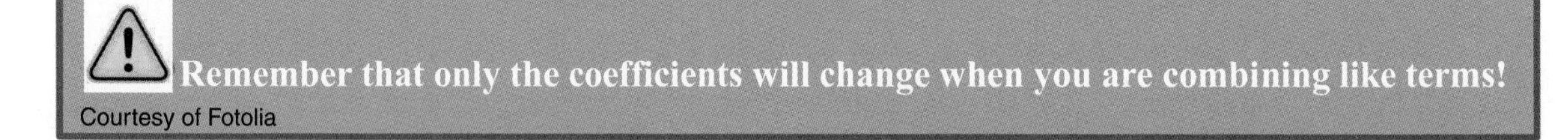

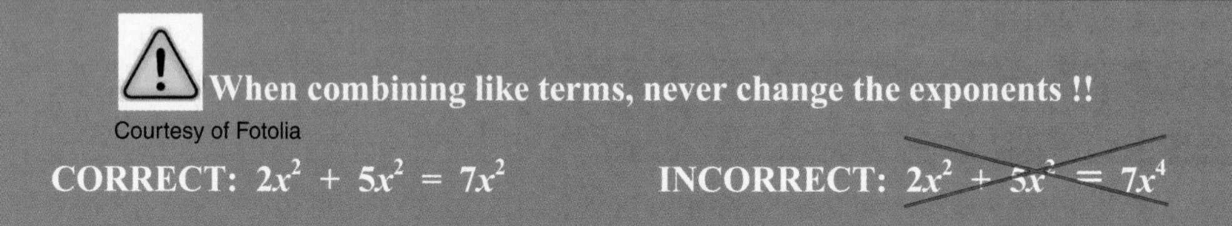

YOUR TURN

Simplify each algebraic expression.

2. $6d - 10 - 4d + 7$

3. $4x - 7x + 3 + 7x - 4$

4. $-4x^2 + 2x^2 - 6$

4.03 Simplifying Expressions 4.03

When simplifying an expression containing parentheses, we will use the distributive property to remove the parentheses and then we will combine like terms.

To Simplify an Expression

1. **Remove parentheses using the Distributive Property.**
2. **Combine like terms.**

EXAMPLE 3

Simplify.

a. $-7(3x-6)+3x$

$-7(3x+-6)+3x$	Rewrite subtraction as addition.
$-7(3x)+-7(-6)+3x$	Multiply each term by -7.
$-21x+42+3x$	
$-21x+3x+42$	Group like terms.
$-18x+42$	Combine like terms.

WATCH OUT! The 2nd # to distribute is –3.

b. $2(4x-4)-3(x-5)$

$8x-8-3x+15$	Apply the distributive property.
$8x-3x-8+15$	Group like terms.
$5x+7$	Combine like terms.

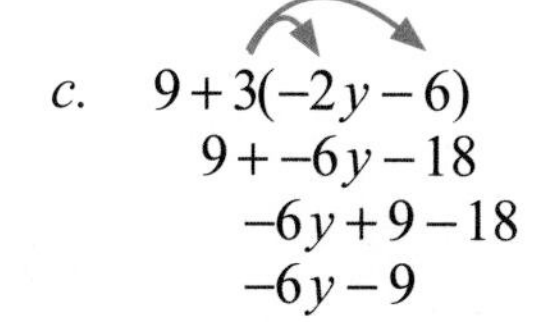

c. $9+3(-2y-6)$

$9+-6y-18$	Apply the distributive property.
$-6y+9-18$	Group like terms.
$-6y-9$	Combine like terms.

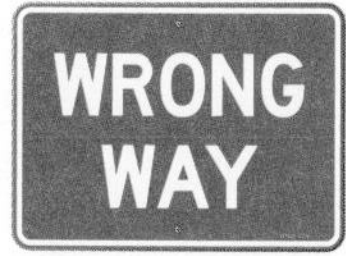

Courtesy of Shutterstock

$9+3(-2y-6)$
$12(-2y-6)$
$-24y-72$

Order of operation tells me that multiplication comes before addition. I cannot add the 9 + 3 before I have multiplied by 3.

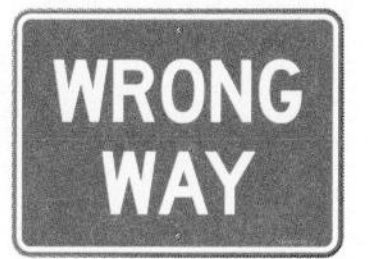

Courtesy of Shutterstock

4.03 Simplifying Expressions **4.03**

EXAMPLE 4

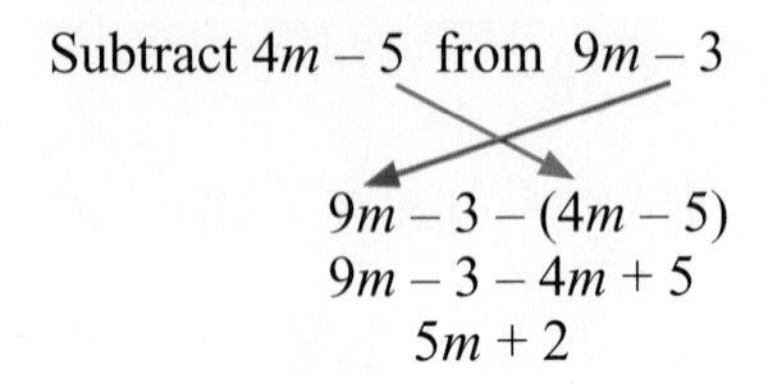

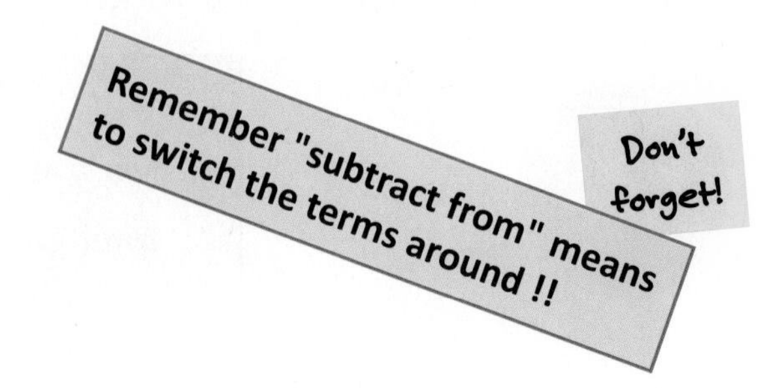

YOUR TURN

Simplify. Watch when distributing negative numbers.

5. $-3(2x-5)-6(x+2)$

6. $2(7w-5)-w$

7. $3(5-2m)-2(m-2)$

8. Subtract $3n+6$ from $2n-3$

TRANSLATING ENGLISH into ALGEBRA, AGAIN !

a. The sum of twice a number and 3. → $2n + 3$ → **"Twice a number" does not require parentheses.**

b. Twice the sum of a number and 3. → $2(n+3)$ → **"Twice the sum" requires parentheses !!!**

c. One-fourth of a number, subtracted from 9. → $9 - \frac{1}{4}n$ → **"Subtracted from" means to switch the order!!**

d. Triple a number, minus the sum of the number and two. → $3n - (n+2)$ → **Keep parentheses around the sum you are subtracting**

e. 44 divided by the sum of a number and 5. → $\frac{44}{n+5}$

YOUR TURN

Translate into an algebraic expression.

9. Three times a number subtracted from 5.

10. Six times the difference of a number and 9.

11. The sum of a number and 6, minus four times a number.

12. Twice a number, increased by three times the number.

13. The sum of twice a number and fourteen.

14. Four times the sum of a number and ten.

YOUR TURN **Answers to Section 4.03.**

1.

Terms	**Coefficients**	**Like or Unlike. Explain why if they are unlike.**
7w , 4w	7,4	Like, same variable & same exponent
5xy, 5x	5,5	Unlike, different variables
2ab, – ab	2,-1	Like, same variables & same exponent
$3x^2$ & $-3y^2$	3,-3	Unlike, different variable but same exponents

2. $2d - 3$ 3. $4x - 1$ 4. $-2x^2 - 6$ 5. $-12x + 3$

6. $13w - 10$ 7. $-8m + 19$ 8. $-n - 9$ 9. $5 - 3n$

10. $6(n - 9)$ 11. $n + 6 - 4n$ 12. $2n + 3n$

13. $2n + 14$ 14. $4(n + 10)$

Complete MyMathLab Section 4.03 Homework in your Homework notebook.

Solving Linear Equations

Courtesy of Shutterstock

We will now begin our study of solving linear equations. This is a very important skill to practice and understand. So, work hard so that you will be able to solve equations with ease.

Equation

An **equation** is a number sentence that says the expressions on either side of the equal sign represent the same number. Equations have "=" signs.

Here are some examples of equations:

$3+2=5$ $\quad$ $14-3=11$ $\quad$ $x+6=13$ $\quad$ $3x-2=5x+6$

Solution of an Equation

Any replacement for the variable that makes an equation true is called a **solution** of the equation. To solve an equation means to find all of its solutions.

UNVEILING THE MYSTERY NUMBERS:

Solving an equation is like a mystery to solve. We will be given a statement such as $\boldsymbol{x}+4=6$. It is our job to find the mystery number, $\boldsymbol{x}$, which will make the statement true. Of course, this one is simple enough that we immediately know that since $\mathbf{2}+4=6$, then $\boldsymbol{x}=2$. Other equations are not as simple so we must develop steps to find the solution.

Courtesy of Shutterstock

4.04 The Addition Property of Equality 4.04

Solving an Equation

Get the **variable** you are solving for alone on **one side** and **everything else** on the **other side** by "undoing" what has been done to the variable. We call this "isolating the variable".

One of the principles that we use in solving equations involves addition. The Addition Property of Equality tells us that , **if two expressions are equal to each other and you add or subtract the exact same thing to both sides, the two sides will remain equal.**

We can relate solving linear equations to a balance scale. If the scales are balanced that indicates that both sides have the same value. In the example below we would know that $a = b$.

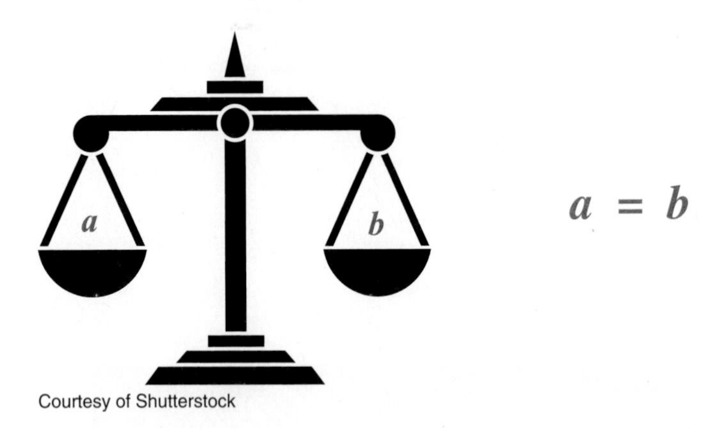

Courtesy of Shutterstock

As long as we add the same value to both sides, we keep the scales balanced! So, if we add ***c*** to both sides, that would tell us that $a + c = b + c$.

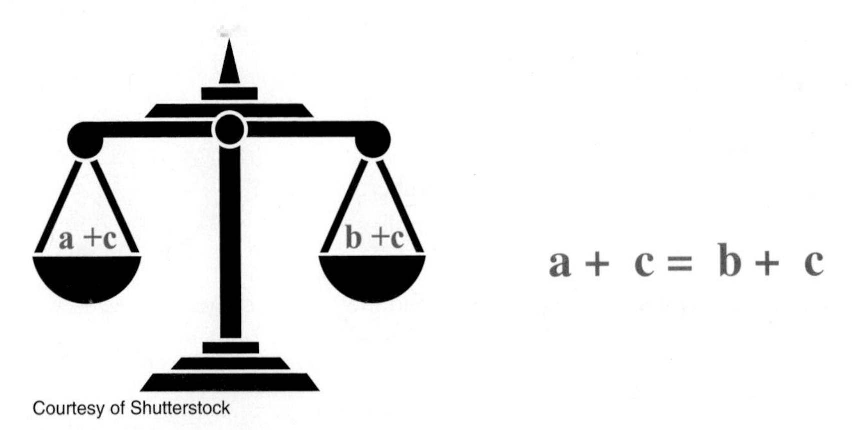

Courtesy of Shutterstock

When we are solving linear equations, we must always keep the equation "balanced". Whatever you do to one side of the equation you must do to the other side of the equation.

4.04 The Addition Property of Equality 4.04

Addition Property of Equality

If a, b, and c are real numbers, and $c \neq 0$, then

$a = b$ and $a + c = b + c$ are equivalent equations.

EXAMPLE 1

Solve $x + 3 = 9$

Remember, when solving a linear equation you need to get the variable you are solving for alone on one side and everything else on the other side.

$$x + 3 = 9$$
$$x + 3 - 3 = 9 - 3$$
$$x + 0 = 6$$
$$x = 6$$

In this equation 3 has been added to the variable, so we should subtract 3 from both sides of the equation.

Subtraction undoes adding.

By subtracting 3 from both sides, we have kept the equation balanced.

CHECK: We should always check our answer to make sure it is a solution. A solution will make the equation true. We will replace x with the number 6 and see if we get a true statement.

$$x + 3 = 9$$
$$6 + 3 = 9$$
$$9 = 9$$

We get a true statement. So, 6 is the solution to $x + 3 = 9$.

EXAMPLE 2

Solve $n - 4 = -5$

$$n - 4 = -5$$
$$n - 4 + 4 = -5 + 4$$
$$n + 0 = -1$$
$$n = -1$$

In this equation 4 has been subtracted from the variable, so we should add 4 to both sides of the equation.

Adding undoes subtracting.

By adding 4 to **both sides**, we have kept the equation balanced.

4.04 The Addition Property of Equality 4.04

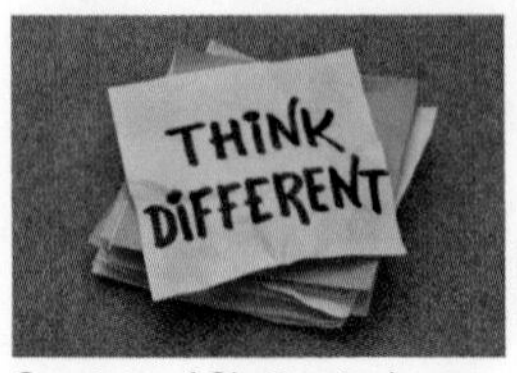

Courtesy of Shutterstock

In the two previous examples the work is written horizontally. Some students will prefer to show their work vertically. Look at the examples below and then decide which way you will choose to show your steps—horizontally or vertically.

horizontal

$$n - 4 = -5$$

$$n - 4 + 4 = -5 + 4$$

$$n = -1$$

vertical

$$\mathrm{n} - 4 = -5$$

$$\begin{array}{rcr} \mathrm{n} - 4 & = & -5 \\ +4 & & +4 \\ \hline & n = -1 & \end{array}$$

EXAMPLE 3

Solve $8 = y + 2$

When you first look at an equation, you should always locate where your variable is first!! That will determine what you do next.

$$\begin{aligned} 8 &= y + 2 \\ 8 - 2 &= y + 2 - 2 \\ 6 &= y + 0 \\ 6 &= y \end{aligned}$$

Notice with this equation that the variable is on the other side of the equal than in our previous examples. It does not matter which side of the "=" the variable is on, our goal is still the same. We need to "isolate" the variable, or in other words, get rid of the number that is on the side with the variable.

Note that equations are **reversible**. That is, if a = b, then b = a is true. In the above example when we solve $8 = y + 2$, we can reverse it and solve $y + 2 = 8$ if we wish. The solution will be the same! Students who feel more comfortable with the variable being on the left side will find this a good option.

$$\begin{aligned} 8 &= y + 2 \\ 8 - 2 &= y + 2 - 2 \\ 6 &= y + 0 \\ 6 &= y \end{aligned}$$

<----- **OR** ----->

$$\begin{aligned} y + 2 &= 8 \\ y + 2 - 2 &= 8 - 2 \\ y + 0 &= 6 \\ y &= 6 \end{aligned}$$

Solving Linear Equations with Variables on Both Sides

Often we will need to solve linear equations where the variable happens to be on both sides of the equal sign. The objective in solving these equations is the same as that of the simpler ones that we have seen – isolate the variable and solve for its value. In order to "isolate" the variable, we will first need to **put all the variables on one side of the equation.**

Variables can be treated just like numbers in an equation. For example, we can add variables to both sides without changing the equation or the values that make it true. We could also subtract a variable from both sides of an equation.

EXAMPLE 4

Solve $4y = 3y + 7$

$$4y = 3y + 7$$

Subtract $3y$ from both sides of the equation.

$$4y - 3y = 3y - 3y + 7$$

$$1y = 7$$

$1y$ is the same as y.

$$y = 7$$

EXAMPLE 5

Solve $2m + 3 = 3m$

$$2m + 3 = 3m$$

$$2m - 2m + 3 = 3m - 2m$$

$$0 + 3 = 1m$$

$$3 = m$$

Remember our goal is to get the variable terms on one side and the constants (numbers) on the other side. Our best move then would be to put the variable terms on the right side since the only constant is on the left. So, we will subtract $2m$ from both sides.

Many times you will encounter equations that have variable terms and constant terms on both sides of the equation. Your goal will be to get all the terms with a variable on one side of the equation and all the terms without a variable on the other side.

EXAMPLE 6

Solve $6n - 7 = 5n - 3$

$$\begin{aligned} 6n - 7 &= 5n - 3 \\ 6n - 5n - 7 &= 5n - 5n - 3 \\ 1n - 7 &= -3 \\ 1n - 7 + 7 &= -3 + 7 \\ n &= 4 \end{aligned}$$

Move the variables to the left by subtracting $5n$ from both sides.

Since we moved variable terms to left, constants must move to the right. So, we will need to move the 7 to the right by adding 7 to both sides.

Sometimes we encounter problems where we need to combine like terms as well. It is always advisable to combine like terms first. Look at examples 7 and 8.

EXAMPLE 7

Solve $2y + 4y = 5y - 6$

$$\begin{aligned} 2y + 4y &= 5y - 6 \\ 6y &= 5y - 6 \\ 6y - 5y &= 5y - 5y - 6 \\ 1y &= -6 \\ y &= -6 \end{aligned}$$

Combine like terms that are on the same side of the equal sign. $2y + 4y = 6y$.

Now solve by getting variables terms to the left.

Subtract $5y$ from both sides.

EXAMPLE 8

Solve $7z + 6 - 5z = 9 + z + 8$

$$\begin{aligned} 2z + 6 &= z + 17 \\ 2z - z + 6 &= z - z + 17 \\ 1z + 6 &= 17 \\ z + 6 - 6 &= 17 - 6 \\ z &= 11 \end{aligned}$$

Combine like terms on each side of the equal sign.

Move the variables to the left by subtracting z from both sides.

Since we moved variable terms to left, constant must move to the right. So, we will need to move the 6 to the right by subtracting 6 from both sides.

Often you will find equations that have parentheses on one or both sides of the equal sign. We will use the distributive property to remove the parentheses. Look at Example 9.

EXAMPLE 9

Solve

$$
\begin{aligned}
-5(n+3) &= -4n - 5 \\
-5n - 15 &= -4n - 5 \\
-5n + 5n - 15 &= -4n + 5n - 5 \\
-15 &= n - 5 \\
-15 + 5 &= n - 5 + 5 \\
-10 &= n
\end{aligned}
$$

Remove ( ) using the distributive property.

Get variable terms on right side by adding 5n.

Get constant terms on left side by adding 5.

Some equations will combine several steps into one equation. Here is the order you should follow in solving more challenging equations.

Step 1

Remove ( ) by using distributive property.

Step 2

Combine like terms on the same side of the = .

Step 3

Get variable terms on one side/numbers on opposite side.

EXAMPLE 10

Solve $-7(w+9)-(7-4w) = -2w$

$$-7(w+9)-(7-4w) = -2w$$
$$-7w - 63 - 7 + 4w = -2w$$
$$-3w - 70 = -2w$$
$$-3w + 3w - 70 = -2w + 3w$$
$$-70 = w$$

Remove ( ) by using Distributive Property.

Combine like terms on each side of the equal.

Add $3w$ to both sides.

YOUR TURN

Solve the following equations. Show all your steps.

1. $-4 = x + 6$

2. $7(y-9) = 8 + 8y$

3. $6x + 2 - 3x = 3 + 2x - 4$

4. $15 - (2 - 3x) = 4x$

5. $-5(w+3) - (4-2w) = -4w$

6. $13x - 9 + 2x - 5 = 12x - 1 + 2x$

WRITING ALGEBRAIC EXPRESSIONS IN APPLICATIONS

We will now practice writing algebraic expressions again to prepare for finding unknown numbers. These algebraic expressions will be used to write equations which we will solve. This is a very important technique to learn when we begin application problems. In general, to solve problems involving sums of quantities, choose a variable to represent one of the unknowns and then **represent the other quantity in terms of the same variable.**

EXAMPLE 11

A 7' rope is cut into 2 pieces. If one piece is "x" feet long, how would you express the length of the other piece?

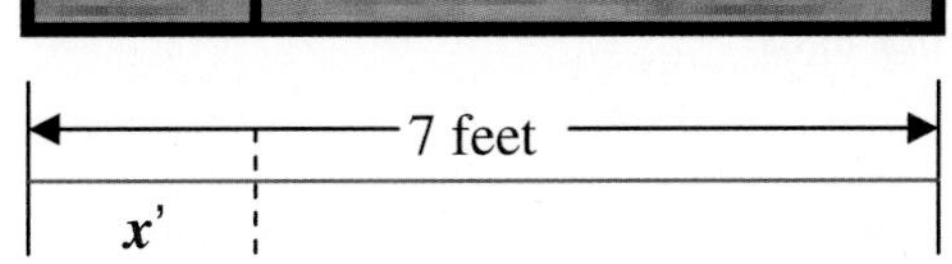

The total length of the rope is 7, then the sum of the two pieces cut would be 7. So, we can think about this as being the sum of two numbers is 7 and one has been given as x.

Therefore, the other unknown would be $7 - x$.

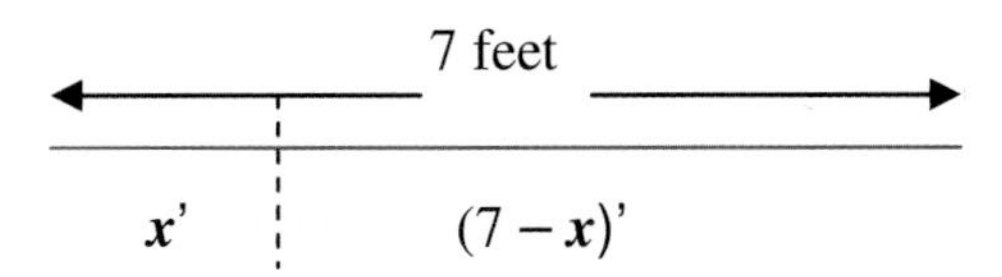

If the sum of two quantities has been given and you are asked to write algebraic expressions for the two quantities, one unknown will always be x, and the other unknown will always be " **the sum** $- x$". Of course, you can use any letter for the variable. For example, if the variable is y, then the other unknown will be "**the sum** $- y$".

EXAMPLE 12

Translate the phrase to an algebraic expression.

Justin had \$40 before spending y dollars on jeans. How much money remains?

Justin has \$40. The sum of the money he spent on jeans and the money he has remaining must add up to \$40. So, this problem can be understood as, the sum of two quantities is \$40. If one unknown is y, find the other unknown.
If the unknown is y, then the other unknown is $40 - y$.

Justin has ($40 - y$) money remaining.

EXAMPLE 13

Two angles add up to 90°. If one angle is x°, how would you express the size of the other angle?

If two angles add up to 90, and
the unknown is x, then
the other unknown is $90 - x$.

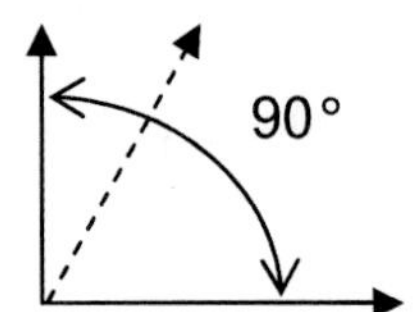

EXAMPLE 14

Mary is n years old. How old will she be in 8 years? How old was she 2 years ago? Write expressions representing her ages.

Courtesy of Shutterstock

Solution:
In 8 years Mary will be 8 more than she is now. If she is n years old , she will be $n + 8$ in 8 years.
Two years ago, Mary was 2 less years than she is now. If she is n years old, two years ago she was $n - 2$.

EXAMPLE 15

Liz rented a truck for a trip. She paid a daily fee of $35 and a mileage fee of 30 cents per mile. Write an expression that represents her total cost when she travels x miles in one day.

Courtesy of Fotolia

Solution:
The cost of the truck is $35 **PLUS** 30 cents per mile. She drives x miles. So, for the total cost we should add $35 to the cost of the mileage. The cost of the mileage would be $0.30 multiplied by the number of miles which is x, or as an expression $0.30x$.
Liz's total cost in terms of x would be $35 + 0.30x$.

EXAMPLE 16

Mike's electricity usage this year decreased by 14% from his electricity use last year, y. Write an expression for his electricity use this year in terms of y.

Solution:
The bill decreased by 14%. His bill this year would be the amount he paid last year decreased by 14 %. The amount he paid last year was y. A decrease would indicate subtraction. To use a percent in an equation, we must convert to a decimal. (14% = 0.14)
Mike's electricity use this year is: bill last year – 14% of bill last year, $y - 0.14y$

4.04 The Addition Property of Equality 4.04

EXAMPLE 17

The price of a rosebush is \$4 less than the price of an azalea. If x represents the price of an azalea, complete the following parts.

a. Represent the price of a rosebush in term of x.

b. Write and simplify an expression that represents the price of 7 rosebushes and 3 azaleas.

Solution:

a. The price of a rosebush is 4 less than the price of an azalea.

x = price of an azalea **$x - 4$ = price of a rosebush**

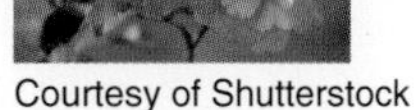

Courtesy of Shutterstock Courtesy of Shutterstock

b. If the cost of one rosebush is $x - 4$, we will need to add 7 of those to find the cost of 7 rosebushes.

$$x-4 + x-4 + x-4 + x-4 + x-4 + x-4 + x-4 = 7x - 28$$

We can do this faster by simply multiplying the cost of one rosebush by 7: $7(x-4) = 7x - 28$

Therefore, the cost of 7 rosebushes is $7x - 28$.

Now we will need to find the cost of 3 azaleas.

If the cost of one azaleas is x, we will need to add 3 of those to find the cost of 3 azaleas.

$$x + x + x = 3x$$

We can do this faster by simply multiplying the cost of one azalea by 3: 3x

Therefore, the cost of 3 azaleas is $3x$.

The total cost can be found by adding the cost of 7 rosebushes and 3 azaleas.

$$7(x-4) + 3x$$
$$7x - 28 + 3x$$
$$10x - 28$$

YOUR TURN

Write each algebraic expression described.

7. A 10-foot cord is cut into two pieces. If one piece is x feet long, express the other length in terms of x.

8. Jane is 6 inches taller than Betty. If Betty's height is represented by x, express Jane's height in terms of x.

9. The cost of a particular Porsche convertible is four times the cost of a Mazda convertible. If the cost of a Mazda convertible is represent by c, express the cost of the Porsche convertible in terms of c.

10. Joe had $90 before spending y dollars on a new sweater. Write an expression in terms of y that represents how much remains?

YOUR TURN **Answers to Section 4.04.**

1. x = -10	2. y = - 71	3. x = - 3	4. x = 13
5. w = 19	6. X = 13	7. 10 – x	8. x + 6
9. 4c	10. 90 - y		

Complete MyMathLab Section 4.04 Homework in your Homework notebook.

Solving an Equation

Multiplication/Division Property of Equality

If a, b, and c are real numbers, and $c \neq 0$, then

$a = b$ *and* $ac = bc$ are equivalent equations.

What this means is that you can multiply or divide both sides of an equation by the same number (a non-zero number) and it will not change the solution. We use this property to solve equations that include multiplication and/or division. You should have all of the variable terms on one side and all of the constant terms on the other side before you begin using this property. Our goal here is to get rid of the coefficient in front of the variable.

EXAMPLE 1

Solve.

$3x = 12$

This equation says that 3 times some number is 12.

$\frac{3x}{3} = \frac{12}{3}$

To find that number we will divide both sides of the equation by 3, the coefficient of the variable.

$1x = 4$

Simplify.

$x = 4$

Now check your answer.

CHECK: $3x = 12$

$3(4) = 12$

$12 = 12$

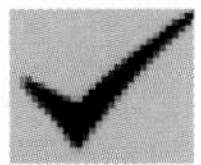

EXAMPLE 2

Solve.

$-5x = 15$

This equation says that -5 times some number is 15.

$\frac{-5x}{-5} = \frac{15}{-5}$

To find that number we will divide both sides of the equation by -5, the coefficient of the variable.

Simplify.

$1x = -3$

Now check your answer.

$x = -3$

4.05 The Multiplication Property of Equality 4.05

Fractional Coefficients

In the next module we will look at a different way to get rid of fractions in an equation. But when we have a one step equation with a fractional coefficient, we will get rid of the fraction by making use of the reciprocal. To get rid of a fractional coefficient, multiply both sides of the equation by the reciprocal of the coefficient. This will give you $1x$, which is just x so you will have completed the solution.

Remember that a number and its reciprocal will have the same sign!

EXAMPLE 3

Solve. $\frac{5}{2}x = 15$

We need the coefficient of x to be 1. Multiply by the reciprocal of the coefficient. The reciprocal of $\frac{5}{2}$ is $\frac{2}{5}$.

$$\frac{2}{5} \cdot \frac{5}{2}x = \frac{2}{5} \cdot 15$$

Simplify.

$$1x = 6$$

EXAMPLE 4

Solve. $-\frac{3}{4} = \frac{2}{5}x$

Multiply both sides by the *reciprocal* of the coefficient (number in front of x). The coefficient of x is $\frac{2}{5}$, whose reciprocal is $\frac{5}{2}$.

$$\frac{5}{2}\left(-\frac{3}{4}\right) = \frac{5}{2}\left(\frac{2}{5}x\right)$$

$$-\frac{15}{8} = 1x$$

$$-\frac{15}{8} = x$$

EXAMPLE 5

Solve. $\frac{x}{3} = -20$

This time we have x divided by 3 so we must multiply both sides by 3.

$$3 \cdot \frac{x}{3} = 3 \cdot -20$$

Simplify.

$$\frac{3}{1} \cdot \frac{x}{3} = 3 \cdot -20$$

$$x = -60$$

Now check your answer.

It is usually easier to **multiply each side if the coefficient of the variable is a fraction**, and **divide on each side if the coefficient is an integer**. For example to solve $-\frac{3}{4}x = 12$, it is easier to multiply by the reciprocal of $-\frac{3}{4}$ which would be $-\frac{4}{3}$ than to divide by $-\frac{3}{4}$.

On the other hand, to solve $-5x = 20$, it is easier to divide by -5 than to multiply by $-\frac{1}{5}$.

EXAMPLE 6

Solve. $5m - 16m = 33$ — Combine like terms on the same side of the equal.

$-11m = 33$ — To isolate the variable, divide by the coefficient -11.

$\frac{-11m}{-11} = \frac{33}{-11}$ — Simplify.

$m = -3$

Avoiding Common Mistakes!

$-x = 20$ Is this finished???

There are many times when solving an equation you will get to a point similar to this, where you have $-x$ equals a number. A common mistake is to think you are finished. No way! We are solving for x, not $-x$. So how do we finish it? Since $-x$ is the same as $-1x$, we can rewrite the equation using $-1x$ instead of $-x$. Divide both sides by -1. You should get $x = -20$.

4.05 The Multiplication Property of Equality **4.05**

YOUR TURN

Solve. Show all your steps.

1. $-5m = -10$	2. $-y = 12$
3. $-5 = \frac{n}{2}$	4. $-\frac{5}{8}x = -40$
5. $-25 = 6m - m$	6. $\frac{3}{2}x = -\frac{7}{8}$

Solving Equations: Using the Addition & Multiplication Properties Together

We will now solve equations that will require applying more than one property. We will always use the Addition Property first to remove addition/subtraction and then we will use the Multiplication Property to isolate the variable.

EXAMPLE 7

Solve. $5x - 3x + 3 = -7 - 2$ Combine like terms on same side of equal.

$$2x + 3 = -9$$
$$\underline{\quad -3 \quad -3}$$
Move the constants to the right hand side by subtracting 3 from both sides.

$$2x = -12$$
Simplify.

$$\frac{2x}{2} = \frac{-12}{2}$$
Isolate the variable by dividing both sides by 2.

$$x = -6$$
Simplify.

EXAMPLE 8

Solve. $-7(x + 2) = 5x - 14$ Remove parentheses by using the Distributive Property.

$$-7x - 14 = 5x - 14$$
$$\underline{-5x \qquad -5x}$$
Move variables on the same side of the = by subtracting $5x$ from both sides.

$$-12x - 14 = -14$$
Simplify.

$$-12x - 14 = -14$$
$$\underline{\quad +14 \quad +14}$$
$$-12x = 0$$
Move the constants to the opposite side of the = by adding 14 to both sides. Simplify.

$$\frac{-12x}{-12} = \frac{0}{-12}$$
$$x = 0$$
Isolate the variable by dividing both sides by – 12. Simplify.

If an equation has parentheses, don't forget to use the distributive property to remove them. Then combine any like terms that are on the same side of the equal sign.

EXAMPLE 9

Solve. $7x - 3 = 8x + 2$

Plan A

$$\begin{array}{rcl} 7x - 3 &=& 8x + 2 \\ -8x & & -8x \\ \hline -1x - 3 &=& 2 \end{array}$$

Move the variable to the left by subtracting 8x from both sides.

Simplify.

$$\begin{array}{rcl} -1x - 3 &=& 2 \\ +3 & & +3 \\ \hline -1x &=& 5 \end{array}$$

Move the constants to the right by adding 3 to both sides.

Simplify.

$$\frac{-1x}{-1} = \frac{5}{-1}$$

Isolate the variable by dividing by the coefficient which is -1.

$$x = -5$$

Solve. $7x - 3 = 8x + 2$

Plan B

$$\begin{array}{rcl} 7x - 3 &=& 8x + 2 \\ -7x & & -7x \\ \hline -3 &=& x + 2 \end{array}$$

Move the variable to the right by subtracting $7x$ from both sides. This move keeps the variable positive.

Simplify.

$$\begin{array}{rcl} -3 &=& x + 2 \\ -2 & & -2 \\ \hline -5 &=& x \end{array}$$

Move the constants to the right by subtracting 2 from both sides.

Look carefully at the two examples above. These two examples illustrate that you always have two options when solving equations.

- In the first example, we moved the variables to the left and the constants to the right.
- In the second example, we moved the variables to the right and the constants to the left.

Two plans but the same answers. Some students always like to move the variable to the left while other students move the variable to keep it positive. Either way the equation will have the same solution!!!!!

YOUR TURN

Solve. Show all your steps.

7. $6x - 4 = -2x - 10$	8. $4a + 1 + a - 11 = 0$
9. $3(2x + 5) = 8x - 5$	10. $z - 5z = 7z - 9 - z$

Writing Algebraic Expressions Again!

In the following examples we want to look at a particular kind of word problem you will be asked to work. These problems involve consecutive integers. Remember, an integer cannot be a fraction or a decimal. Integers consist of {…-3, -2, -1, 0, 1, 2, 3…}.

CONSECUTIVE INTEGERS

Consecutive Integers are integers that occur in order, one after the other.

-7, -6, -5 2, 3, 4

Page numbers are consecutive integers. Look at this workbook. The pages are numbered 1, 2, 3, 4, 5… If you look at house numbers down a local street you will notice that one side of the street has consecutive odd house numbers such as 345, 347, 349 while the other side has consecutive even house numbers such as 346, 348, 350.

EXAMPLE 10

Is there a relationship between consecutive numbers which we can represent using algebraic expressions?

Consecutive integers: 17 , 18 , 19 , 20

+1 +1 +1

Consecutive integers are integers that follow in sequence, each number being 1 more than the previous number. If you look at the example above, we begin with 17. The second number is 17 + 1 = 18. The third number is 17 + 1 + 1 = 19. The fourth number is 17 + 1 + 1 + 1 = 20.

Suppose the first number is given as x

Next consecutive integer $= x + 1$

Third Consecutive integer $= x + 1 + 1 = x + 2$

Fourth Consecutive Integer $= x + 1 + 1 + 1 = x + 3$

This pattern would continue into infinity with each consecutive number being one more than the previous.

4.05 The Multiplication Property of Equality 4.05

For each word problem that mentions " **Consecutive Integers**" you will always use the following definitions to begin your problem.

CONSECUTIVE INTEGERS		
First Integer	=	x
Second Integer	=	$x+1$
Third integer	=	$x+2$

EXAMPLE 11

If x is the first of three consecutive integers, express the sum of the three integers in terms of x.

Key words: three consecutive integers

sum = add together

CONSECUTIVE INTEGERS		
First Integer	=	x
Second Integer	=	$x+1$
Third integer	=	$x+2$

Translate: **first integer** + **second integer** + **third integer**

$x + x+1 + x+2$

$3x + 3$

EXAMPLE 12

If x is the first of three consecutive integers, express the sum of the first and the third integers in terms of x.

Key words: three consecutive integers

sum = add together first and third

CONSECUTIVE INTEGERS		
First Integer	=	x
Second Integer	=	$x+1$
Third integer	=	$x+2$

Translate: **first integer** + **third integer**

$x + x + 2$

$2x + 2$

CONSECUTIVE EVEN INTEGERS

EXAMPLE 13

Consecutive Even Integers: 18 , 20 , 22 , 24 (each +2, +2, +2)

Consecutive EVEN integers are integers that follow in sequence, each number being 2 more than the previous number. If you look at the example above, we begin with 18. The second number is 18+ 2 = 20. The third number is 18 + 2 + 2 = 22. The fourth number is 18 + 2 + 2 + 2 = 24.

Suppose the first number is given by x.

Next consecutive even integer = $x + 2$

Third Consecutive even integer = $x + 2 + 2 = x + 4$

Fourth Consecutive even integer = $x + 2 + 2 + 2 = x + 6$

For each word problem that mentions "**Consecutive Even Integers**" you will always use the following definitions to begin your problem.

CONSECUTIVE EVEN INTEGERS

First Even Integer = x

Second Even Integer = $x + 2$

Third Even Integer = $x + 4$

EXAMPLE 14

If x is the first of three consecutive even integers, express the sum of the three integers in terms of x.

Key words: three consecutive even integers

sum = add together

CONSECUTIVE EVEN INTEGERS

First Integer = x

Second Integer = $x + 2$

Third integer = $x + 4$

Translate:	**first integer**	+	**second integer**	+	**third integer**
	x	+	$x + 2$	+	$x + 4$

$3x + 6$

EXAMPLE 15

If x is the first of three even consecutive integers, express the sum of the first and second integers in terms of x.

CONSECUTIVE EVEN INTEGERS
First Integer $= x$
Second Integer $= x + 2$
Third integer $= x + 4$

Key words: three even consecutive integers

sum = add together first and second

Translate: **first integer** + **second integer**

$x + x + 2$

$2x + 2$

CONSECUTIVE ODD INTEGERS

Consecutive Odd Integers: 13 , 15, 17 , 19

+2 +2 +2

Consecutive ODD integers are integers that follow in sequence, each number being 2 more than the previous number. If you look at the example above, we begin with 13. The second number is 13+ 2 = 15. The third number is 13 + 2 + 2 = 17. The fourth number is 13 + 2 + 2 + 2 = 19.

Suppose the first number is given by x .

Next consecutive even integer $= x + 2$

Third Consecutive even integer $= x + 2 + 2 = x + 4$

Fourth Consecutive even integer $= x + 2 + 2 + 2 = x + 6$

For each word problem that mentions " **Consecutive Odd Integers**" you will always use the following definitions to begin your problem.

CONSECUTIVE ODD INTEGERS
First Even Integer $= x$
Second Even Integer $= x + 2$
Third Even Integer $= x + 4$

EXAMPLE 16

If x is the first of three consecutive odd integers, express the sum of the three integers in terms of x.

Key words: three consecutive odd integers

sum = add together

CONSECUTIVE ODD INTEGERS
First Integer $= x$
Second Integer $= x+2$
Third integer $= x+4$

Translate: **first integer + second integer + third integer**

$x + x+2 + x+4$

$3x + 6$

EXAMPLE 17

If x is the first of three consecutive odd integers, express the sum of the first and second integers in terms of x.

Key words: three consecutive odd integers

sum = add together first and second

CONSECUTIVE ODD INTEGERS
First Integer $= x$
Second Integer $= x+2$
Third integer $= x+4$

Translate: **first integer + second integer**

$x + x + 2$

$2x + 2$

With consecutive integers, consecutive even integers, and consecutive odd integers, the first integer will always be referred to as x. This first integer will also always be the smallest integer.

AVOIDING COMMON MISTAKES

Students often think that just because the word "odd" is used that the definitions should include odd numbers. It is very tempting to use $x, x + 1, x + 3, x + 5,\ldots$ as the starting point for consecutive odd integers . This is incorrect!!! If you look at a list of odd numbers [3, 5, 7] , you should notice that each odd number is **2 more** than the previous odd number.

So as previously stated, the definitions for consecutive odd integers will be $x, x + 2, x + 4, \ldots$

4.05 The Multiplication Property of Equality 4.05

Refer to this page when you are working on consecutive integer word problems.

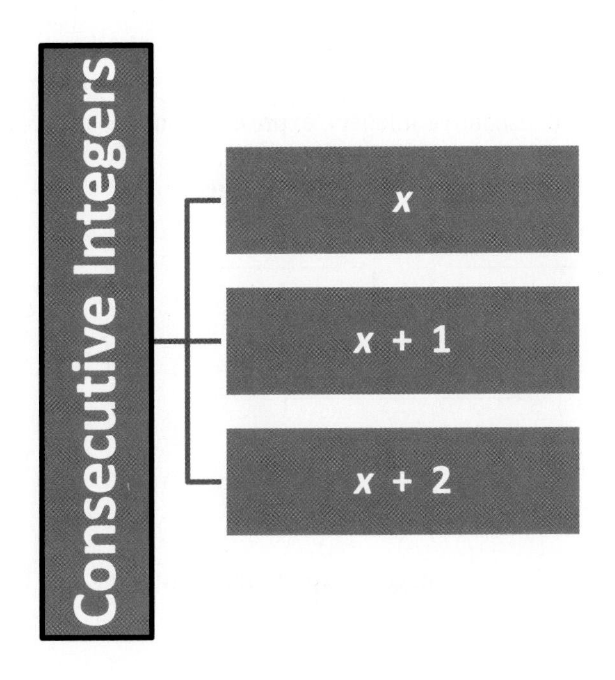

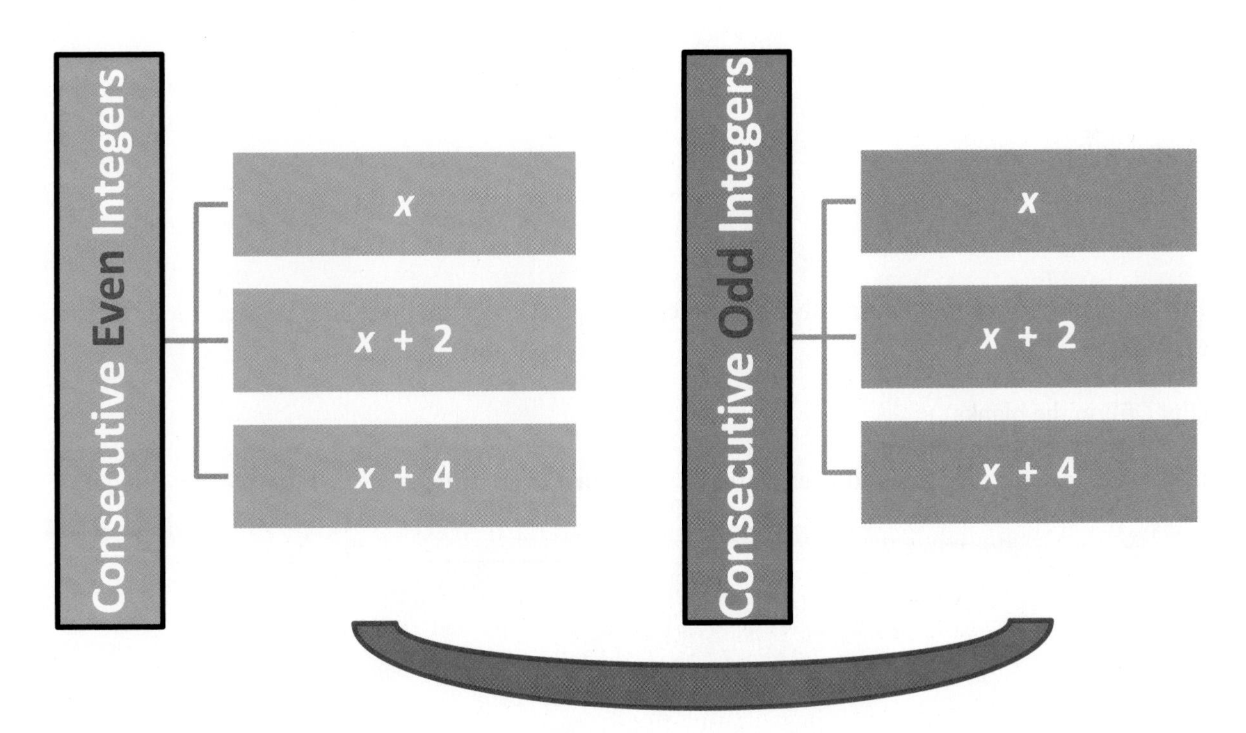

These are the SAME!

YOUR TURN

Fill in the blanks below.

11. If x represents the first of two consecutive integers, express the sum of the two integers in terms of x.

first integer + second integer

________ + __________

____

12. If x is the first of three consecutive odd integers, express the sum of the first odd integer and the second odd integer as an algebraic expression.

Hint (fill in the blanks) : **first odd integer + second odd integer**

________ + __________

____

13. If x is the first of four consecutive even integers express the sum of the four integers as an algebraic expression.

Hint (fill in the blanks) :

first even integer + second even integer + third even integer + fourth even integer

__________ + __________ + __________ + __________

____

YOUR TURN **Answers to Section 4.05.**

1. $m = 2$
2. $y = -12$
3. $n = -10$
4. $x = 64$
5. $m = -5$
6. $x = -\frac{7}{12}$
7. $x = -\frac{3}{4}$
8. $a = 2$
9. $x = 10$
10. $z = \frac{9}{10}$
11. $2x + 1$
12. $2x + 2$
13. $4x + 12$

Complete MyMathLab Section 4.05 Homework in your Homework notebook.

Solving Equations: List of Steps

Remember that when we solve an equation, our goal is to isolate the variable on one side of the equation.

Before we look at some examples, let's make a list of the steps to solving a linear equation.

Step 1: If an equation contains fractions, multiply by the LCD to clear the equation of fractions.

Step 2: If an equation contains parentheses, use the distributive property to remove them.

Step 3: Simplify each side of the equation by combining like terms.

Step 4: Get all variable terms on one side and all constants on the other side by using addition property of equality.

Step 5: Isolate the variable by dividing both sides by the coefficient of the variable.

We will use the steps to help us solve the equations in examples 1 – 6.

EXAMPLE 1

Solve. $5(3x-1)+2 = 12x+6$

The equation has no fraction so we begin with Step 2.

$$5(3x-1)+2 = 12x+6$$

$15x-5+2 = 12x+6$ — *Step* 2: Use the Distributive Property.

$15x-3 = 12x+6$ — *Step* 3: Combine like terms.

$-12x \quad -12x$ — *Step* 4: Move the variable to one side by adding -12 x to both sides.

$3x-3 = 6$ — Simplify.

$+3 \quad +3$ — Add 3 to both sides.

$3x = 9$ — Simplify.

$\frac{3x}{3} = \frac{9}{3}$ — *Step* 5: Isolate the variable by dividing by 3.

$x = 3$ — Simplify.

EXAMPLE 2

Solve. $4 - 5(n+2) = 3(-2+n)$

There are no fractions so we will begin with Step 2.

$$4 - 5(n+2) = 3(-2+n)$$

$$4 - 5n - 10 = -6 + 3n$$ *Step* 2: Use the distributive property.

$$-6 - 5n = -6 + 3n$$ *Step* 3: Combine like terms.

$$-6 - 5n + 5n = -6 + 3n + 5n$$ *Step* 4: Get the variable terms on one side of equation by adding $5n$.

$$-6 = -6 + 8n$$

$$-6 + 6 = -6 + 6 + 8n$$ Get the numbers on the opposite side by adding 6 to both sides.

$$0 = 8n$$

$$\frac{0}{8} = \frac{8n}{8}$$ *Step* 5: Isolate the variable by dividing by 8.

$$0 = n$$

Remember that the variable can be isolated on either side of the equation.

In Example 2 in **Step 3**, the variable was moved to the right by adding $5n$ to both sides. We could have worked the same problem and moved the variable to the left by subtracting $3n$ from both sides as shown below.

$$4 - 5(n+2) = 3(-2+n)$$

$$4 - 5n - 10 = -6 + 3n$$ *Step* 2: Use the distributive property.

$$-6 - 5n = -6 + 3n$$ *Step* 3: Combine like terms.

$$-6 - 5n - 3n = -6 + 3n - 3n$$ *Step* 4: Get the variable term on one side by subtracting $3n$.

$$-6 - 8n = -6$$

$$6 - 6 - 8n = -6 + 6$$ Get the numbers on the opposite side by adding 6 to both sides.

$$-8n = 0$$

$$\frac{-8n}{-8} = \frac{0}{-8}$$ *Step* 5: Isolate the variable by dividing by -8.

$$n = 0$$

EXAMPLE 3

Solve $8a - (3 - 2a) = 2a + 1 + 5a$

$8a - (3 - 2a) = 2a + 1 + 5a$	
$8a - 3 + 2a = 2a + 1 + 5a$	Use the distributive property to remove ( ).
$10a - 3 = 7a + 1$	Combine like terms.
$10a - 3 - 7a = 7a + 1 - 7a$	Move the variable term to left by subtracting $7a$.
$3a - 3 = 1$	Combine like terms.
$3a - 3 + 3 = 1 + 3$	Move the number to right by adding 3.
$3a = 4$	Combine like terms.
$\frac{3a}{3} = \frac{4}{3}$	Isolate the variable by dividing by 3.
$a = \frac{4}{3}$	

YOUR TURN

Solve. Show all your steps.

1. $4(k - 3) - k = k - 6$	2. $4(8 - 3m) = 32 - 8(m + 2)$
3. $7y - (2y - 9) = 39 - y$	4. $9(5 - x) = -3x$

Solving Equations with Fractions:

EQUATIONS WITH FRACTIONS: LET'S CLEAR THEM OUT!

We clear an equation of fractions by multiplying each side of the equation by the least common multiple of all the denominators in the equation.

EXAMPLE 4

Solve $\frac{3}{10}x - \frac{1}{5} = -2$

We will begin by clearing out fractions. Look at all the numbers in the denominators and choose the least common multiple (LCM). The LCM between 10 and 5 is 10. So, we will begin by multiplying both sides of the equation by 10.

Solve $(10)\left(\frac{3}{10}x\right) - (10)\left(\frac{1}{5}\right) = (10)(-2)$ Multiply both sides by LCM which is 10.

$\left(\frac{\cancel{10}}{1}\right)\left(\frac{3}{\cancel{10}}x\right) - \left(\frac{\cancel{10}^{\,2}}{1}\right)\left(\frac{1}{\cancel{5}}\right) = (10)(-2)$ Multiply.

$$3x - 2 = -20$$

$$3x - 2 + 2 = -20 + 2$$ Move constants to right side by adding 2.

$$3x = -18$$

$$\frac{3x}{3} = \frac{-18}{3}$$ Divide by 3 to "isolate the variable."

$$x = -6$$

EXAMPLE 5

Solve. $\frac{x}{2} - 1 = \frac{2}{3}x - 3$

We will begin by clearing fractions. Since the denominators are 2 and 3, we will choose 6 as the least common multiple. So, we will begin solving by multiplying both sides of the equation by 6.

$$\frac{x}{2} - 1 = \frac{2}{3}x - 3$$

$$6\left(\frac{x}{2} - 1\right) = 6\left(\frac{2}{3}x - 3\right)$$ *Step* 1: Multiply both sides by the LCM which is 6.

$$6\left(\frac{x}{2}\right) - 6(1) = 6\left(\frac{2}{3}x\right) - 6(3)$$ *Step* 2: Apply the Distributive Property.

$$3x - 6 = 4x - 18$$ Multiply.

$$3x - 6 - 3x = 4x - 18 - 3x$$ *Step* 3: Get the variable on one side by subtracting $3x$.

$$-6 = x - 18$$ Combine like terms.

$$-6 + 18 = x - 18 + 18$$ Get the number on the opposite side by adding 18.

$$12 = x$$ Combine like terms.

CHECK: $\frac{x}{2} - 1 = \frac{2}{3}x - 3$

$$\frac{12}{2} - 1 = \frac{2}{3}(12) - 3$$

$$6 - 1 = 8 - 3$$

$$5 = 5$$

EXAMPLE 6

Solve. $\frac{3(m-2)}{5} = 3m + 6$

$$\not{5} \cdot \frac{3(m-2)}{\not{5}} = 5(3m + 6)$$ Clear the fraction by multiplying both sides by 5.

$$3(m-2) = 5(3m+6)$$ Simplify.

$$3m - 6 = 15m + 30$$ Apply the Distributive Property.

$$3m - 6 - 3m = 15m + 30 - 3m$$ Subtract $3m$ from both sides.

$$-6 = 12m + 30$$ Simplify.

$$-6 - 30 = 12m + 30 - 30$$ Subtract 30 from both sides.

$$-36 = 12m$$ Simplify.

$$\frac{-36}{12} = \frac{12m}{12}$$ Isolate the variable by dividing by 12.

$$-3 = m$$ Simplify.

EXAMPLE 7

Solve. $\dfrac{5(x-1)}{4} = \dfrac{3(x+1)}{2}$

$\dfrac{5(x-1)}{4} = \dfrac{3(x+1)}{2}$ Use the Distributive Property to remove ( ).

$\dfrac{5x-5}{4} = \dfrac{3x+3}{2}$ Multiply by LCD which is 4.

$^{1}\cancel{4}\left(\dfrac{5x-5}{\cancel{4}}\right) = {}^{2}\cancel{4}\left(\dfrac{3x+3}{\cancel{2}}\right)$

$1(5x-5) = 2(3x+3)$ Simplify.

$5x-5 = 6x+6$ Use the Distributive Property to remove ( ).

$5x-5-5x = 6x+6-5x$ Subtract $5x$.

$-5 = x+6$ Simplify.

$-5-6 = x+6-6$ Subtract 6.

$-11 = x$

OR

Here's another way to solve the same equation!!!

Solve. $\dfrac{5(x-1)}{4} = \dfrac{3(x+1)}{2}$

$\dfrac{5(x-1)}{4} = \dfrac{3(x+1)}{2}$ Use the Distributive Property to remove ( ).

$\dfrac{5x-5}{4} = \dfrac{3x+3}{2}$ These are proportions so we can cross multiply.

$2(5x-5) = 4(3x+3)$ Use the Distributive Property to remove ( ).

$10x-10 = 12x+12$

$10x-10-10x = 12x+12-10x$ Subtract $10x$.

$-10 = 2x+12$ Simplify.

$-10-12 = 2x+12-12$ Subtract 12.

$-22 = 2x$ Simplify.

$\dfrac{-22}{2} = \dfrac{2x}{2}$ Isolate the variable by dividing by 2.

$-11 = x$

YOUR TURN

Solve. Show all your steps.

5. $\frac{5}{2}x - 1 = \frac{3}{2}x - 4$	6. $4x + \frac{1}{3} = 2x - \frac{2}{3}$
7. $\frac{2(m+3)}{3} = 6m + 2$	8. $\frac{3(x-5)}{2} = \frac{2(x+5)}{3}$

Solving Linear Equations with Decimals

Sometimes our equations will involve decimals. A student has two options when solving equations with decimals. A student can choose to simply work with the decimals or the student can choose to clear the equation of decimals before solving. You will need to decide which way you think is easier and follow that example to work the problems.

EXAMPLE 8

Plan A: Use your calculator and just work with the decimals in the equation.

Solve. $0.9x - 4.7 = 0.7$

$0.9x - 4.7 + 4.7 = 0.7 + 4.7$ Add 4.7 to both sides.

$0.9x = 5.4$ Combine like terms.

$\dfrac{0.9x}{0.9} = \dfrac{5.4}{0.9}$ Isolate the variable by dividing by 0.9.

$x = 6$

Plan B: Multiply by the appropriate power of 10 to remove the decimals

Solve. $0.9x - 4.7 = 0.7$

We will clear equation of decimals by multiplying both sides of the equation by 10. We choose 10 because all of the numbers in the equations have only 1 decimal place.

Reminder

- Multiplying by 10 moves the decimal point one place to the right.
- Multiplying by 100 moves the decimal point two places to the right.
- Multiplying by 1000 moves the decimal point three places to the right.

Solve. $0.9x - 4.7 = 0.7$

$10(0.9x - 4.7) = 10(0.7)$ Multiply both sides by 10.

$9x - 47 = 7$

$9x - 47 + 47 = 7 + 47$ Add 47 to both sides.

$9x = 54$

$\dfrac{9x}{9} = \dfrac{54}{9}$ Isolate the variable by dividing by 6.

$x = 6$

EXAMPLE 9

Solve. $0.25x + 0.10(x-3) = 1.1$

We will begin by clearing the decimals. Be careful here! Notice that two of the decimals have 2 decimals places while one decimal only has 1 decimal place. When this occurs, you must choose the power of 10 to clear the largest number of decimal places present. So, in this equation we will multiply both sides by 100.

Courtesy of Fotolia

A word of caution in equations like this one. We cannot multiply one side of the equation by 100 and the other side by 10. This violates the Multiplication Property of Equality. We must always multiply both sides by the same number !!!!!

$0.25x + 0.10(x-3) = 1.1$ Multiply both sides by 100.

$0.25x + 0.10(x-3) = 1.1$ Moves decimal 2 places right.

$25x + 10(x-3) = 110$ Apply the Distributive Property.

$25x + 10x - 30 = 110$ Combine like terms.

$35x - 30 = 110$ Add 30 to both sides.

$35x - 30 + 30 = 110 + 30$ Combine like terms.

$35x = 140$

$\dfrac{35x}{35} = \dfrac{140}{35}$ Isolate the variable by dividing by 35.

$x = 4$

CHECK: $0.25x + 0.10(x-3) = 1.1$

$0.25(4) + 0.10(4-3) = 1.1$

$1 + 0.10(1) = 1.1$

$1 + 0.10 = 1.1$

$1.1 = 1.1$

We could also have solved Example 9 by using our calculator and working with the decimal.

Solve. $0.25x + 0.10(x-3) = 1.1$	Use the distributive property to remove ( ).
$0.25x + 0.10x - 0.30 = 1.1$	Combine like terms.
$0.35x - 0.30 = 1.1$	
$0.35x - 0.30 + 0.30 = 1.1 + 0.30$	Add 0.30 to both sides.
$0.35x = 1.40$	Combine like terms.
$\dfrac{0.35x}{0.35} = \dfrac{1.40}{0.35}$	Isolate the variable by dividing by 0.35.
$x = 4$	

YOUR TURN

Solve. Show all your steps. Write your solution as an integer or decimal to the nearest tenth.

9. $0.9x - 4.1 = 0.4$	10. $0.06 - 0.10(x+1) = -0.2$

Solutions for SPECIAL Linear Equations

So far, all the equations we have solved have had one solution. Such an equation, which is true for a particular value of the variable, is called a **conditional equation**. A conditional equation is true only under certain conditions. In the equation $\boldsymbol{x + 3 = 5}$, the equation is true on the condition that $\boldsymbol{x = 2}$. The equation is false for any other value other than 2.

We will now look at some "special" equations other than conditional equations. There are two other types of equations we need to look at.

- One is called the **Identity** and has an infinite number of solutions.
- The other is called a **Contradiction** and has no solution.

IDENTITY: An equation which is true regardless of what values are substituted for the variable.

Eq. solves to TRUE STATEMENT → *Infinite Number of Solutions* → All Real Numbers

Consider the equation $x = x$. So if $x = 5$, then this statement would say $5 = 5$.

This equation is obviously true for every possible value of x. This is, of course, a ridiculously simple example, but it makes the point. Equations that have this property are called ***identities***. Some examples of identities would be

$$3x = 2x + x$$
$$3 = 3$$
$$0 = 0$$
$$2x - 5 = 2x - 5$$

All of these equations are true for any value of x or in other words, ***the solution would be all real numbers***. The second example, $3 = 3$, is interesting because it does not even contain an x, so obviously its truthfulness cannot depend on the value of x! When you are attempting to solve an equation algebraically and you end up with an obvious identity (like $0 = 0$), then you know that the original equation must also be an **identity**, and therefore it has an infinite number of solutions.

EXAMPLE 10

Solve. $3(x-4) = 2x + x - 12$

$$3(x-4) = 2x + x - 12$$
$$3x - 12 = 3x - 12$$

If we recognize that we have the exact same expression on both sides of the equation, we can stop now and declare this equation an Identity. Every real number may be substituted for $\boldsymbol{x}$ and a true statement will result. **The solution is:** *all real numbers*

If we do not realize we have an identity at that point, we can continue solving and arrive at the same answer.

$$\begin{aligned} 3(x-4) &= 2x + x - 12 \\ 3x - 12 &= 3x - 12 \\ 3x - 12 - 3x &= 3x - 12 - 3x \\ -12 &= -12 \end{aligned}$$

Notice the variable terms with an x have added out. We are left with a true statement. *This means the solution is all real numbers.*

The final equation contains no variable however we realize that - 12 = - 12 regardless of the value of x. Therefore, the equation has an infinite number of solutions or in other words, the solution set would be all real numbers. We can also conclude that if the variable " disappears" and we end up with a **TRUE** statement, then the equation is an **IDENTITY**.

CONTRADICTION: An equation which can never be true.

Eq. solves to FALSE STATEMENT ⟶ *No Solutions*

Now consider the equation $x + 4 = x + 3$.

There is no possible value for x that could make this true. If you take a number and add 4 to it, it will never be the same as if you take the same number and add 3 to it. Such an equation is called a ***contradiction*** because it cannot ever be true.

If you are attempting to solve such an equation, you will end up with an extremely obvious contradiction such as $1 = 2$. This indicates that the original equation is a contradiction, and has no solution.

We can also conclude that if the variable "disappears" and we end up with a **FALSE** statement, then the equation is a **CONTRADICTION**.

EXAMPLE 11

Solve.
$$\begin{aligned} 4(3x+2) &= 12x - 5 \\ 12x + 8 &= 12x - 5 \\ 12x + 8 - 12x &= 12x - 5 - 12x \\ 8 &= -5 \end{aligned}$$

no solution

The final equation contains no variable and the result is a false statement. This means that there is no value of x that makes $\mathbf{8 = -5}$ Therefore, this is a contradiction and has no solution.

Linear Equations – A Summary

Conditional One Solution	$2x + 3 = 15$ $2x = 12$ $x = 6$
Identity Infinite Solutions	$2x + 3 = 2x + 3$ $3 = 3$ $0 = 0$ True Statement → All real numbers.
Contradiction No Solution	$2x + 3 = 2x + 5$ $3 = 5$ False Statement

YOUR TURN

Solve. Show all your steps.

11. Solve. $5(x-6) = 3x - 30 + 2x$	12. Solve. $5(x-6) = 2(x-15) + 7x$

4.06 Further Solving Linear Equations 4.06

YOUR TURN

Solve. Show all your steps.

13. Solve. $5(2-x)+8x = 3(x-6)$	14. Solve. $2-3(2+6z) = 4(z+1)+18$

YOUR TURN Answers to Section 4.06.

1. k = 3	2. m = 4	3. y = 5
4. $x = \frac{15}{2}$	5. x = - 3	6. $x = -\frac{1}{2}$
7. m = 0	8. x = 13	9. x = 5
10. x = 1.6	11. All real numbers	12. x = 0
13. no solution	14. $z = -\frac{13}{11}$	

Complete MyMathLab Section 4.06 Homework in your Homework notebook.

Problem Solving Plan

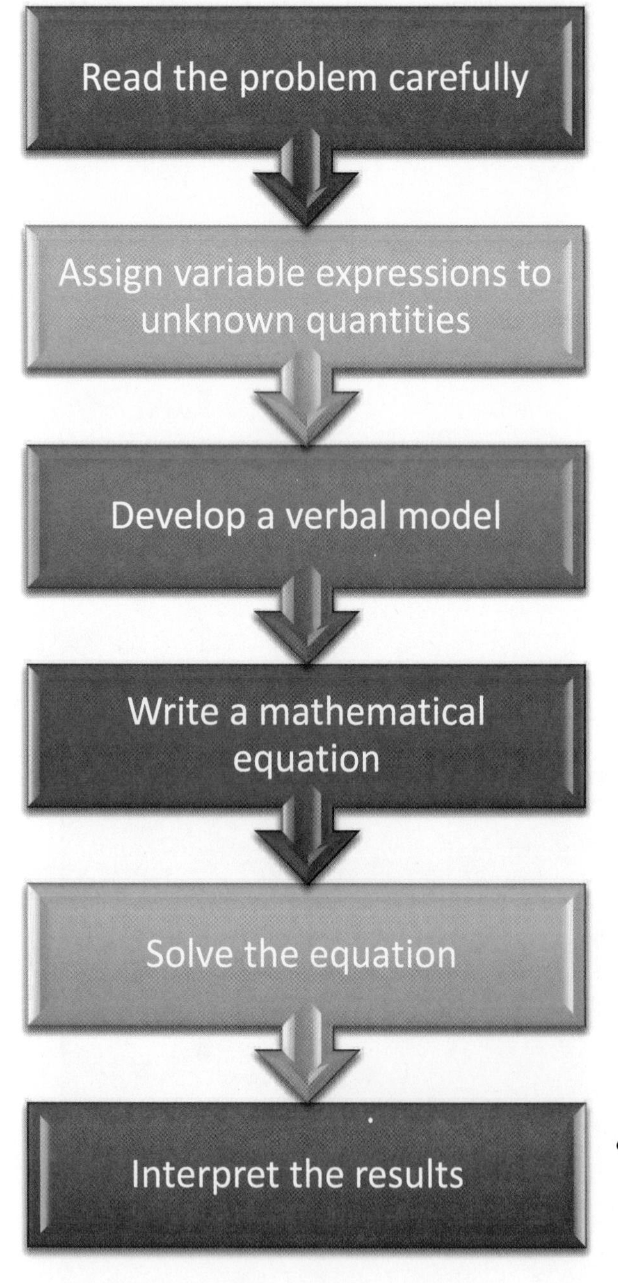

- Read and reread the problem until you understand what it is asking you to do.
- Identify the unknown quantity or quantities. Let ***x*** represent one of the unknowns. Draw a picture and write down any relevant formulas.
- Write an equation in words.
- Replace the verbal model with a mathematical equation using ***x*** or another variable.
- Solve for the variable using the steps you learned in the previous section.
- Once you've solved the equation, look back and see what it represented in the problem. Then use this value to determine any other unknowns in the problems.

4.07 An Introduction to Problem Solving 4.07

Here's that list of key words you've seen before to help translate word problems.

Addition	Subtraction	Multiplication	Division	Equality
sum plus increased by more than total added to longer than	difference less subtracted from less than decreased by minus fewer than	product times multiply of twice double/triple	quotient divided by divide equally per	equals is equal to is/was yields

Let's practice some simple translations before we move on to word problems. Study the example below to make sure you understand each of the translations. You will find this very helpful when you begin to work the problems. Refer to these examples from time to time to refresh your memory.

EXAMPLE 1

Translate into an algebraic expression.

a number y	The sum of twice the number and 7 $2y+7$	Twice the sum of the number and 7 $2(y+7)$
Judy' age x	Judy's age three years from now $x + 3$	Judy's age two years ago $x - 2$
Board length y	Twice the board length $2y$	4 inches longer than seven times the board length $7y + 4$
First consecutive integer x	Second consecutive integer $x+1$	Third consecutive integer $x+2$
First consecutive odd integer x	Second consecutive odd integer $x+2$	Third consecutive odd integer $x+4$
First consecutive even integer x	Second consecutive even integer $x+2$	Third consecutive even integer $x+4$
Measure of smaller angle y	Eight less than three times smaller angle $3y - 8$	Three times the smaller angle $3y$

YOUR TURN

Fill in the table.

1. a number y	The difference of twice the number and 9	Twice the difference of the number and 9
2. Sue's age x	Sue's age ten years from now	Sue's age four years ago
3. string length y	Three times the string length	2 inches longer than five times the string length
4. Measure of smaller angle y	Twice the smaller angle	Three less than twice the smaller angle

There are several kinds of word problems we will look at in this section. Don't panic...... You **CAN** do them! Many students find solving algebra word problems difficult. Learning to solve word problems is like learning to play the piano. First, you are shown how; then you must practice and practice and practice. The more you practice the more confident you will become!

Start every problem with " Let x = **something**". Let x equal what you are trying to find - the unknown. Make sure and label what x stands for as we did on the previous page in the chart. If you have more than one quantity, you will need another definition in terms of x for each unknown.

Many students will become lazy and skip this part of the process - defining your unknowns. However, this step is the key to success in word problems. So, begin every word problem with " Let x = **something**"!!

Application Problems - One Unknown Number

The following examples have only one unknown. We will begin every problem with "Let x = **something**" according to what the unknown value is. The unknown could simply be a number or it could represent miles driven or hours worked just to name a few. The key is to read the question – usually found at the end of the problem. This is where you find the unknown quantity you have been asked to find.

EXAMPLE 2

Four times a number, minus 7, is equal to three times the number plus 9. Find the number.

Solution:
The last sentence says " Find the number". So the unknown is " a number".

x = **a number**

To write the equation, we will reread the problem substituting x every time we read the word **a number.**

Four times a number,	**minus 7,**	**is equal to**	**three times the number**	**plus 9.**
$4x$	-7	$=$	$3x$	$+9$

EXAMPLE 3

A car rental agency advertised renting a car for $28.95 per day and $0.32 per mile. If Kevin rents this car for 2 days, how many whole miles can he drive on a $250 budget? Round to the nearest whole mile.

Define the unknown value:
x = number of miles driven

Write the equation and solve. Then answer the question.

Cost for 2 days	**+**	**cost per mile**	**=**	**total cost**
2 (28.95)	**+**	**0.32x**	**=**	**250**

Solve.

$$2(28.95) + 0.32x = 250$$
$$57.90 + 0.32x = 250$$
$$57.90 + 0.32x - 57.90 = 250 - 57.90$$
$$0.32x = 192.10$$
$$\frac{0.32x}{0.32} = \frac{192.10}{0.32}$$
$$x \approx 600.3125$$

Round down.

Kevin can drive 600 miles.

EXAMPLE 4

Mary's financial aid stipulates that her tuition not exceed $2000. If her college charges a $55 registration fee plus $475 per course, what is the greatest number of courses for which Mary can register?

Define the unknown value:
x = number of courses

Write the equation and solve. Then answer the question.

registration fee + cost per course = total cost
55.00 + 475x = 2000

$$
\begin{aligned}
\textit{Solve}: \quad 55 + 475x &= 2000 \\
55 - 55 + 475x &= 2000 - 55 \\
475x &= 1945 \\
x &= 4.09
\end{aligned}
$$

Mary can register for at most 4 courses.

YOUR TURN

Solve the following application problems by filling in the blanks. Make sure and show your work!

5. Twice the difference of a number and 4 is equal to three times the sum of the number and 2. Find the number.	6. A student takes a job in the college's student center so that she can buy books that cost $178.50. How many hours must she work to make this amount if she earns $8.50 per hour?
a. Define your unknown: x = ____________	a. Define your unknown: x = ____________
b. Write an equation: ________________	b. Write an equation: ________________
c. Solve:	c. Solve:
d. Answer: The number is ____________.	d. Answer: The student must work ________ hrs.

4.07 An Introduction to Problem Solving 4.07

YOUR TURN

Solve the following application problems by filling in the blanks. Make sure and show your work!

7. In one U.S. city, the taxi cost is $2 plus $0.90 per mile. If you are traveling from the airport, there is an additional charge of $4.00 for tolls. How far can you travel by taxi for $46.50 ?

a. Define your unknown: x = ________________

b. Write an equation: ________________

c. Solve:

d. Answer : You can travel ___________ miles.

Courtesy of Shutterstock

Are you ready for the challenge? Remember, you CAN do this !!

In the previous examples, we were asked to find one unknown. Now, we are ready to approach more challenging problems. In the following examples, we will be asked to find more than one quantity or in other words, our application problems will involve two or more unknowns. We may be asked to find "a smaller number **and** a larger number ", "Sally's age **and** Jane's age", or " the length of a short piece of string **and** the length of a longer piece of string" just to name a few examples. So, we will now have more than one definition for the values which are not known. We will need to use a variable expression, in terms of x, for each unknown.

Application Problems - More than One Unknown

EXAMPLE 5

The larger of two numbers is 16 less than twice the smaller number. The sum of the two numbers is 35. Find the **two numbers**.

Define your unknowns: The larger of two numbers is 16 less than twice the smaller number.

x = smaller number

$2x - 16$ = larger number

Equation: The sum of the two numbers is 35. We need to add together the two numbers.

$$\text{smaller number} + \text{larger number} = 35$$
$$x + 2x - 16 = 35$$
$$3x - 16 = 35$$
$$3x - 16 + 16 = 35 + 16$$
$$3x = 51$$
$$\frac{3x}{3} = \frac{51}{3}$$
$$x = 17$$

We have solved the equation and found the value of x, which is the ***smaller number***. The **smaller number is 17**. We will now use this value to find the larger number. According to our definition, $2x - 16$ = larger number. We will plug in 17 for x and find the value of the larger number.

$2x - 16 = 2(17) - 16 = 34 - 16 = 18$ → **larger number.**

The smaller number is 17 and the larger number is 18.

EXAMPLE 6

A 10-foot board is to cut into two pieces so that the length of the longer piece is 4 times the length of the shorter. Find the length of each piece.

Begin by drawing a picture and assign variable expression for each unknown.

10-foot board → two pieces → the length of the longer piece is 4 times the length of the shorter.

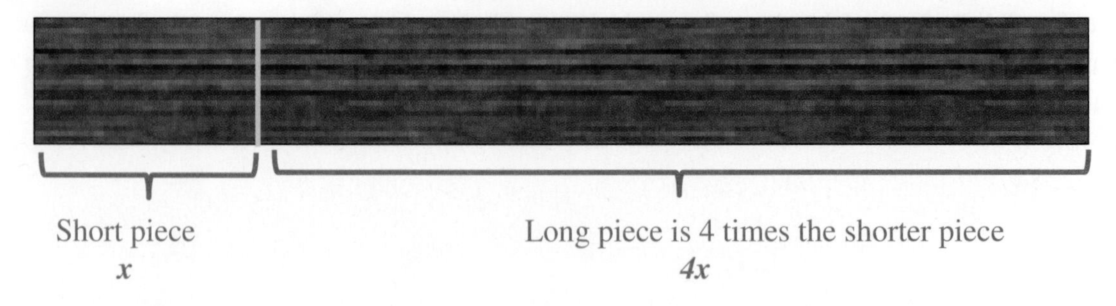

The entire length of the board is 10 ft. The length of the short piece plus the length of the long piece equals the entire length of the board.

Write an equation.

Length of short piece + length of long piece = total length of board

$$x \quad + \quad 4x \quad = \quad 10$$

Solve.
$$x + 4x = 10$$
$$5x = 10$$
$$\frac{5x}{5} = \frac{10}{5}$$
$$x = 2$$

The length of the short piece is 2 feet and the length of the long piece is 8 feet.

Length of short piece = x = 2 feet
Length of long piece = $4x$ = 4 (2) = 8 feet

EXAMPLE 7

College students at a local university were asked whether they preferred chocolate or vanilla ice cream. Out of 500 college students surveyed, there were 200 more students who preferred chocolate over vanilla. How many student preferred chocolate?

Assign variable expressions to each unknown:

number of student who prefer vanilla = x
number of students who prefer chocolate = $x + 200$

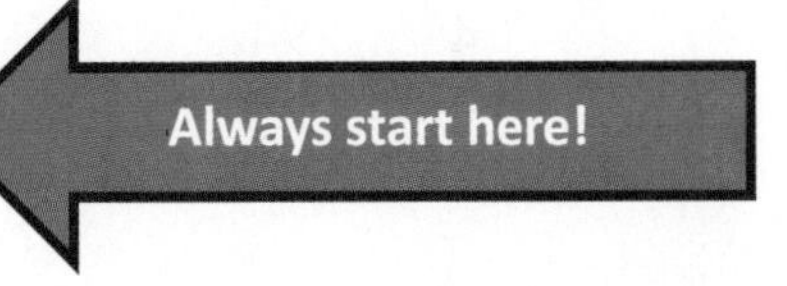

Write an equation.

Students who prefer vanilla + students who prefer chocolate = total number of students

$$x \quad + \quad (x + 200) \quad = \quad 500$$

Solve.
$$x + (x + 200) = 500$$
$$2x + 200 = 500$$
$$2x + 200 - 200 = 500 - 200$$
$$2x = 300$$
$$\frac{2x}{2} = \frac{300}{2}$$
$$x = 150$$

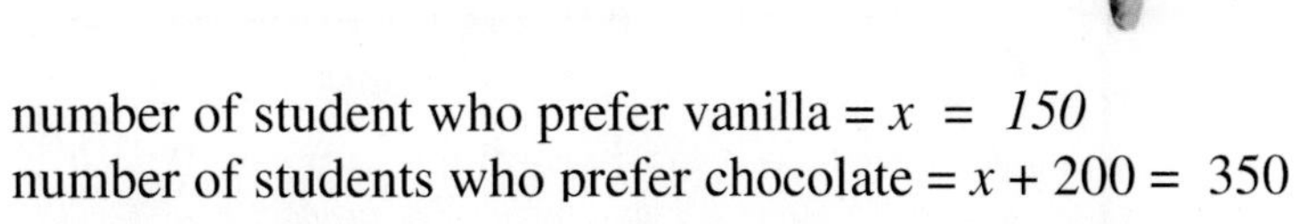

number of student who prefer vanilla = x = *150*
number of students who prefer chocolate = $x + 200$ = 350

There were 150 students who preferred vanilla ice cream and 350 students who preferred chocolate.

EXAMPLE 8

Two angles are supplementary if their sum is 180°. The larger angle measures eight degrees more than three times the measure of the smaller angle. If x represents the measure of the smaller angle and these two angles are supplementary, find the measure of each angle.

Two angles are supplementary if their sum is 180°.

measure of smaller angle + measure of larger angle = 180

$$x + 3x + 8 = 180$$

Solve.

$$x + 3x + 8 = 180$$
$$4x + 8 = 180$$
$$4x + 8 - 8 = 180 - 8$$
$$4x = 172$$
$$\frac{4x}{4} = \frac{172}{4}$$
$$x = 43$$

measure of smaller angle = x = 43
measure of larger angle = $3x + 8 = 3(43) + 8 = 129 + 8 = 137^{\circ}$

The two angles measure 43° and 137 °.

Once you have reached a solution to a word problem, verify that it is reasonable in the context of the problem. For example, in this problem we were asked to find two angles that were supplementary. Were the two angles we found supplementary???? Did they add up to 180 ????

$$4^{\circ} + 137^{\circ} = 180^{\circ}$$

YOUR TURN

Define your unknown(s), write an equation and solve.

8. A 30 inch board is to be cut into three pieces so that the second piece is 4 times as long as the first piece and the third piece is 5 times as long as the first piece. If x represents the length of the first piece, find the lengths of all three pieces.

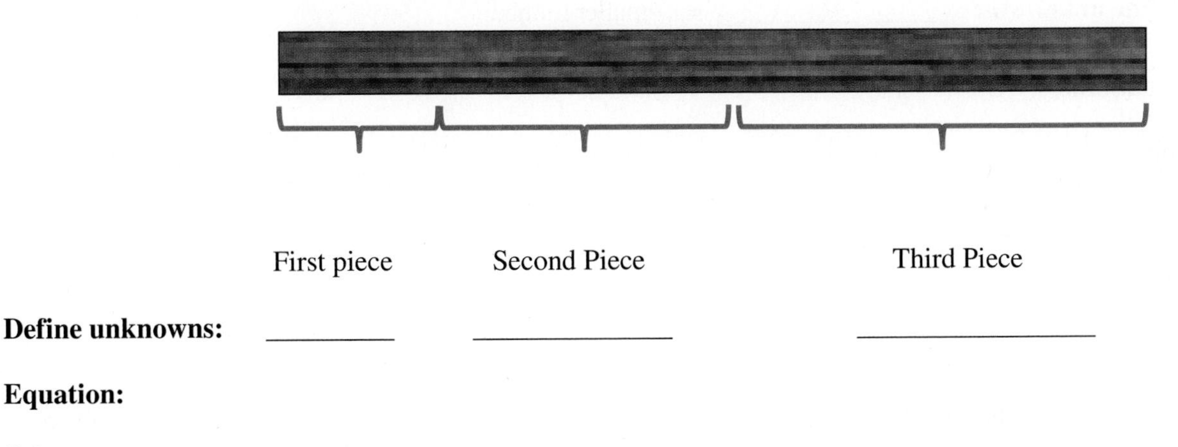

Define unknowns: __________ ______________ ________________

Equation:

Solve:

Answer:

9. Two minor league baseball players got a total of 312 hits. Perez had 20 more hits than Smith. Find the number of hits for each player.

Define unknowns: ______________ = number of hits for Smith
______________ = number of hits for Perez

Equation:

Solve:

Answer:

YOUR TURN

Define your unknown(s), write an equation and solve.

10. The larger of two numbers is three less than four times the smaller number. If the sum of the two numbers is 57, find the two numbers.

Define unknowns: _______________ = smaller number
_______________ = larger number

Equation:

Solve:

Answer:

11. Two angles are supplementary if their sum is 180 degrees. The larger angle measures five degrees more than four times the measure of a smaller angle. Given these two angles are supplementary, find the measure of each angle.

Define unknowns: _______________ = smaller angle

_______________ = larger angle

Equation:

Solve:

Answer:

Important

Once you have reached a solution to a word problem, verify that it is reasonable in the context of the problem.

Application Problems - CONSECUTIVE INTEGER

Recall the definitions we learned in a previous section for consecutive integers.

CONSECUTIVE INTEGERS
First Integer = x
Second Integer = $x+1$
Third Integer = $x+2$

CONSECUTIVE EVEN INTEGERS
First Integer = x
Second Integer = $x+2$
Third Integer = $x+4$

CONSECUTIVE ODD INTEGERS
First Integer = x
Second Integer = $x+2$
Third Integer = $x+4$

EXAMPLE 9

The left and right page numbers of an open book are two consecutive integers whose sum is 27. Find these page numbers.

Look at the definitions above and pick the ones that match your problems. This problems deals with consecutive integers. So we will choose the following definitions.

CONSECUTIVE INTEGERS
First Integer = x
Second Integer = $x+1$

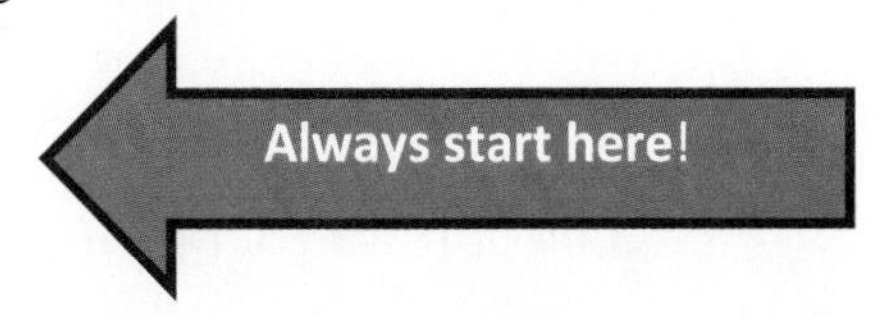

The sentence states that the sum of the two consecutive integers is 27. Sum means to add together.

first consecutive integer	**+**	**second consecutive integer**	**=**	**27**
x	+	$(x+1)$	=	27

Solve. $x + (x+1) = 27$

$$2x + 1 = 27$$

$$2x + 1 - 1 = 27 - 1$$

$$2x = 26$$

$$\frac{2x}{2} = \frac{26}{2}$$

$$x = 13$$

CONSECUTIVE INTEGERS
First Integer = x = 13
Second Integer = $x+1$ = 13 + 1 = 14

The two page numbers are 13 and 14.

EXAMPLE 10

Find three consecutive odd integers such that the sum of the first two odd integers is 19 more than the third integer.

CONSECUTIVE ODD INTEGERS
First Integer = x
Second Integer = $x+2$
Third integer = $x+4$

Sum of the first two odd integers is 19 more than the third integer.

first integer + second integer = third integer + 19

$$x + (x+2) = (x+4) + 19$$

Solve.
$$x + (x+2) = (x+4) + 19$$
$$2x + 2 = x + 23$$
$$2x + 2 - x = x + 23 - x$$
$$x + 2 = 23$$
$$x + 2 - 2 = 23 - 2$$
$$x = 21$$

CONSECUTIVE ODD INTEGERS
First Integer = x = 21
Second Integer = $x+2$ = 21 + 2 = 23
Third integer = $x+4$ = 21 + 4 = 25

The consecutive odd integers are 21, 23, and 25.

Courtesy of Fotolia

Once you have reached a solution to a word problem, verify that it is reasonable in the context of the problem. For example, in this problem we were asked to find three consecutive odd integers. Are the answers we arrived at three consecutive odd integers?? YES!!!

YOUR TURN

Define your unknown(s), write an equation and solve. Then reread and answer the question.

12. The code to unlock a student's combination lock happens to be three consecutive even integers whose sum is 132. Find the integers.

Define unknowns: ______________ = 1st integer

______________ = 2nd integer

______________ = 3rd integer

Equation:

Solve:

Answer:

13. Find three consecutive odd integers such that the sum of the first two is 69 more than the third.

Define unknowns: ______________ = 1st integer

______________ = 2nd integer

______________ = 3rd integer

Equation:

Solve:

Answer:

4.07 An Introduction to Problem Solving 4.07

YOUR TURN

14. When the smaller of two consecutive integers is added to three times the larger, the result is 31. Find the integer.

Define unknowns: _______________ = 1st integer

_______________ = 2nd integer

Equation:

Solve:

Answer:

YOUR TURN **Answers to Section 4.07.**

1. a number y	The difference of twice the number and 9, $2y - 9$	Twice the difference of the number and 9 $2(y - 9)$
2. Sue's age x	Sue's age ten years from now $x + 10$	Sue's age four years ago $x - 4$
3. string length y	Three times the string length $3y$	2 inches longer than five times the string length $5y + 2$
4. Measure of smaller angle y	Twice the smaller angle $2y$	Three less than twice the smaller angle $2y - 3$

5. -14 6. 21 hours 7. 45 miles

8. The short piece is 3 inches, second piece is 12 inches, third piece is 15 inches.

9. Smith: 146 hits, Perez: 166 hits 10. 12, 45 11. 35°, 145°

12. The integers are 42, 44, and 46.

13. 71, 73, 75 14. 7, 8

Complete MyMathLab Section 4.07 Homework in your Homework notebook.

4.08 Formulas and Problem Solving 4.08

A formula is an equation in which variables are used to describe a relationship. We use formulas to find area and perimeter or to calculate interest. Since a formula is an equation we can solve the formula for a particular variable.

Substituting Using Formulas

EXAMPLE 1

Substitute the given values into the formula and solve for the unknown variable.

a. **A = bh; A = 33 , b = 11**

This formula gives the area of a parallelogram with base, ***b,*** and height, ***h***. Substitute the given values into the formula.

$$A = bh$$

$$33 = 11h \qquad \text{Let } A = 33 \text{ and } b = 11.$$

$$\frac{33}{11} = \frac{11h}{11} \qquad \text{Divide by 11.}$$

$$3 = h$$

The height is 3.

b. **P = a + b + c; P = 28 , a = 9 , b = 8**

This is a formula for the perimeter of a triangle with sides a, b, and c.

$$P = a + b + c$$

$$28 = 9 + 8 + c \qquad \text{Let } P = 28\text{, } a = 9\text{, and } b = 8.$$

$$28 = 17 + c$$

$$28 - 17 = 17 + c - 17 \qquad \text{Subtract 17 from both sides.}$$

$$11 = c$$

The length of the third side is 11.

YOUR TURN

Substitute the given values into the formula and solve for the unknown variable.

1. $p = 2l + 2w$; $p = 48$ and $l = 6$	2. $A = \frac{1}{2}bh$; $A = 60$ *and* $b = 8$

Using Formulas in Applications

We can also use formulas to solve application problems. When you read the problem, your first decision will be to decide what formula to use. Is the problem asking about perimeter, area, or distance? You must choose the correct formula to apply to your problem. Here are some formulas you may need in this section.

Perimeter of Rectangle: $P = 2L + 2W$

Area of Rectangle: $A = L\,W$

Perimeter of Triangle: $P = a + b + c$

Area of Triangle: $A = \frac{1}{2} bh$

Circumference of Circle: $C = 2\pi r$ or $C = \pi d$

Area of Circle : $A = \pi r^2$

Distance : $D = rt$

Volume of Rectangular Prism: $V = lwh$

EXAMPLE 2

If the area of a rectangular billboard is approximately 4425 sq ft, and the width is approximately 59 ft, what is the height of the billboard?

Courtesy of Shutterstock

1. Choose the correct formula: Area of a rectangle = lw

2. Plug-in what you know: Area is 4425 sq ft and width is 59 ft

$$A = l \cdot w$$
$$4425 = l \cdot 59$$
$$\frac{4425}{59} = \frac{59l}{59}$$
$$75 = l$$

Note that the length in the formula corresponds to the height of the billboard.

The height of the billboard is 75 ft.

EXAMPLE 3

For the purpose of purchasing new baseboard and carpet,
a. Find the area and perimeter of the room below.
b. Identify whether baseboard has to do with area or perimeter.
c. Identify whether carpet has to do with area or perimeter.

11 ft.

8 ft.

Formulas needed:
Perimeter: $p = 2l + 2w$
Area = lw

a. Perimeter: $p = 2l + 2w$
$= 2(11) + 2(8)$
$= 22 + 16$
$\mathbf{p = 38}$ **ft.**

Area = lw
$= (11)(8)$
A = 88 ft.^{2}

b. Baseboard goes around the outside edges of the room. This would be the perimeter. **So, we would need 38 feet of baseboard for this room.**

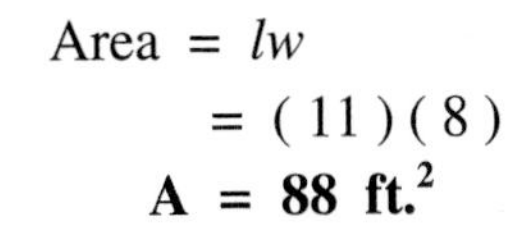

c. Carpet covers the entire floor of the room. This would be area. **So, we would need 88 square feet of carpet for this room.**

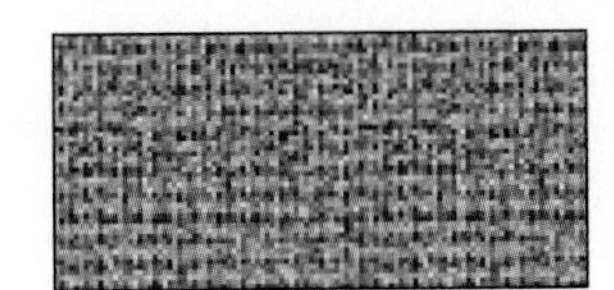

EXAMPLE 4

Patsy's Pizza sells one 14-inch pizza or two 9-inch pizzas for \$11.99. Determine which size gives more pizza.

First we need to decide whether to find the perimeter or area of these two pizzas. Do we want to eat around the outside edges of the pizza or eat the entire pizza? Unless you're a picky eater that only eats the crust, we would prefer to eat the entire pizza. So, we need to find the **area** of these two pizzas and compare.

Formula for Area of Circle: $A = \pi r^2$

Area of 14-inch pizza	Area of two 9- inch pizza
	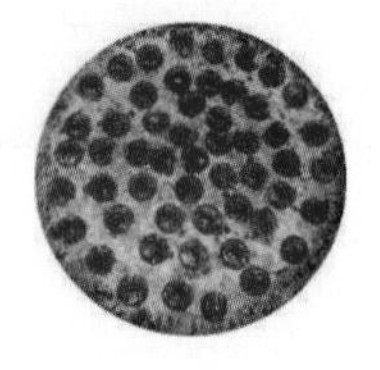 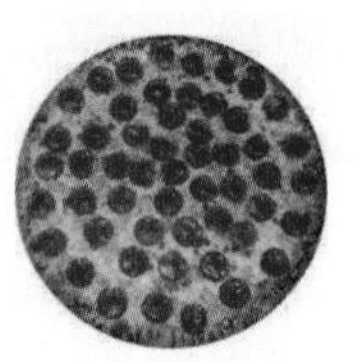
If the diameter of the pizza is 14 inches, then the radius of the pizza is 7 inches.	If the diameter of the pizza is 9 inches, then the radius of the pizza is 4.5 inches.
$A = \pi r^2$ $= (3.14)(7)^2$ $= (3.14)(49)$ $= 153.86$ *square inches*	$A = \pi r^2$ $= (3.14)(4.5)^2$ $= (3.14)(20.25)$ $= 63.585$ *square inches* This is the area for one 9-inch pizza. We will double this answer to get the area for two 9-inch pizzas. $2(63.585) = 127.17$ *square inches*

The 14-inch pizza gives more pizza!!

EXAMPLE 5

A certain species of fish requires 1.5 cubic feet of water per fish to maintain a healthy environment. Find the maximum number of fish you could put in a tank measuring 8 feet by 5 feet by 6 feet.

In this problem, we will need to find the capacity of the aquarium which indicates Volume.

Requires 1.5 cubic feet of water

$$V = lwh$$
$$= (8)(5)(6)$$
$$= 240 \ ft^3$$

If each fish requires 1.5 cubic feet of water and the tank holds 240 cubic feet of water, then we need to know how many "1.5 cubic feet" can fit inside "240 cubic feet". This indicates division!!

$$\frac{240 \ ft^3}{1.5 \ ft^3} = 160$$

You could put 160 fish in the tank.

EXAMPLE 6

A space plane skims the edge of space at 2000 miles per hour. Neglecting altitude, if the circumference of the planet is approximately 20,000 miles, how long will it take for the plane to travel around the planet?

This word problem deals with distance, rate, and time. Our formula to relate those is: d = rt which tells us that distance is equal to rate multiplied by time. Let's reread the problem and see what we are given and what we are asked to find.

Given: rate of the plane: 200 mph
Distance around planet: 20,000 miles
Formula: $d = rt$

Unknown: time it takes to travel around planet

$$d = rt$$

Substitute in formula given d = 20,000 and r = 2000

$$20{,}000 = 2000\,t$$
$$\frac{20{,}000}{2000} = \frac{2000t}{2000}$$
$$10 = t$$

(Reread the question. Then answer.)

It would take the space plane 10 hours to travel around the planet.

YOUR TURN

Solve each application problem. Make sure and show your work.

3. Alice is remodeling her living room and wants to replace the molding around the top of the room and also replace the carpet. The living room measures 12 ft. by 14 ft.

   a. Find the area and perimeter of the room.
   b. How much molding will she need?
   c. How much carpet will she need?

12 ft

14 ft

a. Area = ____________	Perimeter = ____________	b. How much molding? c. How much carpet?

4. Mario's pizza sells one 16-inch pizza or two 10-inch pizzas for $11.99. How much pizza is in each size? Determine which size gives more pizza for the money.

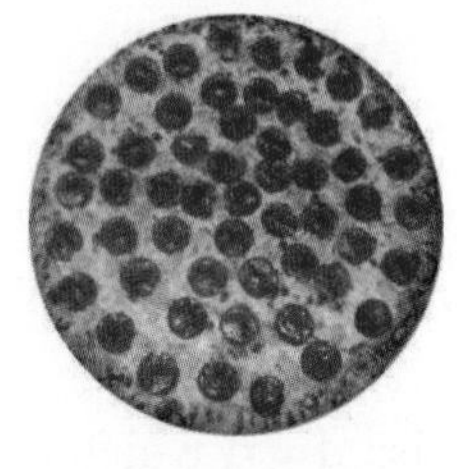

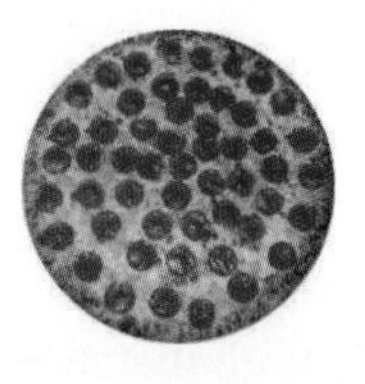

Area of 16-inch pizza:

Area of two 10-inch pizzas:

5. Find the maximum number of goldfish you can put in a cylindrical tank whose diameter is 8 meters and whose height is 3 meters, if each goldfish needs 2 cubic meter of water.

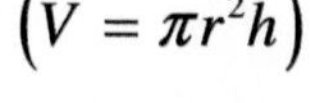

$\left(V = \pi r^2 h\right)$

6. Find how long it takes Louis to drive 550 miles on I-40 if she merges onto I-40 at 7 pm and drives nonstop with his cruise control set on 55 mph.

4.08 Formulas and Problem Solving 4.08

Courtesy of Shutterstock

Using Formulas in Algebraic Application Problems

Formulas can also be used in solving algebraic word problems similar to those we studied in 4.07. The next examples we want to look at involve geometry formulas. It is always a good idea to draw a sketch when a geometric figure is involved.

Remember when solving algebraic word problems, your first step is always to **define your unknowns**. Once again, you will need to write a variable expression for each unknown.

EXAMPLE 7

The length of a rectangular road sign is 2 feet less than three times its width. Find the dimensions if the perimeter is 28 feet.

We have two unknowns: length and width. We need a variable expression for each unknown.

width $= w$
length $= 3w - 2$

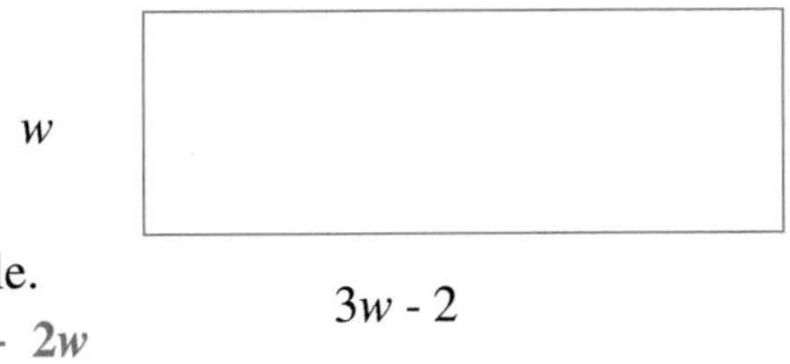

This problem gives us the perimeter of the rectangle.
Formula for the perimeter of a rectangle: $p = 2l + 2w$

Formula: $\mathbf{p = 2l + 2w}$

Substitute: $28 = 2(3w - 2) + 2w$

Solve.	$28 = 2(3w-2) + 2w$	Use distributive property to remove ( ).
	$28 = 6w - 4 + 2w$	Combine like terms on same side of =.
	$28 = 8w - 4$	
	$28 + 4 = 8w - 4 + 4$	Add 4 to both sides.
	$32 = 8w$	
	$\frac{32}{8} = \frac{8w}{8}$	Isolate variable by dividing by 8.
	$4 = w$	

The width of the rectangle is 4 feet and the length of the rectangle is 10 feet.

(Reread the question.)

width $= w = 4$ ft

length $= 3w - 2 = 3(4) - 2 = 12 - 2 = 10$ ft.

EXAMPLE 8

A flower bed is in the shape of a triangle with one side twice the length of the shortest side and the third side is 18 feet more than the length of the shortest side. Find the dimensions if the perimeter is 182 feet.

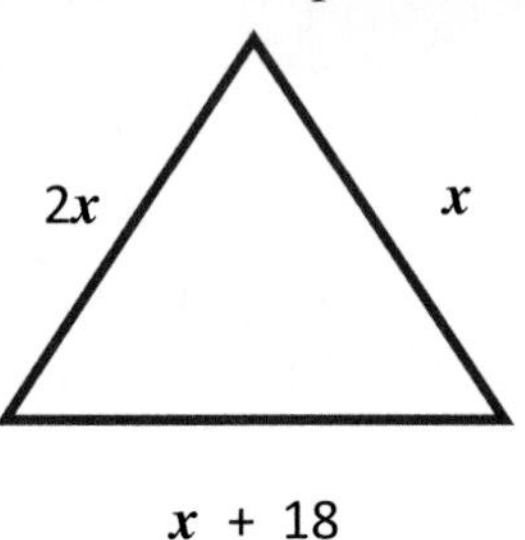

Define the unknown lengths

x = length of shortest side
$2x$ = length of second side
$x + 18$ = length of third side

To find the perimeter of the triangle, we will add the three sides.

Write the equation.

$$\begin{aligned}
P &= \text{short side} + \text{second side} + \text{third side} \\
182 &= x + 2x + x + 18 \\
182 &= 4x + 18 \\
182 - 18 &= 4x + 18 - 18 \\
164 &= 4x \\
\frac{164}{4} &= \frac{4x}{4} \\
41 &= x
\end{aligned}$$

Write the equation only after defining the unknown lengths above.

Now that we know the value of x, we need to find the length of the other two sides. Refer back to those definitions we began with.

x = length of shortest side ⟶ 41 feet
$2x$ = length of second side ⟶ 82 feet
$x + 18$ = length of third side ⟶ 59 feet

4.08 Formulas and Problem Solving 4.08

YOUR TURN

Choose the correct formula. Define the unknown lengths. Write an equation. Solve and answer the question.

7. The length of a rectangular parking lot is 3 feet less than two times its width. Find the dimensions if the perimeter is 51 ft.

Variable Expressions:
width = __________
length = __________

Formula: ________________

Substitute into Formula:

Answer: Length _______________
Width _______________

8. Two sides of a triangle have the same length. The third side measures 2 m less than twice the common length. The perimeter of the triangle is 18 m. What are the lengths of the three sides.

__________ = length of shortest side
__________ = length of second side
__________ = length of third side

Formula: ________________

Substitute into Formula:

Answer:
Length of short side: ________
Length of 2nd side: _________
Length of 3rd side: _________

Solving Literal Equations

When we look at a formula we notice that it is solved for a single variable. The formula, A = bh, is solved for A. The formula P = a + b + c is solved for P. In the next examples, we want to look at how we can manipulate a formula that is solved for one variable but we need to solve it for a different variable. We are going to be solving **literal equations**; this means that we will be solving a formula for a given variable.

We are going to use all of the rules that we've learned for solving equations to solve **literal equations**. You will need to perform "opposite operations" and whatever you do to one side of the equation you must do to the other side of the equation!

At first glance, these exercises appear to be more difficult than your usual solving exercises, but they really aren't that bad. You pretty much do what you've done all along for solving linear equations. The only real difference is that, due to all the variables, you won't be able to simplify your answers as much as you usually do. Here's how "solving literal equations" works:

Look at the equation on the right and the formula on the left. Do you see how they are really the same ?

EXAMPLE 9

• Solve A = bh for b	• Solve 8 = 2b for b
$A = bh$	$8 = 2b$
$\frac{A}{h} = \frac{bh}{h}$ Divide both sides by *h*.	$\frac{8}{2} = \frac{2b}{2}$ Divide both sides by 2.
$\frac{A}{h} = b$	$4 = b$ Simplify

We solved the formula just like we did the linear equation. The only difference was we could actually do the arithmetic in the last step when we divided 8 by 2.

So, to solve a literal equation we will "undo" the side that has the specified variable on it. If a variable has been added, we will subtract to move it to the other side. If a variable is being multiplied, we will divide. Let's look at another example.

EXAMPLE 10

Solve $p = 2l + 2w$ for l.

$$p = 2l + 2w$$

$$p - 2w = 2l + 2w - 2w$$ Subtract $2w$ from both sides.

$$p - 2w = 2l$$

$$\frac{p - 2w}{2} = \frac{\cancel{2}l}{\cancel{2}}$$ Isolate the variable by dividing by 2.

EXAMPLE 11

Solve $x + 6y = 17$ for y.

$$x + 6y = 17$$

$$x + 6y - x = -x + 17$$ Subtract x from both sides.

$$6y = -x + 17$$

$$\frac{6y}{6} = \frac{-x + 17}{6}$$ Isolate y by dividing by –6.

$$y = \frac{-x + 17}{6} \text{ or } y = \frac{1}{6}x + \frac{17}{6}$$ Simplify.

EXAMPLE 12

Solve $w = x + xyz$ for y.

$$w = x + xyz$$

$$w - x = x + xyz - x$$ Subtract x from both sides.

$$w - x = xyz$$

$$\frac{w-x}{xz} = \frac{xyz}{xz}$$ Isolate y by dividing by xz.

$$\frac{w-x}{xz} = y$$ Simplify.

YOUR TURN

Solve each formula for the specified variable.

9. $V = lwh$ *for* h	10. $P = a + b + c$ *for* c
11. $3x + 2y = -6$ *for* y	12. $A = \frac{1}{2}bh$ *for* h (hint: multiply both sides by 2 first.)

YOUR TURN **Answers to Section 4.08.**

1. $w = 18$

2. $h = 15$

3a. P = 52 ft., A = 168 ft^2 3b. perimeter, 52 ft 3c. Area, 168 ft^2 ,

4. 200.96 in^2, 157 in^2, 16 inch pizza

5. 75 goldfish 6. 10 hours

7. width = 9.5 ft., length = 16 ft 8. 5 m , 5 m, 8 m

9. $h = \dfrac{V}{lw}$ 10. $c = P - a - b$ 11. $y = \dfrac{-3x-6}{2}$ *or* $y = \dfrac{-3}{2}x - 3$

12. $h = \dfrac{2A}{b}$

Complete MyMathLab Section 4.08 Homework in your Homework notebook.

4.09 Percent Problem Solving **4.09**

In the previous module we studied percent and percent applications. This section is a review with a few extra problems added. If you need additional help with percent, refer to Module 3.

Finding Percent Increase and Percent Decrease

Percent increase and percent decrease are measures of **percent change**, which is the extent to which something gains or loses value.

When a quantity shrinks (gets smaller), then we can compute its PERCENT DECREASE.

When a quantity grows (gets bigger), then we can compute its PERCENT INCREASE.

To Find Percent Increase or Decrease: Decimal form

$$\textit{percent increase} = \frac{\textit{amount of increase } \{\textit{difference}\}}{\textit{original amount}}$$

$$\textit{percent decrease} = \frac{\textit{amount of decrease } \{\textit{difference}\}}{\textit{original amount}}$$

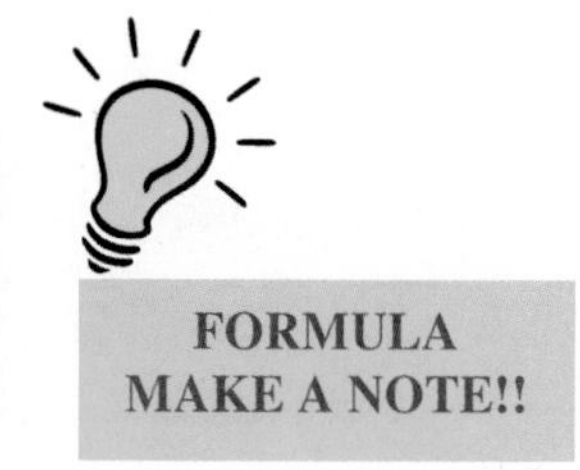

FORMULA
MAKE A NOTE!!

The answer from the formula above will be a decimal. Don't forget to change your answer to a percent by moving the decimal 2 places to the right.

We can write this formula with a 100% at the end to emphasize that since it is a percent increase/decrease, it should be reported as a percent.

EXAMPLE 1

Find percent increase.

Burning Man is a play on Broadway. One of its characters is a mechanical spider. On opening night in 2008, the play saw 49,500 people walk through its doors. In 2010, the play saw its numbers rise to 51,454 attendees. Find the percent increase in attendance. Round your answer to the nearest whole percent.

$$\text{percent increase} = \frac{\text{amount of increase \{difference\}}}{\text{original amount}} \cdot 100\%$$

$$\text{percent increase} = \frac{(51{,}454 - 49{,}500)}{49{,}500} \cdot 100\%$$

$$\text{percent increase} = \frac{1954}{49{,}500} \cdot 100\%$$

$$\text{percent increase} = 0.04 \cdot 100\%$$

$$\text{percent increase} = 4\%$$

Courtesy of Fotolia

If you are going to do this all in one step in your calculator, be sure to keep the numerator in parentheses.

Note that when you compute percent increase or decrease, you always compare how much a quantity has changed from the original amount.

EXAMPLE 2

There are 110 calories in a cup of whole milk and only 64 calories in a cup of skim milk. In switching to skim milk, find the percent decrease in number of calories. Round to the nearest tenth of a percent, if needed.

$$\text{percent decrease} = \frac{\text{amount of decrease \{difference\}}}{\text{original amount}} \cdot 100\%$$

$$\text{percent decrease} = \frac{(110 - 64)}{110} \cdot 100\%$$

$$\text{percent decrease} = \frac{46}{110} \cdot 100\%$$

$$\text{percent decrease} = 0.418 \cdot 100\%$$

$$\text{percent decrease} = 41.8\%$$

Courtesy of Fotolia

4.09 Percent Problem Solving 4.09

EXAMPLE 3

Las Vegas, Nevada increased in population 19% from the 2000 census to the 2010 census. In 2000, the population of Las Vegas was about 478,500.

a. Find the increase in population from 2000 to 2010.

b. Find the population for Las Vegas in 2010.

What number is 19% of 478,500?

$$n = 19\% \cdot 478{,}500$$
$$n = 0.19 \cdot 478{,}500$$
$$n = 90{,}915$$

a. The population increased by 90,915 people between 2000 and 2010.

b. To find the population for Las Vegas in 2010, we will need to add 90,915 to the original population in 2000.

$$\textit{new population} = 90{,}915 + 478{,}500$$
$$\textit{new population} = 569{,}415$$

Mark-up and New Price

$$\textit{amount of mark-up} = \textit{mark-up rate} \cdot \textit{original price}$$
$$\textit{new price} = \textit{original price} + \textit{amount of mark-up}$$

EXAMPLE 4

A student at a university makes money by buying and selling used cars. Charles bought a used car and later sold it for a 15% profit. If he sold it for \$4623, how much did Charles pay for the car?

Notice this time we are given the price after the increase. So, \$4623 is the price of the car plus a 15% profit on the car. Since we don't know the price of the car we will represent it with the variable x.

Price of car + 15 % of price of car = 4623

$$x + 0.15x = 4623$$
$$1x + 0.15x = 4623 \quad \text{The coefficient of } x \text{ is 1.}$$
$$1.15x = 4623 \quad \text{Add like terms.}$$
$$\frac{1.15x}{1.15} = \frac{4623}{1.15}$$
$$x = 4020$$

The original price of the car is \$4020.

4.09 Percent Problem Solving **4.09**

Suppose a television that normally sells for $600 is on sale at 25% off. This means the item's original price of $600 is discounted by 25% of $600, or $150. 25% is known as the discount rate. $150 is the amount of discount {in dollars}. And the sale price is $600 – $150 or $450.

Discount and Sale Price

$$\textit{amount of discount} = \textit{discount rate} \cdot \textit{original price}$$

$$\textit{sale price} = \textit{original price} - \textit{amount of discount}$$

EXAMPLE 5

A Sony DVD Player was advertised on sale for 15% off the regular price of $70. Find the *discount* and the *sale price*.

$\textit{amount of discount} = \textit{discount rate} \cdot \textit{original price}$ Start with the formula.

$\textit{amount of discount} = 0.15 \cdot 70$ Plug in what you know, changing the % to a decimal.

$\textit{amount of discount} = \$10.50$ Multiply.

$\textit{sale price} = \textit{original price} - \textit{amount of discount}$ Start with the formula.

$\textit{sale price} = \$70 - \$10.50$ Plug in what you know.

$\textit{sale price} = \$59.50$ Subtract.

4.09 Percent Problem Solving 4.09

YOUR TURN

Solve. Show the formula used to solve and all steps. Think about your solutions. "Ask, does this answer make sense?"

1. Frosty Fizz increased the price of a $1.65 cola by 15%. Find the increase in price and the new price. Round to the nearest cent.

2. A stereo normally priced for $379 is on sale for 35% off. Find the discount and the sale price.

3. The number of fraud complaints rose from 250,000 in 2000 to 410,000 in 2001. Find the percent increase.

4. A birthday celebration is $47.20 including tax. Find the total cost after a 10% tip is added to the bill.

YOUR TURN **Answers to Section 4.09.**

1. Increase: 0.25, new price: $1.90
2. discount: $132.65 , sale price: $246.35
3. 64%
4. $51.92

Complete MyMathLab Section 4.09 Homework in your Homework notebook.

4.10 Linear Inequalities and Problem Solving 4.10

So far in this module, we have studied expressions and equations. Remember, expressions can only be simplified not solved. Equations have an "=" and can be solved.

INEQUALITIES:

We will now begin our study of inequalities. **Inequalities** are algebraic expressions related by one of four inequality symbols below:

$<$ *"is less than"* $\leq$ *"is less than or equal to"*

$>$ *"is greater than"* $\geq$ *"is greater than or equal to"*

Make sure you can tell the difference between an expression, an equation, and an inequality.

Expressions	Equations	Inequalities
$3x - 2$	$3x - 2 = 7$	$3x - 2 < 8$
$-2x - 4(x+3)$	$-2x - 4(x+3) = 12$	$-2x - 4(x+3) \geq 12$
$\frac{x+3}{7}$	$\frac{x+3}{7} = \frac{3}{5}$	$\frac{x+3}{7} > \frac{3}{5}$

As stated before, we solve equations. Most of the equations we have solved previously have had one solution. If given the equation $x + 3 = 7$, we know the solution is $x = 4$. There is only one number that will make the equation true. The number 4 is the only number we can add to 3 and get exactly 7. All other numbers make the equation false.

However, solutions for inequalities include many more numbers. If the statement reads $x \geq 3$, that means that any number greater than <u>or</u> equal to 3 would make the statement true. How many numbers are greater than or equal to 3??? We could list a few, such as 3, 4, 5, 6, 7, 8, ... but that would not include <u>all</u> the numbers greater than or equal to three. There are many more – too many to list!! Mathematically, we call this an infinite number of solutions.

Let's look at this inequality: $x < 5$

This statement tells us that all the numbers less than 5 would make this equation true. Notice that 5 would not make this equation true. The number 5 is not less than 5. We could list some of the solutions: - 6, 0 , 1, 2.6 , ... but again this would not include all the solutions.

4.10 Linear Inequalities and Problem Solving 4.10

Because we can't list all the numbers in a solution for an inequality, we will be asked to graph the solution set on a number line.

Graph $x \geq 3$

To graph all the numbers greater than or equal to 3, we will shade all the numbers to the right of 3 since they are greater than 3. Then we place a closed circle on the point representing 3 which indicates that 3 is also included in the solution set.

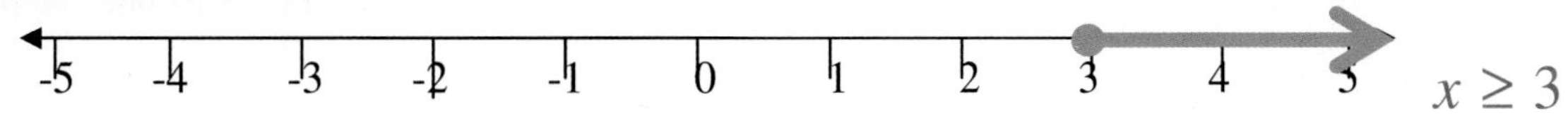

Graph x < 3

To graph all the numbers less than 3, we will shade all the numbers to the left of 3 since they are less than 3. Then we place a open circle on the point representing 3 which indicates that <u>3 is not included</u> in the solution set. 3 is not less than 3.

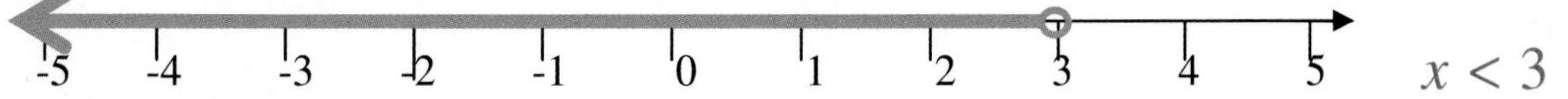

EXAMPLE 1

Graph. $x \leq -\frac{7}{2}$

Rewrite $-\frac{7}{2}$ as the mixed number – 3 ½ to place on number line.

To graph, place a closed circle at **-3 ½** since the inequality is $\leq$. Then we will shade to the left since all the numbers "less than" are to the left on the number line.

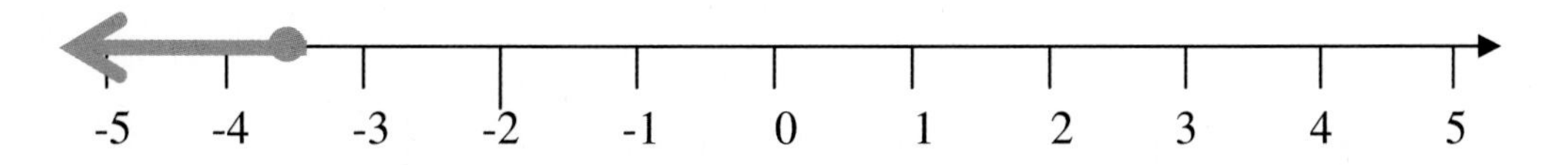

EXAMPLE 2

Graph. $1 < m$

To graph a solution set on the number line, we need the variable to be on the left hand side so that we can read the inequality correctly. Recall, from a previous module that we can read inequalities from left to right or right to left. So, the statement $1 < m$ means the same thing as $m > 1$. We will graph $m > 1$.

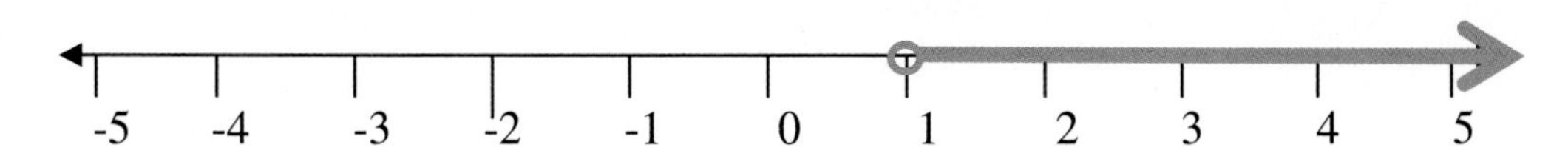

EXAMPLE 3

Graph. $0 < y \leq 5$

This is a ***compound inequality***. You will notice it has two inequality symbols. This statement is telling us that there are two conditions placed on *y* to be in the solution set. If we were to separate the two inequalities we would have: $0 < y$ *and* $y \leq 5$. To read easily, let's change the first one and put *y* first. $y > 0$ *and* $y \leq 5$.

The solution set is all the numbers that are **greater than 0 AND less than or equal to 5**.

We place an open circle at 0 since the symbol is > and a closed circle at 5 since the symbol is $\leq$ and shade between.

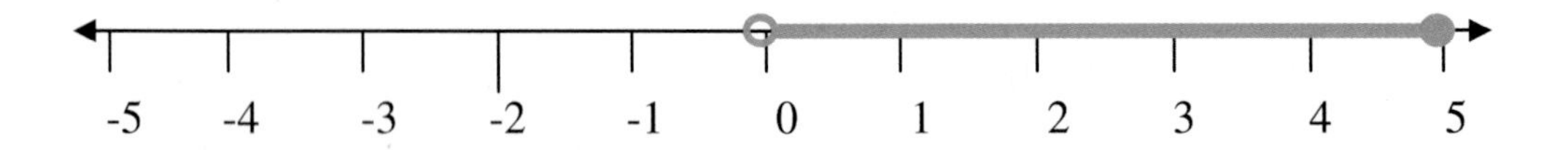

Make a note

To graph an inequality always make sure your variable is on the left side of the inequality. Then follow the guidelines below.

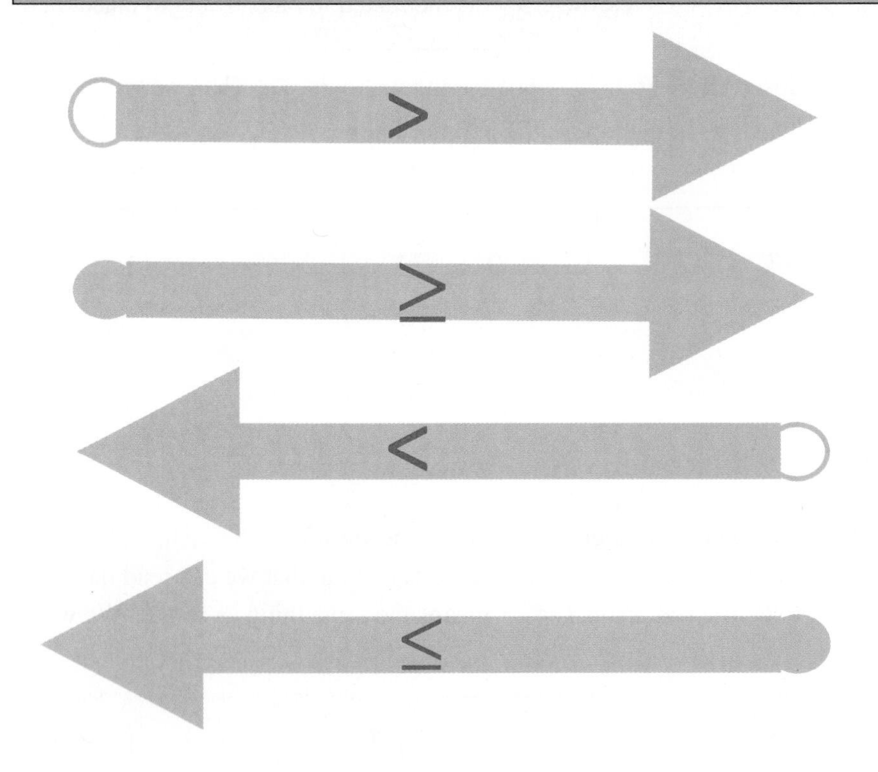

YOUR TURN

1. Graph. $x > -2$

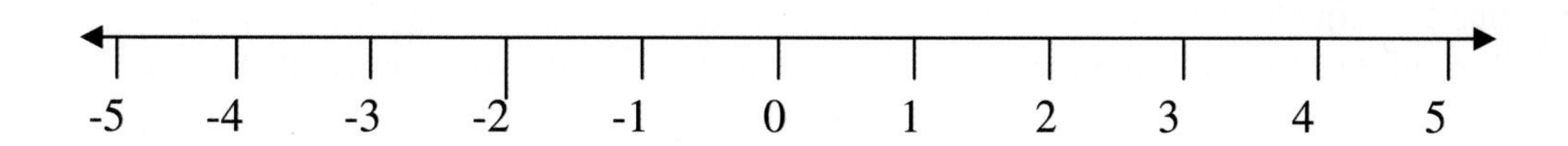

2. Graph. $x \leq \frac{3}{2}$

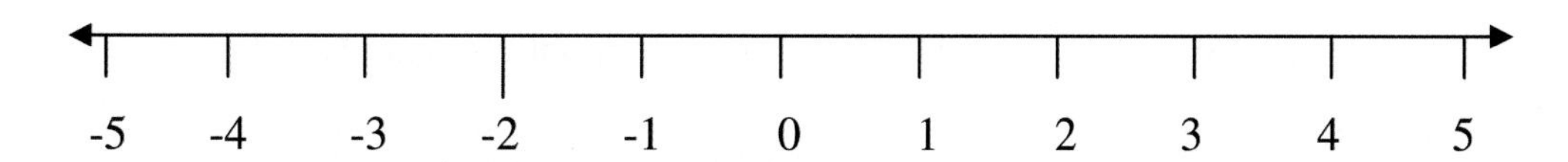

3. Graph. $-4 < m$

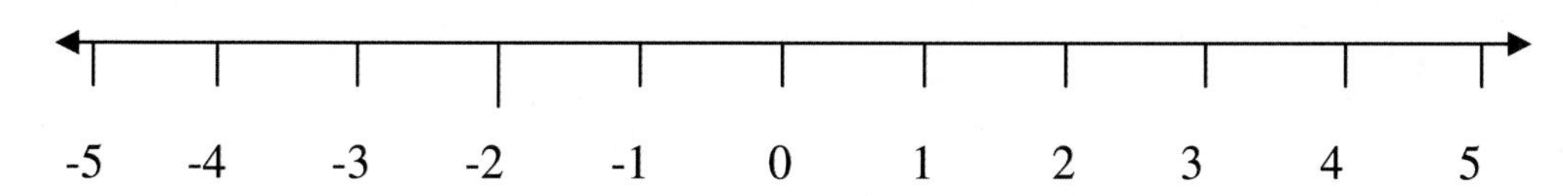

4. Graph. $-1 \leq x < 4$

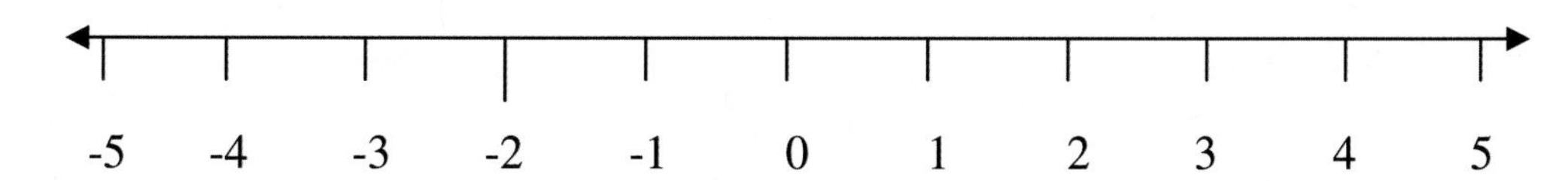

4.10 Linear Inequalities and Problem Solving 4.10

Solving Linear Inequalities

Solving an inequality is similar to solving an equality. Our goal is to get the variable alone on one side of the inequality. When we solved linear equations we were able to add or subtract from both sides and multiply and divide both sides to "undo" the equation. So, our first question will be : Will we be able to do the same to inequalities?

Can we add or subtract the same number from both sides and still get a true equation? Let's test and see.

Begin with a True Inequaltiy

$$3 < 7$$

Add 2 to both sides	Subtract 1 from both sides
$3 < 7$ $3+2<7+2$ $5<9$ **Inequality is still TRUE!!**	$3 < 7$ $3-1 < 7-1$ $2 < 6$ **Inequality is still TRUE!!**

Adding and subtracting the same quantity from both sides of an inequality does not change the inequality.

What about multiply and divide??? Can we multiply or divide both sides of the inequality by the same number and not change the inequality? Let's test again and see.

.

Begin with a True Inequaltiy

$$8 > 2$$

Multiply both sides by 2	Multiply both sides by -2
$8 > 2$ $8(2) > 2(2)$ $16 > 4$ **Inequality is still TRUE!!**	$8 > 2$ $8(-2) > 2(-2)$ $-16 > -4$ **Inequality is FALSE!!**

Divide both sides by 2	Divide both sides by -2
$8 > 2$ $\frac{8}{2} > \frac{2}{2}$ $4 > 1$ **Inequality is still TRUE!!**	$8 > 2$ $\frac{8}{-2} > \frac{2}{-2}$ $-4 > -1$ **Inequality is FALSE!!**

4.10 Linear Inequalities and Problem Solving 4.10

These examples illustrate that an inequality does not remain true if we multiply or divide by a negative number. So, we will need to make some adjustments when this happens. If we look at the two false statements from above, $-16 > -4$ **and** $-4 > -1$, we realize that they would be true if we reversed the inequality symbol. $-16 < 4$ **and** $-4 < -1$.

If you multiply or divide both sides of an inequality by a negative number, the direction of the inequality symbol must be reversed.

The following steps may be helpful when solving inequalities in one variable. These steps are similar to those given to solve a linear equality.

Step 1: If an inequality contains fractions, multiply by the LCD to clear the equation of fractions.

Step 2: If an inequality contains parentheses, use the distributive property to remove them.

Step 3: Simplify each side of the inequality by combining like terms.

Step 4: Get all variable terms on one side and all numbers on the other side by using addition property of equality.

Step 5: Isolate the variable by dividing both sides by the coefficient of the variable. **Remember if you divide by a negative number you must reverse the direction of the inequality symbol.**

EXAMPLE 4

Solve. $-3x > 9$ Divide both sides by -3 and reverse the inequality symbol.

$\frac{-3x}{-3} > \frac{9}{-3}$ Simplify.

Don't forget to reverse symbol when you divide by a negative number!!!

$x < -3$

The graph of the solution set $\{x|\ x < -3\}$ is shown.

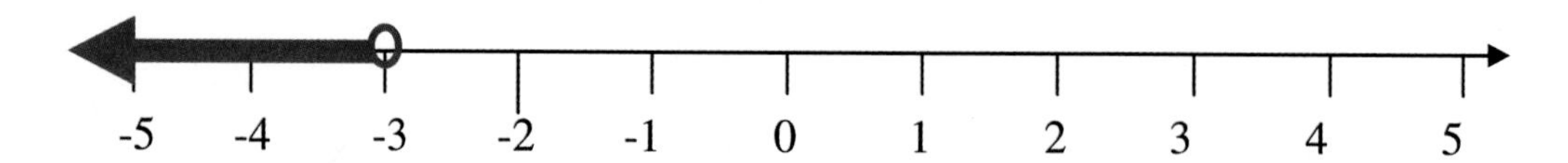

EXAMPLE 5

Solve and graph the solution set: $2x - 1 \geq 5x - 4x$

$2x - 1 \geq 5x - 4x$ Combine like terms on same side of equal.

$2x - 1 \geq x$

$2x - 1 - x \geq x - x$ Subtract x from both sides.

$x - 1 \geq 0$

$x - 1 + 1 \geq 0 + 1$ Add 1 to both sides.

$x \geq 1$

The graph of the solution set $\{x|\ x \geq 1\}$ is shown.

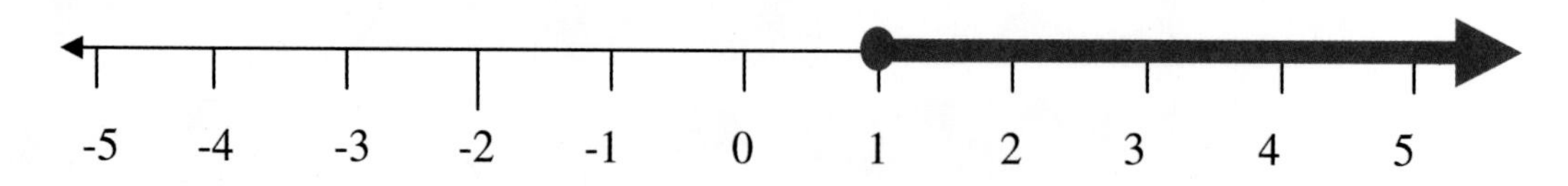

EXAMPLE 6

Solve and graph the solution set: $-8x + 4 \geq 6(4 - x)$

$-8x + 4 \geq 6(4 - x)$

$-8x + 4 \geq 24 - 6x$ Use distributive property to remove ( ).

$-8x + 4 + 6x \geq 24 - 6x + 6x$ Add 6*x* to both sides.

$-2x + 4 \geq 24$ Simplify.

$-2x + 4 - 4 \geq 24 - 4$ Subtract 4 from both sides.

$-2x \geq 20$

$\frac{-2x}{-2} \geq \frac{20}{-2}$ Divide both sides by -2 and reverse inequality.

$x \leq -10$

Don't forget to reverse symbol when you divide by a negative number!!!

The graph of the solution set $\{x \mid x \leq -10\}$ is shown.

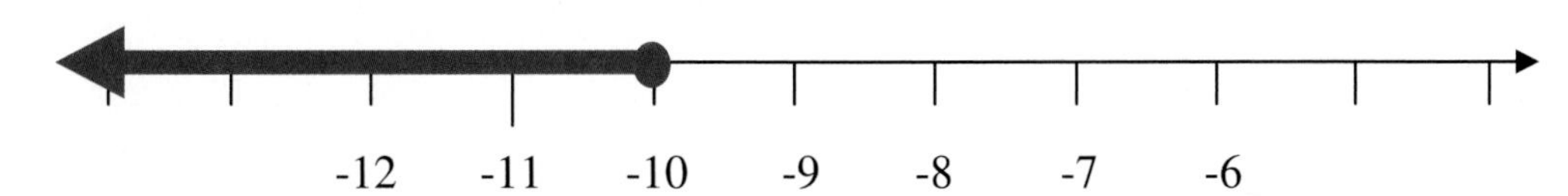

EXAMPLE 7

Solve and graph the solution set: $10 > 5 - x$

$10 > 5 - x$

$10 - 5 > 5 - x - 5$ Subtract 5 from both sides.

$5 > -x$

$\frac{5}{-1} > \frac{-x}{-1}$ Divide both sides by -1 and reverse the inequality symbol.

$-5 < x$

To graph, remember that we need the variable on the left side of the inequality. So, let's rewrite the solution with the variable to the left.

$-5 < x$ *means the same as* $x > -5$

The graph of the solution set $\{x \mid x > -5\}$ is shown.

YOUR TURN

Solve and graph the solution set.

5. $4x - 1 < 19$

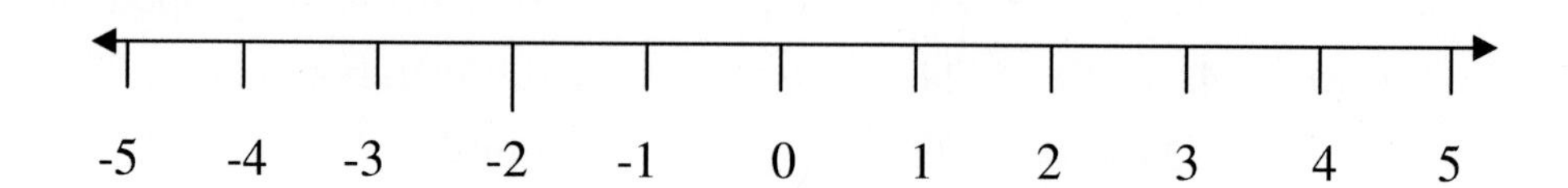

6. $-9 \leq 4x + 7$

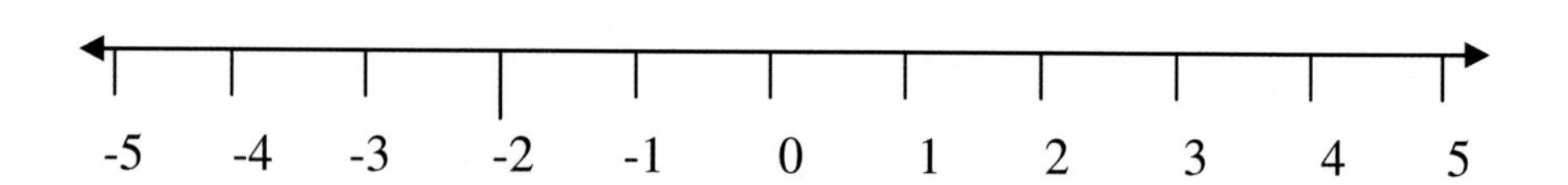

7. $-5x + 7 \geq 2(x - 3) - 1$

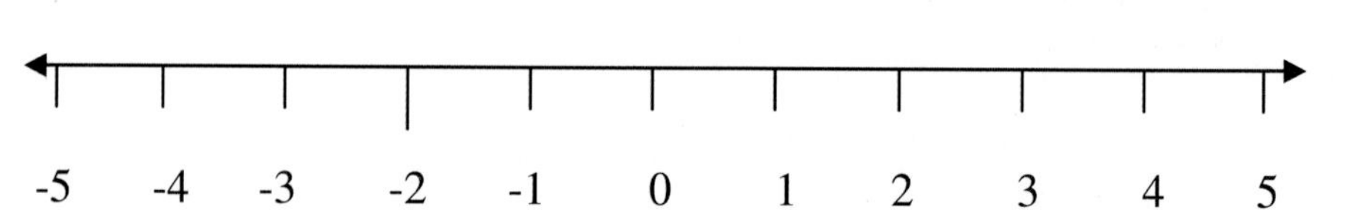

8. $3x - 1 > 7x - 6x$

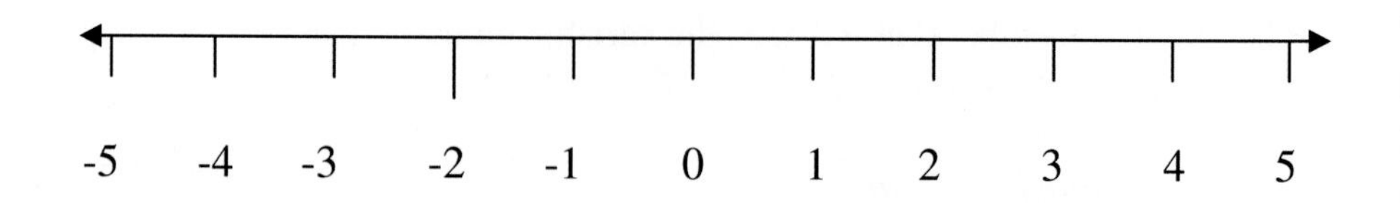

4.10 Linear Inequalities and Problem Solving 4.10

Word problems containing words such as “at least” or “ at most” usually indicate that an inequality should be solved instead of an equation. The table below shows some of the common phrases that suggest inequality.

APPLICATIONS - INEQUALITIES

Inequality Translations			
$>$	$\geq$	$<$	$\leq$
is greater than	at least	is less than	at most
exceeds	no less than		no more than

The next example uses the idea of finding the average of a number of scores. Recall, to find the average of a group of numbers, add the numbers and divide by how many numbers you added together.

EXAMPLE 8

Brent has grades of 86, 88, and 78 on his first three tests in geometry. If he wants an average of at least 80 after his fourth test, what must he score on that test?

Let x = Brent’s score on his fourth test

To find his average after four tests, add the test scores and divide by 4.

$$\frac{86 + 88 + 78 + x}{4} \geq 80$$

$$\frac{252 + x}{4} \geq 80 \qquad \text{Add the known scores.}$$

$$4\left(\frac{252 + x}{4}\right) \geq 4(80) \qquad \text{Multiply by 4.}$$

$$252 + x \geq 320$$

$$252 + x - 252 \geq 320 - 252 \qquad \text{Subtract 252.}$$

$$x \geq 68$$

Brent must score 68 or more on the fourth test.

EXAMPLE 9

A bank offers two checking-account plans. Advantage Checking charges 30 cents per check whereas their Premium Plan cost \$1.17 per month plus 21 cents per check. For what number of checks per month will the Premium Plan cost less?

Let x = number of checks

This problem is asking when the cost of the Premium Plan is less than the cost of Advantage Checking.

cost of the Premium Plan is less than the cost of Advantage Checking

cost of Premium Plan < **cost of Advantage Checking**

\$1.17 plus \$0.21 for every check < **\$0.30 for every check**

$$1.17 + 0.21x < 0.30x$$

$$117 + 21x < 30x$$ Multiply by 100 to remove decimals.

$$117 + 21x - 21x < 30x - 21x$$ Subtract $21x$ from both sides.

$$117 < 9x$$

$$\frac{117}{9} < \frac{9x}{9}$$ Divide by 9.

$$13 < x$$

$$x > 13$$

The Premium Plan will cost less if more than 13 checks are written per month.

The solution is $\{x \mid x > 13\}$

EXAMPLE 10

The width of a rectangle is fixed at 7 cm. Determine (in terms of an inequality) those lengths for which the area will be less than 168 cm^2.

Area is less than 168 ⟶ **A < 168**

Area of a rectangle is length times width. ⟶ $l \cdot w < 168$

7 $A = l \cdot w$

$$l \cdot 7 < 168$$

$$7l < 168$$

$$\frac{7l}{7} < \frac{168}{7}$$

$$l < 24$$

The length must be less than 24 cm.

The solution is $\{l \mid l < 24\ cm\}$

EXAMPLE 11

Elle has $1000 in a savings account at the beginning of the summer. She wants to have at least $250 in the account by the end of the summer to pay her part of a beach house rental. She withdraws $50 each week for food, clothes, and entertainment. How many weeks can Elle withdraw from her account and have at least $250 left?

Solution:
This example is asking us to find the number of weeks Elle can withdraw $50 and still have at least $250 remaining at the end of the summer.
Let's define the unknowns:
S = represents the amount of money left in Elle's savings account and
t = the number of weeks that have passed since the beginning of summer.

Courtesy of Shutterstock

Elle begins with $1000 in checking. She withdraws $50 per week which would be subtracted from her account. If she withdraws $50 per week for t weeks, she would withdraw ***50t*** from her account. Her account balance would be 1000 minus the money she withdraws.

$$S = 1000 - 50t$$

Let's review the question again: How many weeks can Elle withdraw from her account and have at least $250 left? Elle wants her balance to be **at least $250.** ⟶ $S \geq 250$.

$S \geq 250$ ⟶ Plug in for S. ($S = 1000 - 50t$)

$1000 - 50t \geq 250$ or if we prefer we can reverse the inequality as $250 \leq 1000 - 50t$

Now we can solve either of the inequalities: $1000 - 50t \geq 250$ OR $250 \leq 1000 - 50t$.

$$1000 - 50t \geq 250$$

$$1000 - 50t - 1000 \geq 250 - 1000 \quad \text{Subtract 1000.}$$

$$-50t \geq -750$$

$$\frac{-50t}{-50} \geq \frac{-750}{-50} \quad \text{Divide by } -50 \text{ and reverse inequality.}$$

$$t \leq 15$$

Elle could withdraw money each week for 15 weeks or less and still have at least $250 in her account.

YOUR TURN

Write and inequality and solve.

9. Brent has grades of 74, 81, and 82 on his first three tests in geometry. What score must he get on his final exam to receive a B in the class? The final exam counts as 2 tests , and a B is received if the final average is greater than or equal to 80. (Hint: there will be 5 grades to average)

Write an inequality:

Solve.

10. Dennis and Nancy Woods are celebrating their 50th wedding anniversary by having a reception at a local hotel. They have budgeted $3000 for their reception. If the hotel charges a $50.00 cleanup fee plus $34 per person, find the greatest number of people that they may invite and still stay within their budget.

Write an inequality:

Solve.

4.10 Linear Inequalities and Problem Solving 4.10

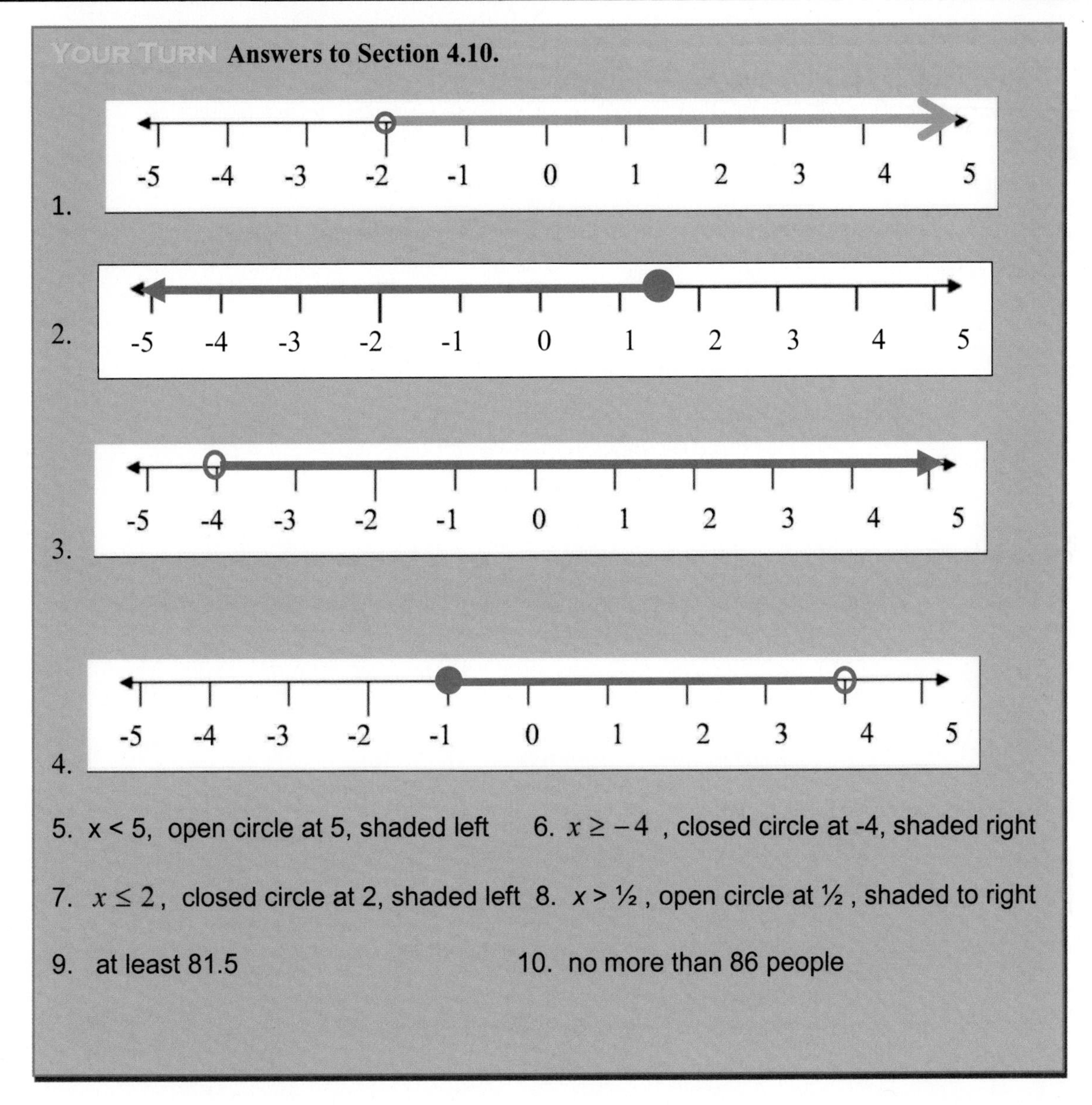

Complete MyMathLab Section 4.10 Homework in your Homework notebook.

DMA 040 TEST: Review your notes, key concepts, formulas, and MML work. ☺

Student Name: ______________________________ **Date:** ____________

Instructor Signature: ______________________________

DMA 050

DMA 050

GRAPHS/EQUATIONS OF LINES

This course provides a conceptual study of problems involving graphic and algebraic representations of lines. Topics include slope, equations of lines, interpretation of basic graphs, and linear modeling. Upon completion, students should be able to solve contextual application problems and represent real-world situations as linear equations in two variables.

You will learn how to

- Generate a table of values given an equation in two variables and plot in Cartesian plane to graph a line.
- Demonstrate an understanding of the concept of slope as a rate of change in real world situations using the slope formula.
- Find and interpret the x- and y-intercepts of linear models in real world situations.
- Graph linear equations using a variety of strategies.
- Represent real world situations in tabular, graphical, and algebraic equation form using two variables.
- Given a contextual application, write a linear equation and use the equation to make predictions.
- Demonstrate a conceptual understanding of horizontal and vertical lines in terms of slope and graphically.
- Demonstrate a conceptual understanding of the concept of an algebraic function.
- Analyze and interpret basic graphs to solve problems.

Graphing Linear Equations:

Slope-Intercept Equation $y = mx + b$

Standard Form Equation $Ax + By = C$

Writing Equations: $y - y_1 = m(x - x_1)$

Special Lines

Vertical Lines $x = \#$

Horizontal Lines $y = \#$

Parallel Lines

Perpendicular Lines

Courtesy of Fotolia

Professions That Use Linear Equations

Management
Management occupations
- Advertising, marketing, promotions, public relations, and sales managers
- Computer and information systems managers
- Construction managers
- Engineering and natural sciences managers
- Farmers, ranchers, and agricultural managers
- Financial managers
- Funeral directors
- Human resources, training, and labor relations managers and specialists
- Industrial production managers
- Medical and health services managers
- Property, real estate, and community association managers
- Purchasing managers, buyers, and purchasing agents

Business and financial operations occupations
- Accountants and auditors
- Budget analysts
- Insurance underwriters
- Loan officers
- Management analysts
- Meeting and convention planners

Healthcare support occupations
- Medical assistants
- Nursing, psychiatric, and home health aides

Protective service occupations
- Correctional officers

Farming
- Agricultural workers
- Forest, conservation, and logging workers

Installation
Electrical and electronic equipment mechanics, installers, and repairers
- Electrical and electronics installers and repairers
- Electronic home entertainment equipment installers and repairers

Vehicle and mobile equipment mechanics, installers, and repairers
- Automotive service technicians and mechanics

Professional
Computer and mathematical occupations
- Actuaries
- Computer programmers
- Computer software engineers
- Computer support specialists and systems administrators
- Computer systems analysts
- Mathematicians
- Statisticians

Architects, surveyors, and cartographers
- Architects, except landscape and naval
- Surveyors, cartographers, photogrammetrists, and surveying technicians

Engineers
- Aerospace engineers
- Biomedical engineers
- Chemical engineers
- Civil engineers
- Computer hardware engineers
- Electrical engineers
- Environmental engineers
- Industrial engineers
- Materials engineers
- Mechanical engineers
- Nuclear engineers
- Petroleum engineers

Drafters and engineering technicians
- Drafters
- Engineering technicians

Life scientists
- Biological scientists
- Conservation scientists and foresters
- Medical scientists

Physical scientists
- Atmospheric scientists
- Chemists and materials scientists
- Environmental scientists and hydrologists
- Physicists and astronomers

Social scientists and related occupations
- Economists
- Market and survey researchers
- Social scientists, other
- Urban and regional planners

Legal occupations
- Judges, magistrates, and other judicial workers
- Paralegals and legal assistants

Education, training, library, and museum occupations
- Librarians
- Teacher assistants
- Teachers-adult literacy and remedial and self-enrichment education
- Teachers-postsecondary
- Teachers-preschool, kindergarten, elementary, middle, and secondary
- Teachers-special education

Art and design occupations
- Artists and related workers

Health diagnosing and treating occupations
- Optometrists
- Pharmacists
- Physical therapists
- Physicians and surgeons
- Registered nurses
- Veterinarians

Health technologists and technicians
- Clinical laboratory technologists and technicians
- Licensed practical and licensed vocational nurses
- Medical records and health information technicians
- Nuclear medicine technologists
- Veterinary technologists and technicians

Administrative

Information and record clerks
- Human resources assistants, except payroll and timekeeping

Material recording, scheduling, dispatching, and distributing occupations
- Stock clerks and order fillers

Other office and administrative support occupations
- Data entry and information processing workers

Metal workers and plastic workers
- Computer control programmers and operators
- Machinists

Printing occupations
- Prepress technicians and workers

Construction
- Carpenters
- Electricians
- Glaziers

http://www.xpmath.com/careers/topicsresult.php?subjectID=2&topicID=2

5.01 Reading Pictographs, Bar Graphs, Histograms, and Line Graphs 5.01

In this section, you will practice reading graphs which contain data (information).

PICTOGRAPHS

Pictograph: a graph that uses pictures or symbols. To read a pictograph, you count the number of pictures.

★ *Make certain you look at the key, do not assume one picture = 1.*

EXAMPLE 1

The pictograph shows last year's fruit production by the top fruit-producing regions. Use the pictograph to answer the following questions. Notice that each apple represents 25 million bushels.

Annual Fruit Production in Top Producing Regions	
Mountain Region	12.5 apples
Central Region	10.5 apples
Southern Region	2 apples
Coastal Region	1.5 apples
Lake Region	5 apples
Northern Region	4 apples

(Key: one apple = 25 million bushels)

1. **How many bushels of fruit were produced in the Southern Region?**
   There are 2 apples next to the Southern Region. Each apple represents 25 million bushels.
   The Southern Region will produce 2 (25 million) = 50 million bushels of fruit.

2. **How many bushels of fruit were produced in the Coastal Region?**
   There are 1.5 apples next to the Coastal Region. Each apple represents 25 million bushels.
   The Coastal Region will produce 1.5 (25 million) = 37. 5 million bushels of fruit.

3. **How many more bushels did the Mountain Region produce than the Central Region?**
   The Mountain Region has 12.5 apples next to it. Central has 10.5 apples. There is a difference of 2 apples which represent 2 (25) = 50 million bushels of fruit.

   The Mountain Region produced 50 million more bushels of fruit than the Central Region.

5.01 Reading Pictographs, Bar Graphs, Histograms, and Line Graphs 5.01

BAR GRAPHS

Bar graph: ***vertical or horizontal*** bars are used to represent data.
Let's look at the parts of a bar graph to better understand how to read a bar graph.

The bar graph shows the number of calories burned per hour during each activity.

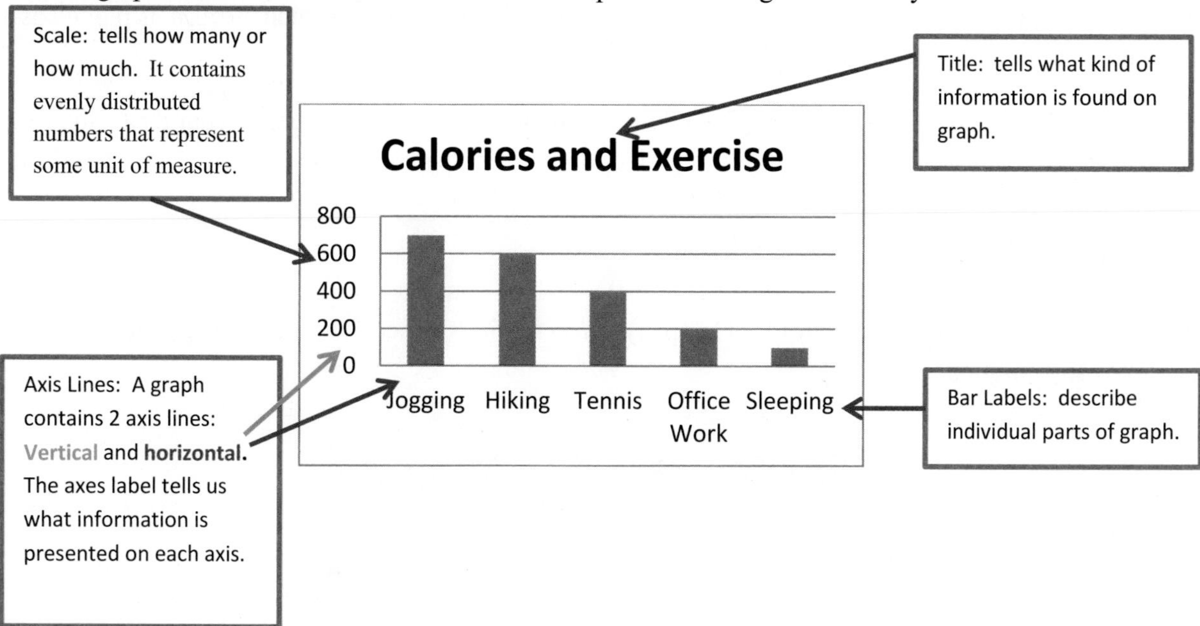

Now that we understand all the pieces that go into a bar graph we are ready to analyze a bar graph

EXAMPLE 2

The bar graph above shows the number of calories burned per hour during each activity by a person weighing 165 pounds. Use the bar graph to answer the following questions. Notice the vertical scale is scaled by increments of 200.

1. **What activity burns the most calories? How many?**
   The tallest bar is jogging. The jogging bar is approximately halfway between 600 and 800. Since the scale is in increments of 200, halfway between 600 and 800 would be 700. Jogging burns the most calories at 700 calories/hr.

2. **How many more calories are burned during hiking than office work?**
   Hiking burns 600 calories/hr . Office work burns only 200 calories/hr. The difference is $600 - 200 = 400$ calories. Hiking burns 400 calories/hr more than office work.

3. **How many fewer calories are burned playing tennis than hiking?**
   Hiking burns 600 calories/hr. Tennis only burns 400 calories/hr.
   Tennis burns 200 fewer calories/hr than hiking.

LINE GRAPHS

A line graph is another way to give a visual representation of the relationship of data. Line graphs are particularly useful for showing change over time and are great tools to track and trend data. Similar to most other graphs, line graphs have a vertical axis and a horizontal axis. If you are plotting changes in data over time, time is plotted along the horizontal or x-axis. Your other data, such as sales, is plotted as individual points along the vertical or y-axis.

When the individual data points are connected by lines, they clearly show changes in your data - such as how sales have increased or decreased over a period of time. You can use these changes to find trends in your data and possibly to predict future results.

EXAMPLE 3

The line graph shows estimated sales for Company A (in millions) since 1996. Use the graph to answer the following question

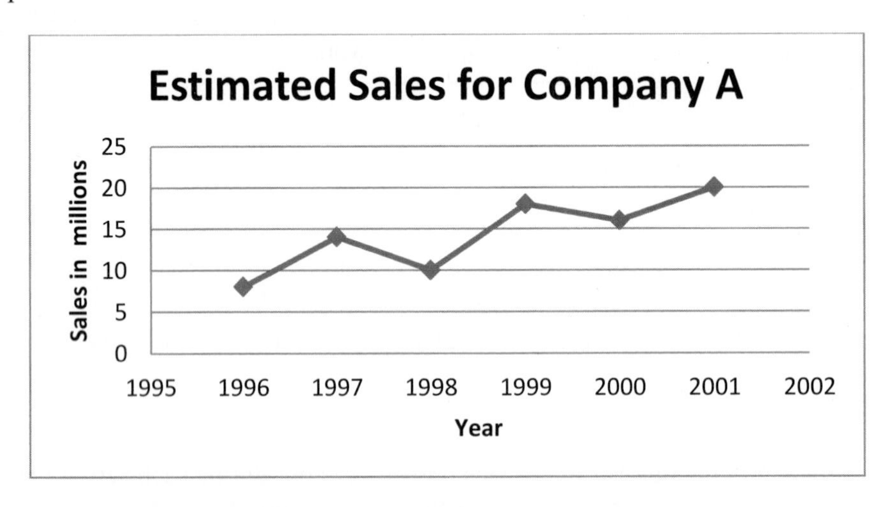

1. **In what year did Company A have the most sales? Least sales?**

   The highest point on the graph occurs in 2001 which represents 20 million in sales.

   The lowest point on the graph occurs in 1996 which represents approximately 8 million in sales each year.

2. **Between what years what was the greatest difference in sales?**
   The steepest part of the line occurs between 1998 and 1999. So, between these two years is the greatest difference in the amount of sales.

3. **How much more did the company sell in 2001 than in 1998?**
   The point for 1998 is at 10 million while the point for 2001 is at 20 million. The company sales were 10 million more in 2001.

HISTOGRAMS

A histogram is a special bar graph. The width of each bar represents a range of numbers called an interval. The height of each bar corresponds to how many times a number in the interval occurs. The bars in a histogram lie side by side with no space between them.

EXAMPLE 4

The test scores of 25 students are summarized in the table below.

Student Scores	Frequency (number of students)
40 – 49	**1**
50 – 59	**2**
60 – 69	**2**
70 – 79	**6**
80- 89	**10**
90 – 99	**4**

We can represent this data in a histogram. The width of the bar will represent each range of the test scores. The height of the bar will represent how many students scored in that interval. Below is the histogram for the above date. Use this histogram to answer the following questions.

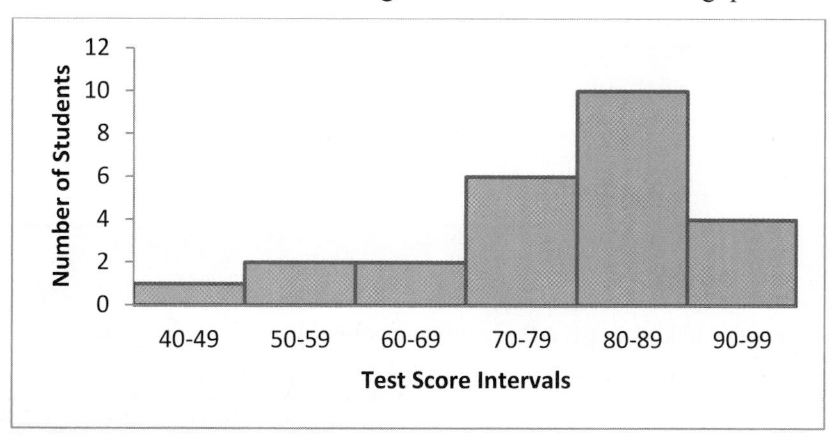

1. **How many students scored between 60 – 69?**
   We find the bar representing 60-69. The height of this bar is 2, which means 2 students scored between 60 – 69 on the test.
2. **What was the most common test interval?**
   The tallest bar is the 80 – 89 bar. The most students scored between 80 and 89.
3. **What is the ratio of students who scored between 90 – 99 to the total number of students?**
   There were 4 students who scored between 90 – 99. There was a total of 25 students who took the test. The ratio of 4 to 25 would be $\frac{4}{25}$.

5.01 Reading Pictographs, Bar Graphs, Histograms, and Line Graphs 5.01

EXAMPLE 5

The list shows the golf scores for an amateur golfer. Use this list to complete the frequency distribution table.

92	82	85	88	90
73	75	83	94	83
87	98	91	83	96

Courtesy of Shutterstock

The data in this list has not been organized. One way to organize the data is to place them in a frequency distribution table. To complete a frequency distribution table, you place a tally mark for every time a number falls in a certain interval and then total the tally marks.

Interval	Tally	Frequency
70-79	II	2
80-89	~~IIII~~ II	7
90-99	~~IIII~~ I	6

- 2 scores fall in the interval 70 – 79 ⟶ bar should be 2 tall.
- 7 scores fall in the interval 80 – 89 ⟶ bar should be 7 tall.
- 6 scores fall in the interval 90 – 99 ⟶ bar should be 6 tall.

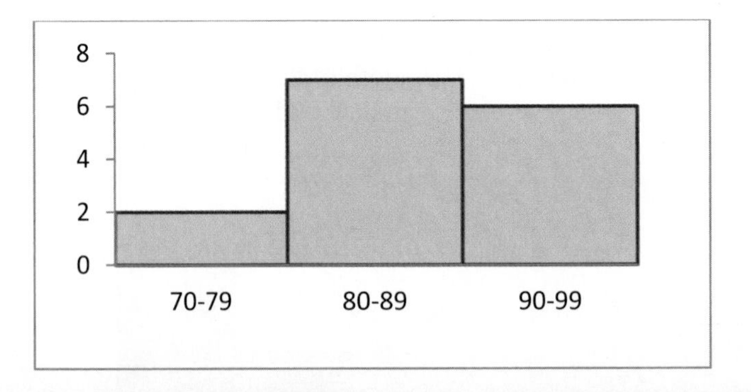

YOUR TURN

1. **Twenty five people were asked to give their current checking account balances. Use the balances shown in the following list to complete the frequency distribution table to the right.**

$53	$105	$443	$109	$162
$468	$47	$261	$315	$229
$207	$357	$15	$301	$75
$86	$77	$512	$219	$101
$192	$288	$353	$169	$295

Intervals Account Balances	Tally	Frequency Number of People
$0-$99		
$100-$199		
$200-$299		
$300-$399		
$400-$499		
$500-$599		

YOUR TURN

2. Use the frequency distribution table from above to construct a histogram.

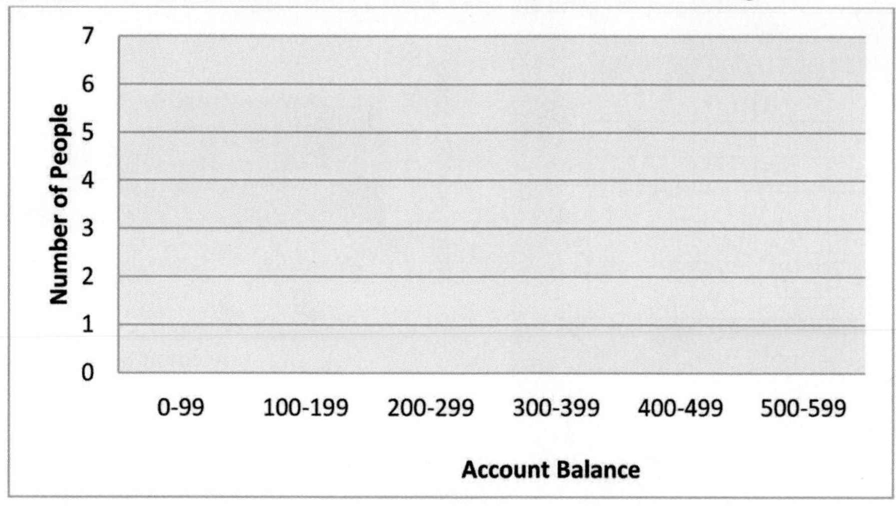

3. The following bar graph show the caffeine content of certain foods for one serving. Use the graph to answer the following questions.

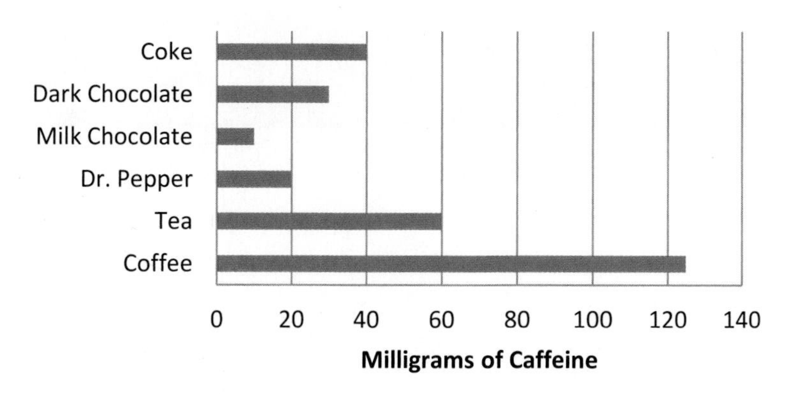

a. How many milligrams of caffeine does Coke have per serving?

c. How many more milligrams of caffeine does tea have than Dr. Pepper?

b. What food has the least amount of caffeine per serving?

d. What is the ratio of the milligrams of caffeine in Coke to milligrams of caffeine in tea?

5.01 Reading Pictographs, Bar Graphs, Histograms, and Line Graphs 5.01

YOUR TURN **Answers to Section 5.01.**

1.

Intervals Account Balances	Tally	Frequency Number of People
$0-$99	111111	6
$100-$199	111111	6
$200-$299	111111	6
$300-$399	1111	4
$400-$499	11	2
$500-$599	1	1

2.

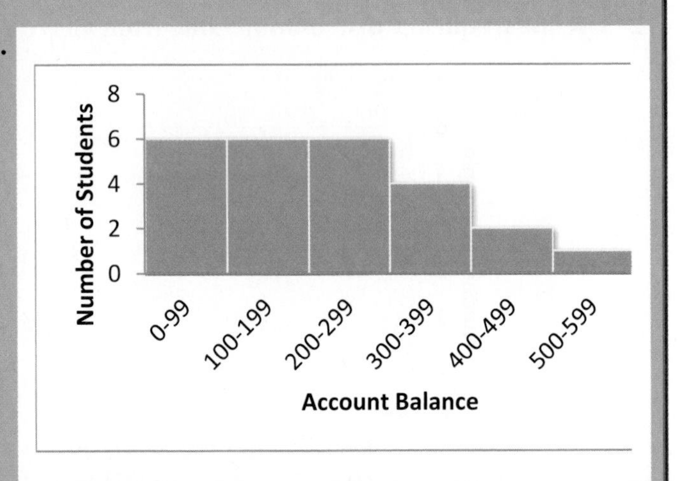

3. a. 40 milligrams

b. milk chocolate

c. 40 milligrams

d. $\frac{40}{60} = \frac{2}{3}$

Complete MyMathLab Section 5.01 Homework in your Homework notebook.

CIRCLE GRAPHS

A circle graph is a graph shaped like a circle that is broken up into sections called sectors. Each sector is shaped like a piece of pie. It shows a particular category and its relative size to the whole. Using a circle graph to display categories of information can provide a quick and easy way to display information about the relationship of **parts to whole**.

EXAMPLE 1

The circle graph shows the favorite sport of 100 people surveyed in the United States. Use this graph to answer the following questions.

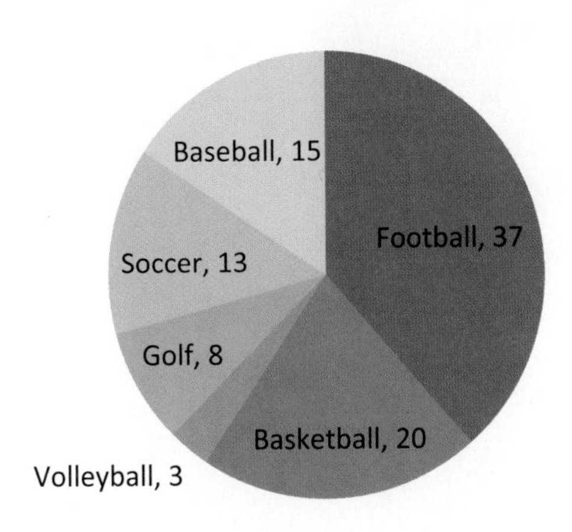

a. What was the favorite sport ?

The biggest piece of the pie represents football. So, football is the favorite sport.

b. What was the least favorite sport?

The smallest piece of pie represents volleyball. So, volleyball is the least favorite sport.

c. What is the ratio of people who preferred baseball to people who preferred basketball.

15 people preferred baseball.
20 people preferred basketball. ⟶ The ratio of 15 to 20 is $\frac{15}{20} = \frac{3}{4}$

d. What is the ratio of people who preferred soccer or golf to the total number of people surveyed?

We need to combine soccer/golf: $13 + 8 = 23$
Total people surveyed was 100. ⟶ The ratio of 23 to 100 is $\frac{23}{100}$

5.02 Reading Circle Graphs 5.02

A circle graph is often used to show percents in different categories. The circle makes one pie. Therefore, the entire circle represents one whole or 100% .

EXAMPLE 2

The circle graph below shows music recordings sold by a music store. A music store sells 4000 recordings a month. Use the graph to answer the following questions.

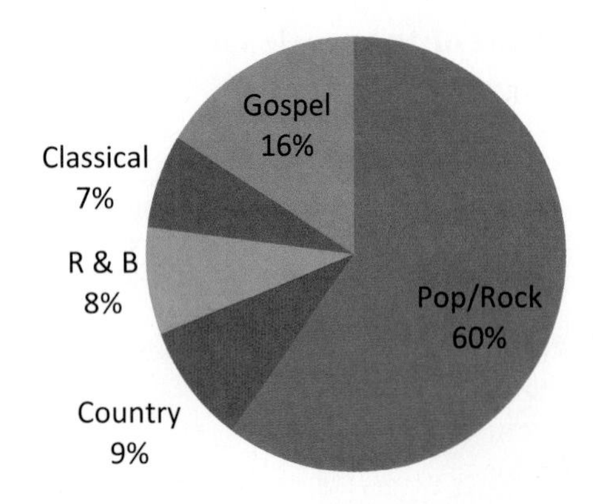

a. What type of recording sold the most records?
The biggest piece is Pop/Rock at 60%.

b. What percent of the records sold were classical or country?
We need to combine classical and country: $7\,\% + 9\,\% = 16\,\%$

c. How many records sold were Pop/Rock ?
Notice we were not asked "What Percent" but "How many".
Pop/Rock accounted for 60% of the records sold. We were told that the pie chart represented a music store that sells 4000 recordings a month. So, we will need to find 60% of 4000.

What is 60 % of 4000?
$n = 0.60\ (4000)$
$= 2400$

There were 2400 pop/rock records sold.

d. **How many records sold were gospel or country?**
Gospel and country accounted for 16% + 9% = 25 % of the records sold.
So, we will need to find 25% of 4000.

What is 25 % of 4000?
$n = 0.25\ (4000)$
$= 1000$

There were 1000 gospel or country records sold.

5.02 Reading Circle Graphs 5.02

EXAMPLE 3

The pie chart shows how federal income tax dollars are used. Sally's taxable income of $45,000 is taxed at an effective rate of 20%. How much of her earnings will go towards defense?

First, we need to calculate how much money Sally will actually pay in taxes. Find 20% of $45,000.

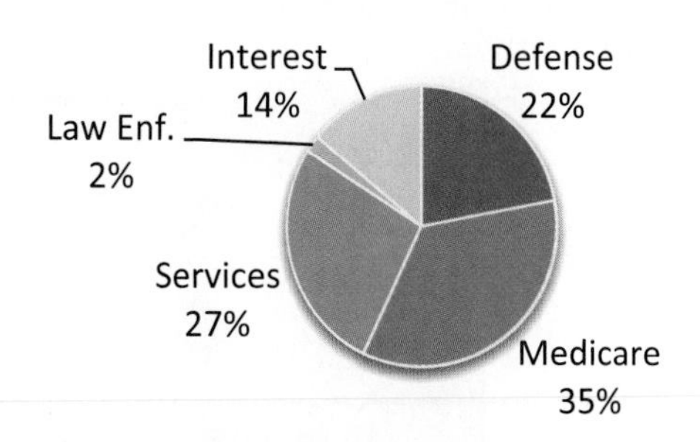

0.20 (45,000) = 9000

Sally will pay $9000 in taxes to the federal government.

Of the $9000 she pays in taxes, 22% of that money will go towards defense. To determine how much of Sally's money will go toward defense, we will need to find 22% of $9000.

0.22 (9000) = $1980 ⟶ $1980 of Sally's earnings will to defense.

YOUR TURN

1. The pie chart in Example 3 above shows how federal income tax dollars are used. David's taxable income of $65,000 is taxed at an effective rate of 15%. Using this pie chart, how much of his earnings will go towards Medicare?

2. The following chart shows the relative sizes of the great oceans. Label the circle graph provided with the appropriate label. Remember the size of the sector must match the size of the percent.

Pacific	49 %
Atlantic	30 %
Indian	16 %
Artic	5 %

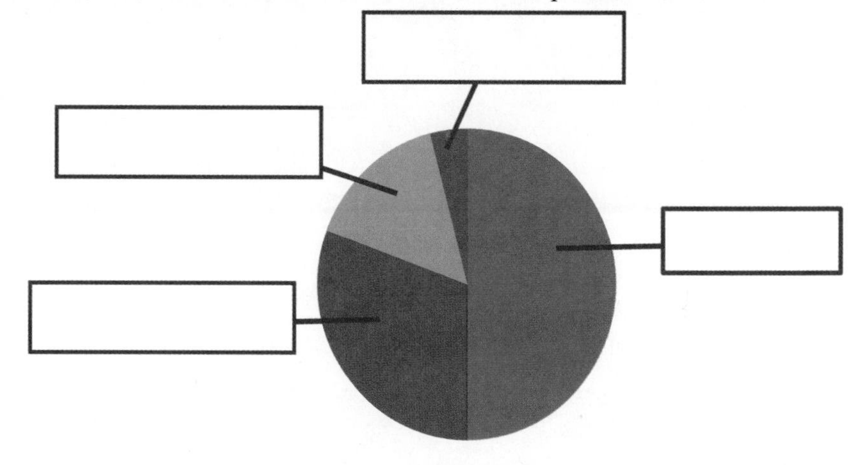

5.02 Reading Circle Graphs 5.02

YOUR TURN

3. The following pie chart shows the distribution of color for M&M's milk chocolate candies. Use the pie chart to answer the following questions.

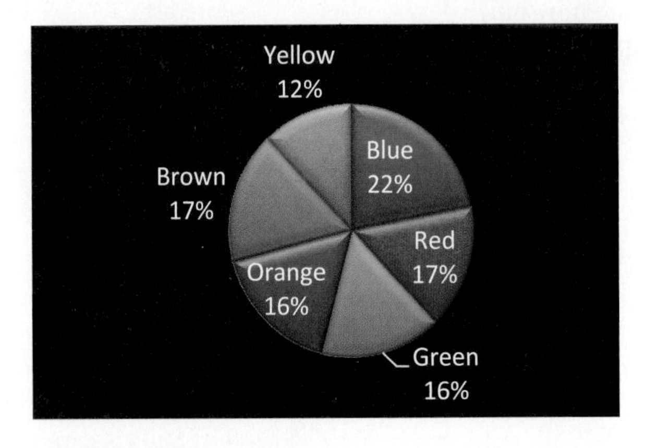

a. What percent of M&M's are either red or brown?

b. How many M&M"s in a bag of 700 would you expect to be orange?

c. How many M&M"s in a bag of 700 would you expect to be blue or red?

YOUR TURN Answers to Section 5.02.

1. $3412.50

2. Blue Sector is Pacific; Red Sector is Atlantic: Green Sector is Indian; Purple Sector is Artic

3. a. 34% b. 112 c. 273

Complete MyMathLab Section 5.02 Homework in your Homework notebook.

5.03 The Rectangular Coordinate System 5.03

The system that is used to describe the location of points in a plane is called the rectangular coordinate system. It consists of two number lines. One is horizontal, called the x-axis, and the other is vertical, called the y-axis. They intersect at the point (0, 0), called the origin, on the number line. This coordinate plane represents two dimensional space, movement left or right and up or down.

The graph is divided into 4 regions called quadrants. Quadrants are always numbered using Roman Numerals counter-clockwise.

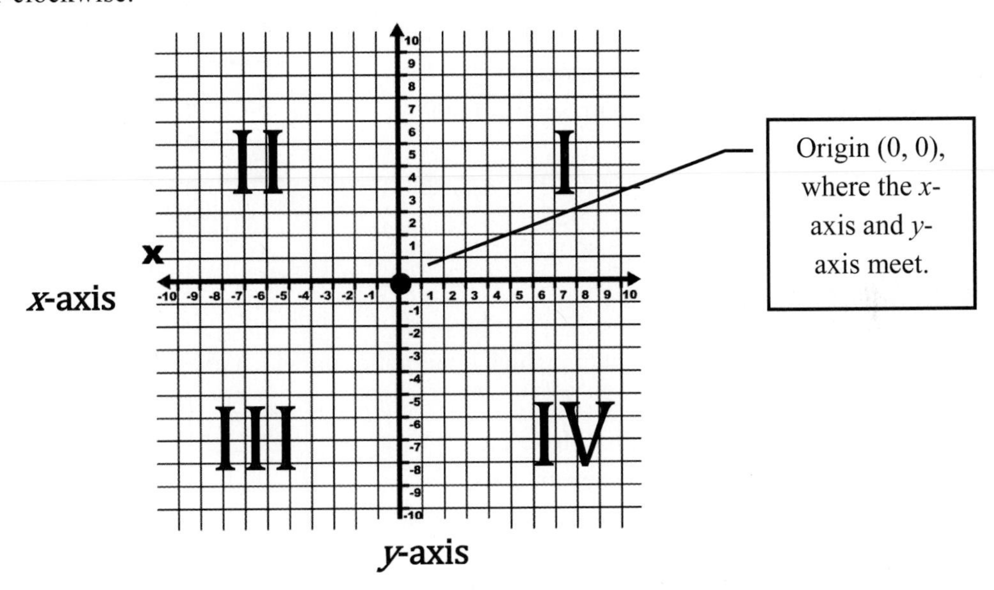

The location of every point on the graph can be determined by two coordinates, written as an ordered pair such as (4, 5). The first number, 4, of an ordered pair is associated with the x-axis and is called the x-coordinate. The second number, 5, is associated with the y-axis and is called the y-coordinate.

To find the **single point** is like driving a car from a beginning location and stopping after a few turns at another. To find the point corresponding to the ordered pair (4, 5), start at the origin. Move 4 units right in the positive direction along the x-axis. From there, move 5 units up in the positive direction along the y-axis. This is called **plotting the point**. Remember, an ordered pair plots as a single point on the grid.

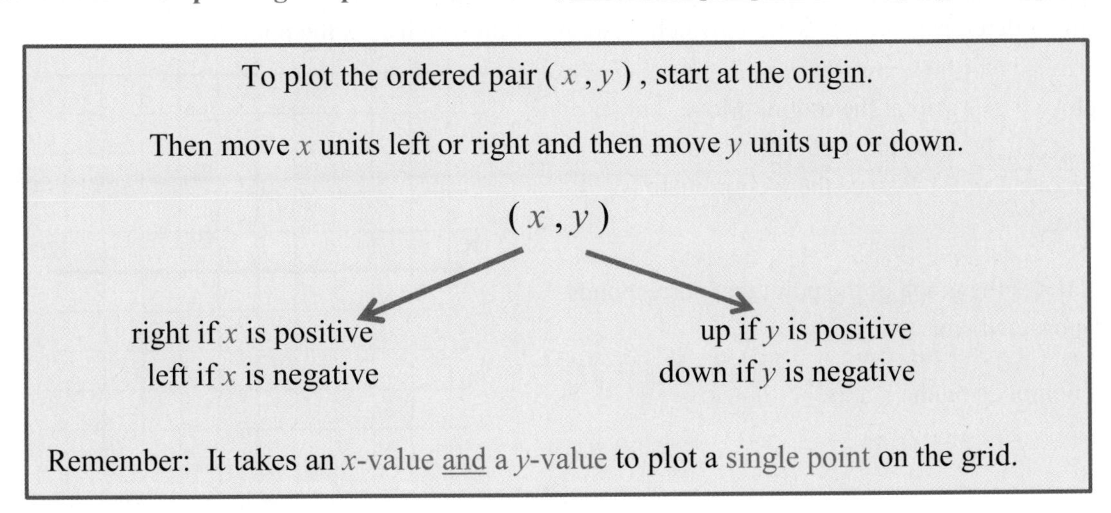

To plot the ordered pair (x, y), start at the origin.

Then move x units left or right and then move y units up or down.

(x, y)

right if x is positive
left if x is negative

up if y is positive
down if y is negative

Remember: It takes an x-value and a y-value to plot a single point on the grid.

5.03 The Rectangular Coordinate System 5.03

EXAMPLE 1

Plot the ordered pair (3, - 4). State in which quadrant or on which axes the point lies.

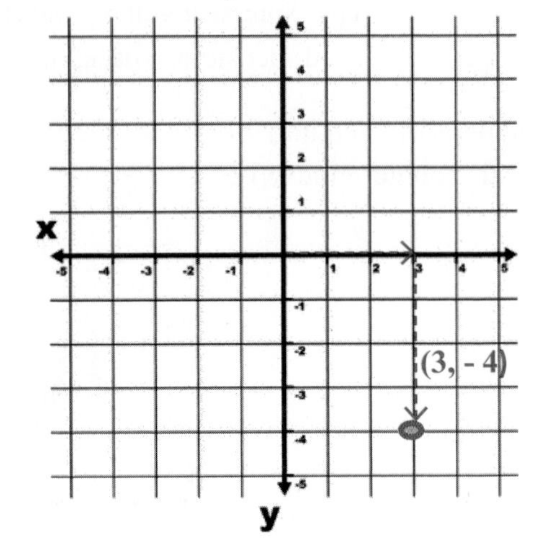

The first number, 3, is the x-coordinate and the second number, - 4, is the y-coordinate.

To plot (3, - 4) start at the origin. Move 3 units right because the x-coordinate is positive and then move **4 units down** since the y-coordinate is **negative**. This dot is the graph of the point that corresponds to the ordered pair (3, - 4) .

This point lies in Quadrant IV.

EXAMPLE 2

Plot the ordered pair (-2, 0). State in which quadrant or on which axes the point lies.

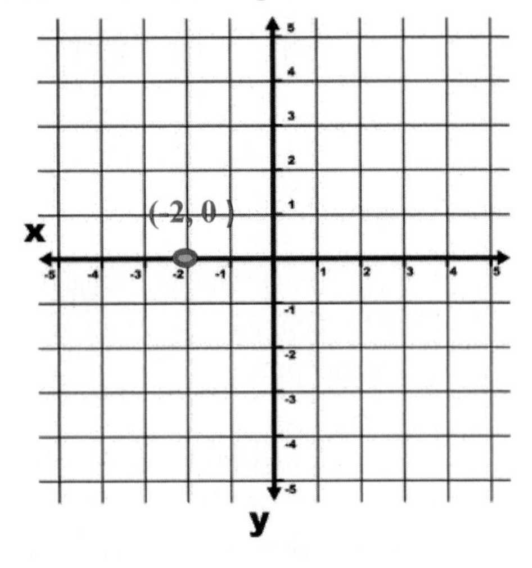

To plot (-2 , 0) start at the origin. Move 2 units LEFT because the x-coordinate is negative and then move 0 units up or down since the y-coordinate is 0.

This dot is the graph of the point that corresponds to the ordered pair (-2 , 0).

This point lies on the x-axis.

EXAMPLE 3

Plot the ordered pair (0, 1). State in which quadrant or on which axes the point lies.

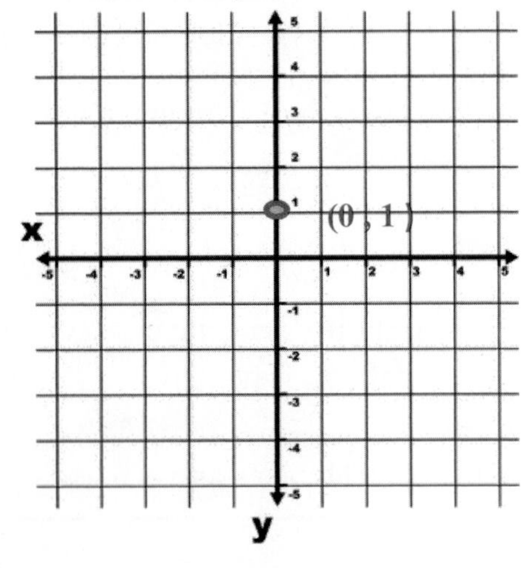

To plot (0 , 1) start at the origin. Move 0 units right or left because the x-coordinate is 0 and then move 1 unit UP since the y-coordinate is positive.

This dot is the graph of the point that corresponds to the ordered pair (0 , 1).

This point lies on the y-axis.

5.03 The Rectangular Coordinate System 5.03

YOUR TURN

1. Plot the given point. Label the point with the appropriate letter.

A: (0, -3)

B: (-5, 0)

C: (0, 2)

D: (5, -4)

E: (2, 3)

F: (-2, 3)

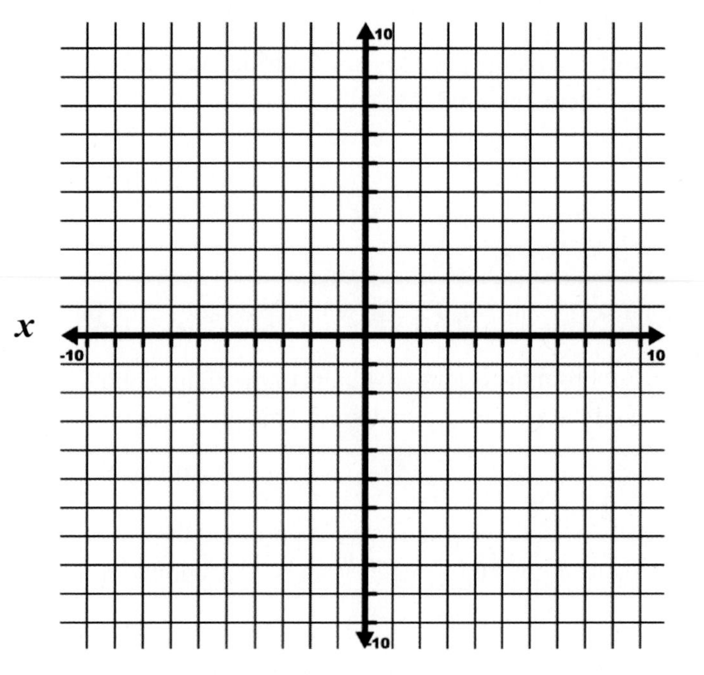

Courtesy of Shutterstock

Remember that each point in the rectangular coordinate system corresponds to exactly one ordered pair and that each ordered pair corresponds to exactly one point.

EXAMPLE 4

Find the *x*- and *y*- coordinate of the labeled points.

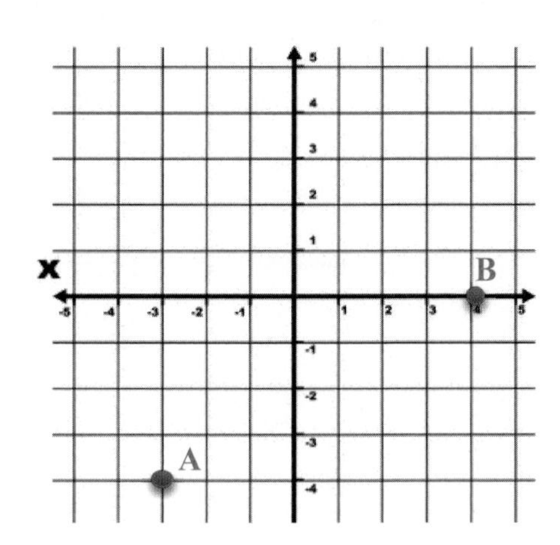

Point A : From the origin we would need to move 3 units to the left along the *x*-axis which tells us the *x*-coordinate is negative and then 4 units down along the *y*-axis which also tells us the *y*-coordinate is negative. The ordered pair would be (- 3, - 4).

Point B: From the origin we would need to move 4 units to the right along the *x*-axis which tells us the *x*-coordinate is positive and then 0 units along the *y*-axis. The ordered pair would be (4, 0).

5.03 The Rectangular Coordinate System 5.03

When you plot an ordered pair, ORDER is the key word. The first number always corresponds to the x-value and the second number always corresponds to the y-value.

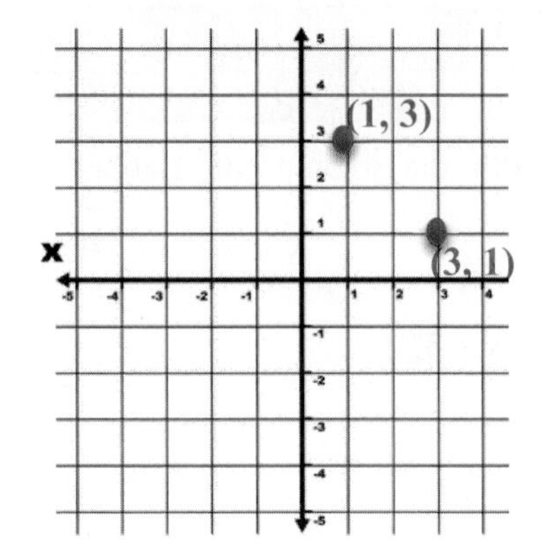

The ordered pair (1, 3) does not describe the same point as the ordered pair (3, 1) .

EXAMPLE 5

Plot (4, 0) and (0, 4). On which axes does each point lie?

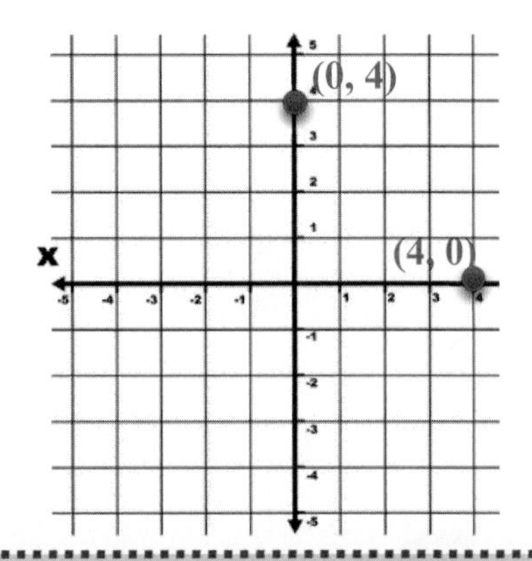

(4, 0) : From the origin go 4 units to the right along the x-axis and then 0 units up or down. This point lies on the x-axis.

(0, 4) : From the origin move 0 units along the x-axis and then move 4 units up along the y-axis. This point lies on the y-axis.

Courtesy of Shutterstock

If an ordered pair has a y-coordinate of 0, its graph lies on the x-axis.

If an ordered pair has a x-coordinate of 0, its graph lies on the y-axis.

YOUR TURN

2. What is the ordered pair represented by point A, B, C, D, E, F on the graph below? Name the quadrant or axes on which the point lies.
Don't forget the parentheses around x and y. (x, y)

A: _______________ _________

B: _______________ _________

C: _______________ _________

D: _______________ _________

E: _______________ _________

F: _______________ _________

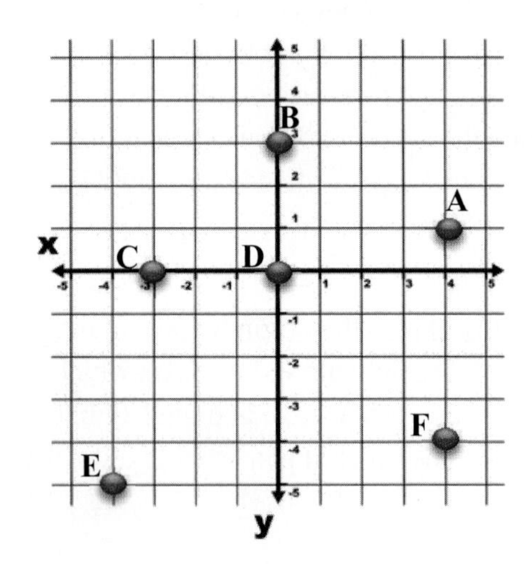

5.03 The Rectangular Coordinate System 5.03

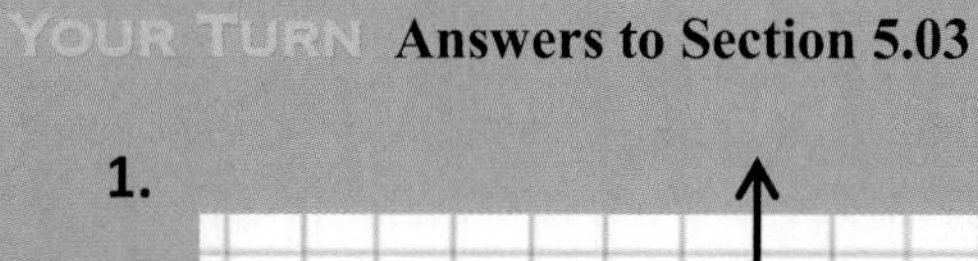

Answers to Section 5.03

1.

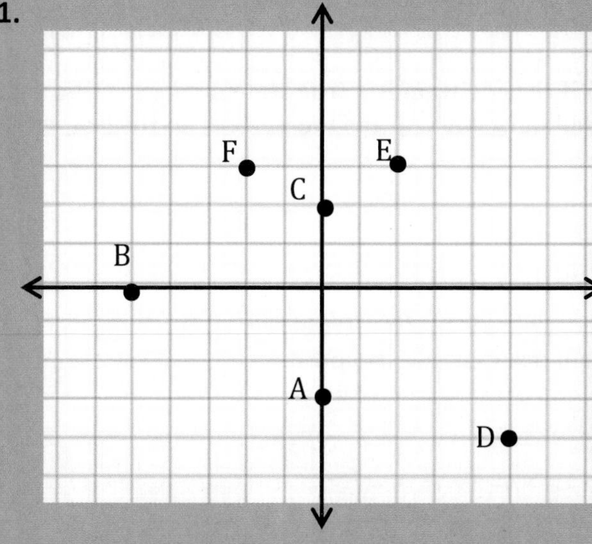

2. A: (4, 1) quadrant I.
   B: (0, 3) on the y axis.
   C: (-3, 0) on the x axis.
   D: (0, 0) is the origin.
   E: (-4, -5) quadrant III.
   F: (4,- 4) quadrant IV.

Complete MyMathLab Section 5.03 Homework in your Homework notebook.

5.04 Graphing Linear Equations 5.04

In this section we will begin to investigate equations in two variables, namely x and y . As you will see, equations in two variables have pairs of numbers for solutions. For equations that have two variables, the solutions are points which come in "ordered pairs". Recall that for an ordered pair (x, y), the first number will always be the x-value and the second number will always be the y-value.

An ordered pair is a solution of an equation if the equation is a true statement when the variables are replaced by the coordinates of the ordered pair. This means the point is on the graph of the equation.

EXAMPLE 1

Is (-1, 6) a solution for the equation $2x + y = 4$?

In the ordered pair (- 1, 6), the x-value is - 1 and the y-value is 6. Replace x with - 1 and y with 6 in the equation $2x + y = 4$ to see if the result makes a true statement.

$$2x + y = 4$$
$$2(-1) + 6 = 4$$
$$-2 + 6 = 4$$
$$4 = 4$$

→ True → (-1,6) is a solution for the equation $2x + y = 4$

Unlike equations with one variable, equations with two variables have more than one solution. In fact, equations with two variables have so many solutions we cannot list them all. We say they have an **infinite** number of solutions.

Let's practice finding ordered pair solutions for these equations. If we are given one coordinate of an ordered pair solution, the other coordinate can be found by replacing the appropriate variable with the known coordinate and then solving for the unknown.

EXAMPLE 2

Complete the ordered pair (3 , ___) for the given equation $y = 2x - 5$.

In the ordered pair (3,____) , the x-value is 3. To find the corresponding y-value, let $x = 3$ in the equation and calculate the value of y.

$$y = 2x - 5$$
$$y = 2(3) - 5$$
$$y = 6 - 5$$
$$y = 1$$

→ The ordered pair solution is (3, 1)

We can check by replacing the x with 3 and the y with 1 in the equation. We should get a true statement.

$y = 2x - 5$

$1 = 2(3) - 5$ Let $x = 3$ and $y = 1$

$1 = 6 - 5$

$1 = 1$ TRUE

5.04 Graphing Linear Equations 5.04

EXAMPLE 3

Complete the ordered pair (___ , - 7) for the given equation $y = 2x - 5$.

In the ordered pair (___ , - 7) , the y-value is -7. To find the corresponding x-value, let $y = -7$ in the equation and calculate the value of x.

$$y = 2x - 5$$
$$-7 = 2x - 5$$
$$-7 + 5 = 2x - 5 + 5$$
$$-2 = 2x$$
$$\frac{-2}{2} = \frac{2x}{2}$$
$$-1 = x$$

→ The ordered pair solution is (-1, - 7)

We can check by replacing the x with -1 and the y with -7 in the equation.

$$y = 2x - 5$$
$$-7 = 2(-1) - 5 \qquad \text{Let } x = -1 \text{ and } y = -7$$
$$-7 = -2 - 5$$
$$-7 = -7 \qquad \text{TRUE}$$

EXAMPLE 4

Complete each ordered pair so that it is a solution for the given linear equation.

$$y = \frac{1}{4}x + 3; \ (20, \_\_\_\_\_) \ , \ (\_\_\_\_\_ , -8)$$

$$y = \frac{1}{4}x + 3 \ ; \ (20, \_\_\_\_\_)$$
$$y = \frac{1}{4}(20) + 3$$
$$y = 5 + 3$$
$$y = 8$$

The ordered pair is $(20, 8)$

$$y = \frac{1}{4}x + 3 \ ; \ (\_\_\_\_\_, -8)$$
$$-8 = \frac{1}{4}x + 3$$
$$-8 - 3 = \frac{1}{4}x + 3 - 3$$
$$-11 = \frac{1}{4}x$$
$$-11 \cdot \frac{4}{1} = \frac{4}{1} \cdot \frac{1}{4}x$$
$$-44 = x$$

The ordered pair is $(-44, -8)$

5.04 Graphing Linear Equations 5.04

Sometimes the equation will not be solved for y. We will use the same process to find solutions.

If given the first coordinate in an ordered pair, the x-value, we will plug-in for x and solve for y.
If given the second coordinate in an ordered pair, the y-value, we will plug-in for y and solve for x.

EXAMPLE 5

Complete each ordered pair so that it is a solution for the given linear equation.

$x - 2y = 4$; (10, _____) , (_____ , −5)

$$\begin{aligned} x - 2y &= 4 \;;\; (10, \_\_\_\_) \\ 10 - 2y &= 4 \\ 10 - 10 - 2y &= 4 - 10 \\ -2y &= -6 \\ \frac{-2y}{-2} &= \frac{-6}{-2} \\ y &= 3 \end{aligned}$$

The ordered pair is (10 , 3)

$$\begin{aligned} x - 2y &= 4 \;;\; (\_\_\_\_, -5) \\ x - 2(-5) &= 4 \\ x + 10 &= 4 \\ x + 10 - 10 &= 4 - 10 \\ x &= -6 \end{aligned}$$

The ordered pair is (- 6 , - 5)

Courtesy of Shutterstock

Given an ordered pair, ORDER is the key word. The **first number** always corresponds to the ***x*-value** and the **second number** always corresponds to the ***y*-value**.

YOUR TURN

1. Is (3, -2) a solution for $3x - 5y = -18$?

5.04 Graphing Linear Equations 5.04

YOUR TURN

2. Complete each ordered pair so that it is a solution for the given linear equation.

$$y = \frac{1}{3}x \; ; \; (-6, \_\_\_\_\_) \; , \; (\_\_\_\_\_, 2)$$

$y = \frac{1}{3}x$; $(-6, \_\_\_\_\_)$

The ordered pair is (-6 ,)

$y = \frac{1}{3}x$; $(\_\_\_\_\_, 2)$

The ordered pair is (, 2)

3. Complete each ordered pair so that it is a solution for the given linear equation.

$$2x - y = 8 \; ; \; (-1, \_\_\_\_\_) \; , \; (\_\_\_\_\_, 2)$$

$2x - y = 8$; $(-1, \_\_\_\_\_)$

The ordered pair is (-1 ,)

$2x - y = 8$; $(\_\_\_\_\_, 2)$

The ordered pair is (, 2)

Courtesy of Shutterstock

Solutions of equations in two variables can be recorded in a table of values to help us stay organized.

The next example is still wanting us to find solutions and complete ordered pairs. However, the information will be presented in a table format.

EXAMPLE 6

Complete the table of ordered pairs for the linear equation:

$$y = -x + 4$$

x	y
0	
	3
- 4	

$y = -x + 4$; $x = 0$

$y = -(0) + 4$

$y = 0 + 4$

$y = 4$

$(0, 4)$

$y = -x + 4$; $y = 3$

$3 = -x + 4$

$3 - 4 = -x + 4 - 4$

$-1 = -x$

$\frac{-1}{-1} = \frac{-x}{-1}$

$1 = x$

$(1, 3)$

$y = -x + 4$; $x = -4$

$y = -(-4) + 4$

$y = 4 + 4$

$y =$

$(-4, 8)$

x	y
0	4
1	3
- 4	8

YOUR TURN

4. Complete the table of ordered pairs for the linear equation:

$$y = 2x - 7$$

x	y
2	
	0
- 3	

$y = 2x - 7$; $x = 2$

$(2, \_\_)$

$y = 2x - 7$; $y = 0$

$(\_\_, 0)$

$y = 2x - 7$; $x = -3$

$(-3, \_\_)$

5.04 Graphing Linear Equations 5.04

Linear equations are used to help us evaluate trends and plan for the future. A few examples are introduced here to help us begin to see how math is used in real life.

EXAMPLE 7

The cost in dollars y of producing x computer desks is given by

$y = 40x + 5000$

a. Complete the table

x	100	200	300
y			

b. Find the number of computer desks that can be produced for $6800.

Courtesy of Shutterstock

a. Complete the table.

First, understand the variables in words:

The # of computer desks: ___x___ ; the cost in $: ___$y$___

x	100	200	300
y	9000	13,000	17,000

$y = 40x + 5000$; $x = 100$
$y = 40(100) + 5000$
$y = 4000 + 5000$
$y = 9000$

It will cost $9000 to produce 100 desks.

$y = 40x + 5000$; $x = 200$
$y = 40(200) + 5000$
$y = 8000 + 5000$
$y = 13{,}000$

It will cost $13,000 to produce 200 desks.

$y = 40x + 5000$; $x = 300$
$y = 40(300) + 5000$
$y = 12000 + 5000$
$y = 17{,}000$

It will cost $17,000 to produce 300 desks.

b. Find the number of computer desks that can be produced for $6800.

Notice the given value has changed. Here we are given $6800, the cost, which is the *y value*. In table above we were given x which is the number of desks.

The **cost in dollars** y of producing **x computer desks** is given by $y = 40x + 5000$.

$$y = 40x + 5000$$
$$6800 = 40x + 5000$$
$$6800 - 5000 = 40x + 5000 - 5000$$
$$1800 = 40x$$
$$\frac{1800}{40} = \frac{40x}{40}$$
$$45 = x$$

The company can produce 45 computer desks for a cost of $6800.

EXAMPLE 8

A man bought a house for \$210,000 in 2012. The area is expected to grow in the coming years. Every year the house is expected to increase in value by \$3,000. Find the value of the house in 5 years and 10 years. Predict what year the house will be valued at \$375,000.

The **value in dollars** y of the house in x **years** will be $y = 3{,}000x + 210{,}000$.

Understand the variables in words:

x-value ⟶ number of years after 2012 y-value ⟶ value of the house in dollars

- **What will be the value of the house in 5 years?**
  We will substitute 5 in for x and find the value of y.

$$y = 3{,}000x + 210{,}000$$
$$y = 3{,}000(5) + 210{,}000$$
$$y = 15{,}000 + 210{,}000$$
$$y = 225{,}000$$

The value of the home in 5 years will be \$225,000.

- **What will be the value of the house in 10 years?**
  We will substitute 10 in for x and find the value of y.

$$y = 3{,}000x + 210{,}000$$
$$y = 3{,}000(10) + 210{,}000$$
$$y = 30{,}000 + 210{,}000$$
$$y = 240{,}000$$

The value of the home in 10 years will be \$240,000.

- **Predict what year the house will be valued at \$375,000.**
  We will substitute \$375,000 in for y and find the value of x.

$$y = 3000x + 210{,}000$$
$$375{,}000 = 3000x + 210{,}000$$
$$375{,}000 - 210{,}000 = 3000x + 210{,}000 - 210{,}000$$
$$165{,}000 = 3000x$$
$$\frac{165{,}000}{3000} = \frac{3000x}{3000}$$
$$55 = x$$

The value of the house will be \$375,000 in 55 years from 2012 which would be 2067.

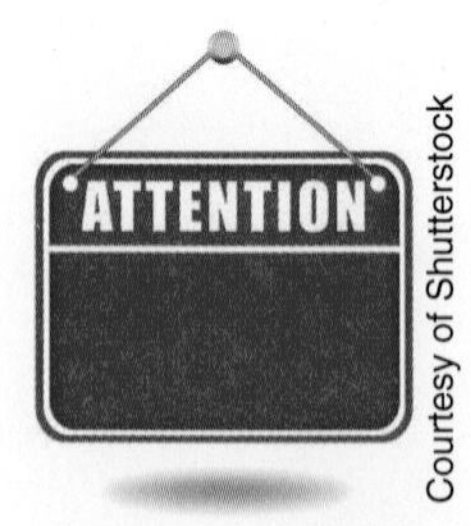

Courtesy of Shutterstock

When working an application problem, it is always important to identify what x and y represent. This will determine whether you are plugging it for x and solving for y **OR** plugging in for y and solving for x. Always reread the problem to make sure you understand what value x represents and what value y represents.

YOUR TURN

5. The hourly wage y of an employee at a local factory is given by $y = 0.25x + 9$ where x is the number of units produced by the employee in an hour.

a. Complete the table.

x	5	10	15
y			

b. Find the number of units that an employee must produce each hour to earn an hourly wage of \$12.25.
(HINT: Find x when $y = 12.25$)

First, understand the variables in words:

x: ______________________________

y: ______________________________

a. Complete the table.

x	5	10	15
y			

$y = 0.25x + 9$; $x = 5$
$y = 0.25(\quad) + 9$

The hourly wage of an employee will be ________ with 5 units produced.

$y = 0.25x + 9$; $x = 10$
$y = 0.25(\quad) + 9$

The hourly wage of an employee will be ________ with 10 units produced.

$y = 0.25x + 9$; $x = 15$
$y = 0.25(\quad) + 9$

The hourly wage of an employee will be ________ with 15 units produced.

b. Find the number of units that an employee must produce each hour to earn an hourly wage of \$12.25. (HINT: Find x when $y = 12.25$)

$y = 0.25x + 9$

The employee must produce ________ units to earn an hourly wage of \$12.25.

5.04 Graphing Linear Equations 5.04

YOUR TURN

6. The average cinema admission price y (in dollars) from 2000 through 2008 is given by $y = 0.2x + 5.39$. In this equation, x represents the number of years after 2000.

First, understand the variables in words:

x: ______________________________

y: ______________________________

a. Complete the table

x	1	3	5
y			

$y = 0.2x + 5.39$; $x = 1$
$y = 0.2(\quad) + 5.39$

The cinema price in one year is ___________.

$y = 0.2x + 5.39$; $x = 3$
$y = 0.2(\quad) + 5.39$

The cinema price in three years is ___________.

$y = 0.2x + 5.39$; $x = 5$
$y = 0.2(\quad) + 5.39$

The cinema price in five years is ___________.

b. In how many years will the average cinema admission price be approximately $8.00. Round to the nearest whole number.

$y = 0.2x + 5.39$; $y = 8.00$

c. Using your answer from above, predict what year the average cinema price might be $8.00.

In the year __________, the average cinema price might be $8.00

5.04 Graphing Linear Equations 5.04

Standard Form of a Linear Equation

$Ax + By = C$ where A, B, and C are real numbers.

The graphs of equations of the form $Ax + By = C$ are straight lines. This is why such equations are called linear. We will look at other ways to write linear equations later.

We have already looked at solutions for linear equations. Recall these two statements we made previously in this section.

- **An ordered pair is a solution of an equation if the equation is a true statement when the variables are replaced by the coordinates of the ordered pair. These ordered pairs are points on the graph.**
- **Unlike equations with one variable, equations with two variables have more than one solution. In fact, equations with two variables have so many solutions we cannot list them all. We say they have an infinite number of solutions.**

EXAMPLE 9

Which of the following ordered pairs are solutions for $y = x + 1$? (2 , 3) , (- 4, - 3) , (0, 1) , (-3, - 2)
(This is asking which points are on the graph of the line!)

(2, 3)	(- 4, - 3)	(0 , 1)	(- 3, - 2)
$x = 2\ ,\ y = 3$	$x = -4\ ,\ y = -3$	$x = 0\ ,\ y = 1$	$x = -3\ ,\ y = -2$
$y = x + 1$	$y = x + 1$	$y = x + 1$	$y = x + 1$
$3 = 2 + 1$	$-3 = -4 + 1$	$1 = 0 + 1$	$-2 = -3 + 1$
$3 = 3$	$-3 = -3$	$1 = 1$	$-2 = -2$
TRUE	TRUE	TRUE	TRUE

All of these ordered pairs are solutions !!! They are all points on the graph of the line $y = x + 1$. How many possible solutions does $y = x + 1$ have? The equation has an unlimited or infinite number of possible solutions, an unlimited number of points on the graph of the line. Since it is not possible to list all the specific solutions, the solutions are illustrated in a graph.

Let's plot all of the ordered pairs on a grid. Notice that they make a straight line when connected. Every point on the line will make the equation true - not just the ordered pairs we found above that were solutions. The line will continue to move up or down, as indicated by the arrows.

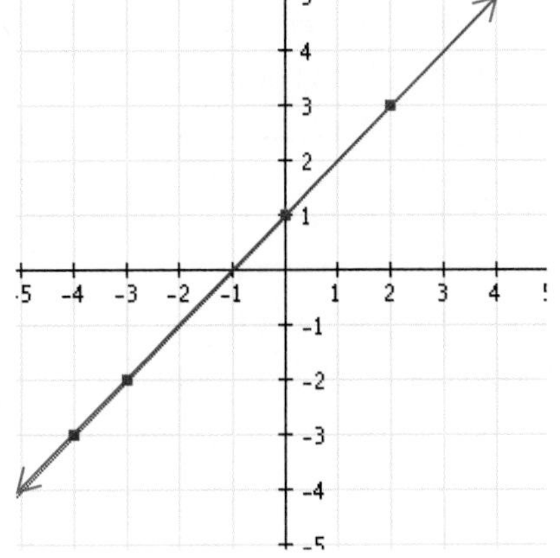

5.04 Graphing Linear Equations 5.04

Courtesy of Shutterstock

A graph of an equation in two variables is an illustration of a set of points whose coordinates are solutions for the equation. To graph an equation, you will need to determine ordered pairs that make the equation true.

How many points are needed to graph a linear equation? A straight line is determined by two points. Since the graph of a linear equation will always be a straight line, then only **two points** are needed to graph a linear equation. However, it is always a good idea to plot at least three points. The third point is considered a check point. In other words, if you use at least three points to plot your graph and they line up in a straight line, you probably have not made a mistake.

EXAMPLE 10

x	*y*
1	
	2
2	

For the given equation, find three ordered pairs by completing the table. Then use the ordered pairs to graph the equation. $x - y = 3$

$$x - y = 3;\ x = 1$$
$$1 - y = 3$$
$$1 - 1 - y = 3 - 1$$
$$0 - y = 2$$
$$-y = 2$$
$$\frac{-1y}{-1} = \frac{2}{-1}$$
$$y = -2$$
$$(1, -2)$$

$$x - y = 3;\ y = 2$$
$$x - 2 = 3$$
$$x - 2 + 2 = 3 + 2$$
$$x = 5$$
$$(5, 2)$$

$$x - y = 3;\ x = 2$$
$$2 - y = 3$$
$$2 - 2 - y = 3 - 2$$
$$0 - y = 1$$
$$-y = 1$$
$$\frac{-1y}{-1} = \frac{1}{-1}$$
$$y = -1$$
$$(2, -1)$$

We **plot the ordered pair solutions and draw a line** through the plotted points. The line is the graph of $x - y = 3$.

Every point on the line represents an ordered pair solution of the equation So, from the graph we can see that (3, 0) , (4, 1) , (6, 3) , and (0, -3) are also solutions. Remember, there are an infinite number of points on a line. So, there are an infinite number of solutions to the equation.

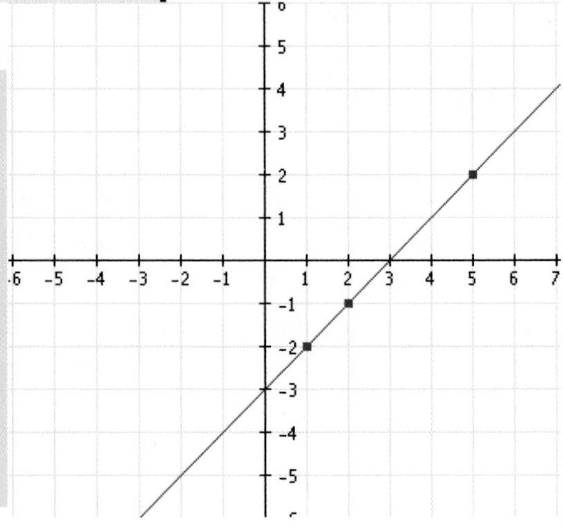

5.04 Graphing Linear Equations 5.04

YOUR TURN

7. For the given equation, find three ordered pairs by completing the table. Then use the ordered pairs to graph the equation. $y = -2x$

x	y
0	
-1	
1	

$y = -2x$; $x = 0$	$y = -2x$; $x = -1$	$y = -2x$; $x = 1$
$y = -2(\quad)$	$y = -2(\quad)$	$y = -2(\quad)$
$y =$	$y =$	$y =$
(0 ,)	(−1 ,)	(1 ,)

8. For the given equation, find three ordered pairs by completing the table. Then use the ordered pairs to graph the equation. $5x - y = 10$

x	y
0	
	0
	-5

$5x - y = 10$; $x = 0$	$5x - y = 10$; $y = 0$	$5x - y = 10$; $y = -5$
(,)	(,)	(,)

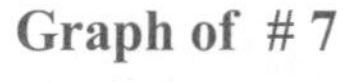

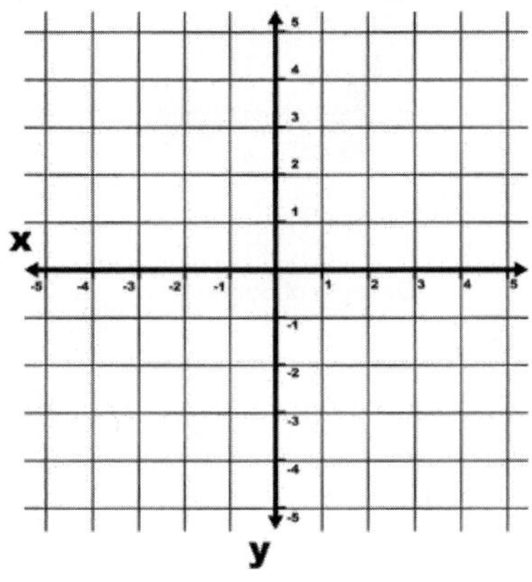

Graph of # 8

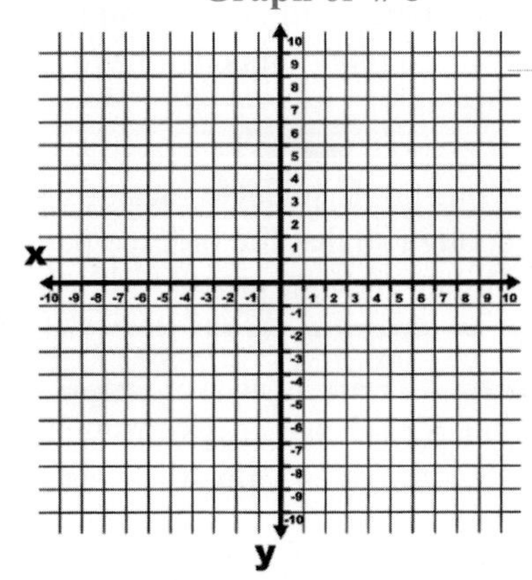

5.04 Graphing Linear Equations 5.04

GRAPHING LINEAR EQUATIONS BY PLOTTING POINTS

Solve the equation for *y*.

Select a value for the variable *x*.

Substitute this value in the equation for *x* and find the corresponding value of *y*.

Repeat Step 2 with two different values of x.

You should now have 3 ordered pairs that are solutions.

Plot the three ordered pairs. The three points should form a straight line. If not, recheck your work for mistakes.

Draw a straight line through the three points.

In step 1 you are asked to solve the equation for y. Although it is not necessary to do this to graph the equation, it can help you select values for x that will make Step 2 much easier. We always want to plot integer values for x and y. That is, you do not want to plot points that have fractions or decimals!

You should select values of x that are small enough so that the ordered pairs obtained can be plotted on your graph.

Since y is often easy to solve for when $x = 0$, 0 is always a great value to select for x.

Courtesy of Shutterstock

5.04 Graphing Linear Equations 5.04

EXAMPLE 11 Graph $y = 2x + 4$.

The equation is already solved for y. We select 3 values for x, substitute them into the equation, and find the corresponding values for y.

Which x-values you pick is totally up to you! And it's perfectly okay if you pick values that are different from the book's choices, or different from your study partner's choices, or different from my choices. Some values may be more useful than others, but the choice is entirely up to you. **No matter what values are chosen the graph of the line will be same!!**

Let's choose the **values -3 , 0 , and 1 for x** . Remember, we could have selected three totally different values and still arrive at the same graph. Let's keep our x- and y-values organized in a table.

x	y
-3	-2
0	4
1	6

$$y = 2x + 4 \; ; \; x = -3$$
$$y = 2(-3) + 4$$
$$y = -6 + 4$$
$$y = -2$$
$$(-3, -2)$$

$$y = 2x + 4 \; ; \; x = 0$$
$$y = 2(0) + 4$$
$$y = 0 + 4$$
$$y = 4$$
$$(0, 4)$$

$$y = 2x + 4 \; ; \; x = 1$$
$$y = 2(1) + 4$$
$$y = 2 + 4$$
$$y = 6$$
$$(1, 6)$$

Plot the three ordered pairs on a graph. Remember, they should be collinear – the points should form a straight line.

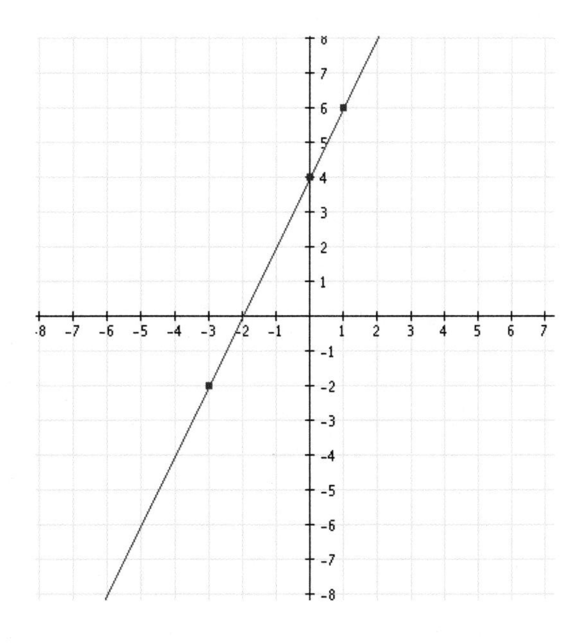

5.04 Graphing Linear Equations 5.04

EXAMPLE 12

Graph $y = \frac{1}{2}x$.

The equation is already solved for y. We select 3 values for x, substitute them into the equation, and find the corresponding values for y.

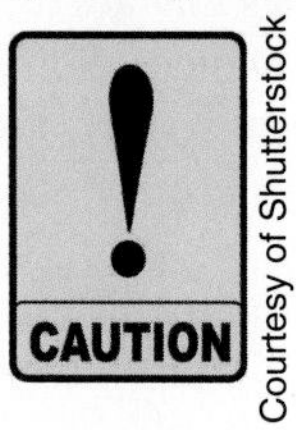

Courtesy of Shutterstock

If there is a fraction in the equation we must be more particular when selecting x-values. Given an equation that has a fraction, we will always **choose x-values that are multiples of the denominator** to avoid fractions in our ordered pairs. Recall that $x = 0$ can always be chosen regardless of the equation!

Since our denominator is 2, we will choose x-values that are multiples of 2. We will select x-values of -2, 0, 2

x	y
-2	-1
0	0
2	1

$$y = \frac{1}{2}x \ ; \ x = -2$$
$$y = \frac{1}{2}(-2)$$
$$y = -1$$
$$(-2, -1)$$

$$y = \frac{1}{2}x \ ; \ x = 0$$
$$y = \frac{1}{2}(0)$$
$$y = 0$$
$$(0, 0)$$

$$y = \frac{1}{2}x \ ; \ x = 2$$
$$y = \frac{1}{2}(2)$$
$$y = 1$$
$$(2, 1)$$

Plot the three ordered pairs on a graph.

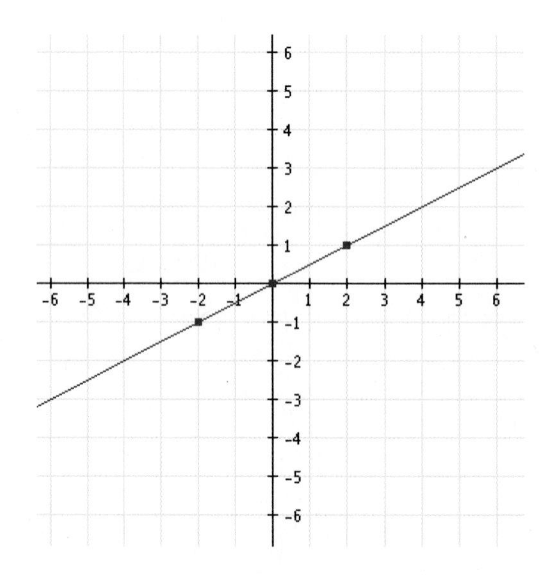

5.04 Graphing Linear Equations

EXAMPLE 13 Graph $2x + 5y = -5$

The equation is not solved for y. So, our first step will be to solve for *y*.

$$2x + 5y = -5$$

$$5y = -2x - 5$$ Move *x*-term to opposite side by changing its sign. Slide the -5 to the right.

$$\frac{5y}{5} = \frac{-2}{5}x - \frac{5}{5}$$ Divide by coefficent of *y* which is 5.

$$y = -\frac{2}{5}x - 1$$ Simplify.

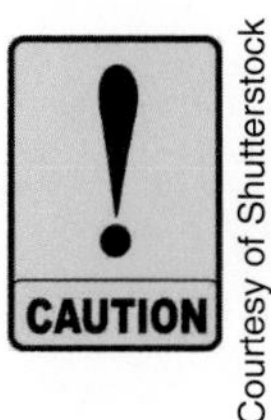

Courtesy of Shutterstock

Given an equation that has a fraction, we will always **choose x-values that are multiples of the denominator**. Since our denominator is 5, we will choose *x*-values that are multiples of 5. We will select *x*-values of -5, 0, 5.

x	*y*
-5	1
0	-1
5	-3

$$y = -\frac{2}{5}x - 1 \;;\; x = -5$$
$$y = -\frac{2}{5}(-5) - 1$$
$$y = 2 - 1$$
$$y = 1$$
$$(-5, 1)$$

$$y = -\frac{2}{5}x - 1 \;;\; x = 0$$
$$y = -\frac{2}{5}(0) - 1$$
$$y = 0 - 1$$
$$y = -1$$
$$(0, -1)$$

$$y = -\frac{2}{5}x - 1 \;;\; x = 5$$
$$y = -\frac{2}{5}(5) - 1$$
$$y = -2 - 1$$
$$y = -3$$
$$(5, -3)$$

Plot the three ordered pairs on a graph.

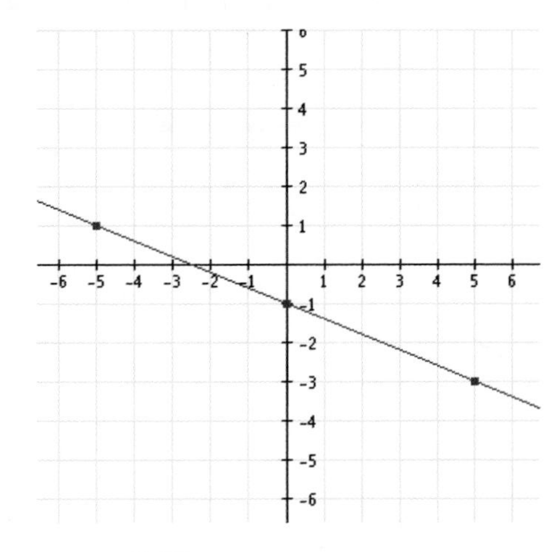

5.04 Graphing Linear Equations 5.04

EXAMPLE 14 Graph $-3x - y = 2$.

The equation is not solved for y So, our first step will be to solve for y.

$-3x - y = 2$

$-y = 3x + 2$ Move x-term to opposite side by changing its sign.

$\frac{-1y}{-1} = \frac{3x}{-1} + \frac{2}{-1}$ Divide by coefficent of y which is -1.

$y = -3x - 2$ Simplify.

x	y
-1	1
0	-2
1	-5

$y = -3x - 2$; $x = -1$
$y = -3(-1) - 2$
$y = 3 - 2$
$y = 1$

$(-1, 1)$

$y = -3x - 2$; $x = 0$
$y = -3(0) - 2$
$y = 0 - 2$
$y = -2$

$(0, -2)$

$y = -3x - 2$; $x = 1$
$y = -3(1) - 2$
$y = -3 - 2$
$y = -5$

$(1, -1)$

Plot the three ordered pairs on a graph.

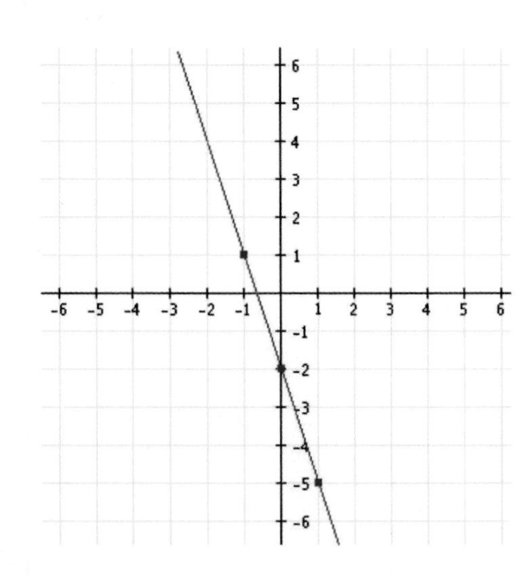

5.04 Graphing Linear Equations 5.04

YOUR TURN

9. Graph $y = -3x + 2$.

x	y

$y = -3x + 2$; $x =$ _____
$y = -3(\quad) + 2$
$y =$
$y =$
(____ , ____)

$y = -3x + 2$; $x =$ _____
$y = -3(\quad) + 2$
$y =$
$y =$
(____ , ____)

$y = -3x + 2$; $x =$ _____
$y = -3(\quad) + 2$
$y =$
$y =$
(____ , ____)

10. Graph $y = \frac{1}{3}x$.

x	y

Since our denominator is 3, make sure and choose multiples of 3 for your x-values.

$y = \frac{1}{3}x$; $x =$ _____
$y = \frac{1}{3}(\quad)$
$y =$
$y =$
(____ , ____)

$y = \frac{1}{3}x$; $x =$ _____
$y = \frac{1}{3}(\quad)$
$y =$
$y =$
(____ , ____)

$y = \frac{1}{3}x$; $x =$ _____
$y = \frac{1}{3}(\quad)$
$y =$
$y =$
(____ , ____)

Graph # 9

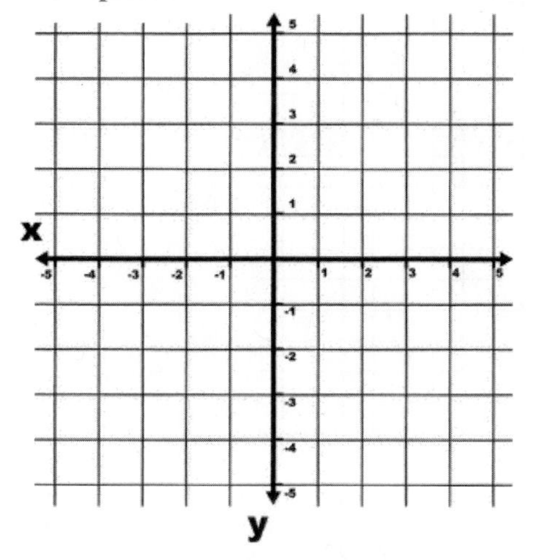

Graph # 10

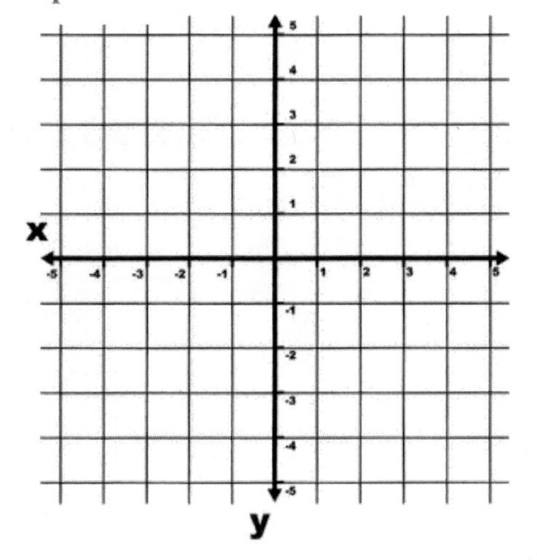

5.04 Graphing Linear Equations 5.04

YOUR TURN

11. Graph $6x - 4y = 24$.

- Solve the equation for y:
- Select x – values.

x	y

- Find corresponding y-values.

(___ , ___) (___ , ___) (___ , ___)

- Plot the three ordered pairs on a graph. Draw the line.

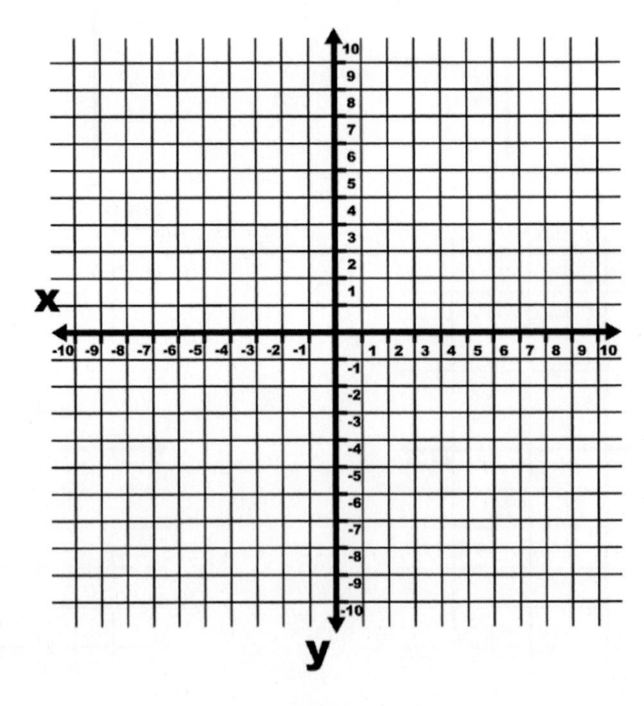

5.04 Graphing Linear Equations 5.04

APPLICATIONS OF LINEAR EQUATIONS

EXAMPLE 15

One American rite of passage is a driver's license. The number of people (in millions) who have a driver's license can be estimated by the linear equation $y = 2.2x + 145$; where x is the number of years after 1990 .

a. Graph the linear equation.
b. Does the point (25 , 200) lie on the line ?
c. Write a sentence explaining the meaning of the ordered pair in b.

a. Graph the linear equation. We will select x-values of 0, 10, and 20.

If $x = 0$ then $y = 2.2(0) + 145 = 145$

If $x = 10$ then $y = 2.2(10) + 145 = 167$

If $x = 20$ then $y = 2.2(20) + 145 = 189$

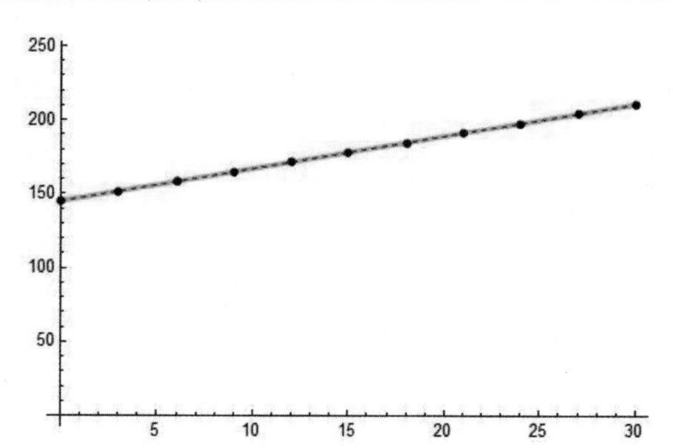

b. Does the point (25 , 200) lie on the line ?

Recall that if a point is on the line it is a solution for the equation. That is, if we plug in the x and y-values we will get a true statement.

(25, 200) ⟶ $x = 25$, $y = 200$

$$y = 2.2x + 145$$
$$200 = 2.2(25) + 145$$
$$200 = 200$$

When we plug-in, we get a true statement. So (25, 200) is a solution and therefore lies on the line.

c. Write a sentence explaining the meaning of the ordered pair in b.

Reread the original statement to determine what the x-value represents and y-value represent.

number of years after 1990 ⟶ x- value
number of people (in millions) who have a driver's license ⟶ y-value

Given the ordered pair (25, 200) , the x-value is 25, the y- value is 200.

number of years after 1990 ⟶ $x = 25$ ⟶ $1990 + 25 = 2015$
number of people (in millions) who have a driver's license ⟶ $y = 200$

In 2015, there will be 200 million people who have a driver's license.

5.04 Graphing Linear Equations 5.04

YOUR TURN

12. College is getting more expensive every year. The average cost for tuition and fees at a public university y from 1978 through 2009 can be approximated by the linear equation $y = 45x + 1089$, where x is the number of years after 1978.

First, understand the variables in words:

x: ______________________________

y: ______________________________

a. Graph the linear equation. (values for x have been selected)

If $x = 0$ then $y = 45(\quad) + 1089 =$ _____

If $x = 10$ then $y = 45(\quad) + 1089 =$ _____

If $x = 20$ then $y = 45(\quad) + 1089 =$ _____

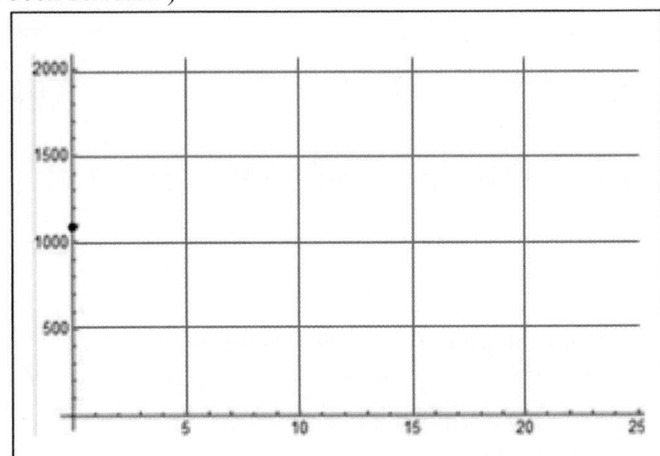

b. Does the point (15, 1764) lie on the line?

c. Write a sentence explaining the meaning of the ordered pair in b.

5.04 Graphing Linear Equations 5.04

YOUR TURN **Answers to Section 5.04.**

1. NO **2. (- 6, - 2) , (6 , 2)** **3. (- 1, - 10) , (5 ,2)** **4.**

x	y
2	-3
7/2	0
-3	-13

5a.

x	5	10	15
y	10.25	11.50	12.75

5b. 13 units

6. a.

x	1	3	5
y	5.59	5.99	6.39

6b. 13 years **6c. 2013**

7.

x	y
0	0
1	-2
-1	2

8.

x	y
0	-10
2	0
1	-5

9.

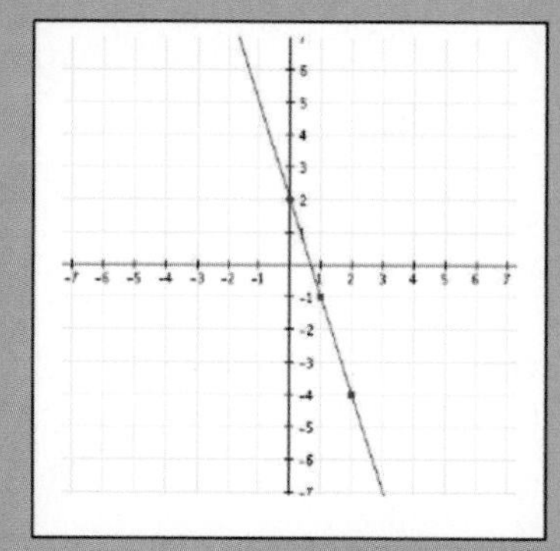

10.

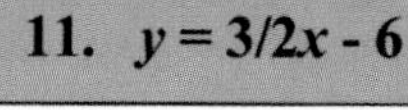

11. $y = 3/2x - 6$

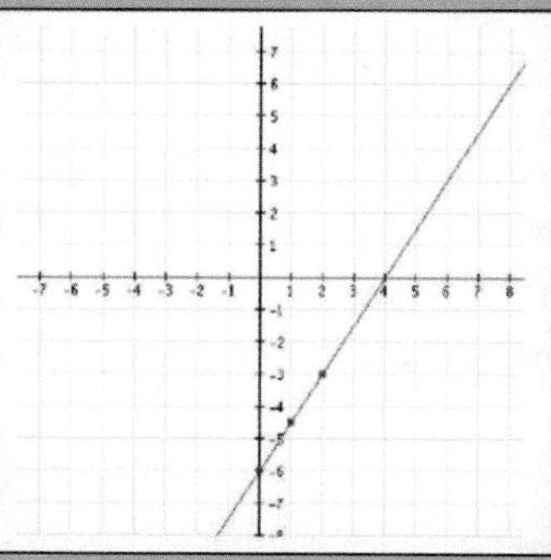

12. a. 1089, 1539, 1989

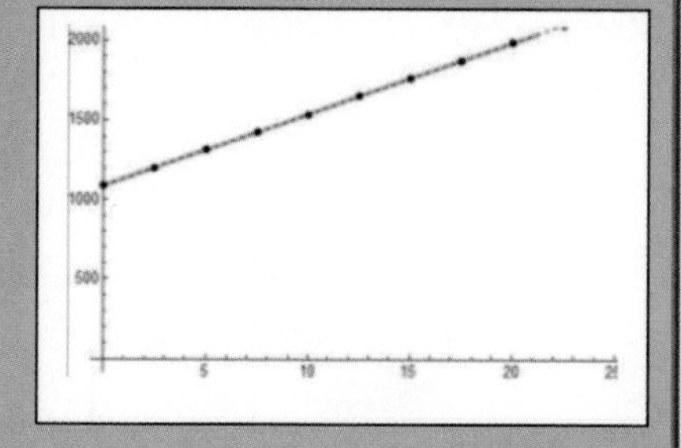

12b. Yes, it is on the line.

12c. The average tuition in 1993 was $1764.

Complete MyMathLab Section 5.04 Homework in your Homework notebook.

5.05 Intercepts 5.05

Think of the English word intersect. What does that mean? Look at the roads in the photo.

Courtesy of Shutterstock

The point at which the roads cross or **INTERSECT** one another is similar to an x or y intercept point.

EXAMPLE 1

The graph of $y = 2x - 4$ is shown below. Notice that the graph crosses the y-axis at the point (0, - 4) . This point is called the **y-intercept**. Similarly, the graph crosses the x-axis at (2, 0) . This point is called the **x-intercept**.

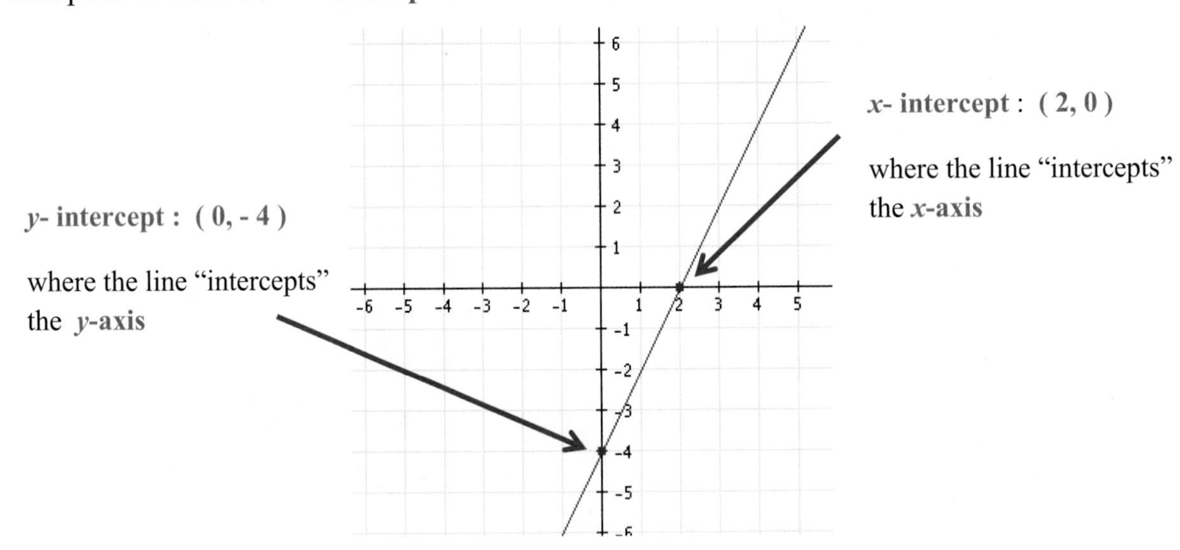

Courtesy of Shutterstock

When you plot an ordered pair, ORDER is the key word . The first number always corresponds to the x-value and the second number always corresponds to the y-value.

The ordered pair (0, 3) does not describe the same point as the ordered pair (3, 0) .

If an ordered pair has a y-coordinate of 0, its graph lies on the x-axis.

If an ordered pair has a x-coordinate of 0, its graph lies on the y-axis.

5.05 Intercepts

The x-intercept of a line is the point at which the line crosses the x axis. (where the y value equals 0)

x-intercept = (x, 0)

The y-intercept of a line is the point at which the line crosses the y axis. (where the x value equals 0)

y-intercept = (0, y)

EXAMPLE 2 Identify the intercepts.

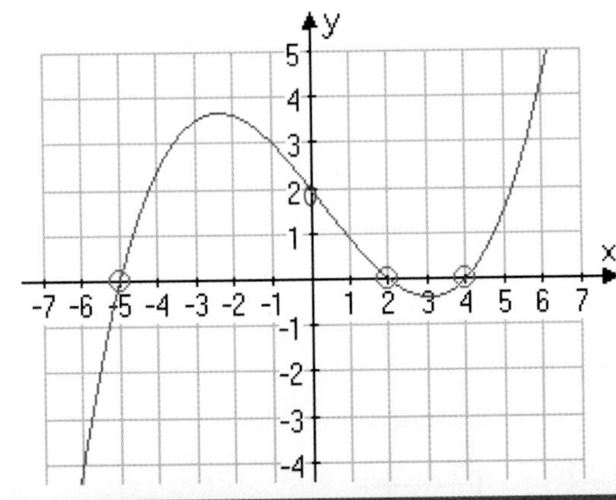

x-intercepts : (-5, 0)
(2, 0)
(4, 0)

y-intercept : (0, 2)

YOUR TURN

Identify the intercepts.

1.

x-intercept: ______

y-intercept: ______

2.

x-intercepts:______ **y -intercept:**______

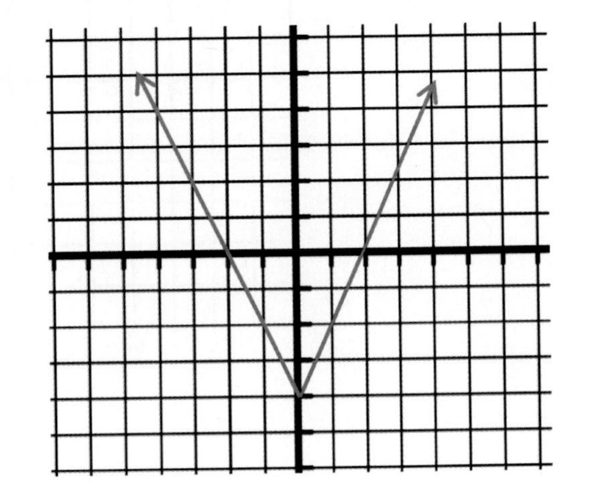

5.05 Intercepts 5.05

To this point, we have identified intercepts from a graph. We can also find intercepts if we are given an equation. We have already stated that an x-intercept will always have $y = 0$. That is, the coordinates of a **x-intercept** are **(x, 0)**. The y-intercept will always have $x = 0$. That is, the coordinates of a **y-intercept** are **(y, 0)**.

To find the x-intercept: Let $y = 0$ and solve for x : (x, 0).

To find the y-intercept: Let $x = 0$ and solve for y : (0 , y).

EXAMPLE 3

Graph $2x + 3y = 6$ using the x and y intercepts.

x-intercept	**y-intercept**
Plug in 0 for y and solve for x.	Plug in 0 for x and solve for y.

$$2x + 3y = 6$$
$$2x + 3(0) = 6$$
$$2x + 0 = 6$$
$$2x = 6$$
$$x = 3$$

x-intercept: (3 , 0)

$$2x + 3y = 6$$
$$2(0) + 3y = 6$$
$$0 + 3y = 6$$
$$3y = 6$$
$$y = 2$$

y-intercept: (0 , 2)

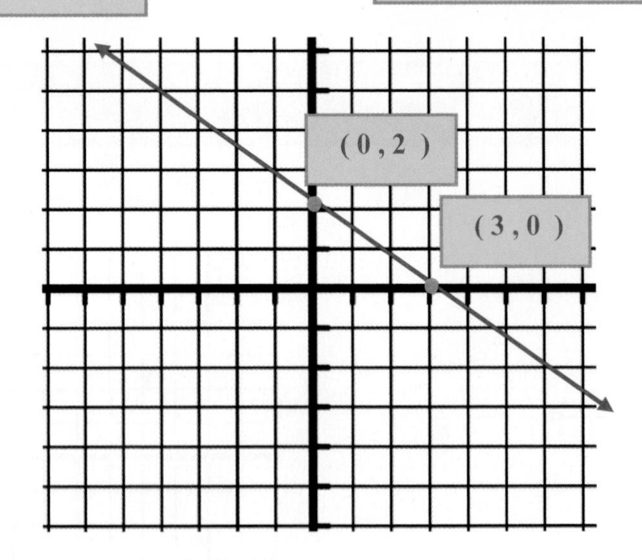

EXAMPLE 4

Graph $x - 5y = 5$ using the x and y intercepts.

***x*-intercept**	***y*-intercept**
Plug in 0 for y and solve for x.	Plug in 0 for x and solve for y.

$$x - 5y = 5$$
$$x - 5(0) = 5$$
$$x - 0 = 5$$
$$x = 5$$

***x*-intercept: (5 , 0)**

$$x - 5y = 5$$
$$0 - 5y = 5$$
$$-5y = 5$$
$$y = -1$$

***y*-intercept: (0 , -1)**

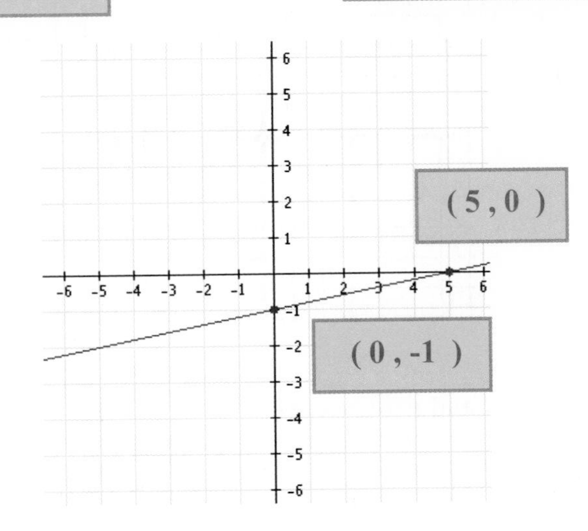

Courtesy of Shutterstock

Graph by intercepts when the equation is in standard form. In other words, the x and y terms are on the same side of the equal. Standard form looks like $Ax + By = C$.

Examples: $x - 2y = 6$, $2x + 5y = -10$, $3x - y = 6$

Graph using the table method when the equation is solved for y.

Examples: $y = 2x$, $y = 3x - 2$, $y = \frac{1}{2}x - 5$

5.05 Intercepts **5.05**

YOUR TURN

3. **Graph** $x - y = 3$ **using the** x **and** y **intercepts.**

x-intercept: Let $y = 0$, solve for x.

$x - y = 3$

y-intercept: Let $x = 0$, solve for y.

$x - y = 3$

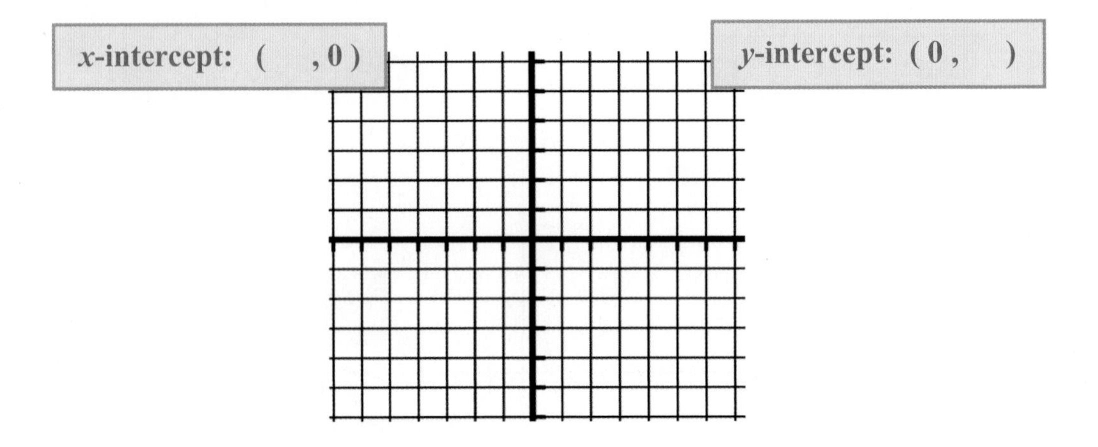

4. **Graph** $2x - 3y = 6$ **using the** x **and** y **intercepts.**

x-intercept: Let $y = 0$, solve for x.

$2x - 3y = 6$

y-intercept: Let $x = 0$, solve for y.

$2x - 3y = 6$

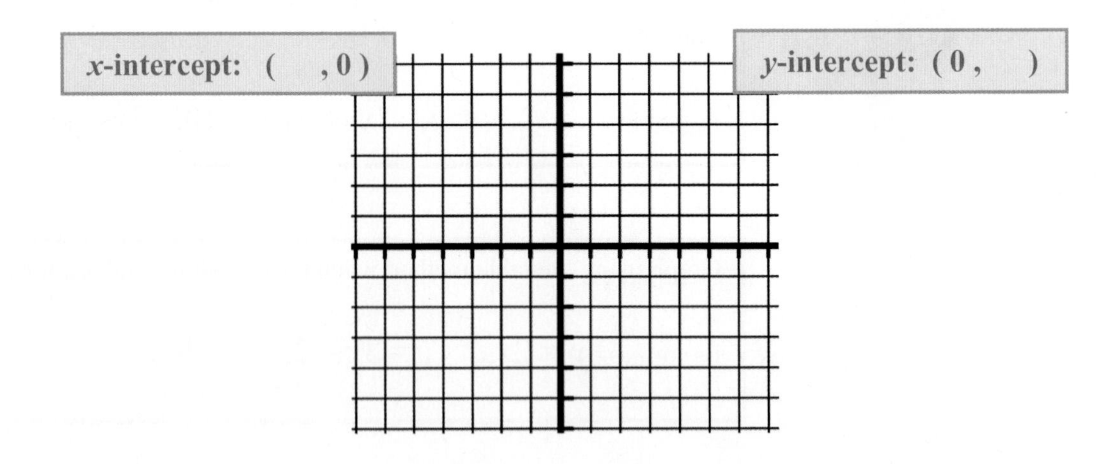

5.05 Intercepts

EXAMPLE 5

The production supervisor at Alexandra's Office Products finds that it takes 3 hours to manufacture a particular office chair and 6 hours to manufacture an office desk. A total of 1200 hours is available to produce office chairs and desks of this style. The linear equation that models this situation is $3x + 6y = 1200$, where x **represents the number of chairs** and y **the number of desks** manufactured.

a. Complete the ordered pair solution (0 ,) for this equation. Describe the manufacturing situation that corresponds to this solution.

$$3x + 2y = 1200;\ x = 0$$
$$3(0) + 2y = 1200$$
$$0 + 2y = 1200$$
$$2y = 1200$$
$$y = 600$$
$$(0, 600)$$

> x **represents the number of chairs** and y **the number of desks**
>
> Our ordered pair is (0, 600) which tells us that $x = 0, y = 600$.
>
> Zero chairs and 600 desks were manufactured.

b. Complete the ordered pair solution (, 0) for this equation. Describe the manufacturing situation that corresponds to this solution.

$$3x + 2y = 1200;\ y = 0$$
$$3x + 2(0) = 1200$$
$$3x + 0 = 1200$$
$$3x = 1200$$
$$x = 400$$
$$(400, 0)$$

> x **represents the number of chairs** and y **the number of desks**
>
> Our ordered pair is (400, 0) which tells us that $x = 400,\ y = 0$.
>
> 400 chairs and zero desks were manufactured.

c. If 50 desks are manufactured, find the greatest number of chairs that can be made.

50 desks ⟶ $x = 50$, find y.

$$3x + 2y = 1200;\ x = 50$$
$$3(50) + 2y = 1200$$
$$150 + 2y = 1200$$
$$150 - 150 + 2y = 1200 - 150$$
$$2y = 1050$$
$$y = 525$$
$$(50, 525)$$

> **The company can produce 50 chairs and 525 desks.**

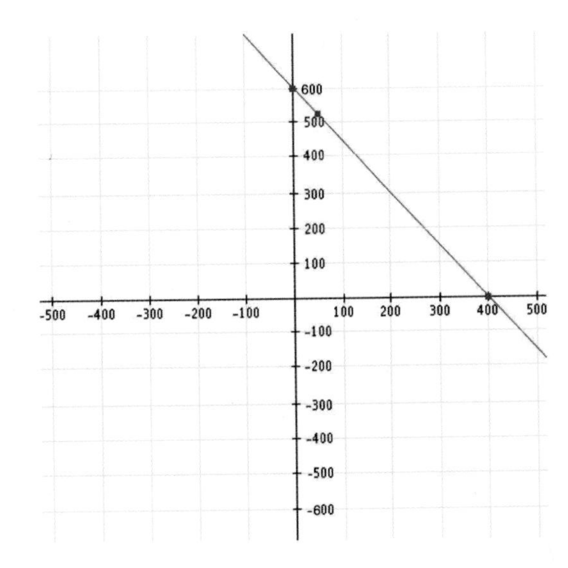

d. Use the ordered pairs found above to graph.

EXAMPLE 6

The average price of a digital camera can be modeled by the linear equation:

$y = -70x + 500$ where x is the number of years after 2000 and y is dollars.

a) Find the y-intercept of this equation. To find the y-intercept, substitute $x = 0$, and solve for y.

$y = -70(x) + 500$

$y = -70(0) + 500$

$y = 500$ y-int. is (0, 500)

b) What does the y-intercept mean?

In the year 2000, the average digital camera was $500.

Remember, the y-intercept is the beginning value when $x = 0$, time has not begun.

c) What does the negative coefficient of "x" mean in the context of this problem?

It means that the camera loses $70 in value every year after it was purchased.

YOUR TURN

5. The number of a certain chain of stores "y" for the years 2003-2007 can be modeled by the equation: $y = -198x + 3991$ where x is the number of years after 2003. (Source: Limited Brands)

a. Find the y intercept of this equation. (0 ,)

b. What does this y-intercept mean?

c. What does the negative coefficient of x tell you in the context of this problem?

5.05 Intercepts 5.05

YOUR TURN

6. In order to prepare for the upcoming snowstorm, Emily went to a local movie rental store. Movie rentals cost \$6 and Wii rentals cost \$3. She plans to spend \$30 to rent movies and games. The linear equation that models this situation is $6x + 3y = 30$ where **x represents the number of movies** rented and **y represents the number of Wii games** rented.

a. Complete the ordered pair solution (0 ,) for this equation. Describe the situation that corresponds to this solution.

$6x + 3y = 30$; $x = 0$

$6(0) + 3y = 30$

Emily rented ________movies and ________Wii games.

(0,   )

b. Complete the ordered pair solution (, 0) for this equation. Describe the situation that corresponds to this solution.

$6x + 3y = 30$; $y = 0$

$6x + 3(0) = 30$

Emily rented ________movies and ________Wii games.

(, 0)

c. If Emily decides to rent only 2 movies, how many Wii games can she rent ?

If Emily rents 2 movies, she can rent ________ Wii games.

d. Use the ordered pairs found above to graph.

1 2 3 4 5 6 7 8 9

Notice we only see Quadrant 1 on the grid.
We can't rent "negative" movies or games!

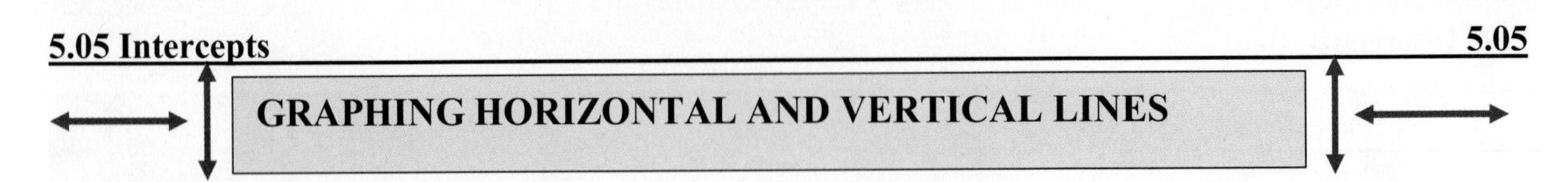

When a linear equation contains only one variable, its graph will be a horizontal or a vertical line.

EXAMPLE 7

Graph $x = 2$.

Notice the y-term is missing. The equation $x = 2$ can be written as $x + 0y = 2$. Choose 3 values for y and find corresponding values of x. We will select -1, 2, and 5 for y.

$x + 0y = 2;\ y = -1$	$x + 0y = 2;\ y = 2$	$x + 0y = 2;\ y = 5$
$x + 0(-1) = 2$	$x + 0(2) = 2$	$x + 0(5) = 2$
$x + 0 = 2$	$x + 0 = 2$	$x + 0 = 2$
$x = 2$	$x = 2$	$x = 2$
$(2,0)$	$(2,2)$	$(2,5)$

No matter what you choose for y, it will get multiplied by 0 in the equation . So, regardless of what you choose we will always get a x-value of 2.

Any ordered pair whose x-coordinate is 2 is a solution of $x + 0y = 2$.

Let's plot the ordered pairs above and see what kind of line we get.

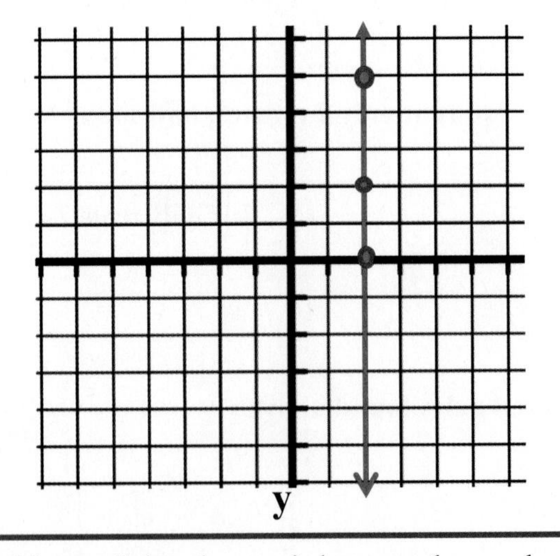

The graph is a vertical line with x-intercept (2 , 0) .

Courtesy of Shutterstock

Notice the graph of $x = 2$ is parallel to the y-axis because the x-value never changes. This will be true for all equations of the form $x = \#$. (y is missing)

The **x-axis is a horizontal** line but the graph of $x = \#$ **is a vertical** line.

EXAMPLE 8

Graph $y = 3$.

Notice the x-term is missing. The equation $y = 2$ can be written as $0x + y = 3$. Choose 3 values for x and find corresponding values of y. We will select -2, 3, and 6 for x.

$0x + y = 3 ; x = -2$	$0x + y = 3 ; x = 3$	$0x + y = 3 ; x = 6$
$0(-2) + y = 3$	$0(3) + y = 3$	$0(6) + y = 3$
$0 + y = 3$	$0 + y = 3$	$0 + y = 3$
$y = 3$	$y = 3$	$y = 3$
$(-2,3)$	$(3,3)$	$(6,3)$

No matter what you choose for x, it will get multiplied by 0 in the equation . So, regardless of what you choose we will always get a y-value of 3.

Any ordered pair whose y-coordinate is 3 is a solution of $0x + y = 3$.

Let's plot the ordered pairs above and see what kind of line we get.

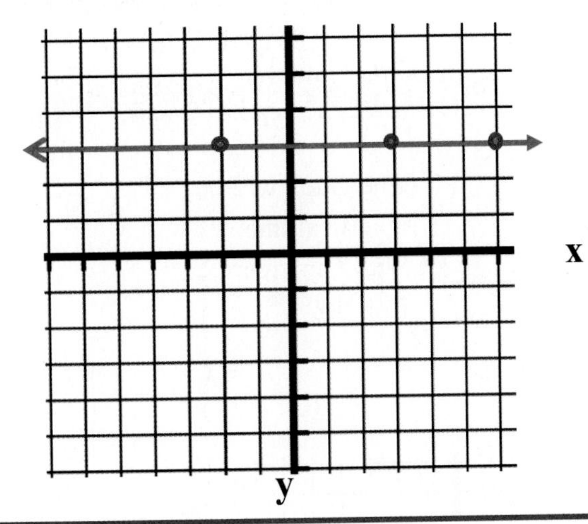

The graph is a horizontal line with y-intercept (0 , 3) .

Courtesy of Shutterstock

Notice the graph of $y = 3$ is parallel to the x-axis because the y-value never changes. This will be true for all equations of the form $y = \#$. (x is missing)

The **y-axis is a vertical** line but the graph of $y = \#$ **is a horizontal** line.

5.05 Intercepts 5.05

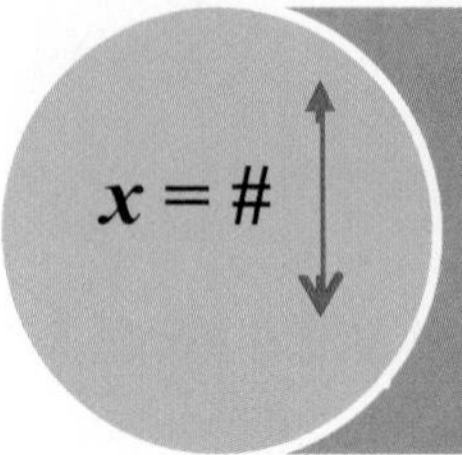

vertical line

x-intercept (# , 0)

Equation contains x but missing y.

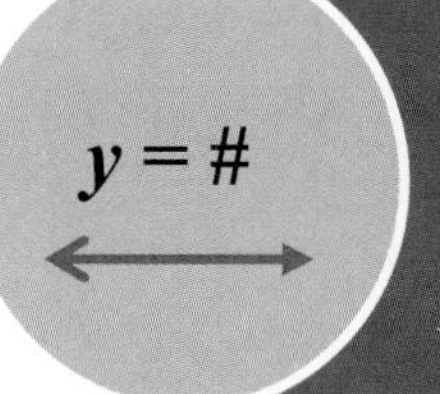

horizontal line

y-intercept (0, #)

Equation contains y but missing x

EXAMPLE 9

Graph: $x + 7 = 4$

The equation is missing y which tells us it is a vertical line of the form $x = \#$. However, we will need to solve for x to see where the vertical line is.

$x + 7 = 4$

$x + 7 - 7 = 4 - 7$

$x = -3$

The graph will be a vertical line with x-intercept (-3, 0).

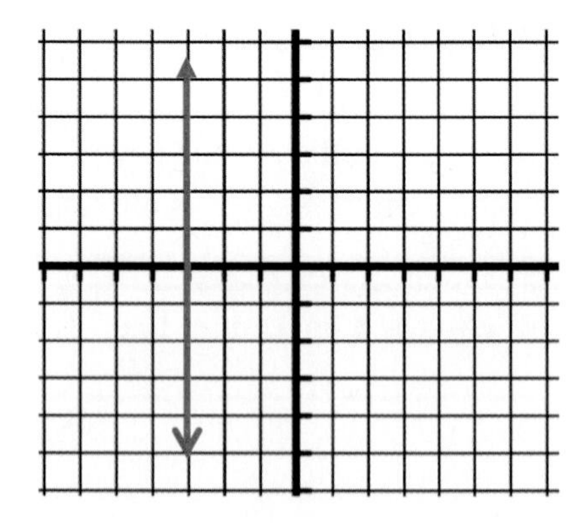

Graph: $3y = -6$

The equation is missing x which tells us it is a horizontal line of the form $y = \#$. However, we will need to solve for y to see where the horizontal line is.

$3y = -6$

$\frac{3y}{3} = \frac{-6}{3}$

$y = -2$

The graph will be a horizontal line with y-intercept (0, -2).

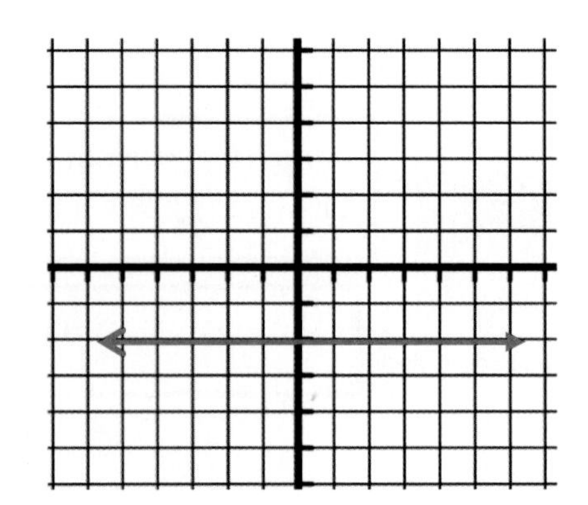

YOUR TURN

7. Graph $x = 4$

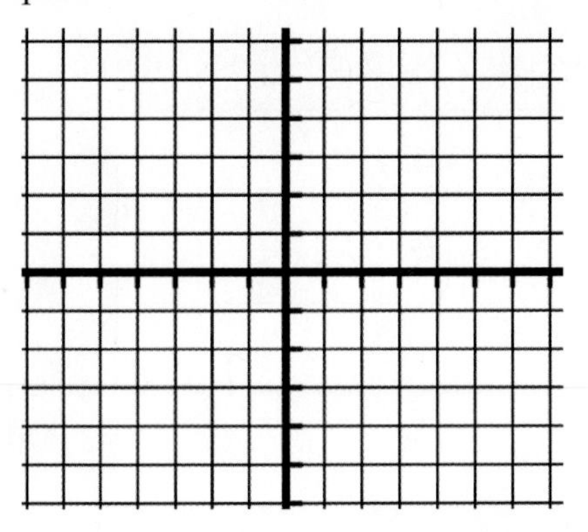

8. Graph $y = -1$

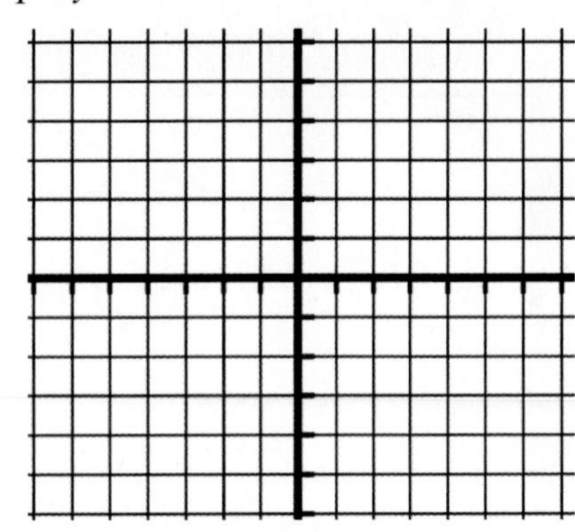

9. Graph $2x = -4$

Equation solve for x : ____________

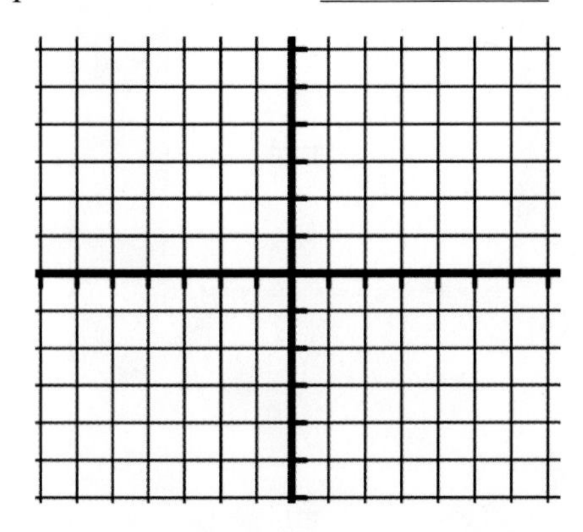

10. Graph $y + 3 = 7$

Equation solve for y : ____________

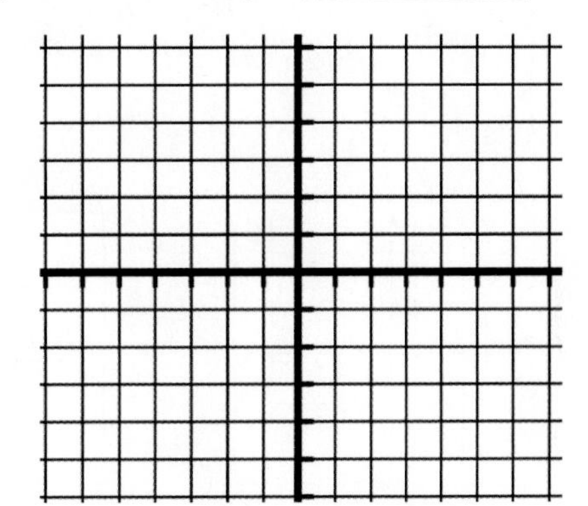

11. Complete the table and graph $y = 2x$. (This question is review.)

x	y
-1	
0	
1	

$y = 2x$; $x = -1$ | $y = 2x$; $x = 0$ | $y = 2x$; $x = 1$

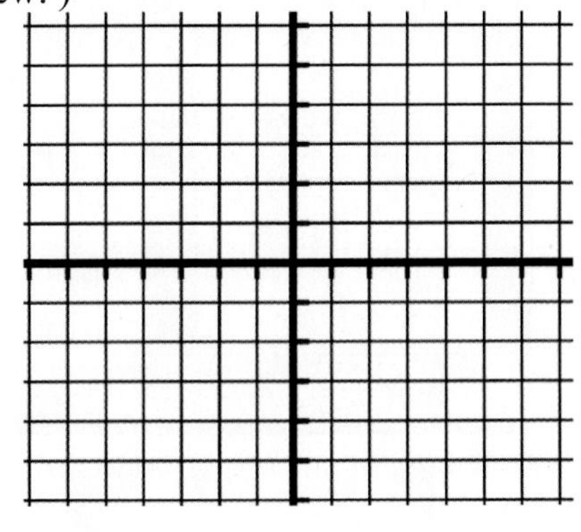

5.05 Intercepts

YOUR TURN **Answers to Section 5.05.**

1. x-intercept (2, 0) ; y-intercept (0, -6)
2. x-intercept (2, 0), (-2, 0) ; y-intercept (0, - 4)

3. x – intercept (3, 0) ; y-intercept (0, - 3)

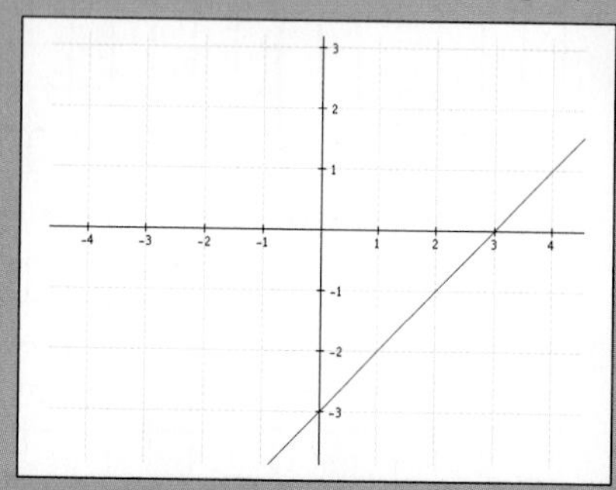

4. x – intercept (3, 0); y-intercept (0, - 2)

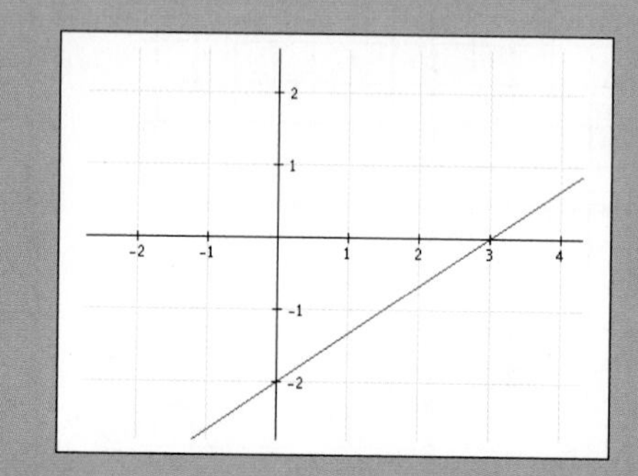

5. a) y-intercept (0,3991)
   b) The y-intercept tells us that in 2003 there were 3,991 stores.
   c) The number of stores is declining by 198 every year.

6. a. (0, 10) ; 0 movies, 10 wii games
   b. (5 , 0) ; 5 movies, 0 wii games
   c. 6 wii games
   d. Graph to right.

7.

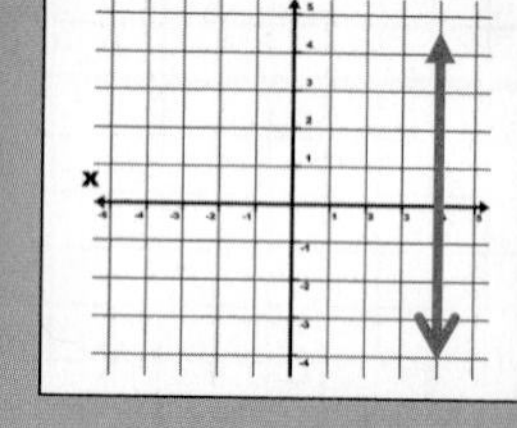

8.

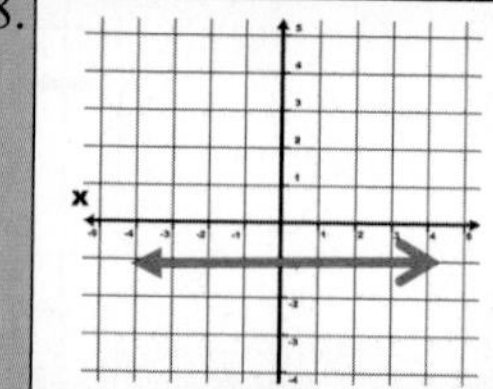

9.

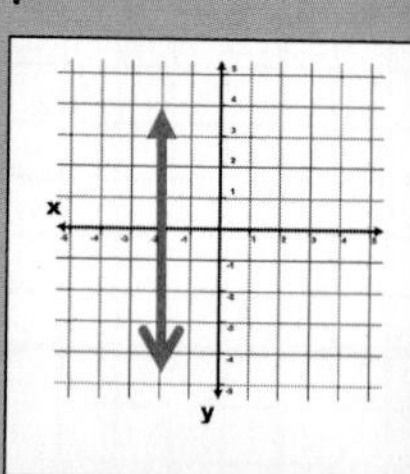

10.

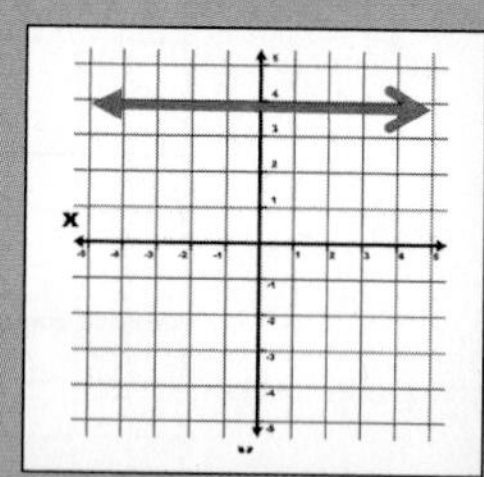

11.

x	y
-1	-2
0	0
1	2

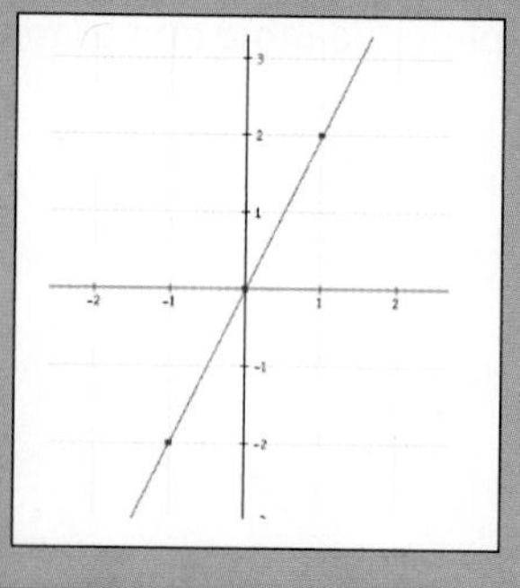

Complete MyMathLab Section 5.05 Homework in your Homework notebook.

5.06 Slope and Rate of Change 5.06

In this section we discuss the *slope* of a line. What is slope?
When you walk up or down a hill, then you are experiencing a real life example of slope.

As you go up hill, you may feel like you are spending lots of energy to get yourself to move.
The steeper the hill, the harder it is for you to keep yourself moving which means the slope is greater.
The slant or steepness of the hill is the *slope*.

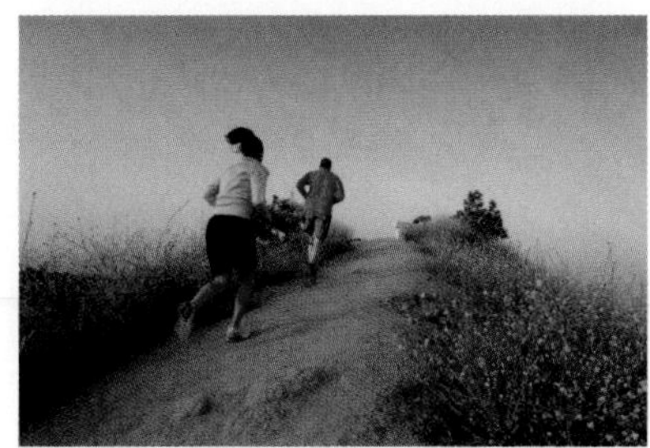

Courtesy of Shutterstock

> **Slope**
> The slope is a measure of steepness which can be found by dividing the vertical change, called the rise, by the horizontal change, called the run.

We always measure slope going from left to right.

Positive slope If you go from left to right and you go up, it is a positive slope	**Negative slope** If you go from left to right and you go down, it is a negative slope
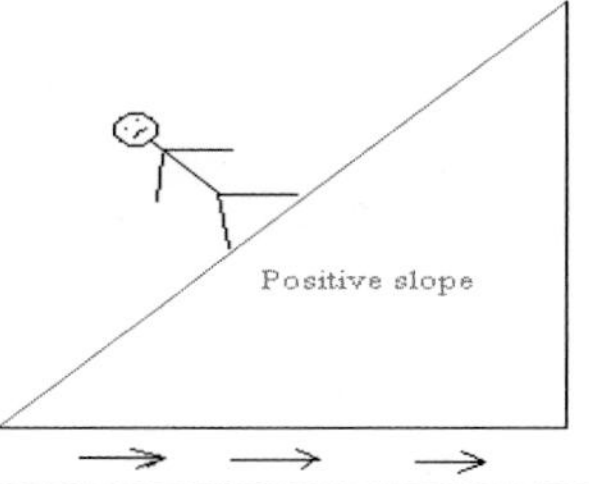	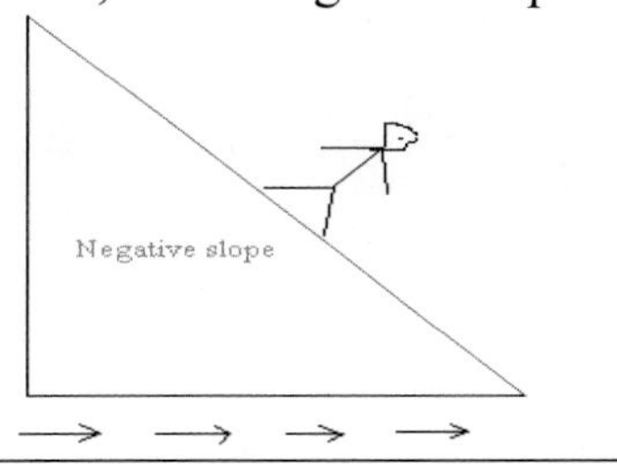

EXAMPLE 1

State whether the slope of the line is positive or negative.

a. 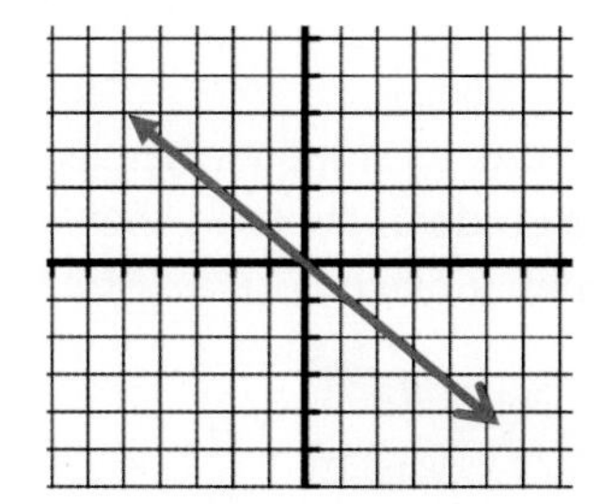

The line is downward sloping from left to right – **negative slope.**

b. 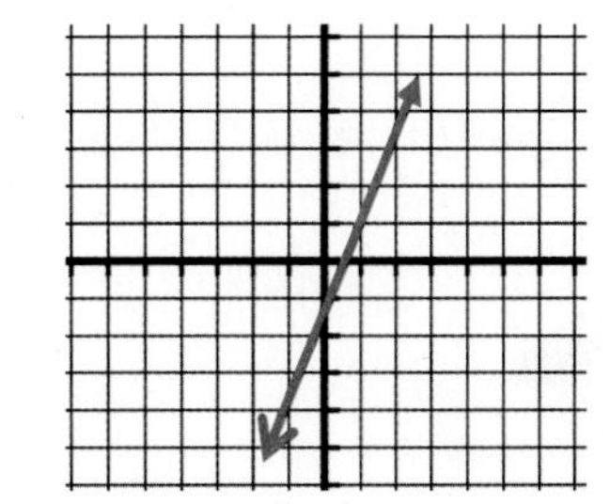

The line is upward sloping from left to right – **positive slope.**

5.06 Slope and Rate of Change 5.06

EXAMPLE 2

Find the slope.

a. Find the pitch (slope) of the roof shown. Recall the slope can be found by dividing the vertical change by the horizontal change.

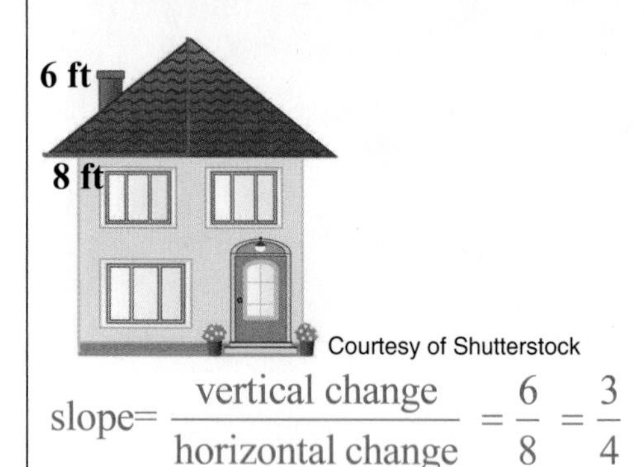

Courtesy of Shutterstock

$$\text{slope} = \frac{\text{vertical change}}{\text{horizontal change}} = \frac{6}{8} = \frac{3}{4}$$

The pitch of the roof is $\frac{3}{4}$

b. Find the slope of the treadmill. Express your answer as a percent.

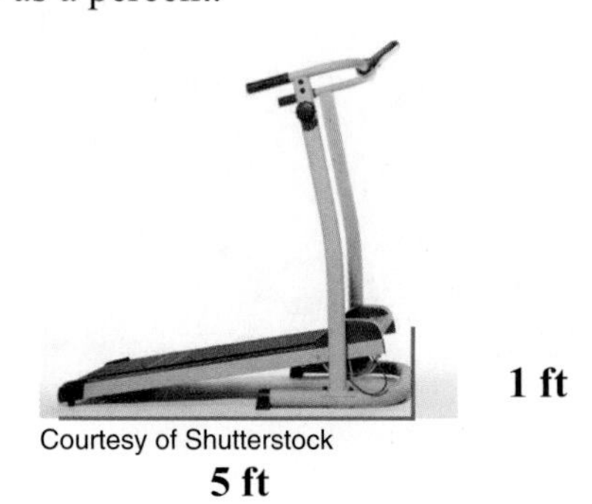

Courtesy of Shutterstock

5 ft

$$\text{slope} = \frac{\text{vertical change}}{\text{horizontal change}} = \frac{1}{5} = 0.2 = 20\%$$

YOUR TURN

State whether the slope of the line is positive or negative.

1.

2.

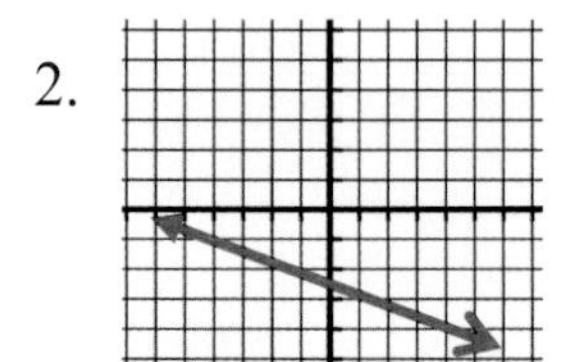

3. Find the pitch (slope) of the roof shown.

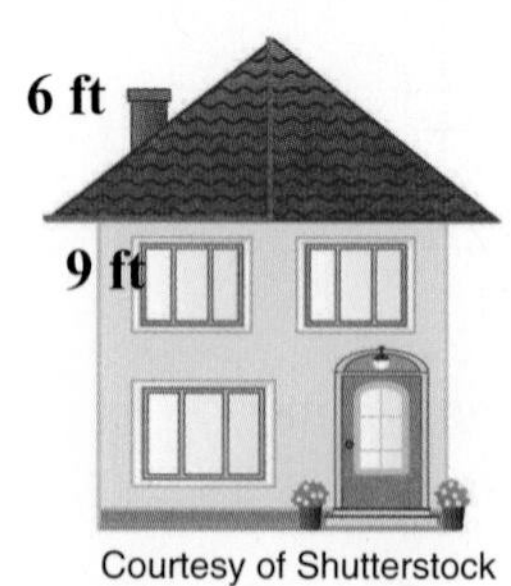

Courtesy of Shutterstock

4. The grade of a road is its slope written as a percent. Find the grade of the road shown. Round to one decimal place.

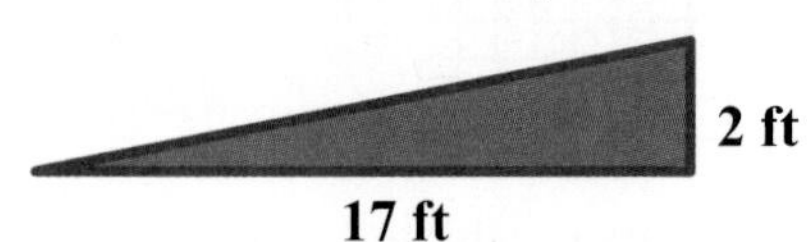

FINDING SLOPE GIVEN TWO POINTS

EXAMPLE 3

The line below contains the points (1, 2) and (4, 6). The vertical change is the change in y-coordinates: 6 – 2 or 4 units. The corresponding horizontal change is the change in *x*-coordinates: 4 – 1 = 3. The ratio of these changes is

$$\text{slope} = \frac{\text{change in } y \text{ - vertical change}}{\text{change in } x \text{ - horizontal change}} = \frac{4}{3}$$

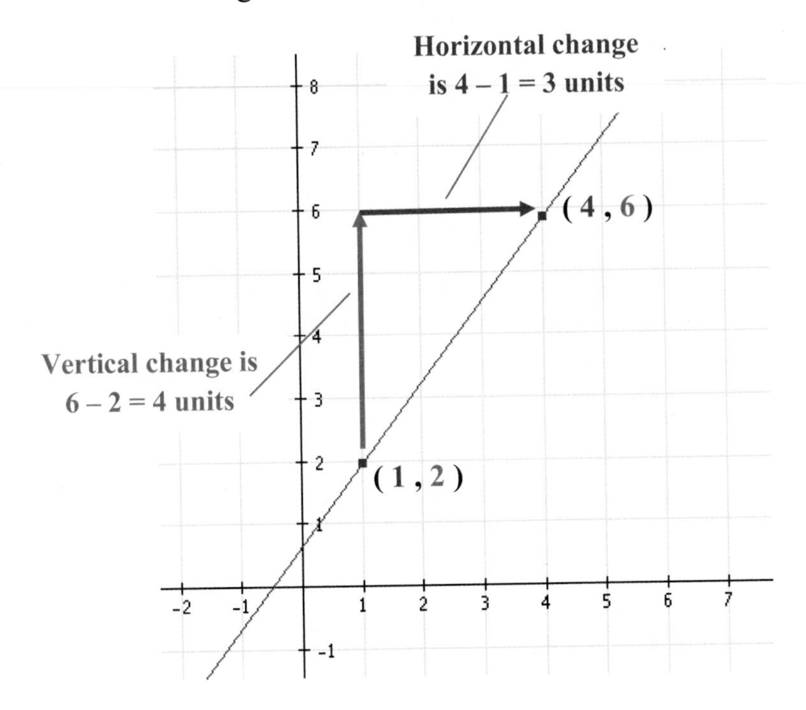

The slope of the line is $\frac{4}{3}$. This tells us that for every 4 units of change in *y*-coordinates, there is a corresponding change of 3 units in *x*-coordinates.

From this example, we can see that if we are given any two points on a line, we can use the same process to find the slope. That is, we can find the **vertical change** by **subtracting the *y*- coordinates** and then find the **horizontal change** by **subtracting the *x*-coordinates**.

We will label two points as (x_1, y_1) and (x_2, y_2).

$$slope = \frac{\text{change in } y \text{ (vertical change)}}{\text{change in } x \text{ (horizontal change)}} = \frac{y_2 - y_1}{x_2 - x_1}$$

5.06 Slope and Rate of Change 5.06

We traditionally use the letter *m* to denote slope.

Find slope given 2 points

$$m = \frac{y_2 - y_1}{x_2 - x_1}$$

EXAMPLE 4

Find the slope of the line that passes thru the points $(-2, 3)$ and $(-4, 6)$, with $(x_1, y_1) = (-2, 3)$ and $(x_2, y_2) = (-4, 6)$.

(Plug the points in the slope formula.)

$$m = \frac{6-3}{-4-(-2)} = \frac{3}{-4+2} = \frac{3}{-2} = -\frac{3}{2}$$

*Be careful with the negative signs!

Courtesy of Shutterstock

When finding slope, it makes no difference which point is identified as (x_1, y_1) and which is identified as (x_2, y_2). Just remember that whatever *y*-value is first in the numerator, its corresponding *x*-value is first in the denominator.

EXAMPLE 5

Find the slope of the line through (7 ,2) and (4, 8) .

(We will find the slope 2 ways illustrating the above statement.)

(x_1, y_1) = **(7 , 2)**, (x_2, y_2) = **(4 , 8)**

$$m = \frac{y_2 - y_1}{x_2 - x_1} = \frac{8-2}{4-7} = \frac{6}{-3} = -2$$

(x_1, y_1) = **(4 , 8)**, (x_2, y_2) = **(7 , 2)**

$$m = \frac{y_2 - y_1}{x_2 - x_1} = \frac{2-8}{7-4} = \frac{-6}{3} = -2$$

SAME SLOPE

YOUR TURN

5. Find the slope of the line.

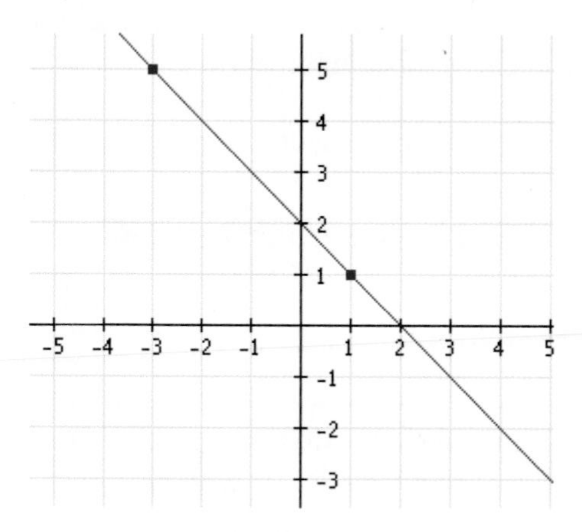

6.

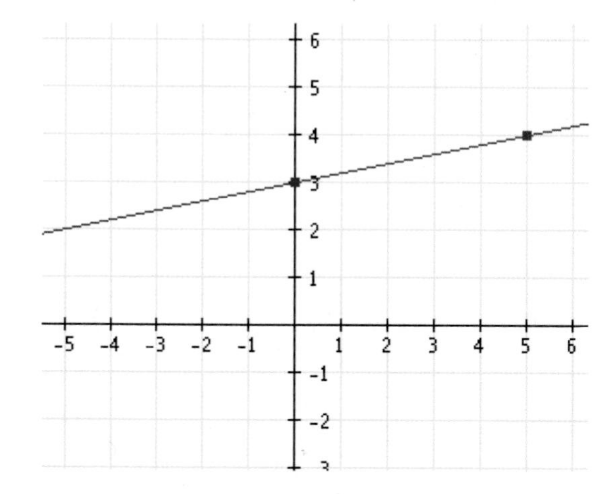

7. Find the slope of the line that passes through the points (1, - 2) and (- 7, -9).

8. Find the slope of the line that passes through the points (- 3, 7) and (5, - 2).

FINDING SLOPE GIVEN AN EQUATION

A very important form of a linear equation is the **slope-intercept form**, $y = mx + b$. We will discuss more about this form and what it tells us in a later section. For now, notice the coefficient of the x. $y = mx + b$. We have seen that letter m before. The letter, m, is used to denote the **slope** of a line.

> **If a linear equation is solved for y,** the coefficient of x is the line's slope.
>
> OR, the slope of the line given by $y = mx + b$ is m, the coefficient of the x.
>
> $y = mx + b$ → m: **slope** **SLOPE-INTERCEPT FORM**

Courtesy of Shutterstock

> **To find the slope given an equation, the equation MUST be solved for y. A common mistake students make is to look at the coefficient of x without making sure the equation is solved for y.**

EXAMPLE 6

Find the slope of each line given the equation.

a. $y = 2x - 1$ Equation is solved for y. The slope is the coefficient of x. $y = 2x - 1$ ⟶ $m = 2$	b. $y = -x + 5$ Equation is solved for y. The slope is the coefficient of x. $y = -x + 5$ $y = -1x + 5$ ⟶ $m = -1$

c. $7x + y = 9$ ⟶ The equation is NOT solved for y. Solve for y first.

$7x + y = 9$

$y = -7x + 9$ Move x-term to opposite side by changing its sign. (add $-7x$ to both sides)

$y = -7x + 9$ ⟶ $m = -7$ *The slope is the # in front of x and does NOT include x.*

EXAMPLE 7 Find the slope of the line $-7x - 8y = -56$.

Our first task is to solve the equation for ***y***.

$$-7x - 8y = -56$$

$$-7x + 7x - 8y = 7x - 56 \qquad \text{Add } 7x \text{ to both sides.}$$

$$-8y = 7x - 56$$

$$\frac{-8y}{-8} = \frac{7x}{-8} - \frac{56}{-8} \qquad \text{Divide both sides by -8.}$$

$$y = -\frac{7}{8}x + 7$$

$$m = -\frac{7}{8}$$

YOUR TURN

Find the slope of the following lines.

9. $y = 3x + 5$	10. $5x + y = 8$ Solve for y first.
11. $-2x - y = 6$ Solve for y first.	12. $-x - 6y = -12$ Solve for y first.

Finding Slope of Horizontal and Vertical Lines

EXAMPLE 8

Find the slope of the line $y = 2$.

Recall that $y = 2$ is a horizontal line where the y-value of each ordered pair is always 2. Two points on the line $y = 2$ would be (3, 2) and (6, 2). Then the slope of the line is

$$m = \frac{y_2 - y_1}{x_2 - x_1} = \frac{2-2}{6-3} = \frac{0}{3} = 0$$

Any two points on the line would yield the same slope 0 since there is no change in y.

Every horizontal line has a slope of 0.

Horizontal > Eq. y = # > slope = 0

EXAMPLE 9

Find the slope of the line $x = 3$.

Recall that $x = 3$ is a vertical line where the x-value of each ordered pair is always 3. Two points on the line $x = 3$ would be (3, 2) and (3, 5). Then the slope of the line is

$$m = \frac{y_2 - y_1}{x_2 - x_1} = \frac{5-2}{3-3} = \frac{3}{0} = \textit{undefined}$$

Recall that division by 0 is undefined.

Any two points on the line would yield the same slope since there is no change in x.

The slope of any vertical line is undefined.

Vertical > Eq. x = # > slope is undefined

5.06 Slope and Rate of Change **5.06**

YOUR TURN

13. **Find the slope of each line.**

a. $y + 3 = 0$ b. $x = 6$ c. $x = 0$

Summary of Slope

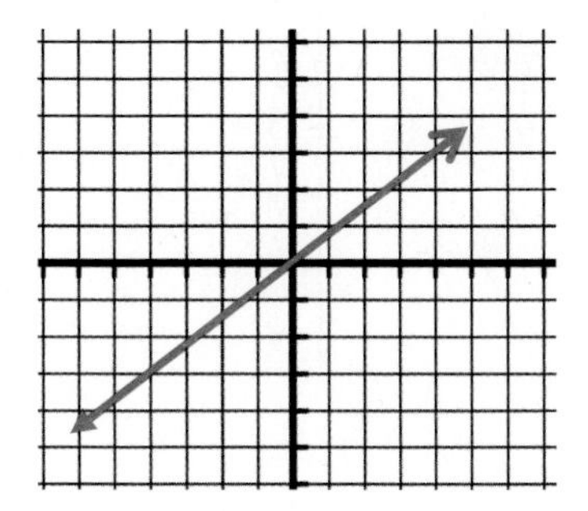

Positive Slope
(rises left to right)

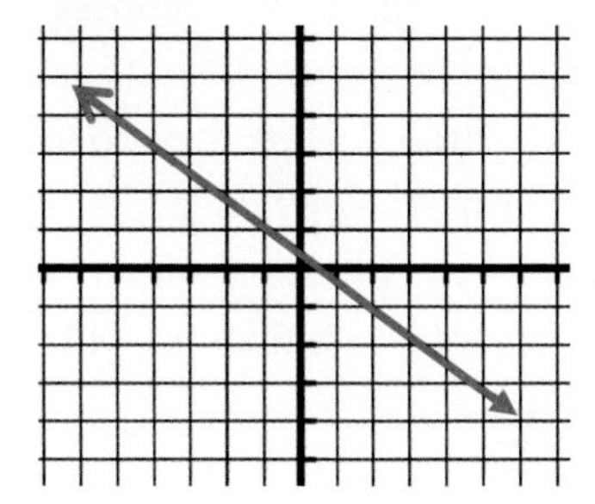

Negative Slope
(falls left to right)

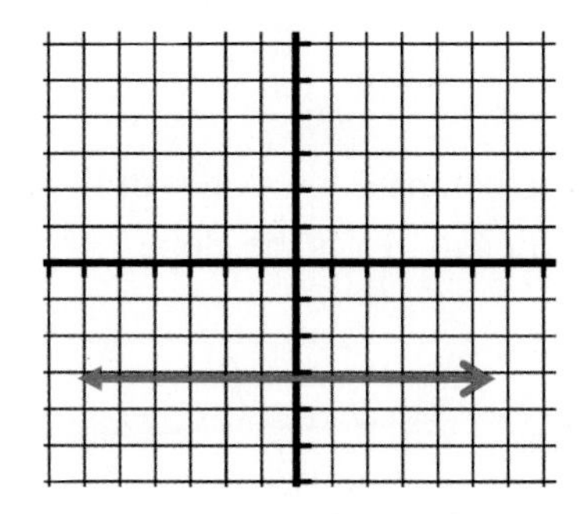

Slope is 0.
(horizontal line)

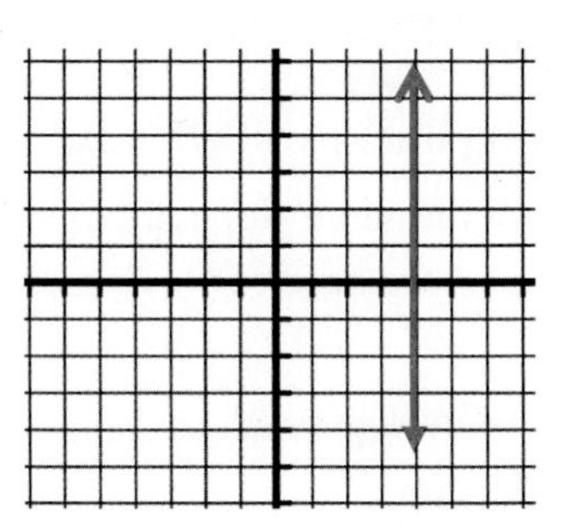

Slope is undefined.
(vertical line)

5.06 Slope and Rate of Change 5.06

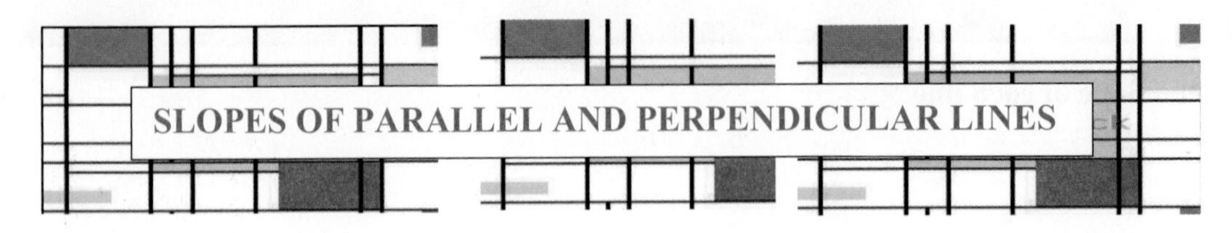

EXAMPLE 10

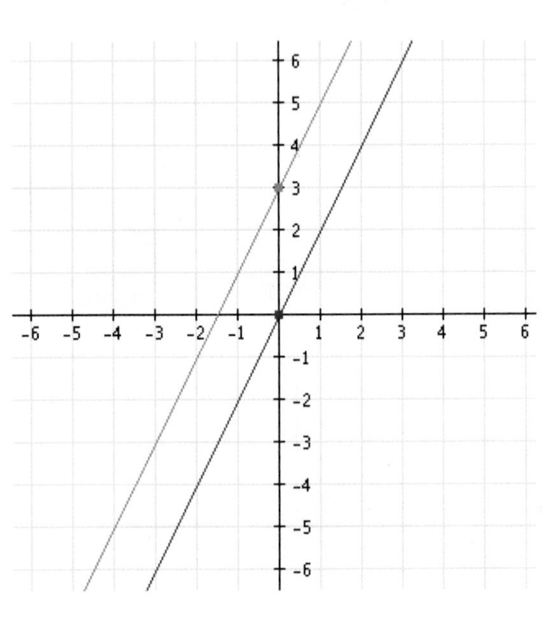

The graphs of $y = 2x$ and $y = 2x + 3$ are shown. By looking at the coefficient of the x, we see that **each line has a slope = 2**. Slopes of lines can help us determine whether lines are parallel. Since parallel lines have the same steepness it follows that they have the same slope.
Two lines are parallel when they do not intersect, no matter how far they are extended.
Note that they have different y-intercepts.
(If the y-intercepts were the same, the lines would be the same)

Parallel Lines → **Same slope**

EXAMPLE 11

Are the following lines parallel? $2x + y = 4$; $-6x - 3y = 6$

If two lines have the same slope, the lines are parallel. So, we need to compare the slopes of these lines. To find the slopes, we need to solve each equation for y and the coefficient of x is the slope.

$2x + y = 4$	$-6x - 3y = 6$
Solve for y to find slope.	Solve for y to find slope.
$2x + y = 4$ $y = -2x + 4$ $m = -2$	$-6x - 3y = 6$ $-3y = 6x + 6$ $y = -2x - 2$ $m = -2$

Same Slopes → Lines are parallel.

5.06 Slope and Rate of Change 5.06

EXAMPLE 12

Two lines are perpendicular if they meet at a 90 degree angle. How do the slopes of perpendicular lines compare? **The product of their slopes is -1.**

The graphs of $y = \frac{3}{2}x + 1$ and $y = -\frac{2}{3}x + 2$ are shown.

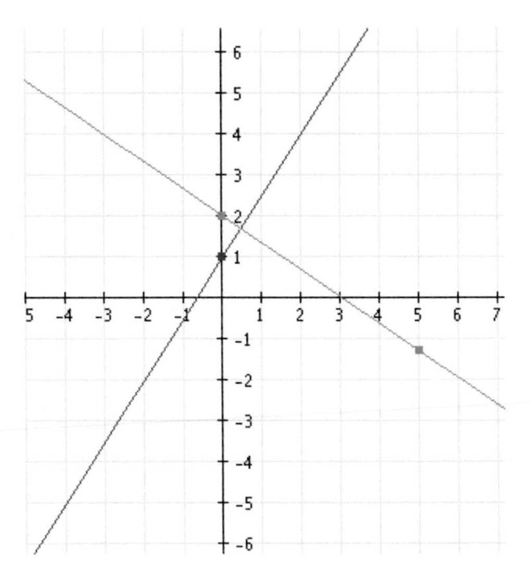

By looking at the coefficient of the x, we see that one line has a slope of $\frac{3}{2}$; the other line has a slope of $-\frac{2}{3}$.

Their product is $\frac{3}{2} \cdot -\frac{2}{3} = -1$

The lines are perpendicular. Notice that the slopes are opposites AND that one slope is the reciprocal of the other slope.

Perpendicular Lines	• The product of their slopes is -1
OR	• The slopes have opposite signs **and** are reciprocals of each other.

EXAMPLE 13

Given a line with slope $\frac{3}{5}$, find the slope of the line parallel and perpendicular to this line.

- Find the slope of the line parallel to this line.

  Parallel lines have the same slope.

  The slope of that line parallel is $m = \frac{3}{5}$.

- Find the slope of the line perpendicular to this line.

  Perpendicular lines have slopes that are opposite reciprocals of each other.

  The slope of the line perpendicular is $m = -\frac{5}{3}$.

5.06 Slope and Rate of Change **5.06**

EXAMPLE 14

Find the slope of the line that is parallel and perpendicular to the line through points (9, 3) and (5, - 6) .

- Find the slope of the line through (9, 3) and (5, -6)

$$m = \frac{y_2 - y_1}{x_2 - x_1} = \frac{-6-3}{5-9} = \frac{-9}{-4} = \frac{9}{4}$$

- Find the slope of the line parallel to this line.

  Parallel lines have the same slope.

  The slope of that line parallel is $m = \frac{9}{4}$.

- Find the slope of the line perpendicular to this line.

  Perpendicular lines have slopes that are opposite reciprocals of each other.

  The slope of the line perpendicular is $m = -\frac{4}{9}$.

EXAMPLE 15

Give the slope of each line and then determine whether the two lines are parallel, perpendicular, or neither parallel nor perpendicular.

$$y = -\frac{1}{5}x + 1; \quad 2x + 10y = 3$$

Recall that to find the slope given an equation, the equation must be solved for y. If the equation is solved for y, the **slope is the coefficient of x.**

$$y = -\frac{1}{5}x + 1$$

Equation is solved for y, so the slope is the coefficient of x.

$$y = -\frac{1}{5}x + 1 \longrightarrow m = -\frac{1}{5}$$

$$2x + 10y = 3$$

Equation is **NOT** solved for y. So, we must solve for y before looking for slope.

$$2x + 10y = 3$$
$$10y = -2x + 3$$
$$\frac{10y}{10} = \frac{-2x}{10} + \frac{3}{10}$$
$$y = -\frac{1}{5}x + \frac{3}{10} \longrightarrow m = -\frac{1}{5}$$

Since the lines have the **SAME SLOPES** they are **PARALLEL.**

EXAMPLE 16

Give the slope of each line and then determine whether the two lines are parallel, perpendicular, or neither parallel nor perpendicular.

$$2x + y = 7; \ -2x + 4y = 6$$

Recall that to find the slope given an equation, the equation must be solved for y. If the equation is solved for y, the **slope is the coefficient of x.** *The slope does not include the variable x.*

$$2x + y = 7$$

Equation is **NOT** solved for y. So, we must solve for y before looking for slope.

$$2x + y = 7$$

$$y = -2x + 7 \rightarrow m = -2$$

$$-2x + 4y = 6$$

Equation is **NOT** solved for y. So, we must solve for y before looking for slope.

$$-2x + 4y = 8$$

$$4y = 2x + 8$$

$$\frac{4y}{4} = \frac{2x}{4} + \frac{8}{4}$$

$$y = \frac{1}{2}x + 2 \rightarrow m = \frac{1}{2}$$

The lines have **OPPOSITE RECIPROCAL SLOPES.**
The lines are **PERPENDICULAR.**

You can also test for perpendicular by multiplying the two slopes with your calculator. Remember the product of their slopes should be -1 if the lines are perpendicular.

$$-2 \bullet \frac{1}{2} = -1$$

Courtesy of Shutterstock

A common mistake students make is to forget that perpendicular lines have slopes that are opposite **AND** slopes that are reciprocals of each other. If a line has a **slope of 3** and another line had a **slope of -3**, these two lines would **NOT** be **perpendicular**. They are opposites but they are not reciprocals of each other.

To be perpendicular the slopes would need to be 3 and $-\frac{1}{3}$.

5.06 Slope and Rate of Change 5.06

YOUR TURN

14. Find the slope of the line through the points (- 4, 2) and (6, 5).

$$m = \frac{y_2 - y_1}{x_2 - x_1} =$$

The slope of the line parallel is __________.

The slope of the line perpendicular is _______.

15. Find the slope of the line through the points (5 , - 6) and (8, -3).

$$m = \frac{y_2 - y_1}{x_2 - x_1} =$$

The slope of the line parallel is __________.

The slope of the line perpendicular is _______.

16. Give the slope of each line and then determine whether the two lines are parallel, perpendicular, or neither parallel nor perpendicular.

$$y = 3x - 9; \quad 9x - 3y = -6$$

$y = 3x - 9$

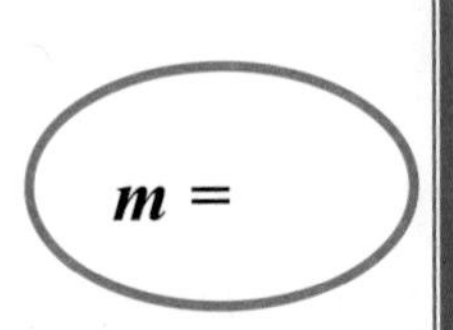

$9x - 3y = -6$

m =

The lines are ________________________.

17. Give the slope of each line and then determine whether the two lines are parallel, perpendicular, or neither parallel nor perpendicular.

$$3y = 4x + 6; \quad 3x + 4y = 8$$

$3y = 4x + 6$

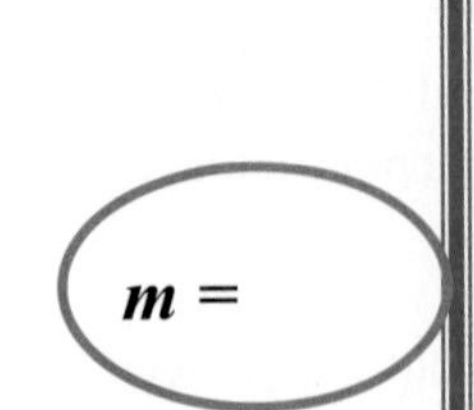

$3x + 4y = 8$

m =

The lines are ________________________.

5.06 Slope and Rate of Change 5.06

APPLICATIONS USING SLOPE

EXAMPLE 17

The maximum grade allowed between two stations in a rapid-transit rail system is 3.5%. Between station A and station B, which are 250 ft. apart, the tracks rise 7.5 ft. What is the grade of the tracks between these two stations? Does this grade meet the rapid-transit standard?

The grade of the tracks (slope) can be found by finding the ratio of the vertical change to the horizontal change. Recall that to simplify a ratio, you divide.

$$m = \frac{\text{vertical change}}{\text{horizontal change}} = \frac{7.5}{250} = 7.5 \div 250 = 0.03 = 3\%$$

The maximum grade allowed between two stations is 3.5%. The grade is 3 % which is less than 3.5% and meets the standard.

EXAMPLE 18

The following graph shows the **cost y (in cents)** of a nationwide long-distance telephone call from Texas with a certain telephone-calling plan, where **x is the length of the call in minutes**. Find the slope of the line and attach the proper units for the rate of change. Then write a sentence explaining the meaning of the slope in this application.

$$m = \frac{y_2 - y_1}{x_2 - x_1} = \frac{62 - 34}{6 - 2}$$

$$= \frac{28}{4} = \frac{7 \text{ cents}}{1 \text{ minute}}$$

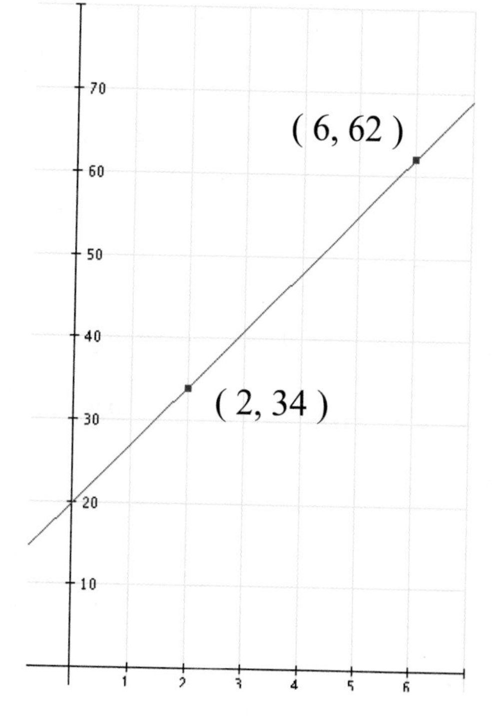

The rate of change of a phone call is 7 cents per 1 minute, or the cost of the phone call is 7 cents per minute.

Courtesy of Shutterstock

5.06 Slope and Rate of Change

YOUR TURN

18. The maximum grade allowed between two stations in a rapid-transit rail system is 3.5%. Between station A and station B, which are 138 ft. apart, the tracks rise 5.5 ft. Round to nearest tenth if necessary. What is the grade of the tracks between these two stations? Does this grade meet the rapid-transit standard?

$$m = \frac{\text{vertical change}}{\text{horizontal change}} = \underline{\qquad}$$

The maximum grade allowed between two stations is 3.5%. The grade is ______ which is ______ than 3.5% and __________ the standard.

19. Mario works at a local supermarket. The following graph shows Mario's earnings y (in dollars) , where x is the number of hours worked.. Find the slope of the line and attach the proper units for the rate of change. Then write a sentence explaining the meaning of the slope in this application.

$$m = \frac{y_2 - y_1}{x_2 - x_1} =$$

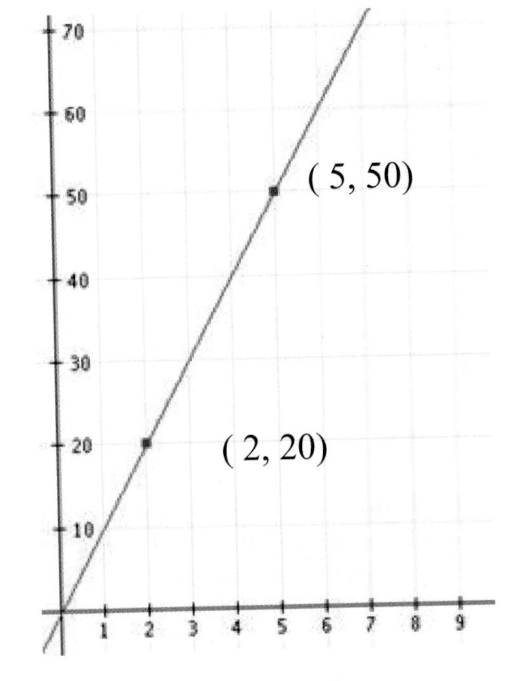

The rate of change of a Mario's earning is is ________ dollars per 1 hour, or Mario's wage is ________ per hour.

5.06 Slope and Rate of Change **5.06**

SUMMARY OF SLOPE

Question	How to do it???	Example
Find the slope given two points	$m = \dfrac{y_2 - y_1}{x_2 - x_1}$	Find the slope of the line through (4, 5) and (- 1 , - 6) $m = \dfrac{y_2 - y_1}{x_2 - x_1} = \dfrac{-6-5}{-1-4} = \dfrac{-11}{-5} = \dfrac{11}{5}$
Find the slope given an equation	Solve the equation for y. Slope is the coefficient of x. $y = mx + b$	Find the slope of $-3x + y = 5$ Solve for y. $-3x + y = 5$ $y = 3x + 5$ $m = 3$
Find the slope of vertical and horizontal lines	Horizontal ⟶ $m = 0$ Vertical ⟶ undefined slope	Find the slope of each line. $x = 5$ vertical line slope is undefined. $y = 6$ horizontal line slope is 0.
Find the slope of parallel and perpendicular lines	Parallel lines ⟶ same slopes Perpendicular lines ⟶ opposite reciprocal slopes Product of slopes is – 1.	Given a line with slope $m = \dfrac{2}{5}$. The slope of the parallel is $\dfrac{2}{5}$. The slope of the line perpendicular is $-\dfrac{5}{2}$.

5.06 Slope and Rate of Change

YOUR TURN **Answers to Section 5.06.**

1. positive
2. negative
3. positive
4. 11.8%
5. -1
6. $m = \frac{1}{5}$
7. $m = \frac{7}{8}$
8. $m = -\frac{9}{8}$
9. $m = 3$
10. $m = -5$
11. $m = -2$
12. $m = -\frac{1}{6}$
13. a. $m = 0$ b. undefined c. undefined
14. parallel : $m = \frac{3}{10}$; perpendicular: $m = -\frac{10}{3}$
15. parallel : $m = 1$; perpendicular: $m = -1$
16. parallel
17. perpendicular
18. The maximum grade allowed between two stations is 3.5%. The grade is 4 % which is more than 3.5% and does not meet the standard.
19. The rate of change of a Mario's earning is 10 dollars per 1 hour, or Mario's wage is $10 per hour.

Complete MyMathLab Section 5.06 Homework in your Homework notebook.

Writing Equations of Lines

Linear Equations are usually written in one of two forms.

Standard Form: $Ax + By = C$

Examples : $x + y = 3$

$2x - y = 8$

$3x + 5y = 10$

x- and y-terms are on same side of equal.
No fractions!

Slope-Intercept Form: $y = mx + b$

Examples : $y = 3x - 7$

$y = -4x$

$y = \frac{2}{3}x - 1$

Equation solved for y.
Fractions are allowed.

The forms are important because as we begin to write equations of lines, you will need to pay close attention to how your answer needs to be expressed. Sometimes you will be asked to write your answer in **standard form** and at other times you will be asked to write your answer in **slope-intercept form**.

So far in our study of linear equation, we have been given the equation and asked to graph the equation or perhaps find the slope. In this section, you will be given information about a line and asked to **find the equation** that corresponds to that information.

Writing Equations of Lines: Given Its Slope and a Point

STARTING PLACE: When the slope of a line and a point on the line are known, we can use the point-slope formula to determine the equation of the line. The point slope formula gets its name because it uses a single point on the graph AND the slope of the line.

Point-Slope Formula

$$y - y_1 = m(x - x_1)$$

m is the slope and (x_1, y_1) is the given point

Courtesy of Shutterstock

To apply this formula, you will have to know the slope of the line and a point on the line.

5.07 Equations of Lines

EXAMPLE 1

Write the equation of a line with a slope of 5 and passing thru the point (4, 2). Write the equation in slope-intercept form.

Since we are given a **point** on the line and the **slope** of the line, we begin by writing the equation using the point-slope formula. The point on the line is **(4, 2)** ; we will use this point for (x_1, y_1) in the formula. We substitute 5 for ***m***.

(4 , 2) $m = 5$

$$y - y_1 = m(x - x_1)$$

$$y - 2 = 5(x - 4)$$

$$y - 2 = 5x - 20$$

$$y - 2 + 2 = 5x - 20 + 2$$

$$y = 5x - 18$$

EXAMPLE 2

Write the equation of the line in slope intercept form through (0 , - 3) with slope $-\frac{5}{7}$.

$$y - y_1 = m(x - x_1)$$

$$y - (-3) = -\frac{5}{7}(x - 0)$$ Substitute into formula.

$$y + 3 = -\frac{5}{7}x$$ Use distributive property.

$$y = -\frac{5}{7}x - 3$$ Subtract 3 from both sides.

EXAMPLE 3

Write the equation of the line with slope $-\frac{1}{4}$ through (2, 3). Write the equation in **standard form**.

$y - y_1 = m(x - x_1)$

$y - 3 = -\frac{1}{4}(x - 2)$ Substitute into point-slope formula

$4(y - 3) = 4 \bullet -\frac{1}{4}(x - 2)$ Multiply both sides by 4.
No fractions in standard form.

$4y - 12 = -1(x - 2)$

$4y - 12 = -1x + 2$ Use distributive property.

$4y = -x + 14$ Add 12 to both sides.

$x + 4y = 14$ Add x to both sides.

YOUR TURN

Write the equation of each line described below. Express your answer in the form asked.

1. $m = -8$ through (- 1 , - 5) ; slope-intercept	2. $m = \frac{3}{2}$ through $(5, -6)$; standard form (Write the equation with a positive coefficient of x.)

Writing Equations of Lines: Given Two Points

EXAMPLE 4

Write the equation of the line through (- 1, 3,) and (- 2, - 5) . Express answer in slope-intercept.

Recall that to use the point-slope formula to write the equation of a line we must have **a point** and the **slope**. In this example, we **DO NOT have the slope** but we do have more than enough points.

First, use the two given points to *find the slope of the line*.

$$m = \frac{y_2 - y_1}{x_2 - x_1} = \frac{-5-3}{-2-(-1)} = \frac{-8}{-1} = 8$$

Next we use the **slope 8** and either one of the given points to write the equation in point-slope form. We will get the same answer regardless of which point we choose. A good suggestion is to pick the ordered pair that has the least amount of negative numbers involved. That will make the arithmetic easier to keep up with.

We will choose **(- 1 , 3)** to use with our **slope of 8**.

$$y - y_1 = m(x - x_1)$$

$y - 3 = 8[x - (-1)]$	Substitute into point-slope formula
$y - 3 = 8[x + 1]$	Simplify.
$y - 3 = 8x + 8$	Use distributive property.
$y = 8x + 11$	Add 3 to both sides.

EXAMPLE 5

Write the equation of the line passing thru points (-1, 6) and (3, 1) in slope-intercept form.

In this problem, we do not have slope, so first we must find the slope. Use the slope formula for 2 points.

$$m = \frac{y_2 - y_1}{x_2 - x_1} \qquad m = \frac{1-6}{3-(-1)} = \frac{-5}{4} = -\frac{5}{4}$$

Now we are ready for Point-slope form.

We have been given two points on the line. We only need to use one of these points to plug into our formula. **So you can choose either point to plug in!!!** Let's pick (3, 1).

A point on the line is given: (x_1, y_1) (3, 1)

The slope of the line is $-\frac{5}{4}$.

$$y-1 = -\frac{5}{4}(x-3)$$

$$y-1 = -\frac{5}{4}x + \frac{15}{4}$$

$$y-1+1 = -\frac{5}{4}x + \frac{15}{4} + 1$$

$$y-1+1 = -\frac{5}{4}x + \frac{15}{4} + \frac{4}{4}$$

$$y = -\frac{5}{4}x + \frac{19}{4}$$

Slope-Intercept Form

> **TIP**
>
> **When deciding which point to plug in, look for a point that has positive values. It will make your math a little easier. Notice we chose the point (3, 1) as opposed to (- 1 , 6).**

Courtesy of Shutterstock

Note: To write an equation in standard form you must clear fractions by multiplying both sides of the equation by the LCM of all denominators.

EXAMPLE 6

Find the equation of the line passing through (7, 10) and (-1 , -1). Write your answer in standard form.

First, use the two given points to find the slope of the line.

$$m = \frac{y_2 - y_1}{x_2 - x_1} = \frac{-1-10}{-1-7} = \frac{-11}{-8} = \frac{11}{8}$$

Next we use $m = \frac{11}{8}$ and **choose** (7, 10) to write the equation in standard form.

5.07 Equations of Lines

$$y - y_1 = m(x - x_1)$$

$$y - 10 = \frac{11}{8}(x - 7)$$ Substitute into point-slope formula.

$$8(y - 10) = 8 \bullet \frac{11}{8}(x - 7)$$ Multiply both sides by 8.

$$8y - 80 = 11(x - 7)$$

$$8y - 80 = 11x - 77$$ Use distributive property.

$$8y = 11x + 3$$ Add 80 to both sides.

$$-11x + 8y = 3$$ Add x to both sides.

OR 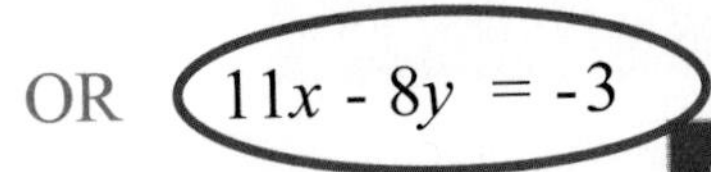

$$11x - 8y = -3$$ Multiply by -1 to make x-term positive.

Standard Form

YOUR TURN

Write the equation of each line below. Express your answer in slope-intercept form.	
3. Passing through (- 1, 3) and (- 2, - 5).	4. Passing through (2, 3) and (0 , 0).

YOUR TURN

Write the equation of each line below. Express your answer in standard form.	
5. Passing through (- 1, 3) and (- 5, 1). (Write the equation with a positive coefficient of x.)	6. Passing through (-6, -2) and (5 , -3). (Write the equation with a positive coefficient of x.)

Writing Equations of Horizontal and Vertical Lines

Remember from a previous section we learned how to graph horizontal and vertical lines.

The graph of $y = \#$ is a horizontal line whose y-intercept is (0, #).
The graph of $x = \#$ is a vertical line whose x-intercept is (#, 0).

If we look closely at those two statements , we can identify what the equation of a horizontal or vertical line should be. Hint: The equations are in red !

Courtesy of Shutterstock

The equation of any horizontal line is $y = \#$.

The equation of any vertical line is $x = \#$.

5.07 Equations of Lines

EXAMPLE 7 **Write the equation of the horizontal line passing through the point (- 6, - 2).**

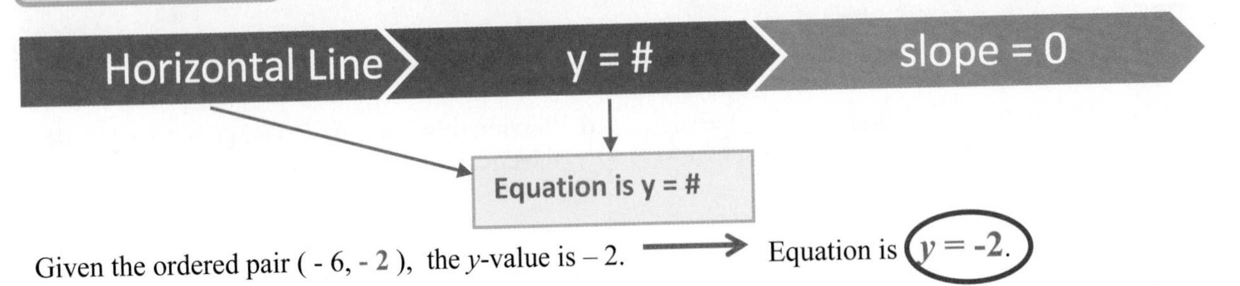

Given the ordered pair (- 6, - 2), the y-value is – 2. ⟶ Equation is $y = -2$.

EXAMPLE 8 **Write the equation of the line with *slope 0* through the point (5, 4).**

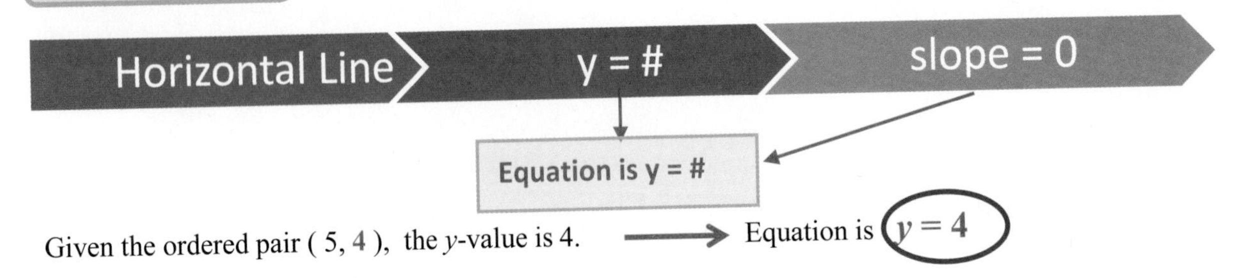

Given the ordered pair (5, 4), the y-value is 4. ⟶ Equation is $y = 4$

EXAMPLE 9 **Write the equation of the vertical line passing through the point (- 7, - 8).**

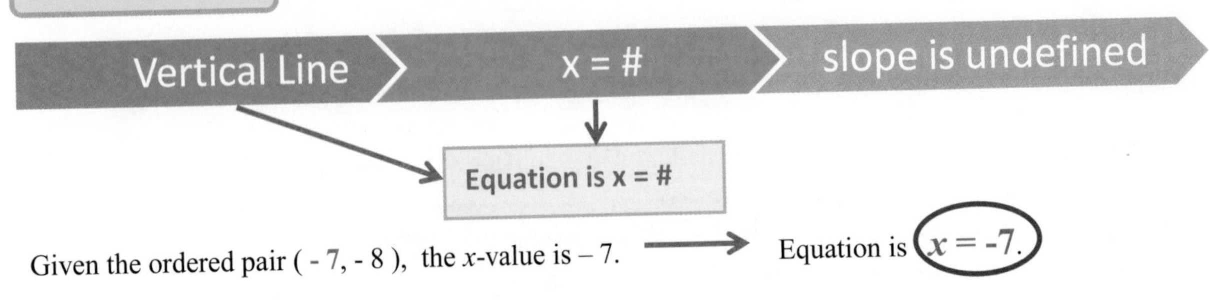

Given the ordered pair (- 7, - 8), the x-value is – 7. ⟶ Equation is $x = -7$.

EXAMPLE 10 **Write the equation of the line with *undefined slope* through (1, - 4) .**

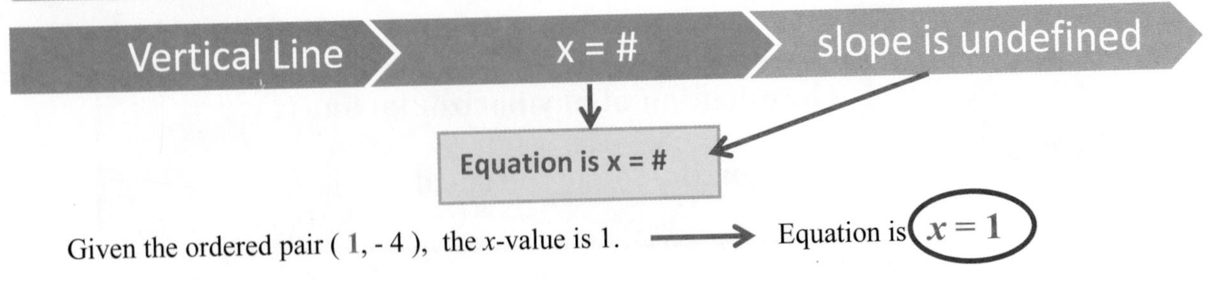

Given the ordered pair (1, - 4), the x-value is 1. ⟶ Equation is $x = 1$

YOUR TURN

Write the equation of each line described below.

7. vertical line through (- 3, 6).	8. line with slope 0 through (2, 5).
9. horizontal line through (- 4, - 6).	10. line with undefined slope through (8, 9).

Application: Writing the Equation of Lines

Problems occurring in many fields can be modeled by linear equations in two variables. The next examples will illustrate how we can use the point-slope form to solve problems.

EXAMPLE 11

The Whammo Company has learned that by pricing a newly released Frisbee at $6, sales will reach 2000 Frisbees per day. Raising the price to $8 will cause the sales to fall to 1500 Frisbees a day.

a. Assume that the relationship between sales price and the number of Frisbees sold is linear and write an equation describing this relationship. Use ordered pairs (sales price, number sold).

b. Predict the daily sales of Frisbees if the sale price is $7.50.

Courtesy of Shutterstock

First, understand the variables in words:

x: price of a Frisbee
y: # of Frisbees sold

An ordered pair in words would be ($Frisbee price, # of Frisbees sold)

a. We use the given information and write two ordered pairs. Our ordered pairs would be (6, 2000) and (8, 1500). To use the point-slope form to write the equation, we find the slope of the line containing these two points.

$$m = \frac{y_2 - y_1}{x_2 - x_1} = \frac{2000 - 1500}{6 - 8} = \frac{500}{-2} = -250$$

Next we use the slope and either of the points to write the equation in point-slope form. We will use (6, 2000) .

$$y - y_1 = m(x - x_1)$$

$$y - 2000 = -250\,(x - 6)$$ Substitute into formula.

$$y - 2000 = -250x + 1500$$ Use distributive property.

$$y = -250x + 3500$$ Add 2000 to both sides.

Equation of line in slope-intercept form is $y = -250x + 3500$

b. To predict the sales if the price is \$7.50, we find y when $x = 7.50$.

$$y = -250x + 3500$$

$$y = -250(7.50) + 3500$$

$$y = -1875 + 3500$$

$$y = 1625$$

If the sales price is \$7.50, sales will reach 1625 Frisbees per day.

EXAMPLE 12

In 2004, there were approximately 83,000 gas-electric hybrid vehicles sold in the United States. In 2007, there were approximately 353,000 such vehicles sold.

Assume the relationship between years after 2004 and the number of vehicles sold is linear .

a. Write an equation describing the relationship between time and the number of gas-electric hybrid vehicles sold. Use ordered pairs of the form (***years after 2004, number of vehicles sold***).

b. Use this equation to predict the number of electric-powered cars in use in 2008.

SOLUTION:

a. We use the given information and write two ordered pairs. Notice that the x-values have been defined as years after 2004. So, the **year 2004** would be $x = 0$. In 2007, 3 years have passed.

(*years after 2004, number of vehicles sold*) ⟶ (0, 83000) and (3, 353000)

To use the point-slope form to write the equation, we find the slope of the line containing these two points.

$$m = \frac{y_2 - y_1}{x_2 - x_1} = \frac{83000 - 353000}{0 - 3} = \frac{-270000}{-3} = 90{,}000$$

The slope tells us that the number of gas-electric hybrid cars increased by 90,000 each year.

Next we use the slope and either of the points to write the equation in point-slope form.
We will use (0, 83000) .

$$y - y_1 = m(x - x_1)$$

$$y - 83{,}000 = 90{,}000\,(x - 0) \qquad \text{Substitute into formula.}$$

$$y - 83{,}000 = 90{,}000\,x + 0 \qquad \text{Use distributive property.}$$

$$y = 90{,}000\,x + 83{,}000 \qquad \text{Add 83,000 to both sides.}$$

Equation of line in slope-intercept form is $y = 90{,}000x + 83{,}000$

b. Use this equation to predict the number of electric-powered cars in use in 2008.

Recall that the x-value is "years after 2004". So, the year 2008 would indicate that $x = 4$.

Find y (number of cars) when $x = 4$.

$$y = 90{,}000x + 83{,}000$$

$$y = 90{,}000(\,4\,) + 83{,}000$$

$$y = 360{,}000 + 83{,}000$$

$$y = 443{,}000$$

Courtesy of Shutterstock

There were 443,000 electric-powered cars in use in 2008.

YOUR TURN

11. In 1998 there were 1508 daily newspapers in the country. By 2005, there were only 1459 daily newspapers.

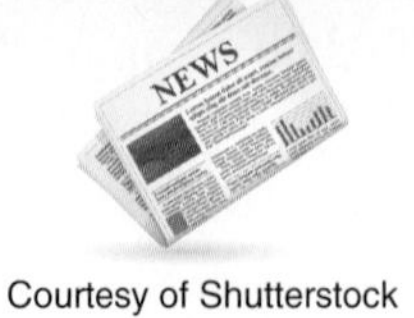

Courtesy of Shutterstock

a. Write two ordered pairs of the form (# years after 1998, # of daily newspaper).

b. The relationship between years after 1998 and numbers of daily newspapers is linear over this period. Use the ordered pairs from part (a) to write an equation for the line relating years after 1998 to numbers of daily newspapers.

c. Use the equation found in (b) to predict the number of daily newspapers in 2003.

12. A rock is dropped from the top of a 500-ft cliff. After 1 second, the rock is traveling 46 feet per second. After 4 seconds, the rock is traveling 142 feet per second.

a. Write two ordered pairs of the form (# of seconds, rate of descent).

b. Use the ordered pairs from part (a) to write an equation that relates time (x) to the rate of descent (y) of the rock.

c. Find the rate of descent 5 seconds after it was dropped.

5.07 Equations of Lines 5.07

Forms of Linear Equations		
Equation	**Description**	**Example**
x = constant	**Vertical Line** Slope is undefined	$x = 3$
y = constant	**Horizontal Line** Slope is 0	$y = 2$
$y = mx + b$	**Slope-Intercept Form** Slope is m *y-intercept* is (0 , b)	$y = 2x - 7$
$Ax + By = C$	**Standard Form**	$3x - 2y = 12$
$m = \dfrac{y_2 - y_1}{x_2 - x_1}$	**Slope Formula** Lines passes through (x_1, y_1) and (x_2, y_2)	Given (4, 8) & (6, 9), $m = \dfrac{9-8}{6-4} = \dfrac{1}{2}$
$y - y_1 = m(x - x_1)$	**Point-Slope Formula** Slope is m Lines passes through (x_1, y_1)	$y - 3 = 4(x-2)$

Parallel and Perpendicular Lines	
Parallel lines have the same slope. The product of the slope of perpendicular lines is -1. Perpendicular lines have slopes that are opposite reciprocals of each other.	Given a slope $= \dfrac{3}{4}$ line parallel has slope = $\dfrac{3}{4}$ line perpendicular has slope = $-\dfrac{4}{3}$

5.07 Equations of Lines

Identifying the Slope and y-intercept of a line

A very important form of a linear equation is the **slope-intercept form, $y = mx + b$.** We have already looked at what the coefficient of the x represents. The letter, m, is used to denote the slope of a line. Now, we will look at what the **b** represents.

If a linear equation is solved for y, the coefficient of x is the line's slope and the point (0, b) is the y-intercept of the line.

$$y = mx + b$$ **SLOPE INTERCEPT FORM**

slope (m)

(0, b) **y-intercept** (b)

This y-intercept is the point where a line crosses the y axis.

EXAMPLE 13

For each equation, identify the slope and y-intercept.

a. $y = 3x + 5$	b. $y = \frac{1}{3}x - 7$	c. $y = -2x$

Each equation is written in slope-intercept form, $y = mx + b$. The slope is the coefficient of x, and the y-intercept is determined by the constant term, b. The y-intercept is the ordered pair (0, b).

a. $y = 3x + 5$	b. $y = \frac{1}{3}x - 7$	c. $y = -2x$
$y = 3x + 5$	$y = \frac{1}{3}x - 7$	$y = -2x + 0$
The slope is 3.	The slope is $\frac{1}{3}$.	The slope is -2 .
The y-intercept is (0, 5).	The y-intercept is (0, -7)	The y-intercept is (0, 0)

EXAMPLE 14

Given the line $3x + 5y = -10$.

a. Write the slope-intercept form of the line.

b. Identify the slope and y-intercept.

a. Write the equation in slope-intercept form, $y = mx + b$, by **solving for y.**

$$3x + 5y = -10$$

$$5y = -3x - 10$$ Move x-term to opposite side by changing its sign.

$$\frac{5y}{5} = \frac{-3x}{5} - \frac{10}{5}$$ Divide by coefficient of y

$$y = -\frac{3}{5}x - 2$$ Slope-intercept form

b. The slope of the line is $-\frac{3}{5}$ and the y-intercept is (0, - 2).

Graphing Lines Using the Slope and the y-intercept

Slope-intercept is a useful tool to graph a line. The y-intercept is a known point on the line. The slope indicates the direction of the line and can be used to find a second point.

EXAMPLE 15

Graph the line $y = \frac{2}{5}x + 1$ by using the slope and y-intercept.

The slope of the line is $\frac{2}{5}$; the y-intercept is (0 , 1).

We begin by plotting the y-intercept (0,1).

The slope $m = \frac{2}{5}$ **can be written as**

$$m = \frac{2}{5} = \frac{\text{the change in } y \text{ is } 2}{\text{the change in } x \text{ is } 5}$$

To find a second point on the line, start at the y-intercept and move **up 2** units and to the **right 5** units.

Draw the line through the two points.

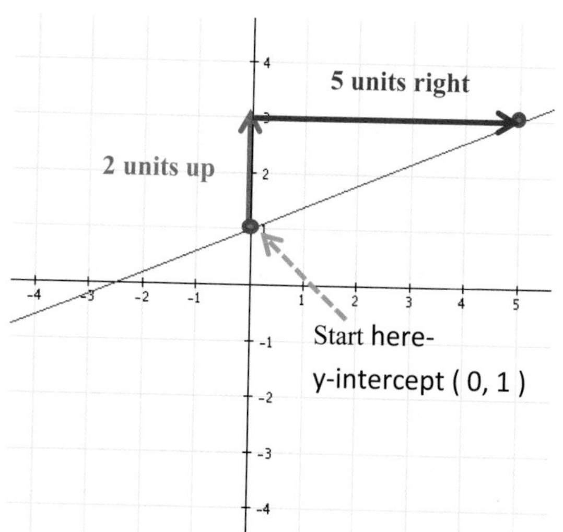

5.07 Equations of Lines

EXAMPLE 16

Graph the line $y = -\frac{5}{3}x - 2$ by using the slope and *y*-intercept.

Slope is $-\frac{5}{3}$; *y*-intercept is (0 , -2) .

We begin by plotting the y-intercept (0,-2) .

The slope $-\frac{5}{3}$ **can be written as**

$$m = \frac{-5}{3} = \frac{\text{the change in } y \text{ is -5}}{\text{the change in } x \text{ is 3}}$$

To find a second point on the line, start at the y-intercept and move **down 5** units and to the **right 3** units.

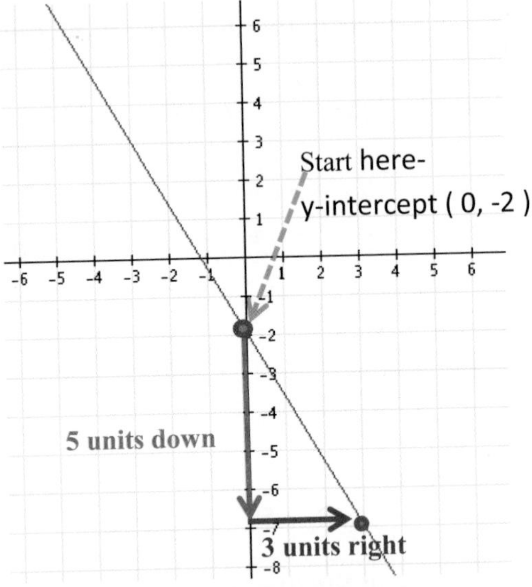

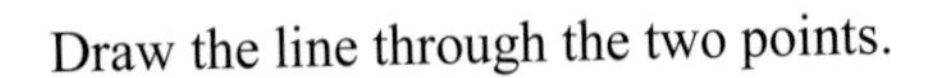

Draw the line through the two points.

If the slope is negative, place the negative sign in the numerator OR denominator.

$$m = -\frac{5}{3} = \frac{-5}{3} = \frac{5}{-3}$$

Similarly, the slope above can be written as

$$m = \frac{5}{-3} = \frac{\text{the change in } y \text{ is 5}}{\text{the change in } x \text{ is -3}}$$

To find a second point on the line, start at the *y*-intercept and move **up 5** units and to the **left 3** units.

Draw the line through the two points.

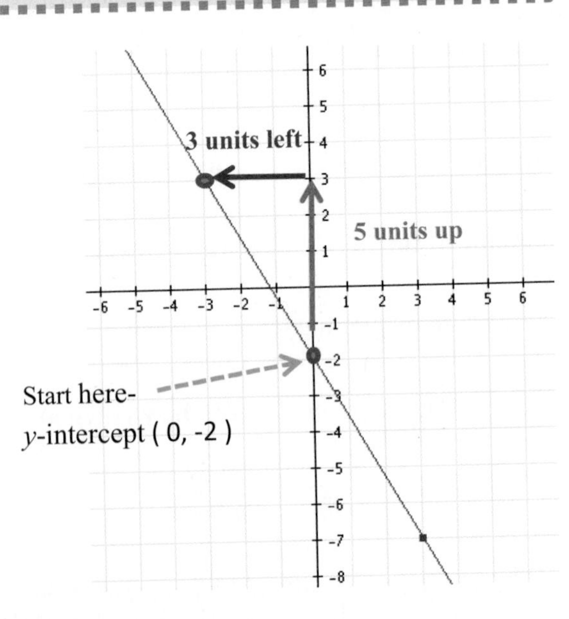

Notice the two graphs above are the same line regardless of whether the negative in the slope was in the numerator OR the denominator.

EXAMPLE 17

Graph the line $y = 4x$ using the slope and the y-intercept.

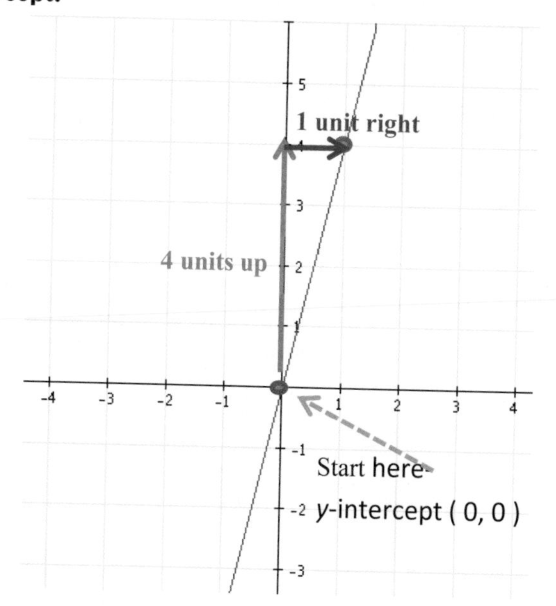

The line can be written as $y = 4x + 0$.

Therefore, we can plot the **y-intercept at (0, 0)** .
The slope $\boldsymbol{m = 4}$ can be written as

$$m = 4 = \frac{4}{1} = \frac{\text{the change in } y \text{ is } 4}{\text{the change in } x \text{ is } 1}$$

To find a second point on the line, start at the y-intercept and move **up 4** units and to the **right 1** unit.

Draw the line through the two points.

Counting Slope

$m = \dfrac{positive}{positve} = \dfrac{up}{right}$ $\quad m = \dfrac{positive}{negative} = \dfrac{up}{left}$ $\quad m = \dfrac{negative}{positve} = \dfrac{down}{right}$ $\quad m = \dfrac{negative}{negative} = \dfrac{down}{left}$

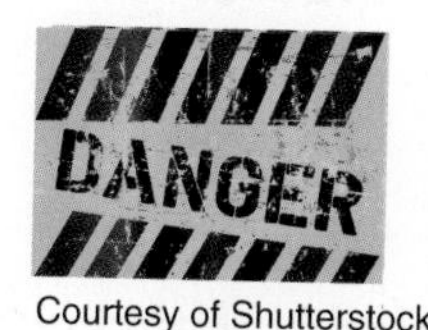

Courtesy of Shutterstock

WATCH OUT FOR THESE COMMON MISTAKES

Courtesy of Shutterstock

To graph using the slope and *y*-intercept, your equation will always have to be solved for *y*.

The *y*-intercept is found on the *y*-axis. So, you will always begin counting slope from a point on the *y*-axis. You should NEVER begin on the *x*-axis.

If the slope is negative, one movement will be positive and the other movement will be negative. To count a negative slope move down/right OR up/left. You CANNOT move down/left. This is two negative moves which is the same as a positive slope!

5.07 Equations of Lines

YOUR TURN

Graph using the slope and the *y*-intercept.

13. $y = 3x + 1$

Slope is ______ ; *y*-intercept is (0 ,   )

Plot *y*-intercept : __________

$m = \frac{\quad}{\quad} = \frac{\text{the change in } y \text{ is}}{\text{the change in } x \text{ is}}$

To find a second point, start at the *y*-intercept and move ____________ units and to the __________ units. Draw line through two points.

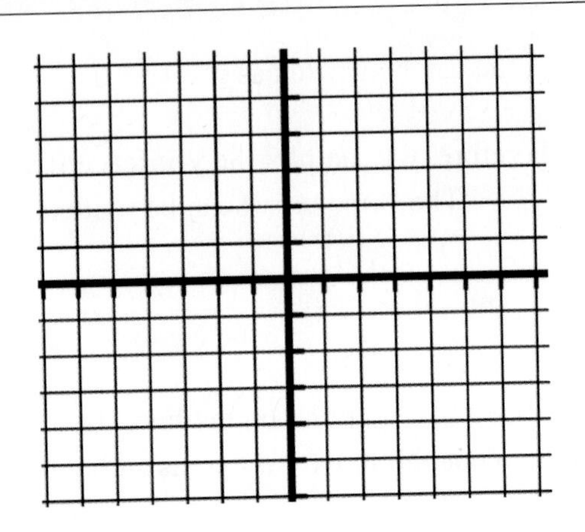

14. $y = -2x$

Slope is ______ ; *y*-intercept is (0 ,   )

Plot *y*-intercept : __________

$m = \frac{\quad}{\quad} = \frac{\text{the change in } y \text{ is}}{\text{the change in } x \text{ is}}$

To find a second point, start at the *y*-intercept and move ____________ units and to the __________ units. Draw line through two points.

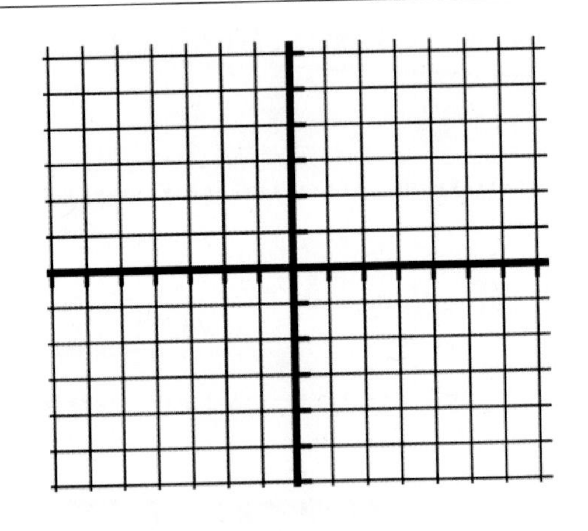

15. $y = -\frac{2}{3}x - 3$

Slope is ______ ; *y*-intercept is (0 ,   )

Plot *y*-intercept : __________

$m = \frac{\quad}{\quad} = \frac{\text{the change in } y \text{ is}}{\text{the change in } x \text{ is}}$

To find a second point, start at the *y*-intercept and move ____________ units and to the __________ units. Draw line through two points.

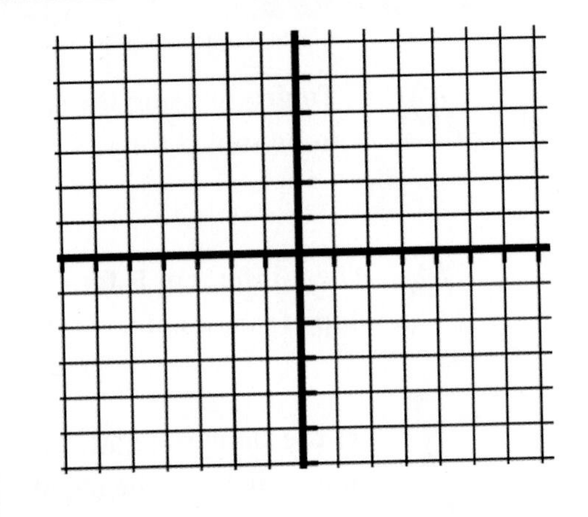

Matching an Equation to its Graph.

To match an equation to its graph, we will need to look at two things. Of course, as you probably expected, these two things are the slope and y-intercept. We will check to see if the y-intercept has been plotted correctly and then we will check the slope of the line.

EXAMPLE 18 Match the following equation to their graphs.

1. $y=-\frac{1}{2}x-1$

slope $=-\frac{1}{2}$; y-intercept $(0,-1)$

First, we will check for the y-intercept. Identify any graphs that **cross the y-axis at (0, -1) .**

Graph d is the only graph that has y-intercept (0, -1).

No need to look any further.

Graph "d" is our answer.

2. $y=-2x+1$

slope $=-2$; y-intercept $(0,1)$

First, we will check for the y-intercept. Identify any graphs that **cross the y-axis at (0, 1) .**

Graph a and Graph c have y-intercept (0, 1).

Next, we will need to check the slope to decide which graph is the correct answer.

Which graph has a slope of -2. From the y-intercept the next point should be down 2 units and right 1 unit.

Graph "a" is our answer.

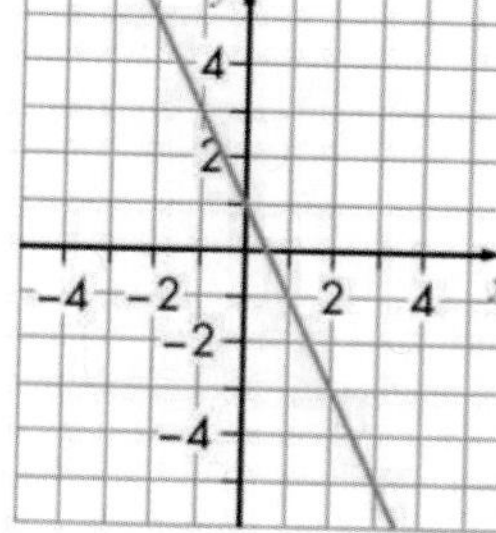

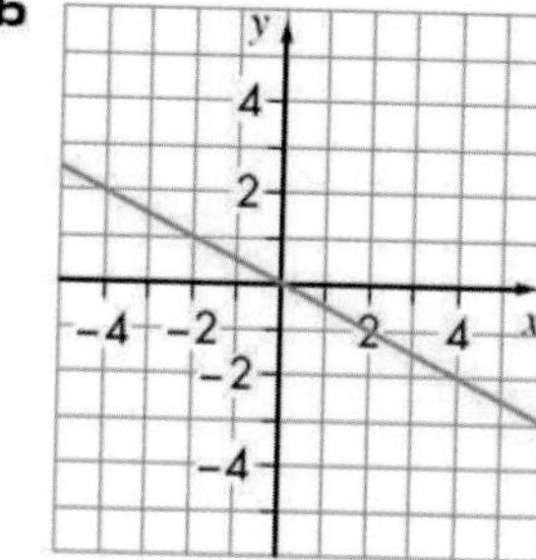

c

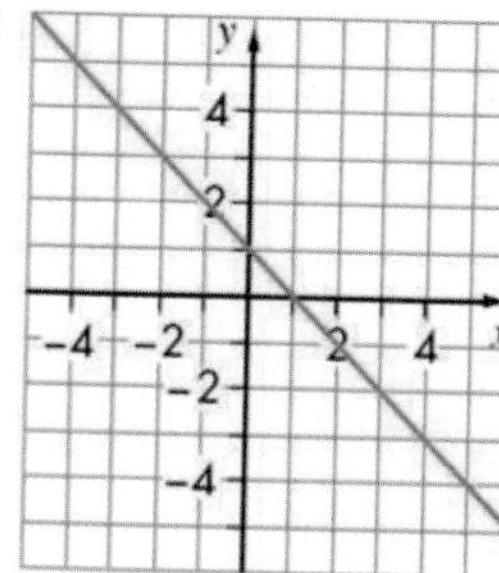

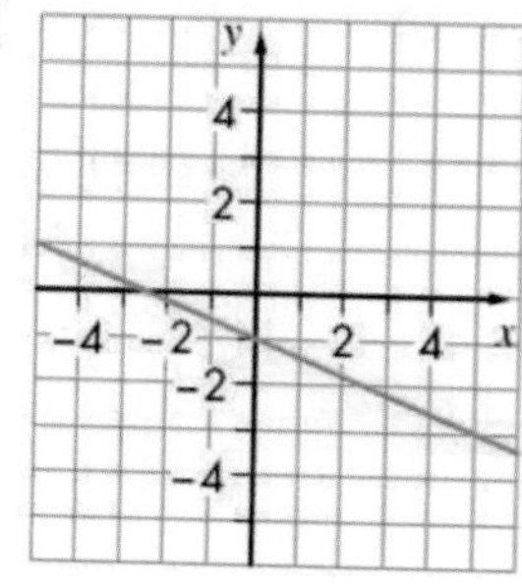

e

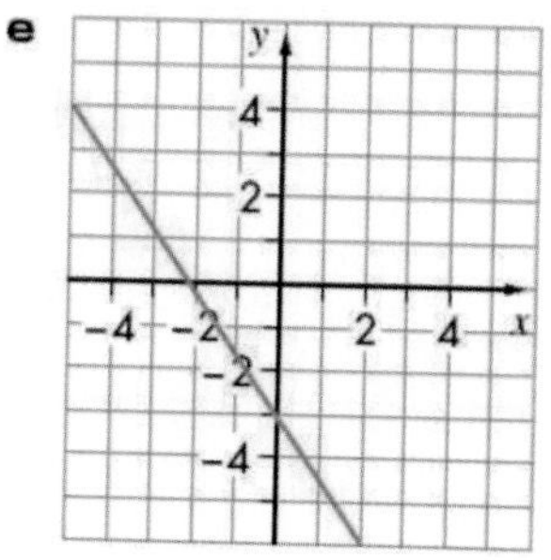

5.07 Equations of Lines

YOUR TURN

Match the following equations to their graphs.

16. $y = -x + 3$

slope = _____ ; *y*-intercept (0,_____)

Graph ______ is the answer.

17. $y = x + 3$

slope = _____ ; *y*-intercept (0,_____)

Graph ______ is the answer.

18. $y = -\frac{2}{3}x + 1$

slope = _____ ; *y*-intercept (0,_____)

Graph ______ is the answer.

19. $y = -\frac{3}{2}x + 1$

slope = _____ ; *y*-intercept (0,_____)

Graph ______ is the answer.

A.

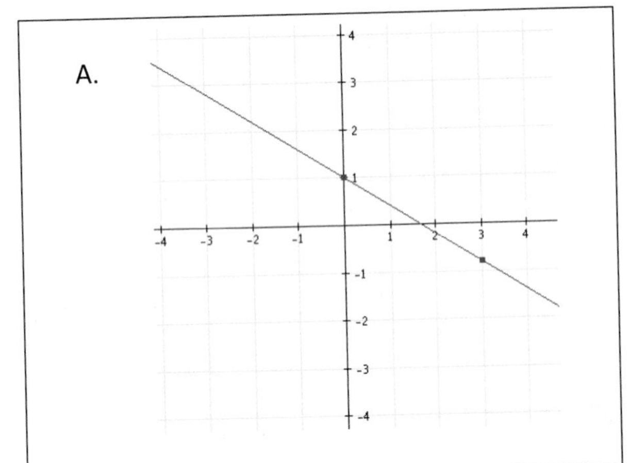

B.

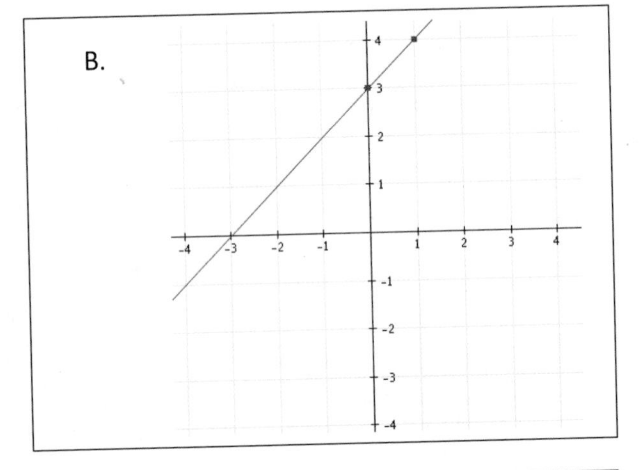

C.

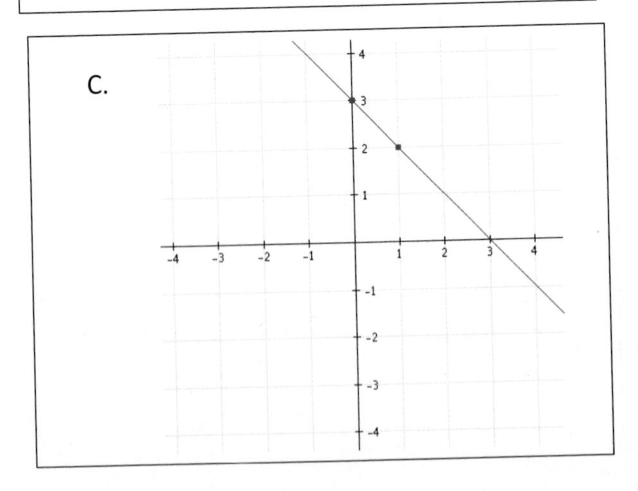

D.

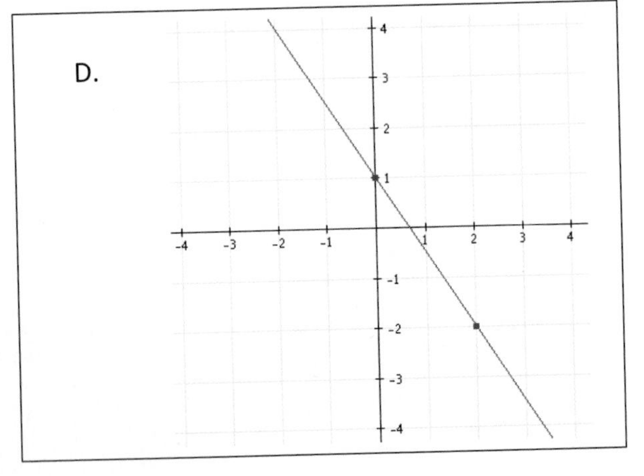

5.07 Equations of Lines

YOUR TURN **Answers to Section 5.07.**

1. $y = -8x - 13$
2. $3x - 2y = 27$
3. $y = 8x + 11$
4. $y = \frac{3}{2}x$
5. $x - 2y = -7$
6. $x + 11y = -28$
7. $x = -3$
8. $y = 5$
9. $y = -6$
10. $x = 8$
11. *a.* $(0,1508), (7,1459)$ *b.* $y = -7x + 1508$ *c.* 1473 newspapers
12. *a.* $(1,46), (4,142)$ *b.* $y = = 32x + 14$ *c.* 174 *feet*

13.

14.

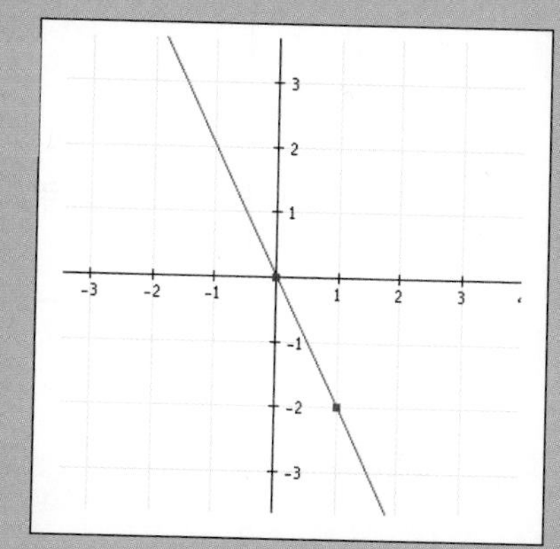

15.

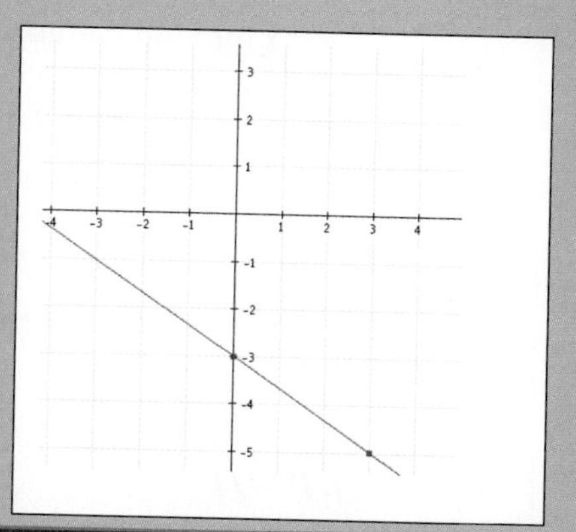

16. Graph C
17. Graph B
18. Graph A
19. Graph D

Complete MyMathLab Section 5.07 Homework in your Homework notebook.

5.09 Functions

Courtesy of Shutterstock

Relating real life situations with mathematical sets of data and equations is a useful tool in today's world. These places include construction and architecture, business management, medical science, engineering and many other areas. It is important to be able to relate how inputting information can result in expected outcomes or output values. The way we discuss and relate starting data **(input values)** with ending data **(output values)** have special names.

FUNCTION NOTATION, $f(x)$

This notation is used when a function is defined by an equation. Rather than using the variable y, use the notation $\boldsymbol{f(x)}$ (read "f of x").

For example, $f(x) = 2x - 4$ is the same as the equation $y = 2x - 4$.

Note: the $f(x)$ notation replaces the variable y. Ordered pairs representing points on the line are written $(x, f(x))$.

For $y = 2x + 3$, the **inputs** are values of x substituted into the equation. The **outputs** are the resulting values of y.

Previous Notation using y	Function Notation
$y = 2x + 3$	$f(x) = 2x + 3$
Find y when $x = 4$. $y = 2x + 3$ $y = 2(4) + 3$ $y = 8 + 3$ $y = 11$	*Find $f(4)$ This means to plug 4 in for x.* $f(4) = 2(4) + 3$ $f(4) = 8 + 3$ $f(4) = 11$ **So when $x = 4$, the function value is 11. (4, 11)**

Finding Function Values

EXAMPLE 1

Given the function $f(x) = 2x - 4$, find $f(-1)$.

This asks the question, *"What is the function value, (the output y value), when x = -1 ? "*

Given $f(x) = 2x - 4$

$f(-1) = 2(-1) - 4$ To find $f(-1)$ means to plug the number in the parentheses, the -1, into the right side of the equation for x and simplify the right side.

$f(-1) = -2 - 4$

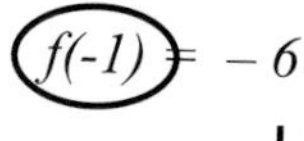

$f(-1) = -6$ Hence, $f(-1)$ is the output value -6.

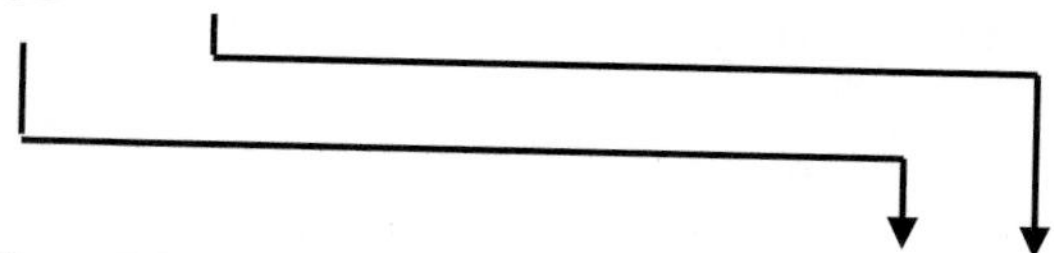

This also means that the ordered pair $(-1, -6)$ is a point on the graph of the function.

Courtesy of Shutterstock

The notation $f(x)$ does not mean "f times x" and should not be read that way.

Functions can be defined using any letter in addition to f. For example, $C(x)$, $h(x)$, $Q(x)$.

EXAMPLE 2

Given the function $h(x) = x^2 - 3x + 5$, find the following.

a. $h(0)$

$h(0) = (0)^2 - 3(0) + 5$

$h(0) = 0^2 - 0 + 5$

$\boldsymbol{h(0) = 5}$

b. $h(-2)$

$h(-2) = (-2)^2 - 3(-2) + 5$

$h(-2) = 4 + 6 + 5$

$\boldsymbol{h(-2) = 15}$

c. $h(3)$

$h(3) = (3)^2 - 3(3) + 5$

$h(3) = 9 - 9 + 5$

$\boldsymbol{h(3) = 5}$

5.09 Functions

YOUR TURN

1. **Given the function,** $g(x) = \frac{1}{3}x + 2$**, find the following.**

   a. $g(0)$

   b. $g(-3)$

2. **Given the function,** $C(x) = 7x^2 - 3$ **, find the following.**

   a. $C(0)$

   b. $C(-2)$

Reading Function Values from a Graph

Remember that functions are graphs of ordered pairs where each x- value is paired with one y- value. Look at the following graphs and name the function values, $f(x)$.
(In other words, state the y value for the given x value using function notation.)

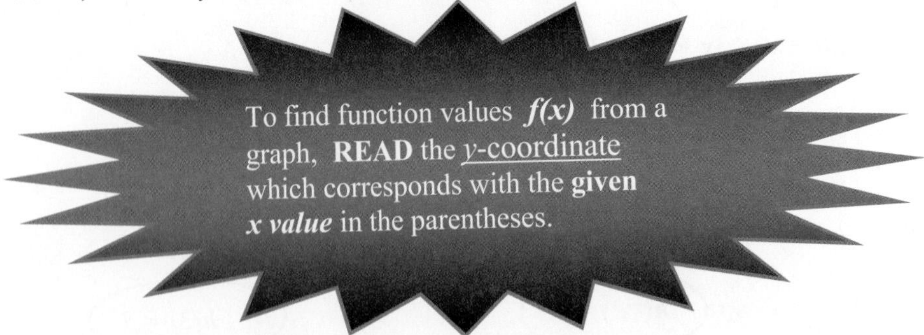

EXAMPLE 3

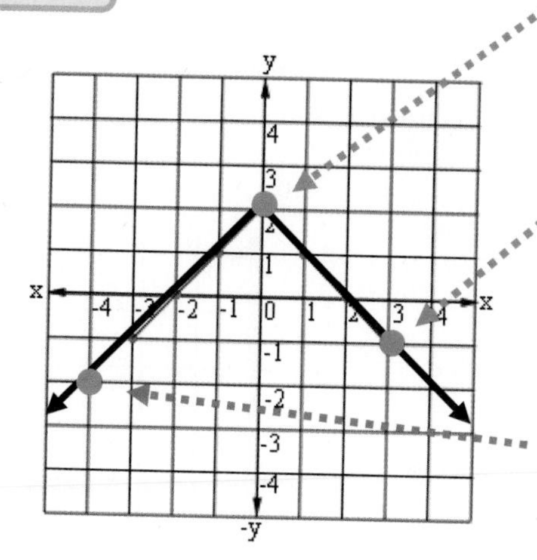

a. Find *f(0)*.
Find x = 0 on the graph, then look up or down.
Read the y value on the graph when $x = 0$. The y coordinate is up 2. Hence, *f(0)= 2*.

b. Find *f(3)*.
Find x = 3 on the graph, then look up or down.
Read the y value on the graph when $x = 3$.The y coordinate is down - 1. Hence, *f(3)= -1*.

c. Find *f(- 4)*.
Find x = - 4 on the graph, then look up or down.
Read the y value when $x = - 4$.
The y coordinate is down - 2. Hence, *f(- 4)= - 2*.

Keep in mind that these values can also be written as ordered pairs.

f (0)= 2	*f (3)*= -1	*f (-4)*= 2
(0 , 2)	(3 , -1)	(-4 , 2)

YOUR TURN

3. Use the graph to find the following values.

   a. $f(2) =$

   (Hint: Look on the graph when $x = 2$ and find the y-value .)

   b. $f(-2) =$

   (Hint: Look on the graph when $x = -2$ and find the y-value)

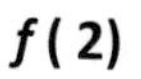

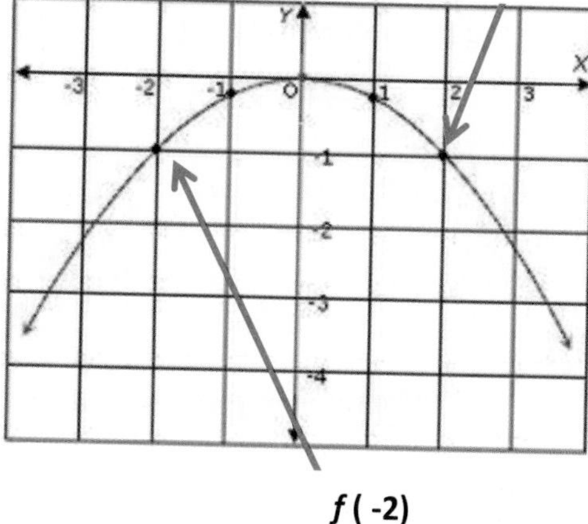

f (-2)

5.09 Functions

EXAMPLE 4

Let C be the function whose graph is given to the right. This graph represents the cost C of using *m* anytime cell phone minutes in a month for a five-person family plan.

a. Determine C(0). Interpret this value.

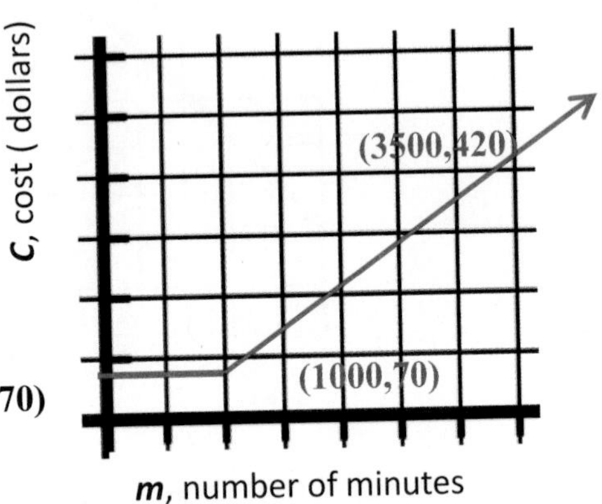

Understand the variables in words:
x-value *m*: # of anytime minutes.
y- value or *C(x)*: the cost in $.
So the ordered pair would match up to (**# minutes, cost $).**

Looking at the graph, we see the ordered pair (0,70). Recall the *x*-value is the input, and the *y*-value is the same as the function value.

C(0) = 70 **(0, 70)**

The ordered pair (0 , 70) would tell us that using 0 minutes costs $70.

b. Determine C(1000) . Interpret.

Looking at the graph, we see the ordered pair (1000, 70). Recall the *x*-value is the input, and the *y*-value is the same as the function value.

C(1000) = 70

The *x*-value corresponds to the number of minutes. The *y*- value or *f(x)* is the cost.

So the ordered pair would match up to **(number of minutes, cost $).** The ordered pair (1000, 70) would tell us that we could use 1000 minutes for a charge of $70.

c. Determine C(3500) . Interpret.

Looking at the graph, we see the ordered pair (3500, 420). Recall the *x*-value is the input, and the *y*-value is the same as the function value.

C(3500) = 420

The *x*-value corresponds to the number of minutes. The *y*- value or *f(x)* is the cost.

So the ordered pair would match up to **(number of minutes, cost).** The ordered pair (3500, 420) would tell us that we could use 3500 minutes for a charge of $420.

5.09 Functions 5.09

Writing Equations of lines Using Function Notation.

EXAMPLE 5

Find the equation of a line through (3, 0) and (– 1, – 2). Write in function notation.

Part 1: Find the slope first → $m=\dfrac{y_2-y_1}{x_2-x_1}$

(x_1, y_1) (x_2, y_2)
$(3, 0)$ $(-1, -2)$

$$m=\frac{-2-0}{-1-3}=\frac{-2}{-4}=\frac{1}{2}$$

$$m=\frac{1}{2}$$

METHOD: Always BEGIN to write an equation using Point–Slope Form

$$y-y_1=m(x-x_1)$$

1st: Find the slope m.
2nd: Use it with one of the points in this formula.
3rd: Change this equation to the form in the question.

Part 2: You can use either ordered pair above, and use the slope (m) from Part 1.

$y-y_1=m(x-x_1)$

$y-0=\frac{1}{2}(x-3)$ Plug in the slope $m=\frac{1}{2}$ and point (3, 0). Then solve for $y=mx+b$.

$y=\frac{1}{2}x-\frac{3}{2}$ Distribute to find Slope–Intercept form.

$f(x)=\frac{1}{2}x-\frac{3}{2}$ Remember to replace the y with $f(x)$ for function notation.

EXAMPLE 6

Write the equation of a line in function notation given the point (6, -8) and $m=-\frac{2}{3}$.

$y-y_1=m(x-x_1)$ Remember, use Point Slope form to write equations.

$y+8=-\frac{2}{3}(x-6)$ Distribute $-\frac{2}{3}$.

$y+8=-\frac{2}{3}x+4$ Subtract 8 from both sides.

$y=-\frac{2}{3}x-4$

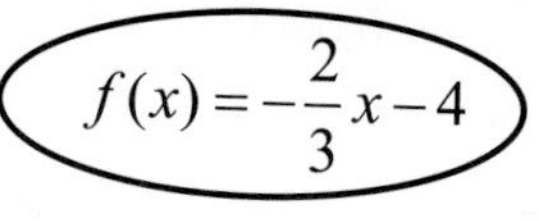

$f(x)=-\frac{2}{3}x-4$ Don't forget to write in function notation by replacing y with $f(x)$.

5.09 Functions

YOUR TURN

4. Find an equation of a line through $(-2, 2)$ and $(1, 3)$. Write the equation using function notation.

5. Write the equation of a line in function notation given point $(0, -1)$ and $m = -2$.

Application of Functions

EXAMPLE 7

The polynomial function $P(x) = 40x - 97{,}000$ models the relationship between the number of briefcases x that a company sells and the profit the company makes, P(x). Find P(4000) , the profit from selling 4000 computer briefcases.

Understand the variables in words:
x-value: # of briefcases.
y- value or P(x): the profit in \$.
So the ordered pair would match up to (**# briefcases, profit \$**).

Courtesy of Shutterstock

To find P(4000), we will substitute 4000 for the x-value in the function.

$$P(x) = 40x - 97{,}000$$

$$= 40(4000) - 97{,}000$$

$$= 160{,}000 - 97{,}000$$

$$= 63{,}000$$

P(4000) = 63,000

The company will make a profit of \$63,000 by selling 4000 briefcases.

5.09 Functions

EXAMPLE 8

Courtesy of Shutterstock

The per capita consumption (in pounds) of all beef in a country is given by the function **C(*x*) = - 0.41*x* + 62.9**, where x is the number of years since 2000.

Understand the variables in words:
x-value: # of years since 2000.

y- value or $C(x)$: the per capita consumption of beef in lbs.
So the ordered pair would match up to **(# years, amount of consumption lbs.).**

a. Find and interpret C(3).

$$\begin{aligned} C(3) &= -0.41(3) + 62.9 \\ &= -1.23 + 62.9 \\ &= 61.67 \end{aligned}$$

The per capita consumption of beef was 61.69 pounds in the year 2003.

b. Estimate the per capita consumption of beef in the country in 2010.

x is the number of years after 2000. So, the year 2010 would indicate $x = 10$.

$$\begin{aligned} C(10) &= -0.41(10) + 62.9 \\ &= -4.1 + 62.9 \\ &= 58.8 \end{aligned}$$

The per capita consumption of beef was 58.8 pounds in the year 2010.

YOUR TURN

6. The total revenue (in dollars) for a company to sell x blank audiocassette tapes per week is given by the polynomial function $P(x) = 8x$. Find the total revenue from selling 20,000 tapes per week.

Understand the variables in words:

x-value: ______________________

$P(x)$: ______________________

The total revenue is $______________/week.

7. Frank plans to take a vacation in the San Diego/Los Angeles area. He determines the cost, C, of the vacation can be estimate using the function $C(n) = 350n + 400$, where n is the number of days spent in the area. Estimate the cost of a 5-day vacation in the area.

Understand the variables in words:

n-value: ______________________

$C(n)$: ______________________

The total cost of a 5-day vacation is $________.

5.09 Functions

YOUR TURN Answers to Section 5.09.

1. a. $g(0) = 2$ b. $g(-3) = 1$

2. a. $c(0) = -3$ b. $c(-2) = 25$

3. a. $f(2) = -1$ b. $f(-2) = -1$

4. $f(x) = \frac{1}{3}x + \frac{8}{3}$

5. $f(x) = -2x - 1$

6. x-value: # audiocassettes/week, $P(x)$: Revenue = $160,000

7. n-value: # days, $C(n)$: Cost of a 5-day vacation = $2,150

Complete MyMathLab Section 5.09 Homework in your Homework notebook.

DMA 050 TEST: Review your notes, key concepts, formulas, and MML work. ☺

Student Name: ______________________________ **Date:** ______________

Instructor Signature: ______________________________

DMA 060

DMA 060

POLYNOMIALS AND QUADRATIC APPLICATIONS

This course provides a conceptual study of problems involving graphic and algebraic representations of quadratics. Topics include basic polynomial operations, factoring polynomials, and solving polynomial equations by means of factoring. Upon completion, students should be able to find algebraic solutions to contextual problems with quadratic applications.

You will learn how to:

- Represent real-world applications as quadratic equations in tabular, graphic, and algebraic forms.
- Apply exponent rules.
- Apply the principles of factoring when solving problems.
- Solve application problems involving polynomial operations and using the zero product property and critique the reasonableness of solutions found.
- Represent contextual applications using function notation.
- Analyze graphs of quadratic functions to solve problems using the graphing calculator to identify and interpret the maximum, minimum, and y-intercept values and the domain and range in terms of the problem.

Quadratic Equation $y = ax^2 + bx + c$

Exponent Rules

$a^m \cdot a^n = a^{m+n}$	$\dfrac{a^m}{a^n} = a^{m-n}$
$(a^m)^n = a^{m \cdot n}$	$a^0 = 1$
$(ab)^m = a^m b^m$	$\left(\dfrac{a}{b}\right)^m = \dfrac{a^m}{b^m}$
$x^{-m} = \dfrac{1}{x^m}$	$\dfrac{1}{x^{-m}} = x^m$

Parabola

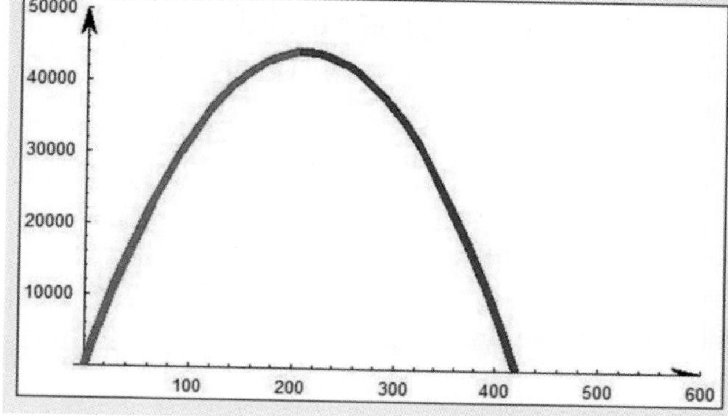

Perfect Square Trinomial

$a^2 \pm 2ab + b^2 = (a \pm b)^2$

Difference of Squares

$a^2 - b^2 = (a + b)(a - b)$

INTRODUCTORY QUADRATIC PROBLEM

We now begin our study of quadratic expressions and functions. Quadratic functions are useful in many professions.

By the end of this study you will have a broad understanding of these quadratic equations. Read through this problem. By the end of this Module, you will be able to analyze this application problem and provide solutions. Give it a try now. Do you remember how to think about this problem?

Remember to come back and answer at the end of your work in this Module.

A baseball player hits a foul ball into the air from a height of 5 feet off the ground. The initial velocity as the ball comes off the bat is 128 feet per second. The motion of the ball can be modeled by the polynomial equation $f(t) = -16t^2 + 128t + 5$, where t is the time in seconds.

1. Write in words what t and $f(t)$ are in this problem.

   t: ____________________

   $f(t)$: ____________________

2. What is the initial height of the ball?

   ____

   In actual life, where would this height be on the ball player? ____________________

   Where is this number in the formula?

   ____

3. How hard is the ball hit? In other words, what is its velocity? Where is this number in the formula?

   ball's velocity: ____________

   where in formula: ______________

   ____

4. What does the term $-16t^2$ reflect in the formula?

   $-16t^2$: ____________________

5. Complete the table of values that shows the height of the ball in terms of time.

Time, *t*	**Height, *h(t)***
seconds	*feet*
0	
1	
2	
4	
6	
8	
10	

6. What is the height of the ball after

   1 second?______ 2 seconds? ______

7. Is the ball rising or falling at 6 seconds?

   ____

8. Do all of the answers in questions 5 - 7 make sense? Use complete sentences to explain your answer.

(Continued on next page.)

INTRODUCTORY QUADRATIC PROBLEM

9. Create a graph that models the path of the ball on the grid below. Carefully label all parts of the graph in words: the horizontal axis, vertical axis, label the units, graphs and label the ordered pairs.

   a. What does the y-intercept represent in words?

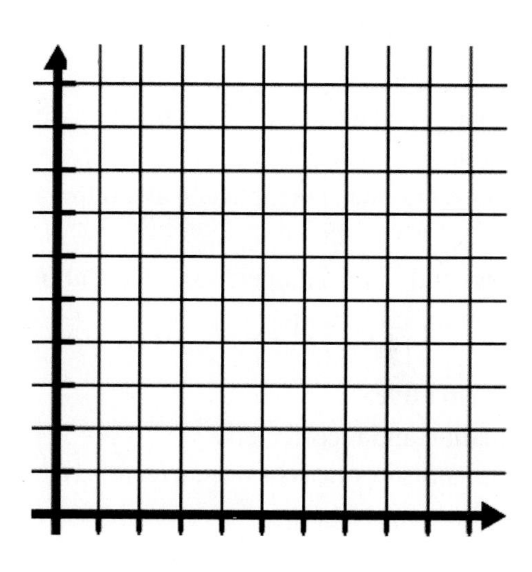

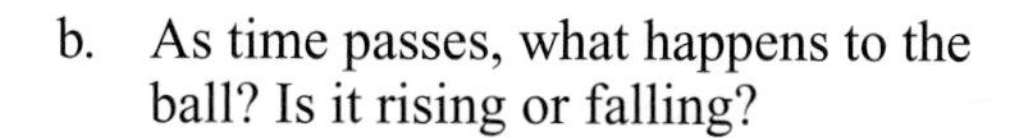

   b. As time passes, what happens to the ball? Is it rising or falling?

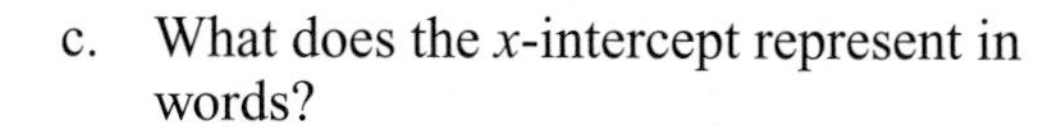

   c. What does the x-intercept represent in words?

10. What is the greatest height of the ball? How long does it take to get to this height? Answer in complete sentences. XMIN = -2 XMAX = 10 YMIN = -50 YMAX = 300 Round to the nearest hundredth.

    greatest height:________________

    how long to reach height:_________

11. What is the height of the ball after 10 seconds? _________

12. How long it will take for the ball to reach 200 feet above the ground?

    ____

13. When or how long will it take the ball to reach 200 ft again before hitting the ground? __________________

14. How long before the ball hits the ground?

    ____

Professions that Use Quadratic Expressions

Management

Management occupations

- ▶ Computer and information systems managers
- ▶ Engineering and natural sciences managers
- ▶ Farmers, ranchers, and agricultural managers
- ▶ Funeral directors
- ▶ Industrial production managers
- ▶ Medical and health services managers
- ▶ Property, real estate, and community association managers
- ▶ Purchasing managers, buyers, and purchasing agents

Business and financial operations occupations

- ▶ Insurance underwriters

Administrative

Information and record clerks

- ▶ Human resources assistants, except payroll and timekeeping

Professional

Computer and mathematical occupations

- ▶ Actuaries
- ▶ Computer software engineers
- ▶ Mathematicians
- ▶ Statisticians

Engineers

- ▶ Aerospace engineers
- ▶ Chemical engineers
- ▶ Civil engineers
- ▶ Electrical engineers
- ▶ Environmental engineers
- ▶ Industrial engineers
- ▶ Nuclear engineers
- ▶ Petroleum engineers

Drafters and engineering technicians

- ▶ Engineering technicians

Life scientists

- ▶ Medical scientists

Physical scientists

- ▶ Chemists and materials scientists
- ▶ Environmental scientists and hydrologists
- ▶ Physicists and astronomers

Social scientists and related occupations

- ▶ Economists
- ▶ Social scientists, other

Education, training, library, and museum occupations

- ▶ Teachers—adult literacy and remedial and self-enrichment education
- ▶ Teachers—postsecondary
- ▶ Teachers—preschool, kindergarten, elementary, middle, and secondary
- ▶ Teachers—special education

Health diagnosing and treating occupations

- ▶ Registered nurses

Health technologists and technicians

- ▶ Medical records and health information technicians

Farming

- ▶ Forest, conservation, and logging workers

Production

Metal workers and plastic workers

- ▶ Computer control programmers and operators

http://www.xpmath.com/careers/topicsresult.php?subjectID=2&topicID=10

GRAPHING CALCULATOR

Polynomial Functions are used in real life in every part of the world. Companies use calculators to make decisions for these mathematical models in this technological age. In this DMA, we will analyze quadratic functions using the some of the graphical functions of the TI-84 calculator.

Use the picture to begin to familiarize yourself with the keys.

THE TI-84 CALCULATOR FEATURES

Y=: Enter equations here. (PLOTS are off, not shaded.)

WINDOW: Set view of max and min values. Then press GRAPH or ZOOM command.

QUIT: (2nd) (MODE)
Returns you to the Home screen.

MODE: Set to (Func) for graphing functions.

X,τ,θ,η : Enters the variable x in an equation in the (Y=) function mode.

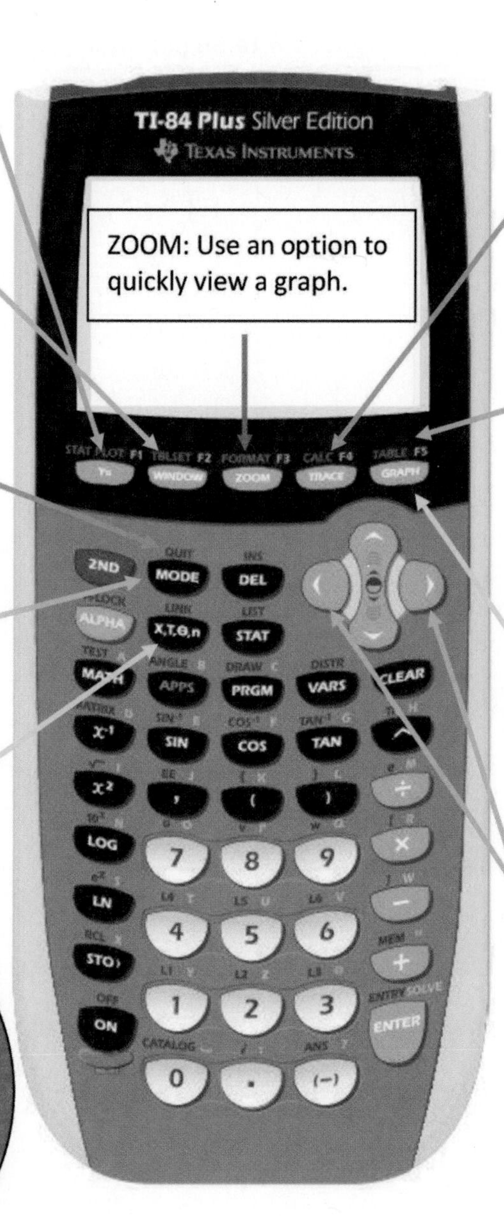

CALC: (2nd) (TRACE)
Allows you to calculate using the functions:
1: value
2: zeros
3: minimum
4: maximum
5: intercept

TABLE: (2nd) (GRAPH)
Shows an x-y chart of ordered pairs for the graph in y=.

GRAPH: Shows a Picture of the graph. You may have to change your WINDOW to get a good view.

SCROLL Keys:
Hint, use <u>ONLY</u> the < left and > right arrows with CALC functions for graphs.

If you get stuck, try (2nd)(QUIT) to get you out of the graphing screen. If this doesn't work, try (2nd)(OFF).

FIRST WE WILL LEARN HOW TO SIMPLIFY POLYNOMIAL EXPRESSIONS.

An exponent is shorthand notation for repeated factors. For example, $3 \cdot 3 \cdot 3 \cdot 3 \cdot 3$ can be written as 3^5 where 3 is called the base and 5 is called the exponent. In an exponential expression, the base is the repeated factor and the exponent is the number of times that the base is used as a factor.

EXAMPLE 1

Evaluate each expression. Watch what happens with no parentheses.

2^4 $2 \bullet 2 \bullet 2 \bullet 2$ 16	$(-5)^3$ $(-5)(-5)(-5)$ -125 * The negative sign is INSIDE the parenthesis.	-5^2 $-(5 \bullet 5)$ -25 * The negative sign is OUTSIDE the parenthesis.	$\left(\frac{1}{2}\right)^3$ $\frac{1}{2} \bullet \frac{1}{2} \bullet \frac{1}{2}$ $\frac{1}{8}$ Cube the top AND the bottom #s.

Avoiding Common Errors: Be careful when deciding what the base is. Pay special attention to parentheses.

$(-4)^2 = (-4)(-4)$ — The repeated factor is – 4.

$-4^2 = -(4 \bullet 4) = -16$ — The repeated factor is 4.

YOUR TURN

Evaluate each expression. Watch the negative sign.

1. -8^2	2. $4 \bullet 5^2$	3. $-(6)^2$	4. $\left(\frac{2}{3}\right)^4$

Exponents may also be used with variables. Remember, variables are standing in for numbers.

$$x^3 = x \cdot x \cdot x$$

$$x^2y^3 = x \cdot x \cdot y \cdot y \cdot y$$

EXAMPLE 2

Evaluate each expression for the given values. Make sure you use parentheses with negative numbers.

$2x^3$; $x = 4$	$-3xy^2$; $x = 2$, $y = -5$
$2 \bullet (4)^3$ $2 \bullet 64$ 128 Evaluate means to plug in the number. Add parenthesis around the number.	$-3(2)(-5)^2$ $-3(2)(25)$ $-6(25)$ -150 Add parenthesis around the number. Watch the negative #.

YOUR TURN

Evaluate the expressions with the given replacement values. Remember to add parentheses.

5. $3x^3y$; $x = 4$ *and* $y = -2$	6. $\frac{9}{x^2}$; $x = -3$

Rules of Exponents

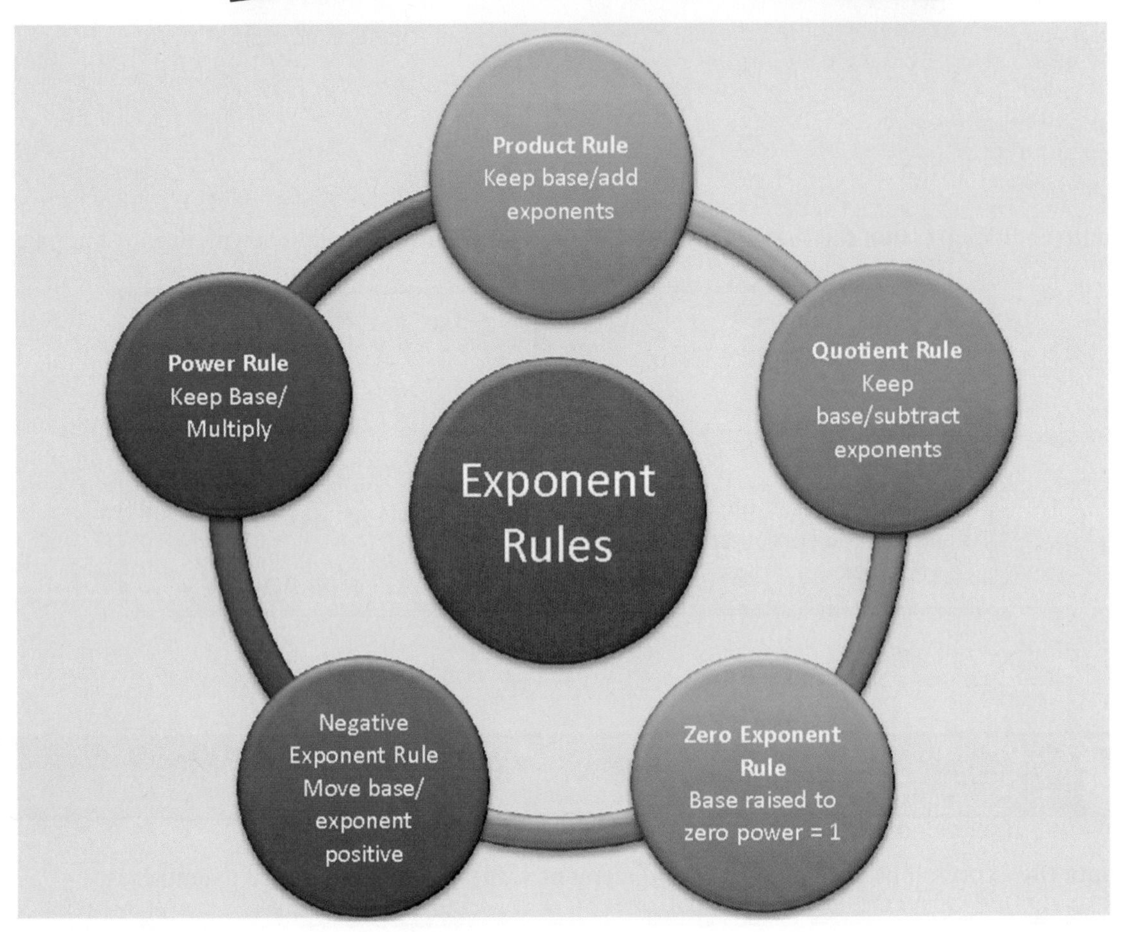

THE EXPONENT RULES:

$$a^m \cdot a^n = a^{m+n} \qquad \frac{a^m}{a^n} = a^{m-n}$$

$$(a^m)^n = a^{m \cdot n} \qquad a^0 = 1$$

$$(ab)^m = a^m b^m \qquad \left(\frac{a}{b}\right)^m = \frac{a^m}{b^m}$$

$$x^{-m} = \frac{1}{x^m}$$

$$\frac{1}{x^{-m}} = x^m$$

Courtesy of Fotolia

Let's zoom in and look at each of the rules individually !

PRODUCT RULE

Exponential expressions can be multiplied together. Let's see if we can develop a shortcut method for multiplying exponential expressions together. Look at the two examples below.

Example: 3 factors 4 factors

$x^3 \cdot x^4 = (x \cdot x \cdot x)(x \cdot x \cdot x \cdot x)$

$= x \cdot x \cdot x \cdot x \cdot x \cdot x \cdot x$

$3 + 4 = 7$ *factors*

$= x^7$

Let's look at another example.

$5^2 \cdot 5^4 = (5 \cdot 5)(5 \cdot 5 \cdot 5 \cdot 5)$

$= 5 \cdot 5 \cdot 5 \cdot 5 \cdot 5 \cdot 5$

$= 5^6$

There are 6 factors of 5.

In both examples, adding the exponents gives the exponent of the product. Thus, we have the product rule for exponents.

> **Product Rule for Exponents – to multiply powers with the same base, keep the base and add the exponents.** $a^m \cdot a^n = a^{m+n}$

EXAMPLE 3

Simplify using the Product Rule.

$3^3 \cdot 3^2$ 3^{3+2} 3^5 or 243	Add exponents.	$x^2 \cdot x^8 \cdot x$ x^{2+8+1} x^{11}	Remember, the last x has an exponent of "1".	$2^3 \cdot 4^3$ $8 \cdot 64$ 512	The product rule does not apply because the bases 2 and 4 are different.
$(x^5y^6)(xy^7)$ $(x^{5+1})(y^{6+7})$ x^6y^{13}	Add exponents of "like" bases.	$(2x^5)(3x^7)$ $(2 \cdot 3)(x^5 \cdot x^7)$ $6x^{12}$	Notice that the coefficients are multiplied but the exponents are added.	$(2a^2b^5)(-5a^6b^9)$ $(2 \cdot (-5))(a^2a^6)(b^5b^9)$ $-10a^8b^{14}$ Put it all together.	

Avoiding Common Errors: when multiplying powers of the same base – DO NOT multiply the bases.

CORRECT	INCORRECT
$3^4 \cdot 3^2 = 3^6$	$3^4 \cdot 3^2 = 9^6$

YOUR TURN

Use the Product Rule to simplify.

7. $8^3 \bullet 8^6$

8. $m \bullet m^3 \bullet m^9$

9. $(-2a^7b^2)(3a^5b)$

Be sure you understand the difference between *adding* and *multiplying* exponential expressions. For example,

$$5x^3 + 8x^3 = 13x^3 \quad \text{but} \quad (5x^3)(8x^3) = 40x^6$$

Addition — Multiplication

Courtesy of Fotolia

Quotient Rule

Exponential expressions can also be divided. Let's see if we can develop a shortcut method for dividing exponential expressions. Look at the example below.

$$\frac{x^5}{x^3} = \frac{x \bullet x \bullet x \bullet x \bullet x}{x \bullet x \bullet x} = \frac{\cancel{x} \bullet \cancel{x} \bullet \cancel{x} \bullet x \bullet x}{\cancel{x} \bullet \cancel{x} \bullet \cancel{x}} = x \bullet x = x^2$$

$$\frac{x^5}{x^3} = x^{5-3} = x^2$$

Notice that the result is exactly the same if we subtract exponents of the common bases!

Quotient Rule for Exponents
To divide powers with the same base,
keep the base and subtract the exponents. $\frac{a^m}{a^n} = a^{m-n}$

EXAMPLE 4

Simplify using the Quotient Rule.

$\frac{x^8}{x^2}$ Think x^{8-2}. x^6 Since there are more x's in the top, the answer has the x's in top.	$\frac{7^9}{7^3}$ Think 7^{9-3}. 7^6 Since there are more 7's in the top, the answer has the 7's in top.	$\frac{5a^9\ b^{10}}{10a\ b^2}$ Think $a^{9-1}\ b^{10-2}$. $\frac{a^8\ b^8}{2}$ Reduce the numbers. Subtract the exponents of the a's and then the b's. Note 2 is in the bottom.

Courtesy of Fotolia

Avoiding Common Errors: when dividing powers of the same base – DO NOT divide the bases.

CORRECT

$\frac{5^8}{5^2} = 5^6$

INCORRECT

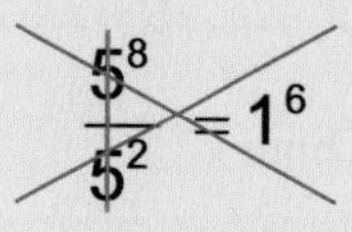

YOUR TURN

Use the Quotient Rule to simplify.

10. $\frac{5^7}{5}$

11. $\frac{-2x^5\ y^2}{14xy}$

12. $\frac{x^4 \bullet x^2}{x^3}$

6.01 Exponents

Courtesy of Fotolia

Power Rule

Exponential expressions can themselves be raised to powers. Let's look at a couple of examples.

$$\left(x^3\right)^4 = \underbrace{\left(x^3\right)\left(x^3\right)\left(x^3\right)\left(x^3\right)} = x^{3+3+3+3} = x^{12}$$

The exponent 4 tells us to multiply the base x^3 four times.

$$\left(x^3\right)^4 = x^{3 \cdot 4} = x^{12}$$

Notice the result is the same if we multiply the exponents.

Power Rule for Exponents
To raise a power to a power,
keep the base and multiply the exponents. $(a^m)^n = a^{m \cdot n}$

EXAMPLE 5

Simplify using the Power Rule.

$\left(x^4\right)^3$ Think $x^{4 \cdot 3}$. x^{12} Multiply the exponents.	$(m)^5$ Think $m^{1 \cdot 5}$. m^5 Remember m has an exponent of "1".	$-\left(\frac{1}{3^4}\right)^2$ Think $-\frac{1^2}{3^{4 \cdot 2}}$. $-\frac{1}{3^8}$

Avoiding Common Errors: Students often confuse the product and power rules. Note the difference carefully.

Product Rule	Power Rule
$x^5 \cdot x^2$	$\left(x^5\right)^2$
x^{5+2}	$x^{5 \cdot 2}$
x^7	x^{10}

Courtesy of Fotolia

YOUR TURN

Use the Power Rule to rewrite the expression with one exponent.

13. $\left(x^5\right)^4$	14. $\left(7^4\right)^3$	15. $\left(\frac{1}{x^2}\right)^5$

As with any other math topic, we need to look at how zero comes into play! Compare these two examples.

$$\frac{x^3}{x^3} = x^{3-3} = x^0$$

Subtract the exponents.

$$\frac{x^3}{x^3} = \frac{x \cdot x \cdot x}{x \cdot x \cdot x} = 1$$

Reduce.

Courtesy of Fotolia

Zero Rule

Zero Rule for Exponents
Any nonzero number raised to the 0 power is 1. $a^0 = 1$

EXAMPLE 6

Use the Zero Exponent Rule to simplify. Ask: "What is being raised to the zero power?"

5^0 1	$(2x)^0$ 1 Watch the parenthesis.	$3x^0 + 4x^0$ $3 \cdot x^0 + 4 \cdot x^0$ $3 \cdot 1 + 4 \cdot 1$ $3 + 4$	The 3& 4 are NOT raised to the zero power.	-7^0 $-1 \cdot 7^0$ $-1 \cdot 1$ -1	Watch the negative.	$(-7)^0$ 1

YOUR TURN

Use the Zero Exponent Rule to simplify.

16. $4^0 + (3x)^0$

17. $-2ab^0$

18. $(2ab)^0$

POWER OF A PRODUCT RULE

Power of Product
Raise each factor inside the parentheses to the power. $(ab)^m = a^m b^m$

EXAMPLE 7

Simplify using the power rule.

$(x\,y^5)^3$ Think $(x^1\,y^5)^3$ $x^{1\cdot 3}\;y^{5\cdot 3}$ $x^3\,y^{15}$ Raise each piece to the power.	$(6\,x\,y^5)^3$ *Think* $6^{1\cdot 3}x^{1\cdot 3}y^{5\cdot 3}$ $6^3x^3y^{15}$ Remember to raise the number $216x^3y^{15}$ to the power.	$(-5x^4)^2$ *Think* $(-5)^2(x^4)^2$ $25x^8$

YOUR TURN

Simplify each expression using the power rule. Remember to raise each piece to the power.

19. $(x^2y^3)^4$

20. $(4ab^3)^3$

21. $(-2x^5yz^2)^4$

POWER OF A QUOTIENT RULE

Power of Quotient

Raise both the numerator and the denominator to the power. $\left(\frac{a}{b}\right)^m = \frac{a^m}{b^m}$

EXAMPLE 8

Simplify each expression.

$\left(\frac{m^2}{n^3}\right)^5$ $\frac{m^{2\cdot5}}{n^{3\cdot5}}$ Raise each piece to the power. $\frac{m^{10}}{n^{15}}$	$\left(\frac{x^3}{3y^7}\right)^4$ $\frac{(x^3)^4}{(3)^4\,(y^7)^4}$ $\frac{x^{12}}{81y^{28}}$	$\left(\frac{-2a^7b^3}{3c}\right)^2$ $\frac{(-2)^2\,(a^7)^2\,(b^3)^2}{(3)^2\,(c)^2}$ $\frac{4a^{14}b^6}{9c^2}$

YOUR TURN

Simplify each expression. Remember to raise each piece to the power.

22. $\left(\frac{a}{b^3}\right)^4$

23. $\left(\frac{xy}{7}\right)^2$

24. $\left(\frac{2xy^8}{-5z^5}\right)^3$

YOUR TURN

Simplify each expression. You will need to decide which rule to use.

25. $\dfrac{3x^{12}}{12x^{3}}$	26. $\left(3x^{2}\right)\left(-2x^{5}\right)$	27. -9^{2}
28. $\left(\dfrac{3}{4}\right)^{3}$	29. $2^{0}+2^{5}$	30. $\left(-3xy^{5}\right)^{2}$
31. $\left(\dfrac{6xy^{5}z^{2}}{-2xyz^{7}}\right)^{3}$	32. $\dfrac{5^{3}\bullet 5^{7}}{5^{2}}$	33. $\left(-2mn^{6}\right)\left(3m^{3}n^{4}\right)^{2}$

6.01 Exponents 6.01

Finding Area

Recall from DMA 040:
To find the area of a square, square the side length. A = s^2
To find the area of a rectangle, multiply the length times the width. A = lw

EXAMPLE 9

Find the area of each figure.

1. Find the area of the rectangle.	2. Find the area of the square.
$A = l\,w$ $A = (6y^5)(5y^4)$ $= 30y^9$ $6y^5$ (length), $5y^4$ (width) 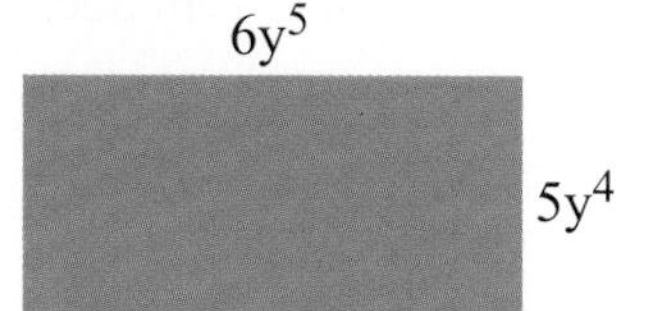	$A = s^2$ $A = (7z^3)^2$ $= 49z^6$ $7z^3$ (side) 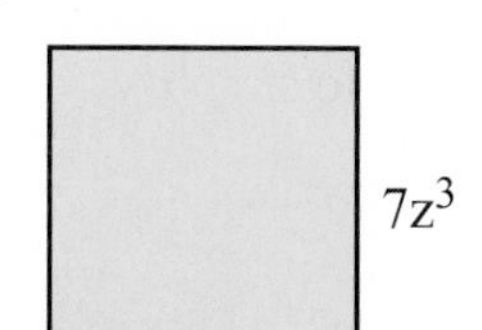

YOUR TURN

34. Find the area of the rectangle.

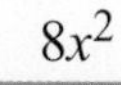

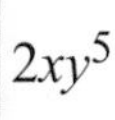

YOUR TURN Answers to Section 6.01.

1. -64	2. 100	3. 36	4. $\frac{16}{81}$	5. -384	6. 1	7. 8^9
8. m^{13}	9. $-6a^{12}b^3$	10. 5^6	11. $-\frac{x^4y}{7}$	12. x^3	13. x^{20}	14. 7^{12}
15. $\frac{1}{x^{10}}$	16. 2	17. -2a	18. 1	19. x^8y^{12}	20. $64a^3b^9$	21. $16x^{20}y^4z^8$
22. $\frac{a^4}{b^{12}}$	23. $\frac{x^2y^2}{49}$	24. $\frac{8x^3y^{24}}{-125z^{15}}$	25. $\frac{x^9}{4}$	26. $-6x^7$	27. -81	28. $\frac{27}{64}$
29. 33	30. $9x^2y^{10}$	31. $-\frac{27y^{12}}{z^{15}}$	32. 5^8	33. $-18m^7n^{14}$	34. $16x^3y^5$	

Complete MyMathLab Section 6.01 Homework in your Homework notebook.

6.02 Negative Exponents

So far in this module, we have looked at expressions that have positive exponents. Now we will need to look at expressions that involve negative exponents, such as x^{-3}.

NEGATIVE EXPONENT RULE

EXAMPLE 1

We can understand negative exponents by looking at the quotient rule from Sec. 6.01. There are two ways to simplify.

$$\frac{x^3}{x^8}$$

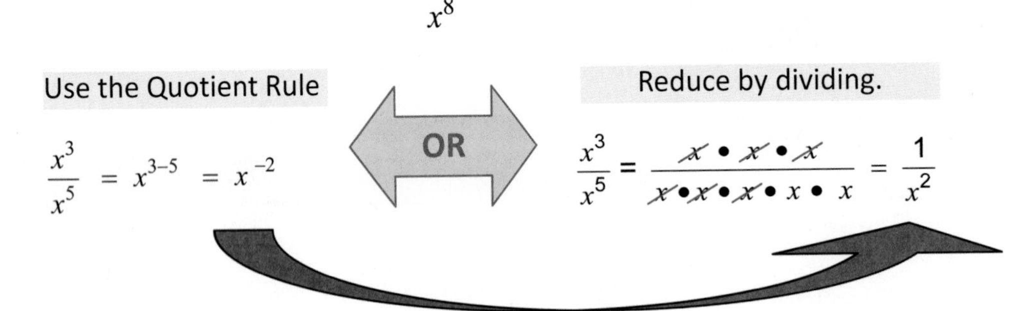

If we work a math problem two different but correct ways, the answer must be the same regardless of how we worked it. Therefore, looking at the example above we can now conclude the following:

$x^{-2} = \frac{1}{x^2}$ This is an example of the negative exponent rule.

Negative Exponent Rule

Move the base to the opposite location and make the exponent positive.

$$x^{-m} = \frac{1}{x^m} \quad or \quad \frac{1}{x^{-m}} = x^m$$

EXAMPLE 2

Simplify by writing each expression with positive exponents.

a. x^{-2}

$x^{-2} = \frac{1}{x^2}$

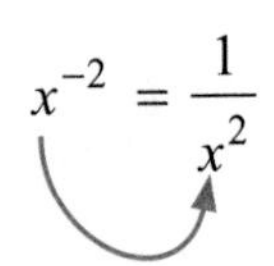

Move the x to the denominator. The exponent 2 is positive.

b. $-m^{-7}$

$-m^{-7} = -\frac{1}{m^7}$

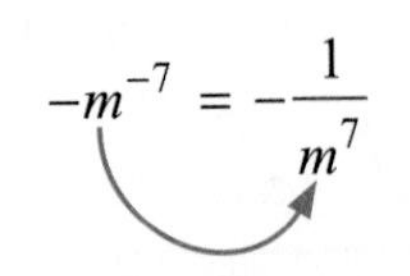

The negative exponent does not affect the negative out front.

c. ab^{-2}

$ab^{-2} = a\left(\frac{1}{b^2}\right) = \frac{a}{b^2}$

ONLY the b^2 moves to the denominator. The a has an exponent of 1.

6.02 Negative Exponents

Avoiding Common Errors :

CORRECT

$3^{-2} = \frac{1}{3^2} = \frac{1}{9}$

INCORRECT

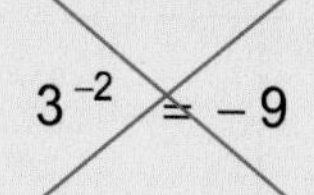

Another way to write negative exponents for fractions is to take its reciprocal and then change the exponent to a positive. Let's be reminded about reciprocals.

Fraction — Its Reciprocal

A negative exponent in the denominator will become a positive exponent in the numerator and a negative exponent in the numerator will become a positive exponent in the denominator.

$$\frac{1}{m^{-5}} = \frac{m^5}{1}$$

$$m^{-3} = \frac{1}{m^3}$$

You can make a negative exponent positive by moving it across the fraction bar!!!!

EXAMPLE 3

Simplify. Write the result using positive exponents only.

$\frac{3}{m^{-7}}$	$\frac{p^{-2}}{q^{-5}}$	$\frac{2x^{-5}y}{3z^{-6}}$
$\frac{3}{m^{-7}} = \frac{3m^7}{1} = 3m^7$ 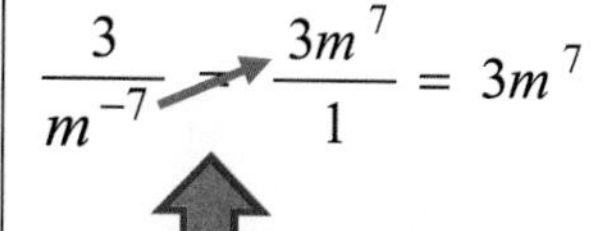	$\frac{p^{-2}}{q^{-5}} = \frac{q^5}{p^2}$ 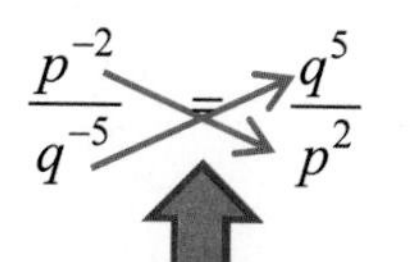	$\frac{2x^{-5}y}{3z^{-6}} = \frac{2yz^6}{3x^5}$
Notice the movement across the fraction bar of all variables/numbers that have negative exponents.		Only variables with negative exponents can move across the fraction bar.

6.02 Negative Exponents 6.02

YOUR TURN

Simplify. Write the result using positive exponents only.

1. 2^{-3}	2. $\frac{2}{y^{-6}}$	3. $\frac{m^{-6}}{m^{-2}}$
4. $\frac{3y^5}{z^{-4}}$	5. $\frac{5x^2y^{-2}}{z^{-7}}$	6. $\frac{-4^{-3}}{3^{-2}}$

Remember all those rules for exponents you learned in Sec. 6.01?
All of those rules apply to negative exponents also!!
Below is a quick review and one example of the exponent rules from 6.01.

EXAMPLE 4

Simplify using the **product rule**. Write each result using positive exponents only.

a. $x^{-5} \bullet x^8$

x^{-5+8} *or* $\frac{x^8}{x^5}$

x^3

b. $x^2 \bullet x^{-8} \bullet x$

x^{2+-8+1}

x^{-5}

$\frac{1}{x^5}$

EXAMPLE 5

Simplify using the **quotient rule**. Write each result using positive exponents only.

a. $\frac{m^7}{m^{-2}}$ *think* $m^{7-(-2)}$

m^7m^2 *or move* m^2

m^9

b. $\frac{12x^{-3}}{6x^4}$ *Think* $\frac{12}{6} \bullet x^{-3-4}$

$\frac{2}{x^4x^3}$ *or reduce #s & move* x^3

$\frac{2}{x^7}$

6.02 Negative Exponents

EXAMPLE 6

Simplify using the **power rule**. Write each result using positive exponents only.

a. $\left(m^{-5}\right)^4$ *Think* $m^{(-5)\cdot(4)}$

m^{-20} Multiply the exponents and then move the base to the bottom.

$\frac{1}{m^{20}}$

b. $\left(x^{-3}\right)^{-6}$

$x^{-3\cdot(-6)}$ Multiply the exponents.

x^{18}

EXAMPLE 7

Simplify. Write results using positive exponents only.

a. $\left(x^{-2}y^{5}\right)^{-3}$ Raise each piece to the power and then move the base y term to the bottom.

x^6y^{-15}

$\frac{x^6}{y^{15}}$

b. $\frac{12m^7n^{-2}}{-4m^{-3}n}$ Move the bases with negative exponents across the fraction bar.

$\frac{12m^7m^3}{-4n\,n^2}$ Then reduce the #s and subtract the exponents.

$\frac{-3m^{10}}{n^3}$

EXAMPLE 8

Simplify. Write results using positive exponents only.

a. $\left(\frac{3m^2}{b}\right)^{-3}$

$=\frac{3^{-3}\left(m^2\right)^{-3}}{b^{-3}}$ Raise each piece to the power.

$=\frac{3^{-3}m^{-6}}{b^{-3}}$ Multiply the exponents of m.

$=\frac{b^3}{3^3m^6}=\frac{b^3}{27m^6}$ Move the bases across the fraction bar.

b. $\left(\frac{2x^3y}{xy^{-1}}\right)^3$

$\frac{2^3\cdot x^9\cdot y^3}{x^3\cdot y^{-3}}$ Raise each piece to the power.

$\frac{2^3x^9y^3y^3}{x^3}$ Move the bases across the fraction bar.

$8x^6y^6$ Subtract the exponents for x and add the exponents for y.

6.02 Negative Exponents 6.02

YOUR TURN

Simplify each expression using the properties of exponents. Write using positive exponents only.

7. $m^{-2} \cdot m^{7}$	8. $\left(m^{-3}\right)^{7}$	9. You're doing great!
10. $\dfrac{3y^{5}}{y^{-4}}$	11. $\left(3m^{-5}\right)\left(2m^{-2}\right)$	12. $\left(5y^{-3}\right)^{2}$
13. $\left(a^{-4}b^{2}\right)^{-5}$	14. $\dfrac{10\,m^{2}\,n}{-5\,m^{3}\,n^{-6}}$	15. $\left(\dfrac{5x^{-3}y^{4}}{xy^{6}}\right)^{-2}$

The End of Negative Exponents!!!!!!

YOUR TURN **Answers to Section 6.02.**

1. $\frac{1}{8}$
2. $2y^6$
3. $\frac{1}{m^4}$
4. $3y^5z^4$
5. $\frac{5x^2z^7}{y^2}$
6. $-\frac{9}{64}$
7. m^5
8. $\frac{1}{m^{21}}$
9. You're doing great!
10. $3y^9$
11. $\frac{6}{m^7}$
12. $\frac{25}{y^6}$
13. $\frac{a^{20}}{b^{10}}$
14. $\frac{-2n^7}{m}$
15. $\frac{x^8y^4}{25}$

Complete MyMathLab Section 6.02 Homework in your Homework notebook.

POLYNOMIALS

We will now look at an important algebraic expression known as a polynomial. At this point, you have seen a variety of *algebraic expressions* like:

$3a^4b^6$, $x + 2y$, $5x^2 + 2x - 7$

Algebraic expressions do NOT have "=".

IDENTIFYING TERMS: The separate pieces that make up algebraic expressions are called terms. Take $x + 2y$ for example. x is a term and $2y$ is a term that together make the expression $x + 2y$.

A term is defined to be a product of numbers and/or variables. Note: A "+" or "–" sign separates terms.

Expression	Identify the Terms
$4x^2 + 3x$	$4x^2, 3x$
$7x^3 - 2x^2 + 3$	$7x^3, -2x^2, 3$
$7y$	$7y$

SPECIAL NAMES OF POLYNOMIALS: If a term is a product of constants and/or variables, it is called a monomial. A polynomial is a monomial or a sum of monomials. For example, $2x^4 - 5x^3 + 2x - 8$ is a polynomial. Some polynomials are given special names depending on how many terms they have.

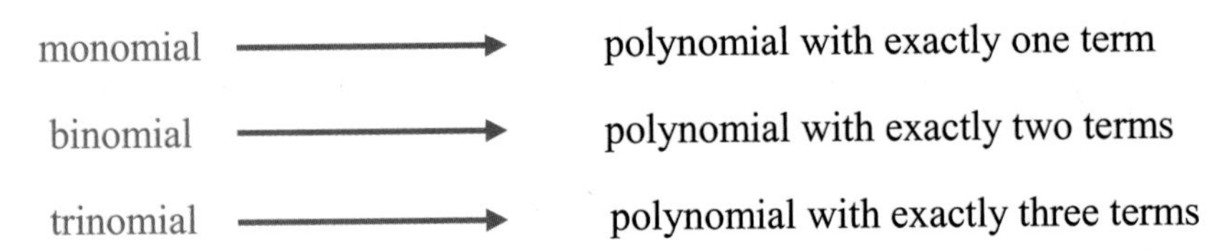

The following are examples of monomials, binomials, trinomials. All of these examples could also be referred to as polynomials.

Monomials – 1 term	Binomials – 2 terms	Trinomials – 3 terms
$2xy$	$3x + y$	$2x^2 - 7x + 6$
-6	$3ab^5 - 2b$	$4x^2 - 2xy + 7y^2$
$2a^2b^5$	$-9x^3 + 3$	$z^6 - 3z^5 + 2z^2$

Polynomials with four or more terms have no special name. We will refer to them as simply polynomials.

6.03 Introduction to Polynomials 6.03

IDENTIFYING COEFFICIENTS: The numerical coefficient of a term is the numerical factor of each term. In the term $3a^4b^6$, the 3 is the numerical coefficient. If no numerical factor appears in the term, such as in x, the coefficient is understood to be 1.

Term	Identify the Coefficient
$-4x^2$	-4
x^5	1
7	7

IDENTIFYING THE DEGREE: Each term of a polynomial has a degree.

The degree of a term is the *sum of the exponents* on the variables contained in a term.

EXAMPLE 1

Find the degree of each term. $-2x^5 \rightarrow$ degree of term is 5.

$6xy^2 \rightarrow$ degree of term is 3, because the exponents $1 + 2 = 3$.

$4 \rightarrow$ degree of term is 0, because $4 = 4x^0$ *or* $4a^0b^0$.

Degree of a Polynomial

The degree of a polynomial is the greatest degree of any term of the polynomial.

EXAMPLE 2

Find the degree of the polynomial.

$$2x^3 - 5x^2 + 7$$

List the degrees of each term.

$2x^3 \longrightarrow$ degree 3.

$-5x^2 \longrightarrow$ degree 2.

$7 \longrightarrow$ degree 0.

The degree of the polynomial is the greatest of these degrees. The degree of this polynomial is 3.

EXAMPLE 3

Find the degree of the polynomial.

$$2x^3y^3 - 5x^2y^5 + xy - 2$$

List the degrees of each term.

$2x^3y^3 \longrightarrow$ degree 6

$-5x^2y^5 \longrightarrow$ degree 7

$xy \longrightarrow$ degree 2

$-2 \longrightarrow$ degree 0

The degree of the polynomial is 7.

6.03 Introduction to Polynomials **6.03**

LEADING COEFFICIENT: To find the leading coefficient of a polynomial, first find the degree of the polynomial. The leading coefficient is the coefficient of the term with the highest degree. It is NOT simply the largest number in the polynomial.

From the previous examples:

EXAMPLE 2A

Find the leading coefficient of the polynomial.

$$2x^3 - 5x^2 + 7$$

The term with the highest degree (3) is the first term, so the leading coefficient is 2.

EXAMPLE 3A

Find the degree of the polynomial.

$$2x^3y^3 - 5x^2y^5 + xy - 2$$

The term with the highest degree (7) is the second term, so the leading coefficient is -5.

YOUR TURN

Find the degree of each polynomial. Also state the leading coefficient.

1. $4xy - 2x^3y + x^5$

Degree of $4xy$ ________

Degree of $-2x^3y$ _______

Degree of x^5 ______

Degree of Polynomial ________

Leading coefficient __________

2. $-3x^3y^2 + 4xy^2 - y^2 + 3x - 2$

Degree of $-3x^3y^2$ ________

Degree of $4xy^2$ _______

Degree of $-y^2$ ______

Degree of $3x$ ______

Degree of -2 ______

Degree of Polynomial ________

Leading coefficient __________

Evaluating Polynomials

Polynomials have different values depending on the values of the variables. When evaluating a polynomial for a value, simply PLUG IN THE NUMBERS. Use parenthesis.

EXAMPLE 4

Evaluate each polynomial when $x = -3$.

a. $-5x + 6$

$-5(-3) + 6$

$15 + 6$

21

Remember, plug the number in for the variable. Use parentheses if needed.

b. $3x^2 - 2x + 1$

$3(-3)^2 - 2(-3) + 1$

$3(9) + 6 + 1$

$27 + 6 + 1$

34

Evaluating Polynomials in Real Life Applications

EXAMPLE 5

The concentration (C), in parts per million, of a certain antibiotic in the bloodstream after t hours is given by the polynomial equation, $C = -0.05t^2 + 2t + 2$. Find the concentration after 2 hours.

First understand what C and t are in words for this problem.

t: is *time in hours* C: is *concentration of antibiotic*

The time given is 2 hours, so let t = 2 and evaluate.

$$
\begin{aligned}
C &= -0.05t^2 + 2t + 2 \\
&= -0.05(2)^2 + 2(2) + 2 \\
&= -0.05(4) + 4 + 2 \\
&= -0.2 + 4 + 2 \\
&= 5.8 \text{ parts per million}
\end{aligned}
$$

Plug t in the formula and solve.

After 2 hours, the concentration is 5.8 parts per million.

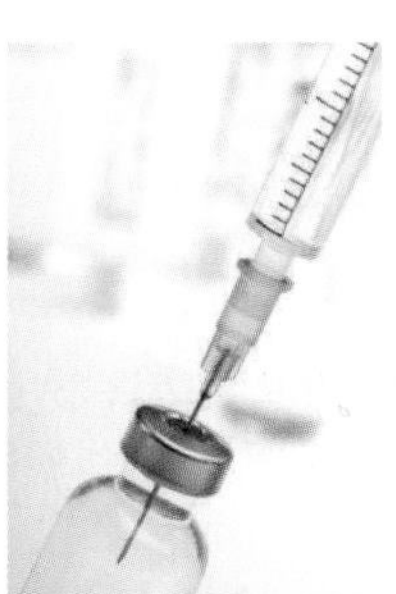

Courtesy of Fotolia

EXAMPLE 6

A rocket is fired upward from the ground with an initial velocity of 200 feet per second. Neglecting air resistance, the height of the rocket at any time, t, is described by the polynomial $h = -16t^2 + 200t$. Find the height of the rocket after 3 seconds.

First understand what h and t are in words for this problem.

t: is *time in seconds* h: is *height in feet.*

The time given is 3 seconds, so let t = 3 and evaluate.

$$
\begin{aligned}
h &= -16t^2 + 200t \\
&-16(3)^2 + 200(3) \\
&-16(9) + 600 \\
&-144 + 600 \\
&456 \textit{ feet}
\end{aligned}
$$

After 3 seconds, the rocket is 456 feet high in the air.

Courtesy of Fotolia

Descending Order

If the polynomial has one variable, arrange the terms so the exponents decrease from left to right. If the polynomial is in more than one variable, arrange the terms so the exponents of the chosen variable decrease from left to right.

EXAMPLE 7

Write each polynomial in descending order.

a. $2x + x^4 \longrightarrow x^4 + 2x$

b. $3m^3 + 7 - 2m \longrightarrow 3m^3 - 2m + 7$

YOUR TURN

Write the following polynomials in descending order.

3. $5x^2 + 7 + 2x^4 - 5x^3$

4. $xy + 3x^2 - 5y^2$

Identify the variables in words and evaluate the polynomial to answer the question.

5. A rocket is fired upward from the ground with an initial velocity of 250 feet per second. Neglecting air resistance, the height of the rocket at any time, t, is described in feet by the polynomial,

$h = -16t^2 + 250t + 20$. Find the height of the rocket after 5 seconds.

In words, identify t and h.

t: ____________ h: __________________

To evaluate, let t = ________

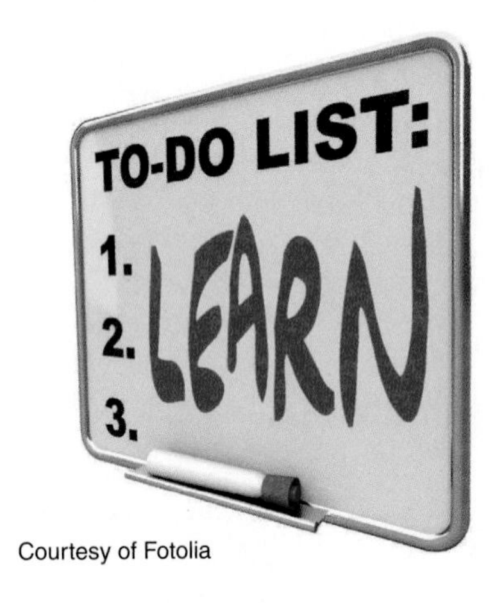

Courtesy of Fotolia

Like terms: terms of a polynomial having the same variables raised to the same powers.

Combine like terms: add or subtract the coefficients of like terms.

EXAMPLE 8

Write an equivalent expression by combining like terms.

$$7a^3 - 5a^2 + 9a^3 + 3a^2 - 1$$

$7a^3 + 9a^3 - 5a^2 + 3a^2 - 1$ ⟶ Rearrange to group like terms.

$(7 + 9)a^3 + (-5 + 3)a^2 - 1$ ⟶ Add or Subtract coefficients.

$$16a^3 - 2a^2 - 1$$

EXAMPLE 9

Write an equivalent expression by combining like terms.

$$4xy - 3x^2 + 5xy - 5y^2 + 12y^2 + 25xy - 8x^2 + 16y^2$$

$-8x^2 - 3x^2 + 4xy + 5xy + 25xy - 5y^2 + 12y^2 + 16y^2$ Rearrange to group like terms.

$-11x^2 + 34xy + 23y^2$ Add or Subtract coefficients.

YOUR TURN

Simplify by combining like terms.

6. $7x^3 + 6x - 3x^2 - x^3 + 6x^2$

7. $\frac{4}{3}x^4 - 5x^3 - \frac{1}{3}x^4 + \frac{7}{10}x^3 + \frac{3}{10}x^3$

Courtesy of Fotolia

Avoiding Common Errors: When combining like terms only add the coefficients . DO NOT ADD the exponents.

CORRECT	INCORRECT
$2x^4 + 5x^4 = 7x^4$	$2x^4 + 5x^4 \neq 7x^8$

USING POLYNOMIALS TO FIND AREA

EXAMPLE 10

Write a polynomial that describes the total area of the squares and rectangles shown below. Then simplify the polynomial.

Recall that the area of a rectangle is length times width. A = LW

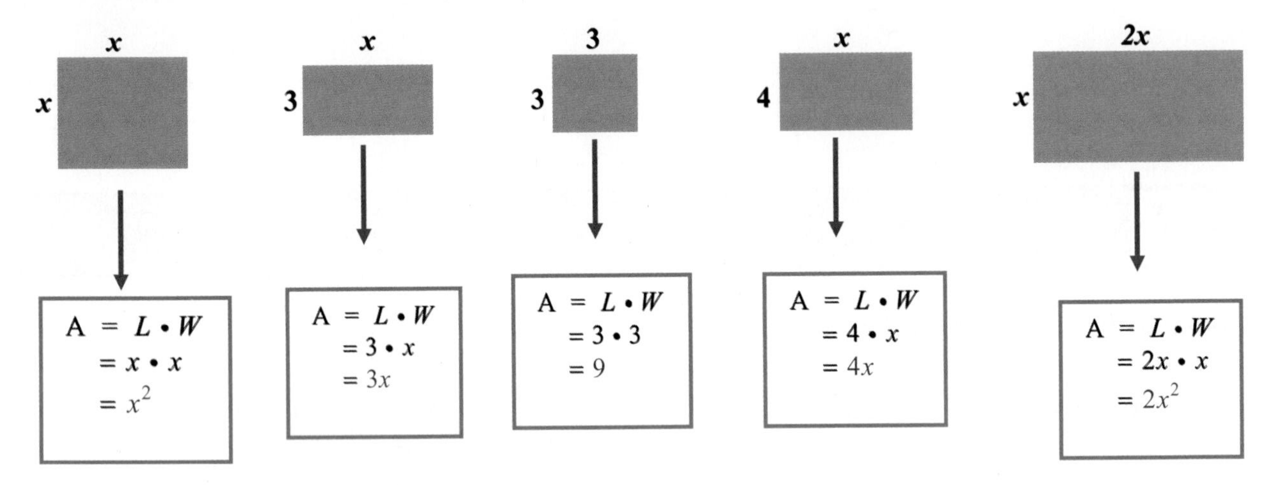

To find the total area we will add the area of each of the rectangles.

$$x^2 + 3x + 9 + 4x + 2x^2$$

$$3x^2 + 7x + 9$$

YOUR TURN

8. Write a polynomial that describes the total area of the squares and rectangles shown below. Then simplify the polynomial.

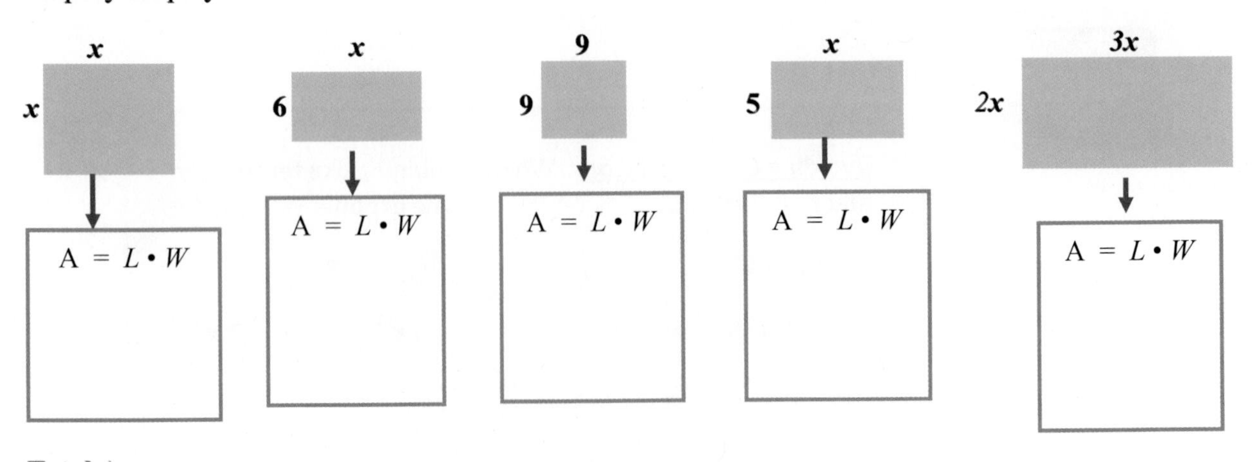

Total Area: ______________________

YOUR TURN **Answers to Section 6.03.**

1. 2, 4, 5, degree 5, leading coefficient 1
2. 5, 3, 2, 1, 0, degree 5, leading coefficient –3
3. $2x^4 - 5x^3 + 5x^2 + 7$
4. $3x^2 + xy - 5y^2$
5. t is time in seconds, h is height in feet. After 5 seconds, the height is 870 feet.
6. $6x^3 + 3x^2 + 6x$
7. $x^4 - 4x^3$
8. $7x^2 + 11x + 81$

Complete MyMathLab Section 6.03 Homework in your Homework notebook.

6.04 Adding and Subtracting Polynomials 6.04

Adding Polynomials

To add polynomials, combine all like terms.

EXAMPLE 1

Add. $\left(4x^2 + 6x + 12\right) + \left(2x^2 - x - 5\right)$

$$\underbrace{4x^2 + 2x^2}_{6x^2} + \underbrace{6x - x}_{5x} + \underbrace{12 - 5}_{7}$$

Remember, *add the coefficients* of like terms and *leave the exponents of the variables alone.*

EXAMPLE 2

Add. $\left(7x^3y - xy^3 + 11\right) + \left(2x^3y - 4\right)$

$$\underbrace{7x^3y + 2x^3y}_{9x^3y} - xy^3 + \underbrace{11 - 4}_{7}$$

Group "like" terms.

Add/subtract the coefficients AND *keep variable parts the same.*

EXAMPLE 3

In the examples above we added polynomials horizontally. You may prefer to add polynomials vertically. To do so, align terms with other like terms in the same column and then add.

Add. $\left(4x^2 + 6x + 12\right) + \left(2x^2 - x - 5\right)$

$$\begin{array}{rrrr} & 4x^2 & + 6x & + 12 \\ + & 2x^2 & - x & - 5 \\ \hline & 6x^2 & + 5x & + 7 \end{array}$$

YOUR TURN

Add.

1. $\left(9x^2 - x + 10\right) + \left(-2x^2 - 7x - 5\right)$

2. $\left(3x^2y - 4xy + y\right) + \left(x^2y + 2xy + 3y\right)$

6.04 Adding and Subtracting Polynomials 6.04

Subtracting Polynomials

To Subtract Polynomials, change all the signs of the second polynomial and add the result to the first polynomial.

EXAMPLE 4

Subtract. $4x^2 - 7x^3 + 2x - (4x^2 + x^3 - 3x + 6)$

$4x^2 - 7x^3 + 2x - (4x^2 + x^3 - 3x + 6)$ Distribute the "-" with all 2nd terms.

$-7x^3 - x^3 + 4x^2 - 4x^2 + 2x + 3x - 6$ Group "like" terms.

Add/subtract the coefficients AND *keep variable parts the same.*

$-8x^3 + 5x - 6$

EXAMPLE 5

Subtract. $(2x^3 - 8x^2y - 6x) - (2x^3 - x^2y + 6)$

$= 2x^3 - 8x^2y - 6x - 2x^3 + x^2y - 6$ Distribute the "-" with all 2nd terms.

$= 2x^3 - 2x^3 - 8x^2y + x^2y - 6x - 6$ Group "like" terms.

$= -7x^2y - 6x - 6$

Add/subtract the coefficients AND *keep variable parts the same.*

EXAMPLE 6

Subtract $(5m - 6)$ from $(7m + 3)$.

$= (7m + 3) - (5m - 6)$

$= 7m + 3 - 5m + 6$

$= 2m + 9$

Courtesy of Fotolia

Don't forget that to translate " subtract from" you must change the order of the two things you are subtracting!!!!

6.04 Adding and Subtracting Polynomials

YOUR TURN

Subtract.

3. $(5x-3)-(2x-11)$	4. $(x^3+2x+6)-(-3x^2-5x+3)$
5. Subtract $(2x^2-7x+3)$ from $(5x^2-12x-9)$ Read this carefully. You will see it again.	6. $(9a^2b^2+6ab-3ab^2)-(5a^2b^2+2ab-9ab^2)$

EXAMPLE 7

Subtract $(5x-3)$ from the sum of (x^2-7x+3) and $(-9x+5)$

Notice that $(5x-3)$ is to be *subtracted from* a sum.

Translation: $\left[(x^2-7x+3)+(-9x+5)\right]-(5x-3)$ Distribute the "-" with all 3rd terms.

$x^2-7x+3-9x+5-5x+3$

$x^2-21x+11$

Add/subtract the coefficients AND *keep variable parts the same.*

6.04 Adding and Subtracting Polynomials **6.04**

POLYNOMIAL APPLICATIONS USING ADDITION AND SUBTRACTION

EXAMPLE 8

Find the perimeter.

To find perimeter, add up the sides.

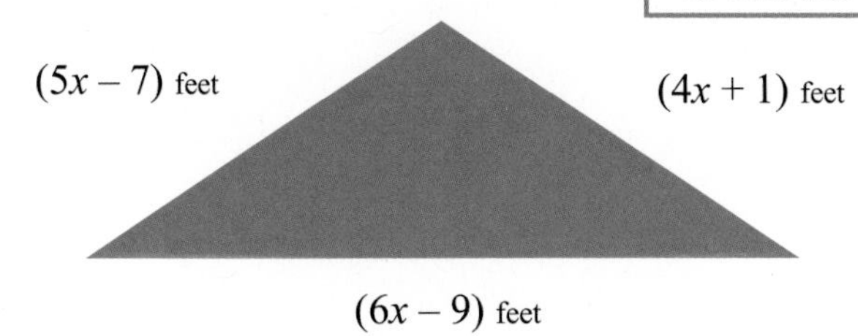

Courtesy of Fotolia

Perimeter: $(5x - 7) + (4x + 1) + (6x - 9)$
$5x - 7 + 4x + 1 + 6x - 9$
$(15x - 15)$ feet

EXAMPLE 9

A wooden beam is $(4y^2 + 4y + 1)$ meters long. If a piece $(y^2 - 10)$ meters is cut off, express the length of the remaining piece of beam as a polynomial in y.

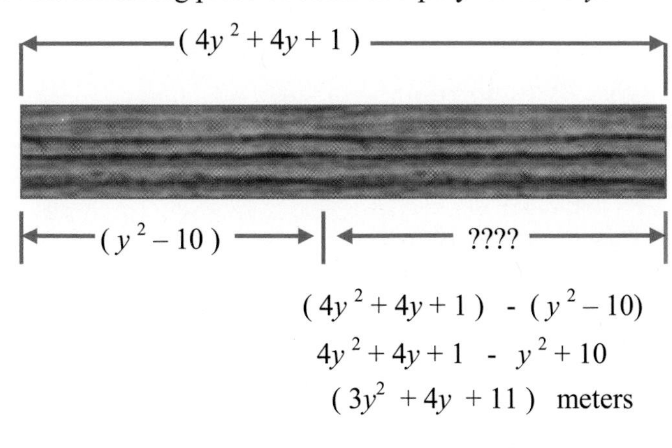

To find the length of the "remaining" piece, we need to subtract the length of the piece we cut off from the total length of the beam.

$(4y^2 + 4y + 1) - (y^2 - 10)$

$4y^2 + 4y + 1 - y^2 + 10$ Distribute the "-" with all 2nd terms.

$(3y^2 + 4y + 11)$ meters Add/subtract the coefficients AND *keep variable parts the same.*

EXAMPLE 10

In developing his business plan, Don wrote an equation to calculate the cost C of producing cupcakes as $C = 250 + 0.78x$ where x is the number of cupcakes. If he sells the cupcakes for \$1.50 each, his revenue R will be modeled by the equation $R = 1.50x$. Profit is caluculated by subtracting cost from revenue. Using what you know, write a new equation to calculate Don's profit for x cupcakes sold.

Profit = *Revenue Equation – Cost Equation*
$= 1.50x - (250 + 0.78x)$
$= 1.50x - 250 - 0.78x$
$= 0.72x - 250$

Courtesy of Fotolia

6.04 Adding and Subtracting Polynomials 6.04

YOUR TURN

Write an expression and then simplify.

7. Subtract $(2x - 3)$ from the sum of $(5x^2 - x + 3)$ and $(-3x + 1)$.

8. Find the perimeter.

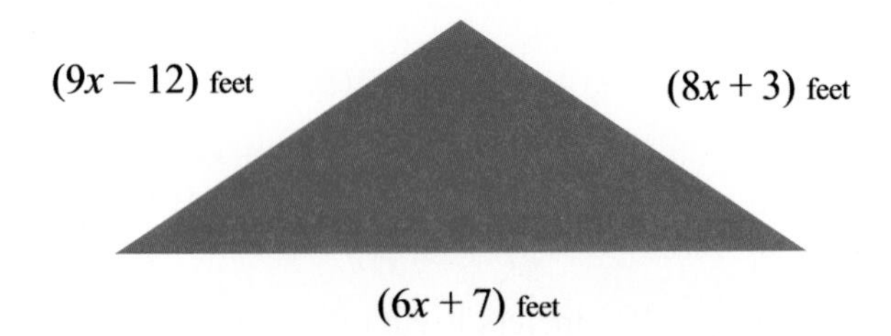

9. A piece of quarter-round molding is ($13x - 7$) inches long. If a piece ($2x + 2$) inches long is removed, express the length of the remaining piece of molding as a polynomial in terms of x.

10. In developing his business plan, Don wrote an equation to calculate the cost C of producing cupcakes as $C = 220 + 0.65x$ where x is the number of cupcakes. If he sells the cupcakes for $2.50 each, his revenue R will be modeled by the equation $R = 2.50x$. If profit is caluculated by subtracting cost from revenue, write a new equation to calculate Don's profit for x cupcakes sold.

Courtesy of Fotolia

Putting the Pieces Together

Courtesy of Fotolia

Simplify each expression by performing the indicated operation.

a. $2m + m = 3m$ Addition ⟶ Add coefficients. Keep base.

b. $2m^3 \bullet 4m^2 = 8m^5$ Multiply ⟶ Multiply coefficients. Add exponents.

c. $-2m - m = -3m$ Subtract ⟶ Add coefficients. Keep base.

d. $(-2m)(-m) = 2m^2$ Multiply ⟶ Multiply coefficients. Add exponents.

YOUR TURN

11. Simplify each expression by performing the indicated operation.

a. $m + m + m$

b. $m \bullet m \bullet m$

c. $-m - m$

d. $(-m)(-m)$

YOUR TURN **Answers to Section 6.04.**

1. $7x^2 - 8x + 5$ 2. $4x^2y - 2xy + 4y$ 3. $3x + 8$

4. $x^3 + 3x^2 + 7x + 3$ 5. $3x^2 - 5x - 12$ 6. $4a^2b^2 + 4ab + 6ab^2$

7. $5x^2 - 6x + 7$ 8. $23x - 2$ feet 9. $11x - 9$ inches 10. $1.85x - 220$

11. *a.* $3m$ *b.* m^3 *c.* $-2m$ *d.* m^2

Complete MyMathLab Section 6.04 Homework in your Homework notebook.

6.05 Multiplying Polynomials 6.05

Multiplying Monomials

To Multiply Monomials, use the product rule for exponents which says to multiply the coefficients and keep the base / add the exponents.

EXAMPLE 1

Multiply.

$(4b^2)(-9b^3)$ — *Think, multiply coefficients*

$(4)(-9) \cdot b^{2+3}$ *add exponents.*

$-36b^5$

EXAMPLE 2

Multiply.

$(5x^2y)(8x^5y^2)$ — Multiply coefficients.

$40x^7y^3$ — Keep the base and add exponents.

YOUR TURN

Multiply.

1. $(2ab^2)(-5a^7b^4)$

2. $(-4x^2y)(-3x^5y^6)$

6.05 Multiplying Polynomials 6.05

Multiplying a Monomial and a Polynomial

To Multiply a Monomial and a Polynomial, use the distributive property and multiply each term of the polynomial by the monomial.

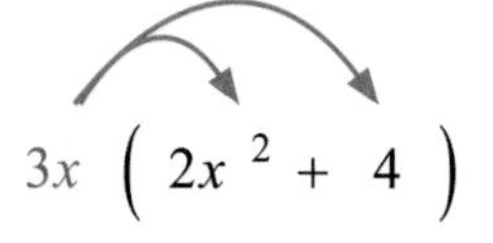

Multiply. $3x(2x^2 + 4)$

$3x\left(2x^2 + 4\right)$

$(3x)\left(2x^2\right) + (3x)(4)$

$6x^3 + 12x$

Distribute the outside term with each term inside the parenthesis.

OR Multiply coefficients and then Add the exponents of like bases.

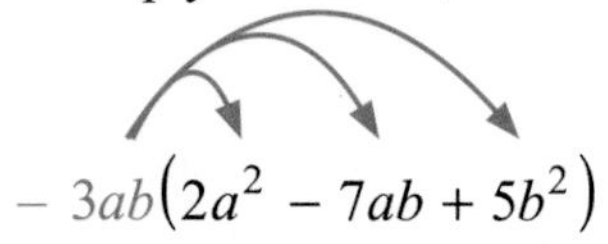

Multiply. $-3ab(2a^2 - 7ab + 5b^2)$

$-3ab(2a^2 - 7ab + 5b^2)$

$(-3ab)(2a^2) + (-3ab)(-7ab) + (-3ab)(5b^2)$

$-6a^3b + 21a^2b^2 - 15ab^3$

YOUR TURN

Multiply. Distribute the outside term.

3. $-3n\left(2n^5 - 5\right)$

4. $2x^2y\left(3x^2 - 5xy + 7y^2\right)$

Multiplying Two Binomials

To Multiply Two Binomials, multiply each term of the first binomial by each term in the second binomial and then combine like terms.

EXAMPLE 5

Multiply. $(3x + 2)(x + 7)$

$(3x + 2)(x + 7)$

$3x(x + 7) + 2(x + 7)$

$3x^2 + 21x + 2x + 14$

$3x^2 + 23x + 14$

Distribute each piece of the 1st term with the 2nd term.

Combine like terms.

EXAMPLE 6

Multiply. $(5x - 3)(2x + 1)$

$(5x - 3)(2x + 1)$

$5x(2x + 1) - 3(2x + 1)$

$10x^2 + 5x - 6x - 3$

$10x^2 - x - 3$

FOILING A POLYNOMIAL

Multiplying polynomials by distributing terms has another name. You can multiply binomials using FOIL. This means you multiply or distribute using this order:

F – Multiply **first** terms
O – Multiply **outer** terms
I – Multiply **inner** terms
L – Multiply **last** terms

EXAMPLE 7

Multiply. $(2x - 3y)(x - 7y)$ (Play the FOIL process below as as movie.)

Multiply *First terms*	*Outer terms*	*Inner terms*	*Last terms*
$(2x - 3y)(x - 7y) \rightarrow$	$(2x - 3y)(x - 7y) \rightarrow$	$(2x - 3y)(x - 7y) \rightarrow$	$(2x - 3y)(x - 7y)$
$2x^2$	$-14xy$	$-3xy$	$+21y^2$

$2x^2 - 17xy + 21y^2$ Finish by combining like terms.

EXAMPLE 8

Multiply using *FOIL.* $(3 - 4m)(2 - 3m)$

Think this step

First *Outer* *Inner* *Last*
$3(2) + (3)(-3m) - 4m(2) - 4m(-3m)$

$6 \quad -9m \quad -8m \quad +12m^2$

$12m^2 - 17m + 6$ Write in descending order.

EXAMPLE 9

Multiply. $(5x^2y - 7xy)^2$

Rewrite first. $(5x^2y - 7xy)(5x^2y - 7xy)$

Multiply coefficients, add the exponents.

$25x^4y^2 - 35x^3y^2 - 35x^3y^2 + 49x^2y^2$

$25x^4y^2 - 70x^3y^2 + 49x^2y^2$

This is squaring a binomial.

6.05 Multiplying Polynomials 6.05

Courtesy of Fotolia

You cannot square a binomial by squaring the terms!!!!!

$$(2x-7)^2 \neq 4x^2+49$$

Let's say it again!!!!!

$$(2x-7)^2 \neq 4x^2+49$$

Or in words:

$(2x-7)^2$ is not equal to and will never be equal to $4x^2+49$.

YOUR TURN

Multiply the following binomials. Use the FOIL method.

5. $(2x+5)(x-1)$	6. $(5x-1)(2x-7)$
7. $(a+5b)(2a-3b)$	8. $(2xy-3y^2)(2xy+3y^2)$
9. $(7-5k)(5+3k)$	10. $(5m-3)^2$

6.05 Multiplying Polynomials 6.05

Now that we can multiply two binomials, we will look at multiplying any two polynomials together. The good news is that we will use the exact same process that we used to multiply two binomials.

Multiplying Two Polynomials

To Multiply Two Polynomials, multiply each term of the first polynomial by each term in the second polynomial and then combine like terms.

EXAMPLE 10

Multiply. $(m-2)(3m^2-7m+1)$

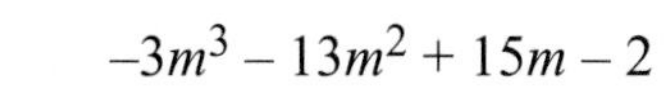

$m(3m^2 - 7m + 1) - 2(3m^2 - 7m + 1)$ Distribute the m and "-2".

$3m^3 - 7m^2 + m - 6m^2 + 14m - 2$ Then combine coefficients of like terms.

$-3m^3 - 13m^2 + 15m - 2$

YOUR TURN

Multiply. Watch when distributing a negative number.

11. $(2a+5)(a^2-3a+2)$	12. $(5a-4)(3a^2-2a+7)$

6.05 Multiplying Polynomials 6.05

Courtesy of Fotolia

Since we're up for a challenge, let's multiply two trinomials together !!!

EXAMPLE 11

Multiply. $\left(x^2 + 5x - 2 \right)\left(3x^2 - 2x + 9\right)$

$$\left(x^2 + 5x - 2 \right)\left(3x^2 - 2x + 9\right)$$

$$x^2\left(3x^2 - 2x + 9\right) + 5x\left(3x^2 - 2x + 9\right) - 2\left(3x^2 - 2x + 9\right)$$

$$3x^4 - 2x^3 + 9x^2 + \quad 15x^3 - 10x^2 + 45x \quad - \quad 6x^2 + 4x - 18$$

$$3x^4 + 13x^3 - 7x^2 + 49x - 18$$

YOUR TURN

13. Multiply. $\left(x^2 - 3x + 2\right)\left(x^2 + x + 1\right)$

POLYNOMIAL APPLICATIONS USING MULTIPLICATION

EXAMPLE 12

Find the area of the rectangular wall mural below.

($2x + 5$) feet

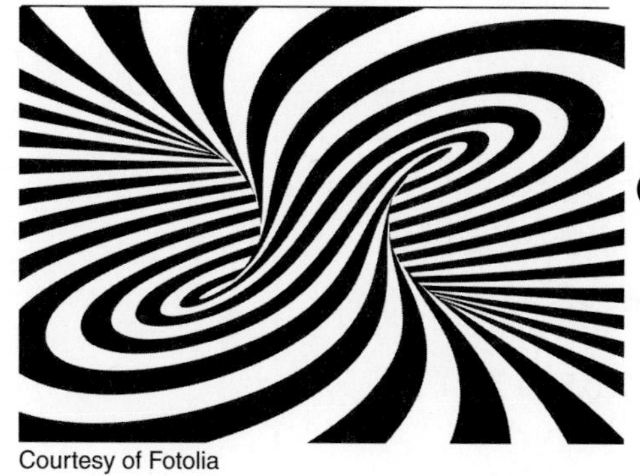

($3x + 1$) feet

Courtesy of Fotolia

To find the area of a rectangle we multiply *length* times *width*.

FOIL.

$(2x + 5)(3x + 1)$

$6x^2 + 2x + 15x + 5$

$6x^2 + 17x + 5$

The area of the wall mural is $(6x^2 + 17x + 5)$ square feet.

YOUR TURN **Answers to Section 6.05.**

1. $-10a^8b^6$ 2. $12x^7y^7$ 3. $-6n^6 + 15n$ 4. $6x^4y - 10x^3y^2 + 14x^2y^3$

5. $2x^2 + 3x - 5$ 6. $10x^2 - 37x + 7$ 7. $2a^2 + 7ab - 15b^2$

8. $4x^2y^2 - 9y^4$ 9. $35 - 4k - 15k^2$ 10. $25m^2 - 30m + 9$

11. $2a^3 - a^2 - 11a + 10$ 12. $15a^3 - 22a^2 + 43a - 28$

13. $x^4 - 2x^3 - x + 2$

Complete MyMathLab Section 6.05 Homework in your Homework notebook.

6.06 Special Products 6.06

The special products in this section are shortcuts for multiplying binomials. Just remember, you can always work out the problems by using the distributive method discussed in the previous section.

SQUARING BINOMIALS

Let's look at two problems and multiply by the method we used in the previous section. Then we will try to look at any patterns that emerge. Remember, when we square something it means to multiply two of those things together. So, $(x+3)^2$ would be the same thing as $(x+3)(x+3)$.

Example 1	Example 2
$(2x+3)^2$	$(5x-3)^2$
$(2x+3)(2x+3)$	$(5x-3)(5x-3)$
$2x(2x+3)+3(2x+3)$	$5x(5x-3)-3(5x-3)$
$4x^2+6x+6x+9$	$25x^2-15x-15x+9$
$4x^2+12x+9$	$25x^2-30x+9$

Let's look a little closer at the two examples above looking for any patterns that we can find.

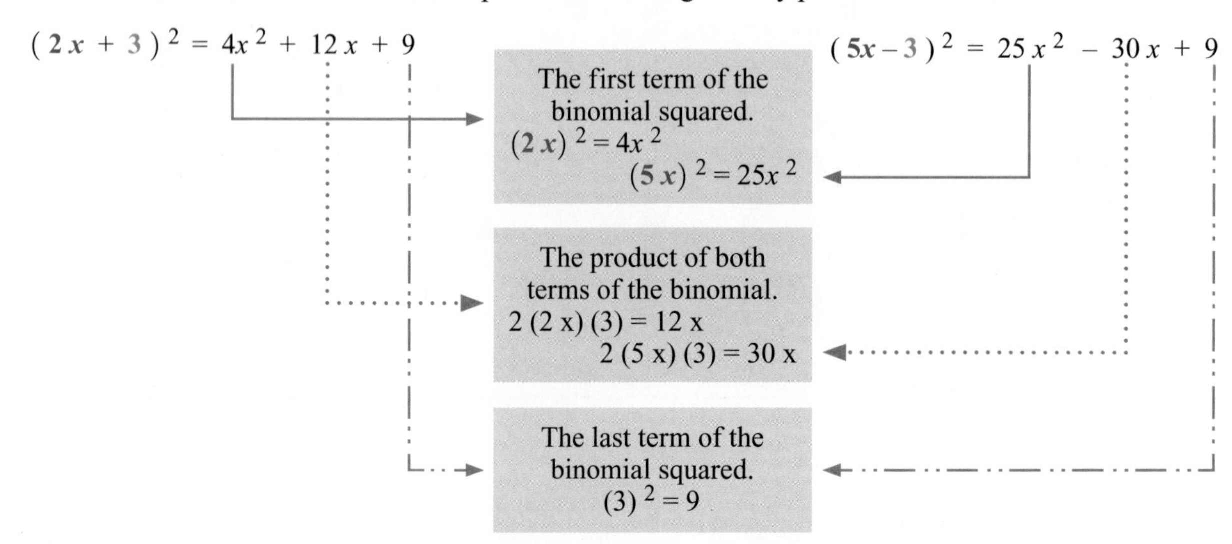

This pattern leads to the following, which can be used when squaring a binomial.

Squaring Binomials

$$(a+b)^2 = a^2 + 2ab + b^2$$

$$(a-b)^2 = a^2 - 2ab + b^2$$

The square of a binomial is a trinomial consisting of the

- square of the first term,
- plus twice the product of the two terms,
- plus the square of the last term of the binomial.

Notice the sign of the middle term is the same as the sign of the binomial.

EXAMPLE 1

Multiply. $(3x+7)^2$

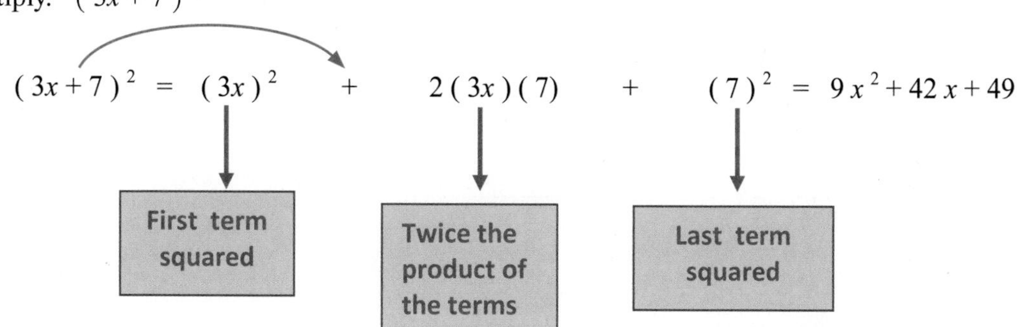

EXAMPLE 2

Multiply. $\left(7x - \frac{2}{9}\right)^2$

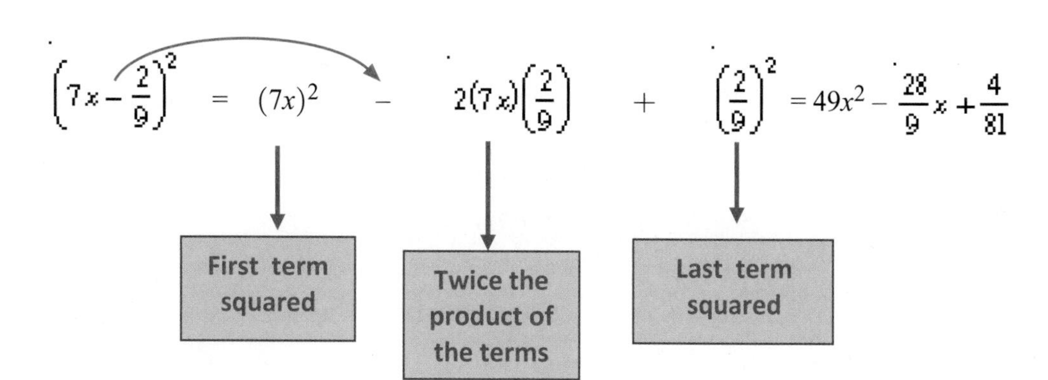

6.06 Special Products 6.06

Now let's see if we can do these steps mentally to make the process even quicker.

EXAMPLE 3

Multiply. $(5x^4 + 7y)^2 = 25x^8 + 70x^4y + 49y^2$

Thinking steps:

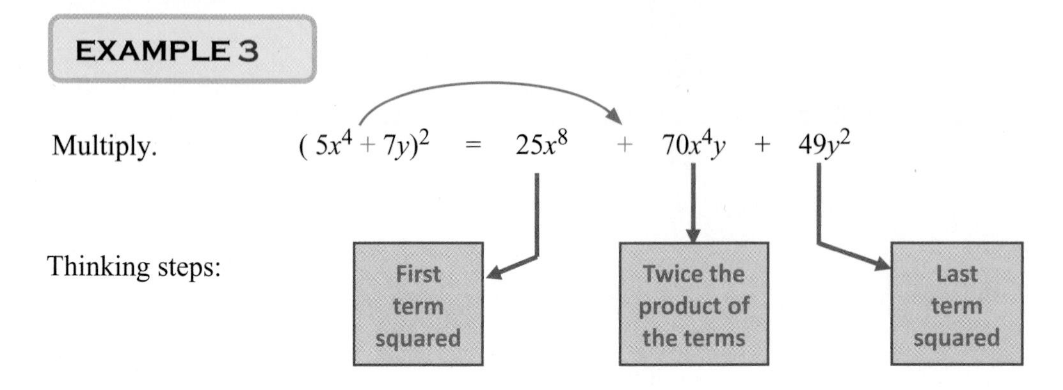

EXAMPLE 4

Multiply. $(5x - 1)^2 = 25x^2 - 10x + 1$

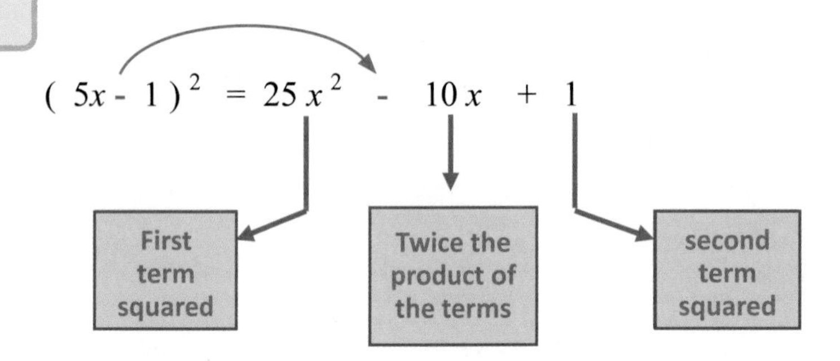

Courtesy of Fotolia

The first and last terms are always positive.
The middle term takes the sign of the binomial.

YOUR TURN

Use the square of a binomial formula to multiply each expression.

1. $(x + 5)^2$	2. $(3x - y)^2$
3. $(5x - 4)^2$	4. $(7m + 3n)^2$

6.06 Special Products 6.06

Courtesy of Fotolia

As a famous poet once said,

"Two roads diverged in a wood, and I took the one less traveled by, and that has made all the difference."

To square a binomial we could travel down 3 roads. You must choose one !

Path 1: Square a Binomial using the Distributive Property

$$
\begin{aligned}
(3x+2)^2 &= (3x+2)(3x+2) \\
&= 3x\,(3x+2) + 2\,(3x+2) \\
&= 9x^2 + 6x + 6x + 4 \\
&= 9x^2 + 12x + 4
\end{aligned}
$$

Path 2: Square a Binomial by using the shortcut pattern.

$$(3x+2)^2 = 9x^2 + 12x + 4$$

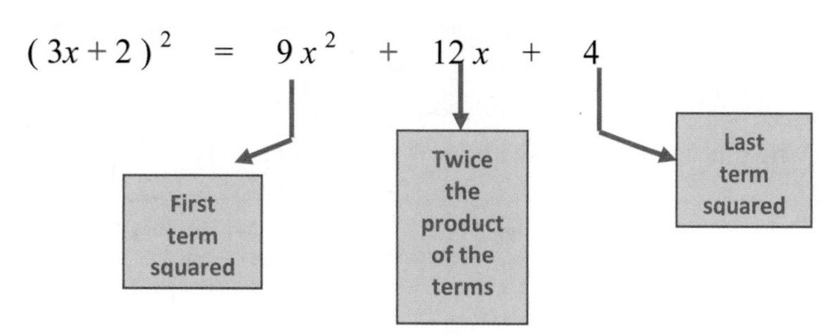

Path 3: Square a Binomial visually by using the square diagram.

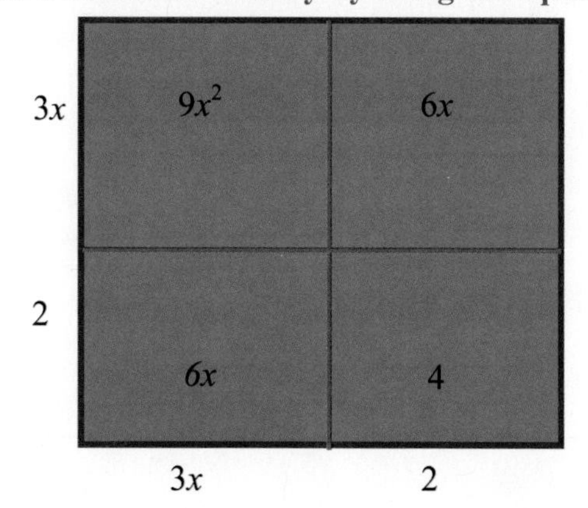

$$
\begin{aligned}
(3x+2)^2 &= 9x^2 + 6x + 6x + 4 \\
&= 9x^2 + 12x + 4
\end{aligned}
$$

6.06 Special Products 6.06

Whichever road you choose, just remember that when you **SQUARE A BINOMIAL** your answer will always be a **TRINOMIAL**!!!

Remember this warning from the previous section.

Courtesy of Fotolia

You cannot square a binomial by squaring the terms!!!!!

$$(2x-7)^2 \neq 4x^2+49$$

Let's say it again!!!!!

$$(2x-7)^2 \neq 4x^2+49$$

Or in words:

$(2x-7)^2$ is not equal to and will never be equal to $4x^2+49$

YOUR TURN

Square the following binomials by the method of your choice.

5. $(7m - 5)^2$	6. $(5x - 9y)^2$

6.06 Special Products 6.06

SUM AND DIFFERENCE OF TWO TERMS

In binomial products of the form $(a+b)(a-b)$, one binomial is the sum of two terms, and the other is the difference of the *same* two terms.

Let's look at two problems and multiply using the distributive property. Then we will try to look at any patterns that emerge.

Problem 1	Problem 2
$(x+5)(x-5)$	$(2x-3)(2x+3)$
$x(x-5)+5(x-5)$	$2x(2x+3)-3(2x+3)$
$x^2-5x+5x-25$	$4x^2+6x-6x-9$
x^2-25	$4x^2-9$

Notice that the middle terms cancel out in both examples. Our answer in both cases is the difference of two squares.

When you multiply two binomials that look exactly the same except one binomial has addition and the other binomial has subtraction , then your answer will be the difference of two squares. The two squares come from squaring the first term of the binomial and squaring the second term of the binomial.

Difference of Two Squares

$$(a+b)(a-b)=a^2-b^2$$

The product of the sum and the difference of two terms is the square of the first term minus the square of the second term.

EXAMPLE 5

Multiply. $(x+4)(x-4) = x^2 - 16$

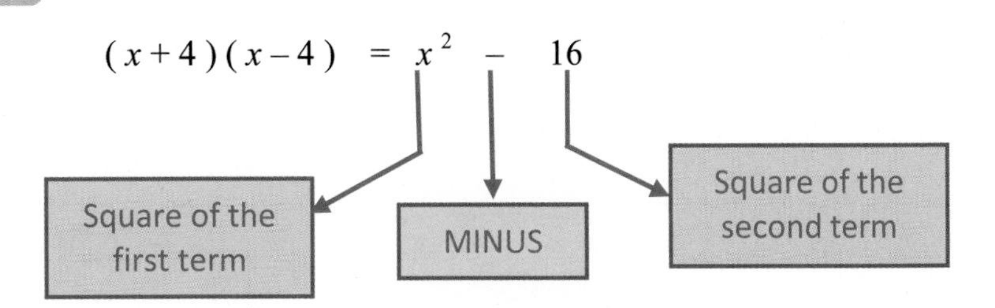

EXAMPLE 6

Multiply. $(5 - 2w)(5 + 2w) = 25 - 4w^2$

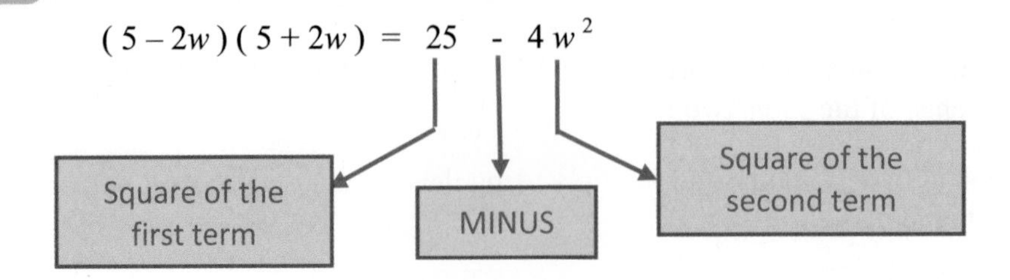

EXAMPLE 7

Multiply. $(3x^2 - 5y)(3x^2 + 5y) = 9x^4 - 25y^2$

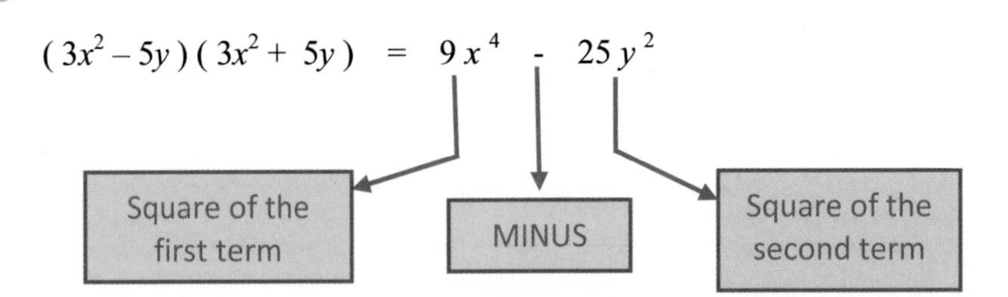

YOUR TURN

Multiply.

7. $(2x + 7)(2x - 7)$	8. $(6 - 5n)(6 + 5n)$
9. $(6x - 2y)(6x + 2y)$	10. $(x^2 - \frac{1}{2}y)(x^2 + \frac{1}{2}y)$

6.06 Special Products 6.06

Here is a summary of the three "special products" we have looked at. Make sure you can tell the difference between each of these products.

Courtesy of Fotolia

$(2x+3)^2 = 4x^2 + 12x + 9$	*Square a sum*
$(2x-3)^2 = 4x^2 - 12x + 9$	*Square a difference*
$(2x+3)(2x-3) = 4x^2 - 9$	*Multiply a sum times a difference*

YOUR TURN

Multiply.

11. $(5x-2)^2$	12. $(5x+2)(5x-2)$	13. $(5x+2)^2$

YOUR TURN Answers to Section 6.06.

1. $x^2 + 10x + 25$
2. $9x^2 - 6xy + y^2$
3. $25x^2 - 40x + 16$
4. $49m^2 + 42mn + 9n^2$
5. $49m^2 - 70m + 25$
6. $25x^2 - 90xy + 81y^2$
7. $4x^2 - 49$
8. $36 - 25n^2$
9. $36x^2 - 4y^2$
10. $x^4 - \frac{1}{4}y^2$
11. $25x^2 - 20x + 4$
12. $25x^2 - 4$
13. $25x^2 + 20x + 4$

Complete MyMathLab Section 6.06 Homework in your Homework notebook.

6.07 Dividing Polynomials 6.07

Now that we know how to add, subtract, and multiply polynomials, we will look at the last operation on polynomials – division.

Recall from Module 2 that when we add two fractions with a common denominator , we keep the denominator and add the numerators. Look at the example below:

$$\frac{3}{7} + \frac{2}{7} = \frac{3+2}{7}$$

If we read the above equation from right to left, we have $\frac{3+2}{7} = \frac{3}{7} + \frac{2}{7}$.

We will use this fact to divide polynomials.

To Divide a Polynomial by a Monomial

- Divide each term of the polynomial by the monomial.

EXAMPLE 1

Divide. $\left(6\text{m}^2 + 2m\right) \div 2m$ — **Split the numerator by dividing each term by the denominator.**

$\frac{6m^2 + 2m}{2m} = \frac{6m^2}{2m} + \frac{2m}{2m}$ — **Use the quotient rule for exponents from 6.01 which tells us to divide the coefficients and subtract the exponents.**

$= 3m + 1$ — **DON'T FORGET THE "1" at the end. Don't throw it away!**

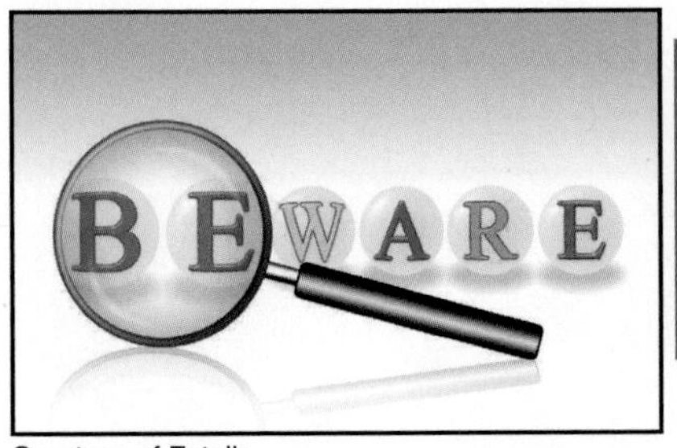

Courtesy of Fotolia

A very common mistake students make is to leave off the 1 at the end. Keep in mind that $3m$ is not the same answer as $3m + 1$. So keep up with those 1's in your answers!!!

EXAMPLE 2

Divide. $\dfrac{9m^5 - 12m^2 + 3m}{3m^2}$

Courtesy of Fotolia

$$\frac{9m^5 - 12m^2 + 3m}{3m^2} = \frac{9m^5}{3m^2} - \frac{12m^2}{3m^2} + \frac{3m}{3m^2}$$

$$= 3m^3 - 4 + \frac{1}{m}$$

Be careful with the last term.

$\dfrac{3m}{3m^2} = \dfrac{1}{m}$ $\qquad \dfrac{3m}{3m^2} \neq m$

EXAMPLE 3

Divide. $10a^5b^4 - 2a^3b^2 + 6a^2b$ *by* $\left(-2a^2b\right)$

$$\frac{10a^5b^4 - 2a^3b^2 + 6a^2b}{\left(-2a^2b\right)} = \frac{10a^5b^4}{-2a^2b} - \frac{2a^3b^2}{-2a^2b} + \frac{6a^2b}{-2a^2b}$$

We divide coefficients and subtract exponents.

$$= -5a^3b^3 + ab - 3$$

Courtesy of Fotolia

Dividing by a negative will change the sign of every term of the polynomial.

EXAMPLE 4

Divide. $\dfrac{6a^2 - 5a + 12}{-2}$

$$= \frac{6a^2}{-2} - \frac{5a}{-2} + \frac{12}{-2} = -3a^2 + \frac{5a}{2} - 6$$

Dividing by a negative will change the sign of every term of the polynomial.

6.07 Dividing Polynomials 6.07

YOUR TURN

Divide.

1. $\dfrac{6m^2 - 9m + 12}{3}$	2. $\dfrac{6a^3 - 4a^2 + 12a}{-2a}$
3. $\dfrac{12x^5y^4 + 9x^3y - 3xy}{3xy}$	4. $\dfrac{-15m^3 - 10m^2 + 6m}{-5m}$

YOUR TURN **Answers to Section 6.07.**

1. $2m^2 - 3m + 4$
2. $-3a^2 + 2a - 6$
3. $4x^4y^3 + 3x^2 - 1$
4. $3m^2 + 2m - \dfrac{6}{5}$

Complete MyMathLab Section 6.07 Homework in your Homework notebook.

6.08 The Greatest Common Factor 6.08

FACTORING A POLYNOMIAL

We have already studied in a previous section how to multiply polynomials. We now turn our attention to a process called **factoring.** To factor means to write as a product.

Factoring is a process that "undoes" multiplying.

We multiply 6 • 2 to get 12, but we factor 12 by writing it as 6 • 2. Look at the following examples. Factoring is the reverse process of multiplication !!

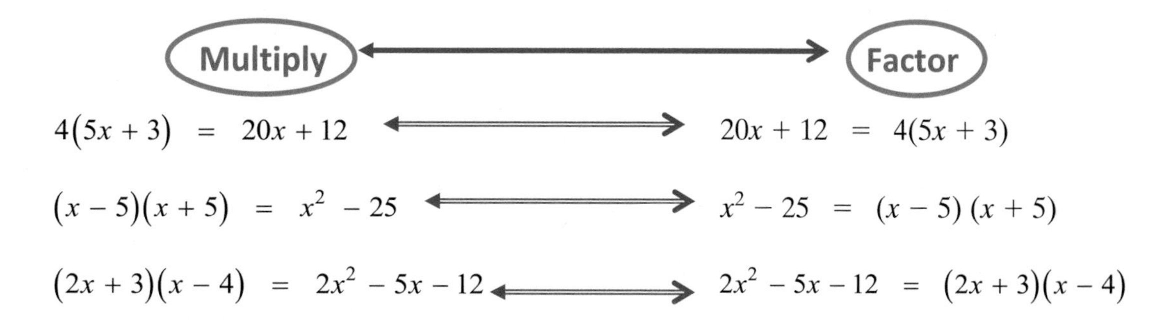

Courtesy of Fotolia

When you are asked to "factor," you have to figure what was multiplied to produce the expression you are given. It can be like trying to find what ingredients went into a cake to make it so delicious. It is sometimes not obvious at all!!!

Courtesy of Fotolia

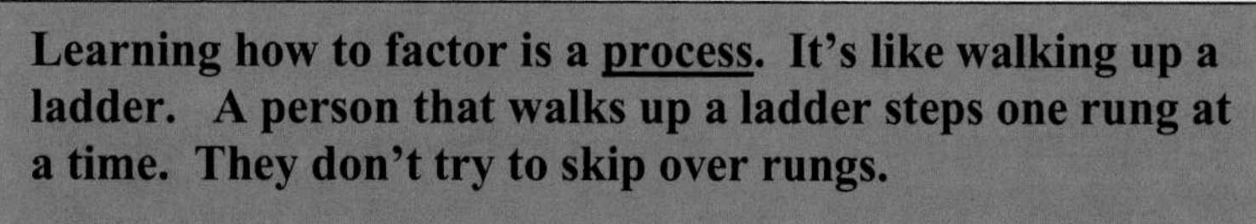

Learning how to factor is a <u>process</u>. It's like walking up a ladder. A person that walks up a ladder steps one rung at a time. They don't try to skip over rungs.

You will be much more successful at factoring if you follow all the steps given and not try to " skip steps" or take " shortcuts."

***Please study each step and make sure you understand before proceeding to the next step. Factoring depends on applying what you have learned in each section.**

6.08 The Greatest Common Factor 6.08

GREATEST COMMON FACTOR: The first step in factoring is to see whether the terms of the polynomial have a common factor. We will always want to factor out the greatest common factor (GCF) first to make problems simpler.

Let's review how to find the greatest common factor.

EXAMPLE 1

Find the greatest common factor of 30 and 45.

First write each number as a product of primes. (This can be found in Module 2 if you need additional help.)

$$30 = 2 \cdot 3 \cdot 5$$

$$45 = 3 \cdot 3 \cdot 5$$

The greatest common factor will be the product of all the numbers each factorization has in common. Looking at the factors of 30 and 45, we see that they both have a 3 and a 5. So, the GCF for 30 and 45 will be the product of 3 and 5 which is 15.

GCF of 30 and 45 is 15.

EXAMPLE 2

Find the greatest common factor of 15, 18 and 66.

First write each number as a product of primes. (This can be found in Module 2 if you need additional help.)

$$15 = 3 \cdot 5$$

$$18 = 2 \cdot 3 \cdot 3$$

$$66 = 2 \cdot 3 \cdot 11$$

The only factor common to all three numbers is 3.

GCF of 15, 18, and 66 is 3.

What about those variables? We need to also look at how to find the GCF if given variables.

EXAMPLE 3

Find the GCF of x^3, x^5, and x^2.

First we will write each as the product of primes just like in the previous examples.

$$x^3 = x \cdot x \cdot x$$

$$x^5 = x \cdot x \cdot x \cdot x \cdot x$$

$$x^3 = x \cdot x$$

There are two common factors, each of which is x. So, the GCF $= x \cdot x = x^2$.

EXAMPLE 4

Find the GCF of x^2y^4, xy^2, *and* x^4y^2.

First we will write each as the product of primes just like in the previous examples.

$$x^2y^4 = x \cdot x \cdot y \cdot y \cdot y \cdot y$$

$$xy^2 = x \cdot y \cdot y$$

$$x^4y^2 = x \cdot x \cdot x \cdot x \cdot y \cdot y$$

The GCF is the product of what each term has in common with the others.

GCF $= x \cdot y \cdot y = xy^2$.

6.08 The Greatest Common Factor 6.08

Courtesy of Fotolia

The greatest common factor (GCF) of a list of terms contains the smallest exponent on each common variable.

The GCF of x^5y^6, x^4y^9 and x^3y^7 is x^3y^6

smallest exponent on x

smallest exponent on y

EXAMPLE 5

Find the greatest common factor of $12x^3$, $6x^2$, *and* $18x^5$.

$$12x^3 = 2 \cdot 2 \cdot 3 \cdot x^3$$
$$6x^2 = 2 \cdot 3 \cdot x^2$$
$$18x^5 = 2 \cdot 3 \cdot 3 \cdot x^5$$

The common factors in the coefficients are $2 \cdot 3$ which is 6.

The least exponent on x is 2.

GCF = $6x^2$

EXAMPLE 6

Find the greatest common factor of $21m^7n$, $18mn^2$, *and* $45m^6$.

$$21m^7n = 3 \cdot 7 \cdot m^7 \cdot n$$
$$18mn^2 = 2 \cdot 3 \cdot 3 \cdot m \cdot n^2$$
$$45m^6 = 3 \cdot 3 \cdot 5 \cdot m^6$$

The common factors in the coefficients is 3. The least exponent on ***m* is 1.** There is no n in the last term, so n will not appear in the GCF.

GCF = $3m$

EXAMPLE 7

Find the greatest common factor of $-18y^2$, $-63y^3$, *and* $27y^4$.

$$-18y^2 = -1 \cdot 2 \cdot 3 \cdot 3 \cdot y^2$$
$$-63y^3 = -1 \cdot 3 \cdot 3 \cdot 7 \cdot y^3$$
$$27y^4 = 3 \cdot 3 \cdot 3 \cdot y^4$$

The common factors in the coefficients are $3 \cdot 3$ which is 9.

The least exponent on ***y* is 2.**

GCF = $9y^2$

6.08 The Greatest Common Factor 6.08

YOUR TURN

Find the greatest common factor of each list of terms by listing out the factors.

1. 14, 24, and 60	2. x^3y, x^2y^2, and x^4y^5	3. $10x^3$, $5x$, *and* $15x^5$
14 = 24 = 60 = GCF = ________	x^3y = x^2y^2 = x^4y^5 = GCF = ________	$10x^3$ = $5x$ = $15x^5$ = GCF = ________
4. $21m^4n^2$, $12mn^2$, *and* $15m^6n$	5. $2x^7y^5$, $8xy^2$, *and* $4x^6$	6. $-18m^3$, $-54m^7$, *and* $24m^4$
$21m^4n^2$ = $12mn^2$ = $15m^6n$ = GCF = ________	$2x^7y^5$ = $8xy^2$ = $4x^6$ = GCF = ________	$-18m^3$ = $-54m^7$ = $24m^4$ = GCF = ________

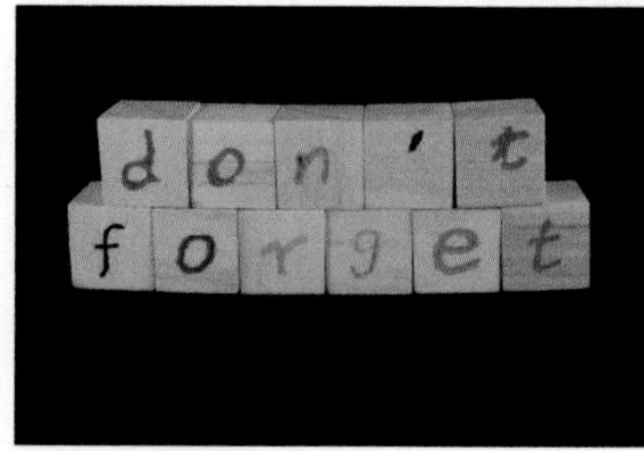

Courtesy of Fotolia

The GCF of the coefficients (numbers) will be the product of all the numbers each factorization has in common.

The GCF of a list of terms contains the smallest exponent on each common variable.

6.08 The Greatest Common Factor 6.08

Now that we know how to find the GCF of a list of terms, let's look at how this will be used to factor a polynomial. The first step in factoring a polynomial is to factor out the greatest common factor (GCF). Begin each and every factoring problem by asking the question, **"Do all the terms have a common factor?"**

Courtesy of Fotolia

Factoring a Polynomial – Step 1

Factor out the greatest common factor of the polynomial.

EXAMPLE 8

Factor. $3m + 12$

$ab + ac = a(b + c)$
To factor is to use the distributive property backwards.

This polynomial has 2 terms: $3m$ and 12.
The greatest common factor of these two terms is 3.

We can rewrite $3m + 12$ as a product with 3 as one of the factors.

$$3m + 12 = 3 \bullet m + 3 \bullet 4$$
$$= 3(m + 4)$$

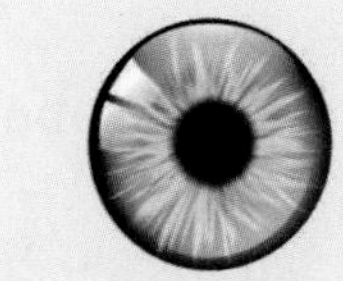

Courtesy of Fotolia

Here's another way to look at it!!!

$$3m + 12 = \frac{3m}{3} + \frac{12}{3}$$ Divide each term by the GCF which is 3.

$$= 3(m + 4)$$ GCF goes outside the ( ).
The answer after you divide goes inside the ( ).

$$3m + 12 = 3(m + 4)$$

This process is called factoring out the greatest common factor (GCF).

EXAMPLE 9

Factor out the greatest common factor. $20m^3 - 10m^2 + 5m$

This polynomial has 3 terms $20m^3, -10m^2,$ *and* $5m$
The greatest common factor of these three terms is $5m$.

Rewrite $20m^3 - 10m^2 + 5m$ as a product with 5m as one of the factors.

$$20m^3 - 10m^2 + 5m = \frac{20m^3}{5m} - \frac{10m^2}{5m} + \frac{5m}{5m}$$

Divide each term by the GCF which is $5m$.

$$= 5m\left(4m^2 - 2m + 1\right)$$

The GCF goes outside the ( ).
The answer when you divide goes inside the ( ).

$$20m^3 - 10m^2 + 5m = 5m\left(4m^2 - 2m + 1\right)$$

This process is called factoring out the GCF.

Courtesy of Fotolia

Be sure to include the 1 in the example above.

CORRECT: $20m^3 - 10m^2 + 5m = 5m\left(4m^2 - 2m + 1\right)$

INCORRECT: $20m^3 - 10m^2 + 5m \neq 5m\left(4m^2 - 2m\right)$

If a term of the polynomial is exactly the same as the GCF, when you divide it by the GCF you are left with 1, NOT 0.

Courtesy of Fotolia

6.08 The Greatest Common Factor 6.08

Remember this statement: **Factoring is a process that " undoes" multiplying.**

This should remind us that we can check any answer by multiplying.

In the last example we **factored** the polynomial : $20m^3 - 10m^2 + 5m = 5m(4m^2 - 2m + 1)$.

Now let's check by **multiplying** .

$$5m(4m^2 - 2m + 1) = 5m \cdot 4m^2 - 5m \cdot 2m + 5m \cdot 1$$

$$= 20m^3 - 10m^2 + 5m$$

This polynomial equals our original problem.

EXAMPLE 10

Factor out the GCF. $4x^5y^3 + 12x^5y^2 - 28x^4y^2 - 4x^3y$

GCF = $4x^3y$

To factor a polynomial completely, we must always factor out the **greatest** common factor and not just a common factor. In this example, we should factor out the GCF which is $4x^3y$ and not just any common factor such as $2x^3y$.

$$4x^5y^3 + 12x^5y^2 - 28x^4y^2 - 4x^3y = \frac{4x^5y^3}{4x^3y} + \frac{12x^5y^2}{4x^3y} - \frac{28x^4y^2}{4x^3y} - \frac{4x^3y}{4x^3y}$$

$$= 4x^3y(x^2y^2 + 3x^2y - 7xy - 1)$$

Don't forget the 1!!!!!

EXAMPLE 11

Sometimes you may be asked to factor out a negative one from a polynomial. Let's look at an example. Factor a "-1" from each polynomial.

a. $-x + 7 = \frac{-x}{-1} + \frac{7}{-1}$

$= -1(x - 7)$

b. $-y - 3 = \frac{-y}{-1} - \frac{3}{-1}$

$= -1(y + 3)$

Notice the change in sign when factoring out a negative factor.

6.08 The Greatest Common Factor 6.08

EXAMPLE 12

Factor out the greatest common factor. $-6x^3 + 3x^2 - 3x$

We can factor out either $3x$ or $-3x$. We will choose to factor out a $-3x$ so that the coefficient of the first term in the trinomial will be positive.

$$-6x^3 + 3x^2 - 3x = \frac{-6x^3}{-3x} + \frac{3x^2}{-3x} - \frac{3x}{-3x}$$

$$= -3x\left(2x^2 - x + 1\right)$$

Don't forget!

Courtesy of Fotolia

Whenever we factor a polynomial in which the coefficient of the first term of the polynomial is *negative*, we will factor out the negative common factor, even if it is just -1.

YOUR TURN

Factor out the greatest common factor.

7. $5m^3 - 15m^2$	8. $12x^3 + 16x^2 - 4x$
9. $4x^2y^3 - 6x^2y + 8xy$	10. Factor a -1 from the polynomial. $-x + 3y$

6.08 The Greatest Common Factor 6.08

Let's look at two more examples of factoring out the GCF.

EXAMPLE 13

Factor out the greatest common factor.
$5(x+3)+y(x+3)$.

The binomial $(x+3)$ is present in both terms and is the greatest common factor. We use the distributive property to factor out $(x+3)$.

$$5(x+3)+y(x+3)$$
$$(5+y)(x+3)$$

EXAMPLE 14

Factor out the greatest common factor.
$x(y^4-2)-4(y^4-2)$

The binomial (y^4-2) is present in both terms and is the greatest common factor. We use the distributive property to factor out (y^4-2).

$$x(y^4-2)-4(y^4-2)$$
$$(x-4)(y^4-2)$$

EXAMPLE 15

Factor out the greatest common factor.
$a(b-3)+(b-3)$.

The binomial $(b-3)$ is present in both terms and is the greatest common factor. We use the distributive property to factor out $(b-3)$.

$$a(b-3)+(b-3)$$
$$a(b-3)+1(b-3)$$
$$(a+1)(b-3)$$

Remember the 1.

YOUR TURN

Factor out the greatest common binomial factor.

11. $a(a+3)-4(a+3)$	12. $2m(m-7)+(m-7)$

FACTORING BY GROUPING: Polynomials with 4 terms.

Once the GCF is factored out, we can proceed to a process called grouping. This is the process we will use to factor polynomials that have 4 terms.

EXAMPLE 16

Factor the four-term polynomial by grouping. $2x + 6 + ax + 3a$

We will " group" the first two terms and the last two terms, since the **first two terms have a common factor of 2** and the **last two terms have a common factor of *a*.**

$2x + 6 + ax + 3a = \underline{2x + 6} + \underline{ax + 3a}$ Group the terms by underlining.

$= 2(x + 3) + a(x + 3)$ Factor out the common factor in each piece.

$= (2 + a)(x + 3)$ Note the common factor. Factor again by pulling out this factor.

In other words, the outside terms make a (new binomial).

The order of the parentheses in the answer does not matter!!

$$(2 + a)(x + 3) = (x + 3)(2 + a)$$

EXAMPLE 17

Factor the four-term polynomial by grouping.

$2a^2 + 5ab + 2a + 5b$

$2a^2 + 5ab + 2a + 5b = \underline{2a^2 + 5ab} + \underline{2a + 5b}$ Group the terms by underlining.

$= a(2a + 5b) + 1(2a + 5b)$ Factor out the common factor in each piece.

$= (2a + 5b)(a + 1)$ Note the common factor. Factor again by pulling out this factor.

Check by multiplying:

$$\begin{aligned}(2a + 5b)(a + 1) &= 2a(a + 1) + 5b(a + 1)\\ &= 2a^2 + 2a + 5ab + 5b\\ &= 2a^2 + 5ab + 2a + 5b\end{aligned}$$

6.08 The Greatest Common Factor 6.08

Courtesy of Fotolia

Factoring a Polynomial – Step 2

If a polynomial has four terms, try factoring by grouping.

Courtesy of Fotolia

Factoring a Polynomial with Four Terms by Grouping

Step 1: **Group Terms.** Collect the terms into two groups so that each group has a common factor.

Step 2: **Factor within groups.** Factor out the greatest common factor from each group.

Step 3: **Factor the entire polynomial.** Factor a common binomial factor from the results of Step 2.

Step 4: **If necessary, rearrange middle terms** and try again.

EXAMPLE 18

Factor by grouping $3x^2 + 4xy - 3x - 4y$

$$3x^2 + 4xy - 3x - 4y = \underline{3x^2 + 4xy} - \underline{3x - 4y}$$
$$= x(3x + 4y) - 1(3x + 4y)$$
$$= (3x + 4y)(x - 1)$$

Make sure and bring down the sign that connects the two underlined groups !

Check by Multiplying: $(3x + 4y)(x - 1)$

$$(3x + 4y)(x - 1) = 3x(x - 1) + 4y(x - 1)$$
$$= 3x^2 - 3x + 4xy - 4y$$
$$= 3x^2 + 4xy - 3x - 4y$$

Courtesy of Fotolia

When the sign between the two underlined groups is **subtraction**, it means you are factoring out a negative factor. **So, the sign in the second group will change to its opposite.**

6.08 The Greatest Common Factor 6.08

We will look at one last example.

EXAMPLE 19

Factor out the GCF from the polynomial. Then factor the four-term polynomial by grouping.

$$4x^4 + 4x^3 - 20x^2 - 20x$$

Step 1: Factor out the GCF. The GCF of all 4 terms is $4x$.

$$4x^4 + 4x^3 - 20x^2 - 20x = 4x\left(x^3 + x^2 - 5x - 5\right)$$

Step 2: Group the polynomial inside the parentheses : $\left(x^3 + x^2 - 5x - 5\right)$

$$\left(x^3 + x^2 - 5x - 5\right) = \underline{x^3 + x^2} - \underline{5x - 5}$$

$$= x^2(x+1) - 5(x+1)$$

$$= (x+1)\left(x^2 - 5\right)$$

Notice the sign change in the second group !!!!

For the final answer, don't forget to include the GCF you factored out in the first step.

$$4x^4 + 4x^3 - 20x^2 - 20x = 4x(x+1)\left(x^2 - 5\right)$$

Courtesy of Fotolia

Use negative signs carefully when grouping as in the previous example, or a sign error will occur. Always check by multiplying.

Courtesy of Fotolia

Factoring A Polynomial

Step 1: Factor out the GCF first.

Step 2: If the polynomial has 4 terms, try factoring by grouping.

6.08 The Greatest Common Factor 6.08

YOUR TURN

Factor completely. (Use the Grouping Method.)

13. $5m + mn + 20 + 4n$	14. $3xy + 9x + y + 3$
15. $7z^2 + 14z - az - 2a$	16. $12x^2y - 42x^2 - 4y + 14$ (Factor out the GCF first and then factor the remaining polynomial.)

YOUR TURN **Answers to Section 6.08.**

1. 2
2. x^2y
3. $5x$
4. $3mn$
5. $2x$
6. $6m^3$
7. $5m^2(m-3)$
8. $4x(3x^2+4x-1)$
9. $2xy(2xy^2-3x+4)$
10. $-1(x-3y)$
11. $(a-4)(a+3)$
12. $(2m+1)(m-7)$
13. $(m+4)(5+n)$
14. $(y+3)(3x+1)$
15. $(7z-a)(z+2)$
16. $2(3x^2-1)(2y-7)$

Complete MyMathLab Section 6.08 Homework in your Homework notebook.

(Stay on our track. We're moving on to Section 6.11 next.)

FACTORING TRINOMIALS BY GROUPING

Recall that factoring is the reverse process of multiplication. So, before we look at how to factor trinomials, let's look at a multiplication problem first to see how trinomials were formed.

Multiply. $(2x+3)(x+5)$

$$(2x+3)(x+5) = 2x(x+5) + 3(x+5)$$
$$= 2x^2 + 10x + 3x + 15$$
$$= 2x^2 + 13x + 15$$

Notice that we combine the two terms in the middle.
$10x + 3x = 13x$

Courtesy of Fotolia

When we **MULTIPLY** two binomials, we end up with TWO center terms which will combine to make the middle term in the answer.

When we **FACTOR**, it makes sense to find these center terms to help us "see" how to factor trinomials correctly.

To factor a trinomial by grouping, the goal is to find a way to split the middle term into two appropriate terms so that the resulting 4-term polynomial can be factored by grouping.

Take the example above. We multiplied $(2x+3)(x+5)$ and the answer was $2x^2+13x+15$.
Now let's look at that process <u>in reverse</u>. In other words, how does a trinomial sum become factors of two binomials?

Courtesy of Fotolia

Factor. $2x^2 + 13x + 15$

$$2x^2 + 13x + 15$$
$$= 2x^2 + 10x + 3x + 15$$
$$= 2x(x+5) + 3(x+5)$$
$$= (2x+3)(x+5)$$

Factoring is changing a sum or difference to a product.

$$2x^2 + 13x + 15 = (2x+3)(x+5)$$

SUM FACTORS

6.11 Factoring Trinomials By Grouping 6.11

FACTORING TRINOMIALS in the FORM

$$ax^2 + bx + c$$

Follow this process to learn to factor trinomials. It'll work EVERY time a quadratic trinomial will factor.

EXAMPLE 1

Factor. $3x^2 + 10x + 8$ (This is a trinomial where $a = 3$, $b = 10$, and $c = 8$.)

Step 1: $3x^2 + 10x + 8$ Multiply the leading coefficient and the constant term.

$$3 \cdot 8 = 24$$

Step 2: List all the factors of 24. (List all the pairs of numbers that will multiply to 24).

Factors of 24
1 • 24
2 • 12
3 • 8
4 • 6

Step 3: From the list of factors, find the one pair that adds to the middle term's coefficient. For this example, we need to find a sum of 10.

$3x^2 + (10x) + 8$

Factors of 24
1 • 24
2 • 12
3 • 8
(4 • 6)

$$4 + 6 = 10$$

Step 4: Rewrite the middle term, forming two terms, using these two values. Order is not important!

$$3x^2 + 10x + 8 = 3x^2 + 4x + 6x + 8$$

$$= x(3x + 4) + 2(3x + 4)$$

$$= (3x + 4)(x + 2)$$

A good check point is that the two factors in the parentheses must be identical.

6.11 Factoring Trinomials By Grouping 6.11

EXAMPLE 2

Factor. $4m^2 + m - 3$ (This is a trinomial where $a = 4$, $b = 1$, and $c = -3$.)

Step 1: $4m^2 + m - 3$ Multiply the leading coefficient and the constant term.

$$4 \cdot -3 = -12$$

Step 2: List all the factors of -12. (List all the pairs of numbers that will multiply together to get -12).

Factors of -12
$1 \cdot -12$ *or* $-1 \cdot 12$
$2 \cdot -6$ *or* $-2 \cdot 6$
$3 \cdot -4$ *or* $-3 \cdot 4$

Step 3: From the list of factors, find the one pair that adds to the middle term's coefficient. For this example, we need to find a sum of 1.

$$4m^2 + m - 3 = 4m^2 + 1m - 3$$

Recall that if you don't see a coefficient, it is 1.

Step 4: Rewrite the middle term, forming two terms, using these two values. The first and last terms will not change. Order is not important!

$$4m^2 + m - 3 = 4m^2 + 4m - 3m - 3$$
$$= 4m(m + 1) - 3(m + 1)$$
$$= (m + 1)(4m - 3)$$

A good check point is that the two factors in the parentheses must be identical.

Just to prove order doesn't matter, let's factor the same trinomial and reverse the order of the two middle terms. It factors into the same two binomials!!!!

$$4m^2 + m - 3 = 4m^2 - 3m + 4m - 3$$
$$= m(4m - 3) + 1(4m - 3)$$
$$= (4m - 3)(m + 1)$$

6.11 Factoring Trinominals By Grouping 6.11

FACTORING TRINOMIALS USING THE "*ac*" METHOD

$$ax^2 + bx + c$$

(Standard Form of a Trinomial)

1) **Arrange** the terms in descending order.
2) Factor out the GCF common to all terms if possible.
3) **Multiply "*a*" times "*c*,"** first & last coefficients. This is your product.
4) Determine all the **factors that multiply to your product**.
5) Pick the **pair of factors that will also add up to your middle** term.
6) Copy down the first term and last term, and put your selected pair of factors in the middle. Remember, order doesn't matter!
7) Now you have 4 terms, so you will finish factoring by **grouping**.

Courtesy of Fotolia

Finding Factors Using Your Calculator.

If the product is a larger number, you may wish to use your calculator to help you generate the table of factors. To make sure you don't miss any factors, you will need to divide your number by every number starting at 1 and continuing until you repeat a factor.

6.11 Factoring Trinomials By Grouping 6.11

YOUR TURN

Complete the steps to factor each trinomial by grouping.

1. $2m^2 + 11m + 12$	2. $6y^2 - 19y + 10$
a. Find two integers whose product is _____•______ = ______ and whose sum is _______.	a. Find two integers whose product is _____•______ = ______ and whose sum is _______.
b. The required integers are _____ and ______.	b. The required integers are _____ and ______.
c. Write the middle terms $11m$ as ________ + __________.	c. Write the middle terms $-19y$ as ________ + __________.
d. Rewrite the given trinomial using four terms.	d. Rewrite the given trinomial using four terms.
e. Factor the polynomial in part d by grouping.	e. Factor the polynomial in part d by grouping.
f. Check by multiplying.	f. Check by multiplying.

(See instructor/tutor for answers.)

6.11 Factoring Trinomials By Grouping 6.11

EXAMPLE 3

Factor. $6x^2 - 2x - 20$ 1. The terms are in order.

$2(3x^2 - x - 10)$ 2. Factor out the GCF of 2.

3. Next notice that $a = 3$, $b = -1$, and $c = -10$ in the resulting trinomial. Find two numbers whose product is $a \cdot c$ or $3 \cdot -10 = -30$ and whose sum is b which is - 1 .

Product = -30 Sum = -1

Finding the factor pair when the product is negative:
→ List the factor pairs of -30.
→ Choose the factor pair: When the product is negative, the pair will have opposite signs, so you will subtract the numbers to make the middle term.
→ Attach the signs to the pair: the larger number in the pair will keep the sign of the middle term.

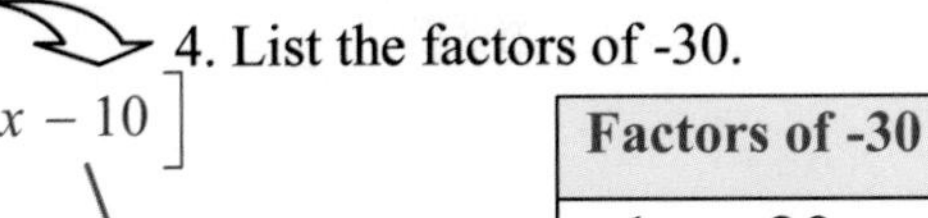

4. List the factors of -30.

Factors of -30
$1 \cdot -30$ *or* $-1 \cdot 30$
$2 \cdot -15$ *or* $-15 \cdot 2$
$3 \cdot -10$ *or* $-3 \cdot 10$
$5 \cdot -6$ *or* $-5 \cdot 6$

$2[3x^2 - x - 10]$

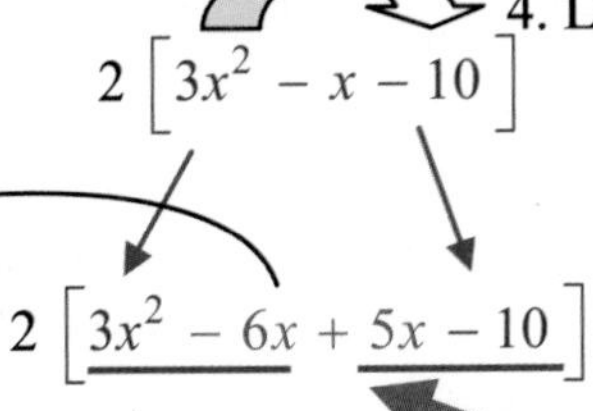

5. The first and last terms stay the same, $3x^2$ and -10.

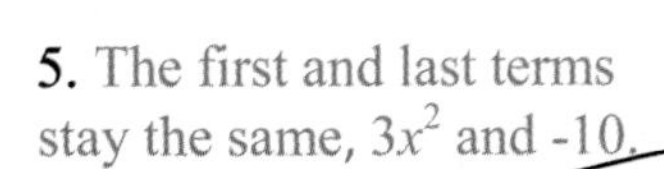

6. Rewrite the middle term using the factor pair from the chart, $5x$ and $-6x$. $2[3x^2 - 6x + 5x - 10]$

7. Factor the 4 terms by grouping. $2[3x(x-2) + 5(x-2)]$

$2[(x-2)(3x+5)]$

$$6x^2 - 2x - 20 = 2(x-2)(3x+5)$$

Don't forget to put the 2 (GCF) back in front for your final answer!!

In example 3, removing the GCF from the originial trinomial produced a new trinomial with smaller coefficients. This makes the factoring process simpler because the product ac is smaller.
By factoring out the GCF first, you will have a completely factored trinomial which is our goal.

Original Trinomial	With the GCF Factored out front
$6x^2 - 2x - 20$	$2(3x^2 - x - 10)$
$a \cdot c = 6 \cdot -20 = -120$	$a \cdot c = 3 \cdot -10 = -30$

6.11 Factoring Trinomials By Grouping 6.11

EXAMPLE 4

Factor. $3x^2 + 10 - 31x$

$3x^2 - 31x + 10$

1. First we will need to rearrange this trinomial so that the exponents are in descending order.

2. There is not a common factor so we get to skip that step.

3. In the trinomial $3x^2 - 31x + 10$, **$a = 3$, $b = -31$, and $c = 10$.** Find two factors which multiply to $a \bullet c = 3 \bullet 10 = 30$ And whose sum is the middle coefficient which is -31.

Product = + 30 Sum = -31.

Finding the factor pair when the product is positive:
→ List the factor pairs of +30.
→ Choose the factor pair: When the product is positive, the pair will have the same signs, so you will add the numbers to make the middle term.
→ Attach the signs to the pair: both numbers in the pair will keep the sign of the middle term.

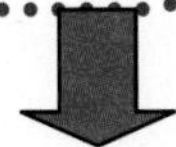

4. List the factors of +30.

Factors of 30
$1 \bullet 30$ or $-1 \bullet -30$
$2 \bullet 15$ or $-15 \bullet -2$
$3 \bullet 10$ or $-3 \bullet -10$
$5 \bullet 6$ or $-5 \bullet -6$

5. Rewrite the middle term using the two factors circled and then factor by grouping.

$3x^2 - 31x + 10$

$3x^2 - 1x - 30x + 10$

$x(3x - 1) - 10(3x - 1)$

$(x - 10)(3x - 1)$

6. Check by multiplying: $(x - 10)(3x - 1) = x(3x - 1) - 10(3x - 1)$

$= 3x^2 - 1x - 30x + 10$

$= 3x^2 - 31x + 10$

6.11 Factoring Trinominals By Grouping 6.11

EXAMPLE 5

Factor. $18y^4 + 21y^3 - 60y^2$

1. The trinomial is in descending order.

$18y^4 + 21y^3 - 60y^2 = 3y^2\left(6y^2 + 7y - 20\right)$ 2. Factor out any GCF . We will factor out $3y^2$.

3. In the trinomial $6y^2 + 7y - 20$,
   a = 6 , b =7 and c = - 20.
   Find two factors which multiply to $a \cdot c = 6 \cdot -20 = -120$.
   And whose sum is the middle coefficient which is 7.
   Product = -120 and Sum = 7.

Finding the factor pair when the product is negative:
→ List the factor pairs of -120.
→ Choose the factor pair: When the product is negative, the pair will have opposite signs, so you will subtract the numbers to make the middle term.
→ Attach the signs to the pair: the larger number in the pair will keep the sign of the middle term.

4. List the factors of – 120.

Factors of - 120		
$1 \cdot -120$	*or*	$-1 \cdot 120$
$2 \cdot -60$	*or*	$-2 \cdot 60$
$3 \cdot -40$	*or*	$-3 \cdot 40$
$4 \cdot -30$	*or*	$-4 \cdot 30$
$5 \cdot -24$	*or*	$-5 \cdot 24$
$6 \cdot -20$	*or*	$-6 \cdot 20$
$8 \cdot -15$	*or*	$-8 \cdot 15$
$10 \cdot -12$	*or*	$-10 \cdot 12$

5. Rewrite the middle term using the two factors circled.

$$3y^2[6y^2 + 7y - 20] = 3y^2[6y^2 - 8y + 15y - 20]$$
$$= 3y^2[2y(3y - 4) + 5(3y - 4)]$$
$$= 3y^2[(3y - 4)(2y + 5)]$$

6. Don't forget to include the GCF you factored out in the first step in your final answer.

$$18y^4 + 21y^3 - 60y^2 = 3y^2(3y - 4)(2y + 5)$$

EXAMPLE 6

Factor. $x^2 + 2xy - 15y^2$

The trinomial is in the correct order and has no GCF. So, we will proceed with finding our product and sum.

$$x^2 + 2xy - 15y^2 = 1x^2 + 2xy - 15y^2$$

$a = 1$, $b = 2$, $c = -15$

$product = 1 \bullet -15 = -15$

$sum = 2$

The numbers are 5 and – 3.

We will rewrite the middle term using 5 and – 3. When we replace the middle term we will make sure we use xy as our variable term.

$$\begin{aligned} x^2 + 2xy - 15y^2 &= x^2 + 5xy - 3xy - 15y^2 \\ &= x(x+5y) - 3y(x+5y) \\ &= (x+5y)(x-3y) \end{aligned}$$

EXAMPLE 7

Factor. $2x^2 - 3x + 4$

1. There is no GCF.
2. $a = 2$, $b = -3$, $c = 4$
   $product = +8 \quad sum = -3$

Factors are: 1, 8
2, 4
There is no pair which will multiply to +8 and add to -3.

Therefore, the polynomial is **PRIME** because it cannot be factored.

Courtesy of Fotolia

Make sure your trinomial is in descending order. If you forget to do this, your values for a,b, and c will be incorrect.

Make sure to check for a GCF before you find your product and sum. Unfortunately, the trinomial will still factor even if you forget. However, the answer you get will not be completely factored and thus will be incorrect.

When we rewrite the middle term using our two factors, we should always use the same variable or variables as the given middle term.

EXAMPLE 8

Factor. $2b^2 - 30b - 648$

The trinomial is in the correct order .

Factor out the GCF of 2.

$$2b^2 - 30b - 648$$
$$2(b^2 - 15b - 324)$$

The resulting trinomial is $b^2 - 15b - 324$

$a = 1, b = -15, c = -324$
product $= 1 \cdot -324 = -324$
sum $= -15$

WOW !! That's a large product ! How will I find all the factors of – 324???

If we depend on the previous method, we will have to use our calculator and begin dividing 324 by 1, divide 324 by 2, divide 324 by 3, divide 324 by 4, etc. !!!

USE THE CALC METHOD BELOW TO FIND FACTORS FAST!!!

The factors are 27 and -12.

To factor example 8, follow this process.

$2b^2 - 30b - 648$

$2[b^2 - 15b - 324] = 2[\underline{b^2 + 12b} - \underline{27b - 324}]$

$= 2[b(b + 12) - 27(b + 12)]$

$= 2[(b + 12)(b + 27)]$

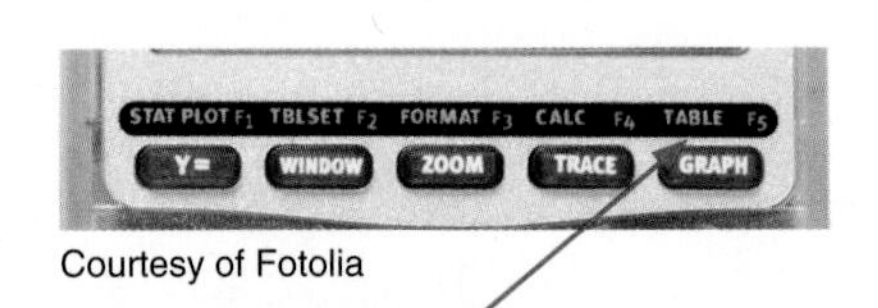

Courtesy of Fotolia

LOOK at FACTORING with your TI-84 Calculator!!!

1. To factor 324, Press the [y =] button on your calculator.
2. Enter the equation: $y_1 = 324/x$ (Product/x).
3. Press the [TABLE] button on your calculator. (This is the [2nd] function above the [GRAPH] button.)
4. In your window, you should see a x-y table with pairs of factors. You will only be interested in the integer factors. Ignore all the decimals!

X	Y_1	
9	36	
10	32.4	
11	29.455	
12	27	
13	23.143	
X = 12		

5. Scroll down through the table of factors until you find the factors whose product is – 324 and whose sum is -15.

The numbers are 27 and – 12. (Assign the signs of the numbers last.)

6.11 Factoring Trinomials By Grouping **6.11**

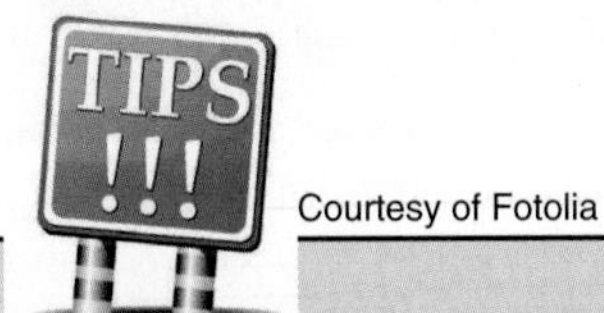

Courtesy of Fotolia

Tips on Finding the Correct Factors

When we are looking for factors that must meet certain conditions, we can get clues from the sign of the **product** and sum.

- If the product is positive, the two factors are both positive OR both negative, *Factor signs are the same*. This tells you to look for factors that will "add" to yield your product.

- If the product is negative, one factor is positive and one factor is negative, *Factor signs are different.* This tells you to look for factors that will "subtract" to yield your product.

NOTE: When you have a *positive and negative factor*, it is very important to assign the negative to the correct factor. Keep in mind that when you add a positive and a negative, the answer takes the sign of the "larger number".

Writing a polynomial in factored form that represents a variable:

Given a figure, how can you determine the area using a polynomial in factored form. These binomials will be the side lengths of the rectangle.

x^2	$5x$
$2x$	10

Write the side lengths which would be used to make the area of each small rectangle. →

	x	5	
x	x^2	$5x$	x
2	$2x$	10	2
	x	5	

To find the total area of the figure, first write the length of each side:

length $= x + 5$ *Then multiply the sides.*

width $= x + 2$ → **Total Area** $= l\,w$

$= (x + 5)(x + 2)$

$= x^2 + 7x + 10$

6.11 Factoring Trinominals By Grouping 6.11

YOUR TURN

Factor each trinomial by grouping.

3. $2x^2 + 7x + 6$	4. $8x^2 + 6x - 9$
5. $16x^2 - 34xy + 18y^2$	6. $15y^3 - 55y^2 + 30y$
7. $4y^2 - 2y - 12$	8. $6 - 11x + 5x^2$

6.11 Factoring Trinominals By Grouping 6.11

YOUR TURN **Answers to Section 6.11.**

1. see instructor/ tutor
2. see instructor/tutor
3. $(2x+3)(x+2)$
4. $(4x-3)(2x+3)$
5. $2(x-y)(8x-9y)$
6. $5y(y-3)(3y-2)$
7. $2(2y+3)(y-2)$
8. $(5x-6)(x-1)$

Complete MyMathLab Section 6.11 Homework in your Homework notebook.

(Stay on our track. We're moving on to Section 6.09 next.)

Section 6.09 Factoring Trinomials of the Form $x^2 + bx + c$

VIEWER DISCRETION IS ADVISED!!!!

THE MATERIAL IN THIS SECTION
MAY NOT BE SUITABLE
FOR CERTAIN STUDENTS!

IN THE FOLLOWING PAGES,
YOU WILL FIND A SHORTCUT
FOR SOME TRINOMIALS.

BY VIEWING THIS SECTION YOU ARE
SAYING THAT YOU ARE AN
EXPERIENCED, GOOD-DECISION MAKING
MATHEMATICIAN!!

6.09 Factoring Trinomials of the Form $x^2 + bx + c$ 6.09

Now that we have your attention, let's begin to look at a special group of trinomials.

The reason for the warning on the front page is that it is very tempting for students to apply a shortcut when one should not be used. So, constantly remind yourself when a shortcut should be used and when it cannot be used!!!

FACTORING TRINOMIALS WITH A LEADING COEFFICIENT OF 1

We will begin by factoring two trinomials using the grouping method we learned in the previous section.

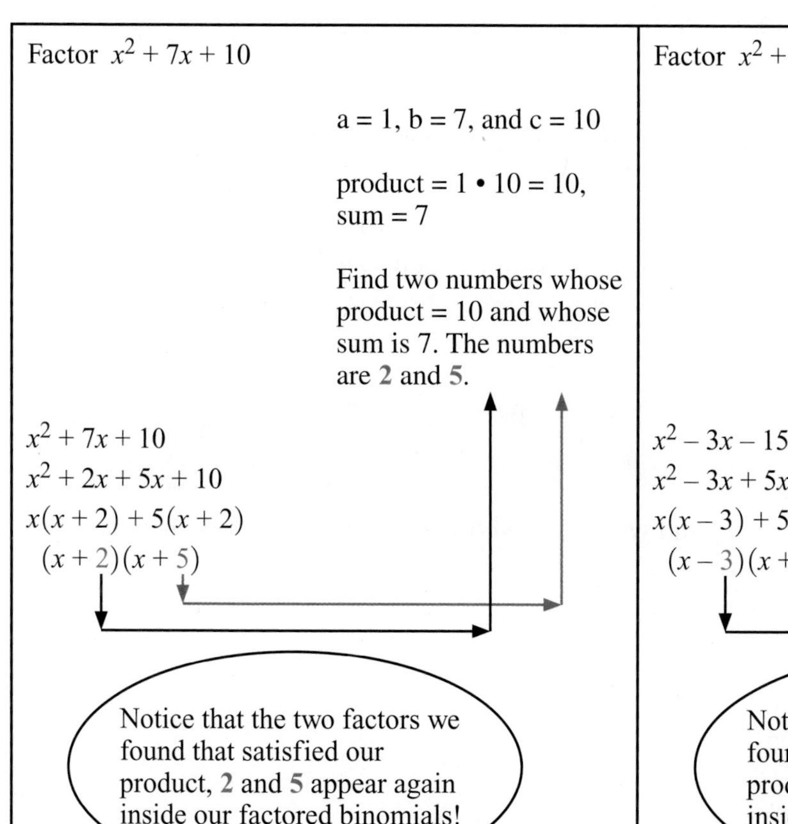

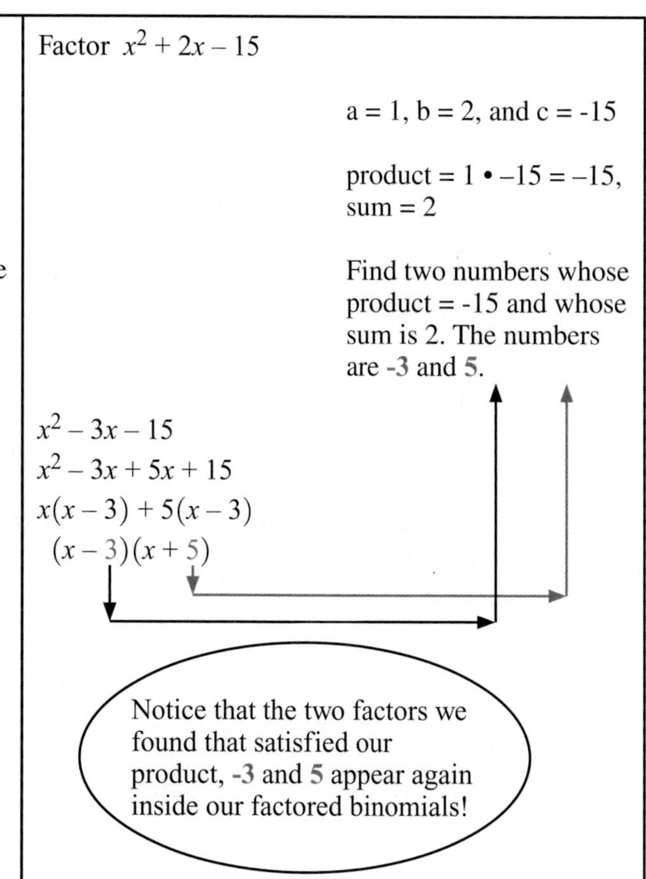

Courtesy of Fotolia

This will always happen if the leading coefficient of the square term is 1!!!!

6.09 Factoring Trinomials of the Form $x^2 + bx + c$ 6.09

EXAMPLE 1

For which of the following trinomials could we use the shortcut? Remember you are looking for a leading coefficient of 1 on the square term.

a. $k^2 - 2k - 6$ → Yes, the coefficient on k^2 is 1.

b. $2m^2 + 3m + 10$ → NO , the coefficient on m^2 is 2.

c. $12 + 10y + y^2$ → Yes, after arranging in descending order the coefficient on y^2 is 1.

d. $3m^2 - 6m + 9$ → Yes, after factoring out the GCF of 3 , the coefficient on m^2 is 1.

YOUR TURN

1. For which of the following trinomials could we use the shortcut? Remember you are looking for a leading coefficient of 1 on the square term.

a. $30 + 11m + m^2$ ________

b. $3m^2 + 2m + 10$ ________

c. $y^2 - y - 20$ ________

d. $2m^2 - 6m + 8$ ________

6.09 Factoring Trinominals of the Form $x^2 + bx + c$ 6.09

Now that we recognize when we can use the shortcut, let's look a little closer at how this formula works.

Courtesy of Fotolia

To factor a Trinomial of the form $x^2 + bx + c$,

$$x^2 + bx + c = \left(x + \square\right)\left(x + \square\right)$$

The two factors you find that satisfy your product and sum go inside the red boxes.

EXAMPLE 2

Factor. $y^2 - 13y + 30$

a = 1 b = -13 c = 30

product = 1 • 30 = 30
sum = –13

The factors are **-10 and -3**.

Notice that a = 1, we can use the shortcut!!

$y^2 - 13y + 30$
$(y + -3)(y + -10)$
$(y - 3)(y - 10)$

EXAMPLE 3

Factor. $m^2 - 4m - 21$

a = 1 b = -4 c = -21

product = 1 • –21 = –21
sum = –4

The factors are **3 and -7.**

Since a = 1, we can use the shortcut!!

$m^2 - 4m - 21$
$(m + 3)(m + -7)$
$(m + 3)(m - 7)$

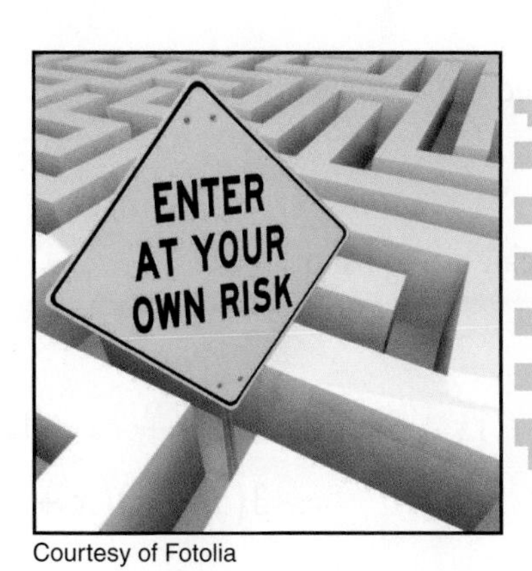

Courtesy of Fotolia

Please remember that you can only use the shortcut if the leading coefficient on the square term is 1.

EXAMPLE 4

Factor. $3x^2 - 31x + 10$

> a = 3 b = -31 c = 10
>
> product $= 3 \bullet 10 = 30$
> sum $= -31$
>
> The factors are **-1 and -30**

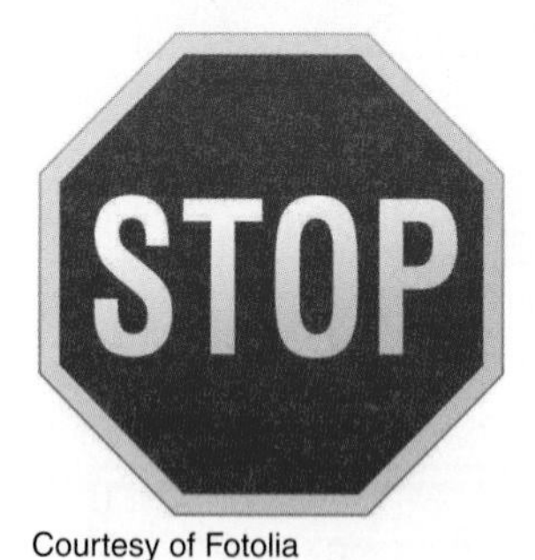

Courtesy of Fotolia

> Notice $\boldsymbol{a \neq 1}$.
>
> We cannot use the shortcut!!!!

We must factor this trinomial as we did in the previous section using grouping.

$$3x^2 - 31x + 10 = 3x^2 - 1x - 30x + 10$$
$$= x(3x - 1) - 10(3x - 1)$$
$$= (x - 10)(3x - 1)$$

EXAMPLE 5

Factor. $3x^2 + 21x + 36$

First, we notice that there is a GCF of 3 which has to be factored out.

$$3x^2 + 21x + 36 = 3(x^2 + 7x + 12)$$

Notice the resulting trinomial now has a coefficient of 1. We can use the shortcut!!

> a = 1 b = 7 c = 12
>
> product $= 1 \bullet 12 = 12$
> sum $= 7$
>
> The factors are **3 and 4.**

$$3x^2 + 21x + 36 = 3(x^2 + 7x + 12)$$
$$= 3(x + 4)(x + 3)$$

Here's the plan for trinomials.

$$a = 1$$

Factor $m^2 - 4m - 21$

a = 1, we can use the shortcut!!

Ask, "What pair of #s can be multiplied to get – 21 and added to get -4?"
The numbers are 3 and -7.

$$m^2 - 4m - 21 = (m + 3)(m + -7)$$
$$= (m + 3)(m - 7)$$

$$a \neq 1$$

Factor $2x^2 + 13x + 15$

a = 2, we cannot use the shortcut !!! Factor using the *ac* method & grouping.

$$2x^2 + 13x + 15$$
$$= 2x^2 + 10x + 3x + 15$$
$$= 2x(x + 5) + 3(x + 5)$$
$$= (2x + 3)(x + 5)$$

30
1, 30
2, 15
3, 10
5, 6

6.09 Factoring Trinomials of the Form $x^2 + bx + c$ 6.09

YOUR TURN

Factor the following trinomials. Be careful when deciding to use the shortcut!!

2. $r^2 - r - 42$	3. $x^2 - 4x - 32$
4. $y^2 + 15y + 36$	5. $2x^2 - 10x - 28$
6. $3y^2 + 13y + 4$	7. $m^2 - 10m + 24$

6.09 Factoring Trinominals of the Form $x^2 + bx + c$ 6.09

YOUR TURN **Answers to Section 6.09.**

1. a. yes b. no c. yes d. yes, after GCF
2. $(r-7)(r+6)$
3. $(x-8)(x+4)$
4. $(y+12)(y+3)$
5. $2(x-7)(x+2)$
6. $(3y+1)(y+4)$
7. $(m-4)(m-6)$

Complete MyMathLab Section 6.09 Homework in your Homework notebook.

(Stay on our track. We're moving on to Section 6.12 next.)

FACTORING PERFECT SQUARE TRINOMIALS

A perfect square trinomial is a trinomial that is the square of a binomial. For example, $x^2 + 8x + 16$ is a perfect square trinomial because it is the square of $(x + 4)$.

$$x^2 + 8x + 16 = (x+4)(x+4)$$
$$= (x+4)^2$$

Courtesy of Fotolia

Keep in mind that not all trinomials are perfect square trinomials. This process is for a select group of trinomials!

We will need to recognize what a perfect square trinomial looks like if we want to use this pattern for factoring.

Here's a reminder of what perfect squares look like. You will find this chart helpful as you factor perfect square trinomials.

PERFECT SQUARES	
$1^2 = 1$	$11^2 = 121$
$2^2 = 4$	$12^2 = 144$
$3^2 = 9$	$13^2 = 169$
$4^2 = 16$	$14^2 = 196$
$5^2 = 25$	$15^2 = 225$
$6^2 = 36$	$16^2 = 256$
$7^2 = 49$	$17^2 = 289$
$8^2 = 64$	$18^2 = 324$
$9^2 = 81$	$19^2 = 361$
$10^2 = 100$	$20^2 = 400$
	$x^2, x^4, x^6, x^8, \ldots$ [Even Exponents]

MEMORIZE: You'll need these again.

One necessary condition for a trinomial to be a perfect square is that *the first and last terms must be perfect squares.* For this reason, $16x^2 + 4x + 15$ is not a perfect square trinomial because only the first term, $16x^2$, is a perfect square but not the last term, 15.

6.12 Factoring By Special Product 6.12

However, even if the two terms are perfect squares, the trinomial still may not be a perfect square trinomial. For example $x^2 + 5x + 36$ has two perfect square terms, x^2 and 36, but it is not a perfect square trinomial either.

The second condition is the *the middle term of a perfect square trinomial* is always *twice the product of the two terms in the squared binomial.*

To Recognize a Perfect Square Trinomial

1. First and last terms are perfect squares which means they have to be positive.
2. The middle term is twice the product of the two terms in the squared binomial. It can be positive or negative.

EXAMPLE 1

Which of the following are perfect square trinomials?

a. $y^2 - 13y + 36$

b. $16x^2 + 40x + 25$

c. $4m^2 - 4m + 1$

d. $25x^2 + 70x - 49$

a. $y^2 - 13y + 36$ is not a perfect square trinomial. Notice the first term is a square $y^2 = (y)^2$, and the last term is a square, $36 = (6)^2$. To be a perfect square the middle term would need to be $2 \bullet 6 \bullet y = 12y$.

b. $16x^2 + 40x + 25$ is a perfect square trinomial. Notice the first term is a square $16x^2 = (4x)^2$, and the last term is a square, $25 = (5)^2$. The middle term is $2 \bullet 5 \bullet 4x = 40x$. This is a perfect square trinomial.

c. $4m^2 - 4m + 1$ is a perfect square trinomial. Notice the first term is a square $4m^2 = (2m)^2$, and the last term is a square, $1 = (1)^2$. The middle term is $2 \bullet 1 \bullet 2m = 4m$. This is a perfect square trinomial.

d. $25x^2 + 70x - 49$ is not a perfect square trinomial. The last term is not positive.

YOUR TURN

1. Tell whether the following trinomials are perfect square trinomials or not. If not, give the reason why.

a. $y^2 - 16y + 64$ b. $4x^2 - 20x + 25$ c. $25m^2 - 10m + 4$ d. $36x^2 + 60x - 25$

6.12 Factoring By Special Product 6.12

Now that we can recognize what a perfect square trinomial looks like we are ready to factor this special group of trinomials. If you have a perfect square trinomial, it will match up with one of the following patterns.

FACTORING PERFECT SQUARE TRINOMIALS

$$a^2 + 2ab + b^2 = (a + b)^2$$

$$a^2 - 2ab + b^2 = (a - b)^2$$

Factor. $25m^2 + 40m + 16$

Notice that the first term is a square: $25m^2 = (5m)^2$, the last term is a square $16 = (4)^2$, and the middle term is $2 \cdot 4 \cdot 5m = 40m$.

This is a perfect square trinomial. The set-up is $(\quad + \quad)^2$ (Use the sign from the middle.)

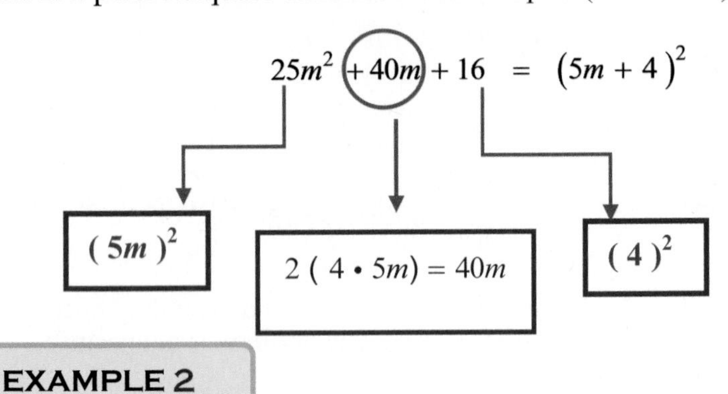

EXAMPLE 2

Factor. $9m^2 - 6m + 1$

Notice that the first term is a square: $9m^2 = (3m)^2$, the last term is a square $1 = (1)^2$, and the middle term is $2 \cdot 1 \cdot 3m = 6m$.

This is a perfect square trinomial .

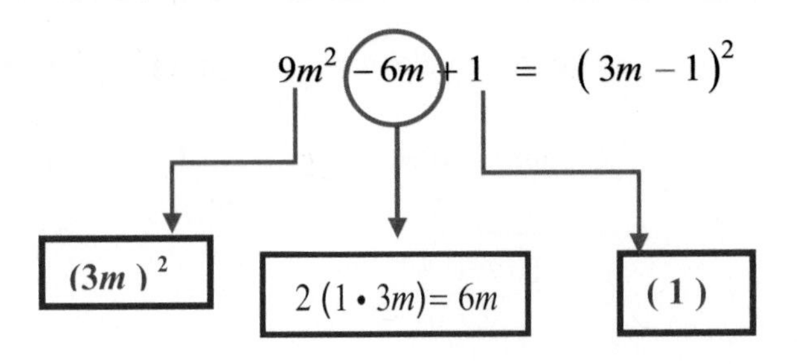

Notice the sign of the middle term!!!

6.12 Factoring By Special Product 6.12

Courtesy of Fotolia

Don't forget to continue looking for any common factors before you begin factoring a trinomial. Also look for the order of your trinomial.

EXAMPLE 3

Factor. $4x^3 - 32x^2y + 64xy^2$

The first thing we should notice is that this trinomial has a common factor of $4x$. So, factor out the common $4x$.

$$4x^3 - 32x^2y + 64xy^2 = 4x\left(x^2 - 8xy + 16y^2\right)$$

Now look at the new trinomial which is $x^2 - 8xy + 16y^2$.

Notice that the first term is a square: $x^2 = (x)^2$, the last term is a square $16y^2 = (4y)^2$, and the middle term is $2 \cdot (x \cdot 4y) = 8xy$.

This is a perfect square trinomial .

$$4x^3 - 32x^2y + 64xy^2 = 4x\left(x^2 - 8xy + 16y^2\right) = 4x(x - 4y)^2$$

Remember to use the sign of the middle term in your factored binomial !!

Courtesy of Fotolia

If you recognize a trinomial as a perfect square trinomial, use the special formula to factor as shown below. However, perfect square trinomials can also be factored using grouping as we learned in section 6.11.

Courtesy of Fotolia

If middle term in perfect square trinomial is positive,

$$a^2 + 2ab + b^2 = (a + b)^2$$

If middle term in perfect square trinomial is negative,

$$a^2 - 2ab + b^2 = (a - b)^2$$

6.12 Factoring By Special Product 6.12

YOUR TURN

Factor using special product rules if possible.

2. $49x^2 + 70x + 25$	3. $49m^2 - 28mn + 4n^2$
4. $3y^2 - 48y + 192$ (Look for GCF first!)	5. $3a^2x - 12abx + 12b^2x$

6.12 Factoring By Special Product 6.12

FACTORING BINOMIALS: A DIFFERENCE OF TWO PERFECT SQUARES

The second special product we will look at concerns binomials. Remember a binomial is a polynomial that has exactly two terms.

In section 6.6 we learned how to multiply the sum and difference of two terms. The examples below should look familiar.

$$\begin{aligned}&(x+3)(x-3)\\&= x(x-3)+3(x-3)\\&= x^2-3x+3x-9\\&= x^2-9\end{aligned}$$

$$\begin{aligned}&(2m-5)(2m+5)\\&= 2m(2m+5)-5(2m+5)\\&= 4m^2+10m-10m-25\\&= 4m^2-25\end{aligned}$$

In section 6.6, we stated that if you multiplied two binomials where one had a sum and the other had a difference, that the result would be the difference of two squares. The general pattern would be

$$(a+b)(a-b) = a^2-b^2$$

Courtesy of Fotolia

By reversing these rules for multiplication of binomials we get rules for factoring. Reversing the rule above leads to the following special factoring rule.

$$a^2-b^2 = (a+b)(a-b)$$

Once again to use this formula , we will need to recognize what the difference of two squares looks like. There are **3 conditions** that must be met to factor using the difference of squares.

Courtesy of Fotolia

Checklist to Factor Using the Difference of Squares

1. **You have two expressions**
2. **Joined by a subtract sign.**
3. **Both expressions are perfect squares.**

EXAMPLE 4

Which of the following are a difference of squares?

a. $x^2 - 49$ 1. You have two expressions. Yes 2. Joined by a subtract sign. Yes 3. Both expressions are perfect squares. Yes This is the difference of two squares.	b. $y^2 + 9$ 1. You have two expressions. Yes 2. Joined by a subtract sign. NO 3. Both expressions are perfect squares. Yes This is NOT the difference of two squares.
c. $2a^2 - 25$ 1. You have two expressions. Yes 2. Joined by a subtract sign. Yes 3. Both expressions are perfect squares. NO This is NOT the difference of two squares.	d. $9m^2 - 1$ 1. You have two expressions. Yes 2. Joined by a subtract sign. Yes 3. Both expressions are perfect squares. Yes This is the difference of two squares.

YOUR TURN

6. Write yes or no in the blanks and then decide if you have a difference of squares.

a. $x^2 + 100$ 1. You have two expressions. ________ 2. Joined by a subtract sign. ________ 3. Both expressions are perfect squares. ________ Is this the difference of two squares?	b. $y^2 - 64$ 1. You have two expressions. ________ 2. Joined by a subtract sign. ________ 3. Both expressions are perfect squares. ________ Is this the difference of two squares?
c. $4a^2 - 35$ 1. You have two expressions. ________ 2. Joined by a subtract sign. ________ 3. Both expressions are perfect squares. ________ Is this the difference of two squares?	d. $9m^2 - 16$ 1. You have two expressions. ________ 2. Joined by a subtract sign. ________ 3. Both expressions are perfect squares. ________ Is this the difference of two squares?

See instructor/tutor to check.

6.12 Factoring By Special Product 6.12

Now that we can recognize the difference of two squares, we will begin to factor using our formula. Here's the formula again!

BINOMIAL: A DIFFERENCE OF TWO PERFECT SQUARES

$$a^2 - b^2 = (a+b)(a-b)$$

EXAMPLE 5

Factor the following binomials.

a. $25m^2 - 9 = (5m)^2 - (3)^2$
$= (5m+3)(5m-3)$

Check for perfect squares. Then set-up (+)(-).

b. $16x^2y^2 - 1 = (4xy)^2 - (1)^2$
$= (4xy+1)(4xy-1)$

Check for perfect squares. Then set-up (+)(-).

c. $49 - 9k^2 = (7)^2 - (3k)^2$
$= (7+3k)(7-3k)$

Check for perfect squares. Then set-up (+)(-).

d. $x^2 - \frac{1}{4} = (x)^2 - \left(\frac{1}{2}\right)^2$
$= \left(x + \frac{1}{2}\right)\left(x - \frac{1}{2}\right)$

Check for perfect squares. Then set-up (+)(-).

Courtesy of Fotolia

Remember to look for the GCF to factor out before you begin!

Always include the common factor in the final factored form.

Remember the order of the parentheses does not matter!

EXAMPLE 6

Factor the following binomials. Factor out the GCF first if there is one present.

a. $50 - 8y^2 = 2\left(25 - 4y^2\right)$

$= 2\left[(5)^2 - (2y)^2\right]$ Check for perfect squares. Then set-up (+)(-).

$= 2\ (5 + 2y)(5 - 2y)$

b. $4x^3 - 49x = x\left(4x^2 - 49\right)$

$= x\left[(2x)^2 - (7)^2\right]$ Check for perfect squares. Then set-up (+)(-).

$= x\ (2x + 7)(2x - 7)$

c. $-25m^2 + 16 = -1\left(25m^2 - 16\right)$

$= -1\left[(5m)^2 - (4)^2\right]$ Check for perfect squares. Then set-up (+)(-).

$= -1\ (5m + 4)(5m - 4)$

YOUR TURN

Factor each binomial.

7. $9 - 4m^2$	8. $25xy^2 - 4x$
9. $49x^2y^2 - 36$	10. $16m^2 + 25$

6.12 Factoring By Special Product 6.12

Your Turn **Answers to Section 6.12.**

1. a. yes b. yes c. no, middle term doesn't match up

 d. no, last term is negative

2. $(7x+5)^2$ 3. $(7m-2n)^2$ 4. $3(y-8)^2$ 5. $3x(a-2b)^2$

6. a. yes, no, yes, no b. yes, yes, yes, yes c. yes, yes, no, no d. yes, yes, yes, yes

7. $(3-2m)(3+2m)$ 8. $x(5y-2)(5y+2)$ 9. $(7xy+6)(7xy-6)$

10. Prime. It does not factor. It is a sum not a difference.

Complete MyMathLab Section 6.12 Homework in your Homework notebook.

Factoring a Polynomial/ Factoring Summary

Are there any common factors? If so, factor out the GCF.

How many terms are in the polynomial?

2 terms

Difference of Squares?

$$a^2 - b^2 = (a + b)(a - b)$$

Perfect Square Trinomial?

$$a^2 + 2ab + b^2 = (a + b)^2$$
$$a^2 - 2ab + b^2 = (a - b)^2$$

If not a perfect square trinomial, factor using grouping.

- **Find two numbers whose product equals ac , and whose sum equals b. Be careful with signs.**
- **Use the two numbers found in step 1 to rewrite the trinomial as a 4 term polynomial**
- **Factor by grouping.**

4 terms

Factor 4 terms by GROUPING.

Polynomials which do not factor are PRIME.

Module 6 Factoring Summary **Module 6**

It is very important in factoring that you recognize what kind of polynomial you have so that you can determine your plan of action. Of course, factoring out the GCF is always the first process you complete. After that, a decision is made based on how many terms you have. As you factor a polynomial, we suggest asking yourself these questions to decide on a suitable factoring technique.

1. *GCF: Is there a common factor (GCF) other than 1?*

If so, factor out the greatest common factor (GCF) from all terms of the polynomial.

2. *# TERMS: How many terms are in the polynomial?*

Two terms: Check to see whether it is a *difference of squares*.

Three terms: *a.* Is it a *perfect square*? OR

b. If not a perfect square, is the *leading coefficient of the x^2 term 1*? If so we can use the shortcut learned in 6.09. OR

c. If the coefficient of the square term is not 1, use the *"ac" grouping method* learned in 6.11 for terms in standard form $ax^2 + bx + c$. ALWAYS WORKS!

Four terms: Factor by *grouping*.

3. CHECK: *Can any factors by factored further?* If so, factor them.

YOUR TURN

Match each polynomial in Column 1 with the method you would use to factor it in Column II. The choice in Column II may be used once, more than once, or not at all. Write the letter of the method you would choose for each polynomial in the blank column.

	I	II
________	1. $12x^2 + 20x + 8$	A. Factor out the GCF; no further factoring is possible.
________	2. $x^2 - 6x - 72$	B. Factor a difference of squares.
________	3. $64a^2 - 25b^2$	C. Factor a perfect square trinomial.
________	4. $36m^2 - 60mn + 25n^2$	D. Factor out the GCF; then factor a trinomial by grouping.
________	5. $2x + 4a + x^2 + 2ax$	E. Factor into two binomials by using the trinomial shortcut method.
________	6. $4w^2 + 9$	F. Factor by grouping.
________	7. $8x^2 - 10x - 3$	G. The polynomial is prime.
________	8. $2m + 10$	
________	9. $100 - x^2y^2$	
________	10. $2m - 15 + m^2$	

See instructor/tutor to check.

SOLVING QUADRATIC EQUATIONS

In a previous module, we solved linear equations in which the power on the variable is 1. An equation where the variable has an exponent of two is called a quadratic equation.

Quadratic Equation in One Variable

If a, b, and c are real numbers such that $a \neq 0$ then a **quadratic equation** is an equation that can be written in the form

$$ax^2 + bx + c = 0$$

The following equations are quadratic because they can be written in the form $ax^2 + bx + c = 0$.

$2x^2 + 3x = 1$	$x(x-2) = 3$	$x^2 = 25$
$2x^2 + 3x - 1 = 0$	$x^2 - 2x = 3$	$x^2 - 25 = 0$
	$x^2 - 2x - 3 = 0$	

Your Turn

1. Which of the following are quadratic equations?

   a. $7x^2 + 3x - 5 = 0$ ______________

   b. $x^2 + 3x = 6$ ______________

   c. $2x + 6 = 10$ ______________

   d. $x^2 = 16$ ______________

One method to solve a quadratic equation is to factor and apply the zero product rule. The zero product rule states that if the product of two numbers is 0, then at least one of the numbers must be 0.

Zero-Factor Property

If a and b are real numbers and if $a \bullet b = 0$, then $a = 0$ or $b = 0$.

This states that one number *must* be 0, but both *may* be 0.

6.13 Solving Quadratic Equations by Factoring 6.13

EXAMPLE 1

Solve. $(x+3)(x-4) = 0$

The product $(x+3)(x-4)$ is equal to 0. By the zero-factor property, the only way that the product of these two factors can be 0 is if at least one of the factors equals 0.
Therefore, either $x+3=0$ or $x-4=0$.

$x+3=0$	$x-4=0$
$x+3-3=0-3$	$x-4+4=0+4$
$x=-3$	$x=4$

The given equation $(x+3)(x-4) = 0$ has two solutions, 4 and -3 .

Check these solutions by substituting 4 for x in the original equation, $(x+3)(x-4) = 0$. Then start over and substitute – 3.

Check If $x=4$, then

$$(x+3)(x-4) = 0$$
$$(4+3)(4-4) = 0$$
$$(7)(0) = 0$$
$$0 = 0$$

Check If $x=-3$

$$(x+3)(x-4) = 0$$
$$(-3+3)(-3-4) = 0$$
$$(0)(-7) = 0$$
$$0 = 0$$

The solutions are 4 and – 3.

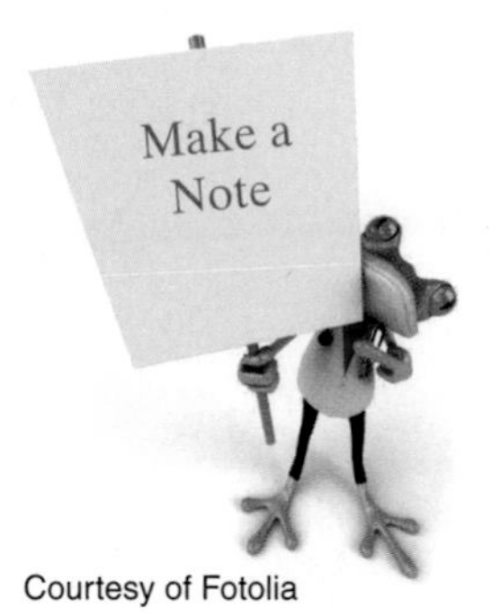

Courtesy of Fotolia

The zero-factor property says that if a product is 0, then a factor is 0.

We can only use this property if one side of our equation is equal to 0. If the equation is not equal to 0 , then we will need to take steps to change the equation so that it is equal to 0.

EXAMPLE 2

Solve. $x(2x-3)=0$

The product $x(2x-3)$ is equal to 0. By the zero-factor property, the only way that the product of these two factors can be 0 is if at least one of the factors equals 0. Therefore, either $x=0$ or $2x-3=0$.

$x=0$ **OR** $2x-3=0$

$$2x-3+3=0+3$$

$$2x=3$$

$$\frac{2x}{2}=\frac{3}{2}$$

$$x=\frac{3}{2}$$

The solutions are 0 and $\frac{3}{2}$.

EXAMPLE 3

Solve. $x^2-3x-18=0$

One side of the equation is 0. However, to use the zero-factor property, one side of the equation must be 0 and the other side must be written as a product (must be factored). So, we must factor the polynomial first.

$x^2-3x-18=0$ For this trinomial, the product = - 18 , and the sum = -3.
The two factors are -6 and 3.
Since a = 1, we can use the shortcut for trinomials!

$$x^2-3x-18=0$$
$$(x-6)(x+3)=0$$

Now we can use the zero-factor property to set each factor = 0 and then solve.

$x-6=0$ **OR** $x+3=0$

$x-6+6=0+6$ $\quad$ $x+3-3=0-3$

$x=6$ $\quad$ $x=-3$

The solutions are 6 and -3 .

6.13 Solving Quadratic Equations by Factoring 6.13

YOUR TURN

Solve each quadratic equation.

2. $m(3m-1)=0$	3. $(y-6)(2y+7)=0$
4. $x^2-5x-24=0$	5. $4x^2-11x-3=0$ (Hint: You cannot use the trinomial shortcut! You will need to rewrite the middle term and group to factor this trinomial.)

6.13 Solving Quadratic Equations by Factoring 6.13

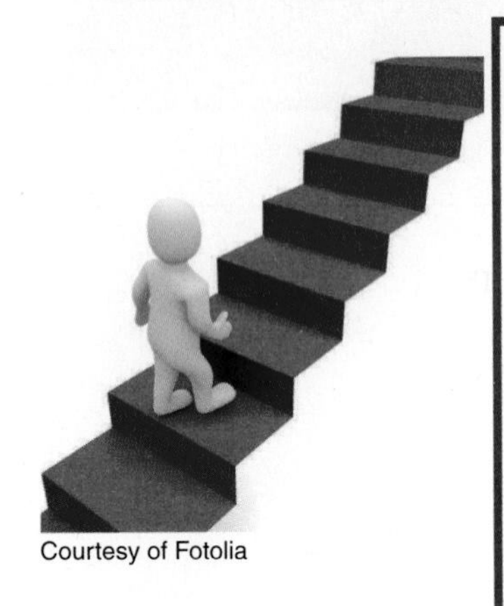
Courtesy of Fotolia

Steps to Solving Quadratic Equations

Step 1: Write the equation in standard form so that one side of the equation is 0. $ax^2 + bx + c = 0$

Step 2: Factor completely.

Step 3: Set each factor with a variable equal to 0.

Step 4: Solve the resulting equation.

Step 5: Check each solution in the original equation.

EXAMPLE 4

Solve. $x^2 - 5x = -6$

Notice that this equation is not in standard form so that one side is 0. We will need to get 0 on one side of this equation before we factor.

$x^2 - 5x = -6$

$x^2 - 5x + 6 = 0$

Move the -6 to the opposite side of the equation by changing its sign. This leaves a 0 on the other side of the =.

Now factor $x^2 - 5x + 6$. Find two numbers whose product is 6 and whose sum is -5. The numbers are -2 and -3 . Since the coefficient of the square term is 1, we can use the shortcut.

$$x^2 - 5x + 6 = 0$$
$$(x-2)(x-3) = 0$$

$x - 2 = 0$ **OR** $x - 3 = 0$

$x = 2$ $\quad$ $x = 3$

The solutions are 2 and 3.

EXAMPLE 5

Solve. $x(2x-7) = 4$

First we write the equation in standard form.

$$x(2x-7) = 4$$
$$2x^2 - 7x - 4 = 0$$

Now, we need to factor.

Notice that in the trinomial $2x^2 - 7x - 4$, a = 2. So , we CANNOT USE THE SHORTCUT!!

Product = - 8
Sum = - 7
The numbers are – 8 and 1.

$$2x^2 - 7x - 4 = 0$$
$$2x^2 - 8x + 1x - 4 = 0$$
$$2x(x-4) + 1(x-4) = 0$$
$$(2x+1)(x-4) = 0$$

The solutions are 4 and $-\frac{1}{2}$

$$2x + 1 = 0 \quad \textbf{OR} \quad x - 4 = 0$$
$$2x = -1 \qquad x = 4$$
$$x = -\frac{1}{2}$$

Courtesy of Fotolia

In solving the equation above, $x(2x-7) = 4$, **do not** set each factor equal to 4. Remember that to apply the zero-factor property, one side of the equation must be 0.

6.13 Solving Quadratic Equations by Factoring 6.13

EXAMPLE 6

Solve. $3x^2 - 12 = 0$

Notice the equation is equal to 0. So, we just need to factor the left hand side.

$$3x^2 - 12 = 0$$
$$3\left(x^2 - 4\right) = 0$$
$$3(x-2)(x+2) = 0$$

$x - 2 = 0$, $x = 2$

$x + 2 = 0$, $x = -2$

The solutions are 2 and -2.

Notice that the first factor 3 does not have a variable. So, this is not part of our solution. Or if we think about this, 3 could never be 0, 3 is always 3. That's a no brainer!! Thus, **3 is not a solution.**

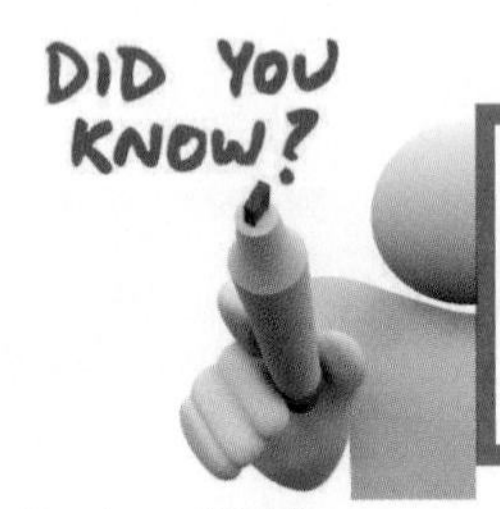

A common error is to include the common factor 3 as a solution. *Only factors containing variables lead to solutions!!!*

Courtesy of Fotolia

EXAMPLE 7

Solve. $3x^3 - 12x = 0$

Notice the equation is equal to 0.

So, we just need to factor the left hand side.

The degree of the polynomial tells how many solutions are possible. This polynomial is 3rd degree x^3, hence 3 solutions.

$$3x^3 - 12x = 0$$
$$3x\left(x^2 - 4\right) = 0$$
$$3x(x-2)(x+2) = 0$$

$3x = 0$, $x = 0$

$x - 2 = 0$, $x = 2$

$x + 2 = 0$, $x = -2$

The solutions are 0, 2 and -2.

Notice in this example that our GCF was $3x$ which has a variable. So, we must also set this factor equal to 0 and solve. This type of equation will have 3 answers!

6.13 Solving Quadratic Equations by Factoring 6.13

YOUR TURN

Solve each quadratic equation. Remember to put the equation in standard for first if needed.
Also, remember the degree of the polynomial tells how many solutions you should have.

6. $5y^2 = 20$	7. $x(x-4) = 5$
8. $x(3x+7) = 6$	9. $2y^3 = 18y$

6.13 Solving Quadratic Equations by Factoring 6.13

YOUR TURN **Answers to p. 6-111 & Section 6.13.**

Sec. 6-13:

1. a. yes b. yes c. no d. yes
2. $0, \frac{1}{3}$
3. $6, -\frac{7}{2}$
4. 8, -3
5. $3, -\frac{1}{4}$
6. 2, -2
7. 5, -1
8. $-3, \frac{2}{3}$
9. 0, 3, -3

Complete MyMathLab Section 6.13 Homework in your Homework notebook.

6.14 Quadratic Equations and Problem Solving 6.14

We can use factoring to solve quadratic equations that appear in applications. We looked at these steps in solving application problems in a previous module. Let's look at them again!

Problem Solving Plan

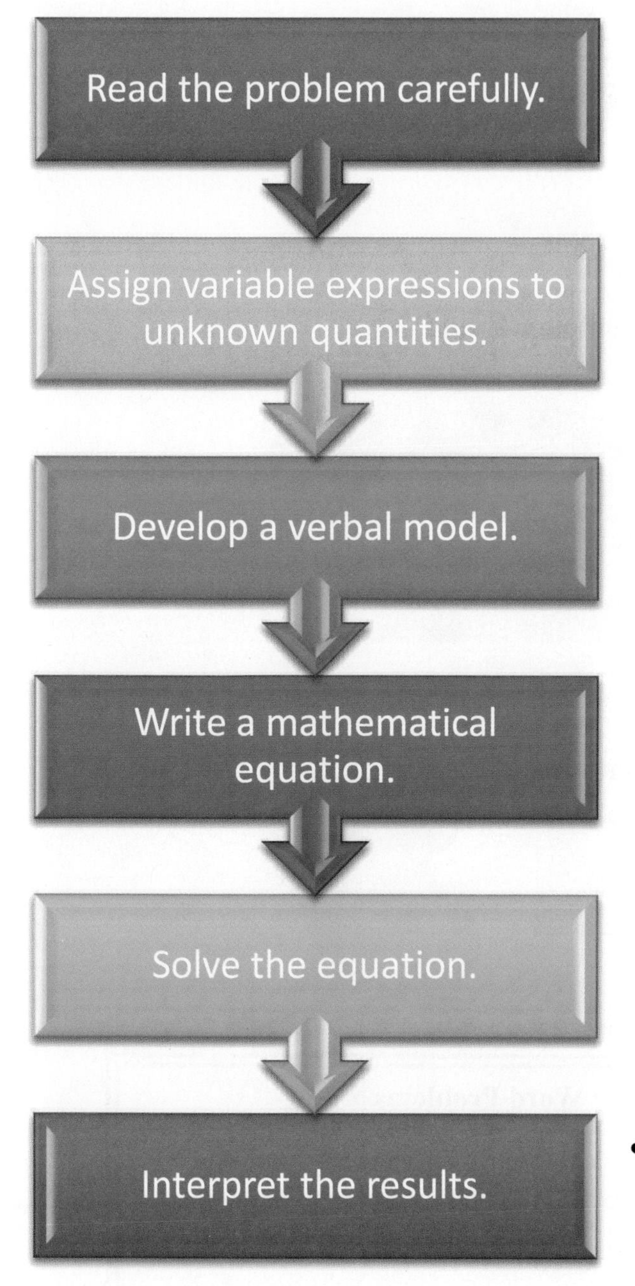

- Read and reread the problem until you understand what it is asking you to do.
- Identify the unknown quantity or quantities. Let x represent one of the unknowns. Draw a picture and write down any relevant formulas.
- Write an equation in words.
- Replace the verbal model with a mathematical equation using x or another variable.
- Solve for the variable using the steps you learned in the previous section.
- Once you've solved the equation, look back and see what it represented in the problem. Then use this value to determine any other unknowns in the problems.

6.14 Quadratic Equations and Problem Solving 6.14

In this section we will look at a variety of application problems. Here is a list of the different kinds of problems and what examples illustrate that type of problem. You can use this outline to assist you when you are working your homework. When you read the problem and the instructions say "solve," first find the topic that matches here in your workbook and then study the corresponding example.

SOLVING QUADRATIC APPLICATIONS

Finding a Number	Example 1
Geometric Figures	Examples 2 and 3
Consecutive Integers	Examples 4 and 5
Pythagorean Theorem	Examples 6 and 7
Quadratic Models	Examples 8,9,10

Keep reminding yourself throughout this section that

WORD PROBLEMS ARE FUN!!

Courtesy of Fotolia

Solving Quadratic Word Problems

- After solving a word problem, ask yourself, "Does my answer make sense?" For example, if you are asked to find an integer, your answer cannot be a fraction or a decimal. If you are asked to find time, your answer cannot be negative.
- Check the answer in the words of the original problem.
- Answer the question you have been asked.

6.14 Quadratic Equations and Problem Solving 6.14

Finding an Unknown Number

Courtesy of Fotolia

EXAMPLE 1

The square of a number plus three times the number is 70. Find the number(s).

READ the problem. We need to find an unknown number.

ASSIGN a variable: Let x = **the number**

WRITE AN EQUATION: We will read the problem again and everywhere we see the word " a number" or "the number" we will substitute x .

The square of a number plus three times the number is 70.

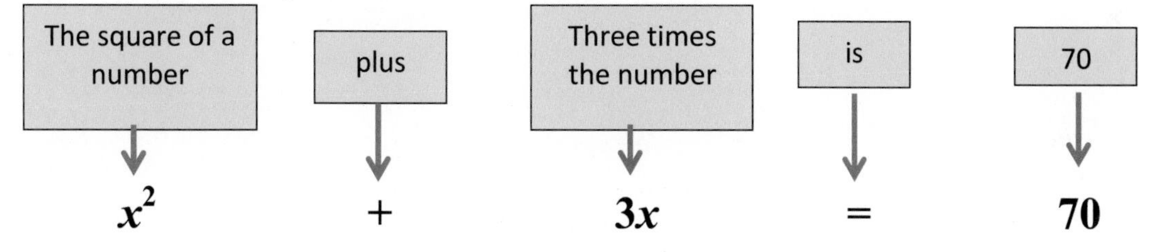

SOLVE.

$$x^2 + 3x = 70$$
$$x^2 + 3x - 70 = 0$$
$$(x+10)(x-7) = 0$$

$$x + 10 = 0 \qquad OR \qquad x - 7 = 0$$
$$x = -10 \qquad\qquad\qquad x = 7$$

The numbers are 7 and -10.

CHECK: **The square of a number plus three times the number is 70.**

Check $x = 7$: The square of 7 is 49. Three times 7 is 21. Then 49 + 21 = 70.

Check $x = -10$: The square of -10 is 100. Three times -10 is -30. Then 100 + -30 = 70.

YOUR TURN

1. The square of a number minus twice the number is 63. Find the number(s).

Assign a variable: *Write an equation and solve:*

Number = __________ ______________________________________

Reread the question and answer: ________________________

2. The sum of a number and its square is 132. Find the number(s).

Assign a variable: *Write an equation and solve:*

Number = __________ ______________________________________

Reread the question and answer: ________________________

6.14 Quadratic Equations and Problem Solving 6.14

Courtesy of Fotolia

Geometric Figures

EXAMPLE 2

The Smith's want to plant a rectangular garden in their yard. The width of the garden will be 4 ft less than its length, and they want it to have an area of 96 ft^2. Find the length and width of the garden.

Assign a variable:

Let x = the length of the garden

$x - 4$ = the width (the width is 4 ft less than the length)

x - 4

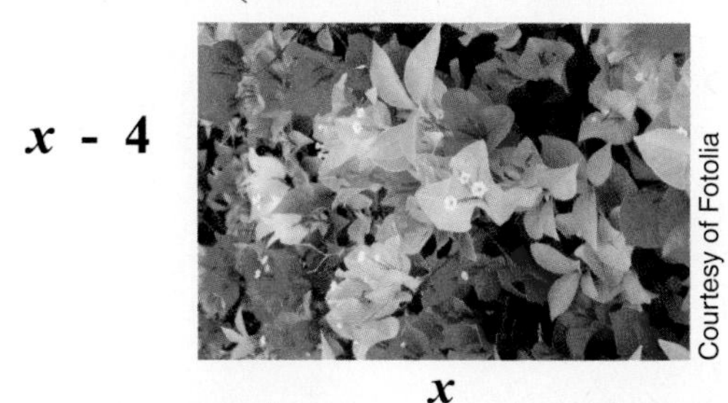

Courtesy of Fotolia

x

The area of a rectangle is given by
Area = length • width

Choose the formula. AREA = LENGTH • WIDTH

Write the equation. $96 = x \cdot x - 4$

Solve.

$$96 = x(x-4)$$
$$96 = x^2 - 4x$$
$$x^2 - 4x - 96 = 0$$
$$(x-12)(x+8) = 0$$

$x - 12 = 0$ *OR* $x + 8 = 0$

$x = 12$ $\qquad$ $x = -8$

The solutions are 12 and -8. A rectangle cannot have a side of negative length, so discard -8. The length of the garden will be 12 feet.

Use the definitions of the variables from above to answer the question.

x = the length of the garden = 12 ft

$x - 4$ = the width = 12 - 4 = 8 ft

The length is 12 ft and the width is 8 ft.

Courtesy of Fotolia

When solving applied problems, always check solutions against physical facts and discard any answers that are not appropriate.

EXAMPLE 3

The length of the base of a triangle is twice its height. If the area of the triangle is 100 square kilometers, find the height.

Assign a variable:

Let x = length of the height

$2x$ = length of the base (twice its height)

The area of a triangle is given by

$$Area = \frac{1}{2} b\, h$$

Choose the formula.

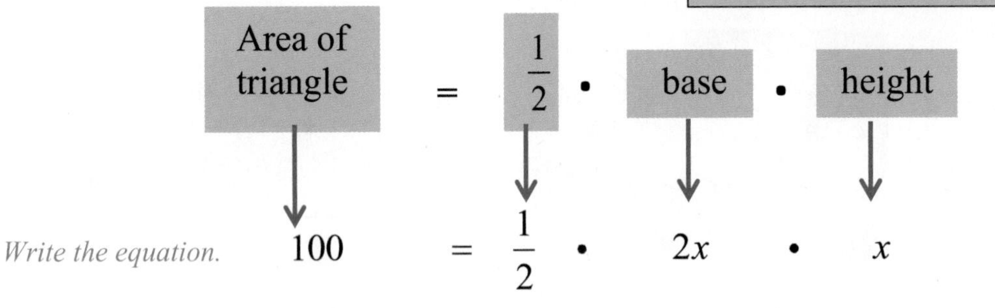

Write the equation.

$$100 = \frac{1}{2} \cdot 2x \cdot x$$

Solve.

$$100 = \frac{1}{2} \cdot 2x \cdot x$$

Multiply $\frac{1}{2} \cdot 2 = 1$

$$100 = 1x \cdot x$$

$$100 = x^2$$

$$0 = x^2 - 100$$

$$0 = (x + 10)(x - 10)$$

$x + 10 = 0$ *OR* $x - 10 = 0$

$x = -10$ $x = 10$

Reread the question and answer.

Since x represents the length of the height, we discard the solution – 10. The height of a triangle cannot be negative. The height is 10 kilometers.

Let x = length of the height = 10 km

$2x$ = length of the base = 20 km

The length of the height is 10 km and the length of the base is 20 km.

6.14 Quadratic Equations and Problem Solving 6.14

YOUR TURN

3. The length of a rectangular rug is 2 feet more than the width. The area of the rug is 48 ft^2. Find the length and width of the room.

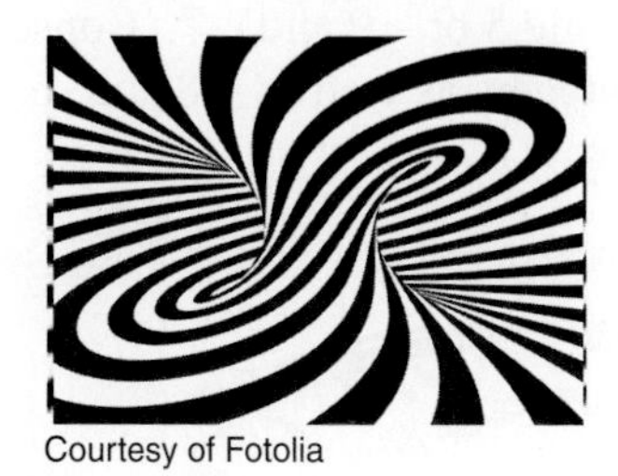
Courtesy of Fotolia

Assign variables:

length: ______________

width: ______________

Choose a formula:

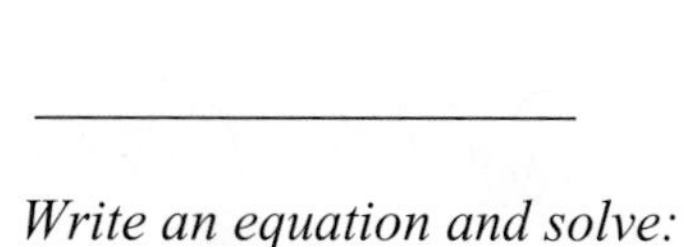

____

Write an equation and solve:

Reread the question and answer:

____

4. The height of a triangular sail is 2 meters less than twice the length of the base. If the sail has an area of 30 square meters, find the length of its base and the height.

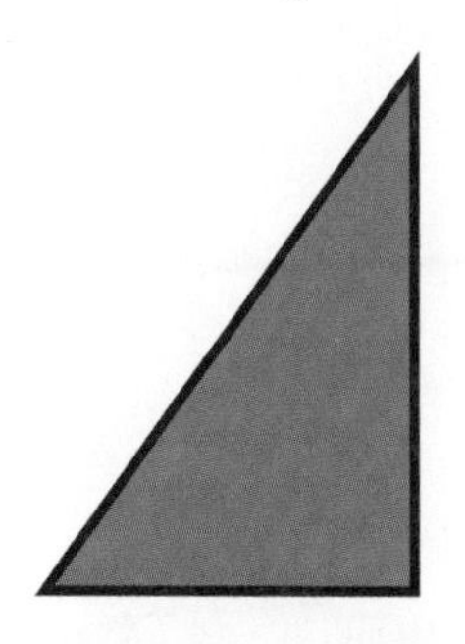

height = ______

base = ______

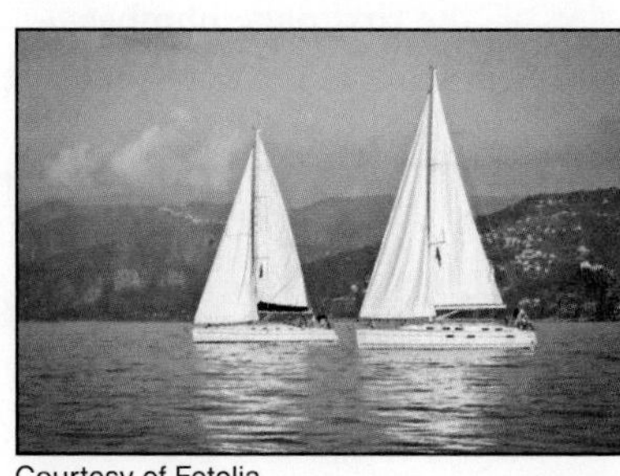
Courtesy of Fotolia

Assign variables:

base: ______________

height: ______________

Choose a formula:

____

Write an equation and solve:

Reread the question and answer:

____

6.14 Quadratic Equations and Problem Solving 6.14

Consecutive Integers

Courtesy of Fotolia

Recall that consecutive integers are integers that are next to each other on a number line, such as 5 and 6 or -10 and – 9 . Consecutive odd integers are odd integers that are next to each other, such as 3 and 5 or – 9 and – 7. Consecutive even integers are defined similarly; 4 and 6 are consecutive even integers as well as – 8 and -6.

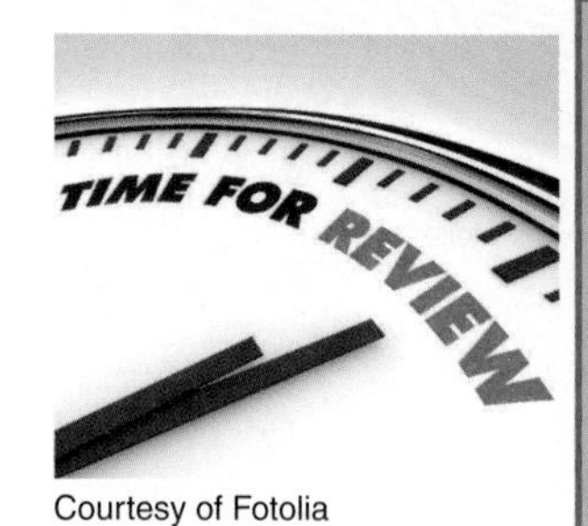

Courtesy of Fotolia

In consecutive integer problems, if x represents the first integer, then for

Two consecutive integers, use $x \, , x+1$

Two consecutive *even* integers, use $x \, , x+2$

Two consecutive *odd* integers, use $x \, , x+2$

EXAMPLE 4

The product of two consecutive page numbers is 420. Find the page numbers.

Assign variables:
x = the first page number
$x+1$ = the next consecutive page number

The product of two consecutive page numbers is 420.

Write an equation. $x \bullet (x+1) = 420$

$$x(x+1) = 420$$
$$x^2 + x = 420$$
$$x^2 + x - 420 = 0$$
$$(x-20)(x+21) = 0$$

$x - 20 = 0$ *OR* $x + 21 = 0$
$x = 20$ $x = -21$

The page numbers are 20 and 21.

Reread the question and answer.
The solutions are -21 and 20. Discard the -21 since a page number cannot be negative.

x = the first page number = 20
$x+1$ = the next consecutive page number = 21

EXAMPLE 5

Find two consecutive odd integers whose **product is 23 more than their sum.**

Assign a variable:

x = first consecutive odd integer

$x + 2$ = second consecutive odd integer

	Product of integers	is	Sum of integers	More than	23
Write an Equation.	$x(x+2)$	$=$	$x + (x+2)$	$+$	23

$$x(x+2) = x + (x+2) + 23$$
$$x^2 + 2x = x + x + 2 + 23$$
$$x^2 + 2x = 2x + 25$$
$$x^2 + 2x - 2x - 25 = 0$$
$$x^2 - 25 = 0$$
$$(x+5)(x-5) = 0$$

$$x + 5 = 0 \quad OR \quad x - 5 = 0$$
$$x = -5 \qquad\qquad x = 5$$

Reread the question and answer.

Since integers can be positive or negative we will need to keep both values for *x*. So, in other words, *x* can be 5 or -5.

If $x = 5$

x = first consecutive odd integer = 5

$x + 2$ = second consecutive odd integer = 7

If $x = -5$

x = first consecutive odd integer = -5

$x + 2$ = second consecutive odd integer = -3

The two consecutive odd integers are 5 and 7 or -5 and -3.

6.14 Quadratic Equations and Problem Solving 6.14

Courtesy of Fotolia

DO NOT USE $x, x+1, x+3$, and so on to represent consecutive odd integers. To see why, let $x = 3$. Then

$$x+1=3+1=4$$
$$x+3=3+3=6$$

3, 4, and 6 are *not* consecutive odd integers.

Courtesy of Fotolia

DO USE $x, x+2, x+4$, and so on to represent consecutive odd integers. To see why, let $x = 3$. Then

$$x+2=3+2=5$$
$$x+4=3+4=7$$

3, 5, and 7 are consecutive odd integers.

YOUR TURN

5. The product of two consecutive even integers is 528. Find the numbers.	6. Find two consecutive odd integers such that their product is 15 more than three times their sum.
1st #: ______________	1st #: ______________
2nd #: ______________	2nd #: ______________
Equation:	Equation:
Answer: ____________________________	Answer: ____________________________

6.14 Quadratic Equations and Problem Solving 6.14

Pythagorean Theorem

Courtesy of Fotolia

The next two examples make use of the ***Pythagorean Theorem***. Recall that a ***right triangle*** is a triangle that contains a 90 degree angle. The ***hypotenuse*** of the right triangle is the side opposite the right angle and is the longest side of the triangle. The ***legs*** of a right triangle are the other sides of the triangle.

Pythagorean Theorem

In a right triangle, the sum of the squares of the lengths of the two legs is equal to the square of the length of the hypotenuse.

$$(leg)^2 + (leg)^2 = (hypotenuse)^2 \qquad or \qquad a^2 + b^2 = c^2$$

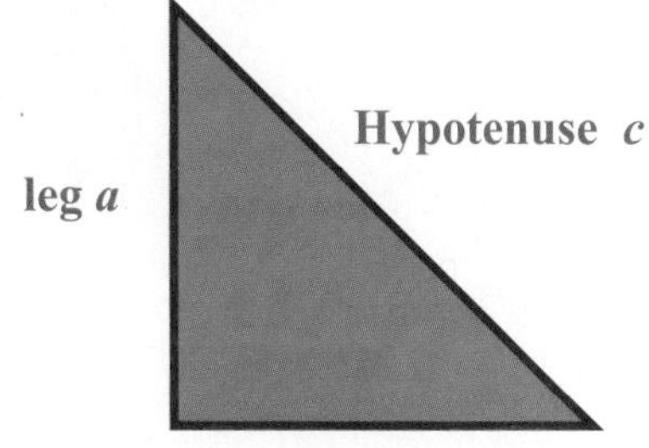

EXAMPLE 6

Find the lengths of the sides of a right triangle if the lengths can be expressed as three consecutive even integers.

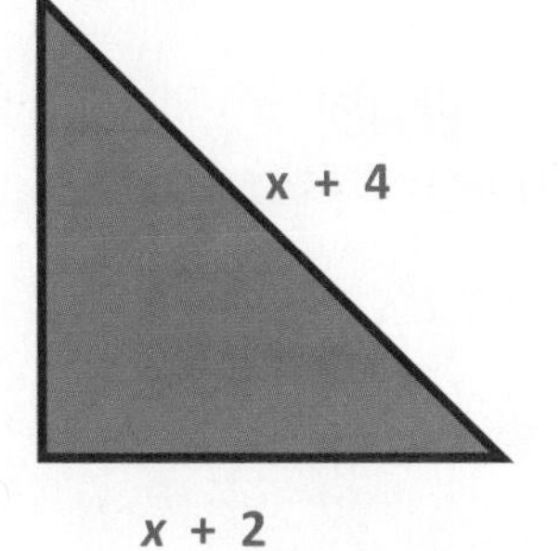

Assign variables.

x = one leg

$x + 2$ = other leg

$x + 4$ = hypotenuse

6.14 Quadratic Equations and Problem Solving 6.14

Use the Pythagorean Theorem to write an equation.

$$(leg)^2 + (leg)^2 = (hypotenuse)^2$$

$$(x)^2 + (x+2)^2 = (x+4)^2$$

Refer to green box at bottom of page to recall how to square a binomial.

$$x^2 + x^2 + 4x + 4 = x^2 + 8x + 16$$

$$2x^2 + 4x + 4 = x^2 + 8x + 16$$

$$2x^2 + 4x + 4 - x^2 - 8x - 16 = 0$$

$$x^2 - 4x - 12 = 0$$

$$(x-6)(x+2) = 0$$

$x - 6 = 0$ *OR* $x + 2 = 0$

$x = 6$ $\qquad$ $x = -2$

Reread the question and answer.

We discard -2 since length cannot be negative.

x = one leg = 6 units

$x + 2$ = other leg = 6 + 2 = 8 units

$x + 4$ = hypotenuse = 6 + 4 = 10 units

The sides have lengths 6 units, 8 units, and 10 units.

Courtesy of Fotolia

Recall that the square of a binomial results in a perfect square trinomial.

$$(a+b)^2 = a^2 + 2ab + b^2$$
$$(a-b)^2 = a^2 - 2ab + b^2$$

Example: $(x+2)^2 = x^2 + 4x + 4$

If you don't feel comfortable using the pattern shortcut then you can always multiply the two binomials using the distributive property.

$$\begin{aligned}(x+2)^2 &= (x+2)(x+2)\\ &= x(x+2) + 2(x+2)\\ &= x^2 + 2x + 2x + 4\\ &= x^2 + 4x + 4\end{aligned}$$

6.14 Quadratic Equations and Problem Solving **6.14**

EXAMPLE 7

The ladder is leaning against a building so that the distance from the ground to the top of the ladder is one foot less than the length of the ladder. Find the length of the ladder if the distance from the bottom of the ladder to the building is 5 feet.

x

$x - 1$

Courtesy of Fotolia

5

Assign variables.

$x - 1$ = one leg
5 = other leg
x = hypotenuse

Use the Pythagorean Theorem to write an equation.

$$(leg)^2 + (leg)^2 = (hypotenuse)^2$$

$$(x-1)^2 + (5)^2 = (x)^2$$
$$x^2 - 2x + 1 + 25 = x^2$$
$$x^2 - 2x + 26 = x^2$$
$$x^2 - 2x + 26 - x^2 = 0$$
$$-2x + 26 = 0$$
$$-2x = -26$$
$$\frac{-2x}{-2} = \frac{-26}{-2}$$
$$x = 13$$

The length of the ladder is 13 ft.

The length of the ladder was defined to be x. So, the ladder is 13 ft in length.

Notice the equation highlighted in yellow above. This is a linear equation, not a quadratic equation. The exponent is not squared and therefore cannot be factored and solved with the zero property rule. We solve by isolating the variable on one side of the equal.

6.14 Quadratic Equations and Problem Solving **6.14**

YOUR TURN

7. The hypotenuse of a right triangle is 1 cm longer than the longer leg. The shorter leg is 7 cm shorter than the longer leg. Find the length of the longer leg of the triangle. Label each side of the triangle and then fill in the remaining definitions. Then substitute into the Pythagorean Theorem.

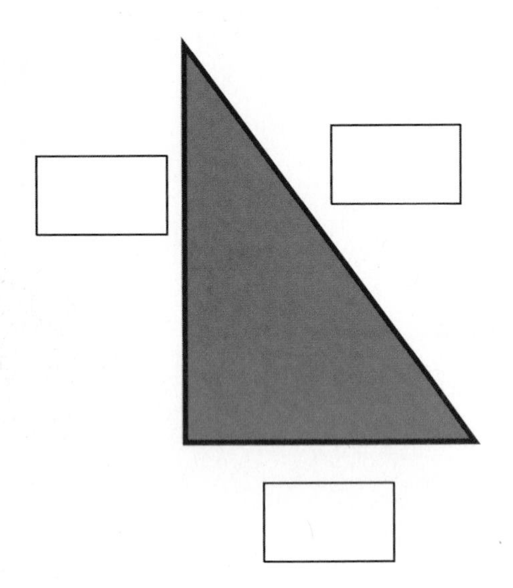

Assign variables:

x = length of longer leg

= length of hypotenuse

= length of shorter leg

Choose the formula:

$$(leg)^2 + (leg)^2 = (hypotenuse)^2$$

Write the equation and solve.
HINT: Remember how to square a binomial. Look at the example for help.

6.14 Quadratic Equations and Problem Solving 6.14

YOUR TURN

8. A 40 foot ladder is leaning against a building so that the distance from the bottom of the ladder to the building is 8 feet more than the distance the ladder reaches up the building. Find how far up the building the ladder reaches.

Fill in the squares with the appropriate value in the diagram below. Read carefully to make sure you assign the right expressions to each side of the triangle.

Courtesy of Fotolia

Assign variables:
one leg = ________

other leg = ________

hypotenuse = ________

Choose the formula: ____________________________

Write an equation and solve.

6.14 Quadratic Equations and Problem Solving

Quadratic Models

Courtesy of Fotolia

In all the previous examples we wrote quadratic equations to model various situations and then solved the equations. In the next examples, we will be given the quadratic models and must use them to determine data.

EXAMPLE 8

Since the 1940s, one of the top tourist attractions in Acapulco, Mexico, is watching cliff divers off La Quebrada. The divers' platform is about 144 feet above the sea. These divers must time their descent just right, since they land in the crashing Pacific, in an inlet that is at most 9.5 feet deep. Neglecting air resistance, the height h in feet of a cliff diver above the ocean after t seconds is given by the quadratic equation $h = -16t^2 + 144$.

Find how long it takes the diver to reach the ocean.

Courtesy of Fotolia

Courtesy of Fotolia

The KEY to every word problem is to understand what the variables represent in words.
time, seconds = t
height, feet = h

When the diver reaches the ocean, he will be 0 feet above sea level. In other words, when the diver reaches the ocean, his height will be 0 ft. So, we will substitute 0 in for the height in the given quadratic equation.

$$h = -16t^2 + 144$$

$h = 0$ feet at sea level

$$0 = -16t^2 + 144$$
$$0 = -16\left(t^2 - 9\right)$$
$$0 = -16(t + 3)(t - 3)$$

$$t + 3 = 0 \quad OR \quad t - 3 = 0$$
$$t = -3 \qquad\qquad t = 3$$

The diver will reach the ocean in 3 seconds.

We discard -3 because time cannot be negative. It will take the diver 3 seconds to reach the ocean.

EXAMPLE 9

A tennis player can hit a ball 180 ft per sec. If she hits a ball directly upward, the height h, of the ball in feet at time t in seconds is modeled by the quadratic equation

$$h = -16t^2 + 180t + 6.$$

When will the ball be 206 feet above the ground?

Courtesy of Fotolia

First remember to understand the variables in words.

t = time in seconds This question is asking for the time, *t, when the ball is 206 feet in the air.*

h = height in feet

A height of 206 ft means $h = 206$. The ball reaches 206 feet on the way up and then again on the way down as it falls back to the ground.

To find the time the ball reaches the height of 206 feet on the way up and the way down, substitute 206 for h in the equation and then solve for t. (Hint: Because this is a quadratic equation, we will get two possible solutions.)

$$206 = -16t^2 + 180t + 6$$

After substituting,

- put all terms on the same side,
- factor out a negative gcf,
- and then factor the remaining trinomial.

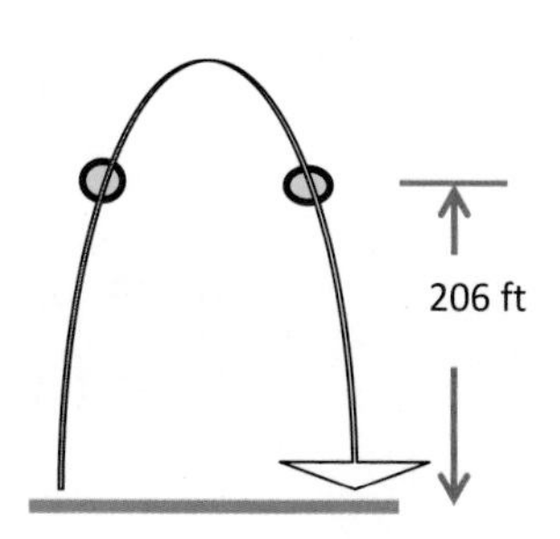

$$206 = -16t^2 + 180t + 6$$

$$-16t^2 + 180t + 6 = 206$$
$$-16t^2 + 180t + 6 - 206 = 0$$
$$-16t^2 + 180t - 200 = 0$$
$$-4\left(4t^2 - 45t + 50\right) = 0$$
$$-4(4t^2 - 40t - 5t + 50) = 0$$
$$-4[4t(t - 10) - 5(t - 10)] = 0$$
$$-4(4t - 5)(t - 10) = 0$$

$4t - 5 = 0$ *OR* $t - 10 = 0$

$4t = 5$ $t = 10$

$t = \frac{5}{4} = 1.25$

The ball will be 206 ft above the ground at 1.25 sec and 10 sec.

Since we found two acceptable answers (neither answer was negative), the ball will be 206 ft above the ground twice – once on its way up and once on its way down.

Courtesy of Fotolia

EXAMPLE 10

If the cost, C, for manufacturing x printer cartridges is given by $C = x^2 - 5x + 40$,
Find the number of printer cartridges manufactured at a cost of $13,840.

First remember to understand the variables in words.
 # of printer cartridge units: x
 Cost, $: C

The problem gives the C value, a cost of $13, 840. It is asking for the number of printer cartridges, the x value.

$$13{,}840 = x^2 - 5x + 40$$ *Substitute the cost, $13,840 in for C.*

$$0 = x^2 - 5x + 40 - 13{,}840$$ *Subtract to put all terms on the same side of the "=".*

$$0 = x^2 - 5x - 13{,}800$$

$$0 = (x - 120)(x + 115)$$ *Factor and solve. Use the "y=" feature of your calculator to find the factors.*

$x - 120 = 0$ *OR* $x + 115 = 0$

$x = 120$ $\qquad$ $x = -115$

120 units can be produced at a cost of $13,840.

Since a company cannot product a negative number of products, we will discard the negative answer.

The answer is 120 printer cartridges.

6.14 Quadratic Equations and Problem Solving 6.14

YOUR TURN

9. An object is thrown upward from the top of an 80-foot building with an initial velocity of 64 feet per second. The height, h of the object after t seconds is given by the quadratic equation $h = -16t^2 + 64t + 80$**. When will the object hit the ground?** (Hint: The height of an object on the ground is 0 ft.)

Understand the variables in words.

Time, seconds: ____

Height, feet: ______ What is h when the object hits the ground? $h =$____

Write an equation and solve:

Courtesy of Fotolia

10. If the cost, C, for manufacturing x units of a certain product is given by $C = x^2 - 15x + 50$,

Find the number of units manufactured at a cost of 9500.

Understand the variables in words.

units: ______

Cost, $: ______ What is the cost? C = ________

Write an equation and solve.

11. The number of impulses I fired after a nerve has been stimulated is modeled by $I = -x^2 + 2x + 60$**, where I is the number of impulses and x is time in seconds. When will 45 impulses occur?**

Understand the variables in words.

Time, seconds = ____

impulses: = _____

How many impulses? I = _____

Write an equation and solve.

6.14 Quadratic Equations and Problem Solving 6.14

YOUR TURN **Answers to Section 6.14.**

1. 9 and -7
2. -12, 11
3. Length = 8 ft , width = 6 ft
4. base = 6 meters, height = 10 meters
5. 22 and 24 or -24 and -22
6. 7 and 9 or -3 and -1
7. longer leg = 12 cm
8. 24 ft
9. 5 seconds
10. 105 units
11. 5 seconds

Complete MyMathLab Section 6.14 Homework in your Homework notebook.

6.15 Introduction to Functions, Domain and Range 6.15

Relating real life situations with mathematical sets of data and equations is a useful tool in today's world. These places include construction and architecture, business management, medical science, engineering and many other areas. It is important to be able to relate how inputting information can result in expected outcomes or output values. The way we discuss and relate starting data (input values) with ending data (output values) have special names

Courtesy of Fotolia

RELATIONS, DOMAIN, AND RANGE

DEFINITIONS:

RELATIONS: A set of ordered pairs with starting values called inputs, x values, paired with a set of ending values called outputs, y values.

Relations may be represented as a set of ordered pairs, a graph, or an equation.

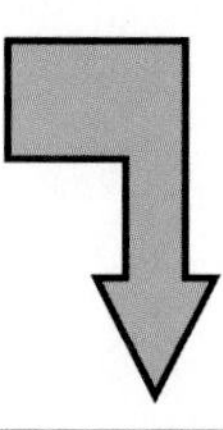

Domain, D:

The set of all first components of ordered pairs, the x-coordinates.

To find a domain given *a set of ordered pairs, list the x values* inside the set symbols { , , } separated by a comma.

Range, R:

The set of all second components of ordered pairs, y-coordinates

To find a range *given a set of ordered pairs, list the y values* inside the set symbols { , , } separated by a comma.

Function, *f(x), g(x) or h(x):*

A *special relation* in which each x-coordinate corresponds to exactly one y-coordinate . . . meaning that there are no points on a graph where an x-value is used more than once; x-values cannot repeat in a function.

6.15 Introduction to Functions, Domain and Range 6.15

Relations can be represented several ways. Sometimes there are a set of points written as ordered pairs. Sometimes they are represented using a picture called a mapping. Other times relations are graphs of points, lines or curves. Let's look at how to determine *the domain and range of a relation* which is a set of ordered pairs or is a mapping. We will also see a few ways to determine if a relation is a *function*, either by looking at a set of ordered pairs, or by looking at a graph .

Type 1: If the Relation is a set of ordered pairs or a mapping, (x, y) or (inputs, outputs), then find the domain and range by listing the coordinates separated by a comma in set notation.
To determine if the relation is a function, look to see that x values are used once, they are not repeated.
(Remember: The set notation symbols "{ , , }" are used to indicate a set of data.)

EXAMPLE 1

a. Find the domain, range and determine if the relation is a function.

This relation is a set of ordered pairs.

Coordinates which are repeated can be listed once.

{ (1, 8), (4, -9), (2, -9), (3, 8) }

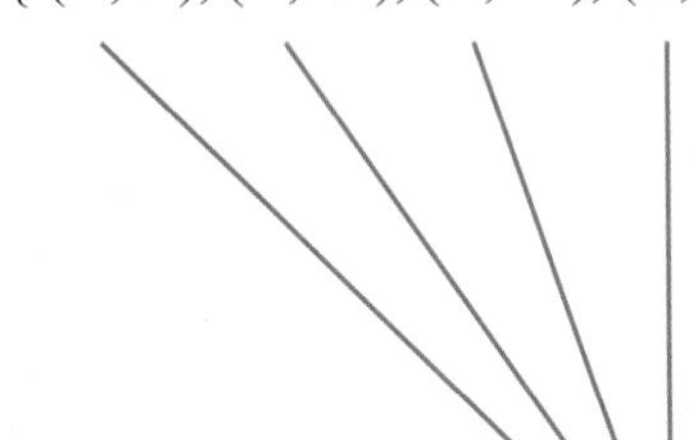

Solution:
Domain: { 1, 2, 3, 4 } List of the *x values* from least to greatest in set notation brackets { , , } separated by a comma.

Range: { -9, 8 } List of the *y values* from least to greatest in set notation brackets { , , } separated by a comma.

Function: Yes Notice that *no x value is repeated*. Therefore this relation IS a function. It does not matter that the y values are repeated.

b. Find the domain, range and determine if the relation is a function.

This relation is a Mapping.

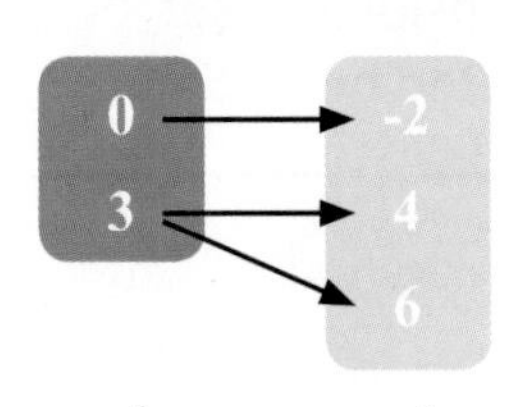

x values
inputs

y values
outputs

Solution:
It will help to list the mapping as a set **{ (0, -2), (3, 4), (3, 6)}.**

To do this, use the arrows to pair an x-coordinate with its corresponding y-coordinate. In other words, the ordered pairs are made by mapping an x to a y.

Domain: { 0, 3 } List of the *x values* from least to greatest in set notation brackets { , , } separated by a comma.

Range: { -2, 4, 6 } List of the *y values* from least to greatest in set notation brackets { , , } separated by a comma.

Function: No Notice that the *x value 3 is repeated*. Therefore this relation is not a function.

6.15 Introduction to Functions, Domain and Range 6.15

Type 2: If the Relation is a graph of points, a curve or line(s), then use the Vertical Line Test to determine if it is a function. We will find the domain and range of these types of relations in the next section.

EXAMPLE 2

This relation is the graph of an ellipse.

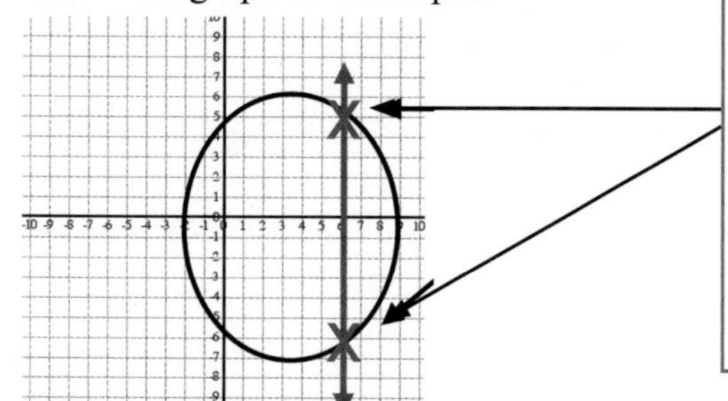

> ***Use the Vertical Line Test:***
>
> **The Vertical Line Test states that if a line can be drawn which crosses the relation or curve in *more than one point*, then the relation is a NOT a function.**
>
> **The Vertical Line Test is for graphs.**

Function: __No__ Notice that many x values can be repeated by drawing a vertical line.

EXAMPLE 3

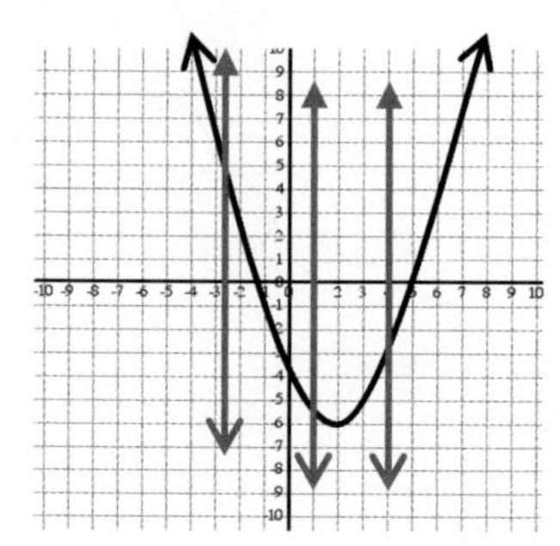

> ***Use the Vertical Line Test:***
>
> **The Vertical Line Test states that if a line cannot be drawn which crosses the relation or curve in *more than one point*, then the relation IS a function.**
>
> **No matter where we draw a line on this graph, it will not cross more than once.**

Function: __Yes__ Notice that no x values can be repeated by drawing a vertical line anywhere on the graph.

EXAMPLE 4

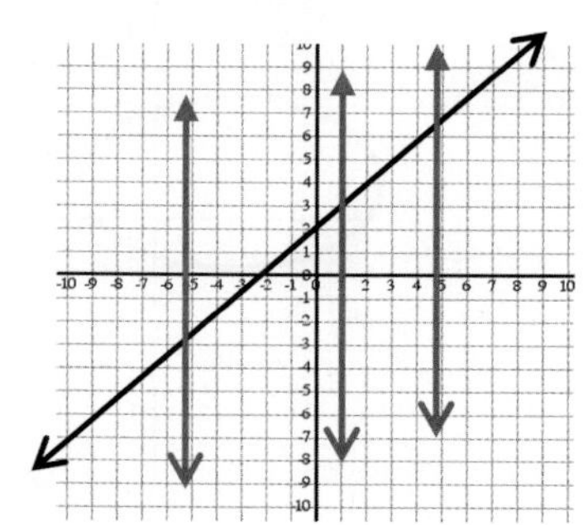

> ***Use the Vertical Line Test:***
>
> **The Vertical Line Test states that if a line cannot be drawn which crosses the relation or line in *more than one point*, then the relation IS a function.**
>
> **No matter where we draw a line on this graph, it will not cross more than once.**

Function: __Yes__ Notice that no x values are used more than once by drawing a vertical line anywhere on the graph.

6.15 Introduction to Functions, Domain and Range 6.15

YOUR TURN

Find the domain and range of each relation. Also determine if the relation is a function

1. { (- 3, -2), (0, -1), (-10, 3), (-3, 4) }

What is the domain? ________________

What is the range? ________________

Does the given relation represent a function?

______

2. { (8, -2), (0, -2), (7, -2), (4, -2) }

What is the domain? ________________

What is the range? ________________

Does the given relation represent a function?

______

Look at the following graphs and determine whether each graph is a function using the vertical line test. Write yes or no in the blank underneath each graph.

3.

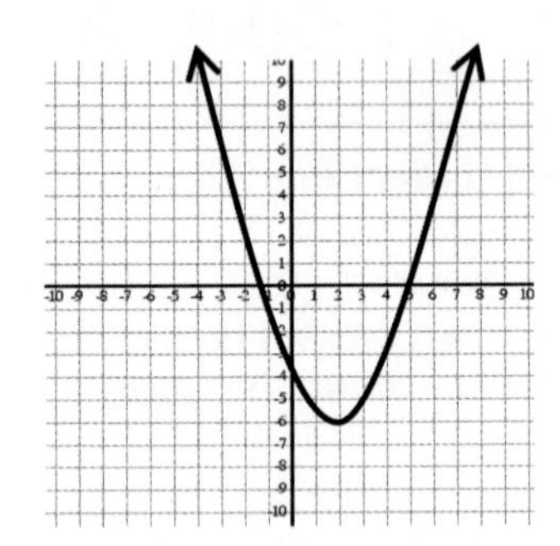

Function: ________________

4.

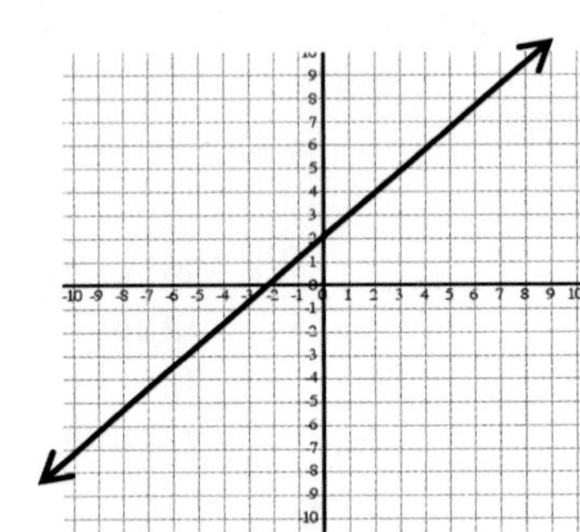

Function: ________________

5.

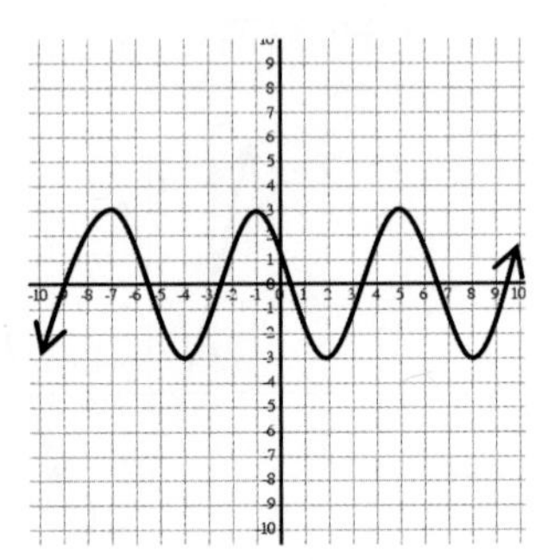

Function: ________________

6.

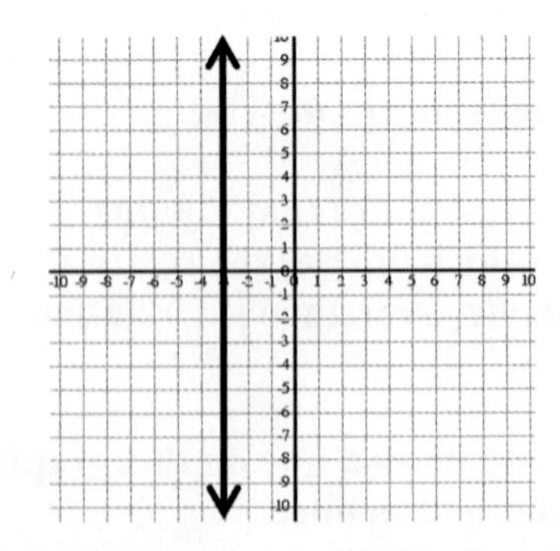

Function: ________________

7.

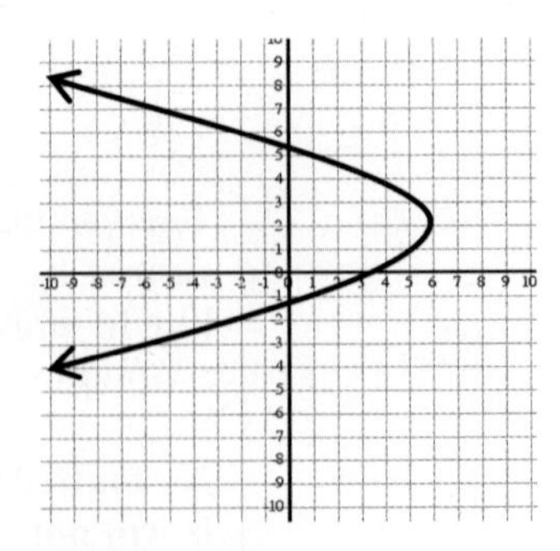

Function: ________________

8.

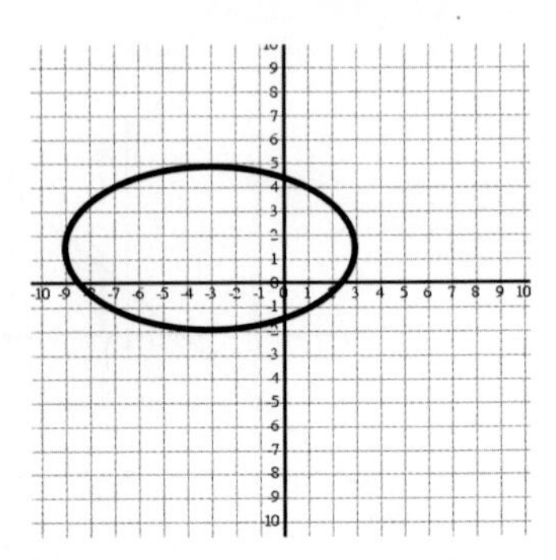

Function: ________________

FINDING FUNCTION VALUES FROM EQUATIONS.

FUNCTION NOTATION, $f(x)$

The symbol $f(x)$ is used when a function is defined by an equation.

The notation $f(x)$ is read *"f of x"*, not *"f times x"*. It is one piece which represents the y value in an equation or on a graph.

For example, $f(x) = 2x - 4$ is the same as the equation $y = 2x - 4$.
Note: The $f(x)$ notation replaces the variable y.

Ordered pairs representing points on the line are written $(x, f(x))$.

To use function notation with the function $y = 4x + 3$, we write $f(x) = 4x + 3$. The notation $f(1)$ means to replace x with 1 and find the resulting y or function value.

Given $f(x) = 4x + 3$, THEN $f(1) = 4(1) + 3 = 4 + 3 = 7$.

The function notation $f(1)$ means that when $x = 1$, the function value $f(x) = 7$. The corresponding ordered pair is (1,7).

EXAMPLE 5

Given the function $f(x) = 2x - 4$, find $f(-1)$.

This asks the question, ***"What is the function value, (the output y value), when x = -1?"***

Given $\boldsymbol{f(x) = 2x - 4}$

$f(-1) = 2(-1) - 4$ — To find $\boldsymbol{f(-1)}$ means to plug the number in the parentheses, the -1, into the right side of the equation for x and simplify the right side.

$f(-1) = -2 - 4$ — *Leave the left side of the equation alone.*

$f(-1) = -6$ — Hence, $f(-1)$ is the output value -6. *This says the function value at -1 is -6.*

This also means that the ordered pair (-1, -6) is a point on the graph of the function.

6.15 Introduction to Functions, Domain and Range 6.15

Courtesy of Fotolia

You can write the input and output of a function as an "ordered pair," such as (4,16). They are called **ordered** pairs because the input always comes first, and the output second:

(input, output) or ($x, f(x)$)

If $f(1) = -10$, the corresponding ordered pair is (1, -10).

If $g(4) = 56$, the corresponding ordered pair is (4, 56).

EXAMPLE 6

Given the function $h(x) = x^2 - 3x + 5$, find the following.

a. $h(0)$ Plug in 0 for x.

$h(0) = (0)^2 - 3(0) + 5$

$h(0) = 0^2 - 0 + 5$

$h(0) = 5$

(0 , 5)

The function value at 0 is 5.

b. $h(-2)$ Plug in -2 for x.

$h(-2) = (-2)^2 - 3(-2) + 5$

$h(-2) = 4 + 6 + 5$

$h(-2) = 15$

(-2, 15)

The function value at -2 is 15.

Courtesy of Fotolia

Note that $f(x)$ is a special symbol in mathematics used to denote a function. The symbol $f(x)$ is read " f of x ".

It **does not** mean $f \cdot x$ (f times x).

YOUR TURN

Given the function $g(x) = -\frac{1}{3}x^2 + 2$, find the following.

9. $g(0)$

10. $g(3)$

11. $g(-6)$

READING FUNCTION VALUES FROM A GRAPH

Remember that functions are graphs of ordered pairs where each x value is paired with one y value. Look at the following graphs and name the function values, $f(x)$.
(In other words, f(x) means to state the y value at the given x value.)

To find function values $f(x)$ from a graph:

- First find the the **given x value(** in the parentheses) on the x axis.
- Locate the point on the graph which has this x coordinate, $(x, \quad)$.
- **READ** the corresponding y-coordinate, $(x, f(x))$

$f(x) = y$ coordinate of a given x value.

EXAMPLE 7

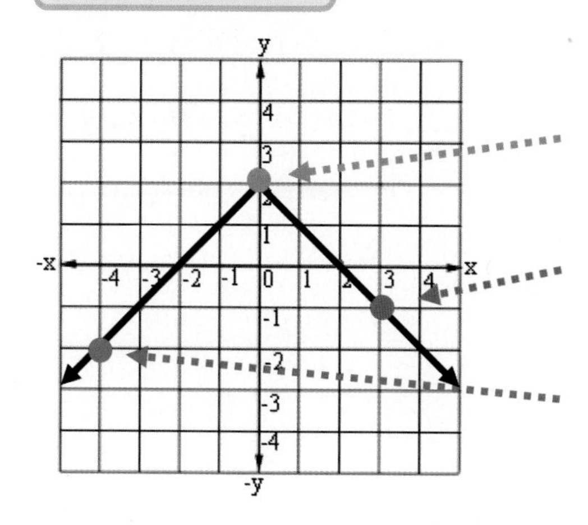

a. Find $f(0)$.
Find $x = 0$ on the graph, then look up or down.
Read the y value on the graph when $x = 0$. The y coordinate is up 2. Hence, $f(0) = 2$.

b. Find $f(3)$.
Find $x = 3$ on the graph, then look up or down.
Read the y value on the graph when $x = 3$.
The y coordinate is down - 1. Hence, $f(3) = -1$.

c. Find $f(-4)$.
Find $x = -4$ on the graph, then look up or down.
Read the y value when $x = -4$.
The y coordinate is down - 2. Hence, $f(-4) = -2$.

6.15 Introduction to Functions, Domain and Range 6.15

We can also find the x-value if given the y-value or function value. We have already stated that $f(x)$ is the same value as y. So, when we see $f(x) = 3$, we are actually saying that **$y = 3$, or the y coordinate is 3.**

In the previous examples, you were given the x value and asked to find the function value. Now, let's try it the other way.

EXAMPLE 8 **REVERSE THE PROCESS.**

Find all x-values such that $f(x) = y$ coordinate which is given.

a. Find *f(x) = 0.*

Find y = 0 on the graph.(Change it up!)
Look left and right. There are two points when y = 0 in this problem.
Read the x values when $y = 0$. The x coordinates are left – 4 and another right 2.
Hence, *$x = -4$ or 2.*

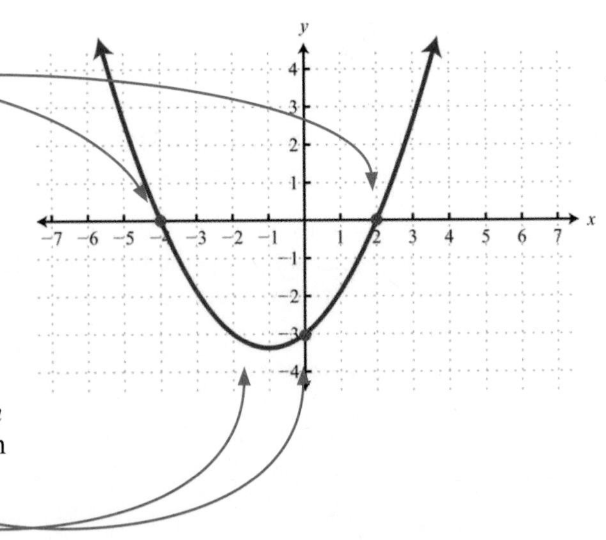

If you were asked to write the points where $f(x) = 0$, they would be (- 4, 0) or (2, 0).

b. Find *f(x) = - 3.*

To find y = - 3 on the graph, go down to -3.
Look left and right. There are two points when y = -3 in this problem. Read the x values when $y = -3$. The x coordinates are left -2 and 0.
Hence, ***x = -2 or 0.***

If you were asked to write the points where $f(x) = -3$, they would be (- 2, - 3) or (0. - 3).

YOUR TURN

Answer each question. Use the graph in Example 8.

12. Use the graph above to find $f(-5)$.

$f(-5) =$ ________

13. Use the graph above to find $f(x) = -2$.

$x =$ __________ (Use a comma to separate answers as needed.)

GRAPHING LINEAR FUNCTIONS

A Review of Slope-Intercept Form with Function Notation

All *linear equations* which can be written in slope intercept form are functions. To emphasize that these equations are functions, we can write them using function notation by replacing the *y* with *f(x)*. Hence,

$$y = mx + b \longrightarrow f(x) = mx + b$$

(All linear equations are functions except vertical lines where $x = \#$ is the equation. Remember, horizontal lines are $y = \#$ and are always functions.)

I. Graphing Lines Using Slope-Intercept Form, $y = mx + b$ or $f(x) = mx + b$.

EXAMPLE 9

Graph $2x + 3y = 3$.

$2x + 3y = 3$
$-2x \quad\quad -2x$

First solve the equation for *y*. (It must be in $y = mx + b$ form.) Subtract $2x$ from both sides.

$\frac{3y}{3} = \frac{-2x}{3} + \frac{3}{3}$

Remember to write the $-2x$ term *in front* of the constant term 3. Then divide both sides by the 3 in front of the *y*.

$y = -\frac{2}{3}x + 1$

This is the same as $f(x) = -\frac{2}{3}x + 1$, *where* $m = -\frac{2}{3}$ *and* $b = 1$.

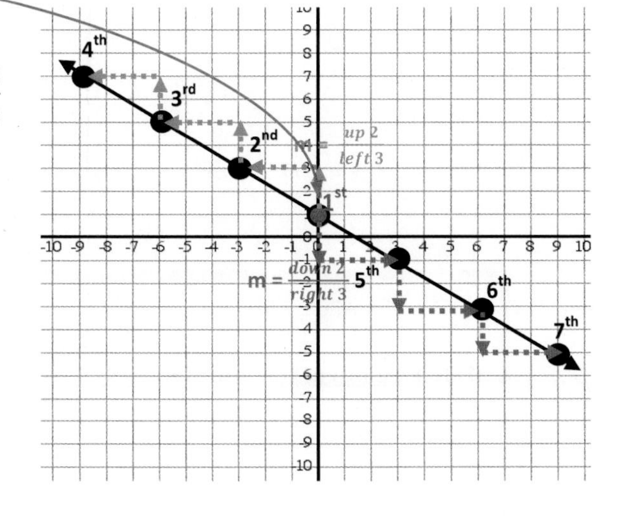

1st: Plot the y-intercept b = 1, point (0, 1).

2nd: Plot other points using the slope,

$m = -\frac{2}{3}$ **which means** $\frac{up\ 2}{left\ 3}$ *or* $\frac{down\ 2}{right\ 3}$ **from the *y*-intercept $b = 1$.**

(Note: The order the points have been plotted are numbered on the graph.)

3rd: Draw a line through the points.

6.15 Introduction to Functions, Domain and Range 6.15

EXAMPLE 10

Graph $f(x) = 4x$.

$f(x) = 4x$ The equation is already in slope-intercept form, $f(x) = mx + b$

$f(x) = mx + b$
$f(x) = 4x + 0$

1st: Graph the y-int. (0, 0)
Why is the y-intercept 0?

$m = \frac{4}{1}, b = 0$

The y-intercept is b = 0 since there is no # following the $4x$ term in the equation.

Use the slope as a fraction to graph.

2nd: $m = 4 = \frac{4}{1}$, count up 4 and right 1 or down 4 and left 1.

3rd: Draw a line through the points.

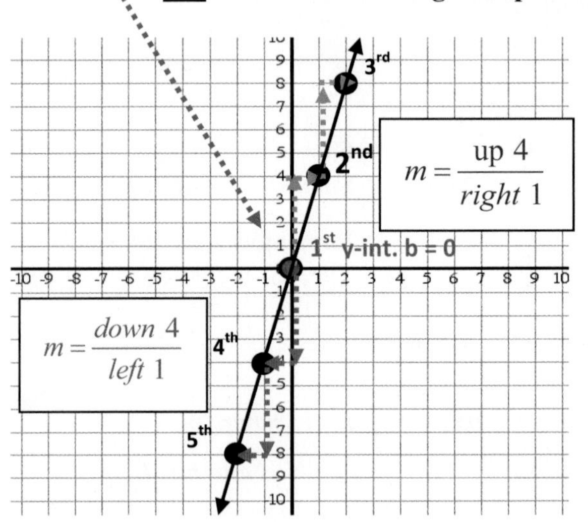

REMINDER ABOUT SLOPES:

If m is positive, the line slants right.

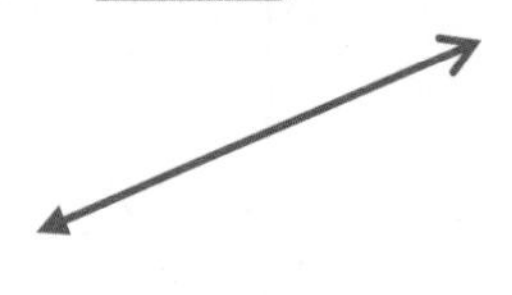

If m is negative, the line slants left.

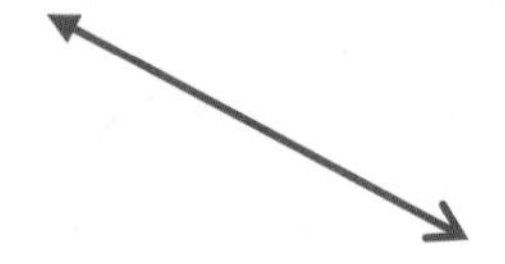

Note: For a review of graphing lines, see DMA 050, Section 5.06 and 5.07.

YOUR TURN

Graph using slope-intercept.

14. $f(x) = \frac{1}{2}x - 3$

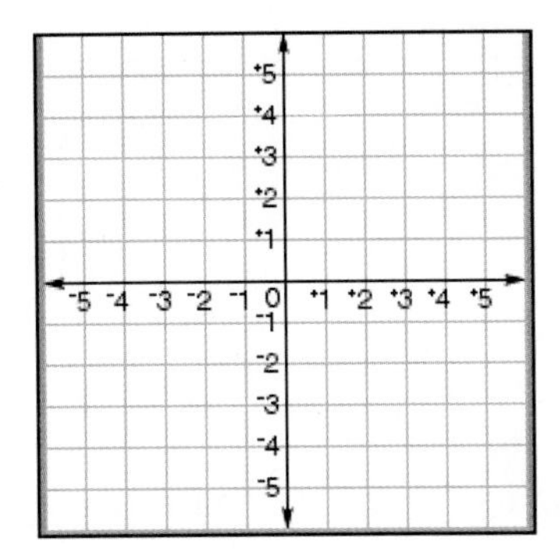

15. $f(x) = -x - 1$

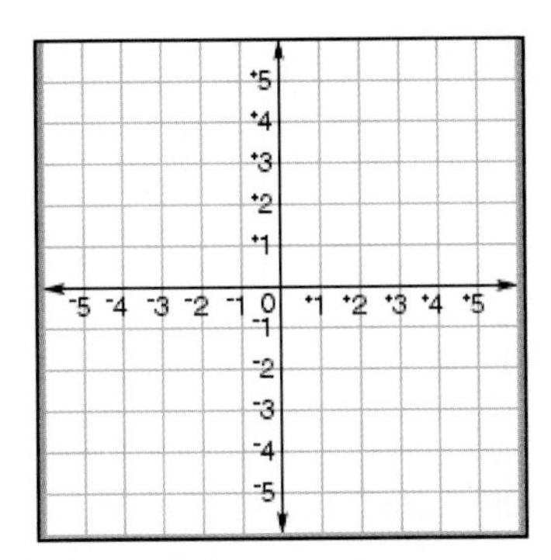

POLYNOMIAL APPLICATION PROBLEMS

Many real-world phenomena are modeled by polynomial functions. If the polynomial function is given, we can often find the solution to a given problem by evaluating the function at a certain value.

EXAMPLE 11

The dosage in milligrams D of Ivermectin, a heartworm preventive, for a dog weighing x pounds is given by $D(x) = \frac{1.36}{25}x$. Find the proper dosage for a dog that weighs 30 pounds.

First, understand the variables in words.

Weight of the dog in lbs. = x

Dosage of Ivermectin in mg = D

Courtesy of Fotolia

$$D(x) = \frac{1.36}{25}x$$

Since the weight of the dog is given, substitute $x = 30$.

$$= \frac{1.36}{25} \cdot 30$$

$$= 1.632\ mg$$ **The dosage of Ivermectin is 1.632 *mg*.**

EXAMPLE 12

A rocket is fired upward from the ground with an initial velocity of 100 feet per second. The height, h, of the rocket at any time t is given by the equation $h(t) = -16t^2 + 100t + 25$.

a. Find the height of the rocket at the given times by filling in the table below.

Time, *t in seconds*	0	1	2	3	4	5	6	7
Height, *h in feet*								

Solve algebraically:

First, understand the variables in words. The time in seconds = t, the height in feet = h.
We will plug in each t into the equation, $h(t) = -16t^2 + 100t + 25$ to find its value. Use your calculator to find these values. The first four values are shown below.

when $t = 0$

$$h(0) = -16t^2 + 100t + 25$$
$$= -16(0)^2 + 100(0) + 25$$
$$= 0 + 0 + 25$$
$$= 25$$

when $t = 1$

$$h(1) = -16t^2 + 100t + 25$$
$$= -16(1)^2 + 100(1) + 25$$
$$= -16 + 100 + 25$$
$$= 109$$

when $t = 2$

$$h(2) = -16t^2 + 100t + 25$$
$$= -16(2)^2 + 100(2) + 25$$
$$= -64 + 200 + 25$$
$$= 161$$

when $t = 3$

$$h(3) = -16t^2 + 100t + 25$$
$$= -16(3)^2 + 100(3) + 25$$
$$= -144 + 300 + 25$$
$$= 181$$

Substitute the other t values into the equation to find the corresponding *h(t)* values in this chart.

Time, *t in seconds*	0	1	2	3	4	5	6	7
Height, *h in feet*	25	109	161	181	169	125	49	-59

6.15 Introduction to Functions, Domain and Range 6.15

Solve graphically:

This sketch is for visual representation only. You do not need to graph.

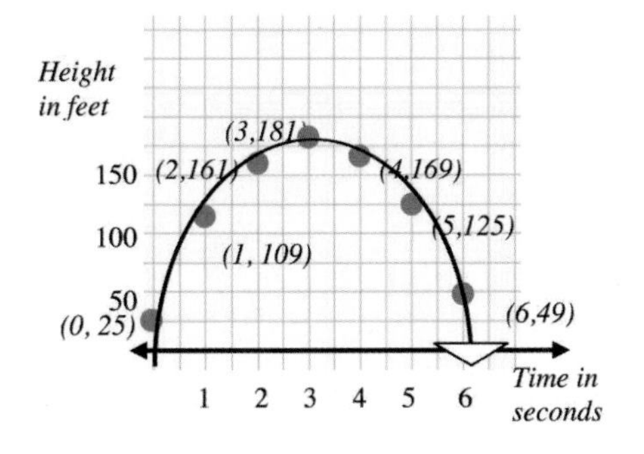

In your calculator, enter the equation in [Y_1=].

Plot1 Plot 2 Plot3
\ Y1= $-16x^2 + 100x + 25$
\ Y2=
\ Y3=

Then go to the [TABLE] by pressing the [2nd] key, and the [GRAPH] key. Read the *y* values from the t-chart.

X	Y_1	
0	25	
1	109	
2	161	
3	181	
4	169	

X =

b. Explain why the height of the rocket increased and decreased.

Height is determined by how far away an object is from the ground. A projectile's height would increase as it was going up, reach its maximum height, and then begin to fall. So, we would expect to see the height increase on its way up and then decrease as the projectile begins its journey back to the ground.

Courtesy of Fotolia

c. Use the table to approximate the maximum height of the rocket.

The rocket reaches its maximum height somewhere near 3 seconds because this is the largest *h* value. The approximate height would be 181 feet.

d. Use the table to determine between what two whole-numbered seconds the rocket strikes the ground.

The rocket hit the ground between 6 and 7 seconds. Notice the height at 6 seconds was 49 feet but the height at 7 seconds is negative. The negative height really doesn't exist but it tells us that somewhere between 6 and 7 seconds the projectile hit the ground.

EXAMPLE 13

The per capita consumption (in pounds) of all beef in a country is given by the function $C(x) = -0.42x + 65.2$, where x is the number of years since 2000.

a. Find and interpret C(7).

b. Estimate the per capita consumption of beef in the country in 2015.

Understand the variables in words.

The years since 2000 = x (This is where x = 0, with 2000 being the initial year.)

The per capita consumption of beef in pounds = C

a. Find C(7). This means that $x = 7$ years from the year 2000. Plug in 7 for the x in the equation.

C(7) $= -0.42(7) + 65.2 =$ **62.26 pounds**.
(Type an integer or a decimal.)

Interpret C(7).

The per capita consumption of beef was **62.26 pounds** in the year **2007**. (Add 7 years to the year 2000.)

b. Estimate the per capita consumption of beef in the country in 2015.

This means that $x = 15$, since the year 2015 is 15 years from the initial year 2000. Plug in 15 for x.

$C(15) = -0.42(15) + 65.2 =$ **58.9 pounds**.

The per capita consumption in **2015** will be **58.9** pounds.
(Type an integer or a decimal.)

YOUR TURN

16. A rocket is fired upward from the ground with an initial velocity of 40 feet per second. The height, $f(x)$, of the rocket at any time x is given by the equation $f(x) = -16x^2 + 90x$.

Courtesy of Fotolia

a. Find the height of the rocket at the given times by filling in the table below.

Time, *x in seconds*	0	1	2	3	4	5	6	7
Height, *y in feet*								

Understand the variables in words:

Time in seconds = ______

Height in feet = ______

Plug *x values* into the equation, $f(x) = -16x^2 + 90x$ to find the height values. Use your calculator to find these values.

b. Use the table to determine between what two whole-numbered seconds the rocket strikes the ground.

c. Use the table to approximate the maximum height of the rocket.

6.15 Introduction to Functions, Domain and Range 6.15

Forensic scientists use the following functions to find the height of a woman if they are given the length of her femur bone (f) or her tibia bone (t) in centimeters.

$H(f) = 2.59f + 47.24$ $H(t) = 2.72t + 61.28$

Understand the variables in words:

Length of the femure bone in cm = ______

Length of the tibia bone in cm = ______

Height of a woman in cm = ________

Courtesy of Fotolia

17. Find the height of a woman whose femur measures 46 centimeters.

Write an equation and solve.

18. Find the height of a woman whose tibia measures 35 centimeters.

Write an equation and solve.

19. The polynomial function $P(x) = 45x - 100{,}000$ models the relationship between the number of computer briefcases x that a company sells and the profit the company makes $P(x)$. Find P(4000), the profit from selling 4000 computer briefcases.

Understand the variables in words:

# of computer briefcases = ______

Profit in $ = ______

Write an equation and solve.

20. The per capita consumption (in pounds) of all beef in a country is given by the function $C(x) = -0.42x + 65.2$, where x is the number of years since 2000.

a. Find and interpret C(9).

b. Estimate the per capita consumption of beef in the country in 2020.

6.15 Introduction to Functions, Domain and Range 6.15

YOUR TURN **Answers to Section 6.15.**

1. D:{ -10, -3, 0 }, R: { -2, -1, 3, 4) } Function: No 2. D: { 0, 4, 7, 8 }, R:{ -2 } Function: Yes

3. Yes 4. Yes 5. Yes 6. No 7. No 8. No

9. $g(0) = 2$ 10. $g(3) = -1$ 11. $g(-6) = -10$

12. $f(-5) = 3$ 13. $x = -3, 1$ 14 & 15. refer to graphs below

16 a. (0, 0), (1, 74), (2, 116), (3, 126), (4, 104), (5, 50), (6, -36), (7, -154)

16 b. The rocket returns to ground level between 5 and 6 seconds.

16 c. Maximum height is around 126 feet 17. 166.38 cm 18. 156.48cm

19. $80,000 profit

20a. C(9) = 61.42, The consumption of beef in 2009 will be 61.42 pounds.

20b. The per capita consumption of beef in 2020 will be 56.8 pounds.

14.

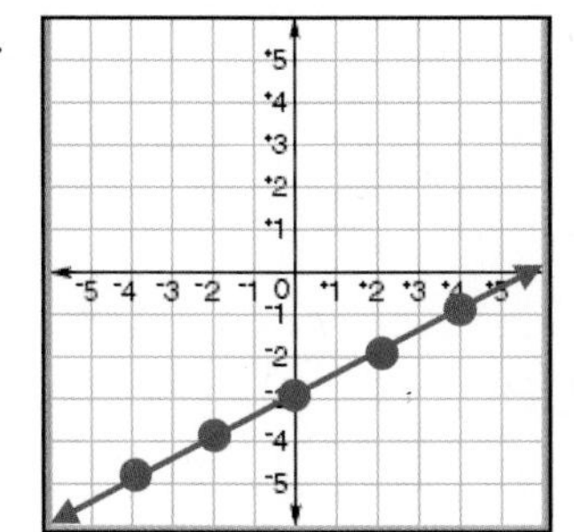

15.

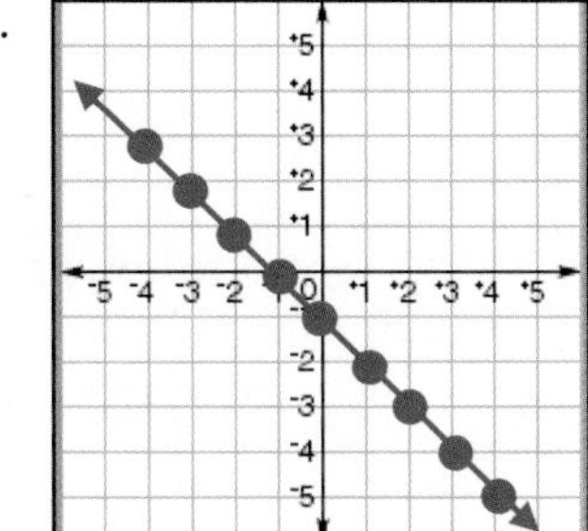

Complete MyMathLab Section 6.15 Homework in your Homework notebook.

6.16 Interval Notation, Finding Domain and Range 6.16

Interval Notation

In a previous module, we solved inequalities. The solution set was a section of the number line. For example, a solution could be $x < 4$. This means that all the numbers less than 4 but not including 4 would be solutions. A solution of $x \geq 1$ would include all the numbers greater than 1 and including 1. Below is how we expressed our solutions on a number line. Notice the use of the *parentheses which excludes the end point* and the *bar which includes the endpoint*.

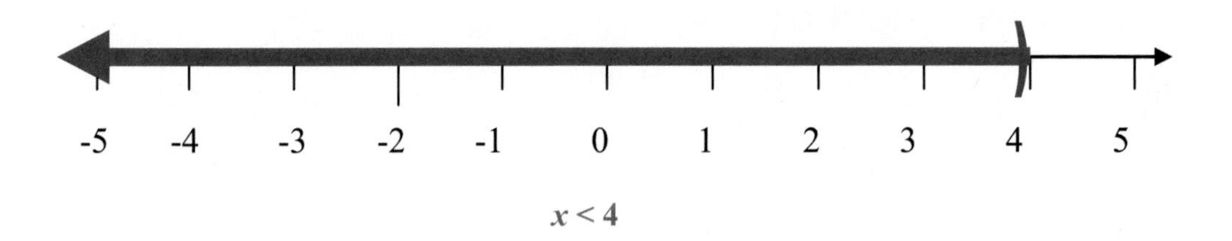

$x < 4$

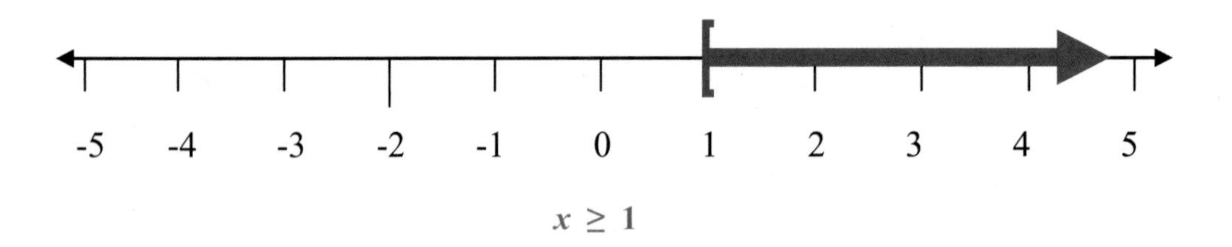

$x \geq 1$

When using interval notation, the symbol	
(	means " not included" or "open"
[	means "included" or "closed"

(smallest #, to largest #)

$x < 4$ expressed in interval notation would be $(-\infty\ ,\ 4)$

$x \geq 1$ expressed in interval notation would be $[1\ ,\ \infty)$

6.16 Interval Notation, Finding Domain and Range 6.16

In interval notation, there are only 5 symbols to know:

- Open Parentheses ( ) used to show the point is not included.
- Closed Bar [ ] used to show the point is included.
- Infinity ∞ solution goes on forever in the positive direction.
- Negative Infinity $-\infty$ solution goes on forever in the negative direction
- Union Sign $\cup$ used to "union" (join) two parts as one answer

Courtesy of Fotolia

Always use parentheses with ∞ or $-\infty$.

EXAMPLE 1

Graph the solution set of each inequality on a number line and then write it in interval notation.

a. $\{ x \mid x > -3 \}$ $(-3, \infty)$

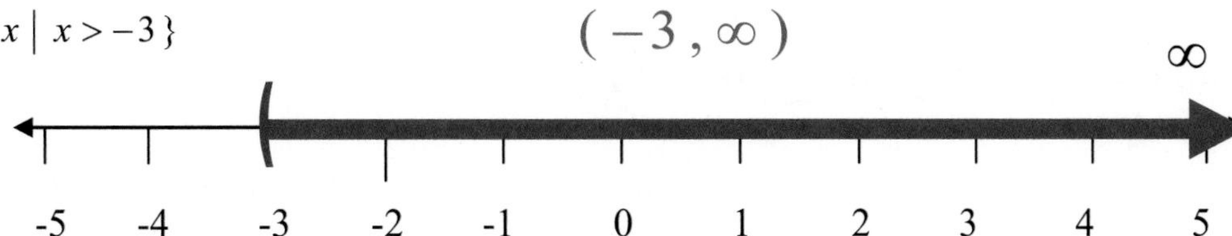

b. $\{ x \mid 2 \geq x \}$ (Rewrite $x \leq 2$.) $(-\infty, 2]$

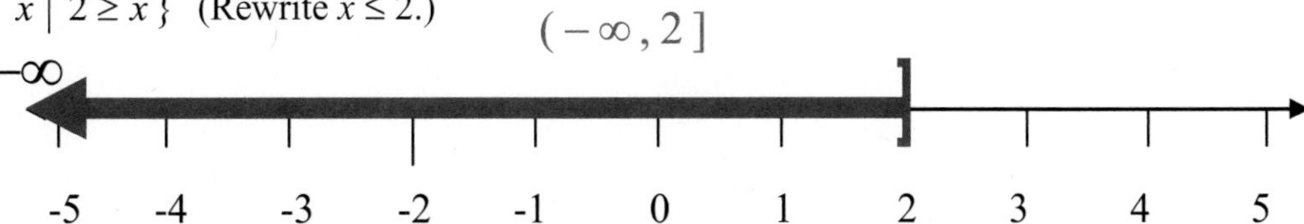

6.16 Interval Notation, Finding Domain and Range 6.16

YOUR TURN

1. Graph the solution set of each inequality on a number line and then write it in interval notation.

a. $\{\ x \mid x < -2\}$ Interval notation: ________________

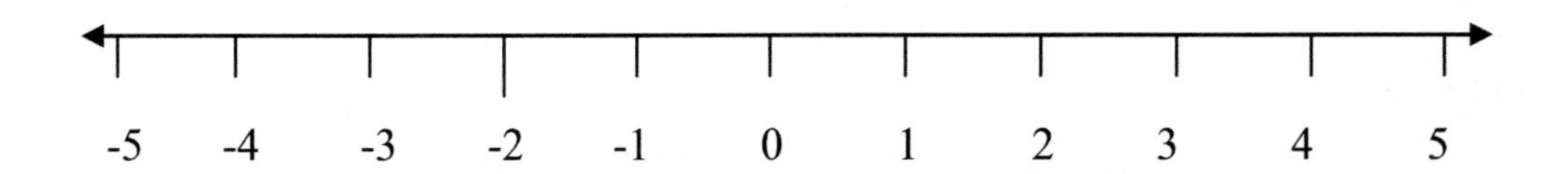

b. $\{\ x \mid 3 \leq x\}$ Interval notation: ________________

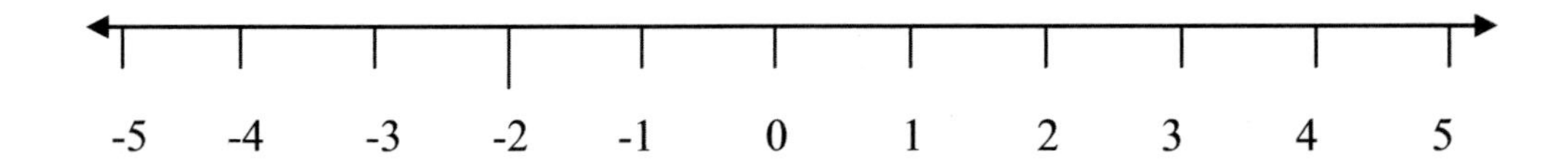

Now we will use interval notation to express domain and range of relations from a graph.

6.16 Interval Notation, Finding Domain and Range 6.16

YOUR TURN

1. Graph the solution set of each inequality on a number line and then write it in interval notation.

a. $\{ x \mid x < -2 \}$ Interval notation: ________________

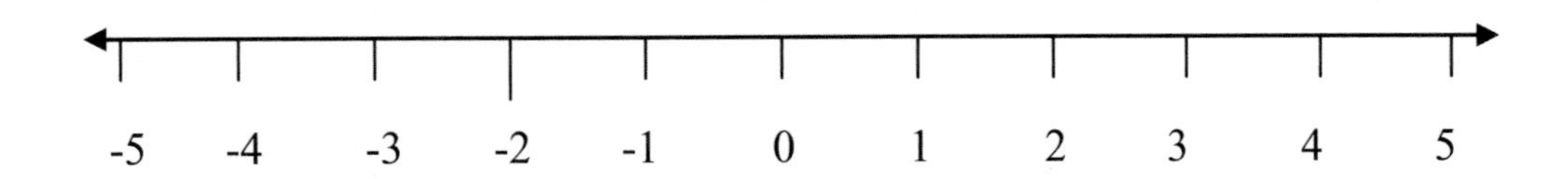

b. $\{ x \mid 3 \leq x \}$ Interval notation: ________________

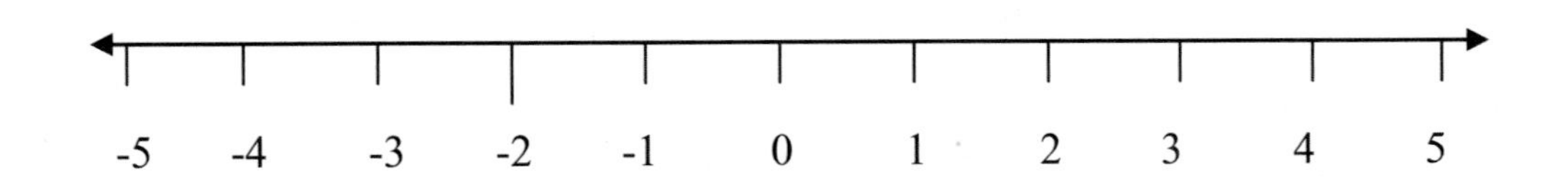

c. $\{ x \mid -1 \leq x < 4 \}$ Interval notation: ________________

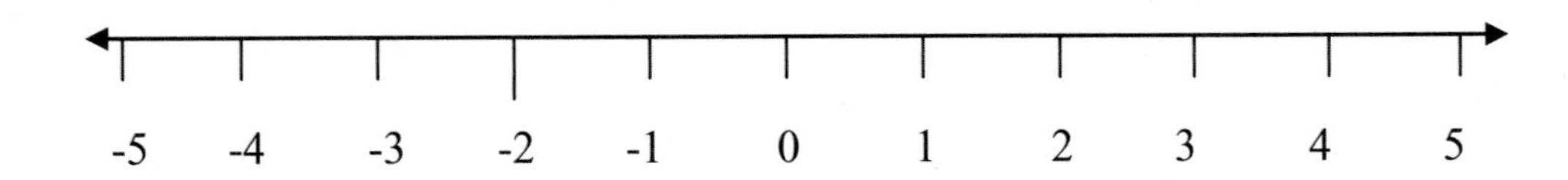

Now we will use interval notation to express domain and range of relations from a graph.

6.16 Interval Notation, Finding Domain and Range 6.16

Domain: is the set of all first components of the ordered pairs of the relation.
Range: is the set of all second components of the ordered pairs of the relation.

If given a graph of a relation which is a curve or a line, we will find its domain and range using interval notation. Remember to use a parentheses to indicate that a number is not part of the domain/range and use a bar to indicate a number is part of the domain/range. Parentheses are always used with infinity or negative infinity.

Domain, D:
The set of x-coordinates of a relation.

Given a graph of a line or curve, use *interval notation to list the x values* from smallest to largest.
(Hint: Read from left to right horizontally →)

Range, R:
The set of y-coordinates of a relation.

Given a graph of a line or curve, use *interval notation to list the y values* from lowest to highest.
(Hint: Read from bottom to top vertically. ↑)

EXAMPLE 2

Find the domain and range of each relation.

a.

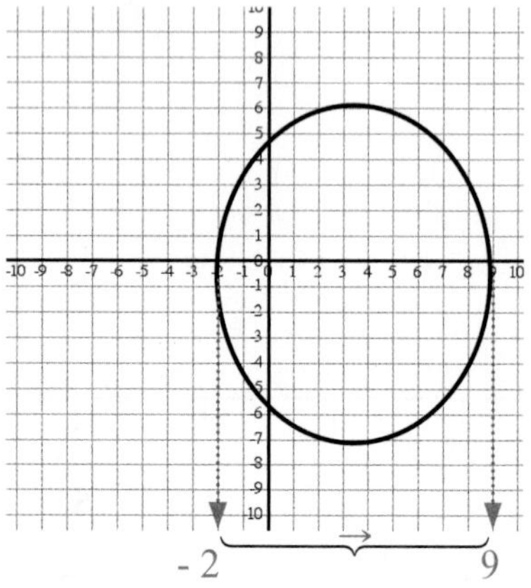

- 2 → 9

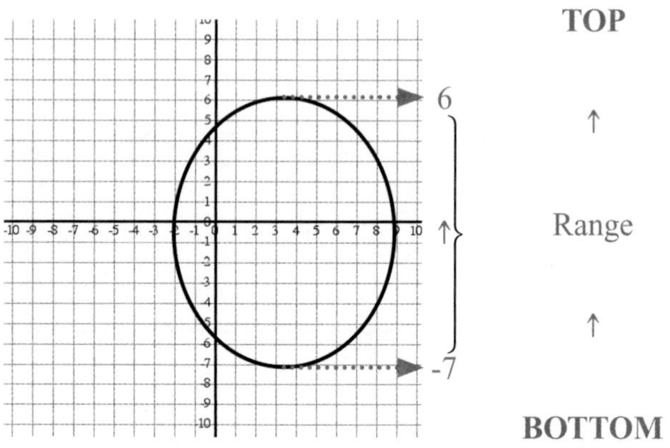

6 ↑ Range ↑ -7

TOP ↑ Range ↑ BOTTOM

Domain reads LEFT → to → RIGHT, **[-2, 9]** Range reads BOTTOM to TOP, **[-7, 6]**

Domain: ___[-2, 9]___ Use interval notation to read the ***x values*** used from left to right, smallest to largest.

Range: ___[-7, 6]___ Use interval notation to read the ***y values*** used from bottom to top, lowest to highest.

b.

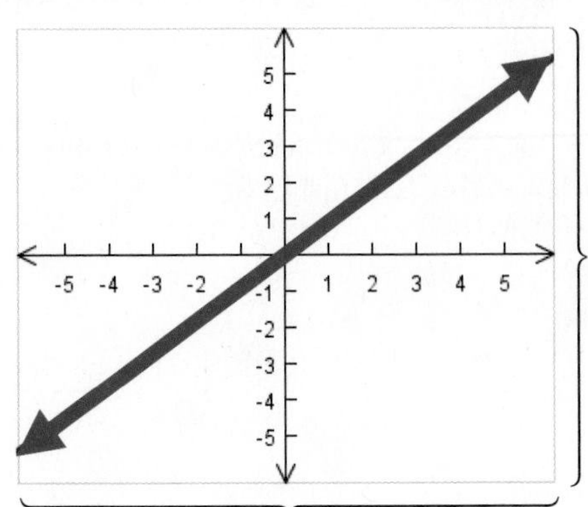

Range goes from negative infinity at the bottom and continues up to positive infinity.

Domain goes from negative infinity on the left and continues right to positive infinity.

Domain: $(-\infty, \infty)$

Range: $(-\infty, \infty)$

c.

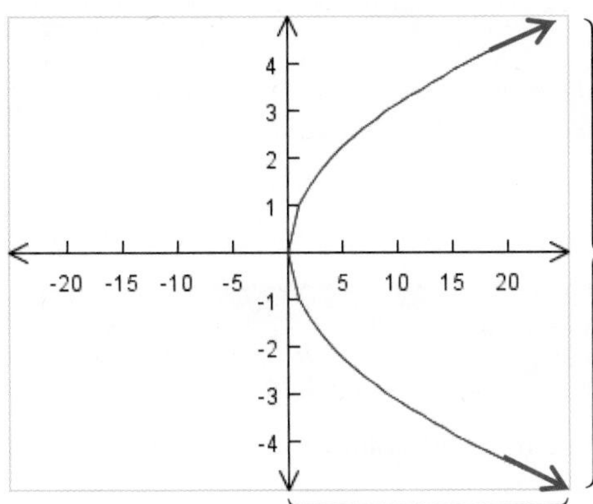

Range goes from negative infinity at the bottom and continues up to positive infinity.

Domain goes from 0 on the left and continues right to positive infinity.

Domain: $[0, \infty)$

Range: $(-\infty, \infty)$

d.

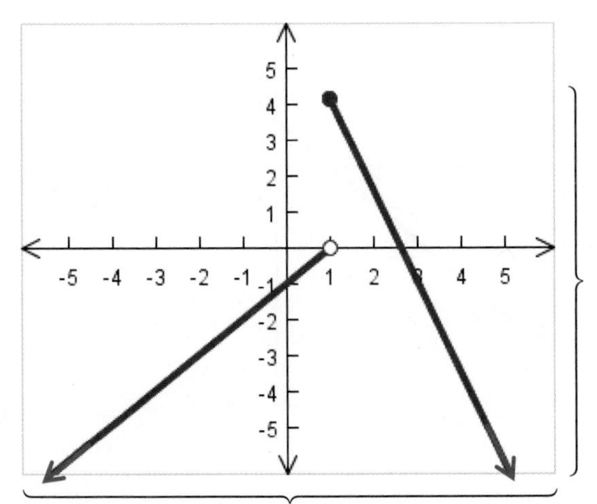

Range goes from negative infinity at the bottom and continues up to positive 4.

Domain goes from negative infinity on the left and continues right to positive infinity.

Domain: $(-\infty, \infty)$

Range: $(-\infty, 4]$

6.16 Interval Notation, Finding Domain and Range 6.16

Using Interval Notation to Write Domain and Range from a Graph

- Use the open parentheses ( ) if the value is not included in the graph. (i.e. the graph is undefined at that point... there's a hole or asymptote, or a jump)
- Use the brackets [ ] if the value is part of the graph.
- If the graph goes on forever to the left, the domain will start with $(-\infty$.
- If the graph travels downward forever, the range will start with $(-\infty$. Similarly, if the graph goes on forever at the right or up, end with $\infty)$
- Whenever there is a break in the graph, write the interval up to the point. Then write another interval for the section of the graph after that part. Put a union sign ∪ between each interval to "join" them together.

YOUR TURN

Find the domain and range of each relation. Remember to use interval notation.

2. 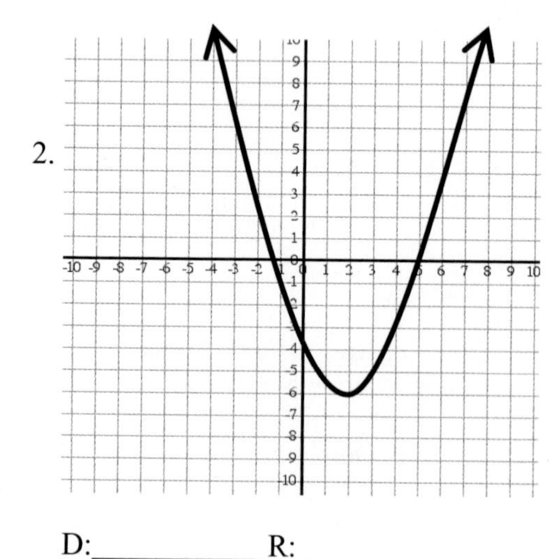

D:__________ R:______________

3. 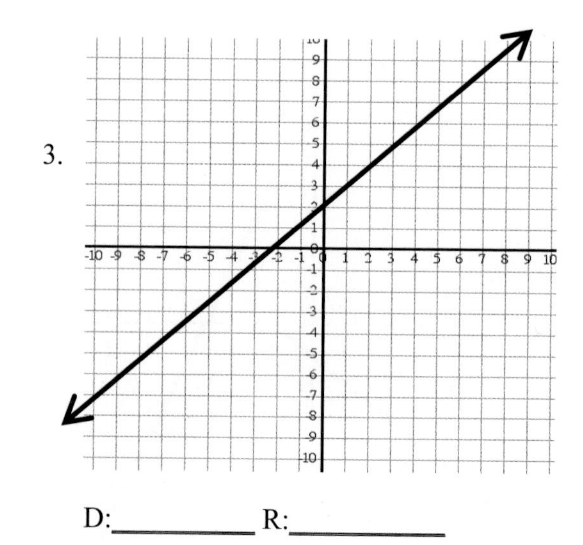

D:__________ R:__________

6.16 Interval Notation, Finding Domain and Range 6.16

YOUR TURN

Find the domain and range of each function.

4.

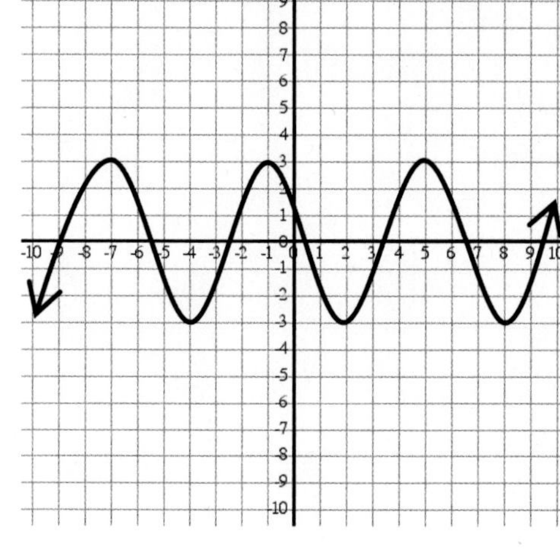

D:__________ R:____________

5.

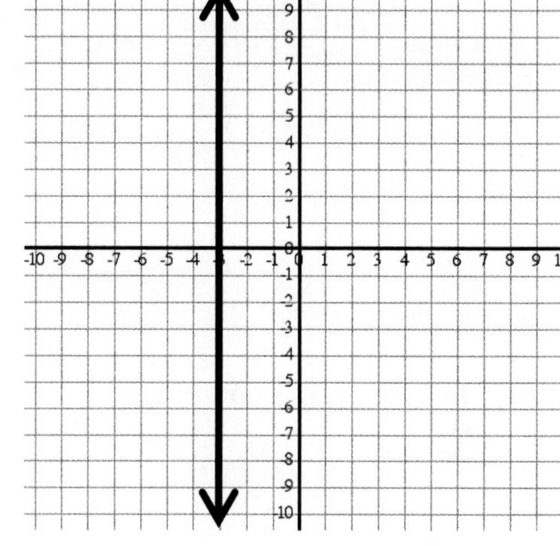

D:__________ R:__________

6.

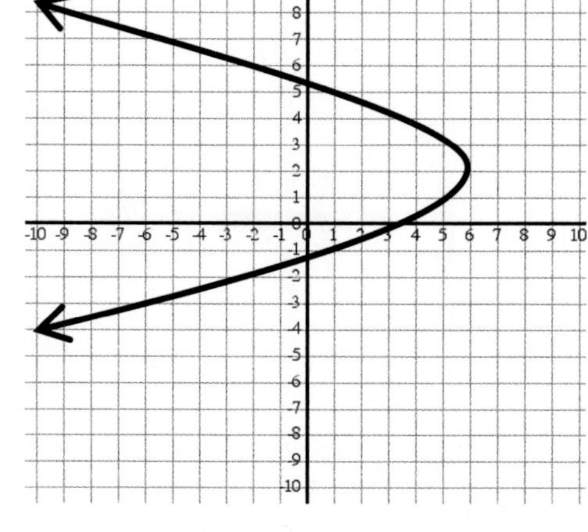

D:__________ R:____________

7.

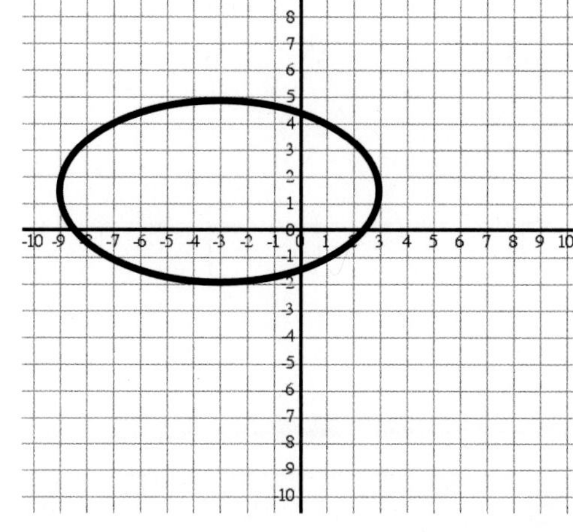

D:__________ R:__________

6.16 Interval Notation, Finding Domain and Range 6.16

YOUR TURN **Answers to Section 6.16.**

1. a. $(-\infty, -2)$ b. $[3, \infty)$
2. Domain: $(-\infty, \infty)$ Range: $[-6, \infty)$
3. Domain: $(-\infty, \infty)$ Range: $(-\infty, \infty)$
4. Domain: $(-\infty, \infty)$ Range: $[-3, 3]$
5. Domain: $\{-3\}$ Range: $(-\infty, \infty)$
6. Domain: $(-\infty, 6]$ Range: $(-\infty, \infty)$
7. Domain: $[-9, 3]$ Range: $[-2, 5]$

1a.

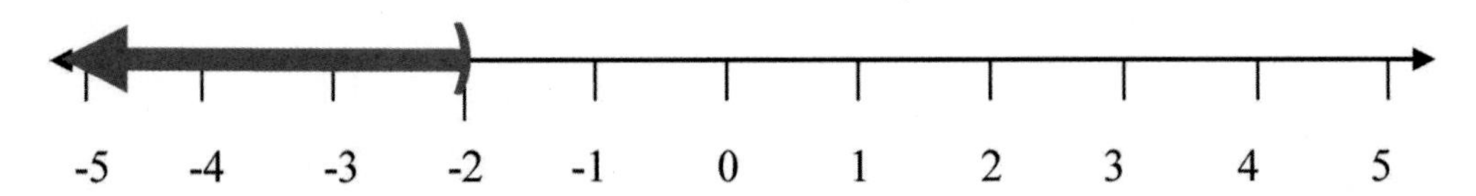

1b.

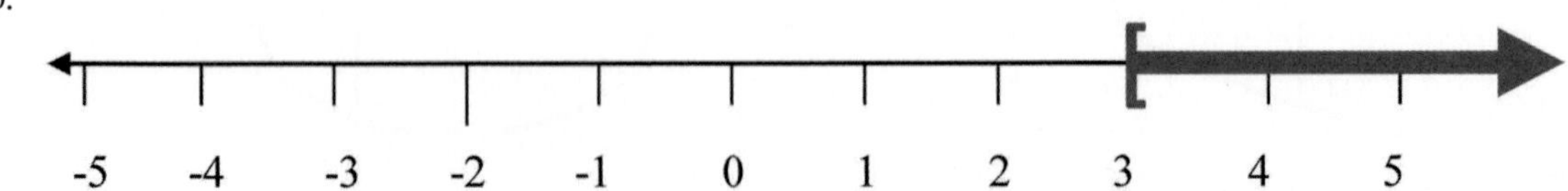

Complete MyMathLab Section 6.16 Homework in your Homework notebook.

(Stay on our track. We're moving on to Section 6.18 next.)

6.18 Further Graphing of Quadratic Functions 6.18

Polynomial Functions are used in real life in every part of the world. Companies use calculators to make decisions for these mathematical models in this techonological age. In this section, we will analyze quadratic functions using the some of the graphical functions of the TI-84 calculator. Some simple functions we will introduce here are finding maximum and minimum values of a curve, an axis of symmetry, x-intercepts and y-intercepts, and other intercepts of a graph.

THE TI-84 CALCULATOR

Y=: Enter equations here. (PLOTS are off, not shaded.)

WINDOW: Set view of max and min values. Then press GRAPH or ZOOM command.

QUIT: (2nd) (MODE)
Returns you to the Home screen.

MODE: Set to (Func) for graphing functions.

X,τ,θ,η : Enters the variable x in an equation in the (Y=) function mode.

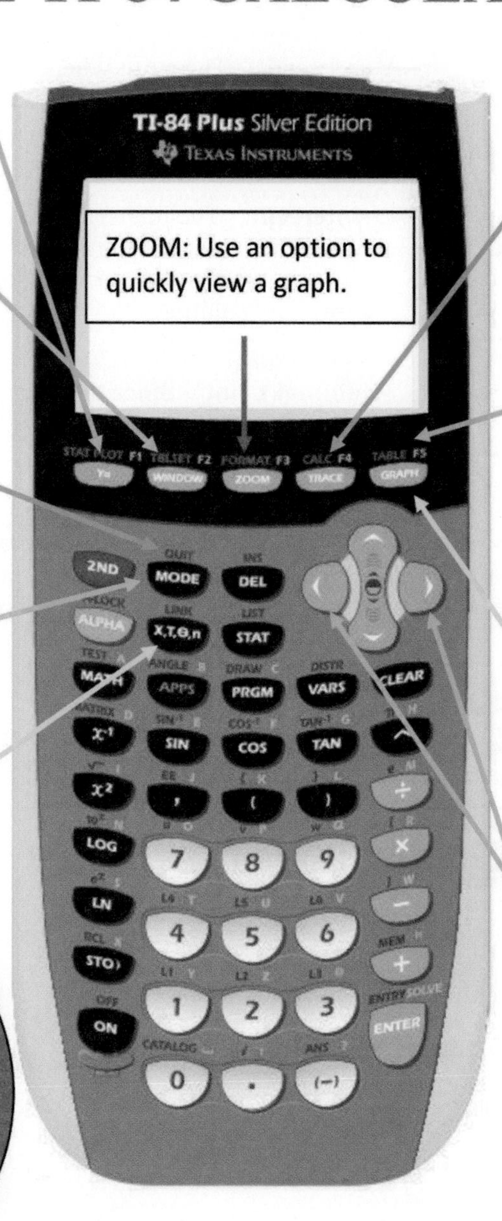

CALC: (2nd) (TRACE)
Allows you to calculate using the functions:
1: value
2: zeros
3: minimum
4: maximum
5: intercept

TABLE: (2nd) (GRAPH)
Shows an x-y chart of ordered pairs for the graph in y=.

GRAPH: Shows a Picture of the graph. You may have to change your WINDOW to get a good view.

SCROLL Keys:
Hint, use <u>ONLY</u> the < left and > right arrows with CALC functions for graphs.

If you get stuck, try (2nd)(QUIT) to get you out of the graphing screen. If this doesn't work, try (2nd)(OFF).

6.18 Further Graphing of Quadratic Functions 6.18

Graphs of Quadratic Functions

Graphs of quadratic functions make the shape of a parabola, a bowl.

The quadratic equation $y = ax^2 + bx + c$ makes a picture similar to the one below. The parabola can curve up or curve down.

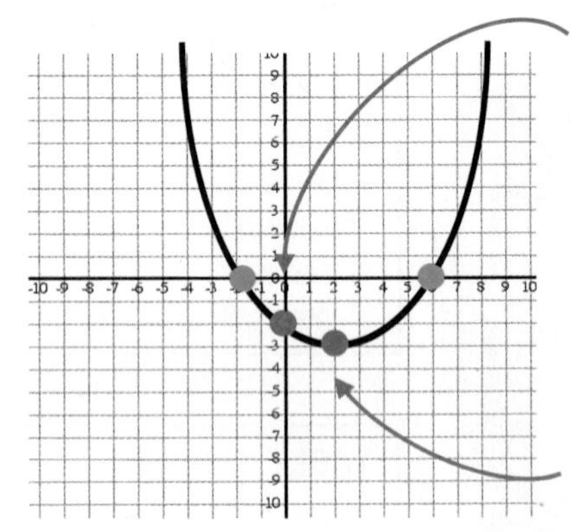

Y-intercept – The point where the graph crosses the y-axis is the y-intercept. In a quadratic equation which is set equal to 0, this is always the "c" value at the end.

***x*-intercept** – The points where the graph crosses the x-axis are the x-intercepts. In a quadratic equation, these are also called "zeros" or "solutions".

Minimum point – Since this graph curves up, the graph has a minimum point, a lowest point.

We will investigate how the numberic values in quadratic equations make these shapes and analyze some special points on these parabolas using the calculator. This section will explain how to use the functions of the CALC feature of your calculator which are listed below.

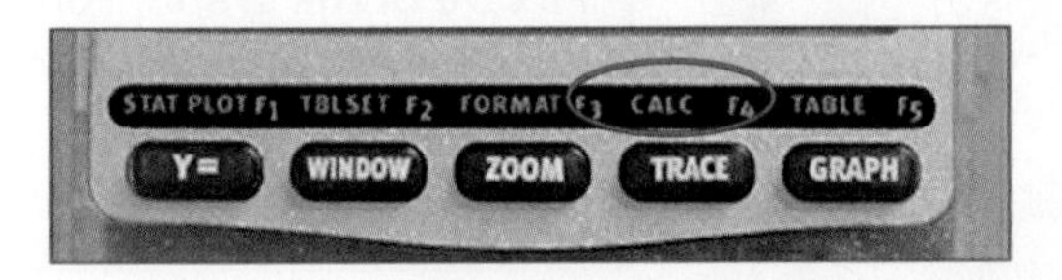

CALC Functions [2nd][CALC] keys

1: value Calculates a function y value for a given x. At $x = 0$, it gives the y-intercept of the graph.

2: zero Finds a zero (x-intercept) of a function, an x-intercept or equation's solution.

3: minimum Finds a minimum of a function, a low point.

4: maximum Finds a maximum of a function, a high point.

5: intersect Finds an intersection of two functions.

FINDING Y-INTERCEPTS

Facts about y-intercepts in an equation:

1. A GRAPH: The y-intercept is the point where the graph crosses the y-axis, (0, #).
2. AN EQUATION: The ending "c" value is the y-intercept in an equation written in standard form, $y = ax^2 + bx + c$.

EXAMPLE 1

Find the y-intercept of the equation $y = 2x^2 - 11x + 5$. (Always make sure the equation is in standard form, $y = ax^2 + bx + c$.)

Press [Y=].
Enter your equation in Y_1=. → Press [WINDOW] → Press [GRAPH] to view the parabola.

Plot1 Plot 2 Plot3
\ Y_1= 2x^2 - 11x +5
\ Y_2=
\ Y_3=

Press (WINDOW):
Enter values for
Xmin = -3
Xmax = 8
Ymin = -16
Ymax = 10

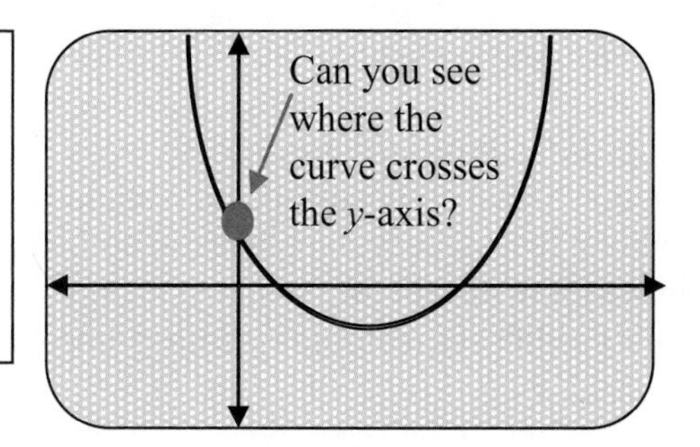

To find the y-intercept, press [2nd][CALC] and choose "1: value". → Then type X= 0 in the bottom left of the [GRAPH] screen. Press [ENTER].

CALCULATE
1: value
2: zero
3: minimum
4: maximum
5: intersect

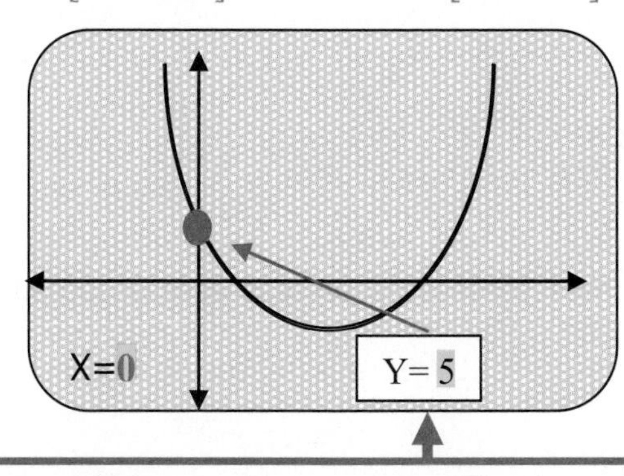

The y-intercept value will appear in the lower right, Y= 5.
Write the y-intercept as a point, (0, 5).

IMPORTANT NOTE: The y-intercept is always the "c" value in an equation in standard form, $y = ax^2 + bx + c$.

APPLICATIONS WITH THE CALCULATOR

APPLICATIONS: FINDING Y-INTERCEPTS

EXAMPLE 2

BEGINNING COST PROBLEM: Suppose an entrepenuer is jump starting a company which will sell sportswear t-shirts. He knows that a good repesentation of the cost of running his company is $C(x) = 2x^2 - 120x + 3{,}000$, where C is the cost of producing the t-shirts and x represents the number of sportswear t-shirts produced. What are his start-up costs going to be so that he can plan to open his business?

First, understand the variables in words:

The # of t-shirts: _x_

The company's costs, $: _C_

Since the start up cost occurs before any t-shirts are made, the solution, the start up cost, is the y-intercept. The x value for any point on the y axis is zero, because no t-shirts will have been made. So we know, $x = 0$.
Let's use the calculator to find the y-intercept.

See an instructor if you need help with setting the WINDOWS.

Go to the TI-83/84 calculator.

Graph the equation using the [Y_1 =] key.
Enter $Y_1 = 2x^2 - 120x + 3{,}000$

Press [WINDOW] and enter these settings.
XMIN = -10 XMAX = 100
YMIN = - 500 YMAX = 5,000

Press [2nd] [CALC]
Choose [1:value]

In the graph screen, "X=" will appear in the bottom left. Enter the number 0, because *for any y-intercept the x valus is always 0.*
Press [ENTER]
The y-intercept will appear in the bottom right, $y = 3000$.

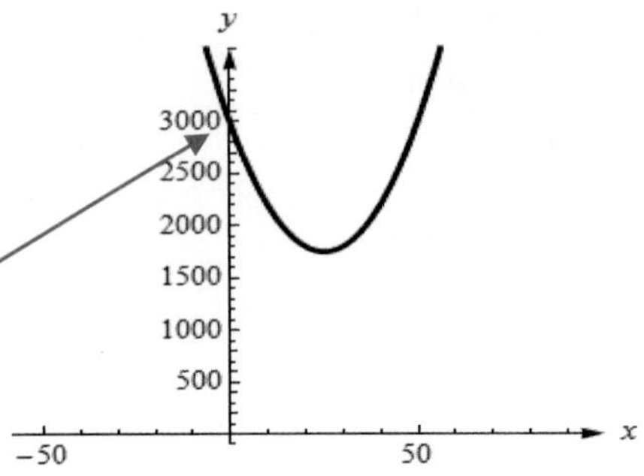

y
3000
2500
2000
1500
1000
500
−50
50
x

The start up cost for this company is $3,000.

(Do you notice this is the constant value at the end of the equation?) $Y_1 = 2x^2 - 120x + 3{,}000$

EXAMPLE 3

INITIAL HEIGHT PROBLEM: If a projectile is fired upward from the ground with an initial speed of 48 feet per second, then its height in feet after *t* seconds is given by the function

$h(t) = -16t^2 + 48t$. **Find the initial height of the projectile.**

WINDOW: Xmin = -3 Xmax = 6 Ymin = -10 Ymax = 50

First, understand the variables in words:

The time in seconds: _t_

The height in feet: _h_

The projectile will go high in the air before coming down, so the Ymax value will need to be increased to 50 feet. We will want to see more *x* or *y* max values than are on the standard graph for a good view.

Your graph will look similar to the one below. Note the labeling of the axes.

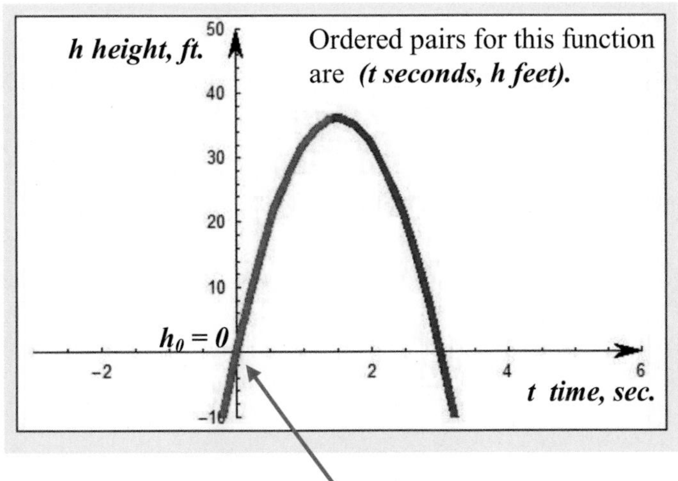

Go to the TI-83/84 calculator.

To find the initial height, use these keys.

- In [Y=], enter the equation $h(t) = -16t^2 + 48t$.
- In the WINDOW, use the settings given above.
- Press [GRAPH].
- Press [2nd][CALC] and select "1:value").
- Enter X = 0 & Press [ENTER].

The *y*-intercept will be Y = 0.

This means the object leaves from the ground, 0 feet.

6.18 Further Graphing of Quadratic Functions 6.18

YOUR TURN

Find the y-intercept using the [1:value] function of the calculator.

1. $y = x^2 + 1.52x + 3.264$

   Y-intercept: _(0, __________)_

2. COST PROBLEM: Suppose a master baker is starting a bakery which will sell cupcakes. She knows that a good model of her complanys costs can be represented by the function $C(x) = 0.5x^2 - 50x + 6{,}000$, where C is the cost of producing the cupcakes and x represents the number of cupcakes produced. What are her start-up costs to open the business?

Understand the variables in words:

The # of ________________: x

The company's costs, $: ______

Go to the TI-83/84 calculator.

Use the WINDOW setttings XMIN = -100 XMAX = 500 YMIN = - 500 YMAX = 10,000.

The start up costs for openning the bakery are _$___________________.

6.18 Further Graphing of Quadratic Functions 6.18

FINDING X-INTERCEPTS

Facts about *x* intercepts:

1. GRAPH: The *x*-intercept is the point where the graph crosses the *x*-axis, (#, 0). In a quadratic equations there are often two places where the parabola will cross the *x*-axis.

2. EQUATION: Factors of an equation $y = ax^2 + bx + c$ give solutions which are the *x*-intercepts. When an equation will not factor but it crosses the graph, we will use the calculator to find the solutions, also called "*x*-intercepts" or "zeros".

EXAMPLE 4

Find the *x*-intercepts of the equation $y = 2x^2 - 11x + 5$. This graph will have two *x*-intercepts.

Press [Y=].
Enter your equation in Y_1=. → Press [WINDOW] → Press [GRAPH] to view the parabola.
$Y_1 = 2x^2 - 11x + 5$ or [ZOOM] [6]. The curve crosses the *x*-axis in two points.

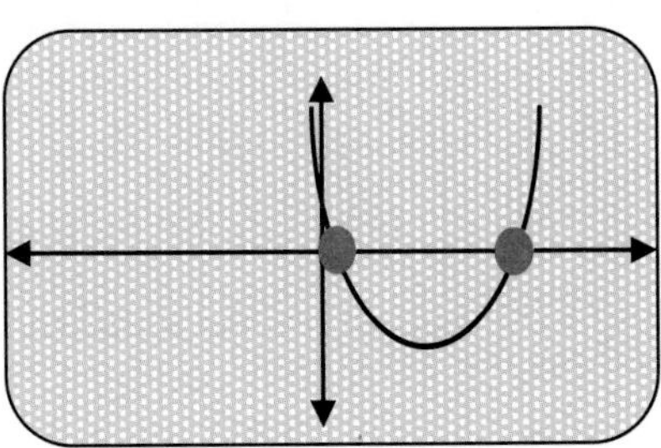

Find the 1st *x*-intercept.
Press [2nd][CALC] and choose "2: zero".
Press [ENTER].

CALCULATE
1: value
2: zero
3: minimum
4: maximum
5: intersect

Use the [<] scroll left key to move the blinking cursor to the left side of the 1st *x*-intercept, just keep pressing until it is left. This is the Left bound of the left *x*- intercept. Press [ENTER].

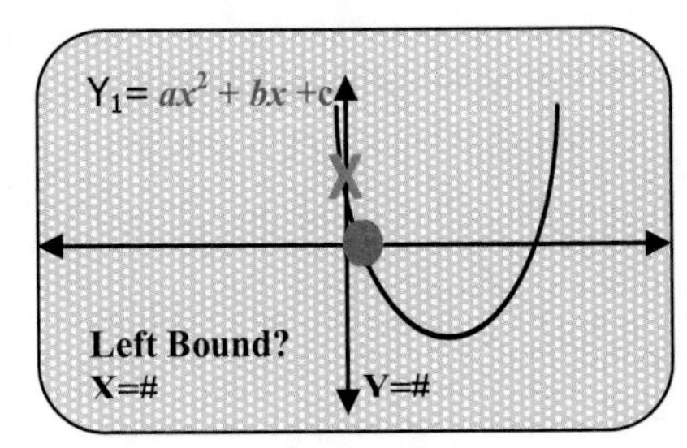

6.18 Further Graphing of Quadratic Functions 6.18

Use the [>] scroll left key to move the blinking cursor to the right side of the 1st x-intercept, just keep pressing until it is right. This is the Right bound of the this x- intercept. Press [ENTER].

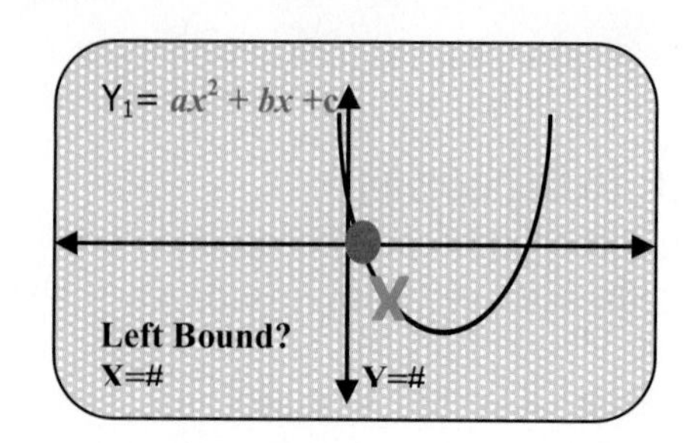

(Don't pass the 2nd x-intercept or you will get an error message.)

Press [ENTER] again to ignore the GUESS screen.

The 1st x-intercept will appear in the bottom of the screen.

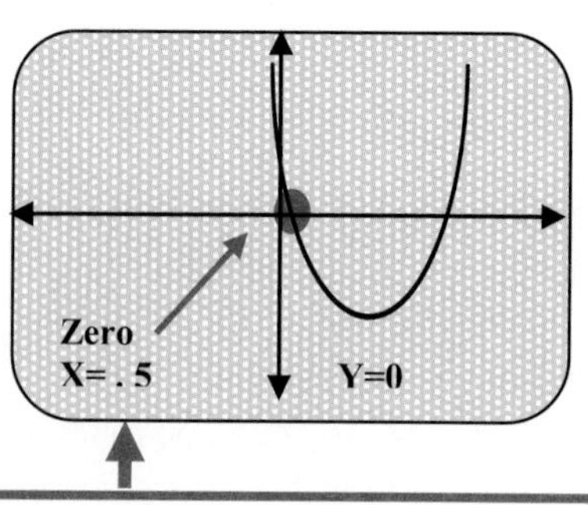

The 1st x-intercept value will appear in the lower part of the screen, X= 0.5 Y=0.

Write the 1st x-intercept (0.5, 0).

Find the 2nd x-intercept.
Press [2nd][CALC] and choose "2: zero".
Press [ENTER].

CALCULATE
1: value
2: zero
3: minimum
4: maximum
5: intersect

Use the [<] scroll left key to move the blinking cursor to the left side of the 2nd x-intercept, just keep pressing until it is left. This is the Left bound of this x-intercept. Press [ENTER].

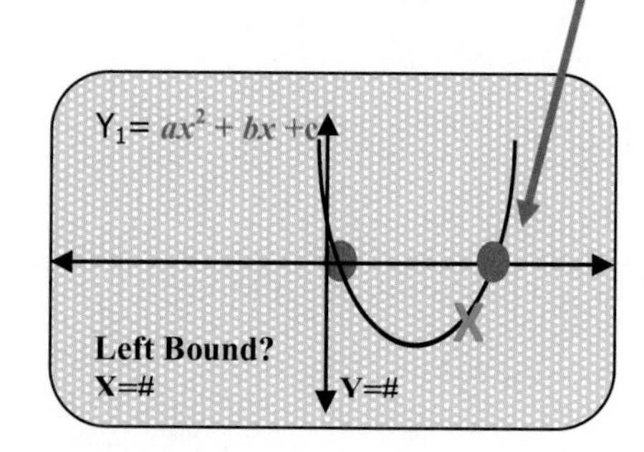

Use the [>] scroll left key to move the blinking cursor to the right side of the 2nd x-intercept, just keep pressing until it is right. This is the Right bound of this x-intercept. Press [ENTER].

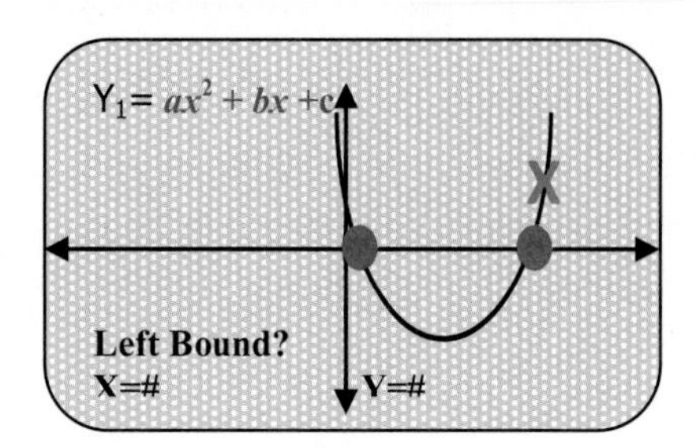

Press [ENTER] again to ignore the GUESS screen.

The 2nd x-intercept will appear in the bottom of the screen.

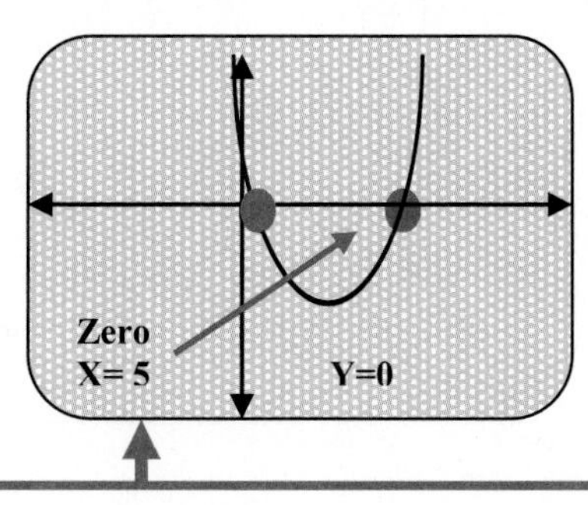

The 2nd ***x*-intercept** value will appear in the lower part of the screen, X= 5 Y=0.

Write the 2nd *x*-intercept (5, 0).

The *x*-intercepts are (0.5, 0) and (5, 0).

6.18 Further Graphing of Quadratic Functions 6.18

FINDING A VERTEX: MAXIMUM OR MINIMUM

The vertex of a parabola is the point where the curve changes directions.

VERTEX OF A PARABOLA

- For a parabola curving upward, ⋃ the *vertex is a minimum* value. It is the *lowest point* on the graph. Use the "minimum" feature of the calculator.
- For a parabola curving downward, ⋂ the *vertex is a maximum* value. It is the *highest point* on the graph.
  - Use the "maximum" feature of the calculator

To find the vertex of a parabola, first look at it's graph. The equation must be in standard form $y = ax^2 + bx + c$ to enter it in your calculator. We will use an example to demonstrate.

EXAMPLE 5

Find the vertex of the graph of $y = x^2 + 5x + 4$. Use a standard WINDOW, (Zoom 6).

Press [Y=].
Enter your equation in Y_1=.
$Y_1 = x^2 + 5x + 4$

Press (GRAPH) to view the parabola.

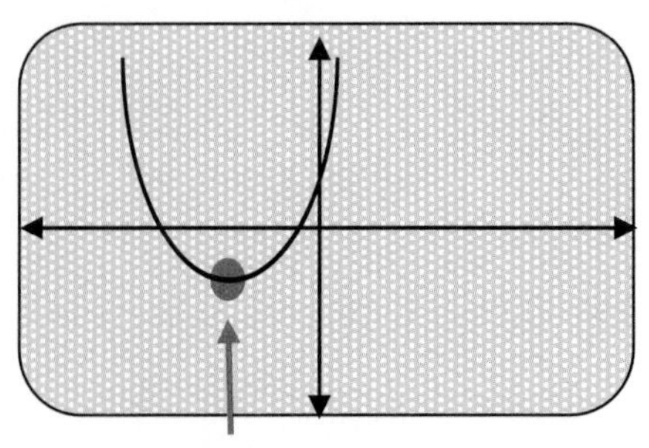

Vertex or Minimum

6.18 Further Graphing of Quadratic Functions 6.18

Because the graph opens upward, the *vertex is a minimum*. We will use the minimum value in the CALC function to find this point.

Press [2nd][CALC]. Choose "3:minimum".

Use the < scroll key to move the blinking cursor to the left side of the minimum. Press [ENTER].	Use the > scroll key to move the blinking cursor to the right side of the minimum. Press [ENTER].	Press [ENTER] to ignore the Guess screen.	The minimum to find the vertex, (x, y).

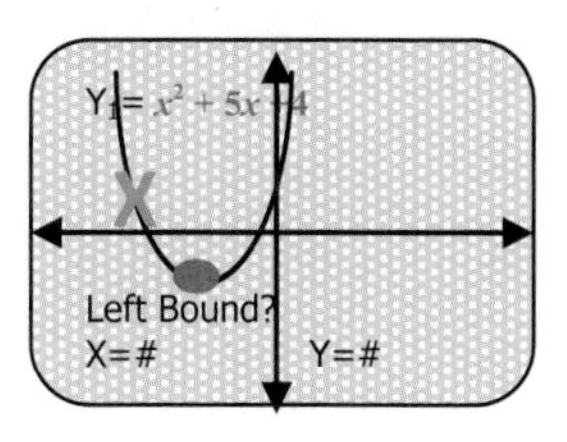

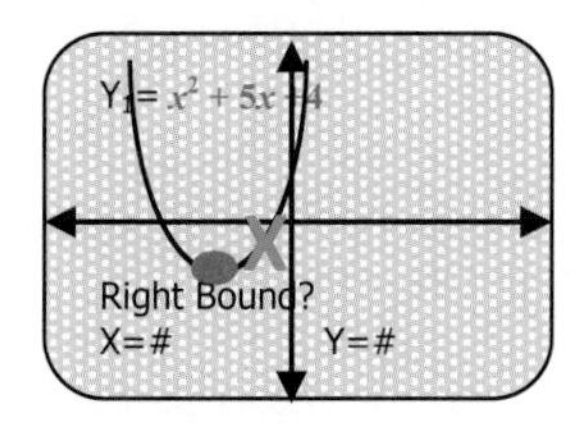

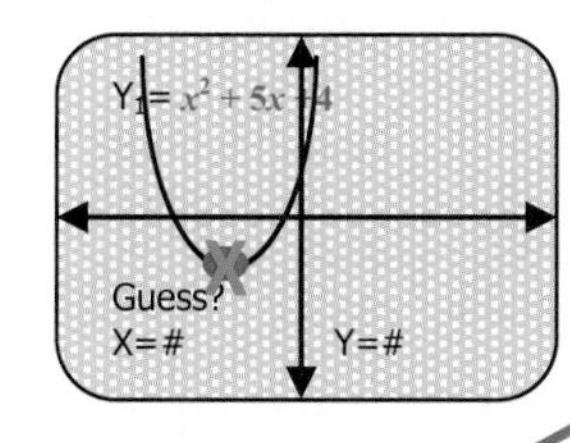

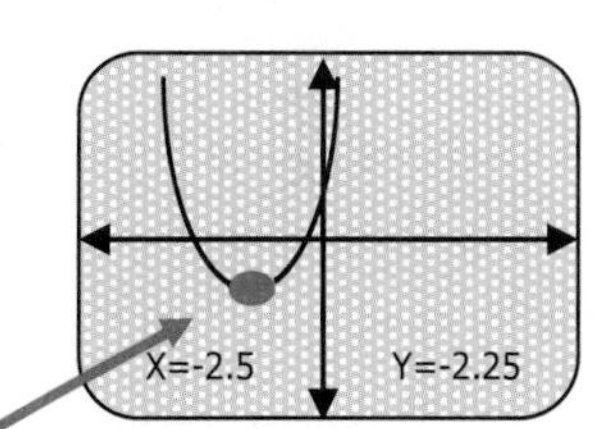

The *vertex or minimum* is (-2.5, -2.25).

HOW CAN YOU TELL IF A PARABOLA WILL OPEN UPWARD OR DOWNWARD WITHOUT LOOKING AT THE GRAPH?

This is easy. LOOK at the value in front of the squared term, x^2, in the equation $\boldsymbol{y = ax^2 + bx + c}$.

- If the value in front of x^2 is positive, the parabola curves upward.
- If the value in front of x^2 is negative, the parabola curves downward.

6.18 Further Graphing of Quadratic Functions 6.18

YOUR TURN

Match each function with its graph. [Use a graphing utility to find the equation's vertex and x and y intercepts. Use a standard WINDOW, (Zoom 6).]

3. $f(x) = x^2 - 4x + 3$

A.
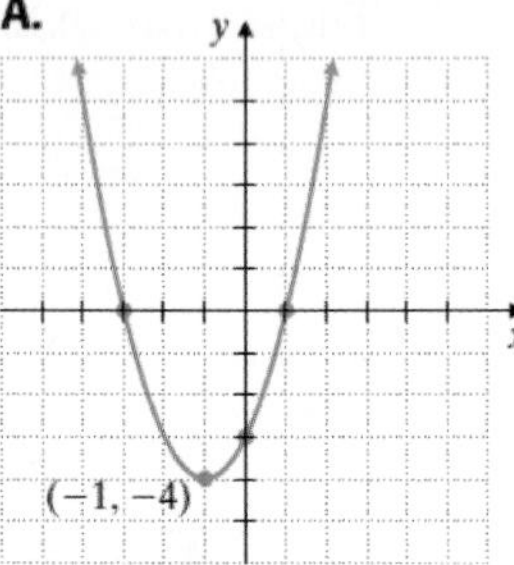

B.
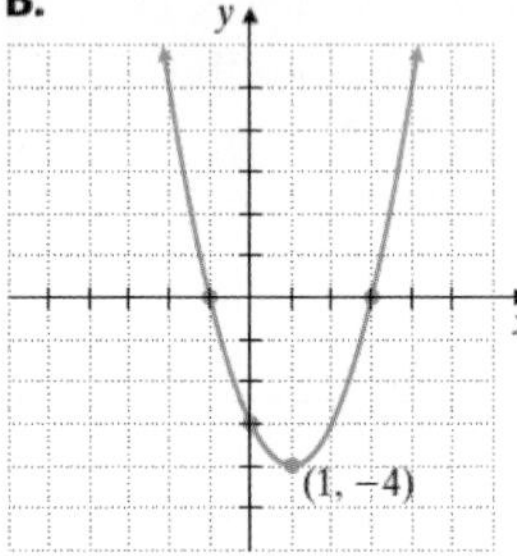

C.
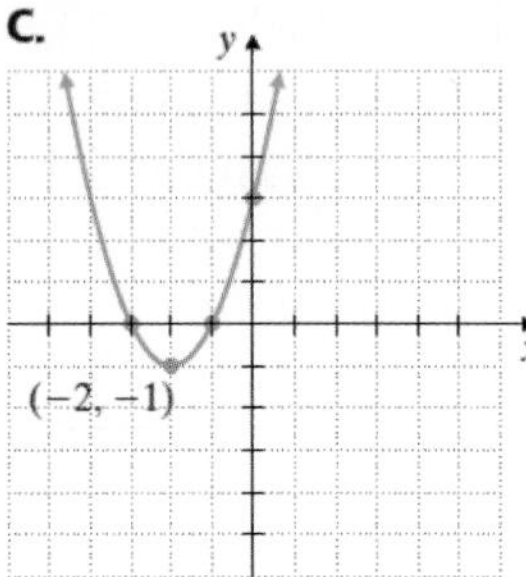

D.
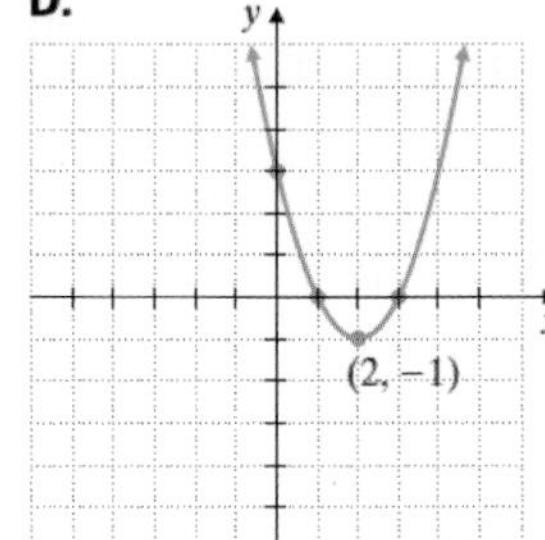

(ebook, page 854.)

Which graph represents the equation? ________

x-int.: __________

y-int.: __________

Determine whether each parabola opens upward or downward and state if the parabola has a maximum or minimum. Use a graphing utility to find the vertex. Use a standard WINDOW, [Zoom 6]. Then state the x & y intercepts, if any.

4. $f(x) = x^2 + 4x - 5$

Opens: ___________________

(Look at the equation 1st.)

Circle: Max or Min

Vertex: ___________________

x-int.: __________

y-int.: __________

EXAMPLE 6

MAXIMUM HEIGHT PROBLEM: If a projectile is fired upward from the ground with an initial speed of 48 feet per second, then its height in feet after *t* seconds is given by the function

$h(t) = -16t^2 + 48t$.

a. **Find the maximum height of the projectile.**
b. **Find when the object hits the ground.**

WINDOW: Xmin = -3 Xmax = 6 Ymin = -10 Ymax = 50

The projectile will go high in the air before coming down, so the Ymax value will need to be increased to 50 feet. We will want to see more *x* or y max values than are on the standard graph for a good view.

Your graph will look similar to the one below. Note the labeling of the axes. The beginning height h_0 of the projectile is 0. In other words, the projectile leaves from the ground.

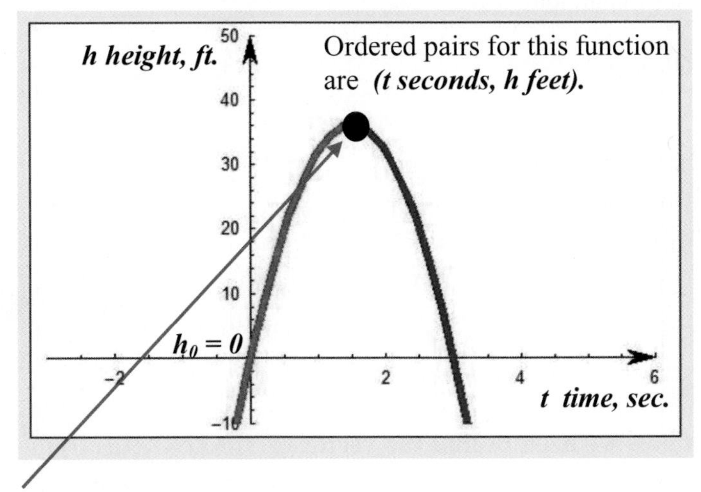

a. To find the maximum height, use these keys.
- In [Y=], enter the equation $h(t) = -16t^2 + 48t$.
- In the WINDOW, use the settings given above.
- Press [GRAPH].
- Press [2nd][CALC] and select "4:maximum".
  Left Bound?: Scroll left of the vertex point. Press [ENTER].
  Right Bound?: Scroll right of the vertex point. Press [ENTER].
  Guess?: Press[ENTER].

The vertex is (1.5, 36).

In terms of this problem, this means it took 1.5 seconds for the projectile to reach a height of 36 feet.

The maximum height of the projectile is 36 feet.

b. To find when the projectile hits the ground, find the x-intercepts. Refer to Example 4 for more assistance.

The projectile returns to the ground after 3 seconds .

EXAMPLE 7

MINIMUM COST PROBLEM: The cost in dollars of manufacturing bicycles at Brown's Production Plant is given by the function
$C(x) = 2x^2 - 800x + 92{,}000$.

a. Find the number of bicycles that must be manufactured to minimize the cost.
b. Find the minimum cost.

WINDOW: Xmin = -10 Xmax = 400 Ymin = -10 Ymax = 100,000

Understand the variable is words:
x: The number of bicycles manufactured.

$C(x)$: Cost of manufacturing a bicycle.

y -int.: $ 92,000 is the start-up cost of the business. Therefore our YMAX must be more than 92,000. We have selected 100,000.

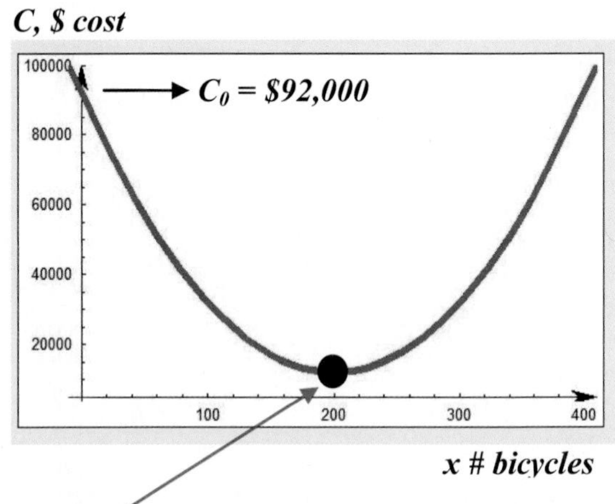

To answer both questions, we need to find the vertex or minimum. Remember, the ordered pair for the vertex in words is (*# bicycles, cost $*).

To find the minimum height, use these keys.

- In [Y=], enter the equation $C(x) = 2x^2 - 800x + 92{,}000$.
- In the WINDOW, use the settings given above.
- Press [GRAPH] to view the parabola.
- Press [2nd][CALC] and select "3:minimum".
  - Left Bound?: Scroll left of the vertex point. Press [ENTER].
  - Right Bound?: Scroll right of the vertex point. Press [ENTER].
  - Guess?: Press[ENTER].
  - The vertex is (200, 12,000).

a. The number of bicycles that must be manufactured to minimize the cost is the x coordinate, 200 bicycles.
b. The minimum cost is the y coordinate, $12,000.

YOUR TURN

5. The Utah Ski Club sells calendars to raise money. The profit, in cents, from selling calendars is given by the function $P(x) = 360x - x^2$.

WINDOW: Xmin = -10 Xmax = 500 Ymin = -10 Ymax = 50,000

Understand the variables in words:

The _______________: x

The profit, cents: ________

Go to the TI-83/84 calculator

a. Find how many calendars must be sold to maximize profit.

b. Find the maximum profit.

BUSINESS PROBLEMS

EXAMPLE 8

BUSINESS PROBLEM: The projected number of Wi-Fi-enabled cell phones in the United States can be modeled by the quadratic function $C(x) = -0.4x^2 + 21x + 35$,where $c(x)$ is the projected number of Wi-Fi-enabled cell phones in millions and x is the number of years after 2009. Round to the nearest whole million. (*Source:* Techcrunchies.com)

a. **Will this function have a maximum or a minimum? How can you tell?**

b. **According to this model, in what year will the number of Wi-Fi-enabled cell phones in the United States be at its maximum or minimum?**

c. **What is the maximum/minimum number of Wi-Fi-enabled cell phones predicted?**

d. **In what year will the number of cell phones reach 200 million?**

WINDOW: Xmin = -10 Xmax = 100 Ymin = -10 Ymax = 500

First, understand the variables x, $C(x)$and y-intercept in the equation $C(x) = -0.4x^2 + 21x + 35$:

x: The number of years after 2009. $x = 0$ is the year 2009, the initial year.
$C(x)$: The number of Wi-Fi-enabled cell phones in millions.

y = intercept: 32 million cell phones is thenumber of Wi-Fi-enabled cell phones in 2009, the initial year.

The ordered pair (*years, # cell phones*) represents the points in words for this function.

6.18 Further Graphing of Quadratic Functions 6.18

a. This function will have a maximum because the value in front of the x^2 term is negative.

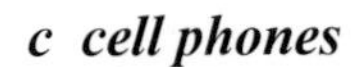

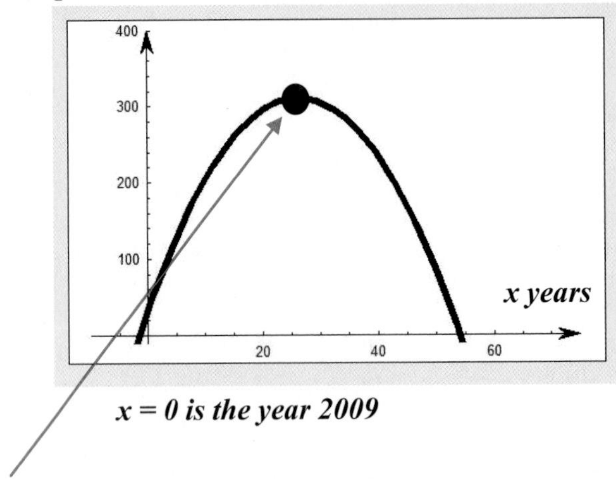

x = 0 is the year 2009

To find the maximum # of cell phones, use these keys.

- In [Y=], enter the equation $C(x) = -0.4x^2 + 21x + 35$.
- In the [WINDOW], use the settings given above.
- Press [GRAPH] to view the parabola.

- Find the vertex or maximum: Press [2nd][CALC] and select "4:maximum".
  - Left Bound?: Scroll left of the vertex point. Press [ENTER].
  - Right Bound?: Scroll right of the vertex point. Press [ENTER].
  - Guess?: Press [ENTER].
  - The vertex is (26.25, 310.625), (*years, # cell phones*).

b. The year the number of W-Fi-enabled cell phones are at a maximum comes from the x coordinate of the vertex.

The number of Wi-Fi-enabled cell phones will be at a maximum in 26 years after 2009 or the year 2035.

c. The maximum number of Wi-Fi-enabled cell phones is the y coordinate of the vertex. (Don't forget your units, millions.)

There will be a maximum of 311 million Wi-Fi-enabled cell phones.

UNSING AN INTERSECTION POINT TO SOLVE PORBLEMS

d. To find the year in which the number of Wi-Fi-enabled cell phones reaches 200 million, we need to use a new feature of the calculator, the INTERSECT feature.

To predict the year the number of cell phones will reach 200 million, use these keys:

- In [Y_1=], enter the equation $C(x) = -0.4x^2 + 21x + 35$.
- In [Y_2=], enter 200. This represents the 200 million cell phones as a new y value.

- In the [WINDOW], use the settings given above.
- Press [GRAPH] to view the parabola.
  Notice that a horizontal line appears on the graph. This is where the y value is 200. Remember, this represents 200 million cell phones.

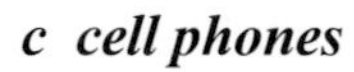

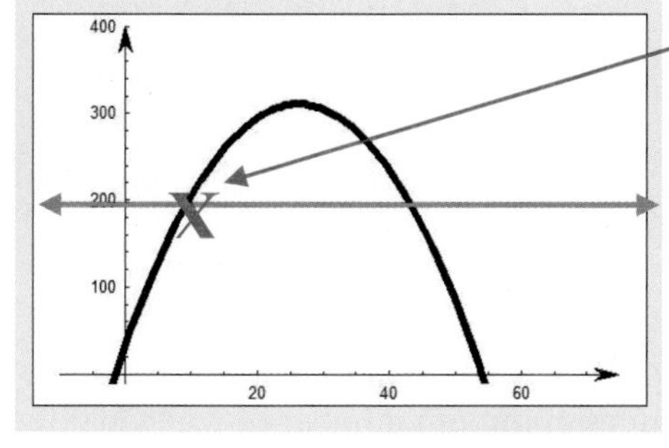

- Find the intersection of the first point to the left of the vertex, the "X" on the graph. Press [2nd][CALC] and select "5:intersect".
  - First curve?: Scroll close to the intersection point where the "X" is located. Remember to use only the left or right scroll keys, not up or down keys. Press [ENTER].
  - Second curve?: Press [ENTER].
  - Guess?: Press [ENTER].
  - The "X" should be on the intersection point. It is (9.62, 200), (*years, # cell phones*).

The year in which the number of Wi-Fi-enabled cell phones reaches 200 million is 9.62 years after 2009. This would be the year 2018.

6.18 Further Graphing of Quadratic Functions 6.18

YOUR TURN

6. The number of McDonald's restaurants worldwide can be modeled by the quadratic equation $f(x) = -96x^2 + 1018x + 28824$, where $f(x)$ is the number of McDonald's restaurants and x is the number of years after 2000. (*Source:* Based on data from McDonald's Corporation)

WINDOW: Xmin = -10 Xmax = 25 Ymin = -10 Ymax = 50,000

Understand the variables in words:

The # of ______________________________: x

The ________________________________: $f(x)$

Go to the TI-83/84 calculator.

a. Will this function have a maximum or minimum? How can you tell?

b. According to this model, in what year was the number of McDonald's restaurants at its maximum/minimum?

c. What is the maximum/minimum number of McDonald's restaurants predicted?

7. If the cost, C, for manufacturing x computers is given by $C = x^2 - 5x + 40$,

a. Find the number of computers manufactured at a cost of **$10,000**. (Use the intersect feature of the calc. Let $Y_2 = 10{,}000$.)

WINDOW: Xmin = -10 Xmax = 200 Ymin = -1000 Ymax = 15,000

Understand the variables in words:

The # of ________________: x

The company's costs, $: __________

Go to the TI-83/84 calculator.

The number of computers is

__________________.

REMEMBER TO GO BACK AND AND ANSWER QUESTIONS IN THE INTRODUCTORY PROBLEM AT THE BEGINNING OF THE MODULE.

6.18 Further Graphing of Quadratic Functions 6.18

YOUR TURN **Answers to Section 6.18.**

1. 9 0, 3.264) 2. The # of cupcakes: *x,* The company's costs, $: *C(x)*, The start-up cost is $6,000.

3. Graph D, *x*-int: (1, 0) and (3, 0) *y*-int: (0, 3)

4. opens upward, minimum,

vertex: (-2, -9)

x-int: (-5, 0) and (1, 0)

y-int: (0, -5)

5. & 6. See below.

7. The number of computers: x, The company's cost: *C(x)*, 102 computers

5. The # of calendars: *x,* The profit, cents: *P(x)*

a. 180 calendars b. maximum profit is 32,400 cents

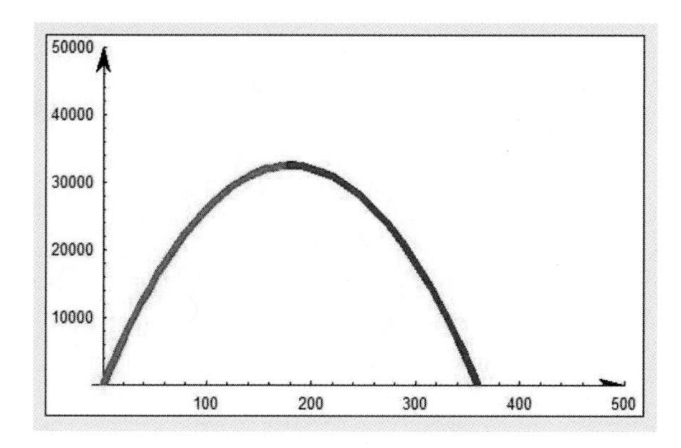

6. The # of years after 2000: *x,* The # of McDonald's restaurants: *f(x)*

a. maximum, Because the value in front of x^2, a = -96, is negative.

b. In 5.3 years or 2005. Add 5 to the beginning year,2000.

c . 31,523 restaurants

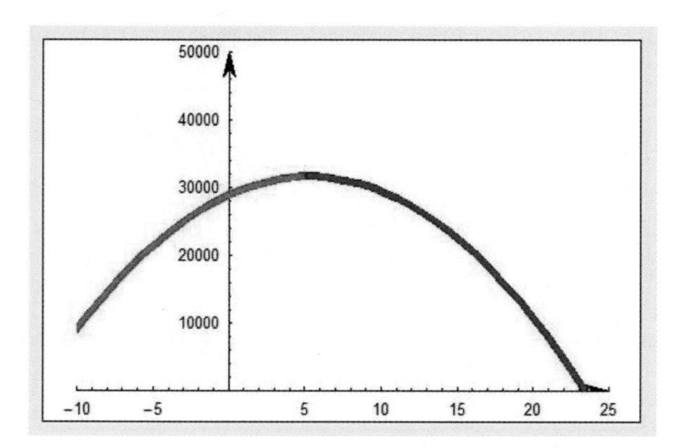

6.18 Further Graphing of Quadratic Functions 6.18

Introductory Problem:

1. t in words is: seconds, $f(t)$ in words is the height of the ball as it rises and then falls.

2. The initial height of the ball (before it is hit) is 5 feet. This would be shoulder height of the baseball player, from where he swings the bat. In the formula, this initial height is at the end. It is the constant term. $f(t) = -16t^2 + 128t + 5$

3. The velocity is 128 ft/sec. This is how hard the ball is hit. It is in coefficient of the second term of the formula. $(t) = -16t^2 + 128t + 5$ When solving the equation with actuall time, it will be multiplied by the time x when used.

4. The term $-16t^2$ reflects the pull of gravity on the ball as time passes. Note that gravity should evenutally pull the ball back toward the earth.

5. (0, 5), (1, 117), (2, 197), (4, 261), (6, 197), (8, 5), (10, -315).

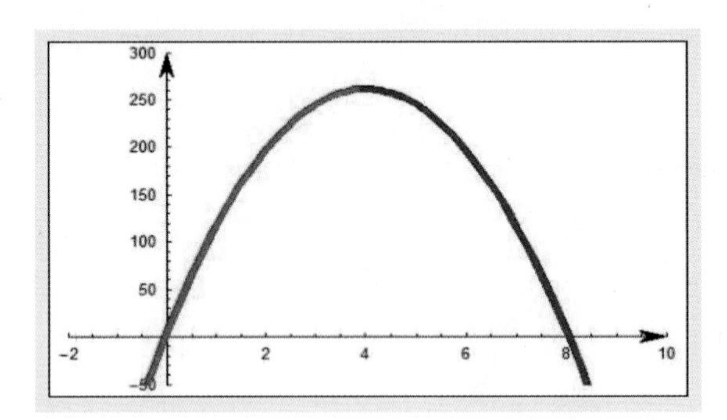

6. 1 second: 117 feet, 2 seconds: 197 feet

7. At 6 seconds the ball is again at 197 feet. The ball has begun to fall back to the ground.

8. The solutions at 1 second and 4 seconds indicate that the ball is rising in the air. This makes sense. At 6 seconds the ball has begun to fall back toward the ground. However, at 10 seconds the ball cannot be at -315 feet unless there is a huge hole in the ground. The ball would land on the ground before 10 seconds.

9. a. The y-intercept is the initial height of the ball before it is hit. It is the point (0, 5).
   b. The ball follows a path upward, reached a maximum height, and then turns to fall toward the ground.
   c. The x-intercept is the time it takes for the ball to hit the ground, where $h = 0$.

10. On the calculator, find the maximum. The greatest height of the ball is 261 feet, the y value. The ball reaches this height after 4 seconds.

11. After 10 seconds, the ball is -315 feet. This makes no sense unless it has fallen in a big hole.

6.18 Further Graphing of Quadratic Functions 6.18

12. On the calculator, enter $y_2 =$ 200 ft. and find the left intersection point. This intersection point is the first one the ball reaches as it would rise into the air, the intersection point on the left. The ball reaches 200 feet after 2.05 seconds.

13. The second intersection point, on the right, is the point when the ball would reach 200 feet on the way back down to the ground. It is 5.95 seconds.

14. On the calculator, find the zero. The ball reaches the ground after 8.04 seconds.

Complete MyMathLab Section 6.18 Homework in your Homework notebook.

DMA 060 TEST: Review your notes, key concepts, formulas, and MML work. ☺

Student Name: ______________________________ **Date:** ____________

Instructor Signature: ______________________________

DMA 070

DMA 070

RATIONAL EXPRESSIONS/EQUATIONS

This course provides a conceptual study of problems involving graphic and algebraic representations of rational equations. Topics include simplifying and performing operations with rational expressions and equations, understanding the domain, and determining the reasonableness of an answer. Upon completion, students should be able to find algebraic solutions to contextual problems with rational applications.

You will learn how to

- Multiply and divide rational expressions.
- Add and subtract rational expressions.
- Solve rational equations.
- Graph rational functions using the graphing calculator to identify and interpret the y-intercept values and domain in terms of the problem.
- Analyze the meaning of asymptotes using a graphing calculator.
- Demonstrate the use of a problem solving strategy to include multiple representations of the situation, organization of the information, and algebraic representation of rational equations.

Courtesy of Fotolia

Application Types

Unknown Number Problems

Work Hour Problems

Distance Problems

Proportions/Similar Triangles

Plane Speed/Wind Speed Problems

Boat Speed/Current Speed Problems

Average Cost Problems

Rational Equations: Equations with FRACTIONS.

$$C(x)=\frac{5x+1250}{x}$$

Courtesy of Fotolia

INTRODUCTORY PROBLEM

We now begin our study of rational expressions and functions.

Here is an example of how rational functions are used. Do as much as you can now and come back to complete it as you work through the module.

Psychologists study memory and learning. In an experiment on memory, students in a language class are asked to memorize 40 vocabulary words in Latin, a language with which the students are not familiar. After studying the words for one day, students are tested each day after to see how many words they remember. The class average is then found. The function, $f(x)=\dfrac{5x+30}{x}$, models the average number of Latin words remembered by the students, after x days.

Courtesy of Fotolia

WINDOW: Xmin = -1 Xmax = 50 Ymin = -5 Ymax = 45

1. What does the point (25, 6.2) mean in the context of this scenario?

2. As time passes, does memory loss slow down or speed up?

3. Find y when $f(11)$. Explain the meaning of the solution in the context of this problem.

4. Use the function to find the number of days it takes for the students to remember half of the original vocabulary list. Check the reasonableness of your answers by referring to the graph.

$f(x)$ average # of Latin words

40
30
20
10

10 20 30 40 50

x # days

Professions that Use Rational Expressions

Management
Management Occupations:
Computer and Information Systems managers
Engineering and natural science managers
Farmers, ranchers, and agricultural managers
Funeral directors
Industrial production managers
Medical and health services managers
Property, real estate, and community association managers
Purchasing managers, buyers, and purchasing agents
Business and Financial Operations Occupations:
Budget analysts
Insurance underwriters

Service
Building and Grounds Cleaning Occupations:
Grounds maintenance workers

Professional
Computer and Mathematical Occupations:
Actuaries
Computer software engineers
Mathematicians
Statisticians

Engineers:
Aerospace engineers
Chemical engineers
Civil engineers
Environmental engineers
Mechanical engineers
Nuclear engineers
Petroleum engineers

Production
Metal workers and Plastic workers:
Computer control programmers and operators

To see more about these professions you can go to the website below:

http://www.xpmath.com/careers/topicsresult.php?subjectID=2&topicID=6

FIRST WE WILL LEARN HOW TO SIMPLIFY RATIONAL EXPRESSIONS.

7.01 Rational Functions and Simplifying Rational Expressions 7.01

Recall that a rational number is also known as a fraction. It can be written as the quotient $\frac{p}{q}$ of two integers p and q as long as q, the denominator $\neq 0$. A rational expression is an expression that can be written in the form $\frac{P}{Q}$ where P and Q are polynomials and the value of Q is not zero.

EXAMPLES OF RATIONAL EXPRESSIONS:

$\frac{2x-7}{6}$ $\frac{3a^2-1}{a+4}$ $\frac{2y+1}{y^2-4y+3}$

Rational expressions can be used to describe functions. Recall, function notation can be written as *f(x)*, meaning "*f* of *x*." It does not mean to multiply *f* times *x*. A function, $f(a) = \frac{2a^2+2}{a+1}$, is a rational function since $\frac{2a^2+2}{a+1}$ is a rational expression.

Pictures or graphs of these rational functions are curves of many shapes. We will not graph these by hand in this module. However, we will graph them using the graphing calculator. This rational function would look something like this. Notice that this rational function is really two curves – one part on the bottom left and another part on the top right. The dotted lines are called asymptotes and are important because you will notice that edges of the curves approach these lines but they will not ever touch them. We will analyze these curves later on. They'll be fun.

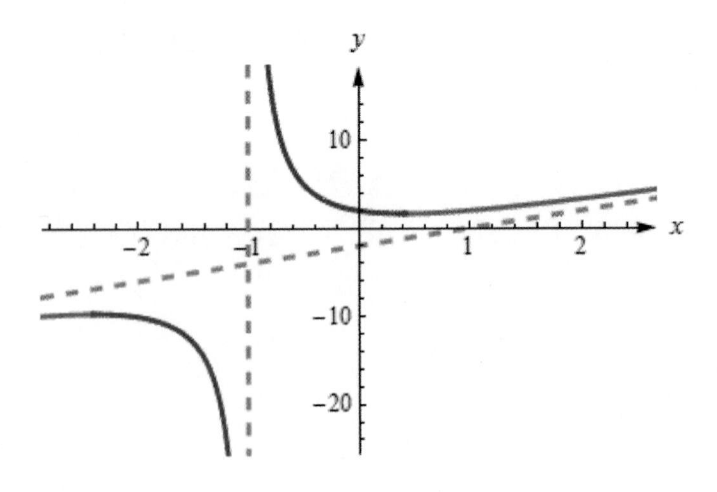

7.01 Rational Functions and Simplifying Rational Expressions 7.01

EVALUATING A RATIONAL EXPRESSION

We evaluate rational expressions the same way that we evaluation any expression. ***Remember that if the denominator is equal to zero, then the expression is undefined!***

EXAMPLE 1

Find the value of $\frac{x+5}{x^2-x}$ when $x = -3$ and $x = 1$. Remember to use parentheses.

when $x = -3$		**when $x = 1$**
$\frac{x+5}{x^2-x} = \frac{(-3)+5}{(-3)^2-(-3)}$	Substitute the given value for ***x***.	$\frac{x+5}{x^2-x} = \frac{(1)+5}{(1)^2-(1)}$
$= \frac{2}{9+3}$	Simplify the numerator. Begin simplifying the denominator.	$= \frac{6}{1-1}$
$= \frac{2}{12} = \frac{1}{6}$	Now simplify the fraction.	$= \frac{1}{0}$ *undefined* ***Since 0 is in the denominator, the expression is undefined when x = 1.***

7.01 Rational Functions and Simplifying Rational Expressions 7.01

YOUR TURN

Find the value of each expression for the given values of the variable.

1. $\dfrac{x^2-5x+6}{x-8}$ when $x=8$	2. The average cost of per t-shirt for Noel and Kelly to produce skateboard t-shirts is given by the formula $C=\dfrac{5x+1250}{x}$ where x is the number of t-shirts produced. (Remember to think about what x and C represent in words before attempting to solve the problem.) a. How much would it cost to produce 1 t-shirt? b. How much is the average cost for producing 100 t-shirts? c. Does the average cost decrease or increase when more t-shirts are produce?

FINDING THE DOMAIN OF A RATIONAL EXPRESSION

The domain of a function tells the x values where a function is defined. Recall, a rational expression is **undefined** if the denominator = 0. If you replace a variable in a rational expression with a number that makes the denominator = 0, this also makes your expression **undefined.** We then say the expression is **undefined** for this value of x.

In other words, as discussed in Module 6, the domain of a function defines the real number values which are represented by the variable x in the expression. For rational functions, some real numbers will not be possible because of variables appearing in the denominator. It is easiest to find the domain of a rational function by identifying the real number values which won't work, the excluded numbers which would make the denominator 0.

7.01 Rational Functions and Simplifying Rational Expressions 7.01

INSTRUCTIONS TO FIND DOMAIN: Look only at the part of the rational function in the denominator.

1. If the denominator has a *variable*, then there will be real numbers which would make the function be UNDEFINED. To find these numbers, set the denominator equal to zero and solve. *This gives you the real numbers to exclude from the domain.* These real numbers would NOT BE in the domain of the rational expression.

2. If the denominator has *no variable*, then the rational function is defined for all real numbers. There are NO excluded numbers which would make the function be undefined.

EXAMPLE 2

Find the domain of $f(a)=\dfrac{3a^2+5}{2a-6}$.

$$f(a)=\frac{3a^2+5}{2a-6}$$

$2a-6$ cannot equal 0 because it is in the denominator!

Remember that variables can represent any real number except those values that make the expression undefined. Let's find those values.

Solve $2a-6=0$ Set denominator = 0 & then solve.

$$2a-6=0$$
$$2a=6$$
$$a=3$$

This means that when a = 3 the denominator is 0, which makes the fraction **undefined**. Therefore we must **exclude** 3 from the domain.

We write the domain of *f* in set builder notation: $\{a \mid a \text{ is a real number and } a \neq 3\}$.

This set notation is read "the set of all real numbers a such that a is a real number and $a \neq 3$.

Since the numerator can be any real number, we do not need to consider it when finding the domain.

DOMAIN: FINDING THE VALUES FOR WHICH THE RATIONAL EXPRESSION IS UNDEFINED

1. Set any denominator **with a variable** = 0.
2. Find the numbers to exclude by solving the equation from step 1.
   [This equation can be linear, 1 answer, or quadratic, 2 answers.]
3. The domain will be all real numbers except these excluded numbers.
   Note: A rational function with only a number in the denominator (no variables) has no restrictions. There are not values for which the expression is undefined. The domain is all real numbers.

7.01 Rational Functions and Simplifying Rational Expressions 7.01

EXAMPLE 3

Find any numbers for which $\frac{x^2+7x+6}{3}$ is undefined.

Since the denominator does not contain a variable, then there are no exclusions!

EXAMPLE 4

Find any numbers for which $\frac{x^2+7x+6}{35x^2-30x}$ is undefined. (Hint: Look only in the denominator for variable terms.)

$35x^2-30x=0$ — Set the denominator equal to 0. Since this is quadratic, we will factor the denominator.

$5x(7x-6)=0$ — Set each factor = 0.

$5x=0$ *or* $7x-6=0$ — Now solve each of the equations.

$\frac{5x}{5}=\frac{0}{5}$ $\qquad$ $7x=6$

$\boxed{x=0}$ $\qquad$ $\boxed{x=\frac{6}{7}}$ — These are the values that must be excluded since each value would make the expression undefined (the denominator would equal 0).

YOUR TURN

Find any numbers for which the rational expression is undefined. This is the same as finding the excluded numbers in the domain.

3. $\frac{-6}{x-8}$	4. $\frac{y+5}{3}$	5. $\frac{t}{2t^2-t-1}$ **Hint:** Factor the denominator 1st!

7.01 Rational Functions and Simplifying Rational Expressions 7.01

Simplifying Rational Expresssions

RECALL: A fraction is in simplest form if both the numerator and denominator have no other number in common other than 1. To simplify a rational expression, we will use the same method we use to simplify a fraction.

$$\text{Simplify: } \frac{12}{216} = \frac{\cancel{2}\cdot\cancel{2}\cdot\cancel{3}}{\cancel{3}\cdot 3\cdot 3\cdot\cancel{2}\cdot\cancel{2}\cdot 2} = \frac{1}{18}$$

Factor and reduce.

We will use this same technique to simplify rational expressions.

First we will look at how two different students simplified the same rational expression. One is correct and the other is wrong. Which method do you think is correct?

STUDENT A	STUDENT B
$\frac{3x+6}{9x+18}$ $\frac{3(x+2)}{9(x+2)}$ $\frac{\cancel{3}(\cancel{x+2})}{\cancel{9}(\cancel{x+2})} = \frac{1}{3}$	$\frac{3x+6}{9x+18}$ $\frac{\cancel{3}\cancel{x}+\cancel{6}}{\cancel{9}\cancel{x}+\cancel{18}} = \frac{1}{9}$

Which one is correct? To help make a good decision, you can plug in 2 (or your favorite number) for the variable and follow the order of operations to see which method works! Show your work above and then circle the correct method. Put a BIG X over the incorrect method.

Check your answer with a tutor.

REMEMBER THAT WHEN WE ARE SIMPLIFYING, WE ARE LOOKING FOR COMMON FACTORS, NOT COMMON TERMS!

7.01 Rational Functions and Simplifying Rational Expressions 7.01

EXAMPLE 5

Simplify. Remember "Simplify" means to FACTOR and REDUCE. $\dfrac{x+2}{x^2+6x+8}$

$$\frac{x+2}{x^2+6x+8}=\frac{(x+2)}{(x+4)(x+2)}$$

Factor the numerator and the denominator. If you cannot factor the polynomial, then just put parentheses around it.

$$\frac{x+2}{x^2+6x+8}=\frac{\overset{1}{\cancel{(x+2)}}}{(x+4)\underset{1}{\cancel{(x+2)}}}=\frac{1}{(x+4)}$$

Divide out [cancel] the common factors.

EXAMPLE 6

Simplify $\dfrac{x^2+3x-28}{2x+14}$ to lowest terms.

1. Factor: $\dfrac{x^2+3x-28}{2x+14}=\dfrac{(x+7)(x-4)}{2(x+7)}$
2. Simplify: $\dfrac{\overset{1}{\cancel{(x+7)}}(x-4)}{2\underset{1}{\cancel{(x+7)}}}=\dfrac{(x-4)}{2}$

Steps to Simplify a Rational Expression

1. **Factor both the numerator and denominator.**
2. **Simplify by dividing out common factors in the numerator and denominator.**
   [cancel out common factors]

Courtesy of Fotolia

Be very careful when you are crossing out common factors. You can only cancel FACTORS (some number times another number). In the example above, it is tempting to cancel 4 and 2 in the final step. But 4 is not a factor; it is being subtracted from x instead of being multiplied by x.

7.01 Rational Functions and Simplifying Rational Expressions 7.01

YOUR TURN

Simplify each rational expression. Be sure your answer is in lowest terms. *Make sure that you are canceling FACTORS, not terms!*

6. $\dfrac{x-5}{x^2-25}$	7. $\dfrac{5y-15}{3y-9}$
8. $\dfrac{x^2+8x-9}{x^2-5x+4}$	9. $\dfrac{6x^2-24}{3x+6}$

Sometimes with rational expressions we need to be able to recognize equivalent factors. For example, we know that $(x + 2)$ is the same factor as $(2 + x)$ because addition is commutative. Let's look at a factor that involves subtraction.

Is subtraction commutative? NO!!! For example $7-5 \neq 5-7$.
So it would follow that $(\boldsymbol{x} - 2)$ is not the same factor as $(2 - \boldsymbol{x})$.

Let's see what would happen if we factor out a negative one.

$$2-x = -1\,(-2+x) = -1(x-2)$$

It is helpful to recognize that (x – 2) and (2 – x) are the opposites of each other!

7.01 Rational Functions and Simplifying Rational Expressions 7.01

So when we notice that two expressions are opposites, we can factor –1 out of one of the factors so that you can then divide out the like factors. Check these out and see if they make sense to you.

$(5-x) = -1(-5+x) = -1(x-5)$

$(3-y) = -1(-3+y) = -1(y-3)$

$25-x^2 = -1\left(-25+x^2\right) = -1\left(x^2-25\right)$

Now let's look at some rational expressions that involve factors that are opposites!

EXAMPLE 7

Simplify. $\dfrac{x-5}{5-x}$

$\dfrac{x-5}{5-x} = \dfrac{(x-5)}{-1(x-5)}$

Since $x - 5$ and $5 - x$ are opposites we need to factor –1 out of one of them. Most students would chose to factor –1 out of the expression in the denominator.

Now divide out the common factors.
Multiply the results.
Finished!

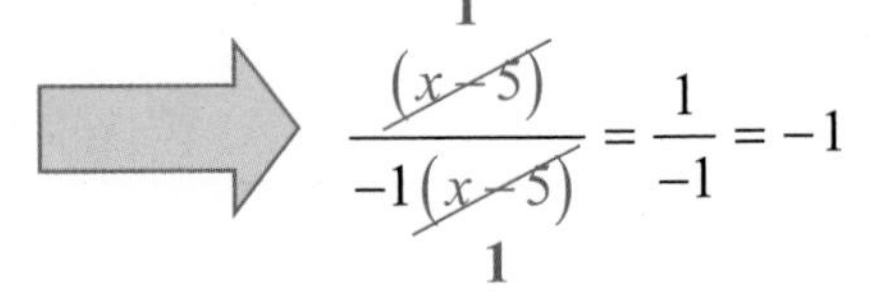

EXAMPLE 8

Simplify. $\dfrac{4-x^2}{x+2}$ One way to simplify this expression is to change the denominator to $2 + x$ and then factor the numerator. Look below for a different way.

$\dfrac{4-x^2}{x+2} = \dfrac{-1\left(x^2-4\right)}{x+2}$

Many students would chose to factor –1 out of the expression in the numerator, changing $4-x^2$ to an expression with the variable first, like the denominator!

Now divide out the common factors.

Multiply the results and simplify.

Finished!

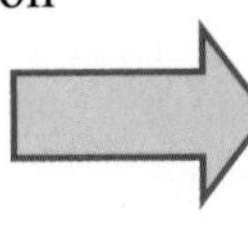

$\dfrac{-1\left(x^2-4\right)}{x+2} = \dfrac{-1(x+2)(x-2)}{(x+2)} = \dfrac{-1(x-2)}{1} = -(x-2) = -x+2$

7.01 Rational Functions and Simplifying Rational Expressions 7.01

Tip: The factor of – 1 may be attached to the numerator or denominator of a fraction. You may also write it out front. It all means the same thing!

$$\frac{-2}{x+1} = \frac{2}{-(x+1)} = -\frac{2}{x+1}$$

YOUR TURN

Simplify the following rational expressions. (In other words, FACTOR and REDUCE.)

10. $\dfrac{x-3}{3-x}$	11. $\dfrac{49-y^2}{y-7}$
12. $\dfrac{2x-8}{20-5x}$	13. $\dfrac{20-5x}{x^2-x-12}$

7.01 Rational Functions and Simplifying Rational Expressions 7.01

YOUR TURN Answers to Section 7.01.

1. $\frac{30}{0} \Rightarrow$ undefined 2a. \$1255 2b. \$17.50 2c. decreases 3. $x = 8$

4. There are no values for which the fraction is undefined. 5. $t = 1, -\frac{1}{2}$ 6. $\frac{1}{x+5}$ 7. $\frac{5}{3}$

8. $\frac{x+9}{x-4}$ 9. $2(x-2)$ 10. -1 11. $-1(y+7) = -y-7$ 12. $-\frac{2}{5}$ 13. $-\frac{5}{x+3}$

Complete MyMathLab Section 7.01 Homework in your Homework notebook.

7.02 Multiplying and Dividing Rational Expressions 7.02

We just looked at simplifying rational expressions in the last section. Simplifying these expressions is the same as simplifying a fraction. We are now going to look at multiplying and dividing rational expressions. As you might recognize, these operations are the same as multiplying and dividing fractions.

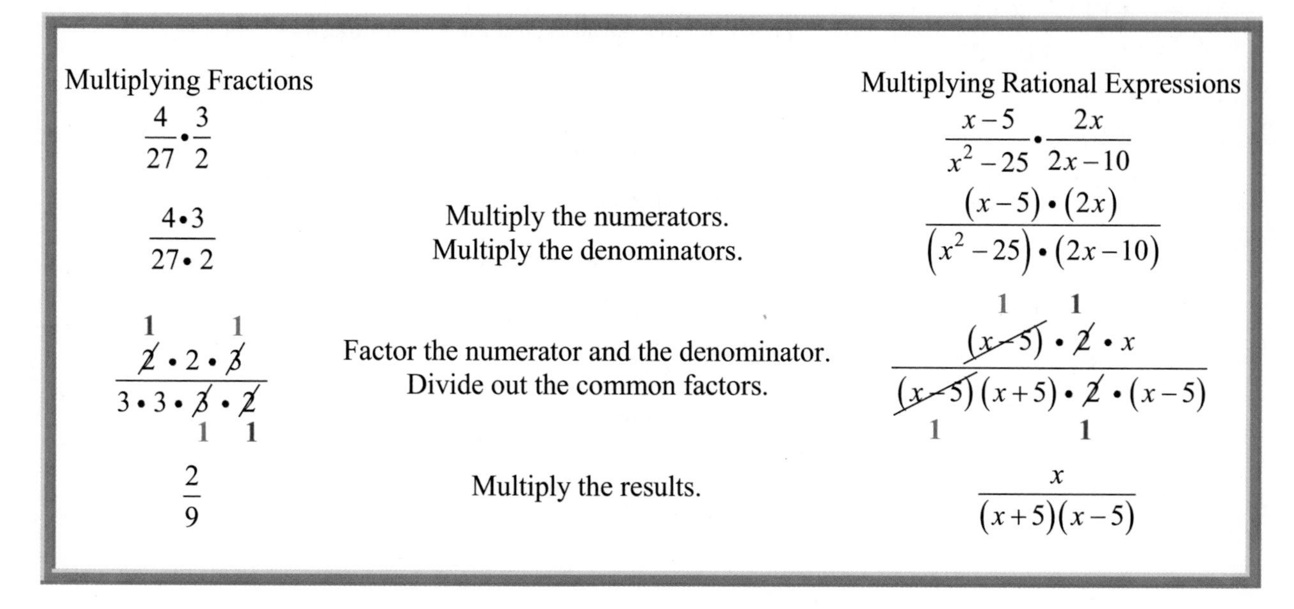

The process is exactly the same for multiplying rational expressions as it is for multiplying fractions!

Steps to Multiply Rational Expressions

1. **Factor each numerator and denominator.**
2. **Simplify by dividing out the common factors.**
3. **Multiply the results.**

To reduce or "cancel", the same factor must be present in a numerator and in a denominator. In this multiplication problem below we cannot divide out the $(x - 5)$ since $(x - 5)$ appears only in the numerator.

$$\frac{3(x-5)}{(x+3)} \cdot \frac{(x-5)}{7}$$

7.02 Multiplying and Dividing Rational Expressions 7.02

EXAMPLE 1

Simplify. (Look at the problem. Simplify here means to factor and reduce.)

$$\frac{3x^2+18x+15}{6x+6} \bullet \frac{x-5}{x^2-25}$$

Factor the numerators and the denominators.

$$\frac{3(x+1)(x+5)}{6(x+1)} \bullet \frac{(x-5)}{(x+5)(x-5)}$$

$3x^2+18x+15=3(x^2+6x+5)=3(x+1)(x+5)$

$6x+6=6(x+1)$

$x-5=(x-5)$ We cannot factor x – 5; put it in parentheses.

$x^2-25=(x+5)(x-5)$

$$\frac{3\cancel{(x+1)}\cancel{(x+5)}}{6\cancel{(x+1)}} \bullet \frac{\cancel{(x-5)}}{\cancel{(x+5)}\cancel{(x-5)}}$$

Now divide out the common factors.

$$\frac{1}{2}$$

Multiply the results.

We handle problems that contain only monomials a little differently.

EXAMPLE 2

Simplify. (This problem can be solved two ways.)

METHOD 1:

$$\frac{5x^2y^2}{15x} \bullet \frac{6x^3}{3xy3}$$

Factor the numerators and the denominators.

$$\frac{\cancel{5}\bullet\cancel{x}\bullet\cancel{x}\bullet\cancel{y}\bullet\cancel{y}}{3\bullet\cancel{5}\bullet\cancel{x}} \bullet \frac{2\bullet\cancel{3}\bullet x\bullet x\bullet x}{\cancel{3}\bullet\cancel{x}\bullet\cancel{y}\bullet\cancel{y}\bullet y}$$

Now divide out the common factors. Look at the colors!

$$\frac{2x^3}{3y}$$

Multiply the results.

EXAMPLE 2 CONTINUED

METHOD 2:

$\frac{5x^2y^2}{15x} \bullet \frac{6x^3}{3xy^3}$ — We can use the rules for exponents.

$\frac{30x^5y^2}{45x^2y^3}$ — Use the product rule to multiply the numerators and then to multiply the denominators. (Multiply the coefficients; add the exponents on the like bases.)

$\frac{2x^3}{3y}$ — Use the quotient rule to continue simplifying. (Divide or reduce the coefficients; subtract the exponents on the like bases.)

YOUR TURN

Multiply and simplify. (This really means factor and reduce.)

1. $\frac{7a^2b}{3ab^2} \bullet \frac{21a^2b^2}{14ab}$	2. $\frac{3m^2+12m}{6} \bullet \frac{9}{2m+8}$
3. $\frac{x^2-25}{x^2-3x-10} \bullet \frac{x+2}{x}$	4. $\frac{x+6}{7-x} \bullet \frac{x^2-9x+14}{x^2+3x-18}$ (Remember what to do with opposites!)

7.02 Multiplying and Dividing Rational Expressions 7.02

We divide rational expressions the same way we divide fractions. To divide a fraction, simply keep the first fraction and multiply by the reciprocal of the second fraction. **KEEP, CHANGE, FLIP!**

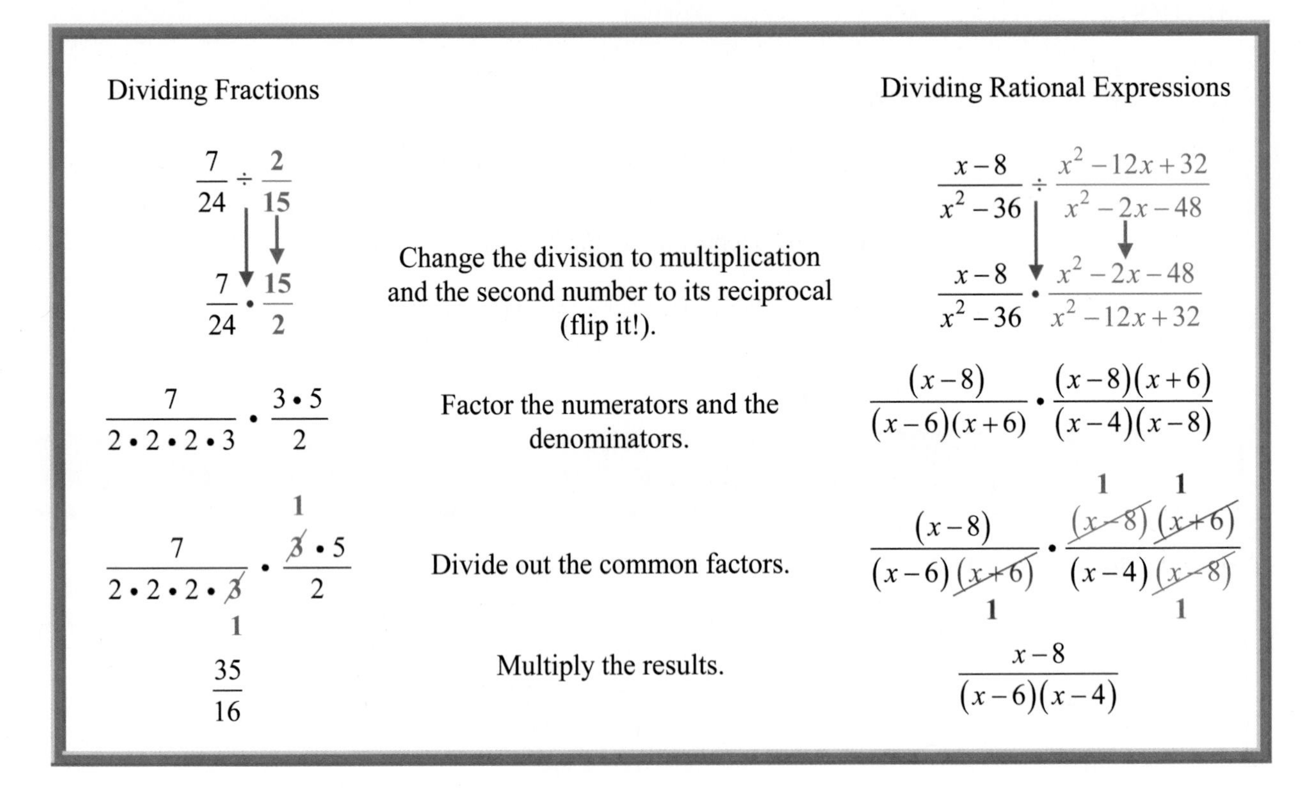

Dividing Fractions		Dividing Rational Expressions
$\frac{7}{24} \div \frac{2}{15}$		$\frac{x-8}{x^2-36} \div \frac{x^2-12x+32}{x^2-2x-48}$
$\frac{7}{24} \bullet \frac{15}{2}$	Change the division to multiplication and the second number to its reciprocal (flip it!).	$\frac{x-8}{x^2-36} \bullet \frac{x^2-2x-48}{x^2-12x+32}$
$\frac{7}{2 \bullet 2 \bullet 2 \bullet 3} \bullet \frac{3 \bullet 5}{2}$	Factor the numerators and the denominators.	$\frac{(x-8)}{(x-6)(x+6)} \bullet \frac{(x-8)(x+6)}{(x-4)(x-8)}$
$\frac{7}{2 \bullet 2 \bullet 2 \bullet \cancel{3}} \bullet \frac{\cancel{3} \bullet 5}{2}$	Divide out the common factors.	$\frac{(x-8)}{(x-6)\cancel{(x+6)}} \bullet \frac{\cancel{(x-8)}\cancel{(x+6)}}{(x-4)\cancel{(x-8)}}$
$\frac{35}{16}$	Multiply the results.	$\frac{x-8}{(x-6)(x-4)}$

So we can see that the process is exactly the same for dividing rational expressions as it is for dividing fractions!

Steps to Divide Rational Expressions

1. **Change the multiplication to division and flip the second fraction to its reciprocal (FLIP IT).**
2. **Factor each numerator and denominator.**
3. **Simplify by dividing out (cancel) the common factors.**
4. **Multiply the results.**

DO YOURSELF A FAVOR AND TAKE CARE OF FLIPPING THAT SECOND FRACTION BEFORE YOU BEGIN FACTORING!

7.02 Multiplying and Dividing Rational Expressions 7.02

YOUR TURN

Divide. Be sure to write your answer in simplest form. (Flip that second fraction!)

5. $\dfrac{9x^2y^3}{6z} \div \dfrac{x^7y^9}{2z}$	6. $\dfrac{3x^5}{x^2-4} \div \dfrac{6x^2}{2-x}$
7. $\dfrac{(m+n)^2}{m+n} \div \dfrac{m^2-mn}{m}$	8. $\dfrac{a^2+7a+12}{a^2+5a+6} \div \dfrac{a^2+5a+4}{a^2+8a+15}$

YOUR TURN **Answers to Section 7.02**

1. $\dfrac{7a^2}{2}$ 2. $\dfrac{9m}{4}$ 3. $\dfrac{x+5}{x}$ 4. $-\dfrac{x-2}{x-3}$

5. $\dfrac{3}{x^5y^6}$ 6. $-\dfrac{x^3}{2(x+2)}$ 7. $\dfrac{m+n}{m-n}$ 8. $\dfrac{(a+5)(a+3)}{(a+2)(a+1)}$

Complete MyMathLab Section 7.02 Homework in your Homework notebook.

7.03 Adding/Subtracting Rational Expressions with Common Denominators 7.03

Addition and subtraction of rational expressions is similar to addition and subtraction of fractions. In this section we will add and subtract rational expressions that already have common denominators.

Adding Fractions

$$\frac{3}{11}+\frac{5}{11}=\frac{3+5}{11}=\frac{8}{11}$$

Recall that when we add fractions we simply ***add the numerators*** and ***keep the denominator***.

Adding Rational Expressions

$$\frac{3}{x+5}+\frac{4}{x+5}=\frac{3+4}{x+5}=\frac{7}{x+5}$$

To add or subtract rational expressions, add or subtract the numerators and place the sum or difference over the common denominator.

EXAMPLE 1

Add and express the answer in simplest form.

$$\frac{5x}{6y}+\frac{x}{6y}=\frac{\textbf{numerator + numerator}}{\textbf{common denominator}}=\frac{5x+x}{6y}=\frac{6x}{6y}=\frac{x}{y}$$

Simplify the answer!

EXAMPLE 2

Subtract and express the answer in simplest form.

$$\frac{7x}{3x+5}-\frac{4x-5}{3x+5}$$

$$\frac{7x-(4x-5)}{3x+5}$$

Insert parentheses around the second numerator so that the entire numerator is being subtracted, not just the first term.

$$\frac{7x-4x+5}{3x+5}$$

Distribute the "-" to $4x$ and -5. NOTE: It changes the sign of BOTH #s.

$$\frac{3x+5}{3x+5}$$

Put parentheses around $3x+5$ since it cannot be factored.

$$\frac{\overset{1}{\cancel{(3x+5)}}}{\underset{1}{\cancel{(3x+5)}}}=\frac{1}{1}=1$$

Simplify (factor numerator and denominator).

7.03 Adding/Subtracting Rational Expressions with Common Denominators 7.03

EXAMPLE 3

Subtract. Simplify the result if possible.

$$\frac{4m}{m^2-2m-63}-\frac{36}{m^2-2m-63}$$

$\frac{4m-36}{m^2-2m-63}$ Subtract the numerators; keep the denominator.

$\frac{4(m-9)}{(m-9)(m+7)}$ Simplify, if possible. Factor numerator; factor denominator.

$\frac{4\overset{1}{\cancel{(m-9)}}}{\underset{1}{\cancel{(m-9)}}(m+7)}$ Divide out the common factors.

$\frac{4}{(m+7)}$ **FINISHED!**

Courtesy of Fotolia

DO NOT FACTOR ANY NUMERATORS UNTIL YOU HAVE FINISHED WITH THE ADDITION AND SUBTRACTION AND ARE READY TO SIMPLIFY THE ANSWER!

YOUR TURN

Add or subtract the rational expressions. Simplify if possible.

1. $\frac{8c}{5d}+\frac{2c}{5d}$	2. $\frac{12x}{3x-5}-\frac{20}{3x-5}$

7.03 Adding/Subtracting Rational Expressions with Common Denominators 7.03

YOUR TURN

3. $\dfrac{2x^2+7x}{x-7}-\dfrac{10x+77}{x-7}$ **Be careful! Distribute the subtraction to ALL of the terms in the 2nd numerator.**	4. $\dfrac{3y}{y^2-3y-10}+\dfrac{6}{y^2-3y-10}$

The rational expressions we have looked at so far had common denominators. Now we will begin to look at the process of what happens when we do not have common denominators. To add or subtract fractions or rational expressions with unlike denominators, we will first have to find a least common denominator (LCD).

FINDING THE LCD

Find the LCD of $\dfrac{1}{72}$ and $\dfrac{9}{60}$

$72 = 2 \cdot 2 \cdot 2 \cdot 3 \cdot 3$ $60 = 2 \cdot 2 \cdot 3 \cdot 5$	Prime factor each of the denominators.
$72 = 2 \cdot 2 \cdot 2 \cdot 3 \cdot 3$ $60 = 2 \cdot 2 \cdot 3 \cdot 5$	The LCD will have all of the factors within each of the denominators. We have one factor of 5. Include 5 in the LCD. Since there are two factors of two in 60 but three factors of two in 72, we must include three factors of 2 in the LCD so that all of the factors of 72 are in the LCD. There also must be two factors of 3 in the LCD in order for all the factors of 72 to be in the LCD.
LCD $= 2 \cdot 2 \cdot 2 \cdot 3 \cdot 3 \cdot 5$	The LCD must include three factors of 2, two factors of 3 and one factor of 5.
LCD = 360	The LCD is 360.

7.03 Adding/Subtracting Rational Expressions with Common Denominators 7.03

You can see from the previous example that we need to include all of the different factors of each denominator in the LCD, without unnecessarily duplicating the factors.

FINDING THE LCD OF RATIONAL EXPRESSIONS

1. Factor the denominators only.
2. The LCD is the product of all the different factors of the denominators, each raised to the highest power of that factor in either of the denominators.

We factor the polynomials and put together the factors in the smallest way possible so that each of the factorizations is contained in the LCD. This means that we include all of the factors to the highest powers that they occur in a single denominator.

EXAMPLE 4

Find the LCD of $\dfrac{2}{3x^5y}$ and $\dfrac{15}{5xy^2}$.

Factor each denominator.

$3x^5y = 3 \bullet x^5 \bullet y$

$5xy^2 = 5 \bullet x \bullet y^2$

You want to use the factor the **maximum** number of times it is used in denominators!

We have 4 different factors: 3, 5, x, y

The greatest power of 3 is one, so we need 3^1.
The greatest power of 5 is one, so we need 5^1.

The greatest power of x is 5, so we need x^5.
The greatest power of y is 2, so we need y^2.

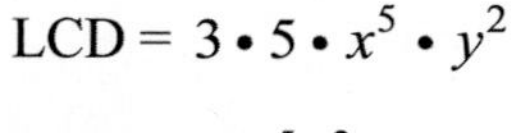

LCD = $3 \bullet 5 \bullet x^5 \bullet y^2$

LCD = $15x^5y^2$

 Make sure that the LCD is divisible by each of the original denominators! 

EXAMPLE 5

Find the LCD of $\dfrac{5}{y^2-4}$, $\dfrac{2y+4}{3y+6}$, $\dfrac{6y}{y^2+5y+6}$.

Factor each denominator.

$y^2-4=(y+2)(y-2)$ $\quad 3y+6=3(y+2)$ $\quad y^2+5y+6=(y+3)(y+2)$

We have these unique (different) factors: 3, $(y+2)$, $(y-2)$ and $(y+3)$.

Each of these factors will be in the LCD. What power will each have?

Since each factor appears only one time in any of the denominators then each factor will have the exponent of 1 in the LCD.

$\text{LCD}=3(y+2)(y-2)(y+3)$.

 Make sure that the LCD is divisible by each of the original denominators! 

EXAMPLE 6

Find the LCD of $\dfrac{3x}{x^2+10x+25}$, $\dfrac{9}{2x+10}$.

Factor the denominators.

$x^2+10x+25=(x+5)(x+5)$

$2x+10=2(x+5)$

$\text{LCD}=2(x+5)(x+5)$

Notice that the factor of (x + 5) occurs two times in the first denominator but only one time in the second denominator. So, the greatest number of times it appears in any one denominator is two times.

The LCD will contain two factors of (x + 5).

 Make sure that the LCD is divisible by each of the original denominators! 

7.03 Adding/Subtracting Rational Expressions with Common Denominators 7.03

EXAMPLE 7

Find the LCD of $\frac{1}{x-7}$ and $\frac{1}{7-x}$.

Recall that these denominators are opposites! $[7-x=-1(x-7)]$ Because of this fact, either $x-7$ or $7-x$ can be the LCD. More about this later!

YOUR TURN

Find the LCD of the following rational expressions.

5. $\frac{2}{7x^3}, \frac{-13}{5x^2}$	6. $\frac{7}{b+4}, \frac{b+12}{b^2-16}$
7. $\frac{2}{3x+6}, \frac{7x}{3x^2+12x+12}$	8. $\frac{x-5}{x^2-10x+21}, \frac{4x}{x^2-4x+3}$

After finding the LCD, it is necessary to change the fraction with the old denominator into an equivalent fraction with the new, chosen LCD.

EXAMPLE 8

Write $\frac{7}{20}$ as an equivalent fraction with the given denominator.

$$\frac{7}{20}=\frac{}{100}$$

$$\frac{7}{20}\bullet\frac{5}{5}=\frac{35}{100}$$

We ask ourselves how we change the denominator 20 into the denominator, 100. Remember that the only number we can multiply a fraction by and not change its value is 1. This time 1 will be in the form of $\frac{5}{5}$ since $100 \div 20 = 5$.

new denominator

old denominator

7.03 Adding/Subtracting Rational Expressions with Common Denominators 7.03

To write an equivalent fraction, multiply the numerator and the denominator of each fraction by the factors needed to change it to the LCD.

EXAMPLE 9

Rewrite the rational expression as an equivalent rational expression with the given denominator.

$$\frac{13}{2y^4} = \frac{\quad}{10y^4z}$$

$$\frac{13}{2y^4} \bullet \frac{5z}{5z} = \frac{65z}{10y^4z}$$

What would we need to multiply by to make $2y^4$ match $10y^4z$? We would need to multiply by $5z$.

$2y^4(5z) = 10y^4z$

EXAMPLE 10

Find the missing numerator.

$\frac{y}{3(y+2)} = \frac{?}{12(y+2)(y-5)}$ $\frac{y}{3(y+2)} = \frac{?}{12(y+2)(y-5)}$ $\frac{y}{3(y+2)} \bullet \frac{4(y-5)}{4(y-5)} = \frac{4y(y-5)}{12(y+2)(y-5)} = \frac{4y^2-20y}{12(y+2)(y-5)}$	We already have a factor of (y + 2) in the new denominator. What else do we need? In other words, what will we multiply by 3 to get 12? Multiply by $\frac{4(y-5)}{4(y-5)}$. Multiply the numerator out. Do not multiply the denominator out!

7.03 Adding/Subtracting Rational Expressions with Common Denominators 7.03

Let's look at opposites again. Recall that $4-x$ and $x-4$ are opposite.
This means $4-x = -x+4 = -1(x-4)$. Look at how we would change a fraction with $-1(x-4)$ as its denominator to a fraction with $x-4$ as its denominator.

EXAMPLE 11

Find the missing numerator.

$$\frac{9}{-1(x-4)} = \frac{?}{x-4}$$

We can make this change by simply multiplying by $\frac{-1}{-1}$!

$$\frac{9}{-1(x-4)} \bullet \frac{-1}{-1} = \frac{-9}{x-4}$$

YOUR TURN

Rewrite the rational expression as an equivalent rational expression whose denominator is the given polynomial.

9. $\frac{5}{2x^3y^4} = \frac{}{8x^6y^4}$	10. $\frac{y}{2(y-6)} = \frac{}{12(y-6)(y+3)}$
11. $\frac{2}{3(2x+1)} = \frac{}{6x(2x+1)(x-3)}$	12. $\frac{3}{(x+5)} = \frac{}{4(x+5)(x+5)}$
13. $\frac{x}{-1(x-5)} = \frac{}{(x-5)}$	14. $\frac{5}{x^2+6x} = \frac{}{x(x+6)(x-6)}$

YOUR TURN Answers to Section 7.03

1. $\frac{2c}{d}$
2. 4
3. $2x+11$
4. $\frac{3}{y-5}$
5. $35x^3$
6. $(b+4)(b-4)$
7. $3(x+2)(x+2)$
8. $(x-7)(x-3)(x-1)$
9. $\frac{20x^3}{8x^6y^4}$
10. $\frac{6y^2+18y}{12(y-6)(y+3)}$
11. $\frac{4x^2-12x}{6x(2x+1)(x-3)}$
12. $\frac{12x+60}{4(x+5)(x+5)}$
13. $\frac{-x}{(x-5)}$
14. $\frac{5x-30}{x(x+6)(x-6)}$

Complete MyMathLab Section 7.03 Homework in your Homework notebook.

7.04 Adding/Subtracting Rational Expressions with Different Denominators 7.04

In the last section, we looked at adding and subtracting rational expressions with like denominators and then learned how to find the LCD. Now we will look at adding and subtracting rational expressions with unlike denominators. As you can guess by now it works much the same as adding and subtracting fractions with unlike denominators. You use the LCD to write equivalent fractions.

Let's look at adding fractions and then at adding rational expressions. It is the same process!

Fractions	Step	Rational expressions
$\frac{1}{4} + \frac{1}{3}$	Find the LCD.	$\frac{4}{5x} + \frac{1}{3x^3}$
$\frac{1}{4}\bullet\frac{3}{3} + \frac{1}{3}\bullet\frac{4}{4}$	Multiply each fraction so that it has the LCD as its denominator.	$\frac{4}{5x}\bullet\frac{3x^2}{3x^2} + \frac{1}{3x^3}\bullet\frac{5}{5}$
$\frac{3}{12} + \frac{4}{12} = \frac{3+4}{12} = \frac{7}{12}$	Add the numerators. Simplify, if possible.	$\frac{12x^2}{15x^3} + \frac{5}{15x^3} = \frac{12x^2+5}{15x^3}$

To Add or Subtract Rational Expressions with Unlike Denominators

1. **Factor the denominators ONLY.**
2. **Find the LCD.**
3. **Multiply the numerator and denominator of each faction by the necessary factors so that the LCD is now the denominator of each of the fractions.**
4. **Keep the denominator factored; multiply out the numerators, where necessary. DO NOT CANCEL here.**
5. **Add or subtract the numerators.**
6. **Simplify. Now you can factor the numerators!**

7.04 Adding/Subtracting Rational Expressions with Different Denominators 7.04

EXAMPLE 1

Simplify.

Notice that the denominators are opposites of each other. By factoring out **–1,** we can make the denominators match!!

$$\frac{6}{x-5} - \frac{7}{5-x}$$

$$\frac{6}{(x-5)} - \frac{7}{-1(x-5)}$$

$$\frac{6}{(x-5)} - \frac{7}{-1(x-5)} \bullet \frac{-1}{-1}$$

Multiply the 2nd fraction by $\frac{-1}{-1}$ so the denominators will be the same.

$$\frac{6}{(x-5)} - \frac{-7}{(x-5)}$$

Perform the subtraction.

$$\frac{6+7}{(x-5)} = \frac{13}{(x-5)}$$

Be careful when subtracting a negative!

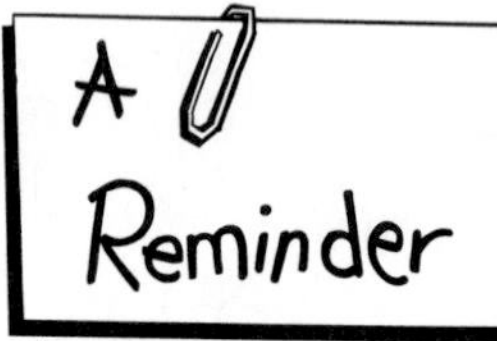

Courtesy of Fotolia

Remember that a fraction can be thought of as division. Here we have a negative divided by a positive and then a positive divided by a negative. Both of these yield a negative quotient. *The negative sign needs to be written outside the fraction.*

$$\frac{-4}{15} = \frac{4}{-15} = -\frac{4}{15}$$

Let's take another look at the last example and apply this thinking.

EXAMPLE 1 REDO!

Simplify.

$$\frac{6}{x-5} - \frac{7}{5-x}$$

$$\frac{6}{(x-5)} - \frac{7}{-1(x-5)}$$

Look at the denominator in the second fraction. We know that $-1(x-5) = -(x-5)$.

$$\frac{6}{(x-5)} - \frac{-7}{(x-5)}$$

So we have a negative in the denominator that we can just move up to the numerator.

$$\frac{6+7}{(x-5)} = \frac{13}{(x-5)}$$

Be careful when subtracting a negative!

You decide which way you understand better and use that process.

7.04 Adding/Subtracting Rational Expressions with Different Denominators 7.04

EXAMPLE 2

Simplify.

$$\frac{9}{x-3}+9$$

$$\frac{9}{x-3}+\frac{9}{1}$$

Change the 2nd number to a fraction and find the LCD. [LCD = (x – 3]

$$\frac{9}{(x-3)}+\frac{9\,(x-3)}{1\,(x-3)}$$

Multiply the fractions by the needed factors. ***No cancelling yet!***

$$\frac{9}{(x-3)}+\frac{9x-27}{(x-3)}$$

Multiply the expressions in the numerator of the 2nd fraction.

$$\frac{9+9x-27}{(x-3)}=\frac{9x-18}{(x-3)}$$

Add the numerators and simplify; factor the numerators and now you can cancel, if possible.

EXAMPLE 2 REDO!

Let's look at that last example again, beginning with the 3rd step.

$$\frac{9}{(x-3)}+\frac{9}{1}\bullet\frac{(x-3)}{(x-3)}$$

Multiply the fractions by the needed factors. ***No cancelling yet!*** To help you to resist the urge to cancel common factors, at this point you can just write **LCD** in the denominator. At the end, we will replace **LCD** with the actual **LCD**.

$$\frac{9}{\textbf{LCD}}+\frac{9x-27}{\textbf{LCD}}$$

$$\frac{9+9x-27}{\textbf{LCD}}$$

Add the numerators and simplify; factor the numerators and now you can cancel, if possible.

$$\frac{9x-18}{(x-3)}=\frac{9(x-2)}{(x-3)}$$

Now that we have combined like terms in the numerator, we will replace **LCD** with the actual **LCD** which is $(x-3)$. Now factor the numerator to make sure that you cannot reduce the fraction.

7.04 Adding/Subtracting Rational Expressions with Different Denominators 7.04

EXAMPLE 3

Simplify.

$$\frac{2a+3}{a^2+a-2}-\frac{5}{3a-3}$$ Factor the denominators.

$$\frac{2a+3}{(a+2)(a-1)}-\frac{5}{3(a-1)}$$ Find the LCD. [LCD = $3(a+2)(a-1)$]

$$\frac{2a+3}{(a+2)(a-1)}\bullet\frac{3}{3}-\frac{5}{3(a-1)}\bullet\frac{(a+2)}{(a+2)}$$ Multiply each fraction so that you have common denominators in each of the fractions.

$$\frac{3(2a+3)}{3(a+2)(a-1)}-\frac{5(a+2)}{3(a-1)(a+2)}$$ **DO NOT CANCEL YET!** Simplify the numerators to remove parentheses. This is the point that you can just use LCD as your denominator!

$$\frac{6a+9}{\textbf{LCD}}-\frac{5a+10}{\textbf{LCD}}$$ Subtract the numerators. $-(5a+10)=-5a-10$

CAUTION! The most common mistake in working this problem is not distributing the negative to both terms in this numerator!

$$\frac{6a+9-5a-10}{\textbf{LCD}}$$ Combine like terms.

$$\frac{a-1}{3(a-1)(a+2)}$$ Simplify (reduce) the answer. Now we can factor the numerator. We cannot factor $a-1$ so we put that expression in parentheses. Now cancel like **factors**!!

$$\frac{\overset{1}{\cancel{(a-1)}}}{3\underset{1}{\cancel{(a-1)}}(a+2)}=\boxed{\frac{1}{3(a+2)}}$$

EXAMPLE 4

Simplify.

$$\frac{2x}{x-7}-\frac{4}{x-2}$$ We cannot factor the denominators this time. Find the LCD. LCD = $(x-7)(x-2)$

$$\frac{2x}{(x-7)}\bullet\frac{(x-2)}{(x-2)}-\frac{x}{(x-2)}\bullet\frac{(x-7)}{(x-7)}$$ Multiply each fraction by the factor(s) needed to form the LCD.

$$\frac{2x^2-4x}{\textbf{LCD}}-\frac{x^2-7x}{\textbf{LCD}}$$ **DO NOT CANCEL YET!** Multiply numerators to remove the parentheses in the numerators.

$$\frac{2x^2-4x-x^2+7x}{(x-7)(x-2)}$$ Be careful subtracting the numerators.

$$\boxed{\frac{x^2+3x}{(x-7)(x-2)}}=\frac{x(x+3)}{(x-7)(x-2)}$$ Combine like terms in the numerator and try to simplify. Factor the numerator and cancel, if possible. This time we cannot simplify the fraction!

7.04 Adding/Subtracting Rational Expressions with Different Denominators 7.04

YOUR TURN

1. Complete the solution by filling in the boxes. $\frac{5a}{3a-12} - \frac{a-7}{a+4}$ $\frac{5a}{\square} - \frac{a-7}{(a+4)}$	Factor the denominators. If you cannot factor the denominator and it is more than one term, then use ( ) around the denominator.
LCD = $\square$	Find the LCD.
$\frac{5a}{3(a-4)} \bullet \frac{\square}{\square} - \frac{(a-7)}{(a+4)} \bullet \frac{\square}{\square}$	Multiply each fraction by the necessary factors so that each has the LCD.
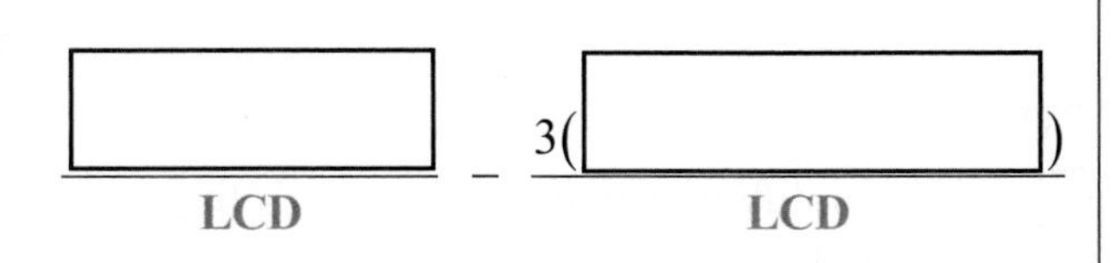 $\frac{\square}{\text{LCD}} - \frac{3(\square)}{\text{LCD}}$	Simplify numerators only. In the second fraction, multiply the two binomials together first. Remember FOIL? Do your multiplication here and then put the product in the box.
$\frac{5a^2+20a}{\text{LCD}} - \frac{\square}{\text{LCD}}$	Now distribute the 3 in the numerator of the second fraction.
$\frac{5a^2 \quad 20a \quad 3a^2 \quad 33a \quad 84}{\text{LCD}}$	Now subtract the numerators. Be sure to distribute the negative to ALL terms in the 2nd fraction! Place the correct signs in the numerator.
$\frac{\square}{\square}$	Simplify the numerator. Now is the time to put the actual LCD back into the fraction. Try to factor the numerator to see if you can simplify the fraction. Don't forget this step! Check your answer in the answer key and make adjustments, if necessary.

7.04 Adding/Subtracting Rational Expressions with Different Denominators 7.04

YOUR TURN

2. Complete the solution by filling in the boxes. $\frac{-18}{2x-3} - 3$ $\frac{-18}{2x-3} - \frac{3}{1}$ $\frac{-18}{\square} - \frac{3}{1}$	Write 3 as $\frac{3}{1}$. Now factor the denominators. This time we are not able to factor the denominators. Put $2x - 3$ in parentheses.
LCD = $\square$	Find the LCD.
$\frac{-18}{(2x-3)} \bullet \frac{\square}{\square} - \frac{3}{1} \bullet \frac{\square}{\square}$	Multiply each fraction by the necessary factors so that each has the LCD.
$\frac{\square}{\text{LCD}} - \frac{\square}{\text{LCD}}$	Simplify numerators only. This is time to replace the denominator with LCD so that you are not tempted to cancel here. Remember that you can only at the end!
$\frac{-18 \quad 6x \quad 9}{\text{LCD}}$	Now subtract the numerators. Be sure to distribute the negative to ALL terms in the 2nd fraction! Place the correct signs in the numerator.
$\frac{\square}{\square}$	Simplify the numerator. Now is the time to put the actual LCD back into the fraction.
	Try to factor the numerator to see if you can simplify the fraction. Don't forget this step!
$\frac{\square}{(2x-3)}$	Check your answer in the answer key and make adjustments, if necessary.

7.04 Adding/Subtracting Rational Expressions with Different Denominators 7.04

YOUR TURN

Simplify. Now you are on your own!

3. $\dfrac{5a}{3a+12} - \dfrac{a-7}{a+4}$	4. $\dfrac{3b}{b-5} + \dfrac{15}{5-b}$
5. $\dfrac{5}{x^2-3x+2} + \dfrac{1}{x-2}$	6. $\dfrac{3}{2x^2+x} - \dfrac{2x}{6x+3}$

YOUR TURN Answers to Section 7.04.

1. $\dfrac{2a^2+53a-84}{3(a-4)(a+4)}$ 2. $\dfrac{-6x-9}{2x-3}$ 3. $\dfrac{2a+21}{3(a+4)}$ 4. 3 5. $\dfrac{x+4}{(x-2)(x-1)}$ 6. $\dfrac{-2x^2+9}{3x(2x+1)}$

Complete MyMathLab Section 7.04 Homework in your Homework notebook.

7.01-7.04 Review of the Steps 7.01-7.04

YOUR TURN

Complete the review by writing the steps for the operations on rational expressions.

Review of Operations on Rational Expressions	
Simplifying	1. ______ 2. ______
Multiplying	1. ______ 2. ______ 3. ______
Dividing	1. ______ 2. ______ 3. ______ 4. ______
Adding & Subtracting	1. ______ 2. ______ 3. ______ 4. ______ 5. ______ 6. ______

7.05 Solving Equations Containing Rational Expressions 7.05

In the previous sections, we worked on techniques to add, subtract, multiply, and divide rational **expressions.** Now we will begin work on solving rational **equations.**

A rational equation is an equation that contains at least one rational expression. Here are two examples of rational equations.

$\frac{3}{y} + \frac{4}{3} = 7$ $\qquad\qquad$ $\frac{x}{x+6} = \frac{72}{x^2 - 36} + 4$

Nobody is crazy about working with fractions (rational expressions), right? So when we have fractions in an equation the first thing we want to do is to get rid of those fractions. You already know how to do this! Let's look at a rational equation that you already know how to solve by clearing the fractions.

EXAMPLE 1

Solve.

$\frac{x}{2} + \frac{8}{3} = \frac{1}{6}$ Find the common denominator of ALL of the fractions in the equation.

$6\left(\frac{x}{2}\right) + 6\left(\frac{8}{3}\right) = 6\left(\frac{1}{6}\right)$ Multiply EVERY term by the LCD. This should clear all of the fractions.

$3x + 16 = 1$ Fractions are gone! Now solve the linear equation.

$3x = -15$

$x = -5$ You can check your answer in the ORIGINAL equation.

WARNING WARNING WARNING

Multiply EVERY term by the LCD of ALL the denominators!!

YOUR TURN

Solve.

1. $\frac{x}{2} + \frac{3}{5} = \frac{1}{10}$

2. $\frac{x+2}{3} - \frac{x-1}{5} = \frac{1}{15}$

Be careful to distribute the subtraction in the second term.

7.05 Solving Equations Containing Rational Expressions 7.05

Recall that a rational expression is defined for all real numbers except those that make the denominator of the expression 0. This means that if a rational equation contains variables in any denominator, we must make sure that the proposed solution(s) does not make the denominator 0.

If replacing the variable with the proposed solution does make any denominator 0, the rational expression is undefined and this proposed solution must be 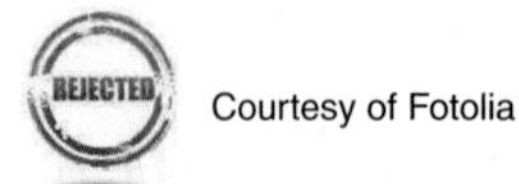

Courtesy of Fotolia

> **TIP:** Examine each denominator of the rational equation that has a variable to determine the domain's exclusions before you begin your work.
> - Reminder, to determine your exclusions, set each denominator that has a variable equal to zero and solve. (Do this on the side.)
> - After solving the rational equation, LOOK back at the excluded values of the domain to throw out any proposed solutions which are not defined. Keep only the solution(s) that have not been excluded.

Consider the equation: $\dfrac{2}{x+2} - \dfrac{x}{x-5} = \dfrac{1}{x}$

Domain: To find the values to exclude, set each denominator = 0 and solve.

$x+2=0 \qquad x-5=0 \qquad x=0$

$x=-2 \qquad x=5$

These values would be excluded from the possible solutions of the equation.

If we put all of this together we come up with these steps for solving rational equations.

STEPS TO SOLVING A RATIONAL EQUATION

Step 1: Factor the denominators. Identify any values of the variable for which an expression is undefined. [Identify any values that make the denominator = 0].

Step 2: Multiply both sides of the equation by the LCD of all denominators.

Step 3: Remove any grouping symbols and solve the resulting equation.
Note: This equation can be linear or quadratic.

Step 4: Check the potential solutions in the original equation.
Note: Any value from step 1 for which the equation is undefined cannot be a solution to the equation.

EXAMPLE 2

Solve. $2+\frac{3}{m-3}=\frac{m}{m-3}$

Step 1: Factor the denominators. Identify any values for which an expression is undefined.

To find the exclusions set each denominator = 0 and solve: $m-3=0$

$m=3$

Step 2: Multiply EVERY TERM of the equation by the LCM of all denominators.

$$2\bullet(m-3)+\frac{3}{(m-3)}\bullet(m-3)=\frac{m}{(m-3)}\bullet(m-3)$$

Some students prefer to express all terms as fractions and cancel.

Remember that you can only *cancel factors, not terms.*

$$\frac{2}{1}\bullet\frac{(m-3)}{1}+\frac{3}{\cancel{(m-3)}_1}\bullet\frac{\cancel{(m-3)}^1}{1}=\frac{m}{\cancel{(m-3)}_1}\bullet\frac{\cancel{(m-3)}^1}{1}$$

Step 3: Remove any grouping symbols and solve the resulting equation.

$$2(m-3)+3=m$$
$$2m-6+3=m$$
$$2m-3=m$$
$$-3=-m$$
$$3=m$$

Note that this is a linear equation.
You will have one proposed solution.

Step 4: Check the potential solutions in the original equation.

Our proposed solution is 3.
However, in step 1, we stated that x cannot be 3 because it makes the denominator equal to 0.
Therefore, this equation has no solution.

7.05 Solving Equations Containing Rational Expressions 7.05

EXAMPLE 3

Solve. * These instructions will not remind you to check your answers using the domain's exclusions.**

Solve. $\dfrac{3x-5}{x^2+5x-14}+\dfrac{2}{x+7}=\dfrac{1}{x-2}$

Step 1: LCD = $(x+7)(x-2)$

Define the domain.

Exclusions: $x+7=0 \qquad x-2=0$

$x=-7 \qquad x=2$

-7 or 2 cannot be solutions

Step 2: Multiply all numerators by the LCD & reduce.

$$(x+7)(x-2)\bullet\frac{3x-5}{(x+7)(x-2)}+(x+7)(x-2)\bullet\frac{2}{(x+7)}=(x+7)(x-2)\bullet\frac{1}{(x-2)}$$

$$\cancel{(x+7)(x-2)}\bullet\frac{3x-5}{\cancel{(x+7)(x-2)}}+\cancel{(x+7)}(x-2)\bullet\frac{2}{\cancel{(x+7)}}=(x+7)\cancel{(x-2)}\bullet\frac{1}{\cancel{(x-2)}}$$

Step 3:

$$3x-5+2(x-2)=1(x+7)$$
$$3x-5+2x-4=x+7$$
$$5x-9=x+7$$
$$4x=16$$
$$x=4$$

Note that this is a ***Linear Equation.*** You will have one proposed solution.

*** LOOK at the domain's exclusions to CHECK your answers! This time we keep the proposed solution.

Step 4: Proposed solution: $x=4$

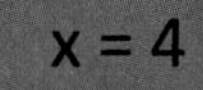

x = 4

YOUR TURN

Solve. * These instructions will not remind you to check your answers using the domain's exclusions.**

3. Solve. $\dfrac{3}{x+3}=\dfrac{12x+19}{x^2+7x+12}-\dfrac{5}{x+4}$

7.05 Solving Equations Containing Rational Expressions 7.05

Sometimes, the equation you get after removing the fractions will be a quadratic equation. Remember that to solve a quadratic equation you will need to get the equation equal to zero, factor the polynomial, and then set each factor = 0 to solve.

EXAMPLE 4

Solve. * These instructions will not remind you to check your answers using the domain's exclusions.**

$$x - \frac{6}{x+3} = \frac{2x}{x+3} + 2$$

Step 1: LCD = $(x+3)$
Define the domain.

Exclusions: $x + 3 \neq 0$

– 3 cannot be a solution

Step 2: Multiply EVERY term by the LCD and reduce.

$$x \bullet (x+3) - (x+3) \bullet \frac{6}{(x+3)} = (x+3) \bullet \frac{2x}{(x+3)} + 2 \bullet (x+3)$$

$$x \bullet (x+3) - \cancel{(x+3)} \bullet \frac{6}{\cancel{(x+3)}} = \cancel{(x+3)} \bullet \frac{2x}{\cancel{(x+3)}} + 2 \bullet (x+3)$$ Reduce.

Step 3:

$x^2 + 3x - 6 = 2x + 2x + 6$ Simplify each side of the equation.

$x^2 + 3x - 6 = 4x + 6$ Combine like terms. Note that this is a ***Quadratic Equation***.

$x^2 - x - 12 = 0$ Write the quadratic equation in standard form, $ax^2 + bx + c = 0$. You will have two proposed solutions. Factor.

$(x-4)(x+3) = 0$ Set each factor equal to zero.

$x - 4 = 0$ or $x + 3 = 0$ Solve.

$x = 4$ $x = -3$ Compare solutions to the exclusions at the top.

Solution is 4

As we have already noted **– 3 cannot be a solution** to the equation. So, the only solution is 4.

7.05 Solving Equations Containing Rational Expressions 7.05

EXAMPLE 5

Solve. * These instructions will not remind you to check your answers using the domain's exclusions.**

$$\frac{x+1}{x+3} = \frac{x^2-11x}{(x+3)(x-2)} - \frac{x-3}{x-2}$$

Step 1: LCD = $(x+3)(x-2)$

Define the domain.

** Exclusions: $x+3 \neq 0$ or $x-2 \neq 0$

so -3 and 2 cannot be solutions.

STEP 2: Multiply EVERY term by the LCD and reduce.

$$\frac{x+1}{x+3}\cdot(x+3)(x-2) = \frac{x^2-11x}{(x+3)(x-2)}\cdot(x+3)(x-2) - \frac{x-3}{x-2}\cdot(x+3)(x-2)$$

$$\frac{x+1}{\cancel{x+3}}\cdot\cancel{(x+3)}(x-2) = \frac{x^2-11x}{\cancel{(x+3)(x-2)}}\cdot\cancel{(x+3)(x-2)} - \frac{x-3}{\cancel{x-2}}\cdot(x+3)\cancel{(x-2)}$$

Cancel out the common factors.

STEP 3:

$(x+1)\cdot(x-2) = x^2-11x-(x-3)\cdot(x+3)$ Multiply the binomials using FOIL.

$x^2-2x+1x-2=x^2-11x-(x^2+3x-3x-9)$ Combine like terms in the parentheses.

$x^2-x-2 = x^2-11x-(x^2-9)$ Distribute to clear the parentheses. ***Be careful, very careful!***

Most common error. Remember to change the sign when distributing the "-".

$x^2-x-2 = x^2-11x-x^2+9$ Notice the $+9$, not -9. Now combine like terms.

$x^2-x-2 = -11x+9$

This is a quadratic equation. Write the quadratic equation in standard form. (Get 0 on one side.) Factor. You will have two proposed solutions. Set each factor equal to zero and solve.

$x^2+10x-11=0$

$(x+11)(x-1)=0$

$x+11=0$ *or* $x-1=0$

$x=-11 \qquad x=1$

Compare solutions to the exclusions.

Neither of these answers needs to be excluded. The solutions are 1 and -11.

7.05 Solving Equations Containing Rational Expressions 7.05

YOUR TURN

4. Solve. $\dfrac{7a}{a-2}-\dfrac{14}{a-2}=8$

Most common error. Remember to change the sign when distributing the "-" in the middle of the problem.

5. Solve. $x+\dfrac{x}{x-5}=\dfrac{5}{x-5}-7$

To avoid the 3 biggest mistakes, make sure to:

- Multiply EVERY term by the LCD of ALL the denominators.
- Be very careful when subtraction is involved in the equation.
- Compare the solutions found to the exclusions.

For each of the following, circle the step with the mistake. Explain what the mistake is and what should have been done instead. Then work the problem correctly.

6. Circle the mistake below and then explain what is wrong and how to work that step properly.	Now work the problem correctly.
$5+\dfrac{12}{a-3}=\dfrac{4a}{a-3}$ $5\bullet\dfrac{(a-3)}{}+\dfrac{12}{a-3}\bullet\dfrac{(a-3)}{}=\dfrac{4a}{a-3}\bullet\dfrac{(a-3)}{}$ $5(a-3)+12=4a$ $5a-15+12=4a$ $5a-3=4a$ $-3=-a$ $3=a$	$5+\dfrac{12}{a-3}=\dfrac{4a}{a-3}$

7.05 Solving Equations Containing Rational Expressions 7.05

YOUR TURN

7. Circle the mistake below and then explain what is wrong and how to work that step properly.

$$\frac{6}{3y}+\frac{3}{2}=1 \quad \textit{Exclusions}: 3y=0$$

$$y=0$$

$$\frac{6}{3y}\bullet\frac{6y}{}+\frac{3}{2}\bullet\frac{6y}{}=1$$

$$12+9y=1$$

$$9y=-11$$

$$y=-\frac{11}{9} \quad \text{Answer is not excluded!}$$

Now work the problem correctly.

$$\frac{6}{3y}+\frac{3}{2}=1$$

8. Circle the mistake and then explain what is wrong and how to work that step properly.

$$\frac{18}{x+8}=\frac{14x+13}{x^2+14x+48}-\frac{3}{x+6}$$

$$\frac{18}{x+8}=\frac{14x+13}{(x+6)(x+8)}-\frac{3}{x+6} \quad \textit{Exclusions}: x=-8;\ x=-6$$

$$\frac{18}{x+8}\bullet\frac{(x+6)(x+8)}{}=\frac{14x+13}{(x+6)(x+8)}\bullet\frac{(x+6)(x+8)}{}-\frac{3}{x+6}\bullet\frac{(x+6)(x+8)}{}$$

$$18x+108=14x+13-3x+24$$

$$18x+108=11x+37$$

$$7x+108=37$$

$$7x=-71$$

$$x=-\frac{71}{7} \quad \text{Answer is not excluded!}$$

Now work the problem correctly.

$$\frac{18}{x+8}=\frac{14x+13}{x^2+14x+48}-\frac{3}{x+6}$$

7.05 Solving Equations Containing Rational Expressions 7.05

Now we will look at solving literal equations (equations with several variables). The goal is still the same. We need to get the specified variable on one side and all other variables and numbers on the other side.

EXAMPLE 6

Solve the equation for the indicated variable: $I = \frac{9a}{b+c}$ for $\boldsymbol{b}$. Compare this to the equation on the left.

You know how to solve this equation for x.		(Compare solving this equation for *b* with the problem on the left which has numbers.)
$5 = \frac{92}{x+10}$	The goal is to get ***b*** on one side and every other variable on the other side. Perform the same operations as you would for problems with numbers.	$I = \frac{9a}{b+c}$ for $\boldsymbol{b}$
	Begin by multiplying both sides of each equation by the LCD.	The goal is to isolate *b*.
$5 \cdot (x+10) = \frac{92}{\cancel{x+10}} \cdot (\cancel{x+10})$	Distribute to remove the parentheses. Cancel common factors.	$I \cdot (b+c) = \frac{9a}{\cancel{b+c}} \cdot (\cancel{b+c})$
$5x + 50 = 92$	Now subtract the term that does NOT contain the variable you are solving for.	$\boldsymbol{b}I + cI = 9a$
$5x + 50 - 50 = 92 - 50$	Simplify as much as you can.	$\boldsymbol{b}I + cI - \boldsymbol{c}\boldsymbol{I} = 9a - \boldsymbol{c}\boldsymbol{I}$
$5x = 42$	Now divide by the coefficient of the variable you are solving for.	$\boldsymbol{b}I = 9a - cI$
$\frac{5x}{5} = \frac{42}{5}$	Simplify as much as you can.	$\frac{\boldsymbol{b}I}{\boldsymbol{I}} = \frac{9a - cI}{\boldsymbol{I}}$
$x = \frac{42}{5}$	***Finished!***	$\boldsymbol{b} = \frac{9a - cI}{I}$

EXAMPLE 7

Solve $\frac{1}{a}+\frac{1}{b}=\frac{1}{c}$ for b.

We have fractions in this equation, so we need to do exactly what we do with any equation that contains fractions - ***clear the fractions***! So let's look at an equation with only one variable and compare it to how to solve the literal equation we have.

Solve the equation for the indicated variable in blue.

$\frac{1}{3}+\frac{1}{x}=\frac{1}{5}$	Begin by multiplying every term by the LCD.	$\frac{1}{a}+\frac{1}{b}=\frac{1}{c}$ Solve for b.
$(15x)\frac{1}{3}+(15x)\frac{1}{x}=(15x)\frac{1}{5}$	Now reduce to clear fractions.	$(abc)\bullet\frac{1}{a}+(abc)\bullet\frac{1}{b}=(abc)\bullet\frac{1}{c}$
$5x+15=3x$	Look at the equation in column 1. By now you know that you need to move $5x$ to the right side of the equation and combine it with the $3x$. In column 3, we have two terms with b in them. We do the same thing with the literal equation that we did with the equation in column 1.	$bc+ac=ab$
$15=3x-5x$	Now factor out the common factor. We don't normally do this with the equation in column 1 but it will help you see what is going on with the literal equation.	$ac=ab-bc$
$15=(3-5)x$	Now divide by the "coefficient" of the variable we are solving for.	$ac=(a-c)b$
$\frac{15}{(3-5)}=\frac{(3-5)x}{(3-5)}$	Simplify by reducing on the right.	$\frac{ac}{(a-c)}=\frac{(a-c)b}{(a-c)}$
$-\frac{15}{2}=x$	Yes, this new literal equation looks worse than the original!	$\frac{ac}{a-c}=b$

SOLVING EQUATIONS FOR A SPECIFIED VARIABLE

1. Multiply both sides by the LCM to clear fractions.
2. Remove parentheses by using Distributive Property.
3. Get all terms with the specified variable alone on one side.
4. Factor out the specified variable if it is in more than one term.
5. Divide to isolate the specified variable.

EXAMPLE 8

Solve $m = \dfrac{y_2 - y_1}{x_2 - x_1}$ for x_2.

$(x_2 - x_1) \bullet m = (x_2 - x_1) \bullet \dfrac{y_2 - y_1}{x_2 - x_1}$	Multiply both sides by the LCD.
$(x_2 - x_1) \bullet m = y_2 - y_1$	Simplify.
$mx_2 - mx_1 = y_2 - y_1$	Use distributive property to remove ( ).
$mx_2 = y_2 - y_1 + mx_1$	Specified variable is x_2. Get all terms with x_2 on the same side.
$x_2 = \dfrac{y_2 - y_1 + mx_1}{m}$	Divide to isolate the variable.

7.05 Solving Equations Containing Rational Expressions 7.05

YOUR TURN

9. Solve $P = \dfrac{R - C}{n}$ for C.	10. Solve $\dfrac{1}{a} + \dfrac{1}{b} = \dfrac{1}{t}$ for ***t***.

YOUR TURN **Answers to Section 7.05.**

1. $x = -1$ 2. $x = -6$ 3. $x = 2$ 4. no solution [exclude 2] 5. $x = -8$ [exclude 5]

6. [1st, didn't exclude solution of 3.] Answer: No solution
7. [1st, didn't multiply left side by LCD $6y$.] Answer: $y = -4$
8. [1st: in 4th line, didn't distribute -3 correctly.] Answer: $x = -17$

9. $C = R - Pn$ or $C = -Pn + R$ 10. $t = \dfrac{ab}{a + b}$

Complete MyMathLab Section 7.05 Homework in your Homework notebook.

The process for adding/subtracting rational expressions is easy to confuse with the process for solving rational equations. Look at these examples and make some notes! Especially when you can cancel and when you cannot.

$$\frac{1}{3x+6}+\frac{1}{2}-5$$

$$\frac{1}{3(x+2)}+\frac{1}{2}-\frac{5}{1}$$

$$\frac{1}{3(x+2)}\bullet\frac{2}{2}+\frac{1}{2}\bullet\frac{3(x+2)}{3(x+2)}-\frac{5}{1}\bullet\frac{6(x+2)}{6(x+2)}$$

$$\frac{2}{6(x+2)}+\frac{3(x+2)}{6(x+2)}-\frac{30(x+2)}{6(x+2)}$$

$$\frac{2+3x+6-30x-60}{6(x+2)}$$

$$\frac{-27x-52}{6(x+2)}$$

$$\frac{1}{3x+6}+\frac{1}{2}=5$$

$$\frac{1}{3(x+2)}+\frac{1}{2}=5$$

$$6(x+2)\bullet\frac{1}{3(x+2)}+6(x+2)\bullet\frac{1}{2}=6(x+2)\bullet 5$$

$$\overset{2}{\cancel{6}}\overset{1}{\cancel{(x+2)}}\bullet\frac{1}{\underset{1}{\cancel{3}}\underset{1}{\cancel{(x+2)}}}+\overset{3}{\cancel{6}}(x+2)\bullet\frac{1}{\underset{1}{\cancel{2}}}=6(x+2)\bullet 5$$

$$2+3(x+2)=30(x+2)$$

$$2+3x+6=30x+60$$

$$-52=27x$$

$$x=\frac{-52}{27}$$

7.06 Proportions and Problem Solving with Rational Equations 7.06

PROPORTIONS

Recall that a ratio is the quotient of two numbers. A proportion is when two ratios are set equal to each other. For example $\frac{1}{2}=\frac{3}{6}$ is a proportion. You already know that cross products are equal. In the above proportion, $1 \bullet 6 = 2 \bullet 3$.

This proportion contains only numbers, but we can use the same process to solve for an unknown number if the proportion only gives us 3 of the 4 numbers.

EXAMPLE 1

Solve. $\frac{2}{5}=\frac{6}{x}$.

Since this is a proportion, we will use cross products to solve.

$$\frac{2}{5} \times \frac{6}{x}$$

$2 \bullet x = 6 \bullet 5$ Set cross products equal.

$2x = 30$

$x = 15$

EXAMPLE 2

Solve. $\frac{x+5}{3}=\frac{x-1}{5}$

$\frac{x+5}{3} \times \frac{x-1}{5}$ Set the cross products equal.

$5(x+5)=3(x-1)$ Use the distributive property to remove ( ).

$5x+25=3x-3$ Solve the linear equation.

$2x+25=-3$

$2x=-28$

$x=-14$

7.06 Proportions and Problem Solving with Rational Equations **7.06**

YOUR TURN

Be careful! This one is quadratic. Since there is an unknown in the denominator, make sure that you check your answers against the domain.

1. Solve. $\frac{x}{x+5}=\frac{1}{x-3}$

Now, let's use proportions to solve a few application problems.

EXAMPLE 3

A 6–ounce Milky Way has 240 calories. How many calories would a 10-ounce Milky Way have?

6–ounce
240 calories

10–ounce
??? calories

$$\frac{\text{ounces}}{\text{calories}}=\frac{6}{240}=\frac{10}{c}$$

It's a good idea to write a ratio in words of the two things you are comparing. Here we are comparing ounces to calories. It does not matter which is the numerator and which is the denominator. Just be consistent.

$$\begin{aligned} 6\bullet c &= 240\bullet 10 \\ 6c &= 2400 \\ c &= 400 \end{aligned}$$

Now solve.

A 10-ounce Milky Way has 400 calories. YIKES!

YOUR TURN

2. If 4 grapefruits sell for 79 cents, how much will 6 grapefruits cost? Round to nearest cent, if necessary. (Watch the decimals in the money!)

7.06 Proportions and Problem Solving with Rational Equations 7.06

We can also use proportions to work with similar triangles.

SIMILAR TRIANGLES

Similar triangles have the following properties:

- They have the same shape but not the same size.
- Each corresponding pair of angles is equal.
- The ratio of any pair of corresponding sides is the same.

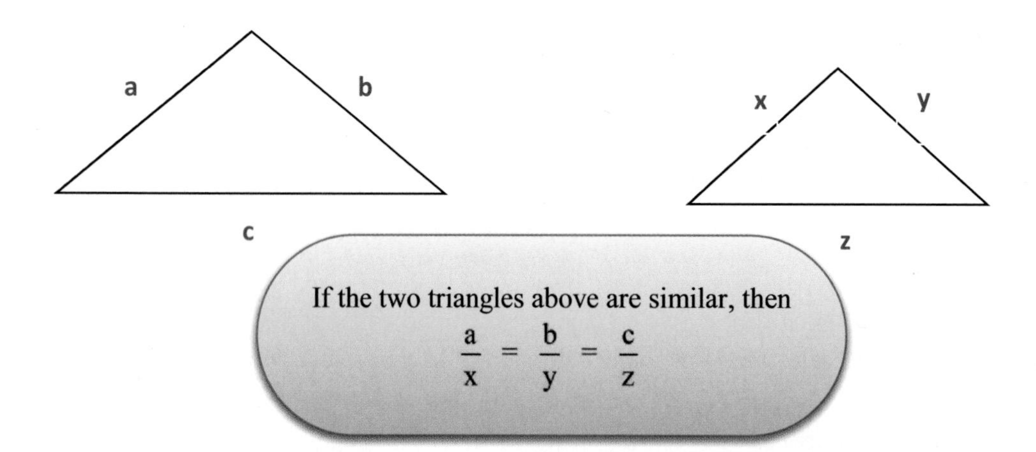

EXAMPLE 4

If the following two triangles are similar, find the missing length.

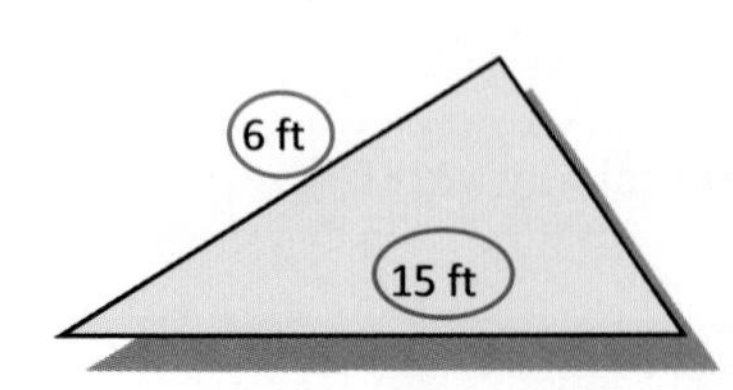

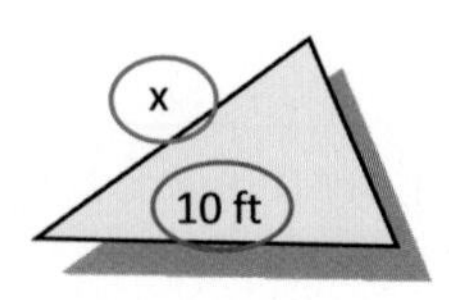

$$\frac{6}{x} = \frac{15}{10}$$

Be careful setting up the proportion. Make sure that you have the corresponding sides in a ratio and that you go from first triangle to second triangle each time.

$$6 \cdot 10 = 15 \cdot x$$

Cross multiply and solve.

$$60 = 15x$$

$$4 = x$$

The missing side has length 4 ft.

YOUR TURN

3. Find the length of the missing side.

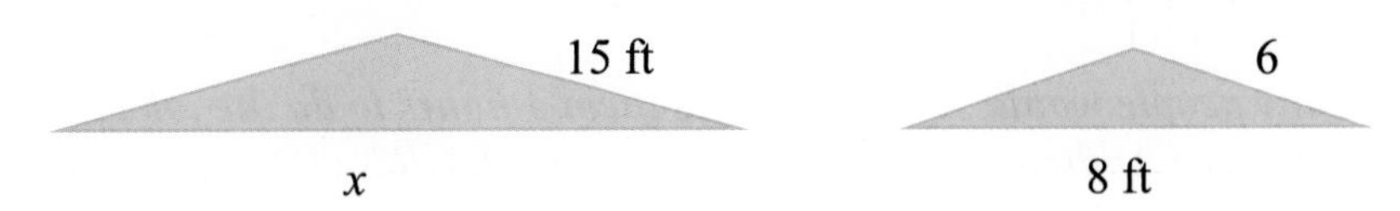

FINDING AN UNKNOWN NUMBER

These problems are straight translations, from an English statement into an algebraic statement.

EXAMPLE 5

The difference of 2 and 8 times the reciprocal of a number is 6.

The difference of 2 and 8 times the reciprocal of a number is 6.

$$2 \; - \; 8 \; \bullet \; \frac{1}{x} \; = 6$$

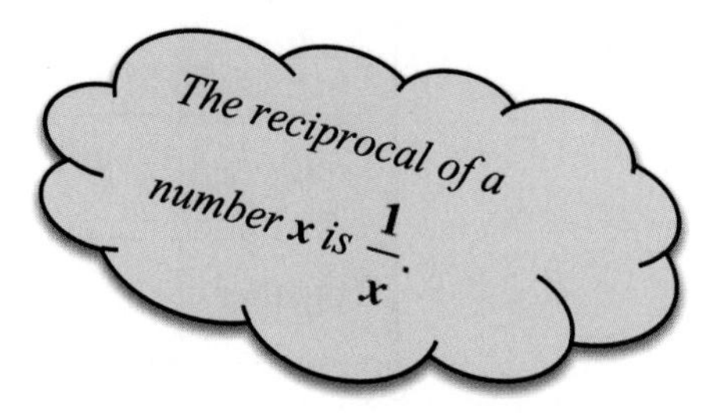

$$2 - 8 \bullet \frac{1}{x} = 6$$

$$2 - \frac{8}{x} = 6$$

$$(x) \bullet 2 - (x) \bullet \frac{8}{x} = (x) \bullet 6$$ Clear the fractions.

$$2x - 8 = 6x$$ Solve the linear equation.

$$-8 = 4x$$

$$-2 = x$$ The number is -2.

YOUR TURN

4. The sum of a number and twice its reciprocal is 3. Find the numbers.

TIP: After clearing fractions, you will have a quadratic equation.

7.06 Proportions and Problem Solving with Rational Equations 7.06

WORK PROBLEMS

Andrea can wallpaper a bathroom in 3 hrs. Janie can wallpaper the same bathroom in 5 hrs. How long would it take them if they worked together?

Many people would say that it takes them 4 hours to do the job together. Does that make sense when it only takes Andrea 3 hours to do the job by herself? NO! Janie would actually be slowing her down!

Notice here the relationship between time *(hours)* it takes to complete the job and the part of the job completed in 1 unit of time *(1 hour)*. For example, if it takes Andrea 3 hours to complete the job, she can complete $\frac{1}{3}$ of the job in 1 hr. Likewise, Janie can complete $\frac{1}{5}$ of the job in 1 hr.

$\boldsymbol{x}$ = time it takes to complete job together ⟶ $\frac{1}{x}$ part of job they complete in 1 hr.

	Hours to Complete Total Job	Part of Job Completed in 1 Hour
Andrea	3	$\frac{1}{3}$
Janie	5	$\frac{1}{5}$
Together	x	$\frac{1}{x}$

$$\begin{pmatrix}\text{Portion of job}\\ \text{Andrea completed}\\ \text{in 1 hour}\end{pmatrix} + \begin{pmatrix}\text{Portion of job}\\ \text{Janie completed}\\ \text{in 1 hour}\end{pmatrix} = \begin{pmatrix}\text{Portion of job}\\ \text{they completed}\\ \text{together in 1 hour}\end{pmatrix}$$

$$\frac{1}{3} \qquad + \qquad \frac{1}{5} \qquad = \qquad \frac{1}{x}$$

$$\frac{1}{3}+\frac{1}{5}=\frac{1}{x}$$

$$(15x)\frac{1}{3}+(15x)\frac{1}{5}=(15x)\frac{1}{x}$$ Multiply ***EVERY TERM*** by the LCD, **15*x*.**

$$5x+3x=15$$

$$8x=15$$

$$x=\frac{15}{8} \quad \text{or } 1\frac{7}{8} \text{ hours}$$

7.06 Proportions and Problem Solving with Rational Equations 7.06

EXAMPLE 6

Mark and Katie volunteer at a recycling center. Together they can sort a batch in about an hour and twelve minutes or $1\frac{1}{5}$ hours. If Katie can sort the batch in 2 hours, how long would it take Mark to sort the batch by himself?

$$\begin{pmatrix}\text{Portion of job}\\\text{Katie completed}\\\text{in 1 hour}\end{pmatrix} + \begin{pmatrix}\text{Portion of job}\\\text{Mark completed}\\\text{in 1 hour}\end{pmatrix} = \begin{pmatrix}\text{Portion of job}\\\text{they completed}\\\text{together in 1 hour}\end{pmatrix}$$

$$\frac{1}{2} + \frac{1}{x} = \frac{1}{1\frac{1}{5}}$$

$$\frac{1}{2}+\frac{1}{x}=\frac{5}{6}$$

$$\frac{(6x)}{}\frac{1}{2}+\frac{(6x)}{}\frac{1}{x}=\frac{(6x)}{}\frac{5}{6}$$

$$3x+6=5x$$

$$6=2x$$

$$3=x$$

It takes Mark 3 hours to sort the batch of recyclables by himself.

This is a complex fraction, since the numerator contains a fraction. So the first thing to do is to simplify the complex fraction.

Remember that the fraction bar means to divide.

$\frac{1}{1\frac{1}{5}} = 1 \div 1\frac{1}{5}$ *Change to* $\div$

$1 \div \frac{6}{5}$ *Change to* fraction

$1 \bullet \frac{5}{6}$ *Flip* 2nd *fraction &*

Change to $\bullet$

$\frac{5}{6}$ *Multiply*

Always make sure that your answer make sense!

YOUR TURN

Define a variable, write an equation, solve and then answer each question.

5. Billie can wax her SUV in 2 hours. Matt can wax the same SUV in 3 hours. If they work, together how long will it take them to wax the SUV?	6. Jasmine proofreads a book in 2 hours. If a second proofreader is employed then the job can be completed in $1\frac{1}{3}$ hours. How long does it take the second proofreader to do the job alone?

7.06 Proportions and Problem Solving with Rational Equations 7.06

YOUR TURN

7. Will and Daniel work for the same company. It takes Will 4 days to paint a house. Daniel needs an additional day to paint the same house. If they work on the job together and the cost of labor for the two of them is $45 an hour, what should the labor cost estimate be?	8. A pipe can fill a vat in 8 hours. A second pipe can fill the vat in 4 hours. If a third pipe is added, the vat is filled in 2 hours. How long would it take for the third pipe to fill the vat?

DISTANCE PROBLEMS

Now we will look at distance problems.
You probably know the basic distance formula (distance = rate x time). This formula allows us to find the distance. Suppose we want to find the rate or the time. We would need to solve the distance formula for rate or for time.

distance = rate • time

$$d = r \cdot t$$
$$\frac{d}{t} = \frac{r \cdot t}{t}$$
$$\frac{d}{t} = r$$
$$\text{rate} = \frac{\text{distance}}{\text{time}}$$

distance = rate • time

$$d = r \cdot t$$
$$\frac{d}{r} = \frac{r \cdot t}{r}$$
$$\frac{d}{r} = t$$
$$\text{time} = \frac{\text{distance}}{\text{rate}}$$

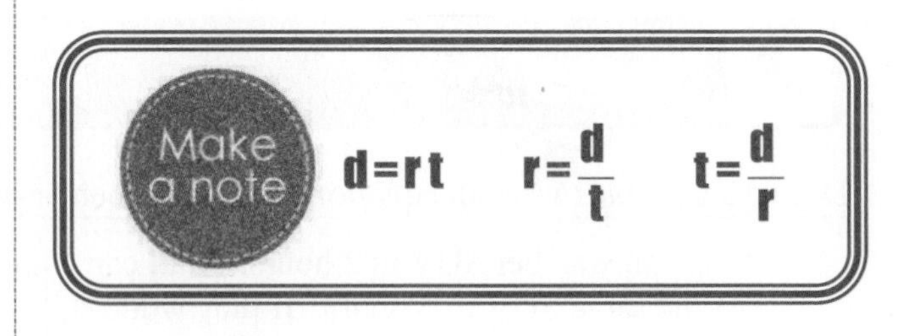

Distance problems are so much easier if you will organize the information in a table.
Ready to give one a try?

EXAMPLE 7

Courtesy of Fotolia

On her road bike, Brandi bikes 15 mph faster than Andy does on his mountain bike. In the time it takes Brandi to travel 80 miles, Andy has traveled 50 miles. Find the speed of each bicyclist.

Let's look at the distances traveled first. Look at the problem again.

"On her road bike . . . . . ***Brandi to travel 80 miles, Andy has traveled 50 miles***. . . . bicyclist."

Put this info in the table.

	Distance	Rate	Time
Brandi	80		
Andy	50		

What else does the problem tell us?

"On her road bike, ***Brandi bikes 15 mph faster than Andy*** does on his mountain bike." This info goes in the rate column.

	Distance	Rate	Time
Brandi	80	$r+15$	
Andy	50	r	

Once you get two of the columns filled with info from the problem, then you use the appropriate formula to fill in the remaining column. We are missing time; we will use the formula solved for time.

$$t = \frac{d}{r}$$

	Distance	Rate	Time
Brandi	80	$r+15$	$\frac{80}{r+15}$
Andy	50	r	$\frac{50}{r}$

Read the problem again to find the relationship between the time for Brandi and the time for Andy.

"On her road bike, Brandi bikes 15 mph faster than Andy does on his mountain bike. ***In the time it takes Brandi to travel 80 miles, Andy has traveled 50 miles***. Find the speed of each bicyclist.

This statement tell us that Brandi's time = Andy's time.

$$\frac{80}{r+15}=\frac{50}{r}$$

Looks like a proportion!

$$\frac{80}{r+15} \nwarrow\nearrow \frac{50}{r}$$

You can solve this by using cross products or by multiplying both sides by the LCD.

$$50(r+15)=80r$$

$$50r+750=80r$$

$$750=30r$$

$$25=r$$

Andy's speed is 25 mph.
Brandi's speed is $r+15=25+15=40$ mph.

YOUR TURN

9. A freight train goes 12 mph slower than a passenger train. If the freight train travels 230 mi in the same time that the passenger train travels 290 miles, find the speed of each train.

	distance	rate	time
Freight Train			
Passenger Train			

Courtesy of Fotolia

EXAMPLE 8

A fisherman on the Hudson River rows 18 miles downstream in the same amount of time that he can row 6 miles upstream. If the current is 4 miles per hour, find the rate of the boat in still water.

There is just one unknown here – the rate (speed) of the boat in still water.

BUT the problem gives us information only about the boat on a river with a current, NOT still water.

Let b = boat's speed in still water (what we are being asked to find!)

Look at the problem again. It talks about the boat traveling upstream and downstream.

A fisherman on the Hudson River ***rows 18 miles downstream in the same amount of time that he can row 6 miles upstream***. If the current is 4 miles per hour, find the rate of the boat in still water.

Downstream means that the current is running the same way that the boat is going. Will the current make the boat speed up, slow down, or have no effect on the boat? ____________________________

CURRENT

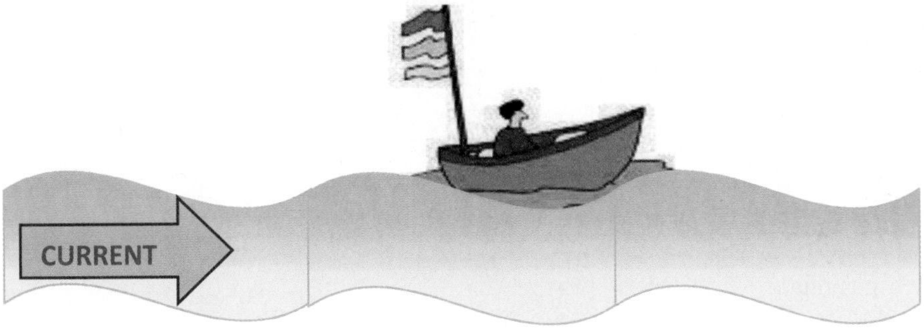

Upstream means that the current is running the opposite way that the boat is going, so will the current make the boat speed up, slow down, or have no effect on the boat?

____

CURRENT

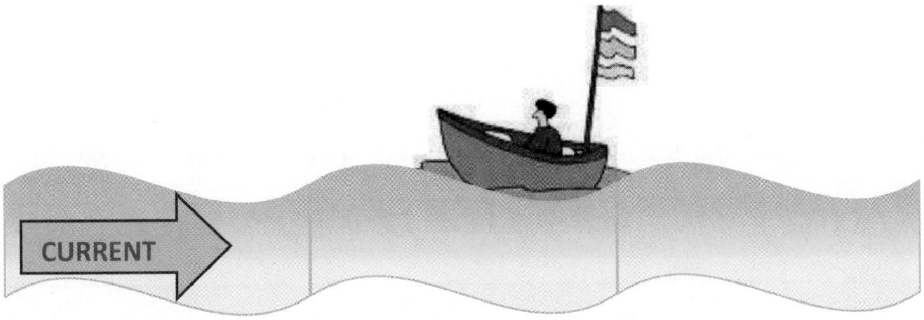

7.06 Proportions and Problem Solving with Rational Equations 7.06

Courtesy of Fotolia

It is so important that you understand this concept. Some students get answers where the current is running faster than the boat is going. Could this ever happen? Would you ever get upstream?

Hopefully, you answered the questions correctly.

Travelling ***downstream increases the speed*** of the boat.
Travelling ***upstream decreases the speed*** of the boat.

Let's read the problem again to fill in the table.

A fisherman on the Hudson River rows 18 miles downstream in the same amount of time that he rows 6 miles upstream. If the current is 4 miles per hour, find the rate of the boat in still water.

$$t=\frac{d}{r}$$

	Distance	Rate	Time
Downstream	18	b + 4	$\frac{18}{b+4}$
Upstream	6	b − 4	$\frac{6}{b-4}$

How do the times relate?

"18 miles downstream in the same amount of time that he row 6 miles upstream"

$$\frac{18}{b+4}=\frac{6}{b-4}$$

Another proportion!

$$18(b-4) = 6(b+4)$$
$$18b-72=6b+24$$
$$12b=96$$
$$b=8$$

The boat is moving at a rate of 8 mph.

TIP: You may be given the rate of the boat and asked for the rate of the current. For example, let's say the problem states that the boat is moving at 10 mph, find the rate of the current. For your rate column, you would then use 10 + c for downstream and 10 – c for upstream.

7.06 Proportions and Problem Solving with Rational Equations 7.06

YOUR TURN

Write an equation and solve. ***Remember the vehicle speed goes first, even if it is the *x*.

10. A boat moves at a rate of 33 mph in still water. It travels 400 miles upstream in the same time that it takes to travel 480 miles downstream. What is the rate of the current?

	Distance	Rate	Time
Downstream			
Upstream			

This same process can be used for working problems with a plane instead of a boat. A plane can fly with the wind or against the wind. The rate going with the wind would be "plane speed + wind speed". The rate going against the wind would be "plane speed – wind speed".

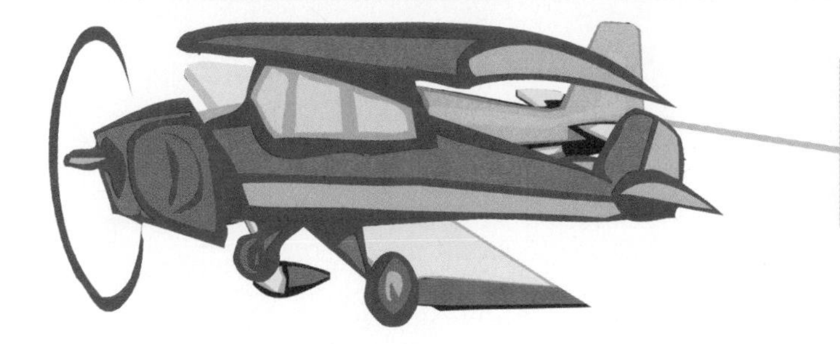

With the wind = plane speed + wind speed

Against the wind = plane speed – wind speed

7.06 Proportions and Problem Solving with Rational Equations 7.06

YOUR TURN

Write an equation and solve. ***Remember the vehicle speed goes first, even if it is the *x*.

11. A plane flies 465 miles with the wind and 345 miles against the wind in the same length of time. If the speed of the wind is 20 mph, find the speed of the plane in still air.

	Distance	Rate	Time
With the wind			
Against the wind			

YOUR TURN Answers to Section 7.06.

1. $x = 5$ or $x = -1$ 2. \$1.19 3. 20 ft. 4. 1 and 2 5. $1\frac{1}{5}$ hours 6. 4 hours

7. \$100 8. 8 hours 9. Freight train: 46 mph; Passenger: 58 mph. 10. 3 mph

11. 135 mph

RECAP OF APPLICATIONS OF RATIONAL FUNCTIONS

Can you work these types of problems without looking back?

- Unknown Number Problems
- Work Hour Problems
- Proportions/Similar Triangle
- Distance Problems
- Plane Speed/Wind Speed Problems
- Boat Speed/Current Speed Problems

Complete MyMathLab Section 7.06 Homework in your Homework notebook.

7.07 Variation and Problem Solving 7.07

We talked about comparisons of two quantities in section 7.06 (proportions). Now we will look at other comparisons of quantities. We will look at three types of relationships:

- ***Direct variation*** – when ***two quantities either both increase or both decrease***. For example, when the radius of a circle increases so does the circumference of the circle.
- ***Inverse variation*** – when ***one quantity increases, the other decreases***. For example, the more checkout lanes open at the grocery store, the less time it takes to get checked out.
- ***Joint variation*** – there are ***more than two quantities in the relationship***. For example, the interest varies jointly as the principle and the time and the rate.

YOUR TURN

Tell whether the following relationships can be modeled by direct variation, inverse variation (sometimes referred to as indirect variation), or joint variation.

1. The amount of gas in your tank and the number of miles you have driven since the last fill-up.

2. The distance traveled and the speed and the time traveled.

3. The amount of time a student studies and the grades the student gets.

4. A person's weekly pay and the number of hours the person worked.

5. The distance from a light and the intensity of the light.

We will limit our study of variation to direct variation only.

DIRECT VARIATION	INVERSE VARIATION	JOINT VARIATION
$y = \mathbf{k}x$	$y = \frac{\mathbf{k}}{x}$	$y = \mathbf{k}xz$
Read "**y varies directly as x.**"	Read "**y varies inversely as x.**"	Read "**y varies jointly as x and z.**"
Both quantities increase together or decrease together.	As one quantity increases, the other quantity decreases.	There are more than 2 quantities related.

7.07 Variation and Problem Solving 7.07

EXAMPLE 1

DIRECT VARIATION

Suppose y **varies directly** as x, and $y = 45$ when $x = 2.5$.

a. Determine the constant of variation (k).
b. Write an equation for this relationship.
c. Use the equation to find the value of y when $x = 4$.

a. Determine the constant of variation (k).

$y = \mathbf{k}x$	Use this equation since it is direct variation.
$45 = \mathbf{k}(2.5)$	Substitute the values for x and y and then solve for k.
$\mathbf{18 = k}$	This is the constant of variation.

b. Write an equation for this relationship.

$y = \mathbf{k}x$	Just plug in the 18 we found for the value of k.
$y = \mathbf{18}x$	This is our equation of variation.

c. Use the equation to find the value of y when $x = 4$.

$y = 18x$	Substitute 4 for x in the equation of variation we just found.
$y = 18(4)$	
$y = 72$	

It is helpful when working an application problem to use variables that are representative of what they stand for. If you are looking for the price of something you might use $\boldsymbol{p}$ to represent the price. Of course you want to keep the **k** for the constant.

EXAMPLE 2

Courtesy of Fotolia

The amount of sales *tax* on a new car **varies directly** to the purchase price of the car. If a \$25,000 car pays \$1750 in sales taxes, what is the purchase price of a new car which has \$3500 sales tax?

Set up the equation and find k.	Write the equation of variation.	Answer the question.
Sales tax = *k* • *purchase price* $t = \mathrm{k}p$ $1750 = \mathrm{k}(25000)$ $\frac{1750}{25000} = \frac{\mathrm{k}(25000)}{25000}$ $\mathbf{0.07 = k}$	$t = \mathrm{k}p$ $t = \mathbf{0.07}p$	$t = \mathbf{0.07}p$ $3500 = 0.07p$ $\frac{3500}{0.07} = \frac{0.07p}{0.07}$ $50{,}000 = p$ **The purchase price is \$50,000.**

EXAMPLE 3

The weight of a ball varies directly with the cube of its radius. A ball that weighs 1.5 pounds has a radius of 3 inches. If a ball has a 10 inch ***diameter***, how much would it weigh? Round the answer to the nearest tenth, if necessary.

Set up the equation and find k.	Write the equation of variation.	Answer the question.
$weight = k \bullet (radius)^3$ $w = k \bullet r^3$ $1.5 = k(3)^3$ $\frac{1.5}{27} = \frac{k(27)}{27}$ $\frac{1}{18} = k$	$w = kr^3$ $w = \frac{1}{18}r^3$	$w = \frac{1}{18}r^3$ $w = \frac{1}{18}(5)^3$ $w = \frac{1}{18}(125)$ $w = 6.9444...$ $w \approx 6.9$ **The ball weighs approximately 6.9 pounds.**

YOUR TURN

6. If y varies directly as the square of x find the constant of variation and the variation equation when $x = 6$ and $y = 9$.

7. The distance traveled varies directly as the speed of the car. If Terri travels 750 miles at 50 mph, how far would she travel if her rate is 60 mph?

YOUR TURN

Answer each question.

8. The distance an object falls from rest varies directly as the square of the time it falls (ignoring air resistance).

   a. If a ball falls 64 feet in 2 seconds, how far will the ball fall in 1.5 seconds?

   b. How long will it take the object to fall 144 feet?

9. Suppose that y varies directly as x. What is the effect on y if x is doubled?

10. Suppose that y varies directly as x squared. What is the effect on y if x is doubled?

YOUR TURN **Answers to Section 7.07.**

1. inverse 2. joint 3. direct 4. direct 5. inverse

6. k = 0.25; $y = 0.25\,x^2$ 7. 900 miles 8a. Variation eq. $d = 16t^2$, $d = 36$ feet

8b. $t = 3$ seconds 9. y is doubled, also 10. y is 4 times what it was originally.

Complete MyMathLab Section 7.07 Homework in your Homework notebook.

7.08 Graphing Rational Functions 7.08

In this last section, we will revisit what a function is and how to find its domain. We will then look at graphs of rational functions and at modeling real world applications with rational functions.

Review of Functions, *f(x), g(x), C(x)* etc):

- A relation in which each *x*-coordinate has exactly one *y*-coordinate.
- Function notation is sometimes used when a function is defined by an equation: ***f(x)***= Eq.***;*** it replaces the *y*.
- The domain of the function is the set of real number values that can be used for the ***x***.
- The domain consists of all the real numbers except for those that make the function undefined.
- Rational expressions are undefined when the denominator equals 0. So are rational functions.

Review: Finding Domains.

EXAMPLE 1

Find the domain of $R(x)=\dfrac{x-2}{x^2-10x-24}$. [Like what we did in 7.01]

Algebraically:

$$x^2-10x-24=0$$
$$(x-12)(x+2)=0$$
$$x-12=0 \;\; or \;\; x+2=0$$
$$x=12 \qquad x=-2$$

We will find the values of the variable to eliminate, (the values that make the denominator equal to 0).

Since it is quadratic we will need to factor and set the factors equal to 0.

Domain: $\{x|x \text{ is a real number } \& \; x \neq -2, \; x \neq 12\}$

Graphically:
Now look at the graph. What do you notice is happening where *x* is -2 and *x* is 12?

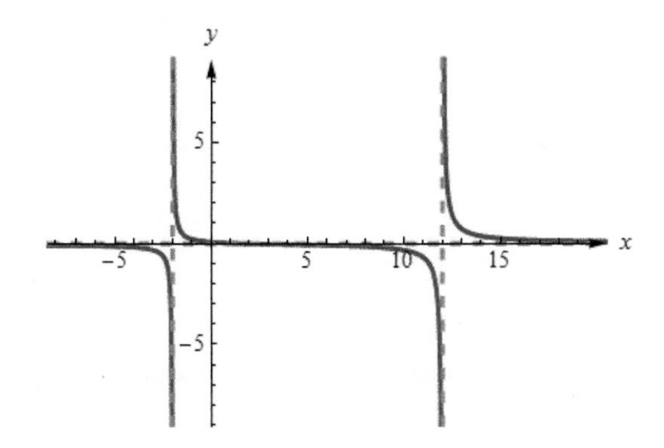

There are vertical lines where *x* = -2 and *x* = 12. The ends of the curve approach these vertical lines, but they never touch them. This is where the rational function is not defined. In other words, the domain for *x* is all other real numbers except for *x* = -2 and *x* = 12

EXAMPLE 2

Find the domain of $H(x) = \dfrac{1800x^2}{x^2+2}$.

Algebraically:

$x^2 + 2 = 0$

We know that when we square a number we get a positive number. If we add a positive number to 2, will we ever get 0?

NO! There are no exclusions.

Domain: $\{x | x \text{ is a real number}\}$

Graphically:

Now look at the graph. What you notice about the curve?

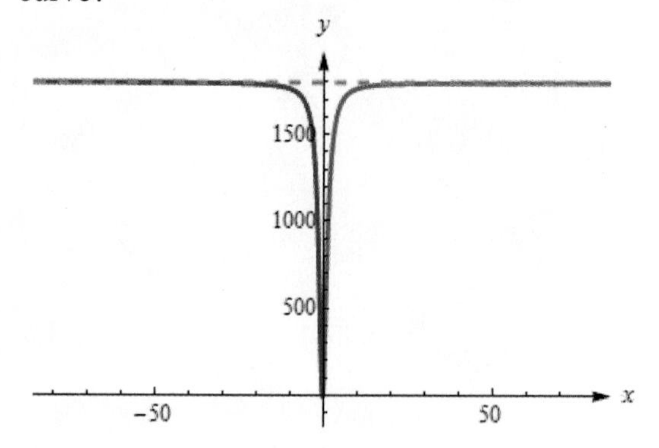

At $x = 0$, there is dip down, but all values of x are used. For the domain, this illustrates that x can be any real number.

7.08 Graphing Rational Functions 7.08

Let's look at evaluating a rational function.

EXAMPLE 3

Evaluate $C(x) = \dfrac{1800x^2}{x^2+4}$ when $x = 1$.

$$C(x) = \frac{1800x^2}{x^2+4}$$

$$C(1) = \frac{1800(1)^2}{(1)^2+4}$$

$$C(1) = \frac{1800 \cdot 1}{1+4}$$

$$C(1) = \frac{1800}{5} = 360$$

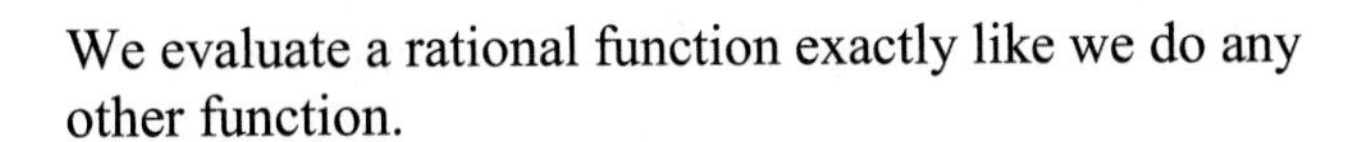

We evaluate a rational function exactly like we do any other function.

Simply plug in the value you are given for the variable and then follow the order of operations.

What value of x would make the function undefined?

The value that makes the function undefined is the same value that makes the denominator equal 0.

So, what number will make this statement true:

$$x^2 + 4 = 0$$

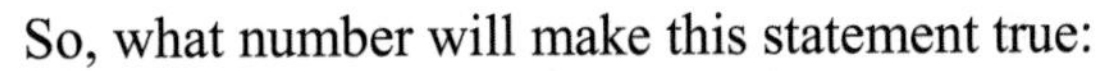

Hopefully, you answered that there is no real number that satisfies the equation above. Thus, there is no real number that will make this function undefined!

Finally something not so tough!

Courtesy of Fotolia

EXAMPLE 4

Given $f(x) = \dfrac{6x+25}{x}$, find x when $f(x) = 8.5$.

$f(x) = \dfrac{6x+25}{x}$ $8.5 = \dfrac{6x+25}{x}$ $8.5 \cdot (x) = \dfrac{6x+25}{x} \cdot (x)$ $8.5x = 6x + 25$ $2.5x = 25$ $x = 10$	Substitute 8.5 for $f(x)$ and solve for x. Remember that in some equations in two variables we can replace the y with function notation. If we needed to graph this function, one of the ordered pairs we would plot would be $(10, 8.5)$.

EXAMPLE 5

The annual cost of a refrigerator can be modeled with the given function, where ***n*** is the number of years you have owned the refrigerator and ***C*** is the cost of owning/operating the refrigerator.

$$C(n) = \frac{1000 + 92n}{n}$$

n = * years frig owned
$C(n)$ = Cost $ of frig.

> Remember, before attempting to answer questions in an application problem, understand what the variables represent in words.

a. Find the cost of owning and operating the refrigerator at the end of the first year.

$C(n) = \frac{1000 + 92n}{n}$ — "***at the end of the first year***" → $n = 1$

$C(1) = \frac{1000 + 92(1)}{1}$ — Substitute 1 in for *n*.

$C(1) = \frac{1000 + 92}{1}$ — Simplify.

$C(1) = 1092$ — The cost of owning and operating the refrigerator for 1 year is **$1,092.**

b. Find the cost of owning and operating the refrigerator at the end of the second year.

$C(n) = \frac{1000 + 92n}{n}$ — "***at the end of the second year***" → $n = 2$

$C(2) = \frac{1000 + 92(2)}{2}$ — Substitute 2 in for *n*.

$C(2) = \frac{1000 + 184}{2}$ — Simplify.

$C(2) = 592$ — The cost of owning and operating the refrigerator for 2 years is **$592.**

c. Find the cost of owning and operating the refrigerator during the second year only.

The cost during the second year only would be the cost for the first year minus the cost for the first two years.

Cost at the end of year 1 – *Cost at the end of year 2* = *Cost for second year only*

$1,092 – $592 = $500

The cost of owning and operating the refrigerator during the second year only is **$500.**

7.08 Graphing Rational Functions 7.08

YOUR TURN

Find the domain of each function.

1. $h(x)=\dfrac{7x}{x+5}$

2. $R(x)=\dfrac{x+9}{x^2-81}$

3. Given $h(x)=\dfrac{7x}{x+5}$, find x when $h(x)=6$.

4. Find $R(3)$ if $R(x)=\dfrac{x+9}{x^2-81}$.

5. The revenue from the sale of an action figure is modeled by $R(x)=\dfrac{1800x^2}{x^2+4}$ where x is the number of years since the creation of the action figure and $R(x)$ is the total revenue in millions of dollars.

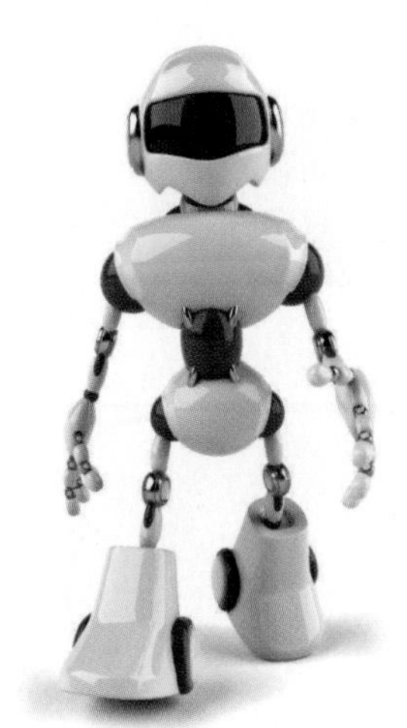

Courtesy of Fotolia

a. Find the total revenue at the end of the first year.__________________

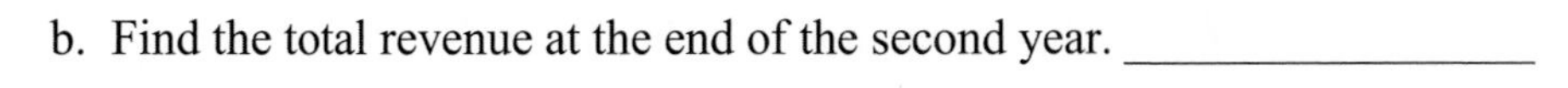

b. Find the total revenue at the end of the second year. __________________

c. Find the revenue during the second year only. __________________

d. Find the domain of *R*.______________________________.

GRAPHING AND ANALYZING RATIONAL FUNCTIONS

EXAMPLE 6

It's a little hard to wrap your mind around what a rational function really is. They say a picture is worth a thousand words so let's look at the picture of a rational function.

As you know, in order to look at the graph (the picture) we need to find some points.

x	0.001	0.01	0.1	0.25	0.5	1	2	4	10	100	1000
$f(x)=\frac{1}{x}$	1000	100	10	4	2	1	0.5	0.25	0.1	0.01	0.001

x	-0.001	-0.01	-0.1	-0.25	-0.5	-1	-2	-4	-10	-100	-1000
$f(x)=\frac{1}{x}$	-1000	-100	-10	-4	-2	-1	-0.5	-0.25	-0.1	-0.01	-0.001

What happens to $f(x)$ as x gets larger?____

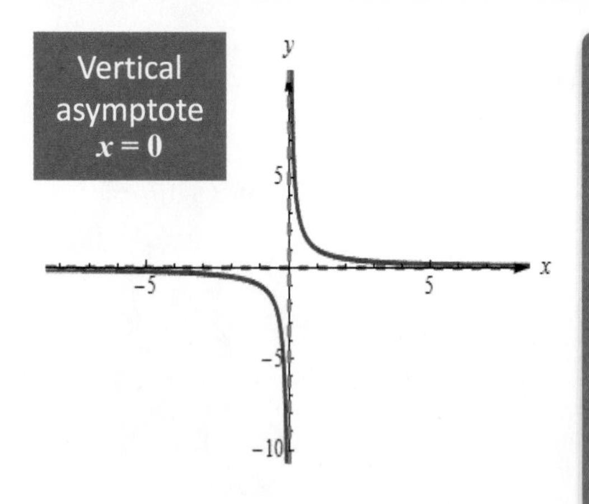

Notice that the graph appears to be approaching the horizontal line y=0 (the x-axis). This function has a horizontal asymptote at y = 0.

ASYMPTOTES:

Look at the curve as it moves from left to right. The curve drops very slowly from -10 until it nears the line where x = 0 where it drops rapidly. It does not touch this line. However, as you continue to move to the right of x = 0, the curve appears and drops from high above and then drops slowly as you approach where x = 10 and beyond.

The vertical part of the curve is getting close to x = 0 (the y-axis).

Will the graph have a point at x = 0?

NO! Think about the domain. The denominator cannot equal 0! Thus $x \neq 0$.

Rational functions have asymptotes. Not all rational functions have the same number of asymptotes.

The asymptote is a line that the graph approaches but does not touch. This function has a vertical asymptote at x = 0.

You will be asked to find horizontal asymptotes in your college algebra or precalculus class, but not in this class!

7.08 Graphing Rational Functions 7.08

Let's look at graphing a rational function on the graphing calculator.

EXAMPLE 7

Graph $f(x) = \dfrac{x}{x+2}$ on your graphing calculator. Use the standard window.

Xmin = –10 Xmax = 10 Xscl = 1 Ymin = –10 Ymax = 10 Yscl=1

Does your graph look like this?

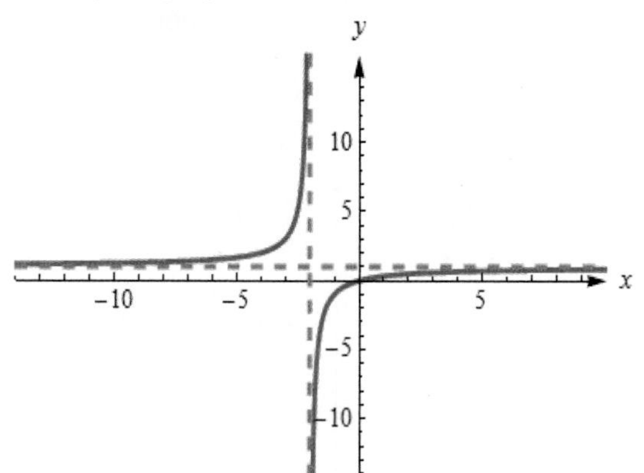

Be very careful when entering the function into the calculator. If the numerator or denominator is more than one term, then group it in parentheses.

In this example you would enter:

$y_1 = x/(x+2)$ Press [Zoom][6:standard window] for quick access to graph.

So what's going on with the dotted lines below? We will only look at the vertical line.

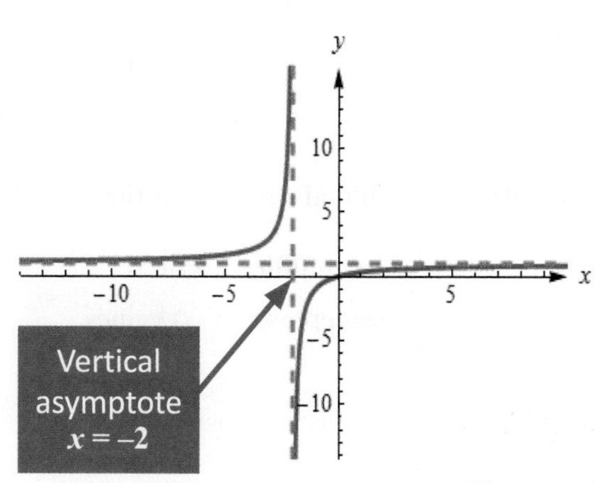

Vertical asymptote $x = -2$

Courtesy of Fotolia

FINDING THE VERTICAL ASYMPTOTE USING THE GRAPHING CALC:
TI-83 calc: When you use the graphing calculator, the vertical asymptote will appear automatically.
TI-84: To see the vertical asymptote, go to the WINDOW settings and change the xres at the bottom of the screen to say xres = 2. This should make the vertical line of the asymptote be visible on the graph.

The asymptote is not a part of the graph.

It just lets you know what number the curve is approaching, $x = -2$. The function is undefined at that point, so it is a good way to check that you determined the domain is correct!

Hint: Always clear your [y_1 =] before entering another equation. Press [Zoom 6] to reset the window.

EXAMPLE 8

Determine the vertical asymptote of $f(x)=\dfrac{x^3+2x^2}{x^2+4x+4}$ both algebraically and graphically.

Xmin = –10 Xmax = 10 Xscl = 1 Ymin = –25 Ymax = 20 Yscl=1

Algebraically

The vertical asymptote occurs for every number that causes the denominator of the function to have a value of 0.

$\dfrac{x^3+2x^2}{x^2+4x+4}$ Simplify the rational expression.

$\dfrac{x^2(x+2)}{(x+2)(x+2)}$ Factor the numerator and denominator.

$\dfrac{x^2}{(x+2)}$ Set the denominator equal to 0.

$x+2=0$ Solve.

$x=-2$ ⬅ Equation of the vertical asymptote.

Clear your calc!

Graphically

Here is the vertical asymptote! Use the [Trace] key to scroll right or left and note where the "X" skips from one curve to the other.

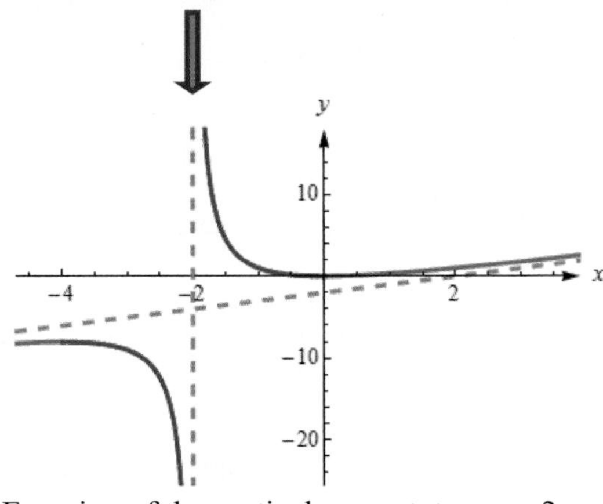

Equation of the vertical asymptote: $x=-2$

EXAMPLE 9

The function $C(t)=\dfrac{5t}{0.01t^2+3.3}$ describes the concentration of a drug in the blood stream over time. In this case, the medication was taken orally. $C(t)$ is measured in micrograms per milliliter and t is measured in minutes. This means the *x*-values represent time *t* and the *y*-values represent C(*t*) values. Round solutions to the nearest hundreth.

1. On the graphing calculator, graph the function over the first two hours after the dose is given. Use Xmin = 0 Xmax = 180 Xscl = 30 Ymin = 0 Ymax = 20 Yscl = 5

2. Why would we limit the graph to the first quadrant only?

3. When is the concentration of the drug at its highest? How much of the drug is in the blood stream at this time?

4. Explain within the context of the problem the shape of the graph between taking the medication orally and the maximum point.

7.08 Graphing Rational Functions 7.08

Solution

1. *Your graph should resemble this graph.*

$C(t)$ concentration of drug, micrograms

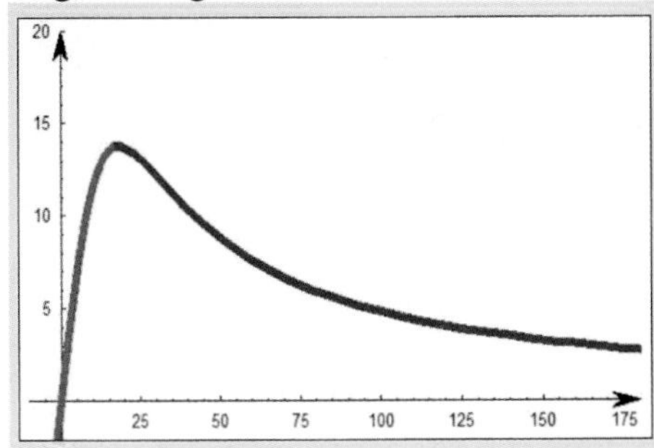

t time

2. *We must limit the graph to the quadrant where x and y are both positive, Quadrant I.* ***Negative time or negative amounts of a drug in one's bloodstream do not make sense!***
3. *Find the maximum on the calculator to answer these questions,* 2nd [Calc → [4:max]. Refer back to module 6.18 if you need help with this. *The maximum point is (18.2, 13.76). The concentration of the drug in the bloodstream is the highest at the maximum point of the function.* ***The maximum concentration in the body is 13.76 mcg (y-value) which occurs at 18.2 minutes (x value).***
4. *The graph shows that the concentration is 0 mcg when time= 0 minutes.* ***Within the first 18.2 minutes, the concentration rises sharply to the maximum at 13.76 mcg.*** *The curve then begins to fall. This means that after reaching the maximum, the kidneys begin cleansing the blood and rapidly remove the drug. Look at the table to see the concentrations at several times. Note that the curve will never reach 0. All of the drug will never be removed from the bloodstream.*

EXAMPLE 10

Courtesy of Fotolia

The rational function $C(x)=\dfrac{100{,}000x}{100-x}$ describes the cost, in dollars, to remove x percent of the air pollutants in the smokestack emission of a utility company that burns coal to generate electricity. Graph the function and answer the following questions.
Use this window for your graph.

Xmin = 0 **Xmax = 140**
Ymin = –100 **Ymax = 2,000,000**

1. What is the vertical asymptote?
2. Graph the function. Remember to identify in words the x values and the y values.
3. Which quadrant should you look at to answer the questions?
4. Find each of the following both graphically and algebraically.
   a. the cost for removing 20% of the pollutants from the air.
   b. the cost for removing 50% of the pollutants from the air.
   c. the cost for removing 90% of the pollutants.
5. What percentage of the pollutants could the company have removed if they can only afford to spend $300,000 on the cleanup?

7.08 Graphing Rational Functions 7.08

Solution to Example 10.

1. To find the vertical asymptote set the denominator equal to 0.

$$100 - x = 0$$
$$-x = -100$$
$$x = 100$$

2. In words, write what the x and $C(x)$ represent.

x = % of air pollutants. C(x) = Cost $ to remove pollutants

Your graph should look like this.

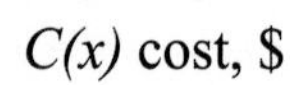

C(x) cost, $

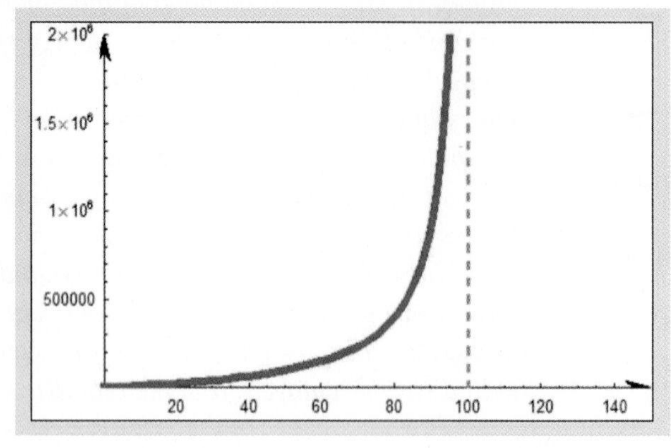

x percent of air pollution

3. Use Quadrant I since percent and the amount of the pollutant would be positive numbers only.

4. *Algebraically:*

a. the cost for removing 20% of the pollutants from the air.

$$C(x) = \frac{100{,}000x}{100 - x}$$
$$C(20) = \frac{100{,}000(20)}{100 - (20)}$$
$$C(20) = \frac{2{,}000{,}000}{80}$$
$$C(20) = 25{,}000$$

b. the cost for removing 50% of the pollutants from the air.

$$C(x) = \frac{100{,}000x}{100 - x}$$
$$C(50) = \frac{100{,}000(50)}{100 - (50)}$$
$$C(50) = \frac{5{,}000{,}000}{50}$$
$$C(50) = 100{,}000$$

c. the cost for removing 90% of the pollutants.

$$C(x) = \frac{100{,}000x}{100 - x}$$
$$C(90) = \frac{100{,}000(90)}{100 - (90)}$$
$$C(90) = \frac{9{,}000{,}000}{10}$$
$$C(90) = 900{,}000$$

7.08 Graphing Rational Functions 7.08

Graphically: use the VALUE feature on the calculator to find the value of y for the given values of x.
2nd → CALC→ "1:value"→ put in the value for x→ enter

5. Substitute 300,000 for $C(x)$

$$C(x)=\frac{100,000x}{100-x}$$

$$300,000=\frac{100,000x}{100-x}$$

$$300,000(100-x)=100,000x$$

$$30,000,000-300,000x=100,000x$$

$$30,000,000=400,000x$$

$$75=x$$

Check your answer by using the INTERSECT feature of your CALC.

Enter Y_1 = (The equation)

Enter Y_2 = 300,000 Then press 2nd CALC.

Select "5:intersect" and press enter three times.

x = % of air pollutants. C(x) = Cost \$(remove pollutants)

x = 75% of air pollutants. C(x) = Cost \$ 300,000

The company can afford to clean up 75% of the pollutants.

YOUR TURN

Answer each question. Practice entering the equations of each previous example in your calculator. Use PARENTHESES!!!!

6. The function $f(x)=\frac{6.5x^2-82.9x+3724}{x^2+576}$ models the pH level, $f(x)$, of the human mouth x minutes after a person eats food containing sugar. Round solutions to the nearest hundredth. In the example, the normal pH level of the human mouth is 6.5 pH.

   Understand what the variables represent in words:

   x = ________________ $f(x)$ = ____________________

   a. Graph the function using the window.

   Xmin = –20 Xmax = 150 Xscl = 1 Ymin = 3 Ymax = 10 Yscl = 1

   Which of these is the graph of this function?

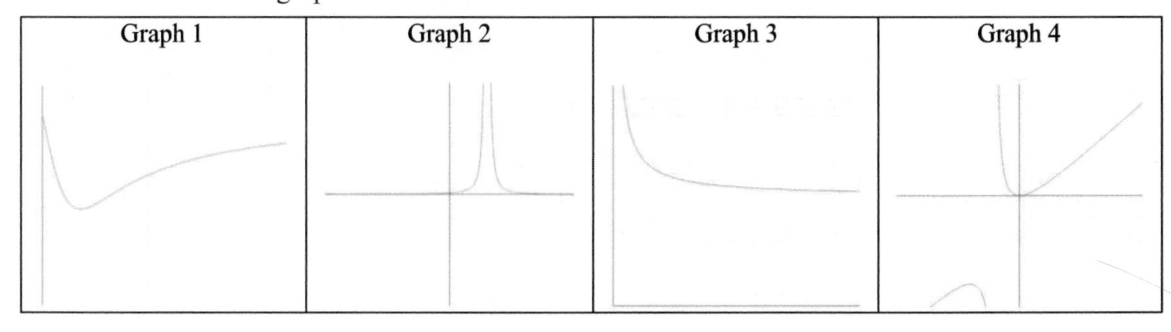

YOUR TURN

b. Does the graph have a vertical asymptote? If not, why not? If it does have a vertical asymptote, what is it?

c. Use the graph to find an estimate to the nearest tenth of the pH level of the human mouth 42 minutes after a person eats food containing sugar. Round to the nearest whole number.

d. Find the minimum point using your calculator. Then use the graph to describe what happens to the PH level during the first hour.

WRITING A RATIONAL FUNCTION

EXAMPLE 11

Kelly and Noel design a cool t-shirt for skateboarders. Everybody they know wanted one so they decided to print and sell the t-shirts to the public. Startup costs amounted to \$1250. They estimate that it will cost \$5 to purchase and print each shirt.

Write a linear function that models the cost of producing ***x*** t-shirts. Call this function ***C(x).***

It cost \$5 for each t-shirt they buy so if they buy ***x*** *shirts it would cost* ***5x*** *for those shirts. We must also include the startup costs.*

$$C(x) = \$5 \text{ per shirt plus the startup costs}$$
$$C(x) = \quad 5x \quad + \quad 1250$$
$$C(x) = 5x + 1250$$

7.08 Graphing Rational Functions 7.08

Complete the table:

x	1	2	10	100
$C(x)$				

*Substitute value for **x** into the function you wrote.*

x	1	2	10	100
$C(x)$	1255	1260	1300	1750

How much does it cost to produce just the 1000th t-shirt?

To answer this question we need to find the ***average cost function,*** $\overline{C}(x)$. *The average cost function tells us the average cost per shirt of producing **x** shirts. To find the average, we will do what we always do: divide the total by the number of shirts.*

$$\overline{C}(x)=\frac{5x+1250}{x}$$

To find the cost of the 1000th t-shirt just substitute 1000 into the average cost function.

$$\overline{C}(x)=\frac{5x+1250}{x}$$

$$\overline{C}(1000)=\frac{5(1000)+1250}{1000}$$

$$\overline{C}(1000)=6.25$$

What is the domain in the context of this problem?

We know that the denominator cannot be 0. Can it be negative? Can it be a fraction?

Since x represents the number of t-shirts the domain includes only positive integers.

Domain: $\{x|x \text{ is an integer and } x>0\}$

What is the vertical asymptote of the average cost function?

Vertical asymptote at x = 0, since x = 0 makes the function undefined.

Graph the average cost function. Remember to look at Quadrant I in the context of this problem. Use the window:

Xmin = –100 Xmax = 1000 Xscl = 100

Ymin = –100 Ymax = 150 Yscl = 10

Your graph should look like this:

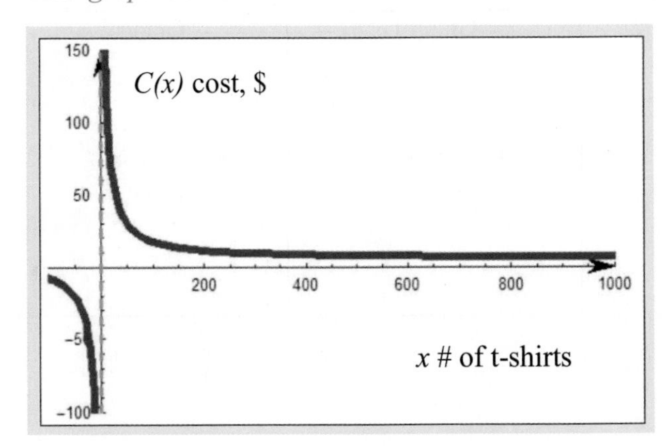

YOUR TURN

7. In New York a premier fitness club membership (the cheaper one!) can cost \$50 per month with a \$100 joining fee. Think, what is the one time fixed amount (y-intercept) and what amount changes per member. Round solutions to the nearest hundredth.

   a. Write a linear function, ***C(x)***, that models the cost of joining and attending the fitness club for x months.

   b. What is the average cost function, $\overline{C}(x)$, that models membership at this fitness club?

   c. What is the domain of $\overline{C}(x)$ in the context of this problem? (Use set notation)

   d. Why should we limit the graph of $\overline{C}(x)$ to the first quadrant?

   e. Graph the average cost function, using the window. Which graph below does your graph look like.

Xmin = 0	**Xmax = 50**	**Xscl = 1**
Ymin = 0	**Ymax = 100**	**Yscl = 1**

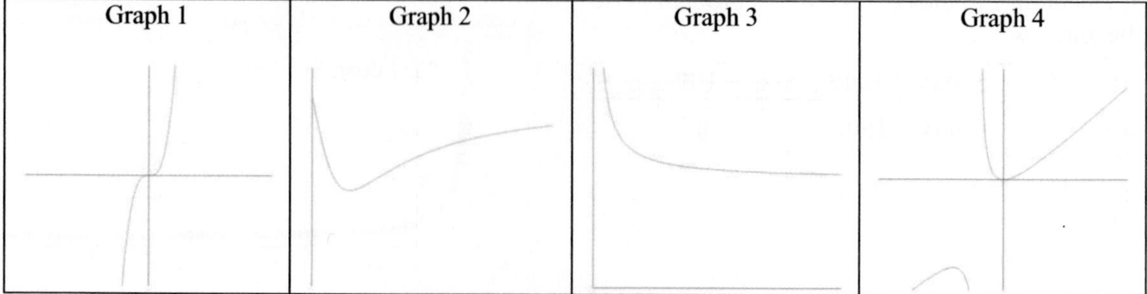

YOUR TURN

f. Find the average cost for attending the gym for 12 months, both algebraically and graphically. Round your answer to the nearest dollar. Do your answers match?

DON'T FORGET TO GO BACK TO THE BEGINNING ANI COMPLETE THE INTRODUCTORY PROBLEM!

7.08 Graphing Rational Functions 7.08

YOUR TURN **Answers to Section 7.08.**

1. $\{x \mid x$ is a real number and $x \neq -5\}$ 2. $\{x \mid x$ is a real number and $x \neq 9, -9\}$ 3. 30 4. $-\frac{1}{6}$

5a. \$360 million 5b. \$900 million 5c. \$540 million 5d. $\{x \mid x$ is a real number$\}$

6. x: minutes, $f(x)$: pH level of mouth 6.a. 1 6b. No, the denominator is a positive number. 6c. 5

6d. The pH level drops sharply immediately after a person eats food containing sugar and then begins to rise again back to almost 6.5. (23.8, 4.76)

7a. $C(x) = 50x + 100$ 7b. $\bar{C}(x) = \frac{50x+100}{x}$ 7c. $\{x \mid x$ is a real number and $x > 0\}$

7d. The number of months nor the amount of money can be negative. 7e. Graph 3

7f. Algebraically: $\bar{C}(x) = \frac{50(12)+100}{12}$ $\bar{C}(x)$ = \$58.00 Graphically: Put the equation in [y=].
Go to [2nd][CALC] and choose "1:value." Let $x = 12$. The y value will be 58.3333. Hence, the average cost per month for 12 months is \$58.33.

Introductory Problem.

1. The x values are days. The $f(x)$ values are the average number of Latin words remembered by students. The point (25, 6.2) represents the fact that after 25 days, students remember an average of 6.2 words or 6 Latin words.
2. As time passes, memory loss slows down.
3. Let $x = 11$. Algebraically: $f(11) = \frac{5(11)+30}{(11)} = 7.7$ Latin words OR

   Graphically: Enter the equation in [y= (5x+30)/x]. Go to [2nd][CALC] and choose "1:value." Let $x = 11$ and press [ENTER]. The y value is 7.7. This means a student will remember about 8 Latin words in 11 days of the original 40 Latin words.
4. Let $f(x) = 20$. Algebraically: $20 = \frac{5x+30}{x}$ $20x = 5x + 30$ $15x = 30$ $x = 2$ *days*

   OR Graphically: Enter the equation in [y_1=] and [y_2]= 20 words, which is half the words. Go to [2nd][CALC] and choose "5:intersect." Press [ENTER] 3 times. The x value is 2 days.

Complete MyMathLab Section 7.08 Homework in your Homework notebook.

DMA 070 TEST: Review your notes, key concepts, formulas, and MML work. ☺

Student Name: ______________________ **Date:** ____________

Instructor Signature: ______________________

DMA 080

DMA 080

RADICAL EXPRESSIONS/EQUATIONS

This course provides a conceptual study of the manipulation of radicals and the application of radical equations to real-world problems. Topics include simplifying and performing operations with radical expressions and rational exponents, solving equations, and determining the reasonableness of an answer. Upon completion, students should be able to find algebraic solutions to contextual problems with radical applications.

You will learn how to:

- Demonstrate the use of a problem solving strategy to include multiple representations of the situation, organization of the information, and algebraic representation of radical equations.
- Correctly use rational exponents to rewrite radical expressions.
- Simplify radical expressions.
- Add, subtract, multiply, and divide radical expressions.
- Solve radical equations.
- Solve quadratic equations using the quadratic formula.
- Graph radical functions using the graphing calculator to identify and interpret the graph in terms of the problem.

Quadratic Formula:

$$x = \frac{-b \pm \sqrt{b^2 - 4ac}}{2a}$$

Radical equations:

$\sqrt{x+3} = x - 9$

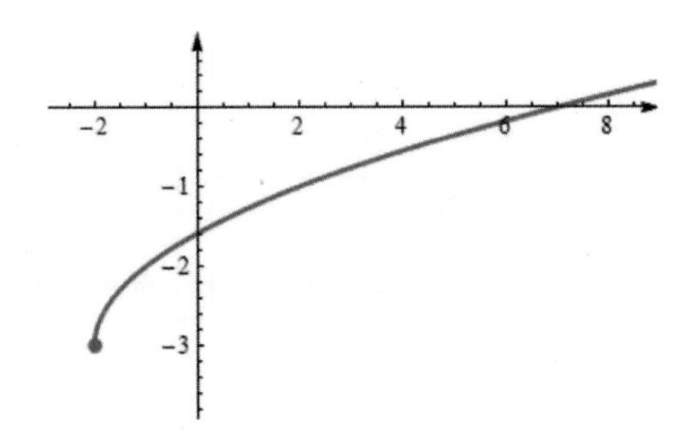

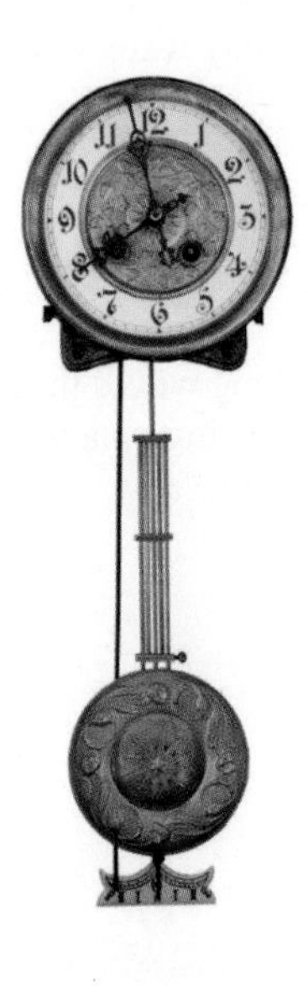

Courtesy of Fotolia

A radical equation is used to find the period of an old clock, the time it takes for a pendulum to swing back and forth.

8.01 Radical Expressions and Equations 8.01

INTRODUCTORY PROBLEM

Here is an example of how radical functions are used. Do as much as you can now and come back to complete it as you work through the module.

Police sometimes use the formula $s = \sqrt{30df}$ to estimate the speed, *s,* in miles per hour, of a car that skidded *d* feet upon braking. The variable *f* is the coefficient of friction determined by the kind of road and the wetness or dryness of the road. The following table gives some values of *f*:

f friction values	On Concrete	On Tar
Wet	0.4	0.5
Dry	0.8	1.0

Answer each question. Round solutions to the nearest whole number.

First, what do the variables represent in words? $s =$ ____________, $d =$ ____________, $f =$ ____________

Round solutions to the nearest whole number.

a) How fast was a car going on a concrete road if the skid marks on a rainy day were 80 feet long?

b) How fast was a car going on a dry tar road if the skid marks were 100 feet long?

c) How long would the skid marks be for a car driving on a wet tar road traveling at 50 mph?

d) Make a table of reasonable values for speed and feet skidded and create a graph for wet tar.

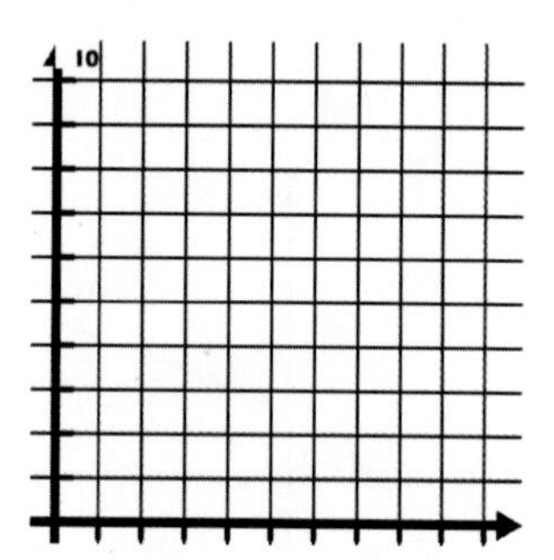

e) What does the domain and range mean within the context of this problem?

f) Use the graphing calculator to determine how long the skid marks would be for a car which had been traveling 35 mph on a dry tar road.

Xmin = –10 Xmax = 250 Ymin = –10 Ymax = 120

8.01 Radical Expressions and Equations 8.01

In this Module, our focus will be on working with radical expressions and equations. Radicals are used in these professions.

Professions that Use Rational Expressions

Management

Management occupations

- Computer and information systems managers
- Engineering and natural sciences managers
- Farmers, ranchers, and agricultural managers
- Funeral directors
- Industrial production managers
- Medical and health services managers
- Property, real estate, and community association managers
- Purchasing managers, buyers, and purchasing agents

Business and financial operations occupations

- Insurance underwriters

Farming

- Forest, conservation, and logging workers

Safety

- Law enforcement

Professional

Computer and mathematical occupations

- Actuaries
- Mathematicians
- Statisticians

Engineers

- Aerospace engineers

Physical scientists

- Chemists and materials scientists
- Physicists and astronomers

Social scientists and related occupations

- Economists

Education, training, library, and museum occupations

- Teachers-adult literacy and remedial and self-enrichment education
- Teachers-postsecondary
- Teachers-preschool, kindergarten, elementary, middle, and secondary
- Teachers-special education

Health diagnosing and treating occupations

- Registered nurses

Production

Metal workers and plastic workers

- Computer control programmers and operators

To see more about these professions you can go to the website below:

http://www.xpmath.com/careers/topicsresult.php?subjectID=2&topicID=9

8.01 Radical Expressions and Equations 8.01

In a previous module we looked at squaring a number. For example, $5^2 = 25$. In this module, we will look at the root of a number. The inverse operation of squaring a number is to find its square root. This is the same as asking the question, "What number squared would give you 25?"

All positive numbers have two square roots, one positive and one negative.

$\sqrt{x}$ ⟶ *positive square root of x* ⟶ $\sqrt{25} = 5$

$-\sqrt{x}$ ⟶ *negative square root of x* ⟶ $-\sqrt{25} = -5$

Remember:

- $\sqrt{0} = 0$
- **The square root of a negative number is not a real number.**

  $\sqrt{-9}$ **is not a real number.**

There can never be a situation where a number multiplied by ITSELF gives us a negative number. Right? Let's try a few. $-4 \bullet -4 = 16$, $4 \bullet 4 = 16$. It is never possible to multiply a number by itself and get a negative answer.

EXAMPLE 1

Simplify. Assume that all variables represent positive numbers.

a) $\sqrt{49}$ *b)* $\sqrt{x^6}$ *c)* $\sqrt{9y^{12}}$ *d)* $\sqrt{-16}$ *e)* $\sqrt{-16}$ *f)* $\sqrt{\frac{1}{25}}$ *g)* $\sqrt{\frac{36x^2}{y^2}}$

SOLUTIONS

To take the square root of a power divide the exponent by 2.

a. $\sqrt{49} = 7$ because $7^2 = 49$.

b. $\sqrt{x^6} = x^3$ because $\left(x^3\right)^2 = x^6$. Remember?

Power to a power means multiply the exponents.

c. $\sqrt{9y^{12}} = 3y^6$ because $(3y^6)^2 = 9y^{12}$.

d. $\sqrt{-16}$ is not a real number.

e. $-\sqrt{16} = -4$. Note the negative is outside the radical. [negative square root.]

f. $\sqrt{\frac{1}{25}} = \frac{1}{5}$ because $\left(\frac{1}{5}\right)^2 = \frac{1}{25}$.

g. $\sqrt{\frac{36x^2}{y^2}} = \frac{6x}{y}$ because $\left(\frac{6x}{y}\right)^2 = \frac{36x^2}{y^2}$.

8.01 Radical Expressions and Equations **8.01**

YOUR TURN

Simplify. Assume that all variables represent positive numbers.

1. $\sqrt{81}$	2. $\sqrt{\frac{16}{49}}$	3. $\sqrt{36b^8}$
4. $-\sqrt{25}$	5. $\sqrt{-64}$	

Finding Cube Roots

Let's look at cube roots. $\sqrt[3]{8}$ is read the cube root of 8. This time the question is "What number would you raise to the third power to get 8?" 2 is the cube root of 8 because $2^3 = 8$.

EXAMPLE 2

Simplify. Assume that all variables represent positive numbers.

a. $\sqrt[3]{64}$ b. $\sqrt[3]{-27}$ c. $\sqrt[3]{x^{12}}$ d. $\sqrt[3]{-64x^9}$ e. $\sqrt[3]{\frac{8}{27}}$

SOLUTIONS

a. $\sqrt[3]{64} = 4$ because $4^3 = 64$

b. $\sqrt[3]{-27} = -3$ because $(-3)^3 = (-3)(-3)(-3) = -27$

c. $\sqrt[3]{x^{12}} = x^4$ because $(x^4)^3 = x^{12}$

d. $\sqrt[3]{-64x^9} = -4x^3$ because $(-4x^3)^3 = -64x^9$

To take the cube root of a power divide the exponent by 3.

e. $\sqrt[3]{\frac{8}{27}} = \frac{2}{3}$ because $\left(\frac{2}{3}\right)^3 = \frac{8}{27}$

8.01 Radical Expressions and Equations **8.01**

Did you notice that unlike square roots, we can take the cube root of a negative number? If you multiply three negative numbers you get a negative number!

$\sqrt{-16}$ is not a real number $\sqrt[3]{-27} = -3$

Finding n^{th} Roots

So far, we have looked at square roots and cube roots. We can raise numbers to powers other than 2 or 3, so we should be able to find roots other than square roots or cube roots! Here's a chart you might find helpful. Of course, you can always use your calculator!

Courtesy of Fotolia

PERFECT SQUARES		PERFECT CUBES	PERFECT 4TH POWERS
$1^2=1$ $2^2=4$ $3^2=9$ $4^2=16$ $5^2=25$ $6^2=36$ $7^2=49$ $8^2=64$ $9^2=81$ $10^2=100$	$11^2=121$ $12^2=144$ $13^2=169$ $14^2=196$ $15^2=225$ $16^2=256$ $17^2=289$ $18^2=324$ $19^2=361$ $20^2=400$ $x^2, x^4, x^6, x^8,$ [Even Exponents]	$1^3=1$ $2^3=8$ $3^3=27$ $4^3=64$ $5^3=125$ $6^3=216$ $x^3, x^6, x^9, \ldots$ [Exponents multiples of 3]	$1^4=1$ $2^4=16$ $3^4=81$ $4^4=256$ $5^4=625$ $x^4, x^8, x^{12} \ldots$ [Exponents multiples of 4] **PERFECT 5TH POWERS** $1^5=1$ $2^5=32$ $3^5=243$ $x^5, x^{10}, x^{15},$ [Exponents multiples of 5]

Before we begin looking at how to work with n^{th} roots , let's look at some words we'll be using.

Index: n^{th} root – tells you what root to take.

The index 2 is omitted for square roots.

$$\sqrt[n]{a}$$

Radicand: expression or number underneath the radical sign.

EXAMPLE 3

Simplify. Assume that all variables represent positive numbers.

a. $\sqrt[5]{32}$ b. $\sqrt[4]{81y^8}$ c. $\sqrt[5]{-32x^{10}y^{15}}$ d. $\sqrt[7]{128x^{21}}$ e. $\sqrt[4]{-16}$

SOLUTIONS:

To take the nth root of a power, divide the exponent by the index.

a. $\sqrt[5]{32} = 2$ because $(2)^5 = 32$

b. $\sqrt[4]{81y^8} = 3y^2$ because $\left(3y^2\right)^4 = 81y^8$

c. $\sqrt[5]{-32x^{10}y^{15}} = -2x^2y^3$ because $\left(-2x^2y^3\right)^5 = -32x^{10}y^{15}$

d. $\sqrt[7]{128x^{21}} = 2x^3$ because $\left(2x^3\right)^7 = 128x^{21}$

e. $\sqrt[4]{-16}$ is not a real number. There is no real number that, when raised to the fourth power is –16.
$(-2)(-2)(-2)(-2) = 16$ not -16.

Finding Cube Roots and Nth Roots on the Calculator

1. Find $\sqrt[3]{64}$

MATH KEY

Press the MATH key.

Select $\sqrt[3]{(}$ and then enter 64). Press enter.

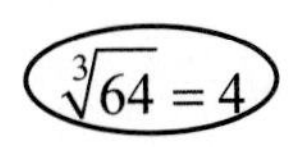

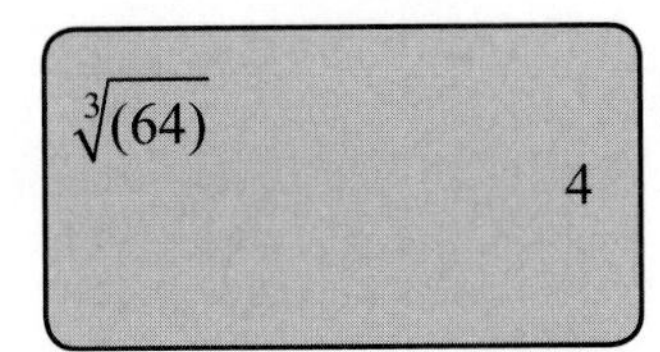

2. Find $\sqrt[5]{(32)}$

Press 5, then the MATH key.

Select $\sqrt{\ }$ and then enter 32). Press enter.

$\sqrt[5]{(32)}$

2

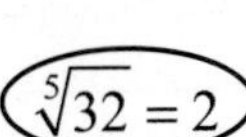

8.01 Radical Expressions and Equations 8.01

Courtesy of Fotolia

When is it OK to have a negative under the radical????

Index is **even** → Radicand cannot be negative → $\sqrt{16}=4$ *but* $\sqrt{-16}$ *is not a real number*; $\sqrt[4]{81}=3$ *but* $\sqrt[4]{-81}$ *is not a real number*

Index is **odd** → Radicand can be positive or negative → $\sqrt[3]{27}=3$ *and* $\sqrt[3]{-27}=-3$; $\sqrt[5]{32}=2$ *and* $\sqrt[5]{-32}=-2$

Your Turn

Simplify. Assume that all variables represent positive numbers.

6. $\sqrt[5]{243m^5}$	7. $\sqrt[4]{16y^{12}}$	8. $\sqrt[5]{-x^{15}y^5}$	9. $\sqrt[3]{-27x^{21}}$	10. $\sqrt{-25}$

8.01 Radical Expressions and Equations **8.01**

Variables Representing Real Numbers

In reading the directions of all the examples in this section, you may have noticed that it always says "assume that all variables represent positive numbers." Let's look a little closer at why that is important.

When there are variables under the radical, we do not know whether they represent a positive or negative number. This fact coupled with the fact that we must have a positive answer if the index is **even** presents a problem! If our index of the radical is even, we can ensure the answer is positive by using absolute value symbols around solutions whose variable exponent is odd. Remember, if the index of the radical is odd, the result can be positive or negative, so no absolute value symbols are needed.

If the variable represents any real number. then check out the index for when to take the absolute value of the variable:

even index → absolute value around variables with odd powers → $\sqrt{4x^2} = 2|x|$, $\sqrt[4]{x^{12}y^4} = |x^3y|$

odd index → do not need absolute value → $\sqrt[3]{8x^3} = 2x$, $\sqrt[5]{32x^5y^{10}} = 2xy^2$

EXAMPLE 4

Simplify. Assume that the variable represents any real number. BE CAREFUL!

1) $\sqrt{(-7)^2} = |-7| = 7$ 2) $\sqrt[4]{(x-3)^4} = |x-3|$ 3) $\sqrt{16x^2} = 4|x|$

Examples 1-3 are all even positive roots. Therefore, they require use of the absolute value bars around the solutions.

4) $\sqrt[3]{(-2)^3} = -2$ 5) $\sqrt[5]{(3x-2)^5} = 3x-2$

Examples 4 & 5 are odd roots. Therefore, they do not need absolute value bars in their answers.

YOUR TURN

Simplify. Assume that the variable represents any real number.

11. $\sqrt{x^6}$	12. $\sqrt[3]{8m^6}$	13. $\sqrt[4]{16y^4}$
14. $\sqrt{25x^2}$	15. $\sqrt{(x-5)^2}$	16. $\sqrt{x^2-20x+100}$ (Factor first!)

If you see the statement "variables represent any real number" and the index of the radical is even, remember to put an absolute value symbol around variables with odd exponents in the solution.

If the directions say "assume variables represent positive real numbers," you will never need to use absolute value!!! The solution is already always positive.

Radical Functions

Radical functions are functions of the form: $f(x) = \sqrt[n]{x}$

Recall that the domain of a function consists of all of the possible values that can be used for the variable. When dealing with radical functions you must consider the index with determining the domain. We will work more with domain when we begin to look at the graphs and applications of radical functions.

Let's look at finding function values.

EXAMPLE 5

If $f(x) = \sqrt{x-2}$ *and* $g(x) = \sqrt[3]{x-5}$, find each function value.

a. $f(11)$ b. $g(13)$ c. $g(4)$ d. $f(1)$

SOLUTIONS: Finding the value of the function is just a matter of substituting the value given for the variable into the function.

a. $f(11) = \sqrt{11-2} = \sqrt{9} = 3$

b. $g(13) = \sqrt[3]{13-5} = \sqrt[3]{8} = 2$

c. $g(4) = \sqrt[3]{4-5} = \sqrt[3]{-1} = -1$

d. $f(1) = \sqrt{1-2} = \sqrt{-1} = \textit{undefined}$

Why is the last one undefined?

There is no real number which when squared makes a -1. The domain of radical functions that involve even roots are limited to values that make the radicand non-negative!

YOUR TURN

Find the function values.

17. If $f(x)=\sqrt{2x+3}$, find $f(11)$

18. If $g(x)=\sqrt{3x-8}$, find $g(2)$.

YOUR TURN Answers to Section 8.01

1. 9	2. $\frac{4}{7}$	3. $6b^4$	4. –5
5. Not a real number		6. $3m$	7. $2y^3$
8. $-x^3y$	9. $-3x^7$	10. Not a real number	
11. $\lvert x^3\rvert$	12. $2m^2$	13. $2\lvert y\rvert$	14. $5\lvert x\rvert$
15. $\lvert x-5\rvert$	16. $\lvert x-10\rvert$	17. 5	18. undefined

Complete MyMathLab Section 8.01 Homework in your Homework notebook.

8.02 Rational Exponents 8.02

Fractional Exponents

In this section we will look at expressions that have fractional exponents. We will begin by looking at a basic definition of $a^{\frac{1}{n}}$.

$$a^{\frac{1}{n}} = \sqrt[n]{a}$$

The denominator of a fractional exponent is equal to the index of the radical.

$8^{\frac{1}{3}}$ - means the cube root of 8 = $\sqrt[3]{8} = 2$

$9^{\frac{1}{2}}$ - means the square root of 9 = $\sqrt{9} = 3$

EXAMPLE 1

Use radical notation to write the following. Simplify if possible. Note: These instructions mean to make the denominator of the exponent the index of the root.

a. $27^{\frac{1}{3}}$	b. $(-8y^9)^{\frac{1}{3}}$	c. $(16x^{12})^{\frac{1}{4}}$	d. $5rs^{\frac{1}{2}}$
$\sqrt[3]{27}$	$\sqrt[3]{(-8y^9)}$	$\sqrt[4]{(16x^{12})}$	*Exponent belongs only to the s.*
3	$-2y^3$	$2x^3$	$5\,r\sqrt{s}$

Courtesy of Fotolia

What happens if the numerator is not 1?

Suppose I have a fractional exponent of $\frac{2}{3}$?

$$a^{\frac{m}{n}} = \sqrt[n]{a^m} = \left(\sqrt[n]{a}\right)^m$$

The denominator of a fractional exponent is equal to the index of the radical
The **numerator** of a fractional exponent is equal to the **power**.

8.02 Rational Exponents **8.02**

Notice with these exponents you have two things to do. Take a root and raise to a power. So, the first question you probably have is "Which one should I do first?" The answer is….. it doesn't matter. Usually it is easier if you take the root first because it makes the number smaller! Then raise that answer to the power.

Look at the example below which we will work both ways.

Raise to Power then evaluate Root		Evaluate Root and then Raise to Power	
$27^{\frac{2}{3}}$		$27^{\frac{2}{3}}$	
$\sqrt[3]{27^2}$	Express in radical notation.	$\left(\sqrt[3]{27}\right)^2$	Express in radical notation
$\sqrt[3]{729}$	Square 27.	3^2	Take cube root of 27 first.
9	Take the cube root.	9	Square 3
		***This is the simplest method.	

Avoiding Common Errors: When we raise a number to a power, only the number beside the exponent is the base. Only that number is raised to the power. If you are to raise a negative number to a power then that negative number must be in parentheses!

Let's compare $-16^{1/4}$ and $(-16)^{1/4}$

$-16^{1/4}$ $-\sqrt[4]{16}$ -2	Only 16 is being raised to the power! Convert to radical notation and then take the 4th root of 16.	$(-16)^{1/4}$ $\sqrt[4]{-16}$ *undefined*	This time -16 is being raised to the power. Convert to radical notation and then take the 4th root of -16. Why is this one undefined? Remember that you can only take *even-numbered roots* (square root, 4th root, etc.) of *positive numbers.*

8.02 Rational Exponents 8.02

YOUR TURN

Use radical notation to write the following. Then simplify if possible. *(Note these instructions.)*

1. $49^{\frac{3}{2}}$	2. $-27^{\frac{2}{3}}$	3. $\left(\frac{1}{9}\right)^{\frac{3}{2}}$	4. $\left(81x^{8}\right)^{\frac{1}{2}}$
5. $(-8)^{\frac{2}{3}}$	6. $2mn^{\frac{1}{3}}$	7. $27^{\frac{4}{3}}$	8. $16^{\frac{3}{4}}$

Courtesy of Fotolia

I hate to ask but . . . . what if the exponents are negative?

A number raised to a *negative* exponent is defined to be the **reciprocal** of that number with a positive exponent. MOVE its LOCATION!

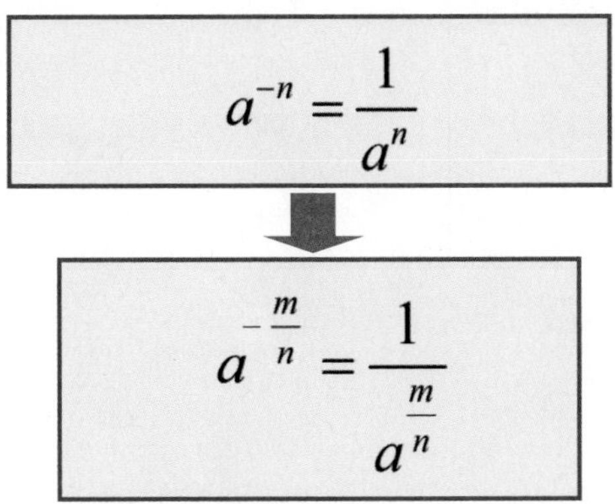

$$a^{-n} = \frac{1}{a^{n}}$$

$$a^{-\frac{m}{n}} = \frac{1}{a^{\frac{m}{n}}}$$

8.02 Rational Exponents 8.02

EXAMPLE 2

Write each expression with a positive exponent, and then simplify. *(Note these instructions.)*

Here's the Plan:

- Rewrite negative exponent as a positive exponent by taking the reciprocal of the base.
- Take the **root** of the base – the **denominator** of the fractional exponent.
- Raise that answer to the **power** – the **numerator** of the fractional exponent.

Follow the directions, write with a positive exponent first.

a.	b.	c.
$16^{-\frac{3}{4}}$ 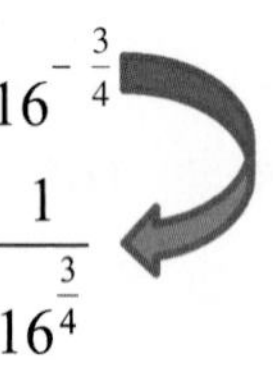 $\frac{1}{16^{\frac{3}{4}}}$ $\frac{1}{\left(\sqrt[4]{16}\right)^3}$ $\frac{1}{(2)^3}=\frac{1}{8}$	$27^{-\frac{2}{3}}$ $\frac{1}{27^{\frac{2}{3}}}$ $\frac{1}{\left(\sqrt[3]{27}\right)^2}$ $\frac{1}{(3)^2}=\frac{1}{9}$	$\frac{1}{(-8)^{-\frac{2}{3}}}$ $(-8)^{\frac{2}{3}}$ $\left(\sqrt[3]{(-8)}\right)^2$ $(-2)^2$ 4

YOUR TURN

Write each expression with a positive exponent, and then simplify. *(Note these instructions.)*

9. $81^{-\frac{3}{4}}$	10. $\frac{1}{8^{-\frac{5}{3}}}$	11. $(-64)^{-\frac{2}{3}}$

8.02 Rational Exponents 8.02

Using Exponent Rules with Rational Exponents

In DMA 060 we learned the rules for exponents. Let's review those.

Product Rule	Multiply the numbers and add the exponents $(2x^3)(3x^5)=6x^8$
Power Rule	Raise numbers to power and multiply the exponents $(3x^2)^3=27x^6$
Quotient Rule	Divide numbers and subtract exponents $\frac{12x^9}{3x^3}=4x^6$

EXAMPLE 3

Use the exponent rules to simplify. Write answers with positive exponents only. *Read the instructions carefully.*

1. $b^{\frac{1}{3}}\cdot b^{\frac{5}{3}}$ Product Rule for Exponents $b^{\frac{1}{3}+\frac{5}{3}}$ $b^{\frac{6}{3}}$ Add the exponents. b^2 Reduce the fractional exponent.	2. $\frac{6a^{\frac{1}{2}}}{a^{\frac{5}{2}}}$ Quotient Rule $6a^{\frac{1}{2}-\frac{5}{2}}$ Subtract Exponents $6a^{-2}$ $\frac{6}{a^2}$ Positive Exponents only
3. $\left(\frac{s^{\frac{1}{2}}t^2}{w^{\frac{3}{4}}}\right)^4$ Power Rule for Exponents $\frac{\left(s^{\frac{1}{2}}\right)^4(t^2)^4}{\left(w^{\frac{3}{4}}\right)^4}$ Multiply the Exponents $\frac{s^2t^8}{w^3}$	4. $\frac{x^{\frac{3}{5}}\bullet x^{\frac{3}{5}}}{x^{\frac{1}{5}}}$ $\frac{x^{\frac{3}{5}+\frac{3}{5}}}{x^{\frac{1}{5}}}=\frac{x^{\frac{6}{5}}}{x^{\frac{1}{5}}}$ Product Rule $x^{\frac{5}{5}}$ Quotient Rule x Simplify exponent

YOUR TURN

Use the exponent rules to simplify. Write answers with positive exponents only. *(Note these instructions.)*

12. $a^{\frac{2}{3}}a^{\frac{5}{3}}$	13. $\dfrac{b^{\frac{1}{2}}b^{\frac{3}{4}}}{-b^{\frac{1}{4}}}$
14. $\dfrac{\left(y^3z\right)^3}{\left(y^{\frac{1}{2}}y^{\frac{1}{3}}\right)^6}$	15. $\dfrac{\left(3x^{\frac{1}{4}}y^{\frac{1}{2}}\right)^4}{x^4y}$

8.02 Rational Exponents 8.02

Use rational exponents to simplify radical expressions.

We can simplify radical expressions by first writing the expression with rational exponents. We would then use our exponent rules to simplify and then convert back to radical notation.

THE PLAN

- Write the radical with rational exponents. (fractions as exponents)
- Use exponent rules to simplify.
- Convert back to radical notation.

EXAMPLE 4

Use rational exponent to write as a single radical expression. Read the instructions carefully. *(Note these instructions. They are your hint about how to start the problem.)*

1. $\sqrt[9]{x^6 y^3}$	2. $\dfrac{\sqrt[3]{b^2}}{\sqrt[4]{b}}$
$x^{\frac{6}{9}} y^{\frac{3}{9}}$ Rewrite expression using rational exponents.	$\dfrac{b^{\frac{2}{3}}}{b^{\frac{1}{4}}}$ Rewrite expression using rational exponents. Then apply the Quotient Rule.
$x^{\frac{2}{3}} y^{\frac{1}{3}}$ Simplify fractions.	$b^{\frac{2}{3}-\frac{1}{4}} = b^{\frac{2}{3}\cdot\frac{4}{4}-\frac{1}{4}\cdot\frac{3}{3}} = b^{\frac{8}{12}-\frac{3}{12}}$ Make a common denominator.
$\sqrt[3]{x^2 y}$ Rewrite as a radical expression.	$b^{\frac{5}{12}}$
	$\sqrt[12]{b^5}$ Rewrite as a radical expression

In the above examples the bases were the same. Remember that to use the exponent rules, we must have the same base. If you look at an example of the product rule notice the bases: $y^2 \cdot y^3 = y^5$. For these examples, the product rule says " keep the like base and add the exponents".

8.02 Rational Exponents 8.02

Let's look at an example where we have different bases. If we have different bases, we cannot use the exponent rules. We need another plan!!

EXAMPLE 5

Use rational exponents to write as a single radical expression. Read the instructions carefully.

$\sqrt[3]{2}\cdot\sqrt{5}$

$2^{\frac{1}{3}}\cdot 5^{\frac{1}{2}}$ Note the radicands become bases which are not like terms so we can't add the exponents. ➡ NEW PLAN: Write the exponents so they have the *same denominator.*

$2^{\frac{1}{3}\cdot\frac{2}{2}}\cdot 5^{\frac{1}{2}\cdot\frac{3}{3}}$ Find the common denominator of the exponents.

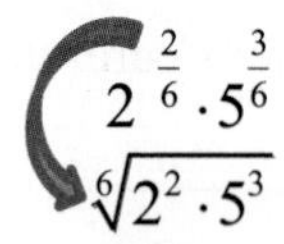

$2^{\frac{2}{6}}\cdot 5^{\frac{3}{6}}$ The common denominator becomes the index.

$\sqrt[6]{2^2\cdot 5^3}$ Exponent's numerators become the power on the base.

$\sqrt[6]{4\cdot 125}$ BE CAREFUL! Raise the base to the power first, then multiply the results.

$\sqrt[6]{500}$

COMMON MISTAKE TO AVOID!!!

$2^{\frac{1}{3}}\bullet 5^{\frac{1}{2}} \neq 10^{\frac{5}{6}}$

You cannot use the product rule if the bases are different!!!

YOUR TURN

Use rational exponents to write as a single radical expression. *Read the instructions carefully.*

16. $\sqrt[8]{x^2y^6}$	17. $\sqrt[8]{x^4}$	18. $\dfrac{\sqrt[5]{m^4}}{\sqrt[3]{m^2}}$
19. $\sqrt{2}\cdot\sqrt[5]{3}$	20. $\sqrt{x}\cdot\sqrt[4]{x}$	

8.02 Rational Exponents 8.02

YOUR TURN Answers to Section 8.02

1. 343
2. –9
3. $\frac{1}{27}$
4. $9x^4$
5. 4
6. $2m\sqrt[3]{n}$
7. 81
8. 8
9. $\frac{1}{27}$
10. 32
11. $\frac{1}{16}$
12. $a^{\frac{7}{3}}$
13. $-b$
14. y^4z^3
15. $\frac{81y}{x^3}$
16. $\sqrt[4]{xy^3}$
17. $\sqrt{x}$
18. $\sqrt[15]{m^2}$
19. $\sqrt[10]{288}$
20. $\sqrt[4]{x^3}$

Complete MyMathLab Section 8.02 Homework in your Homework notebook.

MULTIPLYING and DIVIDING RADICALS with LIKE ROOTS:

Product Rule for Radicals

$$\sqrt[n]{a} \cdot \sqrt[n]{b} = \sqrt[n]{ab}$$

Notice that to use the product rule, the INDEX has to be the same!

EXAMPLE 1

Simplify.

a. $\sqrt{5} \cdot \sqrt{2}$

$\sqrt{10}$

b. $\sqrt[3]{2} \cdot \sqrt[3]{4}$

$\sqrt[3]{8} = 2$

c. $\sqrt[4]{3x^2} \cdot \sqrt[3]{2x}$

Since indices are different we cannot use the product rule to simplify.

d. $\sqrt{7} \cdot \sqrt{7}$

$\sqrt{49} = 7$

Quotient Rule for Radicals

$$\sqrt[n]{\frac{a}{b}} = \frac{\sqrt[n]{a}}{\sqrt[n]{b}}$$

This rule can be used from right to left or left to right. In other words, if you have a single fraction underneath a radical, you can separate into two radicals to simplify. OR, if you have two radicals with the same index, you can put them together underneath one radical to simplify.

$$\sqrt[n]{\frac{a}{b}} = \frac{\sqrt[n]{a}}{\sqrt[n]{b}}$$

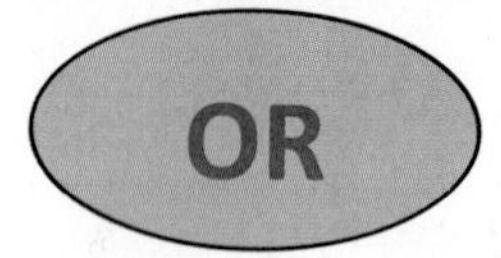

$$\frac{\sqrt[n]{a}}{\sqrt[n]{b}} = \sqrt[n]{\frac{a}{b}}$$

Begin with a single fraction and separate into two radicals to simplify.

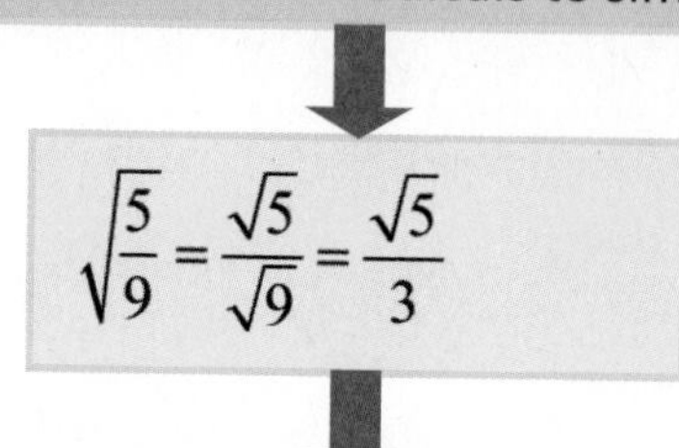

$$\sqrt[3]{\frac{4}{27x^3}} = \frac{\sqrt[3]{4}}{\sqrt[3]{27x^3}} = \frac{\sqrt[3]{4}}{3x}$$

Begin with two radicals and combine in one radical to simplify.

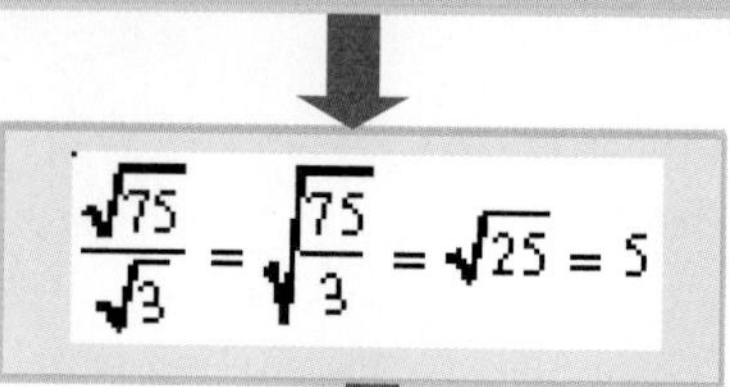

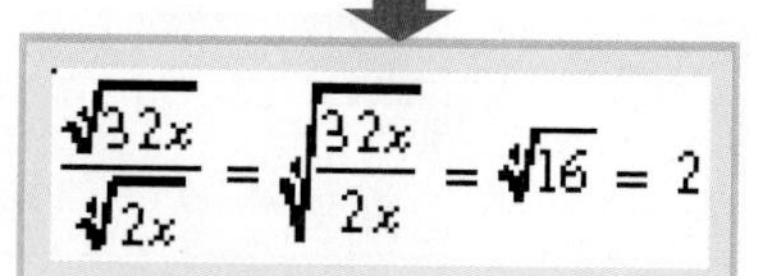

8.03 Simplifying Radical Expressions 8.03

YOUR TURN

Simplify.

1. $\sqrt{\frac{x}{9}}$	**2.** $\sqrt{2}\cdot\sqrt{8}$
3. $\sqrt[3]{4}\cdot\sqrt[3]{16}$	4. $\frac{\sqrt[5]{64}}{\sqrt[5]{2}}$ Remember, reducing is NOT the same as taking the root.
5. $\sqrt[4]{\frac{5}{16y^4}}$	6. $\sqrt{3}\cdot\sqrt{5x}$
7. $\sqrt[3]{\frac{2}{x}}\cdot\sqrt[3]{\frac{9}{y}}$	Radicals are fun!

8.03 Simplifying Radical Expressions 8.03

Simplifying Radicals with Imperfect Roots

We can use the product rule to simplify radicals!

Let's begin by looking at square roots.

PERFECT SQUARES	
$1^2 = 1$	$11^2 = 121$
$2^2 = 4$	$12^2 = 144$
$3^2 = 9$	$13^2 = 169$
$4^2 = 16$	$14^2 = 196$
$5^2 = 25$	$15^2 = 225$
$6^2 = 36$	$16^2 = 256$
$7^2 = 49$	$17^2 = 289$
$8^2 = 64$	$18^2 = 324$
$9^2 = 81$	$19^2 = 361$
$10^2 = 100$	$20^2 = 400$
$x^2, x^4, x^6, x^8, \ldots$	[Even Exponents]

Courtesy of Fotolia

THE PLAN

- Find the largest perfect square from the chart on the right that will divide evenly into the radicand.
- Write the radicand as a product of this perfect square.
- Remove the perfect square from the radical.

If the radical is not a perfect square, $\sqrt{radical}$

Make/split into a $\sqrt{perfect\ square\ factor} \cdot \sqrt{leftover\ radical}$

Take the $root \cdot \sqrt{leftover\ radical}$

Hint: remove the radical symbol after you take the root.

EXAMPLE 2

Simplify.

$\sqrt{18}$	Find the **largest perfect square** factor that will divide evenly into 18. This is 9.
$\sqrt{9 \cdot 2}$	Rewrite 18 as the product of $9 \bullet 2$.
$\sqrt{9} \cdot \sqrt{2}$	Use the product rule to split the radical. $\sqrt{\textit{perfect square}\ \text{factor}} \cdot \sqrt{\textit{leftover radical}}$ Remember, when you take the root of the 1st piece, the radical sign goes away.
$3\sqrt{2}$	Simplify the **"perfect square radical"**. $\textit{root} \cdot \sqrt{\textit{leftover radical}}$
$3\sqrt{2}$	

EXAMPLE 3

Simplify.

$\sqrt{48}$	Find the **largest perfect square** factor that will divide evenly into 48. This is 16.
$\sqrt{16 \cdot 3}$	Rewrite 48 as the product of $16 \bullet 3$.
$\sqrt{16} \cdot \sqrt{3}$	Use the product rule to split the radical. $\sqrt{\textit{perfect square}\ \text{factor}} \cdot \sqrt{\textit{leftover radical}}$
$4 \cdot \sqrt{3}$	Simplify the **"perfect square radical"**. $\textit{root} \cdot \sqrt{\textit{leftover radical}}$
$4\sqrt{3}$	

EXAMPLE 4

What if there is a number outside the radical?

$5\sqrt{18}$	The number outside the radical is being multiplied by the radical.
$5 \cdot \sqrt{18}$	Find the **largest perfect square** factor that will divide evenly into 18.
$5 \cdot \sqrt{9 \cdot 2}$	Use the product rule to split the radical. $\sqrt{\textit{perfect square factor}} \cdot \sqrt{\textit{leftover radical}}$
$5 \cdot \sqrt{9} \cdot \sqrt{2}$	Simplify the **"perfect square radical"**. $\textit{root} \cdot \sqrt{\textit{leftover radical}}$
$5 \cdot 3 \cdot \sqrt{2}$	Finally multiply the numbers outside the radical.
$15\sqrt{2}$	Finished!

8.03 Simplifying Radical Expressions 8.03

What if our radicand has variables and exponents? Remember that to be a perfect root, the exponent is divisible by the index. So, if we have a square root with an index of 2, all perfect square variables must have even exponents.

EXAMPLE 5

Simplify.

$\sqrt{50x^7}$ The largest perfect square that will divide evenly into 50 is 25.

$\sqrt{25 \bullet 2 \bullet x^6 \bullet x}$ Separate the exponents into the largest even number.

$\sqrt{25 \bullet x^6} \bullet \sqrt{2 \bullet x}$ Use product rule to split into a **"perfect square radical"**.

$5 \bullet x^3 \bullet \sqrt{2 \bullet x}$ Simplify the "perfect square radical".

$5x^3\sqrt{2x}$

EXAMPLE 6

Simplify.

$\sqrt{128x^8y^{11}}$ The largest perfect square that will divide evenly into 128 is 64.

$\sqrt{64 \bullet 2 \bullet x^8 \bullet y^{10} \bullet y}$ Separate the exponents into the largest even number.

$\sqrt{64 \bullet x^8 \bullet y^{10}} \bullet \sqrt{2 \bullet y}$ Use product rule to split into a **"perfect square radical"**.

$8 \bullet x^4 \bullet y^5 \bullet \sqrt{2 \bullet y}$ Simplify the "perfect square radical".

$8x^4y^5\sqrt{2y}$

YOUR TURN

Simplify.

8. $\sqrt{27x^5}$	9. $\sqrt{108xy^7}$
10. $3\sqrt{96}$	11. $5\sqrt{12x^3y^{16}}$

8.03 Simplifying Radical Expressions 8.03

Now let's look at **CUBE ROOTS**. The process is exactly the same, except you use the perfect cube root chart as opposed to the perfect square roots. Since we are working with cube roots, we will be putting the variables raised to powers that are multiples of three in the "perfect cube factor."

EXAMPLE 7

Simplify. $\sqrt[3]{16} = \sqrt[3]{8 \cdot 2} = \sqrt[3]{8} \cdot \sqrt[3]{2} = 2\sqrt[3]{2}$

$\sqrt{\textit{perfect cube factor}} \cdot \sqrt{\textit{leftover radical}}$

The largest perfect cube that will divide evenly into 16 is 8.

EXAMPLE 8

Simplify.

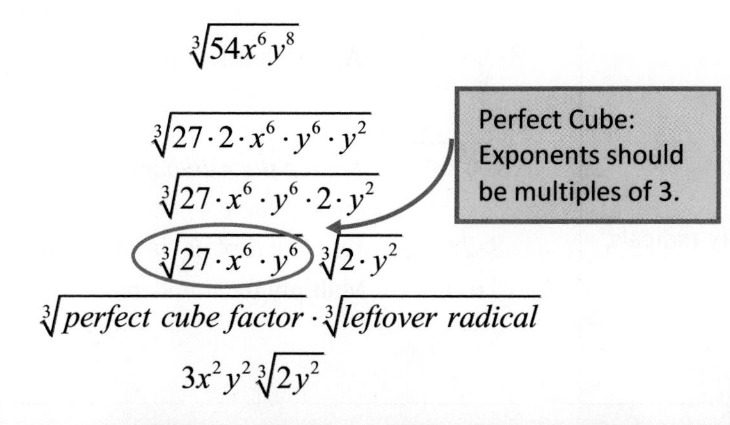

Here is the "perfect cube" list.

PERFECT CUBES	
$1^3 = 1$ $2^3 = 8$ $3^3 = 27$ $4^3 = 64$ $5^3 = 125$ $6^3 = 216$ $7^3 = 343$ $8^3 = 512$	x^3, x^6, x^9, x^{12}... *Exponents are multiples of three.*

YOUR TURN

Simplify.

12. $\sqrt[3]{24a^3b^{10}}$	13. $\sqrt[3]{125x^3}$

8.03 Simplifying Radical Expressions

Quotient Rule for Radicals

$$\sqrt[n]{\frac{a}{b}} = \frac{\sqrt[n]{a}}{\sqrt[n]{b}}$$

Using the Quotient Rule to Simplify

EXAMPLE 9

Simplify.

a. $\dfrac{\sqrt{x^7y^{12}}}{\sqrt{x^3y^8}}$

$\sqrt{\dfrac{x^7y^{12}}{x^3y^8}}$ Apply Quotient Rule.

$\sqrt{x^4y^4}$ Subtract the exponents.

x^2y^2 Take the root or simplify radicals.

b. $\dfrac{8\sqrt[3]{32m^7}}{\sqrt[3]{4m}}$

$8\cdot\sqrt[3]{\dfrac{32m^7}{4m}}$ Apply Quotient Rule.

$8\cdot\sqrt[3]{8m^6}$ ***Reduce the numbers & subtract the exponents.***

$8\cdot 2m^2$ Take the root or simplify radicals.

$16m^2$ Multiply the numbers.

c. $\dfrac{\sqrt[4]{96a^{10}b^6}}{\sqrt[4]{3a^2b^6}}$ In this example, it is best to divide the radicands and then simplify. *Always divide before your simplify, if you can.*

$\sqrt[4]{\dfrac{96a^{10}b^6}{3a^2b^6}}$ Apply the quotient rule.

$\sqrt[4]{32a^8}$ Divide the coefficients and subtract the exponents on the like bases.

$\sqrt[4]{16\cdot 2\cdot a^8}$ Factor the radicand so that you have factored the perfect fourth root of 32.

$\sqrt[4]{16a^8}\cdot\sqrt[4]{2}$ Now use the product rule to split into a perfect square radical.

$2a^2\sqrt[4]{2}$ Simplify the perfect fourth root.

8.03 Simplifying Radical Expressions 8.03

YOUR TURN

Simplify.

14. $\dfrac{\sqrt{98z}}{3\sqrt{z}}$	15. $\dfrac{\sqrt[3]{40x^5y^7}}{\sqrt[3]{5y}}$

YOUR TURN Answers to Section 8.03.

1. $\dfrac{\sqrt{x}}{3}$
2. 4
3. 4
4. 2
5. $\dfrac{\sqrt[4]{5}}{2y}$
6. $\sqrt{15x}$
7. $\sqrt[3]{\dfrac{18}{xy}}$
8. $3x^2\sqrt{3x}$
9. $6y^3\sqrt{3xy}$
10. $12\sqrt{6}$
11. $10xy^8\sqrt{3x}$
12. $2ab^3\sqrt[3]{3b}$
13. 5x
14. $\dfrac{7\sqrt{2}}{3}$
15. $2xy^2\sqrt[3]{x^2}$

Complete MyMathLab Section 8.03 Homework in your Homework notebook.

8.04 Adding, Subtracting, and Multiplying Radical Expressions 8.04

ADD OR SUBTRACT RADICAL EXPRESSIONS

We can **only add and subtract LIKE RADICALS!** Like radicals have the same index and the same radicand.(Remember: The radicand is the number under the radical sign.)

LIKE Radicals: Same index and same radicand	**UNLIKE Radicals:**
$3\sqrt{7}$ and $\sqrt{7}$	$\sqrt{a}$ and $\sqrt[3]{a}$ ⟶ Different index
$\sqrt[3]{a}$ and $2\sqrt[3]{a}$	$\sqrt[3]{b}$ and $\sqrt[3]{a}$ ⟶ Different radicand

Adding and subtracting radicals is **similar to combining like terms**. When we combine like terms such as $2x+3x$, we simply **add/subtract** the coefficients and keep the variable. $2x+3x=5x$

To add/subtract radicals, we add/subtract the coefficients and keep the radical!

EXAMPLE 1

Simplify. $7\sqrt[3]{4} - 2\sqrt[3]{4} + 5\sqrt[3]{4}$

$7\sqrt[3]{4} - 2\sqrt[3]{4} + 5\sqrt[3]{4}$ Notice these radicals have the same indices.

$(7-2+5)\sqrt[3]{4}$ Add/subtract the numbers in front of the radicands.

$10\sqrt[3]{4}$ The radicand stays the same.

Often, at first glance, it appears that radicals are unlike, when in fact, if we simplify one or more of them, they are alike.

EXAMPLE 2

Simplify. $\sqrt{50}+\sqrt{18}$

$\sqrt{25\cdot 2} + \sqrt{9\cdot 2}$ Rewrite radicands using perfect square factors.

$5\sqrt{2}+3\sqrt{2}$ Factor out perfect square roots.

$(5+3)\sqrt{2}$ Add the numbers in front of the radicals.

$8\sqrt{2}$ Simplify.

8.04 Adding, Subtracting, and Multiplying Radical Expressions 8.04

EXAMPLE 3

Simplify. $5\sqrt{20}+\sqrt{72}-3\sqrt{45}$ — Simplify each of the radicals. Look for perfect square factors in each of the radicals.

$5\cdot\sqrt{4\cdot 5}+\sqrt{36\cdot 2}-3\cdot\sqrt{9\cdot 5}$ — Use the product rule to rewrite each radical.

$5\cdot\sqrt{4}\cdot\sqrt{5}+\sqrt{36}\cdot\sqrt{2}-3\cdot\sqrt{9}\cdot\sqrt{5}$ — Take the square root of the perfect squares.

$5\cdot 2\cdot\sqrt{5}+6\cdot\sqrt{2}-3\cdot 3\cdot\sqrt{5}$ — Multiply the numbers outside of each of the radicals.

$10\cdot\sqrt{5}+6\cdot\sqrt{2}-9\cdot\sqrt{5}$ — Now combine the like radicals.

$\boxed{\sqrt{5}+6\sqrt{2}}$

Courtesy of Fotolia

Avoiding Common Errors : When we add or subtract radicals, the end result is that the numerical coefficients are added and the radical factor is unchanged.

CORRECT	INCORRECT
$\sqrt{3}+\sqrt{3}=2\sqrt{3}$	~~$\sqrt{3}+\sqrt{3}=\sqrt{6}$~~

EXAMPLE 4

Simplify. $8\sqrt{x^3y^2}-3y\sqrt{x^3}$ — Radicands are different. Simplify the radicals first.

$8\sqrt{x^2y^2\cdot x}-3y\sqrt{x^2\cdot x}$ — Simplify $\sqrt{x^3y^2}=xy\sqrt{x}$ and $\sqrt{x^3}=x\sqrt{x}$

$8xy\sqrt{x}-3xy\sqrt{x}$ — Subtract coefficients 8 – 3, leave the radical the SAME.

$5xy\sqrt{x}$

8.04 Adding, Subtracting, and Multiplying Radical Expressions 8.04

EXAMPLE 5

Simplify. $\sqrt[3]{\frac{10}{8}}-\sqrt[3]{\frac{10}{125}}$

Radicands are different. Simplify the radicals first.

$$\frac{\sqrt[3]{10}}{\sqrt[3]{8}}=\frac{\sqrt[3]{10}}{2} \quad \text{and} \quad \frac{\sqrt[3]{10}}{\sqrt[3]{125}}=\frac{\sqrt[3]{10}}{5}$$

$\frac{\sqrt[3]{10}}{2}-\frac{\sqrt[3]{10}}{5}$

Remember that to add fractions we must have a common denominator.

$\frac{5\sqrt[3]{10}}{10}-\frac{2\sqrt[3]{10}}{10}$

Can we cancel out the 10's? NO. Remember that in the numerator, 10 is under the radical. However, in the denominator 10 is outside the radical.

$\frac{3\sqrt[3]{10}}{10}$

Subtract, 5 – 2, KEEP the radical the SAME and also KEEP the common denominator.

YOUR TURN

Simplify each expression.

1. $6\sqrt{5}-2\sqrt{5}+8\sqrt{5}$	2. $2\sqrt{24}+4\sqrt{54}+5\sqrt{96}$
3. $\sqrt[3]{8x^4}+3x\sqrt[3]{x}-\sqrt{16x^5}$	

8.04 Adding, Subtracting, and Multiplying Radical Expressions 8.04

MULTIPLY RADICAL EXPRESSIONS

To multiply or divide radicals they must have the same index.

$\sqrt[3]{2} \cdot \sqrt[3]{4}$ Each radical has an index of 3 therefore I multiply the radicands (inside parts) together.
$\sqrt[3]{8} = 2$

$\sqrt[3]{2} \cdot \sqrt{5}$ The first radical has an index of 3 while the second radical has an index of 2.
Since they do not have the same index, I cannot multiply them together.

Don't worry!  There is a way to simplify. Remember rational exponent rules?

Here is the rule for multiplying radicals:

Product Rule for Radicals

$$\sqrt[n]{a} \cdot \sqrt[n]{b} = \sqrt[n]{ab}$$

Multiplying radical expressions is very much like multiplying polynomials. The same methods apply such as the distributive property.

EXAMPLE 6

Simplify. $\sqrt{2}(6+\sqrt{10})$

$\sqrt{2}(6)+\sqrt{2}\left(\sqrt{10}\right)$ Use the distributive property.

$6\sqrt{2}+\sqrt{20}$ Multiply.

$6\sqrt{2}+\sqrt{4\cdot 5}$ Find any perfect square factors for the radicand(s).

$6\sqrt{2}+2\sqrt{5}$ Factor out the perfect square factors.

[We cannot add these two terms because we do not have like radicals!!]

8.04 Adding, Subtracting, and Multiplying Radical Expressions 8.04

EXAMPLE 7

Multiply. $\left(2\sqrt{3}-4\sqrt{2}\right)\left(\sqrt{3}+2\sqrt{2}\right)$ ⟶ Use FOIL to multiply "binomials".

$2\sqrt{3}\bullet\sqrt{3} + 2\sqrt{3}\cdot 2\sqrt{2} - 4\sqrt{2}\cdot\sqrt{3} - 4\sqrt{2}\cdot 2\sqrt{2}$ — In each set, multiply numbers outside and also multiply radicands underneath.

$2\sqrt{9} + 4\sqrt{6} - 4\sqrt{6} - 8\sqrt{4}$ — Product Rule for Radicals

$2\bullet 3 + 4\sqrt{6} - 4\sqrt{6} - 8\bullet 2$ — Take a root of perfect square radicals.

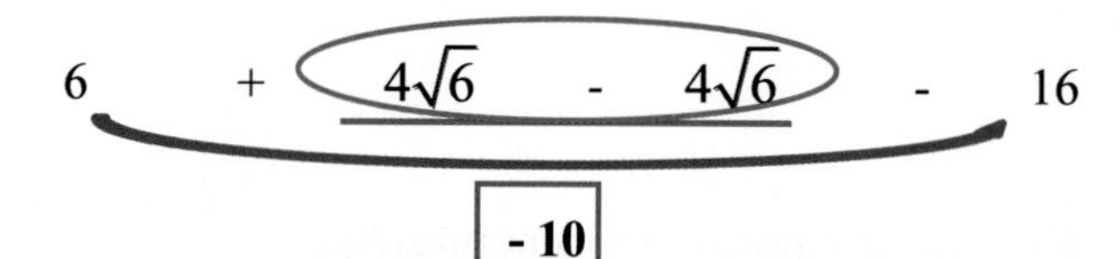

Combine LIKE radicals.

EXAMPLE 8

Multiply.

$\left(5\sqrt{6}+x\right)^2$

$\left(5\sqrt{6}+x\right)\left(5\sqrt{6}+x\right)$ — Rewrite the problem and then FOIL!

$25\sqrt{6}\cdot\sqrt{6} + 5\sqrt{6}\cdot x + 5\sqrt{6}\cdot x + x\cdot x$ — Multiply outsides, then multiply radicands underneath.

$25\sqrt{36} + 10x\sqrt{6}+x^2$

$25\cdot 6+10x\sqrt{6}+x^2$

$150+10x\sqrt{6}+x^2$

EXAMPLE 9

Multiply. $\left(\sqrt[3]{9}-8\right)\left(\sqrt[3]{3}-6\right)$ — Use FOIL to multiply, just like multiplying binomials.

$\sqrt[3]{9}\cdot\sqrt[3]{3}-6\cdot\sqrt[3]{9}-8\cdot\sqrt[3]{3}+8\cdot 6$ — Multiply to remove the parentheses. Remember that you cannot multiply the numbers outside the radical with the numbers inside the radical!

$\sqrt[3]{27}-6\sqrt[3]{9}-8\sqrt[3]{3}+48$ — Take care of the perfect cube, $\sqrt[3]{27}$.

$3-6\sqrt[3]{9}-8\sqrt[3]{3}+48$ — Combine the like terms.

$51-6\sqrt[3]{9}-8\sqrt[3]{3}$ — You cannot combine $6\sqrt[3]{9}-8\sqrt[3]{3}$ because the radicands (numbers in the radical) are not the same.

8.04 Adding, Subtracting, and Multiplying Radical Expressions 8.04

EXAMPLE 10

Multiply. $\left(\sqrt{x-4}+2\right)^2$ Remember what a binomial squared means? Multiply the group by itself.

$$\left(\sqrt{x-4}+2\right)\left(\sqrt{x-4}+2\right)$$

$$\left(\sqrt{x-4}\right)\left(\sqrt{x-4}\right)+2\left(\sqrt{x-4}\right)+2\left(\sqrt{x-4}\right)+4$$

$x-4+4\left(\sqrt{x-4}\right)+4$ When you multiply a square root by itself you get the **radicand** itself.

$x-4+4+4\sqrt{x-4}$ Combine like terms.

$x+4\sqrt{x-4}$

YOUR TURN

Find each product.

4. $\left(\sqrt{5}-\sqrt{2}\right)\left(\sqrt{2}+\sqrt{6}\right)$	5. $\left(1-\sqrt{5}\right)\left(1+\sqrt{5}\right)$
6. $\left(2\sqrt{y}+5\right)\left(2\sqrt{y}-5\right)$	7. $\sqrt[3]{4}\cdot\sqrt[3]{16}$

8.04 Adding, Subtracting, and Multiplying Radical Expressions 8.04

YOUR TURN

Find each product.

8. $\sqrt[3]{5}\cdot\sqrt{5}$	9. $\left(3\sqrt{6}+x\right)^2$
10. $\sqrt{2}\left(\sqrt{6}-3\sqrt{2}\right)$	

YOUR TURN **Answers to Section 8.04.**

1. $12\sqrt{5}$
2. $36\sqrt{6}$
3. $5x\sqrt[3]{x}-4x^2\sqrt{x}$
4. $\sqrt{10}+\sqrt{30}-2\sqrt{3}-2$
5. -4
6. $4y-25$
7. 4
8. Cannot be multiplied.
9. $54+6x\sqrt{6}+x^2$
10. $2\sqrt{3}-6$

Complete MyMathLab Section 8.04 Homework in your Homework notebook.

8.05 Rationalizing Denominators of Radical Expressions 8.05

When working with radical expressions, it is often useful to remove the radical from the denominator. This is called "rationalizing the denominator."

We will need to look at rationalizing the denominator in two situations.

- **Rationalizing a denominator that has one term.**
- **Rationalizing a denominator that has two terms.**

RATIONALIZING A DENOMINATOR – ONE TERM

Remember that the *n*th root of a perfect *n*th power simplifies completely. For example:

$\sqrt{25x^2} = 5x \longrightarrow$ Since $25x^2$ is a perfect square – the square root of a perfect square simplifies.

$\sqrt[3]{8y^3} = 2y \longrightarrow$ Since $8y^3$ is a perfect cube – the cube root of a perfect cube simplifies.

So, the only way to remove a radical is to have a "perfect root."

- If we need to find a square root, we will need to create a perfect square in the denominator.
- If we need to find a cube root, we will need to create a perfect cube in the denominator.
- Keep in mind what makes a perfect root. Here are the charts again!!

PERFECT SQUARES	
$1^2 = 1$	$11^2 = 121$
$2^2 = 4$	$12^2 = 144$
$3^2 = 9$	$13^2 = 169$
$4^2 = 16$	$14^2 = 196$
$5^2 = 25$	$15^2 = 225$
$6^2 = 36$	$16^2 = 256$
$7^2 = 49$	$17^2 = 289$
$8^2 = 64$	$18^2 = 324$
$9^2 = 81$	$19^2 = 361$
$10^2 = 100$	$20^2 = 400$
$x^2, x^4, x^6, x^8, \ldots$	[Even Exponents]

PERFECT CUBES	
$1^3 = 1$	
$2^3 = 8$	$x^3, x^6, x^9, x^{12} \ldots$
$3^3 = 27$	
$4^3 = 64$	Exponents are
$5^3 = 125$	multiples of three
$6^3 = 216$	
$7^3 = 343$	
$8^3 = 512$	

Let's practice making some perfect roots before we begin rationalizing the denominator.

8.05 Rationalizing Denominators of Radical Expressions 8.05

EXAMPLE 1

Instructions: Fill in the missing radicand to make a perfect root.

When you multiply the two radicals together using the product rule, you should have created a perfect square root or a perfect cube root. Use the chart to determine the smallest perfect root you can make using the radical given.

$\sqrt{5} \bullet \sqrt{?} = \sqrt{\textit{perfect square}}$

$\sqrt[3]{4} \bullet \sqrt[3]{?} = \sqrt[3]{\textit{perfect cube}}$

$\sqrt{3x} \bullet \sqrt{?} = \sqrt{\textit{perfect square}}$

$\sqrt[3]{9x^2} \bullet \sqrt[3]{?} = \sqrt[3]{\textit{perfect cube}}$

SOLUTIONS:

$\sqrt{5} \bullet \sqrt{5} = \sqrt{25}$

$\sqrt[3]{4} \bullet \sqrt[3]{2} = \sqrt[3]{8}$

$\sqrt{3x} \bullet \sqrt{3x} = \sqrt{9x^2}$

$\sqrt[3]{9x^2} \bullet \sqrt[3]{3x} = \sqrt[3]{27x^3}$

Keep in mind, that to rationalize we will need to be able to create a "perfect root."

- Look at the radicand you've been given.
- Ask yourself, "What is the smallest square root/cube root I can create from this radicand?" This is when you will use the chart. Square roots – exponents will need to be even.
  Cube roots – exponents will need to be multiples of three.
- Multiply by the appropriate radicand so that when multiplied together, you will have the "perfect root" under the radical.

YOUR TURN

Fill in the missing radicand to make a perfect root.

1. $\sqrt{7} \bullet \sqrt{?} = \sqrt{\textit{perfect square}}$ ________________

2. $\sqrt[3]{2} \bullet \sqrt[3]{?} = \sqrt[3]{\textit{perfect cube}}$ ________________

3. $\sqrt{5y^3} \bullet \sqrt{?} = \sqrt{\textit{perfect square}}$ ________________

4. $\sqrt[3]{3y^2} \bullet \sqrt[3]{?} = \sqrt[3]{\textit{perfect cube}}$ ________________

8.05 Rationalizing Denominators of Radical Expressions 8.05

To rationalize the denominator of a radical expression, multiply the numerator and denominator by an appropriate expression representing "1" to create a perfect root in the denominator. Sometimes it is necessary to simplify the radical before we rationalize.

EXAMPLE 2

Rationalize the denominator. $\frac{\sqrt{6}}{\sqrt{5}}$

$\frac{\sqrt{6}}{\sqrt{5}}$ The denominator is $\sqrt{5}$. We will need to multiply the numerator/denominator by $\frac{\sqrt{5}}{\sqrt{5}}$ to create our perfect square root in the denominator.

$$\frac{\sqrt{6}}{\sqrt{5}} = \frac{\sqrt{6}}{\sqrt{5}} \bullet \frac{\sqrt{5}}{\sqrt{5}} = \frac{\sqrt{30}}{\sqrt{25}} = \frac{\sqrt{30}}{5}$$

Rationalized Denominator

Perfect square root

EXAMPLE 3

Rationalize the denominator. $\frac{9}{\sqrt{28x}}$

$\frac{9}{\sqrt{28x}}$ The denominator is $\sqrt{28x}$. We will need to simplify the radical first.

$$\frac{9}{\sqrt{28x}} = \frac{9}{\sqrt{4 \bullet 7x}} = \frac{9}{\sqrt{4} \bullet \sqrt{7x}} = \frac{9}{2\sqrt{7x}}$$

Simplified Radical

NOW WE WILL RATIONALIZE THE DENOMINATOR. Multiply by factor in red.

$$\frac{9}{2\sqrt{7x}} = \frac{9}{2\sqrt{7x}} \bullet \frac{\sqrt{7x}}{\sqrt{7x}} = \frac{9\sqrt{7x}}{2\sqrt{49x^2}} = \frac{9\sqrt{7x}}{2 \bullet 7x} = \frac{9\sqrt{7x}}{14x}$$

Rationalized Denominator

Perfect square root

Always, always, always simplify the radical before you begin rationalizing!

8.05 Rationalizing Denominators of Radical Expressions 8.05

Now let's look at rationalizing a cube root!!

EXAMPLE 4

Rationalize the denominator. $\frac{\sqrt[3]{7}}{\sqrt[3]{5}}$

The denominator is $\sqrt[3]{5}$. The smallest perfect cube we can create is 125. We will need to multiply the numerator/denominator by $\frac{\sqrt[3]{25}}{\sqrt[3]{25}}$ to create our perfect cube root.

$$\frac{\sqrt[3]{7}}{\sqrt[3]{5}} = \frac{\sqrt[3]{7}}{\sqrt[3]{5}} \cdot \frac{\sqrt[3]{25}}{\sqrt[3]{25}} = \frac{\sqrt[3]{175}}{\sqrt[3]{125}} = \frac{\sqrt[3]{175}}{5}$$

Rationalized Denominator

Perfect cube root

EXAMPLE 5

Rationalize the denominator. $\frac{\sqrt[3]{z^2}}{\sqrt[3]{16y^4}}$

Simplify the radical. $$\frac{\sqrt[3]{z^2}}{\sqrt[3]{16y^4}} = \frac{\sqrt[3]{z^2}}{\sqrt[3]{8 \cdot 2 \cdot y^3 \cdot y}} = \frac{\sqrt[3]{z^2}}{\sqrt[3]{8 \cdot y^3 \cdot 2 \cdot y}} = \frac{\sqrt[3]{z^2}}{\sqrt[3]{8 \cdot y^3} \cdot \sqrt[3]{2 \cdot y}} = \frac{\sqrt[3]{z^2}}{2y\sqrt[3]{2y}}$$

NOW WE WILL RATIONALIZE THE DENOMINATOR

$$\frac{\sqrt[3]{z^2}}{2y\sqrt[3]{2y}} = \frac{\sqrt[3]{z^2}}{2y\sqrt[3]{2y}} \cdot \frac{\sqrt[3]{4y^2}}{\sqrt[3]{4y^2}} = \frac{\sqrt[3]{4y^2z^2}}{2y\sqrt[3]{8y^3}} = \frac{\sqrt[3]{4y^2z^2}}{2y \cdot 2y} = \frac{\sqrt[3]{4y^2z^2}}{4y^2}$$

Perfect cube root

Rationalized Denominator

8.05 Rationalizing Denominators of Radical Expressions 8.05

YOUR TURN

Rationalize the denominators.

5. $\frac{5}{\sqrt{3}}$	6. $\frac{2\sqrt{9}}{\sqrt{20y}}$	7. $\frac{\sqrt[3]{m}}{\sqrt[3]{27n^4}}$

RATIONALIZING A DENOMINATOR – TWO TERMS

To rationalize the denominator that has two terms we must use the conjugates.

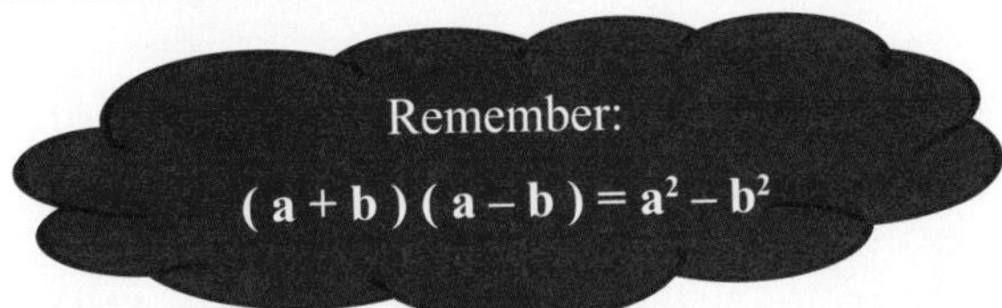

These two expressions $(a + b)(a - b)$ are called conjugates of each other. Conjugates have the **SAME** terms but **DIFFERENT** signs.

Notice what happens when we multiply conjugates that have radicals.

$$\left(2+\sqrt{3}\right)\left(2-\sqrt{3}\right) = 4-2\sqrt{3}+2\sqrt{3}-\sqrt{9}$$

$= 4 - 3$ The two middle terms add out.

$= 1$ We are left with a number with no radical.

So, we can remove a radical with two terms by multiplying by its conjugate.

To rationalize a denominator that is a sum or difference of two terms, we multiply the numerator and the denominator by the conjugate of the denominator.

EXAMPLE 6

Rationalize. $\dfrac{3}{3-\sqrt{5}}$

In order to rationalize the denominator, we must multiply both the numerator and the denominator of the fraction by the conjugate of $3-\sqrt{5}$, which is $3+\sqrt{5}$.

$$\frac{3}{3-\sqrt{5}} = \frac{3}{3-\sqrt{5}} \cdot \frac{(3+\sqrt{5})}{(3+\sqrt{5})}$$ Notice that $\dfrac{3+\sqrt{5}}{3+\sqrt{5}} = 1$

$$\frac{3(3)+3\sqrt{5}}{3^2-\left(\sqrt{5}\right)^2}$$ Distribute the three across the numerator $3+\sqrt{5}$, then multiply the denominators together.

$$\frac{9+3\sqrt{5}}{9-5}$$ Simplify and evaluate radicals.

$$\frac{9+3\sqrt{5}}{4}$$ Simplify the denominator.

**Always remember that the product of the conjugates is a rational number.

EXAMPLE 7

Rationalize the denominator. $\dfrac{2-\sqrt{5}}{3+\sqrt{5}}$

$$\frac{2-\sqrt{5}}{3+\sqrt{5}} = \frac{2-\sqrt{5}}{3+\sqrt{5}} \cdot \frac{3-\sqrt{5}}{3-\sqrt{5}}$$ Multiply the numerator AND denominator by the conjugate of the denominator.

$$\frac{6-2\sqrt{5}-3\sqrt{5}+\left(\sqrt{5}\right)^2}{3^2-\left(\sqrt{5}\right)^2}$$ Use FOIL for the numerator. Multiply the denominators.

$$\frac{6-5\sqrt{5}+5}{9-5}$$ Combine like radicals.

$$\frac{11-5\sqrt{5}}{4}$$ Combine like terms.

8.05 Rationalizing Denominators of Radical Expressions

YOUR TURN

Rationalize the denominators.

8. $\dfrac{6}{\sqrt{8}}$ **(one term)**	9. $\dfrac{12}{\sqrt{3y}}$ **(one term)**
10. $\dfrac{6}{4-\sqrt{3}}$ **(two terms)**	

YOUR TURN Answers to Section 8.01

1. $\sqrt{7}\cdot\sqrt{7} = \sqrt{49} = 7$
2. $\sqrt[3]{2}\cdot\sqrt[3]{4} = \sqrt[3]{8} = 2$
3. $\sqrt{5y^3}\cdot\sqrt{5y} = \sqrt{25y^4} = 5y^2$
4. $\sqrt[3]{3y^2}\cdot\sqrt[3]{9y} = \sqrt[3]{27y^3} = 3y$
5. $\dfrac{5\sqrt{3}}{3}$
6. $\dfrac{3\sqrt{5y}}{5y}$
7. $\dfrac{\sqrt{3mn}}{9n^2}$
8. $\dfrac{3\sqrt{2}}{2}$
9. $\dfrac{4\sqrt{3y}}{y}$
10. $\dfrac{24+6\sqrt{3}}{13}$

Complete MyMathLab Section 8.05 Homework in your Homework notebook.

8.06 Radical Equations and Problem-Solving 8.06

RADICAL EQUATIONS

An equation with one or more radicals containing a variable is called a radical equation.

The equations $\sqrt{x}=5$ and $3+x=\sqrt{x-2}$ are radical equations.

> The basis of solving a radical equation is to eliminate the radical by raising both sides of the equation to a power equal to the index of the radical. Remember $\left(\sqrt[n]{a}\right)^n=a$

To solve the equation $\sqrt[3]{x}=5$, we would cube both sides since the index is 3.

$$\sqrt[3]{x}=5$$
$$\left(\sqrt[3]{x}\right)^3=5^3$$
$$x=125$$

By raising each side of a radical equation to a power equal to the index of the radical, a new equation is produced. However, some of the solutions to this new equation may not be solutions to the original radical equation.

Courtesy of Fotolia

> **All solutions must be checked in the original equation!!!**

The solutions that do not check in the original solution are called **extraneous solutions.**

Extraneous Solutions are Extra Solutions that don't fit the original problem!

Solving a Radical Equation

Step 1: Isolate the radical. If an equation has more than one radical, choose one of the radicals to isolate.
Step 2: Raise each side of the equation to a power equal to the index of the radical.
Step 3: Solve the resulting equation. If the equation still has a radical, repeat steps 1 and 2.
Step 4: Check the potential solutions in the original equation.

8.06 Radical Equations and Problem-Solving 8.06

EXAMPLE 1

Solve and check. $\sqrt{3x-2}=4$ Notice the radical is already by itself. Nothing can come out from underneath until the square root symbol has been removed.

$\left(\sqrt{3x-2}\right)^2=(4)^2$	Square both sides.
$3x-2=16$	Add 2 to both sides.
$3x = 18$	Divide both sides by 3.
$x = 6$	The solution is 6.

All solutions must be checked in the original equation !!!

Check $x = 6$.

$$\sqrt{3x-2}=4$$
$$\sqrt{3(6)-2}=4$$
$$\sqrt{16}=4$$
$$4=4$$

The solution checks so the solution is 6.

EXAMPLE 2

Solve.

$$\sqrt{3x-2}+4=5$$

$\sqrt{3x-2}+4=5$ → Isolate the radical by subtracting 4 from each side.

$$\sqrt{3x-2}=1$$

$\left(\sqrt{3x-2}\right)^2=(1)^2$ → Index is 2 so square both sides.

$3x-2=1$ → Notice this is a linear equation. Solve.

$3x=3$ → Solve the resulting equation for 1 possible solution.

$$x=1$$

Check $x = 1$. $\sqrt{3(1)-2}+4=5$

$$\sqrt{1}+4=5$$
$$1+4=5$$
$$5=5$$

The solution checks so the solution is 1.

EXAMPLE 3

Solve and check. $\sqrt{x-3}+5=x$

$\sqrt{x-3}+5 = x$

$\sqrt{x-3} = x-5$ Isolate the radical by subtracting 5.

$\left(\sqrt{x-3}\right)^2 = (x-5)^2$ Index is two. Square both sides.

$x-3 = x^2-10x+25$ Then, since this is a quadratic equation, get 0 on one side of the equation.

$0 = x^2-11x+28$ Factor the resulting quadratic equation.

$0 = (x-7)(x-4)$

$x-7=0$ $\quad x-4=0$ Set each factor equal to zero and solve.

$x=7$ $\quad x=4$

4 and 7 are potential solutions but must be checked!

Check. $x=7$

$\sqrt{x-3}+5=x$

$\sqrt{7-3}+5=7$

$\sqrt{4}+5=7$

$7=7$ ✓

$x=4$

$\sqrt{x-3}+5=x$

$\sqrt{4-3}+5=4$

$\sqrt{1}+5=4$

$6\neq 4$ 

The solution is 7.

EXAMPLE 4

Solve. $\sqrt[3]{w-1}-2=2$

$\sqrt[3]{w-1}=4$ ⟶ Isolate the radical.

$\left(\sqrt[3]{w-1}\right)^3=(4)^3$ ⟶ Index is 3 – Cube both sides.

$w-1=64$ ⟶ Simplify.

$w=65$

Check: $w = 65$ in the original equation. ⟶ $\sqrt[3]{w-1}-2=2$

$\sqrt[3]{65-1}-2=2$

$\sqrt[3]{64}-2=2$

$4-2=2$

$2=2$ ✓ The solution is 2

EXAMPLE 5

Solve $\sqrt{x+9}=-7$.

Let's think about this for a minute!

This statement says that a square root is equal to a negative number!

Can this be???? . . . .***NO!***

OK then, what is our answer?? NO SOLUTION

Always, always, always work SMART, not HARD!

Courtesy of Fotolia

This example has two radicals!!!

EXAMPLE 6

Solve. $\sqrt{3x+1}-\sqrt{5x}=-1$

$\sqrt{3x+1} = \sqrt{5x}-1$ Isolate a radical by subtracting one radical.

$\left(\sqrt{3x+1}\right)^2 = \left(\sqrt{5x}-1\right)^2$ Index is 2 – Square both sides.

Steps to Isolate second radical!!

$3x+1=5x-2\sqrt{5x}+1$

$3x=5x-2\sqrt{5x}$ Subtract 1 from both sides.

$-2x=-2\sqrt{5x}$ Subtract 5x from both sides.

$x=\sqrt{5x}$ Divide both sides by -2.

$x^2=5x$ Square both sides. Solve.

$x^2-5x=0$ Subtract $5x$ from both sides.

$x(x-5)=0$ Factor.

$x=0$ or $x=5$

Check: Potential solutions are $x=0$ and $x=5$. Now we need to check each of these solutions.

$\sqrt{3(0)+1}-\sqrt{5(0)}=-1$

$\sqrt{1}-\sqrt{0}=1$

$1=-1$ ✗

$\sqrt{3(5)+1}-\sqrt{5(5)}=-1$

$\sqrt{15+1}-\sqrt{25}=-1$

$\sqrt{16}-5=-1$

$4-5=-1$

$-1=-1$ ✓

The solution is 5

8.06 Radical Equations and Problem-Solving 8.06

YOUR TURN

Solve each equation. **Remember to check your solution(s).**

1. $\sqrt{3b-2}+19=24$	2. $\sqrt[3]{2x-3}-2=5$
3. $y+\sqrt{y-2}=8$	4. $\sqrt{8+b}=2+\sqrt{b}$

8.06 Radical Equations and Problem-Solving 8.06

YOUR TURN

5. $\sqrt{4x-8}=\sqrt{8-4x}$	6. $\sqrt{2y+3}=-4$ Work smart, not hard!

APPLICATIONS: Using the Pythagorean Theorem

The Pythagorean Theorem applies to right triangles – triangles that have a right angle. The theorem states that the length of the hypotenuse squared equals the sum of the lengths of each of the legs squared.

Hypotenuse – Slanted Side opposite right angle.
Legs – Two sides that make up right angle.

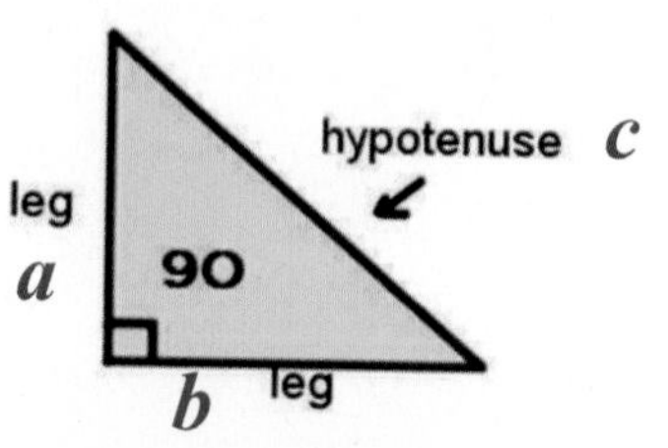

$$a^2 + b^2 = c^2$$

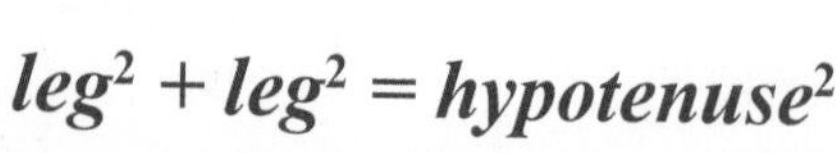

$$leg^2 + leg^2 = hypotenuse^2$$

Avoiding Common Errors: Make sure you plug the side lengths into the formula correctly. Students prefer to plug in for "a" and "b" and solve for "c" which makes the math simpler. However, pay attention to what you have been given. Often the problem will give you "a" and "c" and ask for "b".

8.06 Radical Equations and Problem-Solving

EXAMPLE 7

Find the length of the unknown leg of the right triangle.

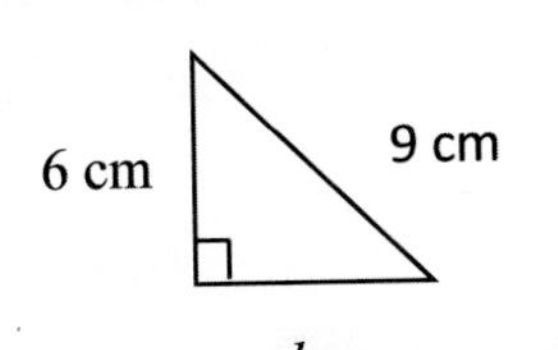

In the formula, $a^2 + b^2 = c^2$, we know that a and b are legs and c is the hypotenuse. Here, $c = 9$ which is the hypotenuse, and $a = 6$. We solve for b.

$a^2 + b^2 = c^2$ Formula for the Pythagorean Theorem.

$6^2 + b^2 = 9^2$ Plug in values for a and c.

$36 + b^2 = 81$ Evaluate exponents.

$b^2 = 45$ Subtract 36 from both sides.

$\sqrt{b^2} = \pm\sqrt{45}$ Take the square root of both sides.

$b = \sqrt{45} = \sqrt{9 \cdot 5}$ Find perfect square factor(s) of radicand. Keep only the positive root.

$b = 3\sqrt{5}$ cm Simplify the radical.

EXAMPLE 8

A ladder 11 meters long rests against a wall with the base of the ladder 4 meters from the base of the wall. How far up the wall does the top of the ladder rest? Round to 3 decimal places if necessary.

$a^2 + b^2 = c^2$ Pythagorean Formula.

$a^2 + 4^2 = 11^2$ Plug in values for variables.

$a^2 + 16 = 121$ Evaluate the exponents.

$a^2 = 105$ Subtract 36 from both sides.

$\sqrt{a^2} = \pm\sqrt{105}$ Take the square root of both sides.

$a = 10.247$ meters Simplify. (Keep only the positive root.)

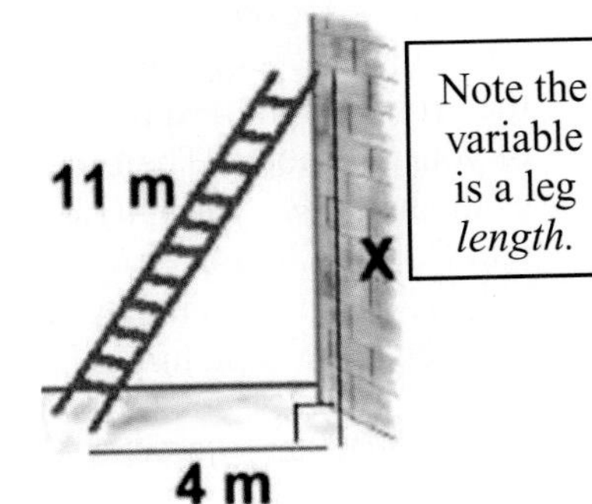

Note the variable is a leg *length.*

EXAMPLE 9

A wire is needed to support a flag pole that is 15 feet high. The wire will be anchored to a stake 8 feet from the base of the pole. How much wire is needed? Round to 3 decimal places.

$a^2 + b^2 = c^2$

$8^2 + 15^2 = c^2$

$64 + 225 = c^2$

$289 = c^2$

$\pm\sqrt{289} = \sqrt{c^2}$

$c = 17$ feet (Keep only the positive root.)

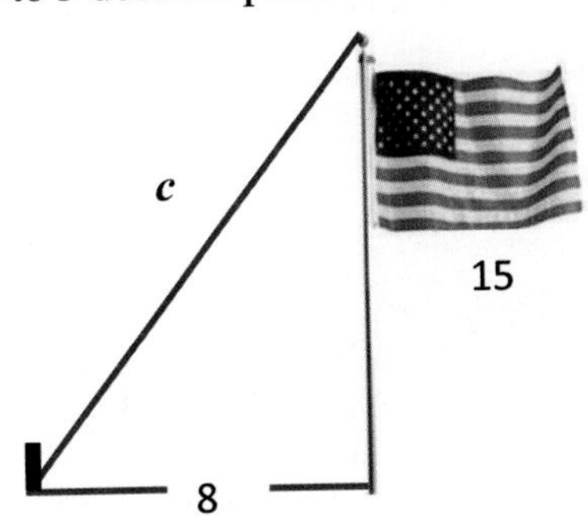

8.06 Radical Equations and Problem-Solving 8.06

EXAMPLE 10

The formula $v = \sqrt{2gh}$ gives the velocity v, in feet per second, of an object when it falls h feet accelerated by gravity g in feet per second squared. If g is approximately 32 feet per second squared, how far will an object fall if its velocity is 80 feet per second?

$v = \sqrt{2gh}$

$g = 32$ and $v = 80$, find h. *Pay attention to the variables and which # values the problem gives.*

$80 = \sqrt{2(32)h}$

$80 = \sqrt{64h}$

$(80)^2 = \left(\sqrt{64h}\right)^2$

$6400 = 64h$

$100 = h$

Object will fall 100 feet.

Courtesy of Fotolia

YOUR TURN

Use the Pythagorean Theorem to solve.

7. You are locked out of your house and the only open window is on the second floor, 25 feet above the ground. You need to borrow a ladder from one of your neighbors. There's a bush along the edge of the house, so you'll have to place the ladder 10 feet from the house. How long will the ladder need to be to reach the window? Round to nearest whole number. Draw a picture.

Courtesy of Fotolia

8. A 50 foot support wire is to be attached to a cell phone antenna. Because of surrounding buildings and sidewalks, the wire must be anchored 20 feet from the base of the antenna. How tall can the antenna be? Round to nearest tenth. Draw a picture.

8.06 Radical Equations and Problem Solving 8.06

EXAMPLE 11

The distance a person can see into the horizon depends on their height from the ground. The distance can be found using the formula $d = 1.5\sqrt{h}$, where d is the distance in miles and h is the height in feet from the ground to the person's eye.

Courtesy of Fotolia

First, understand the variables in words:

h: height from the ground in feet
d: distance a person can see into the horizon in miles.

a. Find the distance someone can see to the horizon if their eye is 6 feet of the ground. Round to the nearest hundredth.

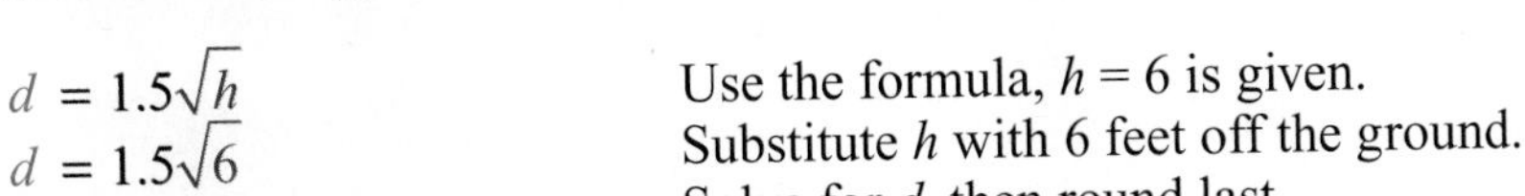

$d = 1.5\sqrt{h}$	Use the formula, $h = 6$ is given.
$d = 1.5\sqrt{6}$	Substitute h with 6 feet off the ground.
$d = 3.6742 \approx 3.67$ miles	Solve for d, then round last.

A person will be able to see approximately 3.67 miles.

b. Find the distance someone can see to the horizon if their eye is 24 feet of the ground. Round to the nearest hundredth.

$d = 1.5\sqrt{h}$	Use the formula, $h = 24$ is given.
$d = 1.5\sqrt{24}$	Substitute h with 24 feet off the ground.
$d = 7.3484 \approx 7.35$ miles	Solve for d, then round last.

A person will be able to see approximately 7.35 miles.

c. A person in an airplane can see 175 miles to the horizon, what is the altitude of the plane? Round to the nearest whole number.

$175 = 1.5\sqrt{h}$ For this problem, the d variable is given, substitute $d = 175$.

$\frac{175}{1.5} = \frac{1.5\sqrt{h}}{1.5}$ To isolate the radical piece first, divide by 1.5. *(WAIT before squaring.)*

$\frac{175}{1.5} = \sqrt{h}$

$\left(\frac{175}{1.5}\right)^2 = \left(\sqrt{h}\right)^2$ Since the index is 2, square both sides to solve for h.

$h = 13{,}611.11 \approx 13{,}611$ Round last.

The height of the person's eye is approximately 13,611 feet.

8.06 Radical Equations and Problem-Solving

YOUR TURN

Answer each question.

9. For a pyramid with a square base, the length of a side of the base b is given by the formula

$$b = \sqrt{\frac{3V}{h}}$$

where V is the volume and h is the height. The Pyramid of the Pharoah at Giza has a square base. If the length of the base of the pyramid is 230.4 m and the height is 146.6 m, what is the volume of the pyramid? Round to the nearest whole square meter.

What are the variables in words:

V: ______________________________

h: ______________________________

b: ______________________________

Substitute values in the equation and solve. Round your solution last.

10. Oceanographers use the model $v = 3\sqrt{d}$ to describe the velocity of a tsunami, a great sea wave produced by underwater earthquakes. In the model, v represents the velocity (speed) in feet per second of a tsunami as it approaches land, and d is the depth of the water in feet. Find the depth of the water if the speed is measured at 66 feet per second.

Courtesy of Fotolia

What are the variables in words:

v: ______________________________

d: ______________________________

Substitute in the equation and solve. Round your solution last.

YOUR TURN Answers to Section 8.06

1. 9 2. 173 3. 6 4. 1 5. 2 6. no solution

7. 27 feet 8. 45.8 feet 9. V: volume, h: height, b: base, $V \approx 2{,}594{,}046 \text{ m}^3$

10. v: velocity, d: water depth, $d \approx 484$ feet

Complete MyMathLab Section 8.06 Homework in your Homework notebook.

8.08 Solving Quadratic Equations using the Quadratic Formula 8.08

The technique of solving quadratic equations by factoring is very limited. In most real-world applications involving quadratic equations, the quadratric is not factorable. When that happens, we use a formula called the **Quadratic Formula**. These solutions are usually decimal values.

First, put your equation in the standard form:

Standard Form of a Quadratic Equation

$y = ax^2 + bx + c$ and $a \neq 0$

Set $y = 0$ and you have $0 = ax^2 + bx + c$. You might remember solving quadratic equations by using the factoring method in DMA 060. We saw that quadratic equations can have one solution, two solutions or no solutions. We found these solutions by factoring. The factoring method does not always work so we will now look at solving quadratic equations by using the quadratic formula.

Quadratic Formula

$$x = \frac{-b \pm \sqrt{b^2 - 4ac}}{2a}$$

What does ± mean?

These a, b & c values are the numbers from the standard form of the quadratic equation above.

We will plug these numbers into the quadratic formula to find the solutions.

What does ± mean? The quadratic equation can have 2 solutions.

One solution is: $x = \dfrac{-b + \sqrt{b^2 - 4ac}}{2a}$

The other solution is: $x = \dfrac{-b - \sqrt{b^2 - 4ac}}{2a}$

EXAMPLE 1

For the quadratic equation $3x^2 = 14$ identify a, b, $\&c$.

Solution:

$3x^2 = 14$	First write in standard form $ax^2 + bx + c$.
$3x^2 + 0x - 14 = 0$	Subtract 14 from both sides. Save a place for the bx term, $b = 0$.
$a = 3 \quad b = 0 \quad c = -14$	The values are now seen easily.

8.08 Solving Quadratic Equations using the Quadratic Formula 8.08

EXAMPLE 2

Solve $x(3x+4)=5$ using the Quadratic Formula. Write exact answers and then approximate answers to the nearest thousandth.

First, write the equation in standard form. $3x^2+4x-5=0$

Second we need to identify the coefficients *a, b, c.*

$a=3$ $b=4$ $c=-5$

Now we substitute the values *a, b, c* into the quadratic formula and simplify.

$x=\dfrac{-b\pm\sqrt{b^2-4ac}}{2a}$ **Quadratic Formula.**

$x=\dfrac{-(4)\pm\sqrt{(4)^2-4(3)(-5)}}{2(3)}$ **Substitute in for *a*, *b*, and *c*.**

$x=\dfrac{-(4)\pm\sqrt{16-(-60)}}{6}$ **Simplify using order of operations**

$x=\dfrac{-4\pm\sqrt{76}}{6}$ **Do not round yet.**

Exact answers: $x=\dfrac{-2+\sqrt{19}}{3}$ *or* $x=\dfrac{-2-\sqrt{19}}{3}$

Approximations: $x=\dfrac{-2+4.358899}{3}\approx 0.7863$ OR $x\approx\dfrac{-2-4.358899}{3}\approx -2.1196$ ROUND LAST!

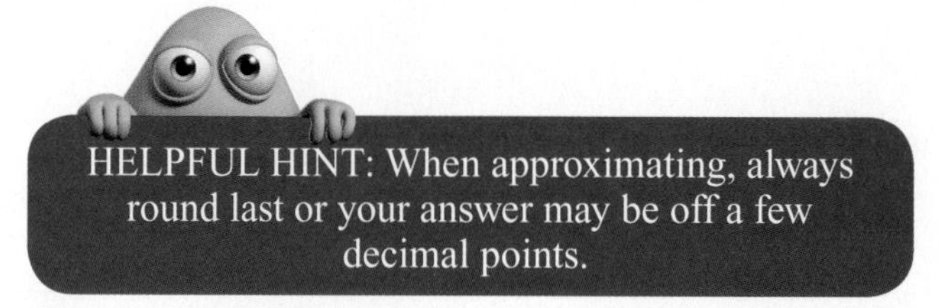

8.08 Solving Quadratic Equations using the Quadratic Formula 8.08

EXAMPLE 3

The following table from the National Health and Nutrition Examination Survey indicates that the number of American adults who are overweight or obese is increasing.

Courtesy of Fotolia

Courtesy of Fotolia

Years since 1960, t	**1**	**12**	**18**	**31**	**39**	**44**	**46**
Percentage of Americans who are overweight or obese, P(t)	**45**	**47**	**47**	**56**	**64.5**	**66.0**	**66.9**

This data can be modeled by the equation: $P(t) = 0.0101t^2 + 0.0530t + 44.537$

First, understand the variable in words:

t: years since 1960
(Pt): percentage of Americans who are overweight

a) Use your graphing calculator to graph a model of the function, $P(t) = 0.0101t^2 + 0.0530t + 44.537$.
Xmin = -1 Xmax = 70 xscl = 1 Ymin = 0 Ymax = 100 yscl = 1

P(t) Percentage of overweight Americans

Remember from Sec. 6.18, the y-intercept is the c value, 44.537.

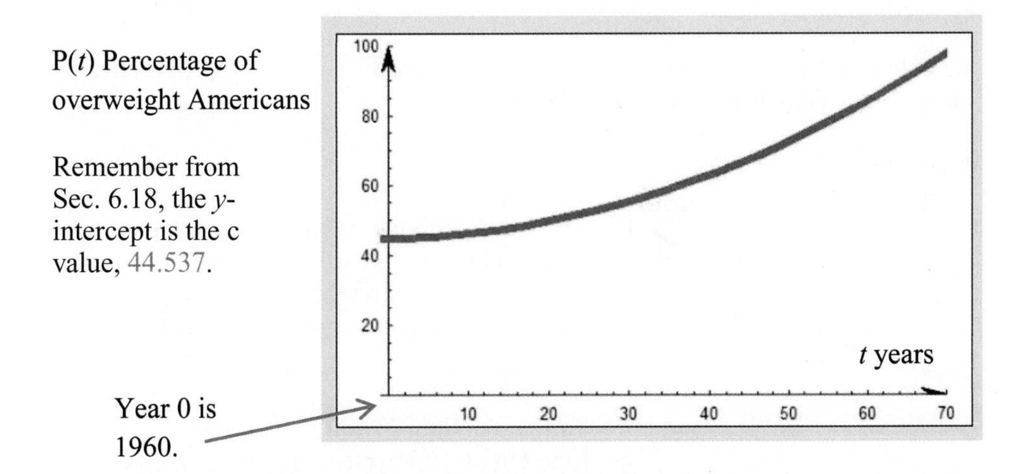

Year 0 is 1960.

Does your graph look like the one above? Only Quadrant I is important. Why?
Because we will only look at the years after 1960.
In looking at the graph at 1960, the percentage of overweight Americans was 44.537%. In the following years after 1960, the percentage of overweight Americans increases.

8.08 Solving Quadratic Equations using the Quadratic Formula 8.08

b) Use the quadratic formula to determine the year when the model predicts the percentage of overweight or obese Americans will first exceed 75%. Round to the nearest year.

Write the equation to make this prediction. $P(t) = 0.0101t^2 + 0.0530t + 44.537$

$75 = 0.0101t^2 + 0.0530t + 44.537$ Since P(t) = 75%, substitute 75 into the equation.

$75 = 0.0101t^2 + 0.0530t + 44.537$ We will use the Quadratic Formula.

$\underline{-75} \qquad \underline{-75}$

$0 = 0.0101t^2 + 0.0530t - 30.463$

$a = 0.0101 \quad b = 0.0503 \quad c = -30.463$

Put the equation in standard form by subtracting 75 from both sides in order to identify *a*, *b*, & *c*. Identify *a, b,* & *c*. Watch the signs.

$$t = \frac{-(0.0530) \pm \sqrt{(0.0530)^2 - 4(0.0101)(-30.463)}}{2(0.0101)}$$

Substitute the values for *a, b,* & *c*, & enter into your calculator.

$$t = \frac{-(0.0530) \pm \sqrt{1.2335142}}{0.0202}$$

Keep more decimals than you need so that you can round last.

$$t = \frac{\left(-0.0530 + \sqrt{1.2335142}\right)}{0.0202} \approx \frac{1.0576368}{.0202} \approx 52.35825 \approx 52.4$$ *Remember to round last.*

$$t = \frac{\left(-0.0530 - \sqrt{1.2335142}\right)}{0.0202} \approx \frac{-1.16364}{.0202} \approx 57.6059 \approx -57.6$$ *Remember to round last.*

Does the negative number make sense in the context of this problem? . . . NO!!!! Do not use –57.6.
SOLUTION: Add 52.4 to 1960, the initial year, 1960 + 52 = 2012.

The predicted year when 75 % of Americans are overweight is 2012.

c) Use the graphing calculator to determine the year when the model predicts the percentage of overweight or obese Americans will first exceed 75%.

P(*t*) % Obese Americans

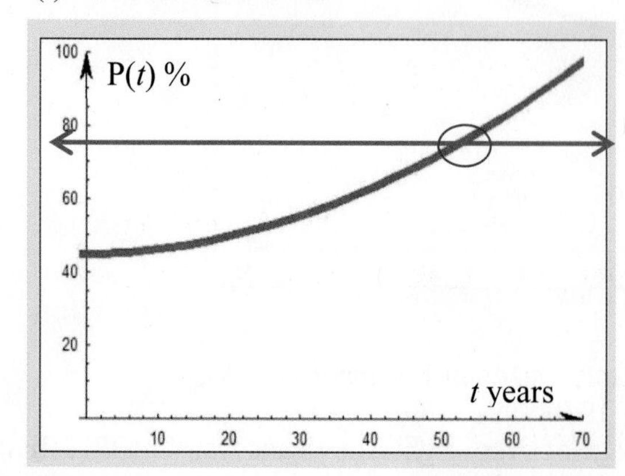

The solution found using the quadratic formula by hand is the same as the *intersection point* on the graph.

Enter $y_1 = 0.0101t^2 + 0.0530t + 44.537$
$y_2 = 75$

To find the intersection point, press:
[2nd CALC, 5:intersection, enter, enter, enter]
X = 52.35826 years

The *x*-intercept of 52.4 determined from the graph is the same as the solution determined by using the quadratic formula. Hence, 52.4 yrs. + 1960 (initial year) = 2012. The solution is the same.

8.08 Solving Quadratic Equations using the Quadratic Formula 8.08

EXAMPLE 4

According to statistics from the U.S. Department of Commerce, the average annual income of each resident of the U.S. from 1960 to 2008 can be modeled by the equation:

$P(t) = 12.6779t^2 + 177.3069t + 1{,}836.8354$ where t is the number of years since 1960.

***Remember:* For 1960, $t = 0$. For 1970, $t = 10$, etc.**

a) **What is the domain that makes sense for this function P?**

The graph begins in 1960 and ends in 2008. Therefore the domain is 0 to 48, [0, 48].

b) **Here is a table that shows some of the values for t. We have rounded to the nearest dollar.**

Year	1960	1970	1980	1990	1995	2000	2005	2008
t	0	10	20	30	35	40	45	48
P(t)	1837	4878	10,454	18,566	23,573	29,214	35,488	39,557

c) **Sketch a graph of the function using your graphing calculator with the given window:**

$P(t) = 12{,}6779t^2 + 177.3069t + 1{,}836.8354$

Xmin = -5 Xmax = 60 Ymin = -100 Ymax = 50,000

d) **Estimate the average income in the year 1989 using the calculator. Round to the nearest cent. (Hint: $t = 29$)**

Use the "VALUE" feature on your calculator to determine the y value when $x = 29$.

Press [2nd][CALC], select :1:value".
For X = type 29. Press ENTER.

Y = 17,640.849 or **P(29) ≈ $17,640.85**

The average income in 1989 is $17,640.85

P(t) $ average income

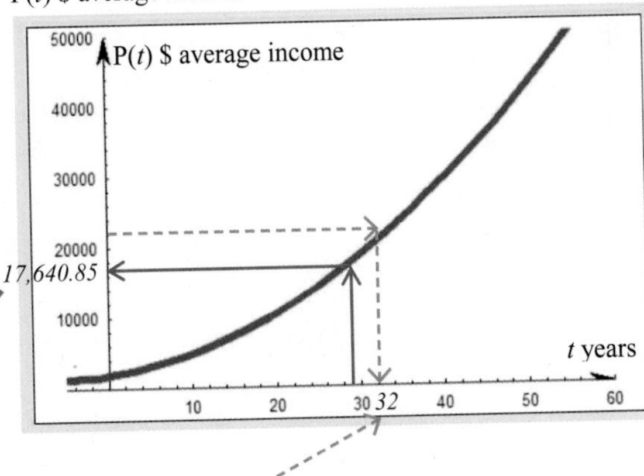

e) **Determine in which year the average personal income reached $20,500. Round to the nearest year.**

Use the INTERSECT feature of your calculator to find the value of t when P(t) = $20,500?

Leave $Y_1 = 12.6779t^2 + 177.3069t + 1{,}836.8654$ & enter $Y_2 = 20{,}500$. Press GRAPH.
Press [2nd][CALC] & select "5:Intersect". Press ENTER 3 times.

X = 32.007256 Y = 20,500 Round $t \approx 32$. Add 32 to 1960, 1960 + 32 = 1992.

The year when the average personal income was P(t) = 20,500 when $t = 32$, the year 1992.

8.08 Solving Quadratic Equations using the Quadratic Formula 8.08

YOUR TURN

Solve the equations using the quadratic formula. Give an exact answer and an approximation to the nearest hundredth.

1. $x^2+6x-3=0$	2. $2x^2-3x=5$
3. $7y=2y^2-4$	4. $(2x-2)(x+2)=1$

8.08 Solving Quadratic Equations using the Quadratic Formula 8.08

YOUR TURN

5. The number n (in millions) of cell phone users in the United States from 2000 to 2006 is given in the following table.

Year	2000	2001	2002	2003	2004	2005	2006
Number of Cell phone Users (in millions)	109.5	128.4	140.8	156.7	182.1	207.9	233.0

This data can be approximated by the quadratic model,

$n(t) = 1.39t^2 + 12.29t + 111.293$ where $t = 0$ corresponds to the year 2000.

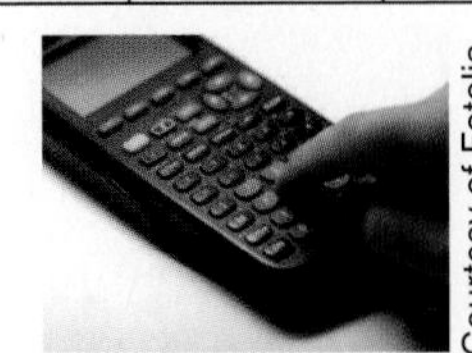
Courtesy of Fotolia

a. **Use your graphing calculator to sketch a graph of the function. Label the axes.** Use the window:

Xmin: -5 Xmax: 15 Ymin: -10 Ymax: 400

Your x-axis begins with 0 and should go for several years past 2008, or 8 years. Your y-axis will go to 400.

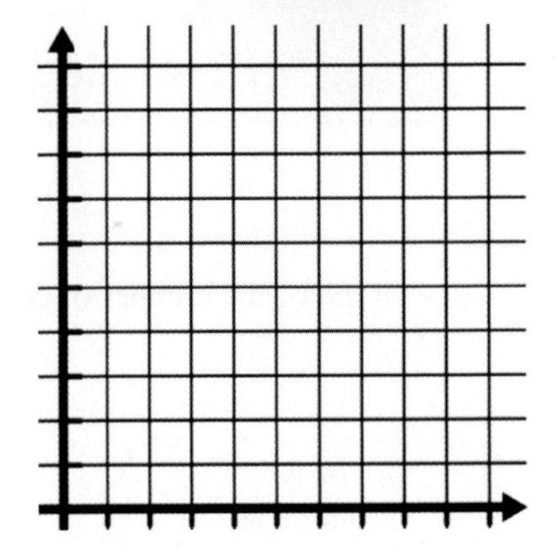

b. **Use the graphing calculator to estimate the year in which there will be 300 million cell-phone subscribers. Round to the nearest year.**
(Use the intersect feature of the calculator.)

CALC Estimate: ____________________

c. **Use the quadratic formula to answer part b.**
How does your answer compare to the estimate you obtained by using a graphical approach? Round to the nearest year.

Quadratic Formula Estimate: $a =$ ________ $b =$ ________ $c =$ ________

Courtesy of Fotolia

$t =$ ____________________

8.08 Solving Quadratic Equations using the Quadratic Formula 8.08

YOUR TURN **Answers to Section 8.08.**

1. $x=-3+2\sqrt{3}$ *or* $-3-2\sqrt{3}$; $x\approx 0.46$ *or* $x\approx -6.46$

2. $x=2.5$ *or* -1 (no approximation needed)

3. $x=4$ *or* $x=-\dfrac{1}{2}=-0.5$ (no approximation needed)

4. $x=\dfrac{-1+\sqrt{11}}{2}$ *or* $x=\dfrac{-1-\sqrt{11}}{2}$; $x\approx\dfrac{(-1+3.316625)}{2}\approx 1.16$ *or* $x\approx\dfrac{(-1-3.3166245)}{2}\approx -2.16$

5. a. See graph below.
   b. $t=$ *about* 8 *years*
   c. $t=8.1$ *or* -16.7 remember to discard the negative answer.

5. Graph of the number of cell phone users.

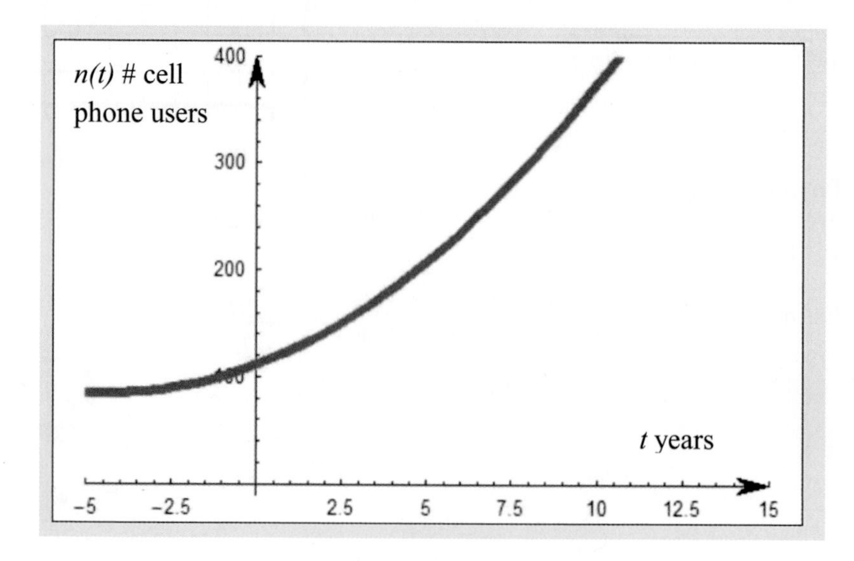

Complete MyMathLab Section 8.08 Homework in your Homework notebook.

8.09 Graphing Radicals

Graphical representations of radical equations can be found the same way we have made graphs of other equations. A graph of any radical equation is a picture of the points of its curve. Let's make a graph by finding points which make a radical equation true. We will use a t-chart of some ordered pairs (x, y).

Look at the graph of $y = \sqrt{x}$.
By choosing x-values and substituting in the equation, we find the corresponding y values of some points on the graph.

x	$y = \sqrt{x}$
0	$y = \sqrt{0} = 0$
.5	$y = \sqrt{.5} = .707$
1	$y = \sqrt{1} = 1$
2	$y = \sqrt{x} = 1.414$
4	$y = \sqrt{4} = 2$
6	$y = \sqrt{6} = 2.449$
-1	$y = \sqrt{-1}$ This is not possible.
-4	$y = \sqrt{-4}$ This is not possible. *The function is undefined for all negative values of x.*

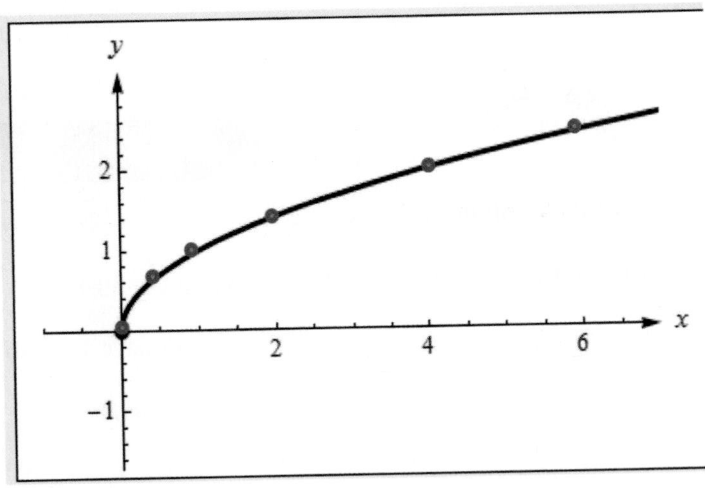

Notice that the curve will continue to move to the right and also move slowly upward.

When you graph a radical equation you have to consider the domain of the equation before you graph. Why? You cannot have a negative number inside a square root.

Let's look at some other graphs of radical functions and compare their curves with the equation above.

$y = \sqrt{x} + 3$

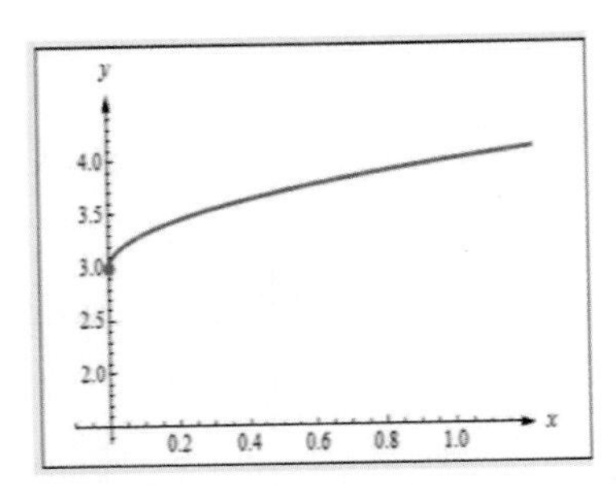

The graph has moved up 3 units.

$y = \sqrt{x-2}$

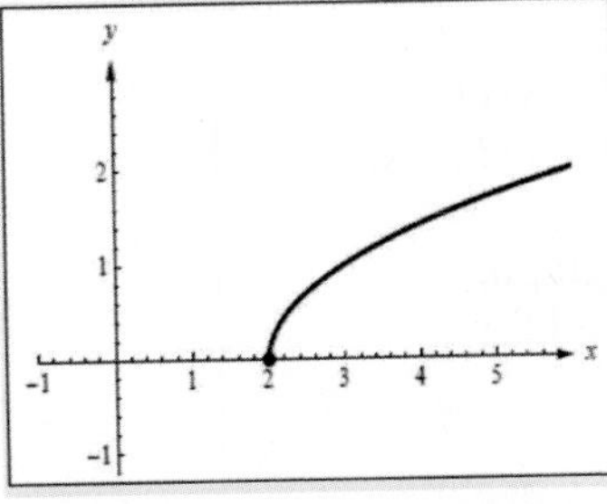

The graph has moved right 2 units.

$y = \sqrt{x-2}$

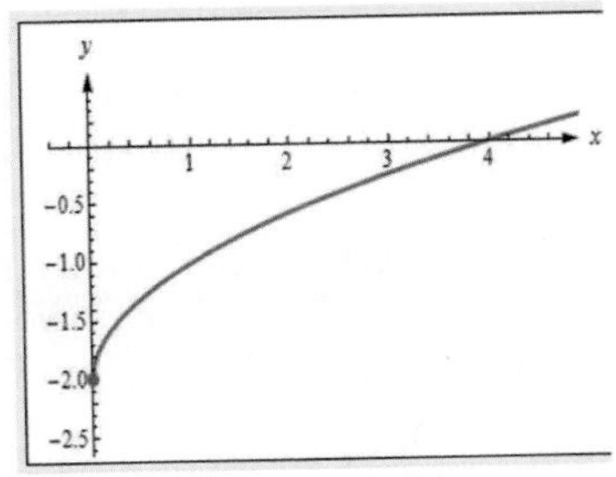

The graph has moved down 2 units.

8.09 Graphing Radicals 8.09

Domain, *D*:

The set of *x*-coordinates of a relation.

Given a graph of a line or curve, use *interval notation to list the x values* from smallest to largest.

(Hint: Read graphs from left to right horizontally.)

There are two ways to find the domain of a radical function. One is algebraically using the equation and the other is by looking at the graph of the equation.

EXAMPLE 1

Here are the steps to solving this.

Find the domain of $y = \sqrt{x+2}$.

To find the domain of a radical function algebraically, let's look under the radical sign.

$y = \sqrt{x+2}$ Remember that you can't have a negative value inside a square root.

$x + 2 \geq 0$ *Set the radicand (inside part) equal to or greater than zero.*

$x \geq -2$ Solve.

The domain in set notation is $\{ x \mid x \geq -2\}$. Remember, this is read "For the set of all real numbers *x*, where *x* is greater than or equal to -2."

(-2, ∞] This is the domain written in Interval Notation.

What does this tell us? → Any number greater than or equal to -2 will make the radicand non-negative.

To find the domain of a radical function graphically, let's look the graph of $y = \sqrt{x+2}$.

1) GRAPH: $y_1 = \sqrt{x+2}$ using the window:

   Xmin = -3 Xmax = 3 Xscl = 1
   Ymin = -1 Ymax = 3 Yscl = 1

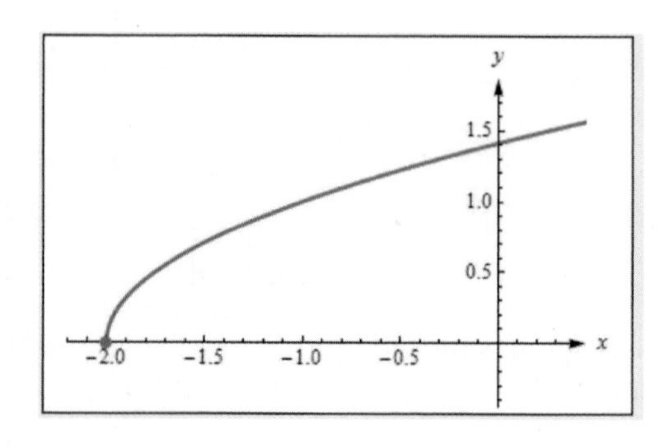

2) As you read the domain from the graph, does it match the domain you solved for algebraically? Yes. [-2, ∞)

   Yes. Any number equal to or greater than -2 will make the radicand non-negative.

EXAMPLE 2

Find the domain of $y_1 = \sqrt{3-x}$.

To find the domain of a radical function algebraically, let's look under the radical sign.

$y_1 = \sqrt{3-x}$ Remember that you can't have a negative number inside a square root.

$3 - x \geq 0$ *Set the radicand (inside part) equal to or greater than zero.*

$3 \geq x$ Solve.

$x \leq 3$ Don't forget to change direction of the inequality sign when you switch places with the number and the variable.

The domain in set notation is $\{ x \mid x \leq 3\}$. $(-\infty, 3]$ This is the domain written in Interval notation.

What does this tell us? → Any number equal to or less than 3 will make the radicand non-negative.

To find the domain of a radical function graphically, let's look the graph of $y_1 = \sqrt{3-x}$.

GRAPH: $y_1 = \sqrt{3-x}$ using the window:

Xmin = -1	Xmax = 5	Xscl=1
Ymin = -1	Ymax = 3	Yscl=1

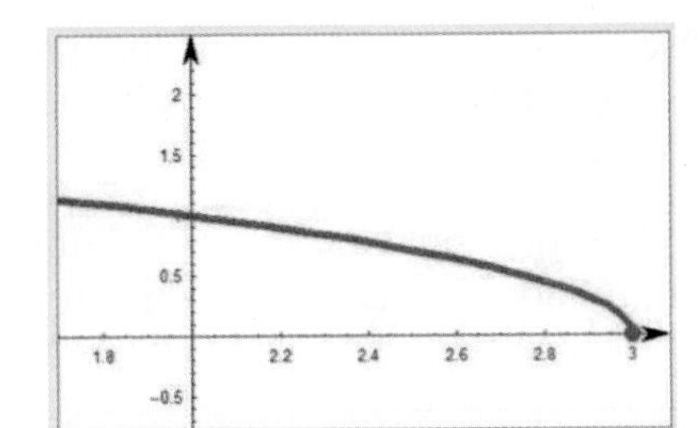

As you read the domain from the graph, does it match the domain you solved for algebraically? $(-\infty, 3]$
Yes. Any number equal to or less than 3 will make the radicand non-negative.

EXAMPLE 3

Find the domain of the function $f(x) = -\sqrt{x+5}$.

a) **Graph the function.** $f(x) = -\sqrt{x+5}$

Use the window:

Xmin = - 6	Xmax = 4	Xscl = 1
Ymin = -4	Ymax = 3	Yscl = 1

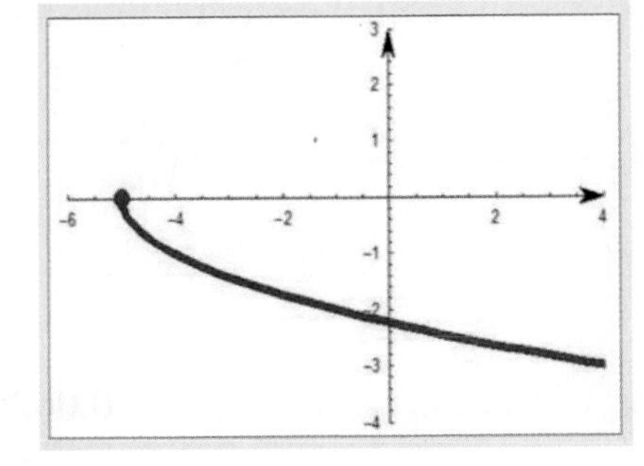

b) **Look at the graph and determine the domain.**

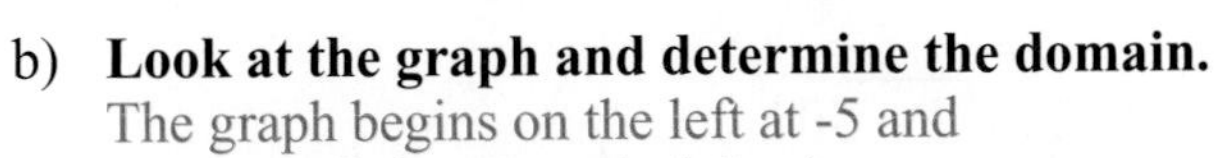

The graph begins on the left at -5 and goes to infinity. Domain $[-5, \infty)$

c) **Find the domain algebraically.**

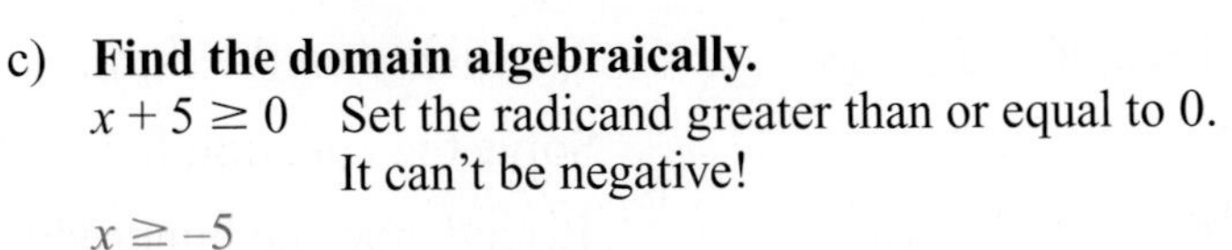

$x + 5 \geq 0$ Set the radicand greater than or equal to 0. It can't be negative!

$x \geq -5$

We can see that the domain is $x \geq -5$ **or** $[-5, \infty)$.

EXAMPLE 4

Suppose two different objects are dropped: a marble and a large beach ball. Because of air resistance, the beach ball will take longer than the marble to fall the same distance. The variable t represents time and d represents the distance the marble falls.

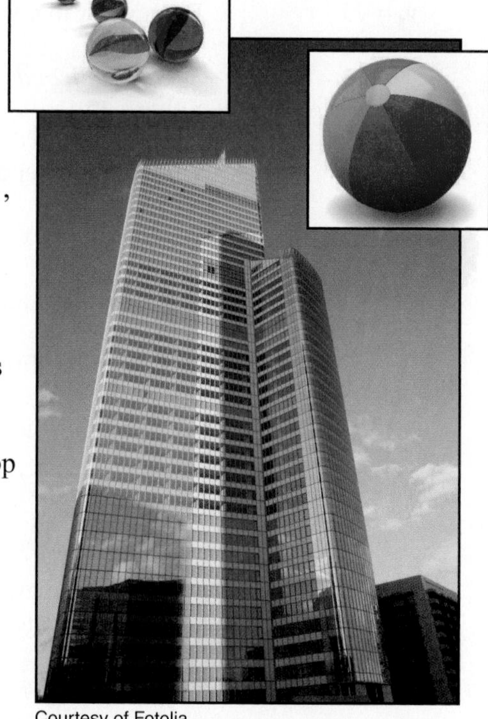
Courtesy of Fotolia

The marble falls according to the equation: $t = 0.25\sqrt{d}$

The beach ball falls according to the equation: $t = 0.26\sqrt{d-k}$,
where k is a positive constant representing the difference in the beginning height of the ball from the marble.

a) **Suppose the beach ball is dropped from a height 50 feet lower than the marble. Find the equation representing this time.**

Use $t = 0.26\sqrt{d-k}$ to find the time for the beach ball to drop d - 50 feet, where d is the height the marble falls. (In other words, $k = 50$ feet.)

So the equation representing the time t it takes for the beach ball to fall in the context of this problem is:

$$t = 0.26\sqrt{d-50}$$

b) **From what height must the marble and the beach ball be dropped so that they will each hit the ground at the same time?**

The time it takes the marble to fall equals the time it takes the beach ball to fall, so set the equations equal and then solve for *d*.

$$0.25\sqrt{d} = 0.26\sqrt{d-50}$$

$$\left(0.25\sqrt{d}\right)^2 = \left(0.26\sqrt{d-50}\right)^2$$

$0.0625d = 0.0676(d-50)$ Square both sides.

$0.0625d = 0.0676d - 3.38$ Distribute on the right.

$-0.0676d \quad -0.0676d$ Solve the equation for d.

$-0.0051d = -3.38$ Rounded to the hundredths.

$d \approx 662.75$ ft.

What does this value of d represent in the context of this problem?

The initial or beginning height of the marble is the *d* distance value.
The marble fell from a height of 662.75 feet.

c) **Find the time it takes for each object to fall from this same height.**

Substitute 662.75 for *d* (distance) in both the marble equation and the beach ball equation.

Marble: $t = 0.25\sqrt{(662.75)} = 6.4$ seconds

Beach ball: $t = 0.26\sqrt{(662.75) - 50} = 6.4$ seconds

If the marble is dropped at 662.75 feet and the beach ball is dropped at 612.75 feet, they will both hit the ground in approximately 6.4 seconds.

d) **Verify your solution by using your graphing calculator. Round to the nearest hundredth.**

Enter in the calculator: $y_1 = 0.25\sqrt{x}$ and $y_2 = 0.26\sqrt{x-50}$ & find the INTERSECTION point.

Use the window: Xmin = -5 Xmax = 800 Ymin = 0 Ymax = 10

Press [2nd][CALC] key and then choose "5:INTERSECT".

Press the "ENTER" key 3 times.

X = 662.7451 Y = 6.435959

You will see the *x* value (which corresponds to distance) = 662.75 feet and the *y* value (which corresponds to time) = 6.44 seconds.

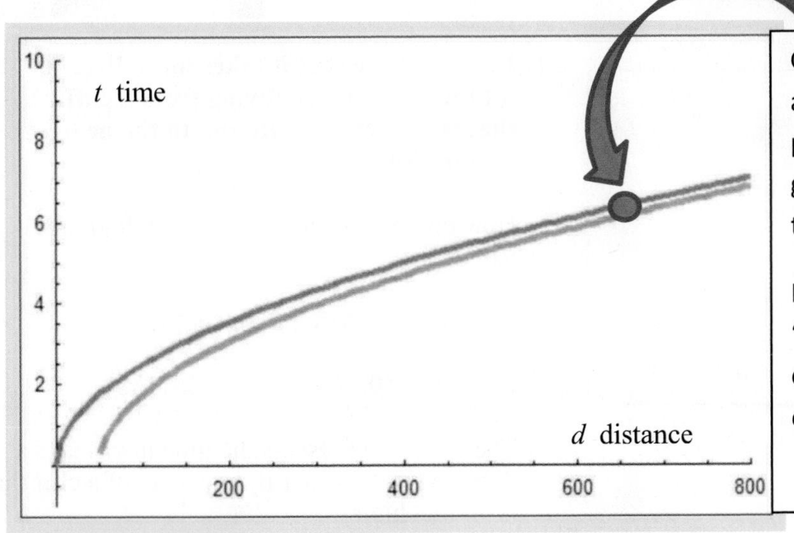

Graphs using a calculator do not always show you accurate pictures. For example, in this graph it is difficult to see that the two curves intersect.

However, when using the "5:INTERSECT" feature of the calculator, you see that the curves intersect at:

x = 662.75 and y = 6.44.

YOUR TURN

Determine the domain of each function. Express your answer in interval notation.

1. $\sqrt{x-5}$	2. $h(x)=\sqrt{6-2x}$
3. $g(x)=\sqrt{3x+2}$	

EXAMPLE 5

The time t (in seconds) that it takes for a cliff diver to reach the water is a function of the height h (in feet) from which he dives: $t=\sqrt{\frac{h}{16}}$.

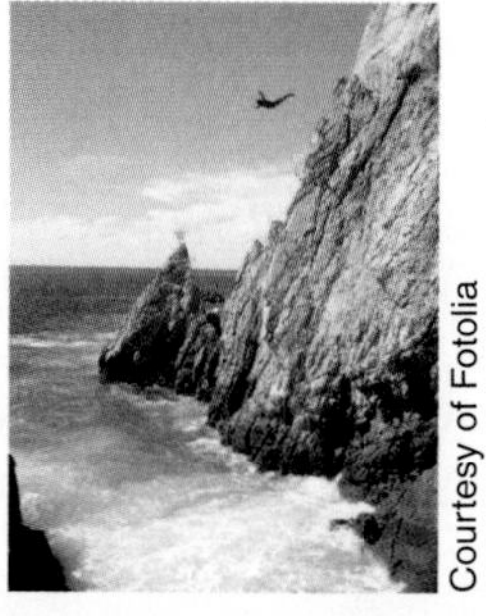
Courtesy of Fotolia

a) Graph the equation on your calculator. Use the window:

Xmin = 0 Xmax = 40 Ymin = 0 Ymax = 5

t time

h height

b) Find the time that it takes for a diver to hit the water when diving from a cliff that is 40 feet high. Round to the nearest hundredth.

How do you use the calculator to find this value?

Press [2nd][CALC] & select "1:value".

Enter X = 40. Notice Y = 1.5811388

The y value represents the time it will take a diver to hit the water if they dive off a cliff that is 40 feet high.

It takes approximately 1.58 seconds.

8.09 Graphing Radicals

YOUR TURN

4. If an object is dropped from a tall building, the time, t, in seconds, it takes for the object to strike the ground is modeled by the equation:

$$t = \frac{\sqrt{d}}{4} \text{ or } t = \frac{1}{4}\sqrt{d}$$ where d is the distance traveled in feet.

Using your graphing calculator, how long will it take an object to fall from the top of the Willis Tower in Chicago, a distance of 1450 feet? Round to the nearest hundredth of a second. Use the window:

Xmin = 0 Xmax = 1750 Xscl = 1 Ymin = -1 Ymax = 11 Yscl = 1

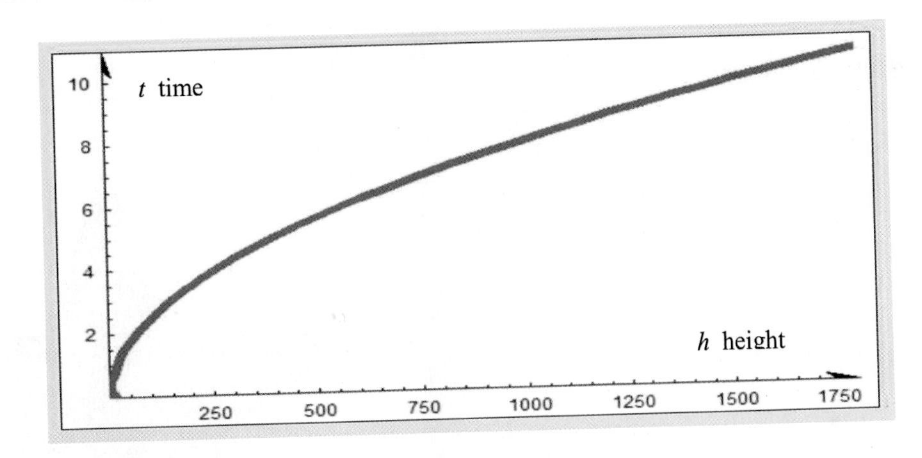

DON'T FORGET TO GO BACK TO THE BEGINNING AND COMPLETE THE INTRODUCTORY PROBLEM!

8.09 Graphing Radicals **8.09**

Answers to Section 8.09

1. $[5, \infty)$ 2. $(-\infty, 3]$ 3. $\left[-\frac{2}{3}, \infty\right)$ 4. Approximately 9.52 seconds.

Introductory Problem: s = speed, d = distance, f = friction

a. $s=\sqrt{30df}$ $s=\sqrt{30(80)(0.4)}$ $s=31\ mph$ b. $s=\sqrt{30(100)(1)}$ $s\approx 55\ mph$

c. $50=\sqrt{30(d)(0.5)}$ $d\approx 167\ feet$ d. See below.

e. range = speed of car

domain = how far a car skids in various conditions.

f. Let $f=1.0$. Enter $y_1=\sqrt{30(d)(1)}$ *and* $y_2=35\ mph$. Change your window. Press [2nd][CALC] and choose "5:intersect." Press enter 3 times.

If a car is traveling 35 mph on dry tar, it will skid approximately 41 feet.

Introductory Problem

d.

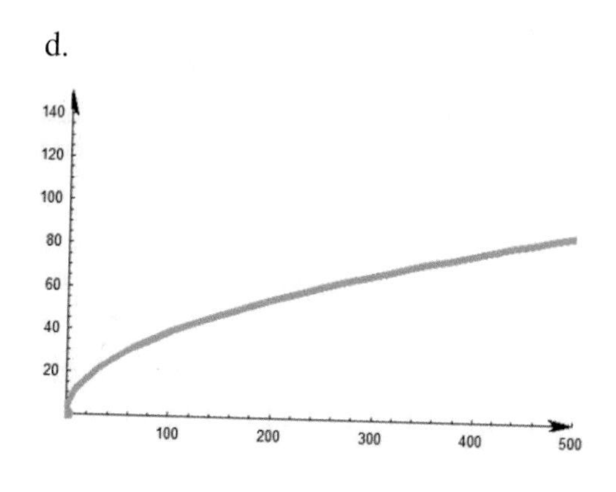

Extra hint for your test: Pay particular attention to the instructions for the problems in Sec. 8.01 & 8.02.

Complete MyMathLab Section 8.09 Homework in your Homework notebook.

DMA 080 TEST: Review your notes, key concepts, formulas, and MML work. ☺

Student Name: ____________________ **Date:** __________

Instructor Signature: ____________________

Index

H

I

J

K

L

M

Q

R

S